AF478330

Nitrogen Fixation:
Global Perspectives

Nitrogen Fixation:
Global Perspectives

Proceedings of the 13th International Congress
on Nitrogen Fixation
Hamilton, Ontario, Canada
2-7 July 2001

Edited by:

Turlough M. Finan, Mark R. O'Brian, David B. Layzell,
J. Kevin Vessey and William Newton

CABI *Publishing*

CABI *Publishing* is a division of CAB *International*

CABI Publishing
CAB International
Wallingford
Oxon OX10 8DE
UK

Tel: +44 (0)1491 832111
Fax: +44 (0)1491 833508
Email: cabi@cabi.org
Web site: www.cabi-publishing.org

CABI Publishing
10 E 40th Street
Suite 3203
New York, NY 10016
USA

Tel: +1 212 481 7018
Fax: +1 212 686 7993
Email: cabi-nao@cabi.org

A catalogue record for this book is available from the British Library, London, UK.

Library of Congress Cataloging-in-Publication Data

International Congress on Nitrogen Fixation (13th : 2001 : Hamilton, Ont.)
  Nitrogen fixation: global perspectives : proceedings of the 13th International Congress on Nitrogen Fixation, Hamilton, Ontario, Canada, July 2-7, 2001 / edited by Turlough M. Finan ... [et al.].
      p. cm.
Includes bibliographical references and index.
  ISBN 0-85199-591-8
  1.   Nitrogen –Fixation--Congresses. I. Finan, Turlough M. II. Title.
  QR89.7.I564 2001
  572'.545--dc21
                              2002000584

ISBN 0 85199 591 8

Printed and bound in the UK by Biddles Ltd, Guildford and King's Lynn, from copy supplied by the editors.

TABLE OF CONTENTS

SECTION 2: BACTERIAL GENOMICS

SECTION 8: STRESSES AND FACTORS LIMITING NITROGEN FIXATION

SECTION 9: REGULATION OF N_2 FIXATION AND METABOLISM

SECTION 15: APPLIED ASPECTS OF NITROGEN FIXATION

POSTER PAPERS

List of Sponsors

McMaster University
United States Department of Agriculture/National Research Institute
National Science Foundation (USA)
Environment Canada
Canadian Forestry Service
The City of Hamilton
Agrium
LiphaTech
Philom Bios Inc.
Alberta Pulse Growers Commission
Agribiotics Inc.
Ag-West Biotech
MicroBio RhizoGen
Urbana Laboratories
Qubit Systems Inc.
Saskatchewan Pulse Growers Association

Local Organizing Committee

Turlough Finan, Department of Biology
McMaster University, Hamilton, Canada

Mark O'Brian, Department of Biochemistry
State University of New York, Buffalo, USA

David Layzell, Department of Biology
Queen's University, Kingston, Canada

J. Kevin Vessey, Department of Plant Science
University of Manitoba, Winnipeg, Canada

National Advisory Committee

H. Antoun, Québec
L. Barran, Québec
T. Charles, Waterloo
B. Driscoll, Montréal
P. Hallenbeck, Montréal

M. Hynes, Calgary
M. Leggett, Saskatoon
L. Nelson, Saskatoon
D. Prevost, Québec
D. Smith, Montréal

International Steering Committee

C. Elmerich, France
H. Hennecke, Switzerland
W.E. Newton, USA
R. Palacios, Mexico

F.O. Pedrosa, Brazil
A. Pühler, Germany
B. Smith, UK
I. Tikhonovich, Russia

International Advisory Committee

C. Atkins, Australia
S. Austin, UK
J. Batut, France
W. Broughton, Switzerland
B. Burgess, USA
F. Dakora, South Africa
F. de Bruijn, France
M. Dilworth, Australia
A. Downie, UK
P. Gresshoff, Australia
G. Hernandez, Mexico
M. Hungria, Brazil
M. Kahn, USA

E. Kondorosi, France
A. Legocki, Poland
P. Ludden, USA
K. Minamisawa, Japan
B. Rolfe, Australia
C. Ronson, New Zealand
T. Ruiz-Argüeso, Spain
L. Seefeldt, USA
G. Stacey, USA
C. Vance, USA
D. Werner, Germany
P. Young, UK

PREFACE

One of the pleasures with working in the area of nitrogen fixation is the fact that while the actual reduction of nitrogen gas (dinitrogen) to ammonia would appear to be a well defined process, many research questions remain and continue to be addressed by diverse groups of scientists including chemists, biochemists, physiologists, molecular biologists, evolutionary biologists, ecologists and applied agricultural scientists. Every two years or so, these groups meet to discuss their most recent findings.

About 300 registrants from 44 countries world-wide attended the thirteenth International Congress on Nitrogen Fixation, which took place in Hamilton, Ontario, Canada on July 2–7, 2001. During the Congress, some 70 oral and approximately 200 poster presentations were made. This volume represents written documentation of these presentations. The very broad participation and wide range of topics justifies our book title, Nitrogen Fixation: Global Prospectives.

The Congress included reports on the continuing progress in understanding the catalytic mechanism of dinitrogen reduction by the conventional molybdenum-containing nitrogenase including the latest ideas on where substrates bind on the biologically unique prosthetic group called FeMo-co. In addition, we heard that homologs of NifHDK which are the structural components of nitrogenase, catalyze steps in chlorophyll synthesis in *Rhodobacter capsulatus*. Further characterization of the *Streptomyces thermoautotrophicus* nitrogenase underlined just how little this enzyme has in common with the conventional nitrogenase.

Genomics was well represented at the Congress. The complete nucleotide sequence and annotation of the *Mesorhizobium loti* and *Sinorhizobium meliloti* genomes were described along with further sequence analyzes of the symbiotic regions from *Bradyrhizobium japonicum* and *M. loti*. It is expected that these genomes will lead to the identification of additional symbiotic genes. Clearly the large numbers of genes with unknown function open up the possibility of much future research.

In the area of plant genomics, we saw that work on the construction of refined maps of the *Lotus japonicus* and *Medicago truncatula* genomes is well underway. One report outlined exciting progress on the identification of a receptor kinase gene which may be the site of a non-nodulating mutation in alfalfa.

The integration and impact of phylogeny on microbiology continues in the area of nitrogen fixation. Of particular note was the identification of rhizobia-like bacteria that induce nitrogen-fixing root nodules on leguminous plants in the β division of the Proteobacteria. That report strengthens the view that the diversity of the rhizobacteria remains to be fully realized.

Overall, from the lively discussions we witnessed in both the auditoria and the hallways, we judged this Congress to be a satisfying and worthwhile experience for the participants. We hope this volume will be a reminder of those good times and a ready reference source of the invaluable information shared in Hamilton. Not all was bright, however, and with sadness and respect, we acknowledge the passing of several of our colleagues from the nitrogen-fixation community.

Finally, we wish to take this opportunity to thank all of the individuals, who helped organize this Congress and to thank the organizations, listed elsewhere, without whose support this Congress would not have been possible. We thank Marlene Mirza for her tireless work in helping to organize the Congress and Vicki Newton for her work and guidance towards getting this book published. We thank Hanna Lindemann, Barbara Reuter, Allyson MacLean, Barbara Finan and Elizabeth Weretilnyk, and most importantly, we thank McMaster University for financial, secretarial, and administrative support throughout the organization of the Congress.

Turlough Finan
David Layzell
Mark O'Brian
Kevin Vessey
William Newton

Hamilton, August 15th, 2001

Johanna Döbereiner Memorial Lecture

JOHANNA DÖBEREINER:

JOHANNA DÖBEREINER: FIFTY YEARS DEDICATED TO THE BIOLOGICAL NITROGEN FIXATION RESEARCH AREA

J.I. Baldani, V.L.D. Baldani, V.M. Reis

Embrapa Agrobiologia, BR465, Km47, 23851-970-Seropédica, Rio de Janeiro, Brazil

Johanna Döbereiner was born in the former Czechoslovakia in 1924, immigrated to Brazil in 1950 and became a Brazilian citizen by option in 1956. She was an extremely enthusiastic person in the field of Biological Nitrogen Fixation (BNF) which began, almost incidentally, when she was employed by the Research Department of the Brazilian Ministry of Agriculture (nowadays Embrapa) to work in the soil microbiology area. Her first publication was in 1951 and referred to the effect of cover vegetables on soil microbe populations. Together with her colleagues she identified in 1958 the first rhizosphere nitrogen-fixing bacterium, named *Beijerinckia fluminensis*, associated with sugarcane plants grown in Rio de Janeiro State. The persistent green color of the grass plants grown on the campus around Embrapa stimulated the young scientist to discover another rhizosphere nitrogen-fixing bacterium. This bacterium was named *Azotobacter paspali* (1966) since its occurrence was predominantly found on the rhizoplane of *Paspalum notatum* cv. *batatais*. Other groups later confirmed this unique association and the amount of fixed nitrogen (around 30 kg N ha^{-1}year^{-1}) was quantified using sophisticated techniques such as the ^{15}N isotopic dilution method.

Despite her important work with grasses, her greatest contribution occurred in the legume field, when together with other Brazilian scientists between the years 1960 and 1970, she convinced the organizers of the Brazilian Soybean breeding program to select soybean varieties dependent only on the BNF process. This approach allowed Brazil to become the World's second largest soybean producer and has saved more than a billion dollars annually in nitrogen fertilizer. It has enabled Brazil to compete successfully in the international market through the production and export of soybean where BNF, biological control and zero tillage are low cost techniques commonly applied. During this period, Johanna trained many students in the legume area investigating the growth of common beans, pastures legumes and legume trees. These former students are now research scientists working in the same field contributing to the development of Brazilian Agriculture with a strong emphasis on sustainability. BNF in legume trees was a special area stimulated by her, which began with experiments on rhizobial inoculation of the legume tree called Sabiá. Later on, one of her first generation students identified new nodule types as well as new modes of infection processes induced by rhizobia in symbioses with legume trees. The group led by Avilio Franco has taken advantage of this symbiosis and, by aggregating mycorrhiza, they developed a technology to rapidly recover degraded areas, including bauxite mining, with legume trees.

The birth of the acetylene reduction technique in the 1970s allowed the detection of very low levels of nitrogenase activity and therefore renewed the interest in BNF associated with graminaceous plants. This period coincided with the energy crisis and The Green Revolution, a system that demanded high amounts of nitrogen fertilizer application and which caused an increase in pollution of the environment. In between 1970 and 1974, Johanna published in collaboration with colleagues at home and abroad many papers measuring nitrogenase activity of *Paspalum notatum* and other pasture grasses as well as sugarcane. With the participation of the late J. Day, she isolated from *Digitaria decumbens*, a micro-aerophilic nitrogen-fixing bacterium resembling *Spirillum lipoferum*, subsequently reclassified as *Azospirillum lipoferum* in collaboration with Dr Noel Krieg. This isolation was made possible by the development of a nitrogen-free semi-solid medium called NFb (Fb stands for Fábio Pedrosa) that reproduced the soil oxygen gradient. Another *Azospirillum* species, named *A. brasilense* was also identified at that time.

Two members of the original BNF grass team, my wife Vera and I, began working with her in late 1976 when the BNF work on graminaceous plants reached its peak. We learnt how to isolate nitrogen-fixing bacteria, not only the known ones but also new species that could be identified based on those "ten commandments" published during the Brazilian BNF Conference, held in Rio de Janeiro, in 1987. These "ten commandments" helped to isolate and identify new nitrogen-fixing bacteria including the diazotrophic endophytes, which colonize the interior of plant tissues. These were the species *Azospirillum amazonense* in 1983, using the semi-solid LGI medium (I stands for Ivo), and the endophytes *Herbaspirillum seropedicae* in 1986, *Gluconacetobacter diazotrophicus* (former *Acetobacter diazotrophicus*) and *Herbaspirillum rubrisubalbicans* in 1996. All these bacteria are known to have an important role in association with graminaceous plants, especially for some sugarcane varieties that are reported to obtain between 30 and 50% of their N from BNF. Johanna's determination was that these new bacteria should be explored as much as possible, therefore, she encouraged many scientists around the world to work with them and today many of these microrganisms are subjects of discussion at this conference.

Johanna's contribution to Science can be measured by the several international prizes that she received, including election to the Vatican Pontifical Academy of Sciences. She was always proud to attend its annual meetings. In addition, she received the title of "Honoris Causa" from the University of Florida (1975) and the Federal Rural University of Rio de Janeiro (1979). She received many other national and international prizes that are listed in her brief biography written by Avilio Franco and Robert Boddey published in Soil Biology and Biochemistry, special issue (29: ix-xi, 1997). She was the most cited female Brazilian scientist by the international community and the seventh most cited Brazilian scientist in a search prepared in 1995 by the Folha de Sao Paulo Newspaper. Johanna has contributed around 190 scientific publications (national and international) in addition to two books, several chapters, more than two hundred abstracts and was an invited speaker at over 50 international meetings. Over a fifty-year scientific career, she averaged 3.7 papers per annum.

Training was another of Johanna's strong characteristics, and today more than one hundred students are learning how to do research at Embrapa. At least three generations of scientists have already trained under her guidance and are now transferring her work philosophy and ideas on the Biological Nitrogen Fixation theme to new students in Brazil and overseas. She was a woman with a very strong personality, sincere and dedicated to her work. Despite her rigorous treatment of scientific aspects, she behaved always like a mother to all students and trainees in our center. In Brazil, she was known as the "soil doctor", although for those that had the privilege to work with her we regarded her as our "Aunt Johanna".

Johanna never agreed that personal names should be given to a bacterium. However, because of her important contribution to the area, two diazotrophs have been named in her honour: *Azospirillum doebereinerae* (Hartmann's group) and *Gluconacetobacter johannae* (Caballero-Mellado's group). For this, and much more, future scientists working in the field of biological nitrogen fixation will remember her name.

Supported by Embrapa, Pronex II, FAPERJ and PADCTII

INORGANIC NITROGEN COMPLEXES: A CANADIAN DISCOVERY

The discovery of Nitrogenpentaamminerutheniumchloride [Ru^{II}(NH$_3$)$_5$N$_2$]Cl$_2$ - the first nitrogen complex by A.D. Allen and C.V. Senoff in 1965

In the early 1960s, A.D. Allen and his coworkers at the University of Toronto were studying the ammine complexes of ruthenium. They routinely prepared these complexes by the reaction between ruthenium trichloride and zinc dust in concentrated ammonia, followed by treatment with hydrochloric acid to give [Ru^{II}(NH$_3$)$_6$]ZnCl$_4$. Mild oxidation and treatment with hydrochloric acid then gave [Ru(NH$_3$)$_5$Cl]Cl$_2$. In early 1965 they became intrigued by the report by some Russian workers of the direct preparation of [Ru^{III}(NH$_3$)$_5$Cl]$^{2+}$ by refluxing an aqueous solution of K$_2$[Ru^{III}Cl$_5$H$_2$O] with hydrazine monohydrochloride. On repeating this preparation but using hydrazine hydrate instead of the hydrochloride, they obtained a different product in the form of a yellow solid which proved very difficult to characterize. Eventually by chemical analysis, infrared spectroscopy of the compound and its fully deuterated form, decomposition with sulfuric acid to give N$_2$, and other experiments they came to the conclusion that it must be [Ru^{II}(NH$_3$)$_5$N$_2$]Cl$_2$ and with some trepidation submitted a paper to this effect to Chemical Communications. The idea that the very unreactive N$_2$ molecule could form a complex with a transition metal was something that at the time nobody had considered might be possible. Finally, at the end of 1966 after having some difficulty obtaining suitable crystals, an X-ray diffraction study was completed and the composition of the compound and its structure were fully confirmed. Many more nitrogen complexes were then rapidly prepared particularly when it was discovered that many could be prepared by the direct reaction of a transition metal complex with N$_2$ under mild conditions. Today many hundreds of inorganic nitrogen complexes of almost all the transition metals are known.

References

Allen AD, Senoff CV (1965) Chem. Commun. 621

Allen AD, Bottomley F (1965) Accts. Chem. Res. Vol. 1, 360

The above summary was generously contributed by:

Professor R.J. Gillespie
Department of Chemistry
McMaster University
1280 Main Street West
Hamilton, Ontario, ON   L8S 4M1
Email: gillespi@mcmaster.ca
Website: http://www.chemistry.mcmaster.ca/faculty/gillespie

KEYNOTE ADDRESS

BIOFIXATION AND NITROGEN IN THE BIOSPHERE AND IN GLOBAL FOOD PRODUCTION

V. Smil

Department of Geography, University of Manitoba, Winnipeg, MB, R3T 2N2 Canada

Meetings such as this one are triumphs of scientific reductionism. In this room, we have dozens of people who know more about some extremely specialized attribute of biofixation or bacterial and plant physiology than does anybody else on this planet of more than six billion people. As a counterweight to this restrictive attention, and following the tradition of keynote addresses (including some given at these biannual meetings) to look further afield, I will offer some broad perspectives on biofixation's peculiar place in the biosphere and on its role in modern food production. I will start with quantifying the remarkable rarity of diazotrophs and end with an appeal for improving efficiencies of existing practices and processes rather than for engaging in a more glamorous search for radically new supplies of reactive N.

I am sure that all of you are aware of the often noted relative scarcity of biofixation among existing species but using the recent reevaluations of global biodiversity makes it even more obvious how rare that capacity is. Of course, we do not know how many species are out there, but the latest consensus points to a much larger total than anticipated just a generation ago. As of the late 1990s, about 1.75 million species were named and described, twice that amount would seem to be a very conservative estimate of actual total biodiversity, and 13.6 million was the most likely total offered by the most comprehensive global assessment (Heywood, Watson 1997). Similarly, we have yet to identify all diazotrophic species. For example, a recent examination of *nifH* gene fragments retrieved from rice roots showed that most of them could not be assigned to any known bacteria (Ladha, Reddy 1999).

Even when assuming that there are no more than 3.5 million species (i.e. twice the known diversity) and that there are 10^2 species of prokaryotes harboring one of the nitrogenases, it means that fewer than 0.0003% of all living organisms can cleave N_2. And when we assume, again, that the eventual grand total of bacteria and archaea will be at least twice the currently identified number of some 5,000 species, 10^2 diazotrophs would still represent only a few percent of all prokaryotes.

What is no less remarkable is the highly conservative evolution of this ability. Although we still cannot reconstruct a satisfactory evolutionary timeline of nitrogenase's emergence and diffusion, it is plausible to speculate that, given the biofixation's undoubted Archaen origins, only a small number of species acquired N-fixing capability during the subsequent 1.5 billion years of late Precambrian evolution and during more than half a billion years of the Phanerozoic eon. The share of diazotrophs among all living organisms peaked, most likely, during the late Archaean era when cyanobacteria ruled the biosphere and when atmospheric oxygen levels were a small fraction of today's concentration. And when the vascular plants conquered the continents some 360 million years ago, fewer than 20,000 of the more than 270,000 species of gymnosperms and angiosperms, that is less than 8% of the total, evolved symbioses or association with N-fixing bacteria.

The single most important reason for this paucity of N-fixing species and N symbioses is the high energy cost of biofixation. Perhaps the best indicator of this high price is the fact that the Haber-Bosch synthesis of NH_3 that takes place in massive steel converters under high pressure (mostly above 10 Mpa) and high temperature (above 350°C) actually requires less energy than the enzymatically-mediated work of N-fixing bacteria operating at ambient conditions. Theoretical calculations indicate that energy cost for the legume nodule is typically between 3-6 g Cg^{-1} N, and actual whole-plant energy costs of symbiotic N_2 fixation can be easily twice as high (Phillips 1980; Hardarson, Lie 1984). Even the lower range is equivalent to roughly 100-200 kJ g^{-1} N. For comparison, the energy requirements of the earliest coal-based Haber-Bosch plants were just over 100 kJ g^{-1} and today's best single-train Kellogg or Topsøe plants need less than 35 kJ g^{-1}, or 1/6 of the energy cost of biofixation (Smil 2001). Of course, there is the essential difference in requisite energies: nonrenewable fossil fuels (now overwhelmingly CH_4) are needed for the Haber-Bosch synthesis while the photosynthetically renewable glucose energizes biofixation.

High energy cost of biofixation is also the best explanation why most plants have eschewed any association with diazotrophs and rely on less energy expensive uptake of nitrates: energy costs of this reduction range from negligible amounts (when done entirely in shoots) to about 1.5 g C g^{-1} N for complete reduction in roots at the expense of respired assimilate (Hardarson, Lie 1984). Although costly in energy terms, biofixation remains the only major provider of reactive N in natural ecosystems (except in those receiving heavy atmospheric N deposition), and recent reevaluations of the global supply of fixed N have only raised its relative importance for the functioning of the biosphere.

Quantifying the contributions of diazotrophs to the biosphere's supply of reactive N remains an uncertain task but the best recent assessment of terrestrial biofixation offers a significantly higher mean than did most of the often cited previous estimates. Appropriately, it uses a range as wide as 100-290 Mt N $year^{-1}$ but it sets the best mean at about 195 Mt N for the Earth's potential natural ecosystems (Cleveland 1999). With no more than 5 Mt N fixed by lightning (Galloway *et al.* 1995), diazotrophs would have provided about 98% of all reactive N available to preagricultural land ecosystems. Today's biofixation adjusted for land use changes would lower the aggregate by about 15%, to almost 170 Mt N $year^{-1}$.

Traditional agricultures expanded their N supply by cultivating leguminous species and this practice became eventually an essential part of intensive cropping particularly in Europe and East and South Asia. My accounts of N inputs in traditional farming show as much as 50% of all available N originating from biofixation by leguminous food, forage and green manure crops in China of the early 1950s (Smil 2001). While the rotations with legumes retain their importance in modern cropping the traditional practice of cultivating leguminous green manure crops has been steadily declining.

I have also prepared the first disaggregated account of N flows in the current global food production (Smil 1999). By far the most accurate part of this exercise, whose results refer to the mid-1990s, is the calculation of N removed by the world's crop harvest: about 60 Mt N in harvested parts and 25 Mt N in crop residues. This total of 85 Mt N includes the nutrient incorporated in all fields crops, be they staple cereals, vegetables or nonfood plants, as well as N in forages grown on arable land and in fruits and plantation species. The only highly accurately known input is the amount of N applied in synthetic fertilizers, close to 80 Mt N during the 1990s (IFA 2001; FAO 2001). Seeds (2 Mt N) and irrigation water (4 Mt N) are minor contributors. Recycling of crop residues and animal manures added more than 30 Mt N (14 and 18, respectively) and atmospheric deposition (all reactive N in precipitation and dry deposition) amounted to some 20 Mt N.

Several factors are responsible for uncertainties in quantifying contributions made by free-living and symbiotic fixers. Fixation rates of all diazotrophs vary widely. Almost all published values for rhizobial fixation in widely cultivated legumes have at least three-fold ranges, much larger spreads are common for some species, and too few fixation rates have been published for many less common legumes (Smil 1997, 1999). Consequently, it is difficult to choose averages needed for national or global calculations. Moreover, unlike for virtually all food and industrial crops, there are no worldwide statistics for the amount of land under leguminous forages and green manures. Previously published totals for the annual global biofixation in agroecosystems were mostly around 40 Mt N (Galloway *et al.* 1995). My best calculation for N fixed by bacteria is about

33 Mt N during the mid-1990s, with the range between 25 to just over 40 Mt N (Smil 1999). Taking the most likely value would mean that diazotrophs now contribute about 20% of all N available to the world's crops.

Their qualitative contribution is even more significant. Bacterially-fixed N is generally much less susceptible to volatilization than are manures and urea, now the world's most important synthetic fertilizer. Biofixed N is also less leaching-prone than the high rates of the nutrient applied in highly soluble nitrates. As a result, more of the biofixed N stays within the soil-plant system to be eventually used by crops. In contrast, losses following applications of N fertilizers have now become the single largest cause of human interference in the global N cycle. They contribute to a variety of undesirable local, regional and even global environmental changes (Galloway *et al.* 1995; Mosier *et al.* 2001; Smil 1995, 2001a). These impacts range from N enrichment of coastal waters of all inhabited continents to high levels of atmospheric nitrate and ammonia deposition on forests and grasslands.

Scientific emphases and goals change with time. A quarter century ago it was widely believed that endowing the nonleguminous staple cereals with the ability to secure their own N is the most important long-term challenge for the biofixation research. At that time optimistic forecasts anticipated that this goal would be accomplished by the century's end (Smil 2001). I believe that the failure to do so is not to be regretted, as there would have been almost certainly a substantial yield cost paid by the new grain- or oil-yielding symbionts. Moreover, with the exception of the sub-Saharan Africa, N currently available to crops worldwide is not in short supply. My global balance shows that crops receive roughly twice as much N as they eventually incorporate in their tissues and hence it is the huge loss (50% global average, close to 70% in rice cropping) of the nutrient that should be, rationally, addressed first (Smil 1999, 2000b, 2001).

Affluent nations have no need for increased food production as their food availability is already 50-70% above the actual average per capita consumption (Smil 2000b). Consequently, they should be actually reducing their overall N needs, and some of them, most notably the Netherlands and Japan, have been doing so for about two decades (Smil 2000b, 2001). In contrast, populous low-income nations will need significantly increased food output in order to accommodate at least another 2-3 billion people during the next two generations and also to improve the overall quantity and quality of average food supply. A key challenge for modern intensive cropping in these countries will be to combine higher food production with minimized throughput of reactive N. Biofixation research could thus make the greatest difference to the world's food production and to the biosphere's integrity by helping to raise the productivity of bacterial diazotrophs in Asia, Africa and Latin America and by improving the efficiency with which the biofixed N is used everywhere by both leguminous symbionts and by nonleguminous crops.

The most rewarding way is to raise the prevailing shares of N derived by leguminous crops from biofixation. Such gains could range from relatively marginal improvements (e.g. from 60% to 70%) that would make a major aggregate difference on the global scale to large increments (e.g. from 20 to 40%) that would be of a great local and regional importance. Soybeans are the best candidates in the first category; many bean cultivars that remain an important source of protein in many poor countries exemplify the second case. Another vigorously pursued strategy should come from a wider use of suitable endosymbionts as their introduction and diffusion does not require drastic genetic modification entailed in transferring symbiotic N fixation to nonleguminous species (Ladha, Reddy 1999). Genetic engineering will eventually produce some N-fixing non-legumes but future adequate food supply that is compatible with sound agroecosystemic management can be maintained without such unprecedented advances.

TABLE OF CONTENTS

Section 1:
Chemistry and Biochemistry of Nitrogenase

CHAIR'S COMMENTS: OVERVIEW OF THE SESSION

B.K. Burgess

Department of Molecular Biology & Biochemistry, University of California, Irvine, Irvine,
CA 92697-3900, USA

This session brought together chemists and biochemists working on nitrogenase and related systems. Four of the talks focused on the conventional molybdenum nitrogenase which is composed of two separately purified proteins. The Fe protein (NifH) is a 60,000 molecular weight dimer of identical subunits connected through a single [4Fe-4S] cluster. It serves as an electron donor for the MoFe protein in a reaction somehow coupled to MgATP hydrolysis. The MoFe protein is often viewed as being composed of two identical halves that do not communicate with each other. Each half has one subunit (NifD) and one subunit (NifK), one [MoFe$_7$S$_9$ homocitrate] cluster designated FeMo-co, one [8Fe-7S] cluster designated the P-cluster, and one binding site for the Fe protein. The P-clusters appear to be involved in electron transfer from the Fe protein to FeMo-co which serves as the site of substrate binding and reduction. Although the structures of both proteins and their metal centers have been available for several years the chemical synthesis of P-cluster and FeMo-co analogs has not yet been accomplished and the specific binding site for substrates on FeMo-co has not been determined.

The first talk (R. Holm) focused on the chemical synthesis problem. The current emphasis is on viewing both the P-clusters and FeMo-co as being composed of two discrete fragments that can be individually synthesized followed by attempts to covalently bridge the fragments. Another recent strategy is to use designed peptides as scaffolds for the construction of the covalent bridges. These strategies have been successful in the synthesis of a carbon monoxide dehydrogenase model that is spectroscopically identical to the active site cluster. In the case of the P-clusters a topological analog has been successfully constructed that includes the hexa-coordinate sulfide and has the composition [Mo$_2$Fe$_6$S$_7$]. The third talk (P. Ludden) approached the synthesis problem from the biological side where progress continues to be made on defining what gene products are required for FeMo-co biosynthesis and on determining the sequence of events. It has been known for many years that the NifH gene product was somehow required for the initial biosynthesis of FeMo-co but this was the first report that an Mo and Fe containing FeMo-co precursor actually accumulates on NifH in an MgATP dependent process. This was also the first report that under certain conditions Mo, as well as Fe, accumulates on NifNE. For many years a protein called gamma was believed to play a critical role in FeMo-co biosynthesis but the gene encoding the protein had not been cloned, precluding detailed characterization of its role in the process. That problem has now been solved and the gene has been isolated and designated *nafY*. The NafY protein is homologous to NifY, it is a dimer both in the presence and absence of FeMo-co and its function is currently proposed to be stabilization of the FeMo-co-deficient protein in a state that can accept FeMo-co.

In R. Thorneley's talk, electron transfer from the [4Fe-4S]$^{1+}$ cluster of the Fe protein to the MoFe protein was examined with emphasis on, different conformational states, the role that MgATP plays in the process and on the MgATP binding site. Three different methods were employed including studies with deoxy ATP where it was shown that removing the H-bond from ribose 2'-OH did not affect the cluster but did influence the rate of electron transfer. In a second series of experiments Asp43, which is H-bonded to H$_2$O coordinated to Mg^{2+} in the complex structure was modified to Glu and Asn, yielding Fe proteins that no longer underwent the normal MgATP induced conformational change and had very little activity. A partially active [4Fe-4Se] form of the Fe protein was also constructed and characterized. The same talk considered the exciting progress that is being made in using stopped-flow FTIR to directly look for intermediates in multi-electron substrate reduction with current efforts focusing on azide reduction. Perhaps the biggest breakthrough in the session came in the talk of D. Dean who used novel approaches to determine where substrates bind on FeMo-co. In earlier studies the authors had used a genetic approach to identify MoFe protein residue Gly[69] as being critical for the interaction of acetylene with the FeMo-co site. In this study, alterations of Gly[69], Val[70], and Arg[96] were characterized, with the latter two residues capping a specific [4Fe-4S] face of FeMo-co, remote from the Mo site. Introducing smaller residues at position 70 allowed efficient reduction of larger substrates than can be efficiently reduced by the native enzyme and led to the elimination of proton reduction during acetylene reduction. These experiments provide compelling evidence in support of the proposed model that all substrates bind to a specific 4Fe-4S face of FeMo-co.

Moving away from conventional nitrogenase, Bauer's presentation considered the fact that *nifHDK* homologs encode proteins with completely different functions. In light-independent chlorophyll/bacteriochlorophyll biosynthesis the double bond of ring D of protochlorophyllide is reduced stereospecifically by an enzyme (DPOR) consisting of three subunits. These three subunits BchL, BchN and BchB are similar to NifH, NifD and NifK respectively. Now for the first time the enzyme has been purified and detailed biochemical characterization has begun. The DPOR catalyzed reaction has the same requirements as the nitrogenase reaction for all three subunits, dithionite and MgATP, however preliminary characterization of the purified BchNB complex and sequence comparisons reveal that it does not have the conserved Cys and His residues required for FeMo-co ligation nor does it appear to contain Mo or V. It appears to resemble NifNE (which has a [4Fe-4S] cluster in place of the P-clusters) more than it does the MoFe protein. Finally, D. Gadkari's talk continued to emphasize biological diversity by considering that one organism, *Streptomyces thermoautotrophicus* fixes atmospheric dinitrogen using an enzyme with no sequence similarity to conventional nitrogenase and with very different biochemical properties. In spite of these differences, new information presented showed that, like conventional nitrogenase the *S. thermoautotrophicus* system was capable of reduction of both azide and cyanide.

LOCALIZATION OF THE NITROGENASE SUBSTRATE BINDING SITE

S.M. Mayer[1], J. Christiansen[1], P.C. Dos Santos[1], W.G. Niehaus[1], P. Benton[2], L.C. Seefeldt[2], D.R. Dean[1]

[1]Department of Biochemistry, Virginia Tech, Blacksburg, Virginia, 24061, USA
[2]Department of Chemistry & Biochemistry, Utah State University, Logan, Utah, 84322, USA

1. Introduction

Nitrogenase is the two-component metalloenzyme that catalyzes the MgATP-dependent reduction of dinitrogen. For the Mo-dependent nitrogenase, the homodimeric Fe protein (component II) delivers electrons to the $\alpha_2\beta_2$ MoFe protein (component I) within which is located two substrate reduction sites (see Christiansen *et al.* 2001; Rees, Howard 2000; and Burgess, Lowe 1996 for recent reviews on the catalytic features of nitrogenase). The path of electron transfer during nitrogenase catalysis is believed to proceed from a [4Fe-4S] cluster contained within the Fe protein to an [8Fe-7S] P cluster bridged between an individual $\alpha\beta$-subunit pair of the MoFe protein and finally to a substrate reduction site located within a MoFe protein α-subunit. One nucleotide-binding site is located within each Fe protein subunit and each of these is separated from an individual substrate reduction site by more than 30 angstroms.

Substrate binding and reduction takes place at a complex metallocluster called FeMo-cofactor. The framework of the FeMo-cofactor is constructed from 4Fe-3S and 3Fe-3S-Mo sub-fragments bridged by three sulfides. An organic constituent, homocitrate, is also attached to the Mo atom through its 2-hydroxy and 2-carboxyl groups. It has not yet been possible to identify the metal-sulfur surface of FeMo-cofactor that provides the substrate interaction site by simple inspection of the FeMo-cofactor structure, nor by kinetic and spectroscopic studies. Nevertheless, a variety of different substrate reduction sites have been proposed. For example, FeMo-cofactor contains three geometrically identical 4Fe-4S faces, and any of these faces is a reasonable candidate for providing a substrate interaction site because each of them contains four coordinately unsaturated Fe atoms. The Mo atom has also been proposed as a site to which substrates might bind during the catalytic cycle.

2. Isolation of an Acetylene-resistant Nitrogenase

In addition to the natural substrate, dinitrogen, nitrogenase is able to reduce a variety of other substrates, including acetylene. The K_m for acetylene reduction (~0.005 atm) is much lower than the K_m for dinitrogen reduction (~0.10 atm). Thus, under conditions of low electron flux, acetylene is a very effective physiological inhibitor of dinitrogen reduction. As an approach to identify the nitrogenase substrate interaction site, we developed a genetic selection for the isolation of mutant strains that are able to effectively reduce dinitrogen, but are significantly impaired in their ability to reduce acetylene. Mutants resistant to acetylene were obtained by plating approximately 10^8 cells of a strain defective in electron flux on a minimal medium containing no fixed nitrogen source and incubating these cells under an ambient atmosphere that also contains 0.025 atm of acetylene. Under these conditions, no immediate growth was observed. However, after about two weeks, a small number of colonies appeared. These colonies were picked and streaked again under a 0.025 acetylene atm to establish that each strain was indeed stably resistant to acetylene. Examination of the growth of these strains without acetylene, showed that they are not impaired in their ability to fix dinitrogen. Subsequent DNA sequence analysis for several independently isolated acetylene-resistant strains revealed that each has the MoFe protein α-subunit 69^{Gly} residue codon substituted by an α-69^{Ser} codon. Site-directed mutagenesis and gene replacement experiments were also used to independently confirm that substitution of α-69^{Gly} by α-69^{Ser} results in the acetylene-resistant phenotype.

An altered nitrogenase α-69^{Ser} MoFe protein was isolated and shown to have the following properties: (1) It is unaffected in its S=3/2 resting state EPR spectrum. (2) It exhibits a 20-fold increase in K_{m} for acetylene reduction but is not affected in its V_{max} for acetylene reduction. (3) It is not significantly impaired in either K_{m} or V_{max} for dinitrogen reduction. (4) Acetylene is converted from a non-competitive to a competitive inhibitor of dinitrogen reduction. (5) The inhibitor CO is converted from a non-competitive inhibitor to a competitive inhibitor of acetylene and dinitrogen reduction. Our interpretation of these results is that there are two functional acetylene-binding sites located within the MoFe protein. One of these is a high-affinity acetylene-binding site and is the one most affected by the α-69^{Ser} substitution. The other is a low-affinity acetylene-binding site that is also the same site as the dinitrogen-binding site. An issue that emerges from this interpretation is the nature of the structural relationship between the high-affinity and low-affinity acetylene-binding sites. We suggest that both of these sites are actually the same with respect to their *general* location within the metal-sulfur surface of FeMo-cofactor but that they are manifested by different redox states of the MoFe protein. The low-affinity acetylene-binding site/state is also proposed to correspond to the dinitrogen-binding site/state.

Where is the actual substrate interaction site located? Inspection of the MoFe protein crystal structure shows that the α-69^{Gly} residue is located on a short helix that extends from the P-cluster coordinating residue α-62^{Cys} to α-70^{Val}. The α-70^{Val} side chain approaches a specific 4Fe-4S face of FeMo-cofactor. Furthermore, the backbone carbonyl oxygen of the α-69^{Gly} residue is located within hydrogen bonding distance to a guanido NH-group of α-96^{Arg}, which also approaches the same 4Fe-4S face. The α-96^{Arg} residue is likewise connected to the P cluster through a short helix extending from the P cluster coordinating α-88^{Cys} residue. Based on these considerations we suggest that redox-dependent structural rearrangements known to occur within the P cluster could result in the communication along the connecting helices and the subsequent movement of either or both α-70^{Val} and α-96^{Arg} during catalysis. Thus, our interpretation of the physiological and kinetic consequences of the α-69^{Ser} substitution is that the redox-dependent movement of either or both α-70^{Val} and α-96^{Arg} is affected.

3. Reduction of Short-chain Alkynes by Altered Nitrogenases

If our model that the 4Fe-4S face of FeMo-cofactor capped by α-70^{Val} and α-96^{Arg} provides the substrate interaction site is correct, then the side chains of either or both of these residues could have an effect on the ability of certain substrates to interact with the substrate binding site. To test this possibility we substituted the α-70^{Val} residue by α-70^{Ala} and by α-70^{Gly}. The effect of these substitutions on the ability of the altered MoFe proteins to catalyze the reduction of short-chain alkynes that are not effectively reduced by the wild type enzyme was then examined. Our initial approach towards this end was to determine if the addition of propargyl alcohol (C_3H_4O) to the growth medium affects the diazotrophic growth of the strain having the α-70^{Ala} substitution. We found that this strain grows well under normal diazotrophic conditions but that addition of 6 mM propargyl alcohol to the growth medium completely inhibits diazotrophic growth. Propargyl alcohol had no effect on diazotrophic growth of the wild type strain and did not affect growth of the α-70^{Ala}-substituted strain if a source of fixed nitrogen was also added to the growth medium. Kinetic characterization of the α-70^{Ala} MoFe protein revealed that propargyl alcohol is a competitive inhibitor of acetylene reduction having a K_{i} of approximately 2 mM.

Because of the difficulty in measuring the product of the reduction of short-chain alkyne alcohols, we turned to an examination of the ability of the α-70^{Ala}- and α-70^{Gly}-substituted MoFe proteins to reduce propyne and butyne. The reduction products of both of these substrates can be easily and accurately quantified by gas chromatography. For the wild type enzyme, propyne reduction is extremely poor and butyne reduction, if it occurs, is below levels that we are able to accurately detect. In contrast, both the α-70^{Ala}- and α-70^{Gly}-substituted MoFe proteins are able to reduce both propyne and butyne, but at different levels depending on the respective substituted residue. For example, both the α-70^{Ala}- and α-70^{Gly}-substituted MoFe proteins are able to reduce both propyne and butyne at readily detectable levels but reduction of these alkynes to yield the corresponding alkene is more effective for the α-70^{Gly} MoFe protein than for the α-70^{Ala} MoFe protein. Also, for both substituted MoFe proteins, propyne reduction is much more effective than butyne reduction. These results establish a direct correlation between a decrease in side-chain length at the α-70 residue position with the capability of the MoFe protein to reduce short-chain alkynes of increasing length. Thus, reduction of short-chain alkynes does occur at the 4Fe-4S face of FeMo-cofactor we have targeted.

Another interesting feature of the α-70^{Ala}- and α-70^{Gly}-substituted MoFe proteins is that acetylene is able to effectively suppress proton reduction catalyzed by these altered MoFe proteins ($\geq$98% suppression by 0.10 atm acetylene). This level of H_2 suppression has previously been predicted on the basis of the extrapolation of kinetic data that suggest acetylene should be able to completely suppress proton reduction. However, it has not been possible to prove this hypothesis experimentally using wild type nitrogenase because acetylene concentrations greater that ~0.20 atm have an adverse effect on electron flux. Our interpretation of the result that 0.1 atm acetylene virtually eliminates proton reduction catalyzed by the α-70^{Ala}- and α-70^{Gly}-substituted MoFe proteins is that proton reduction and acetylene reduction occur at the same 4Fe-4S face. In other words, constraints on acetylene binding in these altered MoFe proteins is sufficiently relaxed so that proton reduction is nearly eliminated when $\geq$ 0.10 atm acetylene is present.

4. The α-96^{Arg} Side-chain Moves to Accommodate Substrate Interaction

A caveat to our suggestion for the substrate interaction site, as well as with every other model that has been proposed, is that there is not sufficient space to accommodate substrate interaction unless certain amino acid side-chains move away from the FeMo-cofactor during catalysis. We have already suggested that the side-chains of either or both α-70^{Val} and α-96^{Arg} move during catalysis. Once again, inspection of the structural model provides clues about how such movement might occur. Close examination of the structural model reveals that there is little room to permit movement of α-70^{Val} unless there are major structural rearrangements within the MoFe protein during catalysis, a possibility we consider unlikely. However, there is sufficient room to accommodate movement of the α-96^{Arg} side-chain without the requirement of other severe structural perturbations. Based on this observation, we are considering a model where the α-96^{Arg} side-chain becomes repositioned during catalysis when compared to its location in the resting state of the MoFe protein. An attractive possibility is that the α-96^{Arg} side-chain moves back and forth during turnover – first to expose a substrate binding site and then, perhaps, to trap bound substrate at an iron-sulfur surface of FeMo-cofactor. Such movement of the α-96^{Arg} side-chain could also serve as a shuttle to facilitate proton transfer. The proposed dynamic movement of the α-96^{Arg} side-chain during catalysis is also in line with the observation that this side-chain occupies three slightly different conformations in each of the crystallographically determined structures for MoFe protein from *Azotobacter vinelandii*, *Clostridium pasteurianum*, and *Klebsiella pneumoniae*.

It does not appear that any substrate or inhibitor can interact with FeMo-cofactor in the as-isolated resting state of the MoFe protein. The basis for this conclusion is that the S=3/2 EPR signal observed for the as-isolated MoFe protein, which originates from FeMo-cofactor, is not perturbed by the addition of any substrate or inhibitor. The usual explanation offered for this situation is that electrons must first be delivered to the active site in order to prime it for substrate interaction and reduction. Although this is almost certainly the case for many substrates and inhibitors, there is no *prima faciae* chemical reason why all substrates and inhibitors should be unable to interact with

FeMo-cofactor contained in the MoFe protein in its resting state, unless these molecules are denied access to the FeMo-cofactor. Indeed, it has been shown that addition of cyanide to isolated FeMo-cofactor, which contains an S=3/2 EPR spectrum very similar to that of the protein-bound species, does result in the appearance of a new S=3/2 EPR signal.

As an approach to investigate whether or not the α-96^{Arg} side-chain might move during catalysis, we placed a leucine substitution at this position and asked if the altered MoFe protein is now able to interact with substrates in the resting state under non-turnover conditions. The logic of the α-96^{Leu} substitution was to shorten the side-chain at this position as a way to expose the substrate interaction site in the resting state. Incubation of acetylene with the α-96^{Leu} MoFe protein in the as-isolated resting state decreased the intensity of the normal S=3/2 FeMo-cofactor signal with appearance of a new EPR signal having inflections at g=4.50 and 3.50. Similarly, incubation of cyanide with the α-96^{Leu} MoFe protein also decreased the normal FeMo-cofactor signal with concomitant appearance of a new EPR signal having an inflection at g=4.06. No such EPR changes were elicited from the wild type MoFe protein under the same conditions or for the α-96^{Leu} MoFe protein when incubated in the presence of dinitrogen. These results support a model where effective interaction of dinitrogen with FeMo-cofactor occurs as a consequence of *both* increased reactivity and *accessibility* of FeMo-cofactor under turnover conditions. For example, the reason that dinitrogen does not interact with the α-96^{Leu} MoFe protein in the resting state, but acetylene does, is that a lower redox state is required for dinitrogen binding.

5. All Substrates are Reduced at the Same 4Fe-4S Face of FeMo-cofactor

We have now suggested that both the high-affinity and low-affinity acetylene-binding sites, as well as the proton reduction site(s), are located at the same 4Fe-4S face of FeMo-cofactor approached by the α-70^{Val} and α-96^{Arg} residue side-chains. Where is the dinitrogen-binding site? If all substrates, including dinitrogen, interact with the same 4Fe-4S face of FeMo-cofactor, it should, in principle, be possible to isolate an altered MoFe protein having a single amino acid substitution which severely impairs the access of all substrates to the reduction site. However, it is experimentally difficult to distinguish effects on substrate reduction that arise from disruption of electron or proton delivery to the active site (flux effects) from alterations that specifically affect the accessibility of substrates to the active site. As a way to circumvent this problem we reasoned that, because protons are so small, it might be possible to isolate an altered MoFe protein that is not altered in proton reduction but is ineffective in the reduction of larger nitrogenase substrates. Namely, an altered MoFe protein that is not altered in its rate of proton reduction is, by definition, not affected in flux.

The next challenge involved selecting the position and choice of amino acid substitution. If there is only one major substrate entry pathway to the 4Fe-4S face we have targeted, then examination of the MoFe protein structure reveals that the most reasonable pathway for access to this site is located between the short arm of homocitrate and the α-69^{Gly} residue. We considered this an attractive possibility because substitution of α-69^{Ser} at this position has a dramatic affect on the ability of the MoFe protein to reduce acetylene. It therefore seemed reasonable to expect that substitution of a residue having an even bulkier side-chain at this position could affect the accessibility of all substrates, with the possible exception of protons. Although we have now placed a number of different amino acid substitutions at this position, the most instructive results obtained so far are with the α-69^{Val} MoFe protein.

Kinetic characterization of the α-69^{Val} MoFe reveals that it is able to reduce protons at the same rate as the wild type protein and, therefore, is not perturbed with respect to electron flux. However, the α-69^{Val} MoFe protein reduces acetylene and dinitrogen at rates of only about 2% when compared to the wild type MoFe protein. In separate experiments, it was also found that the interaction of CO with the α-69^{Val} MoFe protein under turnover conditions is severely impaired when compared to the wild type protein. We, therefore, conclude that most, and perhaps all, nitrogenase substrates and inhibitors interact with the same 4Fe-4S face of FeMo-cofactor and that there is a single major access pathway to that face.

6. Conclusions

Based on the worked described here, we make the following conclusions. First, molybdenum does not provide the site for substrate binding. Second, the 4Fe-4S face of FeMo-cofactor capped by α-70^{Val} and α-96^{Arg} provides the substrate-binding site. Third, all substrates bind at the same 4Fe-4S face of FeMo-cofactor. Fourth, all substrates, with the possible exception of protons, gain access to the substrate-binding site primarily through the same pathway. Fifth, the side-chain of α-96^{Arg} moves during catalysis in order to make the binding site available to substrate. The α-96^{Arg} side-chain might also serve as a proton shuttle during catalysis. Sixth, although we propose that all substrates access the same metal-sulfur surface during catalysis, it is expected that different substrates will interact with this surface in different ways. Seventh, it also seems reasonable to anticipate that an individual substrate might interact with the metal-sulfur surface of FeMo-cofactor in different geometric orientations depending upon the particular redox-state of the enzyme.

We believe that isolation of an altered MoFe protein that interacts with certain substrates in the resting state now makes it possible to crystallographically determine exactly where and how these substrates initially interact with FeMo-cofactor. This approach, as well as the continued mapping of the active site environment by using amino acid substitutions and substrate analogs, will remain the emphasis of our studies aimed at determining the nitrogenase catalytic mechanism.

7. References

Burgess BK, Lowe DJ (1996) Chem. Rev. 96, 2983-3011

Christiansen JC *et al.* (2001) Ann. Rev. Plant Physiol. Plant Mol. Biol. 52, 269-295

Rees DC, Howard JB (2000) Curr. Opin. Chem. Biol. 4, 559-566

BIOSYNTHESIS OF THE IRON-MOLYBDENUM COFACTOR OF NITROGENASE: ROLES OF DINITROGENASE REDUCTASE AND NAFY

P.W. Ludden, P. Rangaraj, L. Rubio, C. Rüttimann-Johnson, D. Dyer, J. Cheng, V.K. Shah

Dept of Biochemistry, University of Wisconsin-Madison, Madison, WI 53706, USA

1. Introduction

The iron molybdenum cofactor (FeMo-co) constitutes the active site of the *nif*-encoded, molybdenum nitrogenase and its analogs, FeV-co and FeFe-co comprise the actives sites of the *vnf*-encoded, vanadium nitrogenase and the *anf*-encoded, iron only nitrogenase, respectively. While the structure for FeMo-co is known (Figure 1), FeV-co and FeFe-co are thought to have very similar structures based on their ability to substitute for FeMo-co in the *nifDK*-encoded nitrogenase protein. Furthermore, all three systems share a requirement for functional *nifB* gene; NifB generates the Fe and S precursor to FeMo-co.

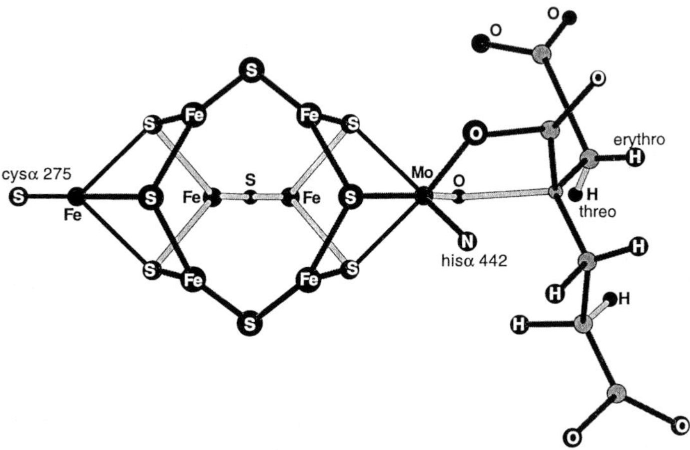

Figure 1. The structure of FeMo-co (adapted from Kim, Rees 1992).

An *in vitro* FeMo-co synthesis system has been developed and each of the components of that system is now available in a highly purified form. The system includes NifB-co (the metabolic product of NifB), NifNE, NifX, NifH, NafY (aka Gamma), homocitrate (the metabolic product of NifV), MgATP, reductant in the form of sodium dithionite, molybdate, and a FeMo-co-deficient form of dinitrogenase (apodinitrogenase in its $\alpha_2\beta_2\gamma_2$ form). In this system, FeMo-co is synthesized and inserted into apodinitrogenase and is capable of substrate reduction. To date, it has not been possible to synthesize FeV-co *in vitro* starting with a source of V that has not been processed biologically. However, it is possible to convert a protein bound form of V into a cofactor that can be inserted into *nif*-encoded apodinitrogenase and function in acetylene reduction (Ruttimann-Johnson *et al.* 2001).

The two main points of this chapter are: (i) that NifH (dinitrogenase reductase) plays a catalytic role in FeMo-co synthesis; and (ii) that there is a family of proteins (the X-proteins) whose members are able to bind FeMo-co/FeV-co and/or their precursors.

2. Dinitrogenase Reductase is a "Moonlighting Protein"

Moonlighting proteins are proteins that have more than one role in the cell (Jeffery 1999). The roles can be related, as they are in the case of dinitrogenase reductase, or seemingly unrelated, as in the case of phosphoglucose isomerase, which serves as both a glycolytic enzyme and an autocrine factor (Jeffery *et al.* 2000). Dinitrogenase reductase is a supermoonlighting protein in that it has at least three essential roles in nitrogen fixation (Figure 2).

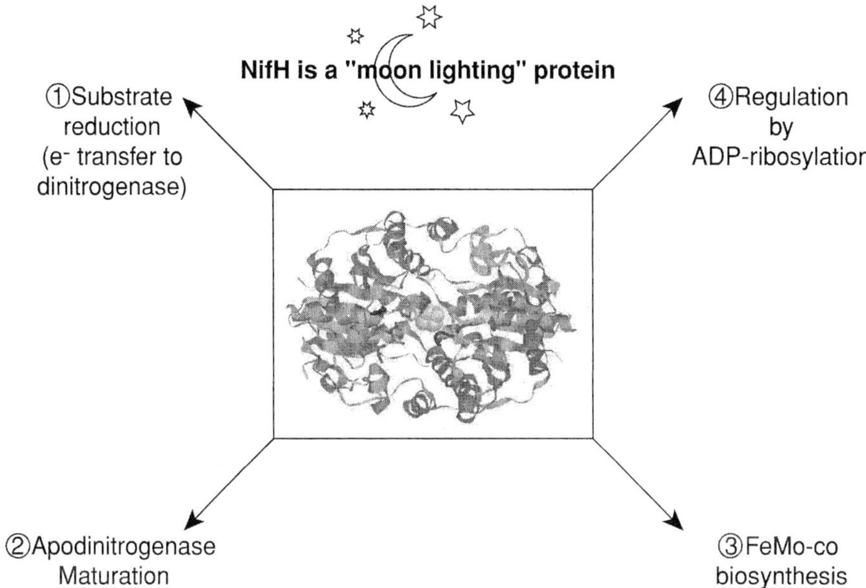

Figure 2. NifH is a "moon lighting" protein

First, dinitrogenase reductase is the unique and specific electron donor to dinitrogenase and was first identified for this role (Bulen, LeComte 1966). No other protein is known to be able to reduce dinitrogenase to a state able to reduce substrates. As the electron donor, it is the target for regulation of this process in some organisms. Dinitrogenase reductase from *Rhodospirillum rubrum* is regulated by reversible ADP-ribosylation at an arginine residue near the Fe_4S_4 cluster that bridges the two subunits (Pope *et al.* 1985).

Second, dinitrogenase reductase is a biosynthetic enzyme. Filler and Smith first noted that non-polar *nifH* mutants of *Klebsiella pneumoniae* accumulated FeMo-co-deficient dinitrogenase (Filler *et al.* 1986). Subsequent experiments showed that dinitrogenase reductase is required for *in vitro* synthesis and that [55]Fe from NifB-co accumulated on dinitrogenase reductase during *in vitro* FeMo-co synthesis (Rangaraj *et al.* 2001). The accumulation of [55]Fe from [55]Fe-NifB-co required NifNE and MgATP. In this presentation, it is demonstrated that [99]Mo can also accumulate on dinitrogenase reductase during FeMo-co synthesis and that this accumulation is NifB-co, NifNE and MgATP-dependent. VnfH can substitute for NifH and can accumulate [99]Mo. The role of dinitrogenase reductase in FeMo-co synthesis is not to provide reductant, because a form of dinitrogenase reductase that is devoid of Fe_4S_4 clusters and the ability to transfer electrons to dinitrogenase is fully functional in FeMo-co synthesis (Rangaraj *et al.* 1997). The amount of [99]Mo accumulated on dinitrogenase reductase is significant, approximately 0.5 moles of Mo per mole of dimeric protein. It is unlikely that Mo is accumulating at the site observed by Georgiadis and Rees in the first crystal structure of dinitrogenase from *Azotobacter vinelandii* (Georgiadis *et al.* 1992) because that molecule of MoO_4^{2-} occupied the binding site for the terminal phosphoryl group of ATP. We observe ^{99}Mo accumulation only in the presence of MgATP when that site would be expected to be occupied.

Dinitrogenase reductase is proposed to accept Fe and S from NifNE and add Mo to the complex. Neither homocitrate nor NifX are required for the accumulation of ^{99}Mo on dinitrogenase reductase, so dinitrogenase reductase is proposed to play its role before the addition of the organic acid ligand to Mo. While dinitrogenase reductase is proposed to serve as the site of entry of the heterometal of the FeMo-co, and by analogy, to FeV-co and FeFe-co, it does not appear to be the site of specificity for heterometal selection. For example, VnfH, the dinitrogenase reductase from the vanadium system, will substitute for NifH and accumulate ^{99}Mo during *in vitro* FeMo-co synthesis (Figure 3).

Finally, dinitrogenase reductase plays an as yet undefined role in FeMo-co insertion that is distinct from its role in FeMo-co synthesis (Allen *et al.* 1993; Robinson *et al.* 1987). Dinitrogenase from cells that lack dinitrogenase reductase accumulate FeMo-co-deficient apodinitrogenase and are unable to insert FeMo-co. Once the FeMo-co-deficient apodinitrogenase has been treated with dinitrogenase reductase, it gains the ability to accept FeMo-co and become active. It also gains the ability to bind NafY (gamma protein); NafY is also a FeMo-co-binding protein (see below). *A. vinelandii* cells with lesions in FeMo-co-biosynthetic genes other than *nifH* accumulate a form of apodinitrogenase with NafY tightly bound. *A. vinelandii nifH* cells accumulate an NafY-free form of dinitrogenase reductase. Now that *nafY* mutants are available, it is known that NafY is not essential for FeMo-co insertion *in vivo*; apodinitrogenase from, for example, *nifBnafY* mutants can be activated by the addition of purified FeMo-co in the absence of NafY. Once again, Fe_4S_4-deficient dinitrogenase reductase is able to perform its role in dinitrogenase maturation required for FeMo-co insertion, thus that role does not require electron transfer to or from dinitrogenase reductase (Rangaraj *et al.* 1997).

It is interesting that the *nifH* gene product plays multiple roles in the overall process of nitrogen fixation and that it is such an amazingly conserved protein across nature. Perhaps the high degree of sequence identity for this protein across the many genera of bacteria and the arachea reflects that fact that it has several domains with different function, each of which must be tightly conserved.

3. NafY is a Member of a Family of FeX-co-Binding Proteins

FeMo-co-deficient dinitrogenase from *A. vinelandii* was discovered to have a third type of subunit (Paustian *et al.* 1990). This protein was termed the "gamma" subunit of apodinitrogenase and it was shown to bind FeMo-co (Homer *et al.* 1995). The gene for this protein has been obtained and designated *nafY* (Nitrogenase Accessory Factor). The deduced sequence for NafY shows that it belongs to a small group of proteins that are known to bind FeMo-co/FeV-co or their precursors. This family of proteins includes NafY, VnfX, NifX, NifY and VnfY (Rüttimann-Johnson *et al.* 2000, 2001). The *nafY* gene is immediately downstream of *rnfGEH* in *A. vinelandii*, but is not co-transcribed with these genes. NafY has been overexpressed and shown to exist as a dimer in both free and FeMo-co-bound forms. NafY binds a single molecule of FeMo-co per dimer with a Kd of 1 μM. The ability to fix nitrogen by *nafY* mutants is affected only at high temperatures and our current model is that NafY plays a role in stabilizing apodinitrogenase. *nafYnifB* mutants accumulate normal levels of apodinitrogenase, but the apodinitrogenase in extracts of these mutants can only be activated to 50% of the level achieved in a *nifB* mutant. The role of the FeMo-co binding site is unknown. NafY will also bind the FeMo-co precursor, NifB-co. Crystals of NafY have been obtained.

Accumulation of ^{99}Mo on VnfH

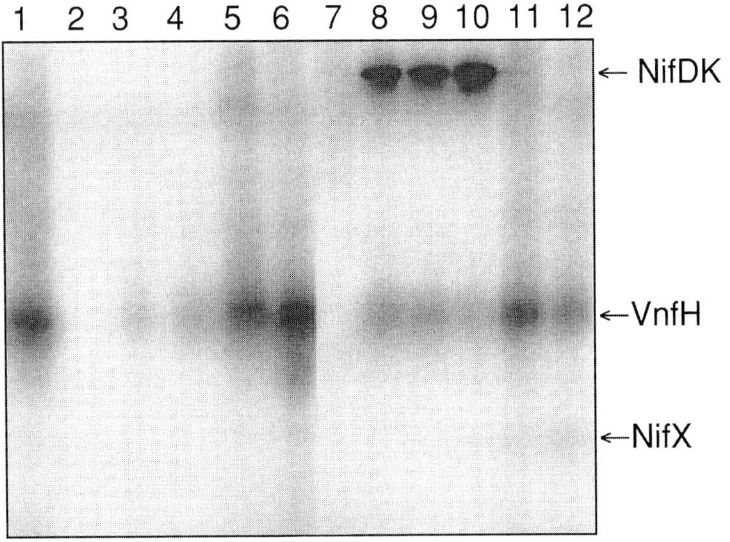

Lanes:
1. Complete system
2. − ATP
3. − NifNE
4. − NifB-co
5. − homocitrate
6. + VnfX (3.0 nmol)
7. − VnfH
8. + Nif-Apodinitrogenase (0.4nmols)
9. + VnfX (3.0 nmol), + Nif-Apodinitrogenase (0.4nmols)
10. + NifX (3.0 nmols) + Nif-Apodinitrogenase (0.4nmols)
11. + NifX (3.0 nmols)
12. + NifX (6.0 nmols)

Phosphorimage of an anoxic native gel of *in vitro* reactions with purified components. The complete system contained 25 mM Tris HCl pH 7.5, MgATP, sodium dithionite, homocitrate, NifB-co, NifNE (0.036 nmols), VnfH (1 nmol) and ^{99}Mo in a final volume of 0.5 ml.

4. Acknowledgements

The authors wish to thank Drs Gary Roberts and Dennis Dean for stimulating discussions. Work from the authors' laboratory was supported by NIGMS grant 35332.

5. References

Allen RM *et al.* (1993) J. Biol. Chem. 268, 23670-23674

Bulen WA, LeComte JR (1966) Proc. Natl. Acad. Sci. USA 56, 979-86

Filler WA *et al.* (1986) Eur. J. Biochem. 160, 371-377

Georgiadis MM *et al.* (1992) Science 257, 1653-1659

Homer MJ *et al.* (1995) J. Biol. Chem. 270, 24745-24752

Jeffery C (1999) Trends in Biochem. Sci. 24, 8-11

Jeffery C *et al.* (2000) Biochem. 39, 955-964

Kim J, Rees DC (1992) Science 257, 1677-1682

Paustian TD *et al.* (1990) Biochem. 29(14), 3515-3522

Pope MR *et al.* (1985) Proc. Natl. Acad. Sci. USA 82, 3173-3177

Rangaraj P *et al.* (2001) J. Biol. Chem. 276, 15968-15974

Rangaraj P *et al.* (1997) Proc. Natl. Acad. Sci. USA 94, 11250-11255

Robinson AC *et al.* (1987) J. Biol. Chem. 262, 14327-14332

Rüttimann-Johnson C *et al.* (2001) J. Biol. Chem. 276, 4522-4526

Rüttimann-Johnson C *et al.* (2000) In Pedrosa FO, Hungria M, Yates MG, Newton WE (ed) Nitrogen Fixation: From Molecules To Crop Productivity, pp. 35-36, Kluwer Academic Publishers, Dordrecht

NITROGENASE MECHANISM: CAN OLD LAGS TEACH US NEW TRICKS?

R.N.F. Thorneley[1], H. Angove[1], G.A. Ashby[1], M.C. Durrant[1], S.A. Fairhurst[1], S.J. George[1], P.C. Hallenbeck[2], A. Sinclair[1], J.D. Tolland[1]

[1]Department of Biol. Chem., John Innes Centre, Norwich, NR4 7UH, UK
[2]Départ. de Micro. et Immunol., Université de Montréal, Montréal, H3C 3J7 Canada

1. Introduction

Our understanding of the mechanism of nitrogenase is still informed by the concepts, experimentally determined rate constants and simulations of Lowe-Thorneley (1984). The Fe-protein cycle describes the minimum number of partial reactions necessary to effect the transfer of an electron from the Fe protein to the MoFe protein coupled to the hydrolysis of two molecules of MgATP. A major challenge is to understand at atomic resolution the energy transducing reactions (electron transfer, hydrolysis of MgATP and associated conformation changes) that only occur within the transient complex formed between the Fe and MoFe proteins. X-ray crystallography (Chiu *et al.* 2001 and references therein) and low angle X-ray scattering (Grossman *et al.* 1997) using AlF_4^-, BeF_3^- and the $\Delta127$-Fe-protein to trap out different conformations of this complex have contributed greatly to our understanding of these events and form the basis for the interpretation of our spectroscopic and kinetic data. We have probed the structural determinants of the kinetic and spectroscopic profiles of reactions occurring in the Fe-protein cycle in three ways: (i) by use of the ATP analog 2'-deoxyATP; (ii) by site-directed mutagenesis of residues that interact with MgATP; and (iii) by replacement of the 4Fe-4S cluster by a 4Fe-4Se cluster in the Fe protein.

In addition, stopped-flow Fourier transform infrared spectroscopy has been used to monitor azide reduction and carbon monoxide inhibition in both the pre-steady state and steady state. These data are beginning to contribute to our understanding of the chemistry that occurs at the FeMo-cofactor consequent on the eight sequential electron and proton transfers that comprise the MoFe-protein cycle of the Lowe-Thorneley model.

2. The ATP Analog, 2'-deoxyATP

The 2'-hydroxy group of the ribose element of ADP is hydrogen bonded to Glu221 in the Fe protein (Figure 1). This hydrogen bond can be easily removed by using 2'-deoxyATP (or ADP) in which the hydroxy-group is replaced by a hydrogen atom. We have studied a number of the partial reactions occurring in the Fe-protein cycle of *Klebsiella pneumoniae* (Kp) nitrogenase using 2'-deoxyATP/ADP and compared the kinetics and EPR data with those obtained with ATP and ADP. The increase in the rate of Fe-chelation from the Fe protein (Kp2) by bathophenanthroline induced by 2'-deoxyATP binding is very similar to that observed with ATP. 2'-DeoxyATP, like ATP, induces the rhombic to axial change in the EPR signal of Kp2. The reduction by SO_2^- of $Kp2ox(MgADP)_2$ and $Kp2ox(Mg2'\text{-}deoxyADP)_2$ occurs at essentially the same rate ($k = 4$ and 5×10^6 $M^{-1}s^{-1}$, respectively).

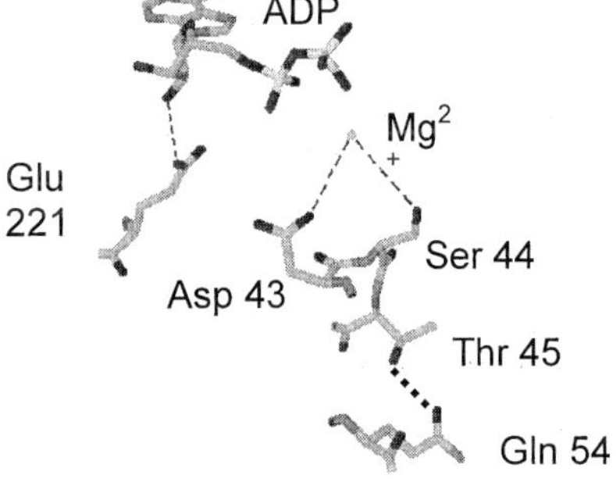

Figure 1. MgADP binding site of Av2 (cartoon by D. Lawson using coordinates of Jang *et al.* 2000).

We conclude that with free Fe protein, the hydrogen bond from the 2'-hydroxy group of the ribose to the carboxylate of Glu221 does not significantly determine the conformation of the protein in the vicinity of the 4Fe-4S cluster. The overlay of the x-ray structures of the *Azotobacter vinelandii* Fe protein (Av2) and Av2(MgADP)$_2$ (Jang *et al.* 2000) shows little or no movement of this hydrogen bond when free Av2 undergoes the ADP-induced conformation change. Reactions occurring within the Fe-MoFe protein complex are, however, significantly affected by the presence of this hydrogen bond as is shown by the kinetic data summarized in Table 1.

Table 1. Comparison of kinetic data for ATP and 2'-deoxyATP with Kp nitrogenase (23°C)

	ATP	2'-DeoxyATP
Electron transfer (k_2, s^{-1})	150	21
K_d nucleotide (mM)	0.43	1.3
Complex dissociation (k_{-3}, s^{-1})	6.4	0.8
Sp.Act. (H$_2$ nmole/mg/min)	842	112
ATP/2e$^-$	4.4	5.4
Reductant independent ATPase (s^{-1})	0.6	1.2

Replacement of ATP by 2'-deoxyATP decreases the first-order rate constants for electron transfer and complex dissociation by a factor of eight. Complex dissociation is still rate-limiting because the specific activity for proton reduction is also decreased by a factor of eight. The affinity for Mg2'-deoxyATP is about four-fold less than for MgATP calculated from the hyperbolic dependence of the electron-transfer rate (k_2) on nucleotide concentration. The rate of MgATP hydrolysis must also be decreased by a factor close to eight since the ATP/2e$^-$ ratio of 5.4 for 2'-deoxyATP is only 25% higher than that for MgATP. Indeed, this difference can be entirely accounted for by an increased rate of reductant-independent ATP hydrolysis catalyzed by the complex of oxidized Kp2 with Kp1, which is about two-fold higher for Mg2'-deoxyATP than for MgATP. These data show that electron transfer, ATP hydrolysis and complex dissociation are co-modulated by changes in Fe protein structure consequent on loss of the hydrogen bond between Glu221 and the 2'-hydroxyl group of ribose element of ATP. Inspection of the overlay of the x-ray structures of Av2(MgADP)$_2$ and Av2(MgADP.AlF$_4$)$_2$ presented by Jang *et al.* (2000) shows that this hydrogen bond is broken in the latter structure. We, therefore, suggest that, since ATP hydrolysis only occurs when the Fe protein is bound to the MoFe protein, this hydrogen bond is sensitive to and contributes to the free energy change on docking the Fe protein onto the MoFe protein prior to electron transfer.

3. Asp43Glu, Asp43Asn and Thr45Ser: Switch 1 Region Mutants of Av2

Figure 1 shows that Asp43 and Ser44 are hydrogen bonded to water molecules coordinated to the Mg^{2+} ion that, in turn, is coordinated to the terminal phosphate group of ADP. Thr45 is hydrogen bonded to Gln54. All three of these residues are part of the signal transduction network in Av2 termed "Switch 1" by analogy to P21ras, which has high structural homology in this region (Jang *et al.* 2000). We have constructed and characterized three Av2 variants, Asp43Glu, Asp43Asn and Thr45Ser.

EPR spectroscopy showed that the Av2-Asp43Glu and Av2-Asp43Asn mutants do not undergo the MgATP-induced rhombic to axial change in form of the g = 1.95 signal, whereas the Av2-Thre45Ser clearly does. An increase in the initial rate of iron chelation by bathophenanthroline is another criterion by which to estimate the extent of MgATP-induced conformation changes in the Fe protein. The data in Table 2 confirm that the Av2-Asp43Glu and Av2-Asp43Asn mutants do not undergo the MgATP-induced conformation change whereas the Av2-Thr45Ser does. The more than three-fold increase in rate for Av2-Thr45Ser (760) relative to wild-type Av2 (220) is quite remarkable and indicates significant structural changes in the protein environment of the 4Fe-4S cluster. However, the effect of these mutations appears to be much less with respect to the reactivity

Table 2. Increase in initial rates of Fe chelation by bathophenanthroline induced by MgATP

Av2-native	220
Av2-Asp43Glu	8
Av2-Asp43Asn	7
Av2-Thr45Ser	760

Table 3. Rate of reduction of oxidized Av2 with MgADP bound by SO_2^- (k_4)

Av2-native	4×10^6 $M^{-1}s^{-1}$
Av2-Asp43Glu	6×10^6 $M^{-1}s^{-1}$
Av2-Asp43Asn	8×10^6 $M^{-1}s^{-1}$
Av2-Thr45Ser	4×10^6 $M^{-1}s^{-1}$

of the oxidized proteins. The rates of reduction of native $Av2_{ox}(MgADP)_2$ and the mutated forms by dithionite (SO_2^-) are all within a factor of two of each other (Table 3). These clearly show the differential effect of these mutations depending on the oxidation levels of Av2 and on whether MgADP or MgATP is bound. This is important since electron transfer and ATP cleavage are early events in the series of reactions occurring within the Fe-MoFe protein complex, and subsequent energy-transducing reactions and rate-limiting complex dissociation, must involve oxidized Fe protein with 2MgADP bound. Native Av2, Av2-Asp43Glu, Av2-Asp43Asn and Av2-Thr45Ser have specific activities of 2360, 18.5 and 1100 nmole H_2 (min mg)$^{-1}$ and primary electron-transfer rates to Av1 of 180, 5.4, not detectable, and 250 s^{-1}, respectively. The significantly higher rate of electron transfer (extrapolated value of k_{obs} at infinite MgATP concentration) for Av2-Thr45Ser (250 s^{-1}) compared to native Av2 (180 s^{-1}) is intriguing and consistent with its increased rate of Fe-chelation induced by MgATP. It appears as though this mutant undergoes a somewhat larger MgATP-induced conformation change than native Av2. However, the rate of complex dissociation involving the oxidized MgADP-bound form is slower (k_{-3} = 3.7 s^{-1}) compared to that of native $Av2_{ox}(MgADP)_2$ (k_{-3} = 6.5 s^{-1}). This results in the two-fold decrease in specific activity for this mutant.

A prediction of the Lowe-Thorneley model is that the Av2-Thr45Ser mutant should be more efficient than native Av2 with respect to the percentage of electron flux going to hydrogen evolution under one atmosphere of nitrogen but be less efficient with respect to the ATP/2e$^-$ ratio. Both of these predictions are based on an increased steady-state concentration of the $Av2ox(MgADP)_2Av1$ complex due to the slower rate of complex dissociation. This suppresses hydrogen evolution but increases the reductant-independent ATPase contribution. The percentage of electron flux into hydrogen evolution under nitrogen for Av2-Thr45Ser and native Av2 (27 ± 2.0 and 32 ± 1.9%, respectively), the ATP/2e ratio under argon (6.1:1 and 4.5:1, respectively), and the k_{cat} for reductant-independent ATPase activity (1.2 and 0.65, respectively) provide some evidence in support of this prediction. Evolution has presumably selected a Thr at position 45 in order to maximize the total rate of ammonia production and minimize the amount of ATP hydrolyzed at the expense of a slightly less efficient enzyme with respect to hydrogen evolution.

4. Replacement of the 4Fe-4S Cluster by a 4Fe-4Se Cluster in the Fe Protein (Kp2)

The 4Fe-4S cluster of Kp2 can be removed, after the addition of MgATP, by chelation of the iron with bathophenanthroline, followed by gel filtration on a Sephadex G-50 under strictly anaerobic conditions in a glove box. The apo-Kp2 can then be reconstituted with a 4Fe-4Se cluster using seleno-D,L-cystine, ferrous ion, dithiothreitol and cysteine-desulphurase (Nif S, supplied by Dr D.R. Dean). The Se-Kp2 protein was purified on DEAE cellulose in Tris-buffer (50 mM, pH 7.4) and eluted with 0.4 M NaCl, 2mM dithionite. The protein contains equivalent amounts of Fe and Se in the range 3 to 5 g atoms per mole of Kp2.

Figure 2 shows the EPR spectra of Se-Kp2 with features at g = 5.2 (S = 7/2) and g = 1.95 (S = 1/2) that are characteristic of 4Fe-4Se clusters (Gaillard *et al.* 1986; Yu *et al.* 1991). Modeling of the 4Fe-4Se cluster into the native Av2 x-ray crystal structure shows that replacing S by Se should barely change the Fe-Fe distance [2.76 Å (S), 2.78 Å (Se)]. The difference between the Fe-S (2.29 Å) and Fe-Se (2.42 Å) bond lengths causes the 4 Se atoms to move outwards with the Fe atoms fixed by their protein cysteinyl sulfur ligands. EXAFS analysis of Se-Kp2 is in progress (with G. George and R. Prince).

Se-Kp2 has a specific activity about 15% of wild type but with an ATP/2e = 17 (cf. 4.4 for native-Av2). The rate of primary electron transfer is decreased to 38 s^{-1} vs. 180 s^{-1} for native Av2, which is associated with a three-fold tighter binding of MgATP (Kd = 0.15 mM vs. 0.44 mM for native Av2).

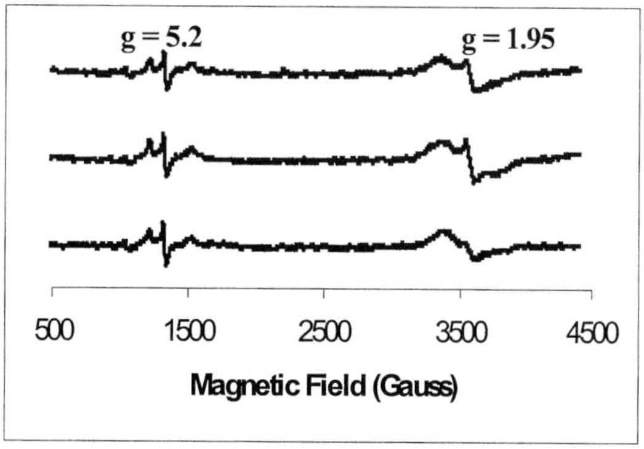

g = 5.2 g = 1.95

500 1500 2500 3500 4500

Magnetic Field (Gauss)

Figure 2. EPR spectra of Se-Kp2 (bottom) with MgADP (middle) and MgATP (top) at 18 K, 10 mW with buffer blank subtracted.

5. Time-Resolved Fourier Transform Infrared Spectroscopy of CO Inhibition of Azide Reduction

The Lowe-Thorneley model predicts a lag phase of ca. 1 s for the eight-electron reduction of azide to ammonia. The reaction can be studied by stopped-flow FTIR using the apparatus previously described (George *et al.* 1997, 2000) by monitoring the loss of azide at 2050 cm^{-1}. A time course is shown (Figure 3), which clearly exhibits the predicted lag phase and an extrapolated, rapid initial loss of 8 µM azide that occurs within the first 200 ms. This amount of azide is stoichiometric with the Mo concentration and indicates that each FeMo-cofactor binds and reduces one equivalent of azide. The reduction of bound azide by eight electrons has to be completed before the second azide binds and a steady state is established at times longer than ca. 1 s. We have also studied the inhibition of this reaction by carbon monoxide under conditions of sub-stoichiometric CO w.r.t Mo that gave ca. 50% inhibition of the rate of azide reduction. No lag phase was observed (<200 ms) for the onset of inhibition. This is intriguing since we have previously monitored the time course for the appearance of an IR band at 1904 cm^{-1} under low CO conditions (George *et al.* 1997). This band, which is the first to appear in the time course, takes ca. seven seconds to reach its maximum intensity and is less than 10% developed at 1 s. These data indicate that the 1904 cm^{-1} species cannot be responsible for the rapid onset of inhibition of azide reduction. We are now seeking, in the technically more difficult region of the IR spectrum where strong protein amide bands absorb

(1500-1700 cm^{-1}), evidence for a new CO species that forms at very short times and is responsible for the inhibition of azide reduction.

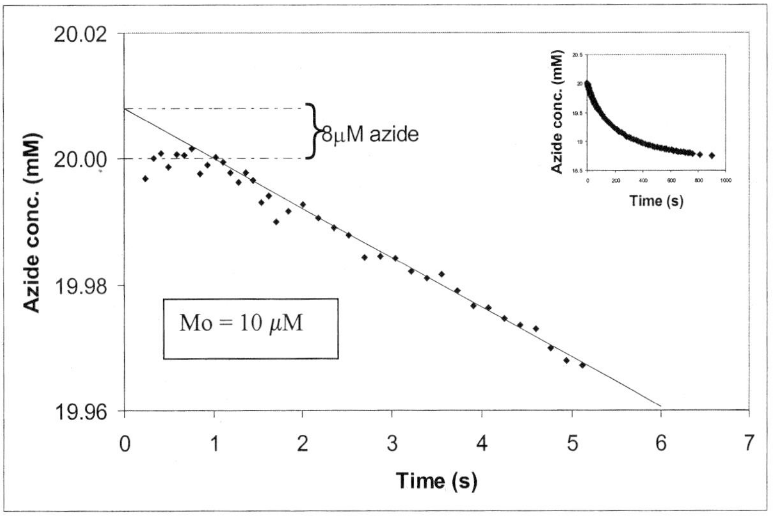

Figure 3. Stopped-flow FTIR time course for azide reduction showing a one-second lag phase (at 23°C, pH 7.4).

6. Conclusions

In this paper, we have shown how three methods of perturbing the structure of the Fe protein, i.e. by ATP analogs, site-directed mutagenesis of ATP-binding site residues, and selenium substitution of sulfide in the 4Fe-4S cluster, can give new insights into how protein structure modulates the partial reactions that comprise the Fe-protein cycle. In addition, substrate-reduction reactions occurring in the MoFe-protein cycle are becoming accessible to study by real time monitoring by stopped-flow FTIR spectroscopy.

7. References

Chiu H-J *et al.* (2001) Biochem. 40, 641-650
Gaillard J *et al.* (1986) Biochem. 25, 464-468
George SJ *et al.* (1997) J. Amer. Chem. Soc. 119, 6450-6451
George SJ *et al.* (2000) J. Biol. Chem. 275, 33231-33237
Grossman JG *et al.* (1997) J. Mol. Biol. 266, 642 -648
Jang SB *et al.* (2000) Biochem. 39, 14745-14752
Lowe DJ, Thorneley RNF (1984) Biochem. J. 224, 877-886
Yu S-B *et al.* (1991) Inorg. Chem. 30, 3476-3485

8. Acknowledgements

RNFT is supported by the UK Biotechnology and Biological Sciences Research Council. HA and JT are grateful to the BBSRC, and AS to the Gatsby Foundation for financial support. PCH was supported on sabbatical leave at the JIC by NSERC and the Royal Society.

AZIDE, CYANIDE AND NITRITE ARE NEW SUBSTRATES OF THE OXYGEN-DEPENDENT NITROGENASE OF THE THERMOPHILIC BACTERIUM *STREPTOMYCES THERMOAUTOTROPHICUS*

D. Gadkari, A. Sarjiya, O. Meyer

Chair of Microbiology, University of Bayreuth, D-95440 Bayreuth, Germany

1. Introduction

Streptomyces thermoautotrophicus is a thermophilic, aerobic and obligately chemolithoautotrophic bacterium (Gadkari *et al.* 1990). It is able to fix dinitrogen with CO or H_2 plus CO_2 as growth substrates (Gadkari *et al.* 1992, Ribbe *et al.* 1997). Nitrogenase of *S. thermoautotrophicus* comprises three enzymes, a heterotrimeric Mo-dinitrogenase (St1), a dimeric Mn-superoxide oxidoreductase (St2) and a heterotrimeric Mo-carbon monoxide dehydrogenase (CODH or St3). These three enzymes are dioxygen insensitive; in fact, dioxygen is an essential intermediate (Ribbe *et al.* 1997). CODH oxidizes CO and transfers the electrons released to oxygen (O_2), thereby producing superoxide (O_2^-). The Mn-superoxide oxidoreductase reoxidizes O_2^- to O_2 and transfers the electrons to a Mo-dinitrogenase which, in turn, reduces dinitrogen to ammonium. In contrast to the electronic coupling in known nitrogenases via ferredoxin/flavodoxin, the *S. thermoautotrophicus* enzyme establishes a molecular coupling via O_2^-. The amino acid sequence of superoxide oxidoreductase, designated SdnO, was highly similar to superoxide dismutases (SOD) of bacilli (Hofmann-Findeklee *et al.* 2000). The amino acid sequences of the three SdnM, SdnS and SdnL polypeptides (Hofmann-Findeklee *et al.* 2000) were very similar to the corresponding subunits of structurally characterized CODH from *Oligotropha carboxidovorans* (Gremer *et al.* 2000).

"Conventional" Mo-containing nitrogenases are versatile and reduce low-molecular-weight compounds containing N-N, N-O, C-N, and C-C double or triple bonds (Burgess 1993) and carbon monoxide (CO) is a non-competitive inhibitor of all substrates except the H^+ (Hardy *et al.* 1965; Rivera-Ortiz, Burris 1975; Pham, Burgess 1993). In contrast, the nitrogenase of *S. thermoautotrophicus* cannot reduce acetylene to either ethylene or ethane (Ribbe *et al.* 1997). Schöllhorn and Burris (1967) showed the reduction of azide and Kelly *et al.* (1967) the reduction of cyanide, which supplies both HCN as a substrate and CN^- as a potent inhibitor. The ratio of these two species is pH dependent (Li *et al.* 1982). Reduction of nitrite to NH_4^+ by nitrogenase has been reported (Vaughn, Burgess 1989), however, very little is known about NO_2^- as a substrate.

Carbon monoxide dehydrogenase (CODH) is the key enzyme of carboxidotrophic metabolism. It is a copper-containing molybdenum-iron-sulfur flavoprotein (Gremer *et al.* 2001), which oxidizes CO to CO_2 using water as oxidant to gain the energy for the carboxidotrophic growth (Meyer, Schlegel 1980). The crystal structure of CODH has been solved (Dobbek *et al.* 1999; Meyer *et al.* 2000). Until now, N_3^- reduction to NH_4^+ by CODH, using dithionite as the electron donor, has not been demonstrated. Here, we report that the purified nitrogenase of *S. thermoautotrophicus* can reduce azide, cyanide, and NO_2^- to NH_4^+, in the presence of MgATP and dithionite, and show the effect of CO on substrate reduction and H_2 formation. In addition, we show that the CODH (St3 enzyme) of *S. thermoautotrophicus* can reduce azide to NH_4^+, which further adds to the analogy between the two enzyme systems.

2. Experimental Procedures

S. thermoautotrophicus UBT1 (DSM 41605, ATCC 49746) was grown chemolitho-autotrophically with CO as a sole source of carbon and energy in mineral medium (Meyer, Schlegel 1980) containing NH_4Cl (28 mM) under a gas mixture of (v/v) 78% air, 13% CO and 9% CO_2 (Ribbe *et al.* 1997).

Crude extracts were prepared by passing bacterial suspensions (50-60 g of cell wet weight suspended in 50-60 ml of 50 mM potassium phosphate buffer, pH 7.7, containing 0.5 mM phenylmethylsulfonyl fluoride and a few crystals of DNase) 5 to 6 times through a French pressure cell at maximum pressure under oxic conditions and then subjected to low spin centrifugation. Cytoplasmic fractions were obtained by ultra centrifugation and then (70-80 mL) loaded on a 15 cm x 2.6 cm anion-exchange Source 30Q (Amersham Pharmacia Biotech) column, equilibrated with 25 mM potassium phosphate buffer (pH 7.7). Elution with 225 mL 25 mM potassium phosphate buffer (pH 7.7) was followed by a linear gradient of 0 to 0.6 M NaCl in phosphate buffer. The St2 protein does not bind to Source 30Q and is eluted in the first 80-160 mL. The proteins, St1 and St3, are eluted close together at about 0.3 M NaCl. Fractions with ammonium-forming activity (St1 protein) were pooled, supplemented with 1.2 M K_2HPO_4/HCl buffer (pH 7.7), and stirred for 15 min. Precipitated protein was removed by low-spin centrifugation. The supernatant (80-100 mL) was loaded onto a hydrophobic interaction chromatography column (13 cm by 2.6 cm; Source 15ISO, Amersham Pharmacia Biotech), and equilibrated with 1 M K_2HPO_4/HCl buffer (pH 7.7). Unbound proteins were eluted with 225 mL of equilibration buffer, followed by 450 mL of a decreasing linear gradient (1.2 to 0 M). The St2 protein was purified as described (Ribbe *et al.* 1997).

Nitrogenase assays were performed as described previously by following NH_4^+ formation from N_2 by the indophenol method (Ribbe *et al.* 1997). CODH activity was measured spectroscopically with iodonitro-tetrazolium chloride (INT) as electron acceptor (Kraut *et al.* 1989). Hydrogen evolution was analyzed by conversion of HgO to mercury in a trace analytical RGD2 reduction gas detector as described (Gadkari *et al.* 1990). Dinitrogen was analyzed by gas chromatograph employing standard methods. Solutions of KCN, NaN_3 and KNO_2 were prepared in 50 mM KH_2PO_4/NaOH buffer (pH 7.5).

3. Azide as a Substrate

Under appropriate assay conditions with dithionite as electron donor, the purified nitrogenase components, St1 and St2, are capable of catalyzing the reduction of azide to ammonium (Figure 1). Maximum NH_4^+ formation activities of 1.29 µmole NH_4^+ h^{-1} x mg^{-1} was at 25 mM azide. Concentrations above 25 mM neither increased nor inhibited activity.

Figure 1 shows the time course of ammonium formation by purified *S. thermoautotrophicus* nitrogenase in the presence or absence of N_2 or azide. After 1 h of incubation at 65°C, the ammonium formed in the control was 0.10 mM, whereas ammonium formation with azide under either nitrogen or helium atmosphere was 0.46 mM and 0.51 mM, respectively. After 6 h of incubation, the activity with azide under dinitrogen or helium increased 3.9-fold and 4.6-fold, respectively, compared to the control. Apparently, azide is a substrate and it produces only N_2 and NH_4^+ in a molar ratio 1.4. Azide reduction to NH_4^+ had a K_m of 2.38 mM and a V_{max} of 26.6 nmole of NH_4^+ formed (min mg)$^{-1}$.

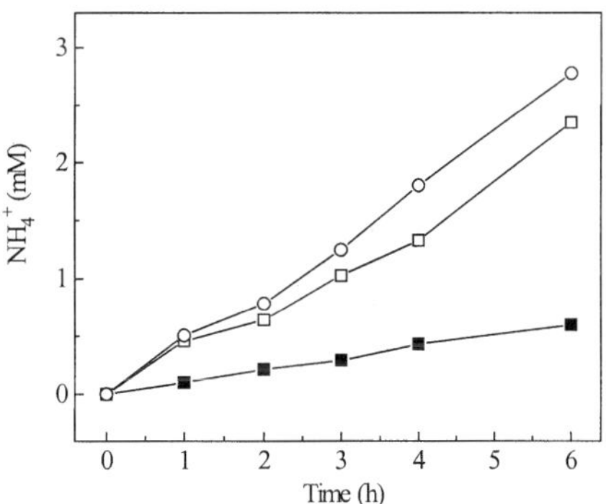

Figure 1. Ammonium formation from azide. Reaction mixtures contained 2.3 mL KH_2PO_4/NaOH buffer (50 mM, pH 7.5), 2.5 mM ATP, 5.0 mM $MgCl_2$, 10.0 mM dithionite, 0.82 mg St1, 0.13 mg St2 and 25 mM NaN_3 where indicated. Assays were performed at 65° C. ■, without azide under 100% N_2; □, with azide under 100% N_2; and O, with azide under 100% helium.

3.1. Influence of carbon monoxide on azide reduction. CO is a powerful non-competitive inhibitor of all substrates except protons of "conventional" nitrogenases (Hardy *et al.* 1965; Bulen *et al.* 1965; Hwang *et al.* 1973). It specifically influences the substrate reduction sites without interfering in the enzyme-catalyzed hydrolysis of MgATP.

Figure 2 shows that, with increasing CO, the reduction of azide (25 mM) correspondingly decreased. With azide at 65°C, NH_4^+ formation without CO was 1.61 µmole h^{-1} x mg^{-1}. With 15% CO, however, it was only 0.19 µmole h^{-1} x mg^{-1}, i.e. only 11.8% of that without CO. In an assay with N_2 (control) in the absence of CO, the activity was 0.28 µmole h^{-1} x mg^{-1} and, in the presence of 15% CO, it was only 0.061 µmole h^{-1} x mg^{-1}, which was 21.8% of that without CO. In both assays (with and without azide) with CO present, there was an increase in H_2 evolved, which corresponded to the CO concentration. After 4 h incubation, the ratio of NH_4^+ to H_2 in the control assay without CO was 1:0.18, which changed with increasing CO. At 15% CO, the ratio was 1:1.6 (Figure 2), indicating that most of the electrons have been transferred to protons. A similar change in this ratio was observed with added azide from 1:0.015 to 1:0.45 without and with CO (15%), respectively. Although these results show a similar pattern of CO inhibition as for the "conventional" nitrogenase, the amount of CO required for inhibition was distinctly higher.

4. Cyanide as a Substrate

Cyanide can be also reduced by the St1 and St2 components (Figure 3). The maximum concentration of cyanide, which is tolerated by the St1 and St2 components, is 20 mM. Above this concentration, activity declined and, at 25 mM, ceased entirely.

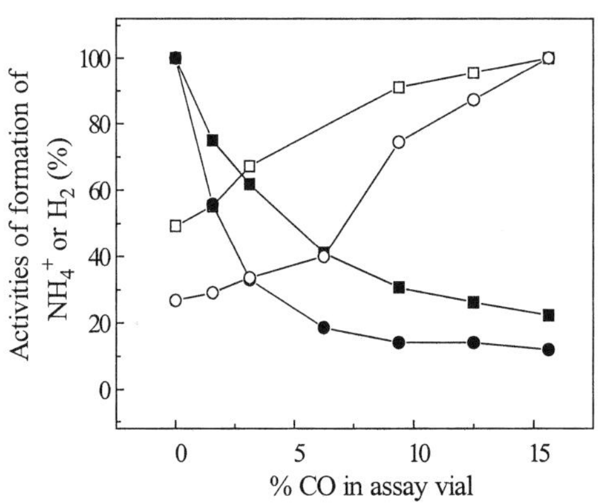

Figure 2. Effect of CO on NH_4^+ and H_2 formation from azide by St1 and St2. Reaction mixtures were as in Figure 1, but with 25 mM NaN_3 and various CO concentrations. NH_4^+ was analyzed after 4 h at 65°C under N_2. ■, NH_4^+ formed without azide; □, H_2 formed without azide; ●, NH_4^+ formed with azide; O, H_2 formed with azide. For NH_4^+, 1.6 µmol $(h.mg)^{-1}$ and 0.28 µmol $(h.mg)^{-1}$ were set as 100% with and without azide, respectively. For H_2, 0.13 µmol $(h.mg)^{-1}$ and 0.43 µmol $(h.mg)^{-1}$ were set as 100% with and without azide, respectively.

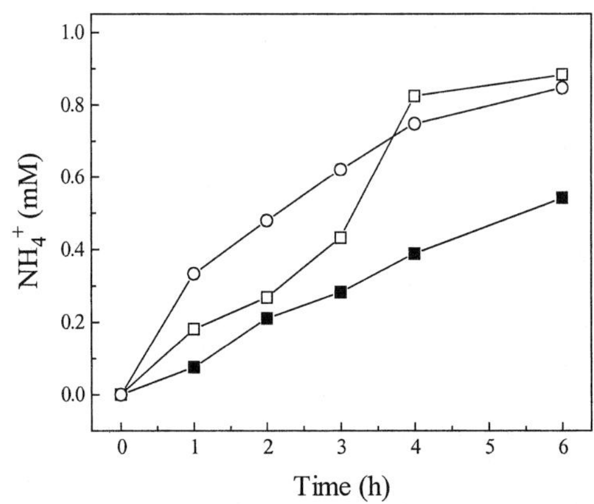

Figure 3. Ammonium formation from cyanide catalyzed by purified St1 and St2 components. Reaction mixtures were as Figure 1 plus 10 mM KCN where indicated. Assays at 65°C under 100% N_2 or He: ■, without cyanide under N_2; □, with cyanide under N_2; O, with cyanide under 100% He.

At 10 mM KCN, the maximum activity was 0.42 μmole NH_4^+ h^{-1} x mg^{-1}. Figure 3 shows ammonium formation in the presence of either KCN or N_2. The reaction occurred with similar efficiency in the presence and in the absence of N_2 (Figure 3). After 6 h of incubation at 65°C, the total NH_4^+ formation with KCN was 0.88 mM and with N_2 (control) it was 0.45 mM, suggesting that cyanide may be preferred. KCN was reduced to NH_4^+ with a K_m of 5.21 mM and a V_{max} of 16.7 nmole of NH_4^+ formed (min^{-1} x mg^{-1}). These values fall into the range (K_m of 1.1 mM to 4.5 and V_{max} 113 to 158 nmole min^{-1} x mg^{-1}) reported for "conventional" nitrogenases (Li *et al.* 1982; Shen *et al.* 1997).

4.1. Influence of carbon monoxide on cyanide reduction.

Increasing amounts of CO (up to 15%) hardly affected KCN reduction (Figure 4). In the absence of CO, NH_4^+ formation was 0.67 μmol $(h.mg)^{-1}$, whereas with 15% CO, it was 0.57 μmol $(h.mg)^{-1}$. Moreover, even 47% CO did not affect NH_4^+ formation (results not shown), however, H_2 formation increased; 4-fold with 15% CO.

5. Nitrite as a Substrate

Under appropriate assay conditions with dithionite as the electron donor, the nitrogenase components St1 and St2 are able to catalyze the reduction of NO_2^- to NH_4^+ linearly up to 9 mM NO_2^-. At 65°C, the specific activity of a control without NO_2^- was 0.82 μmole NH_4^+ h^{-1} x mg^{-1}, whereas with 9mM NO_2^-, it was 1.60 μmole NH_4^+ h^{-1} x mg^{-1}. Higher concentrations were inhibitory and, with 35 mM NO_2^-, ammonium formation was 27% less than the control without NO_2^-. Vaughn and Burgess (1989) observed that, at 60 mM, nitrite is an inhibitor of the Fe protein. In our experiments in the presence of MgATP and dithionite, the St2 protein alone was able to reduce NO_2^- slowly (0.3 μmole NH_4^+ h^{-1} x

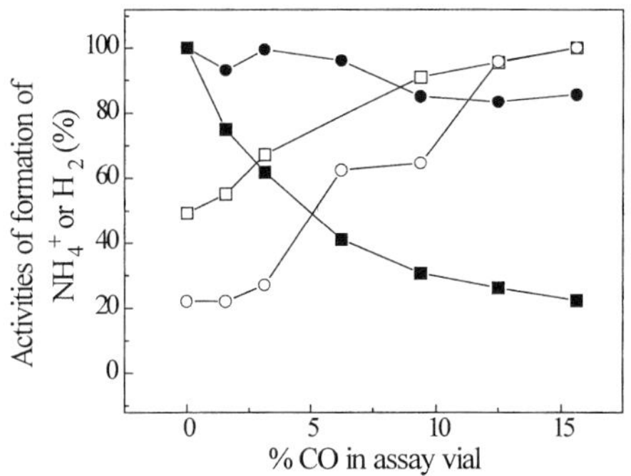

Figure 4. Influence of CO on NH_4^+ and H_2 formation with cyanide present by the St1 and St2 components under N_2. Reaction mixtures were as in Figure 1 with 10 mM KCN and CO as indicated. NH_4^+ was analyzed after 4 h at 65° C. ■, NH_4^+ formed without KCN; □, H_2 formed without KCN; ●, NH_4^+ formed with KCN; O, H_2 formed with KCN. For NH_4^+, 0.67 μmole $(h.mg)^{-1}$ and 0.28 μmol $(h.mg)^{-1}$ were set as 100% with and without KCN, respectively. For H_2, 0.11 μmol $(h.mg)^{-1}$ and 0.43 μmol $(h.mg)^{-1}$ were set as 100% with and without KCN, respectively.

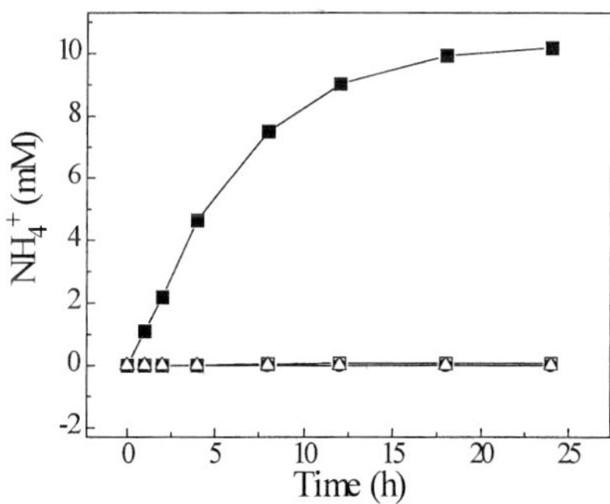

Figure 5. Ammonium formed from azide (10 mM) catalyzed by CODH. Reaction mixtures contained 2 mL of KH_2PO_4/NaOH buffer (50 mM, pH 7.5), 10 mM dithionite and 2.8 mg CODH. Assays were under 100% N_2 65°C. ■, CODH + dithionite + N_3^-; □, CODH + N_3^-; ●, CODH + dithionite; Δ, dithionite + N_3^-.

mg^{-1}), even at a concentration of 35 mM NO_2^-, suggesting that 35 mM NO_2^- inhibited only the activity of the St1 protein. NO_2^- was reduced to NH_4^+ with a K_m of 2.79 mM and a V_{max} of 27.9 nmole of NH_4^+ formed $(min.g)^{-1}$.

6. Reduction of Azide by Carbon Monoxide Dehydrogenase (St3 protein)

Using appropriate conditions with dithionite as electron donor, the purified CODH of *S. thermoautotrophicus* is capable of catalyzing the reduction of azide to NH_4^+ and N_2 (Figure 5). After 25 h at 65°C, total NH_4^+ measured was 10.18 mM. In control vials, where only CODH or CODH plus dithionite or CODH plus N_3^- were present, neither NH_4^+ nor N_2 was formed.

During azide (5 mM) reduction, 5.16 mM NH_4^+ and 5.20 mM N_2 were formed. According to Dilworth and Thorneley (1981), azide can be reduced by one of the following three equations:

$$N_3^- + 3\,H^+ + 2\,e^- \;\rightarrow\; N_2 + NH_3 \qquad (1)$$

$$N_3^- + 7\,H^+ + 3\,e^- \;\rightarrow\; N_2H_4 + NH_3 \qquad (2)$$

$$N_3^- + 9\,H^+ + 8\,e^- \;\rightarrow\; 3\,NH_3 \qquad (3)$$

Because NH_4^+ and N_2 were formed in a stoichiometric ratio, this enzyme appears to operate in accord with reaction (1).

7. Discussion

Many alternative substrates are known for the "conventional" nitrogenases. The structural genes and cofactors of all these nitrogenases are highly similar. The oxygen-insensitive nitrogenase of *S. thermoautotrophicus* (St-nitrogenase) is genetically and structurally different from "conventional" nitrogenases (Hofmann-Findeklee *et al.* 2000). The primary sequence of St1 shows high homology to molybdenum-containing hydroxylases, especially with CODH from *Oligotropha carboxidovorans*, *Hydrogenophaga pseudoflava* and *Pseudomonas thermocarboxydovorans* (Hofmann-Findeklee *et al.* 2000). St2 is highly homologous to Mn-containing SODs (Hofmann-Findeklee *et al.* 2000). A functional difference between the St1/St2 system and the "conventional" nitrogenase is the inability of the former to reduce acetylene. Therefore, it was not obvious that the St-nitrogenase would reduce substrates of the "conventional" nitrogenases, but azide, cyanide and nitrite will indeed serve as substrates for St-nitrogenase. However, there are some differences.

First, the CO concentration (15%) required for inhibition of azide reduction was much higher than for the "conventional" nitrogenases (0.5-2.0%). Second, even with 47% CO, the reduction of cyanide was hardly inhibited. Third, nitrite did not inhibit St2 activity; on the contrary, St2 is able to reduce nitrite to NH_4^+. Fourth, the results obtained with azide were more or less consistent with the results known from other nitrogenases (Figure 1, Figure 2). Fifth, CO inhibits the reduction of azide to NH_4^+ by St-nitrogenase and also simultaneously stimulates the production of H_2. However, the CO concentrations required for inhibition were much higher than for "conventional" nitrogenases, indicating that, for St-nitrogenase, CO appears to act not as a non-competitive inhibitor, but as a competitive inhibitor. Inhibition by CO is dependent on the ratio of azide and CO concentrations. Sixth, CO is a non-competitive inhibitor of cyanide reduction by "conventional" nitrogenase, however, 47% CO hardly affected cyanide reduction by St-nitrogenase. It appears that CO probably does not function as an inhibitor of St-nitrogenase but does stimulate H_2 evolution (Figure 4).

CO can produce many different effects. An altered *A. vinelandii* MoFe protein, where αGly^{69} was replaced by serine, suffers competitive inhibition of substrate reduction rather than non-competitive by CO (Christiansen *et al.* 2000). Also, a mutant of *A. vinelandii*, where the MoFe protein α-glutamine-191 residue was replaced by lysine, exhibited 50% inhibition of proton reduction by CO (Kim *et al.* 1995). Furthermore, at low concentrations, CO acts as a stimulant for V-nitrogenase of *A. vinelandii* (Cameron, Hales 1996). In contrast, *S. thermoautotrophicus* is a carboxidotrophic bacterium that grows under 45-50% CO and, as expected, CO does not exert much, if any, inhibitory effect on St-nitrogenase as exemplified by the high CO concentration required for inhibition of azide reduction. Thus, azide and cyanide may not bind with similar affinity to the St1 component. Nitrite reduction by St1 is consistent with a previous report (Vaughn, Burgess 1989), however, although nitrite inhibits the Fe-protein activity of "conventional" nitrogenase by affecting its 4Fe-4S center, the St2 component of St-nitrogenase, which does not contain Fe-S centers, is not inhibited by nitrite. On the contrary, the St2 protein alone was able to reduce nitrite to NH_4^+.

The CODH (St3 protein) can also reduce azide to NH_4^+. The difference between St3 and St1 is that, for substrate reduction, St1 requires St2 plus MgATP and dithionite, whereas St3 requires only dithionite for azide reduction. Because the structural genes of St1 and St3 are highly homologous, it is not surprising that St3 can also reduce azide to NH_4^+. Possibly many other compounds, which serve as substrates for nitrogenases, could be reduced by St-CODH. However, in general, CODHs of aerobic bacteria have oxidizing characters whereas nitrogenases are reductive. The St1 component is a heterotrimeric monomer and resembles at least one of the CODH species occurring in *S. thermoautotrophicus*. Similar to the "conventional" nitrogenases, St1 can reduce N_2, azide, cyanide, and nitrite. The St-nitrogenase is entirely different from the "conventional" nitrogenases (Ribbe *et al.* 1997) and it remains of interest to know why these two enzymes, in spite of these differences, react more or less similarly with respect to reduction of these substrates.

8. References

Bulen WA *et al.* (1965) In San Pietro (ed) Non-heme Iron Protein: Role In Energy Conversion, pp. 107-112, Antioch Press, Yellow Springs, OH

Burgess B, Lowe DJ (1996) Chem. Rev. 96, 2983-3011

Cameron LM, Hales BJ (1996) J. Am. Chem. Soc. 118, 279-280

Christiansen J *et al.* (2000) J. Biol. Chem. 275, 36104-36107

Dilworth MJ, Thorneley RNF (1981) Biochem. J. 193, 971-983

Dobbek *et al.* (1999) Proc. Natl. Acad. Sci. USA 96, 8884-8889

Gadkari D *et al.* (1990) Appl. Environ. Microbiol. 56, 3227-3234

Gadkari D *et al.* (1992) J. Bacteriol. 174, 26627-26633

Gremer L *et al.* (2000) J. Biol. Chem. 275, 1864-1872

Gremer L *et al.* (2001) Biospectrum

Hardy RWF *et al.* (1965) Biochem. Biophys. Res. Commun. 20, 539-544

Hofmann-Findeklee C *et al.* (2000) In Pedrosa FO, Hungaria M, Yates G, Newton WE (ed) Nitrogen Fixation: From Molecules to Crop Productivity, pp. 23-30, Kluwer, Dordrecht

Hwang JC *et al.* (1973) Biochem. Biophys. Acta 292, 256-270

Kelly M *et al.* (1967) Biochem. J. 102, 1c-3c

Kim CH *et al.* (1995) Biochem. 34, 2798-2808

Kraut M *et al.* (1989) Arch. Microbiol. 152, 335-341

Li JG *et al.* (1982) Biochem. 21, 4393-4402

Meyer OM, Schlegel GH (1980) J. Bacteriol. 141, 74-80

Meyer OM *et al.* (1993) In Murrell JC, Kelly DP (ed) Microbial Growth on C_1 Compounds, pp. 83-90, American Society for Microbiology, Washington, DC

Meyer OM *et al.* (2000) Biol. Chem. 381, 865-876

Pham DN, Burgess BK (1993) Biochem. 32, 13725-13731

Ribbe M *et al.* (1997) J. Biol. Chem. 272, 26627-26633

Rivera-Ortiz JM, Burris RH (1975) J. Bacteriol. 123, 537-545

Schöllhorn R, Burris RH (1967) Proc. Natl. Acad. Sci. USA 57, 1317-1323

Shen J *et al.* (1997) Biochem. 36, 4884-4894

Vaughn SA, Burgess BK (1989) Biochem. 28, 419-424

Section 2: Bacterial Genomics

CHAIR'S COMMENTS: THE WHOLE CHROMOSOME SEQUENCE OF *SINORHIZOBIUM MELILOTI* STRAIN 1021

D. Capela[1,2], F. Barloy-Hubler[2], J. Gouzy[1], G. Bothe[3], F. Ampe[1], J. Batut[1], P. Boistard[1], A. Becker[4], M. Boutry[5], E. Cadieu[2], S. Dréano[2], S. Gloux[2], T. Godrie[6], A. Goffeau[5], D. Kahn[1], E. Kiss[1], V. Lelaure[2], D. Masuy[5], T. Pohl[3], D. Portetelle[6], A. Pühler[4], B. Purnelle[5], U. Ramsperger[3], C. Renard[1], P. Thébault[1], M. Vandenbol[6], S. Weidner[4], F. Galibert[2]

[1]Laboratoire de Biologie Moléculaire des Relations Plantes-Microorganismes, UMR215-CNRS-INRA, Chemin de Borde Rouge, BP27, F-31326 Castanet Tolosan Cedex, France
[2]Laboratoire de Génétique et Développement UMR6061-CNRS, Faculté de Médecine, 2 avenue du Pr. Léon Bernard, F-35043 Rennes Cedex, France
[3]GATC GmbH, Fritz-Arnold-str. 23, D-78467 Konstanz, Germany
[4]Universität Bielefeld, Biologie IV (Genetik), Universitätstr. 25, D-33615 Bielefeld, Germany
[5]Unité de Biochimie physiologique, Université Catholique de Louvain, Place Croix du Sud 2, Bte 20, B-1348 Louvain-la-Neuve, Belgium
[6]Unité de Microbiologie, Faculté des Sciences Agronomiques de Gembloux, Avenue Maréchal Juin 6, B-5030 Gembloux, Belgium

1. Introduction

Sinorhizobium meliloti strain 1021 is a free-living, gram-negative soil bacterium, and symbiont of alfalfa (*Medicago sativa*). Its genome consists of three large replicons – a chromosome and two megaplasmids – that have been entirely sequenced by an international consortium (Galibert 2001). The pair *M. truncatula* and *S. meliloti* has been chosen by many international groups and emerges as *the* model for symbiosis and nitrogen fixation studies worldwide (along with *Lotus japonicus* and *Mesorhizobium loti*).

2. Results and Discussion

The entire double-strand nucleotide sequence of the *S. meliloti* strain 1021 chromosome has been determined by shotgun sequencing of 50 ordered recombinant BAC clones with a sufficient redundancy (over four-fold) and a sequence quality estimated to correspond to less than one error per 100,000 nucleotides. The average GC content is 62.7% (as in the *M. loti* chromosome), although six large regions with lower GC content have been found – three of these correspond to the *rrn* operons and the others to regions of putatively external origin. For instance, 0.5% of the 2.2% of the total chromosome sequence with transposon-related functions are located in one of these regions. In addition, except for the three *rrn* operons, only *tuf* and *purU* are found duplicated on this replicon, with more than 90% nucleotide sequence identity.

Coding regions, including protein-coding genes and RNA-coding genes, represent 86.4% of the total chromosome sequence and their organization includes frequent changes in polarity. A total of 3341 protein-encoding genes are predicted, with a mean length of 938 bp. The longest identified chromosomal gene, *ndvB*, is 8496 nucleotide in size. Putative functions were assigned on the basis of homology to 59% of the chromosomal genes and 5% of the protein-coding ORFs are orphans, with no sequence analogy.

The *S. meliloti* chromosome carries all 57 genes involved in DNA metabolism and replication, while several genes involved in primosome assembly in *E. coli* are missing. Six of the nine genes required for cytokinesis in *E. coli* (*ftsA, ftsI, ftsK, ftsQ, ftsW* and two *ftsZ* genes) are found on the *S. meliloti* chromosome, while *ftsL, ftsN* and *zipA* are apparently missing. We also identified new genes for septum formation (*maf*), chromosome partitioning (*smc* and *parAB*) as well as *ctrA*, a member of the two-component signal transduction family involved in the control of a number of cell cycle-regulated genes.

Fifty-one tRNA genes have also been detected. These are evenly distributed on the chromosome and correspond to 43 different tRNA acceptors, which by wobble pairing can translate all but one codon. The missing essential tRNA, corresponding to an arginine codon, is encoded by the pSymB megaplasmid.

With the notable exception of asparagine synthase, whose two *asn* genes are on pSymB, the complete pathways for amino acid biosynthesis have all been found. Interestingly, *S. meliloti* possesses two different pathways for methionine biosynthesis, the classical *metABC*, as in *E. coli*, and *metZ*, essential for symbiosis in *R. etli*. All the essential genes responsible for the *de novo* synthesis of purines and pyrimidines as well as glycolysis and gluconeogenesis are encoded by chromosomal genes, with the exception of the gluconeogenic fructose-1,6-biphosphatase (cbbF), found on pSymB. Regarding glycolysis, *S. meliloti* lacks the classical ATP-dependent phosphofructokinase (*pfk*) but possesses a complete Entner-Doudoroff pathway, which makes this the main route for glucose utilization. A total of 10.8% of the chromosomal proteins are involved in transport, 33% of which belong to the ABC (ATP Binding Cassette) and 18% to MF (Major Facilitator) superfamilies.

S. meliloti seems well equipped to face a large variety of stress conditions, including osmotic shock, heat and cold shock as well as oxidative stresses. Oxygen protection might be essential for efficient infection if rhizobia, like many pathogens, induce an oxidative burst upon plant cell infection. The eleven *gst* (Glutathione-S transferase) and the three *rpoE* (sigma factor 24) identified on the chromosome might contribute to protection against oxygen or other reactive molecular species. Six additional *gst* and five additional *rpoE* have been recognized on the megaplasmids. In addition to cell surface components, the chromosome sequence reveals a number of genes putatively involved in virulence, including an ortholog of the *acvB* virulence gene of *Agrobacterium tumefaciens*. Finally, regulatory functions have been assigned to 7.2% of the chromosomal genes, incuding an unusually high number of nucleotide cyclases (26 genes). The role of these cyclases and the signal transduction pathways in which they participate are unknown in *S. meliloti*.

3. Conclusions

The *S. meliloti* chromosome sequence carries not only housekeeping genes, but also genetic information for mobility and chemotaxis processes, plant interaction, putative virulence as well as stress responses. However, since up to 41% of the *S. meliloti* chromosomal genes still have unassigned functions many other functions are likely to be encoded by this replicon. Transcriptomic and proteomic analyzes have been initiated to elucidate these.

4. References

Galibert F *et al.* (2001) Science 293, 668-672

THE 1683 KB REPLICON OF *SINORHIZOBIUM MELILOTI*: PAST AND PRESENT INVESTIGATIONS INTO THE NATURE OF A VERY LARGE BACTERIAL PLASMID

S.R. MacLellan, C.D. Sibley, B. Golding, T.M. Finan

Department of Biology, McMaster University, 1280 Main St. West, Hamilton, ON, L8S 4K1 Canada

1. Introduction

Many of the bacteria that employ endosymbiotic nitrogen-fixing lifestyles in association with plant hosts possess large non-chromosomal DNA replicons. *Sinorhizobium meliloti*, for example, possesses a 1400 kb and a 1700 kb unit- or low-copy number plasmid in addition to a 3500 kb chromosome. The plasmids, or megaplasmids, in *S. meliloti* have garnered interest since their detection because numerous genes that influence N_2-fixation and endosymbiosis map to these sequences. Their large size (comparable to some of the smaller bacterial chromosomes) have generated speculation as to their genetic content and the biological role they play in the free-living or bacteroid forms of *S. meliloti* cells. Many laboratories have contributed to our increased understanding of the biology of the megaplasmids. Here, we consider some of the past and present research initiatives in which our laboratories have played some role in investigating the nature of the largest *S. meliloti* megaplasmid, pSymB.

We begin with the first detection of pSymB (originally called pRmeSU47b and now also called pExo), its genetic mapping, and the results of a large-scale deletion analysis of the megaplasmid. pSymB is a member of a growing and well-conserved family of plasmids (based on their replicator regions) called *repABC*-type replicons. We discuss research initiatives that have and will enhance our knowledge of the replication and segregation mechanisms that ensure faithful maintenance of these replicons in cell populations. Finally, we consider some of the general features of the nucleotide sequence of pSymB that have been deduced as the result of the completion of the *S. meliloti* genome sequencing effort.

2. The Discovery of pSymB in *S. meliloti*

The megaplasmid pSymB was originally detected (Finan *et al.* 1986) during attempts to map the location of transposon insertions generating a class of mutants that induced the formation of atypical nodules that were Fix⁻ and devoid of infection threads. These mutants were unable to synthesize an exopolysaccharide now known to be required for effective nodulation. Eckhardt gels used to visualize high molecular weight genomic DNA would often display a doublet character at the band that corresponded to the previously identified megaplasmid, the so-called nod-nif plasmid. Southern hybridization analyses using radiolabeled Tn5 probe showed hybridization to the upper band of the doublet in the *exo* mutant strains while labeled probes hybridized with the lower band in strains containing a nifH::Tn5 insert. This and other data demonstrated the existence of a second megaplasmid similarly sized though somewhat larger than the nod-nif (pSymA) megaplasmid in *S. meliloti*. Today both plasmids are often referred to as symbiotic plasmids since both contain exopolysaccharide synthesis loci (and other genes) that are required for effective nodulation.

3. Genetic Mapping and Deletion Analysis of pSymB

To facilitate the characterization of pSymB, a genetic map of the replicon was constructed by employing transposon insertions with alternating antibiotic resistance genes as genetic markers (Charles, Finan 1990). These markers, linked by transduction, for the first time allowed the unambiguous mapping of known phenotypic loci on the plasmid. This development led to the first genomic scale analysis of any replicon in *S. meliloti* because it was realized that these same transposon markers could be used, via homologous recombination between adjacent or distant IS50 elements, to generate large-scale deletions in pSymB. Accordingly, a series of mutant derivatives containing 120-600 kb deletions were constructed (Charles, Finan 1991). The ensuing phenotypic analysis revealed previously unknown loci required for the utilization of dulcitol, melibiose, raffinose, β-hydroxybutyrate, acetoacetate, protocatechuate and quinate as well as previously unidentified loci required for effective nodulation and exopolysaccharide synthesis. Collectively, nearly 90% of pSymB was demonstrated to be non-essential for viability. Some regions of the plasmid were not represented amongst the deletion derivatives raising the possibility that these sequences contained loci that were essential for viability (see later).

Indirectly, the construction of the linkage map for pSymB played a significant role in the development of a technique to clone out very large regions of plasmids that could then be used as a source of DNA for genome sequencing and also led to the discovery of the pSymB origin of replication (Chain *et al.* 2000).

4. Controlling the Replication and Segregation of a Megaplasmid

Early attempts to isolate active origins of replication from the *S. meliloti* genome were largely unsuccessful. Our initial involvement in the pSymB nucleotide sequencing project concentrated on a 60 kb region of DNA that was not represented amongst the deletion derivative strains described by Charles and Finan (1991). This region appeared to replicate autonomously in *A. tumefaciens*. Within this region, we identified three ORFs (Chain *et al.* 2000) with a high degree of sequence similarity to the *repA*, *repB*, and *repC* genes that had previously been isolated and genetically characterized in plasmids from *Rhizobium leguminosarum* (Turner, Young 1995), *Rhizobium etli* (Ramirez-Romero *et al.* 2000), several strains of *Agrobacterium tumefaciens* (Tabata *et al.* 1989; Suzuki *et al.* 1998; Li, Farrand 2000), *A. rhizogenes* (Nishiguchi *et al.* 1987) and from the soil bacterium *Paracoccus versutus* (Bartosik *et al.* 1998). It is therefore clear that pSymB is a member of a large and growing family of plasmids called *repABC*-type replicons – so called because their replication and segregation processes are dependent upon the *repABC* gene products. A 780-bp sequence previously isolated from pSymB that appears to be capable of autonomous replication (Margolin, Long 1993) is located approximately 650 kb from the *repABC* locus. The biological significance of this apparently non-essential sequence is not known, however recent experiments in our laboratory suggest that its replication is not dependent on *repABC*. The *S. meliloti* genome sequencing project has revealed the presence of a second set of *repAB* genes on pSymB (131 kb distant from *repABC*) that are more closely related to genes in other rhizobia rather than those in *S. meliloti* (Finan *et al.* 2001). The sequencing project has also revealed that the 1400 kb pSymA replicon in *S. meliloti* is a *repABC*-type replicon.

Since the first description of a *repABC* replicator region (Nishiguchi *et al.* 1987) several genetic analyses have demonstrated the following general features of the region. Deletions in the *repA* or *repB* genes do not prevent replication but render the replicon (either the native plasmid or a mini-derivative of the plasmid) unstable. Such a plasmid is rapidly lost from a growing culture under non-selective conditions in contrast to the non-mutated replicon that is reportedly quite stable under the same conditions (Ramirez-Romero *et al.* 2000). Thus the RepA and RepB proteins appear to play a role in plasmid segregation during cell division. The predicted amino acid sequences of these proteins demonstrate their relatedness to a rather large family of proteins from many different species of bacteria. Where studied, these proteins have been demonstrated to positively influence plasmid segregation in the cell. This sequence similarity extends to the SopAB and ParAB proteins from the *Escherichia coli* F plasmid and P1 phage, respectively. Several apparent motifs, including a putative ATPase domain in RepA, are strongly conserved amongst the RepAB proteins and the polypeptides from the aforementioned replicons.

Deletions, insertions, or frameshift mutations in *repC* eliminate replication (Tabata *et al.* 1989; Bartosik *et al.* 1998; Ramirez-Romero *et al.* 2000). RepC is often presumed to initiate replication by initiating the melting of duplex DNA at or adjacent to the origin of replication much in the way that RepE performs the same function at the F plasmid origin. Unlike the case for RepA and RepB, sequence database searches do not detect proteins with significant similarity to RepC, except for those coded by genes that are resident in other *repABC*-type replicons.

A non-coding region of nucleotide sequence between the *repB* and *repC* ORFs exerts an incompatibility effect against a replicon possessing the parental *repABC* replicator region. This effect has been consistently recognized in every replicator region examined although a mechanistic explanation for the observation has not been forthcoming. Interestingly, Ramirez-Romero *et al.* (2000) reported that a deletion in the *repA* gene of the Rhizobium etli p42d plasmid relieves this incompatibility effect – an observation that our lab has also made with a pSymB replicator region derivative possessing a frameshift mutation in *repA*. Finally, the *repB-C* intergenic regions from rhizobial plasmids display a rather high level of nucleotide sequence conservation that is not maintained throughout the entire *repABC* region. This may ultimately reflect a conservation of *cis*-acting biochemical function amongst these related replicator regions.

Discussed above are the most obvious features of a typical *repABC*-type replicator region. Previous genetic analyzes however, have detected other features that are too numerous to mention here. One striking aspect of these regions is their overt similarity in terms of overall organization and sequence similarity. That said, we also note some differences amongst the behavior of these replicator regions. For example, other investigators have indicated the presence of a second incompatibility region that is essential for replication downstream of the *repC* gene (Ramirez-Romero *et al.* 2000). In the case of pSymB, we can detect no obvious incompatibility effect exerted from this region. Furthermore, Bartosik *et al.* (1998) have indicated that the *repC* gene itself (including its ribosome binding site, but without additional upstream and downstream nucleotide sequence) can confer autonomous replication upon an otherwise non-replicating vector. Our experiments using pSymB derivatives have indicated that replication is dependent on having nucleotide sequences both upstream and downstream of the *repC* gene. These differences are significant in so much as they place limitations upon the location of the actual origin of replication and possibly other *cis*-acting sites within the region. Further research should determine whether such differences are largely artifactual due to slightly differing experimental systems or whether they represent some level of functional divergence within an otherwise well-conserved biochemical and genetic framework.

The study of *repABC* replicator regions is essentially in its infancy and if other plasmid systems are any indication, such regions are liable to be very complex from a mechanistic point of view. To date, even in the best studied plasmid replicator regions, processes such as incompatibility and segregation are poorly understood. With respect to *repABC* systems, there are several avenues of investigation that should improve our understanding of the region. Among these, where is the actual origin of replication located and what is the precise role of *RepC*? How is the expression of the proteins regulated and what other proteins play a role in the initiation of replication and plasmid segregation? Are there any cross-interactions between the *cis* and *trans* acting components of different *repABC* systems in the same cell? Finally, will these replicator regions yield insight into the evolutionary history of the very large plasmids, like pSymB, that are typical of the Rhizobeaceae? Some plasmids contain more than one replicator region and it is possible that these represent vestigial replication and segregation control regions for what were once smaller, independent plasmids.

5. The Complete Nucleotide Sequence of pSymB

S. meliloti is perhaps the best characterized of the N_2-fixing endosymbiotic bacteria and given its agricultural and ecological importance and its rather unusual physiology amongst the bacteria, it was a natural choice as a target for complete genome sequencing. It is perhaps fitting that the publication of these Proceedings of the 13th International Congress on Nitrogen Fixation nearly exactly coincides with the publication of the complete genome sequence of *Sinorhizobium meliloti* (Galibert *et al.* 2001).

With regard to the pSymB replicon, its determined length is 1,683,333 bp (Finan *et al.* 2001) that is in good agreement with previous physical and genetic estimations. It has an overall GC content of 62.4% that is rather similar to that of the *S. meliloti* chromosome. It has a gene density similar to that of other bacterial chromosomes (about 90% protein coding) and a predicted 1,570 ORFs.

Eleven gene clusters (totaling 223 kb) that encode for cell surface (lipo-, capsular-, and extracellular-) polysaccharide synthetic machinery are found on pSymB and nine of these were previously unknown. Seventeen percent of the coding capacity of pSymB is devoted to ABC transporter systems with half of these predicted to be sugar-specific. pSymB encodes just over half (235) of all of the ABC transporter systems predicted in the entire genome. There is a predicted 134 ORFs encoding transcriptional regulators on pSymB including four ECF (extracytoplasmic function) sigma factors. In addition, a number of genes predicted to be involved in amino acid catabolism, nucleotide scavenging, aromatic compound degradation, and plant-derived metabolite degradation/utilization were predicted.

A number of potentially essential genes were discovered on pSymB. These include the only copies of the minCDE genes and the only copy of an Arg-tRNA gene that encodes the cognate RNA for the second most frequently used arginine codon, CCG. Both of these loci lie within regions that are not represented in the previously discussed pSymB deletion mutant library. Another region not represented in the library contains only two loci that might obviously be essential. One is *fusA2*, or elongation factor G, but experiments in our laboratory suggest that this gene is not essential (P. Aneja, unpublished data). The other locus is, as previously discussed, the replicator region *repABC* gene cluster. We suspect that these genes, or the *cis*-acting origin of replication within the region, are essential in so much as they are required for pSymB replication and pSymB is essential for cellular viability.

6. Conclusions

Since its initial detection, research involving the largest megaplasmid in *S. meliloti* has advanced considerably. We now know the sequence of the entire pSymB plasmid and the unambiguous location of all known and predicted genes. Oddly, what might seem the ultimate compendium of biological information, the genome sequence really represents a new starting point. It is the current point from which research moves ahead into the so-called post-genomic era, but is also a point from which techniques that pre-date the genomic era must be brought to bear. Understanding the mechanisms of replication and segregation of the megaplasmids in *S. meliloti*, resolving the vast transcriptional regulatory networks within the free-living and endosymbiotic cell, and determining why pSymB possesses so many varied solute uptake systems are but three of untold numbers of issues that wait to be addressed.

7. References

Bartosik D *et al.* (1998) Microbiol. 144, 3149-3157

Chain PSG *et al.* (2000) J. Bacteriol. 182, 5486-5494

Charles TC, Finan TM (1990) J. Bacteriol. 172, 2469-2476

Charles TC, Finan TM (1991) Genet. 127, 5-20

Finan TM *et al.* (1986) J. Bacteriol. 167, 66-72
Finan TM *et al.* (2001) Proc. Natl. Acad. Sci. USA 98, 9889-9894
Galibert F *et al.* (2001) Science 293, 668-672
Li P, Farrand SK (2000) J. Bacteriol. 182, 179-188
Margolin W, Long SR (1993) J. Bacteriol. 175, 6553-6561
Nishiguchi R *et al.* (1987) Mol. Gen. Genet. 206, 1-8
Ramirez-Romero MA *et al.* (2000) J. Bacteriol. 182, 3117-3124
Suzuki K *et al.* (1998) Biochim. Biophys. Acta 1396, 1-7
Tabata S *et al.* (1989) J. Bacteriol. 171, 1665-1672
Turner SL, Young JPW (1995) FEMS Microbiol. Lett. 133, 53-58

SEQUENCING AND ANNOTATION OF THE MEGAPLASMID pSYMB OF THE NITROGEN-FIXING ENDOSYMBIONT *SINORHIZOBIUM MELILOTI*

S. Weidner, J. Buhrmester, L. Sharypova, F.-J. Vorhölter, A. Becker, A. Pühler

Universität Bielefeld, Fakultät für Biologie, Lehrstuhl für Genetik, Universitätsstr. 25, D-33615 Bielefeld, Germany

1. Introduction

All bacteria known as rhizobia induce N_2-fixing root nodules on leguminous plants. In particular, *Sinorhizobium meliloti* nodulates species of the genera *Medicago, Melilotus* and *Trigonella*. The interaction of *S. meliloti* and *Medicago truncatula* has been studied for a long time by numerous international groups because this symbiosis represents a model system in the field of plant-microbe interaction.

Like many other bacteria belonging to alpha-proteobacteria, *S. meliloti* possesses a complex genome composed of a chromosome and two so-called megaplasmids, namely pSymA and pSymB. An international consortium sequenced the complete genome of *S. meliloti* strain 1021 (Galibert *et al.* 2001; Capela *et al.* 2001; Barnett *et al.* 2001; Finan *et al.* 2001). The sequence of the megaplasmid pSymB was established as a collaborative project of the Bielefeld group and in the group of T.M. Finan from the McMaster University (Hamilton, Ontario, Canada). In this article, the contribution of the Bielefeld group is presented.

2. Procedure

The sequencing of the pSymB megaplasmid was based on a BAC clone contig consisting of 24 BAC clones making use of the vector pBeloBAC11 (Barloy-Hubler *et al.* 2000). Later during the sequencing phase of the project, two additional BAC clones covering minor contig gaps were added to the original contig. By restriction analyses and Southern hybridization experiments to genomic DNA of *S. meliloti,* the nativeness of all BAC clone inserts was verified. To determine a suitable sequencing strategy, the sizes of overlaps between individual BAC clone inserts were estimated by restriction analyses and Southern hybridization. Shotgun clone libraries with insert sizes of 1.5 kb and 3.0 kb were generated for all BAC clones. Every second BAC clone of the minimal set was sequenced applying the shotgun sequencing strategy using M13 forward and reverse primers. By sequencing both ends of the remaining BAC clone inserts, the exact size of overlaps was determined. In the case of small overlaps to neighboring BAC clones, the shotgun sequencing strategy was employed again. For BAC clones with larger overlaps, shotgun clones were sequenced from one end first and, in case of a localization in a non-overlapping region, also from the other end.

Sequencing was continued until a 7.5-fold coverage was obtained. Closing gaps and polishing the sequence was carried out by primer walking on BAC insert DNA using custom primers. Sequence assembly was performed using the phred/phrap and Staden (gap4) packages (Ewing, Green 1998; Ewing *et al.* 1998; Staden 1996). Custom primers were designed by PRIDE (Haas *et al.* 1998). The annotation of the pSymB sequence was carried out in collaboration by the teams of McMaster University and the University of Bielefeld. In Bielefeld, the GEN-db annotation database environment, a recently developed program package, was used.

3. Results

As already mentioned, the complete megaplasmid pSymB could be covered by a minimal set of 26 BAC clones. The arrangement of the BAC clone inserts along the pSymB megaplasmid is illustrated in Figure 1. The total length of the *S. meliloti* pSymB megaplasmid turned out to be 1,683,333 bp. The G+C content was found to be 62.4%, which is almost identical to the chromosome 62.7% (Capela *et al.* 2001). In total, 1570 open reading frames have been predicted which means that 90% of the pSymB sequence can be considered as protein-coding.

4. Unexpected Genes Located on pSymB

Several genes annotated on pSymB were not expected to be situated on this replicon, e.g. *minCDE* and *ftsK2* (Figure 1). For *E. coli,* it was shown that the proteins MinCDE play a role in cell division. Surprisingly, the only copy of the *S. meliloti minCDE* genes is located on the pSymB megaplasmid. Another protein involved in the *E. coli* septa formation is FtsK. In *S. meliloti,* one of the two copies of *ftsK* is encoded on pSymB while the other copy is located on the chromosome. Another example of unexpected genes located on megaplasmid pSymB is the only copy of an arginine tRNA gene. The gene product provides the second most frequently used codon CCG (Figure 1).

Beside these individual genes, two more classes of genes dominate on the pSymB megaplasmid: (i) genes proposed to be involved in the biosynthesis of polysaccharides; and (ii) genes coding for transport systems of the ABC-type.

5. Polysaccharide-biosynthesis Gene Clusters Located on pSymB

Surface polysaccharides of *S. meliloti* are essential for successful nodule invasion. From previous studies, it was known that the *exo/exs* cluster (Becker *et al.* 1993; Glucksmann *et al.* 1993), directing the biosynthesis of succinoglycan (EPS I), and the *exp* cluster (Gazebrook, Walter 1989; Becker *et al.* 1997) involved in the biosynthesis of galactoglucan (EPS II), are located on the pSymB replicon. Many more genes whose products are proposed to be involved in the biosynthesis of polysaccharides have been identified during sequence annotation. Together with the *exo/exs* and the *exp* cluster, 11 gene clusters containing 188 predicted genes with a total size of 223 kb have been identified. They comprise over 12% of the genes located on pSymB (Figure 1).

Similarly to the *exo/exs* and *exp* cluster, the cryptic polysaccharide clusters encode diverse biochemical functions required for the production of polysaccharides, including the biosynthesis of sugar precursors (BSP), glycosyltransferase activities (GT) and the export/polymerization (Exp/Pol) machinery. Thus, the genes identified in the two largest cryptic clusters can be grouped into the following functional classes: (i) 7 BSP, 6 GT and 3 Exp/Pol; (ii) 10 BSP, 7 GT and 3 Exp/Pol. The gene products of the other clusters seem to cover only a few steps of a hypothetical polysaccharide biosynthetic pathway. Even the most complete gene clusters do not reveal such a compact and comprehensive organization as the *exo/exs* and *exp* clusters. Genes of the cryptic clusters do not form large transcriptional units and are often separated by genes which are unlikely to be involved in polysaccharide biosynthesis. Probably, the scattered and interrupted organization of these gene clusters resulted from genomic rearrangements and horizontal gene transfer. Consequently, it is possible that some cryptic clusters are silent.

Since *S. meliloti* 1021 was never reported to produce any extracellular polysaccharide in addition to EPS I and EPS II, the cryptic clusters are most probably involved in the synthesis of cell surface antigens, like K antigens (capsular polysaccharide, KPS) and O-antigens of lipopolysaccharides (LPS). In contrast to the enteric bacteria, *Rhizobium leguminosarum* and *R. etli,* the antigenic specificity of *S. meliloti* strains is determined by K antigens rather than O-antigens of LPS (Reuhs *et al.* 1998). This means that nearly every *S. meliloti* strain produces a different K antigen. Chemical structures of K antigen and LPS produced by Rm1021 are unknown. Therefore, pathways leading to the biosynthesis of these polymers cannot be predicted and linked to the functional annotation of the polysaccharide gene clusters. To elucidate functions of the cryptic clusters, knock-out mutants were generated by a plasmid integration approach. The phenotypes of the resulting mutants are under investigation.

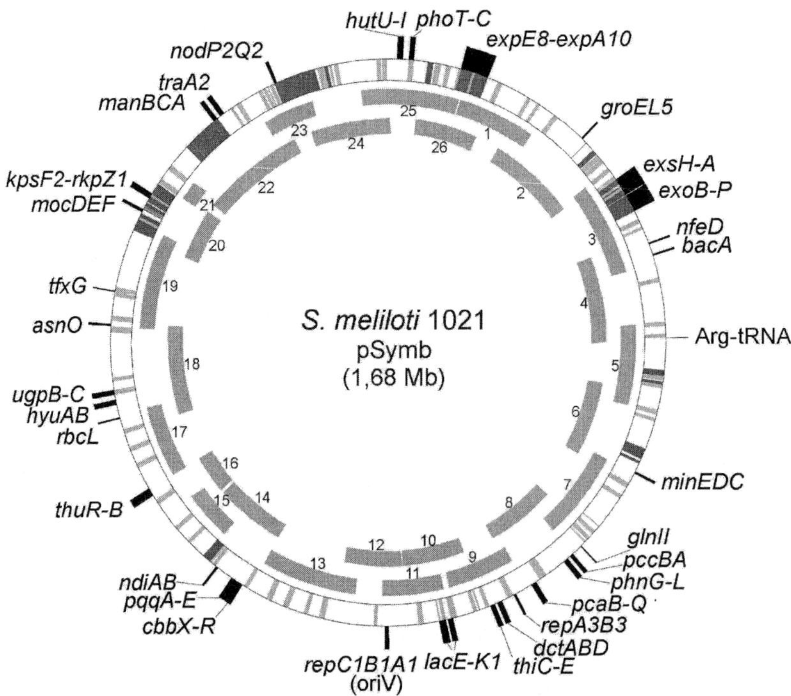

1. Genomic map of the pSymB megaplasmid of *Sinorhizobium meliloti* strain 1021. The positions of
pecific genes are indicated. The circle displays genes and gene clusters involved in polysaccharide
hesis (dark gray) and ABC transporter systems (light gray). The 26 BAC clone inserts forming the
one contig used for sequencing are displayed inside the circle.

C Transporter Gene Clusters Located on pSymB

ort of specific molecules across the cytoplasmic membrane is mediated by proteins associated
e membrane and belonging to different families. The largest and most diverse family of these
ort proteins is the ABC superfamily. ABC transporters consist of one or two integral
rane proteins (permeases), an ATP-binding protein (ATPase), and, in case of uptake systems,
lasmic solute-binding protein. In the whole genome of *S. meliloti,* 430 genes coding for ABC
orter systems have been predicted; 235 of them coding for 65 ABC transport systems are
d on the pSymB megaplasmid (Figure 1). Most of them are import systems, which are
ed to transport sugars (58%), amino acids and peptides (11%), iron (8%),
idine/putrescine (4%), and other solutes (19%). In accordance with these solute-import
s, numerous genes proposed to encode catabolic activities are located on the pSymB
lasmid.

In the symbiotic interaction, there is a great demand for iron, which is used, e.g. for the
enase complex, ferrodoxin and other iron proteins. In the genome of *S. meliloti,* eight iron
transporters have been annotated; three are encoded by the chromosome, two by megaplasmid
A and four by the megaplasmid pSymB. The four iron ABC transporters encoded by pSymB
analyzed in detail. All four ATPases contain the two conserved motifs, Walker$_A$ and Walker$_B$,
are involved in the binding of ATP. A characteristic signature sequence, the function of

which is unknown, could be identified for all four ATPases. Phylogenetic analyses demonstrated that all ATPases clustered into this group of importers. One of the ABC systems belongs to the siderophore/heme/vitamin B_{12} type. Two systems cluster into the group of ABC transporters of the ferric-iron type. The affiliation of the fourth iron ABC transport system is unclear. To clarify their specific role and regulation, knock-out mutants in several genes which encode the iron ABC transporters and putative iron regulators on the three replicons of *S. meliloti,* are under investigation.

7. Conclusions

There are several reasons that justify the view that the megaplasmid pSymB is a second chromosome in *S. meliloti*. First, the pSymB is comparable in size. Second, the gene regions of pSymB coding for the arginine-tRNA and MinCDE are essential for the growth of the *S. meliloti*. Third, the G+C content of pSymB and of the chromosome of *S. meliloti* are almost identical.

The large number of solute transport systems and numerous genes that are proposed to encode catabolic activities lead to the assumption that pSymB provides *S. meliloti* with the ability to utilize many different compounds from the soil and rhizosphere environment and enhances its metabolic flexibility. Similarly to the metabolic adaptation to different habitats, the large amount of polysaccharide gene clusters located on megaplasmid pSymB may extend the surface variability of *S. meliloti* and thereby enable the bacteria to cope with the different conditions and environments it encounters in the soil, rhizosphere and the legume nodule.

8. References

Barloy-Hubler F *et al.* (2000) Curr. Microbiol. 41, 109-113
Barnett MJ *et al.* (2001) Proc. Natl. Acad. Sci. USA 98, 9883-9888
Becker A *et al.* (1993) Mol. Gen. Genet. 241, 367-379
Becker A *et al.* (1997) J. Bacteriol. 179, 1375-1384
Capela D *et al.* (2001) Proc. Natl. Acad. Sci. USA 98, 9877-9882
Ewing B, Green P (1998) Nat. Genet. 25, 232-234
Ewing B *et al.* (1998) Genome Res. 8, 175-185
Finan TM *et al.* (2001) Proc. Natl. Acad. Sci USA 98, 9889-9894
Galibert F *et al.* (2001) Science 293, 668-672
Glazebrook J, Walker GC (1989) Cell 56, 661-672
Glucksmann MA *et al.* (1993) J. Bacteriol. 175, 7045-7055
Haas S *et al.* (1998) Nucleic Acids Res. 26, 3006-3012
Reuhs BL *et al.* (1998) Appl. Environ. Microbiol. 64, 4930-4938
Staden R (1996) Mol. Biotechnol. 5, 233-241

9. Acknowledgements

We thank F. Barloy-Hubler, D. Capela and F. Galibert for providing the pSymB BAC clones. This work was supported by a grant from the Bundesministerium für Forschung und Technologie (0311752) to A.P.

PROTEOME ANALYSIS OF *SINORHIZOBIUM MELILOTI*

M.A. Djordjevic, S.H. Natera, H.-C. Chen, G. Weiller, C. Menzel, G. Van Noorden,
J. Weinman, S. Taylor, K. Guo, B.G. Rolfe

Genomic Interactions Group, Research School of Biological Sciences, Australian National
University, GPO Box 475, Canberra, ACT Australia, 2601

1. Introduction

With the completion of the *Sinorhizobium meliloti* strain 1021 (Galibert *et al.* 2001) and
Mesorhizobium loti genomes (Kaneko *et al.* 2000) and with the knowledge of the NGR234 pSym
(Freiberg *et al.* 1997) and *Bradyrhizobium* symbiosis regions (Gottfert *et al.* 2001), we have now
entered the post-genome era for the analysis of these microsymbionts. The assembly of the strain
1021 genome has enabled the *Sinorhizobium* research community to be in the most advantageous
position yet to understand the complexities of this model organism. Until recently, there was a
frequent implication that knowledge of genome sequences alone would be sufficient to understand
biological systems. It is now well recognized that the genome provides a huge resource that will
assist understanding gene function. However, in isolation, the genome sequence is unable to predict
the following:

(i)   if and when mRNA species are translated;
(ii)   the relative concentrations of the proteins *in vivo*;
(iii)   the extent and types of post-translational modifications of proteins;
(iv)   the cellular or sub-cellular locations of proteins;
(v)   the unexpected pleiotropic effects of mutation or overexpression upon protein levels;
(vi)   the occurrence of small ORFs that are often overlooked by sequence annotation programs;
        and
(vii)   whether start sites for ORFs have been assigned correctly in all cases.

Nevertheless, armed with the genome sequence, there will be an increasing emphasis to use this
extensive resource to undertake a functional genomic analysis of *S. meliloti*. There remain gaps in
our knowledge, for example, in how the microsymbiont (a) escapes the full attention of the host
defence system (Djordjevic *et al.* 1987), (b) establishes the symbiosis with the legume host and (c)
survives in nutrient depleted environments. High throughput methodologies that establish patterns
of transcription (transcriptomics), translation (proteomics) and metabolic profiles (metabolomics)
will be combined with systematic mutagenesis initiatives in the near future to further define new
genes of interest and address points (i) through (vii). A combination of these approaches will
provide a foundation for the identification or elucidation of function using traditional and newly
evolving biochemical strategies.

In this paper we will give an overview of methodological considerations used in proteomic
studies and how proteomics has already contributed to generating new knowledge. Here, the
proteome is defined as the total protein output encoded by a genome and includes proteins that arise
from a single gene due to post-translational modification or cleavage. In the case of eukaryotes, the
proteome would also include the protein products that result from differential splicing.

2. Procedure

Two dimensional gel electrophoresis (2-DGE), is the most powerful technique available to separate
complex mixtures of proteins and it remains the method of choice that underpins proteome analysis.
2-DGE allows the separation of extremely complex mixtures of proteins and has a minimum of 20-

50 fold more resolving ability than reversed phase HPLC. Although approximately 3000 *S. meliloti* proteins can be resolved and visualized that have pIs in the range of pH 4-10 (Guerreiro *et al.* 1999; unpublished results), this technique is not without its drawbacks. Proteins of high hydrophobicity (integral membrane proteins of greater than seven transmembrane domains), and/or of extreme pI (greater than ten or less than 3.5), and/or of high molecular mass (greater than 100 kDa), and/or of low relative abundance (transcriptional regulators comprising a few molecules per cell), remain refractive to this technique. Alternative approaches are needed to address these problematic proteins. Figure 1 summarizes the window of proteins that can be isolated using proteome analysis as predicted from the *S. meliloti* chromosome.

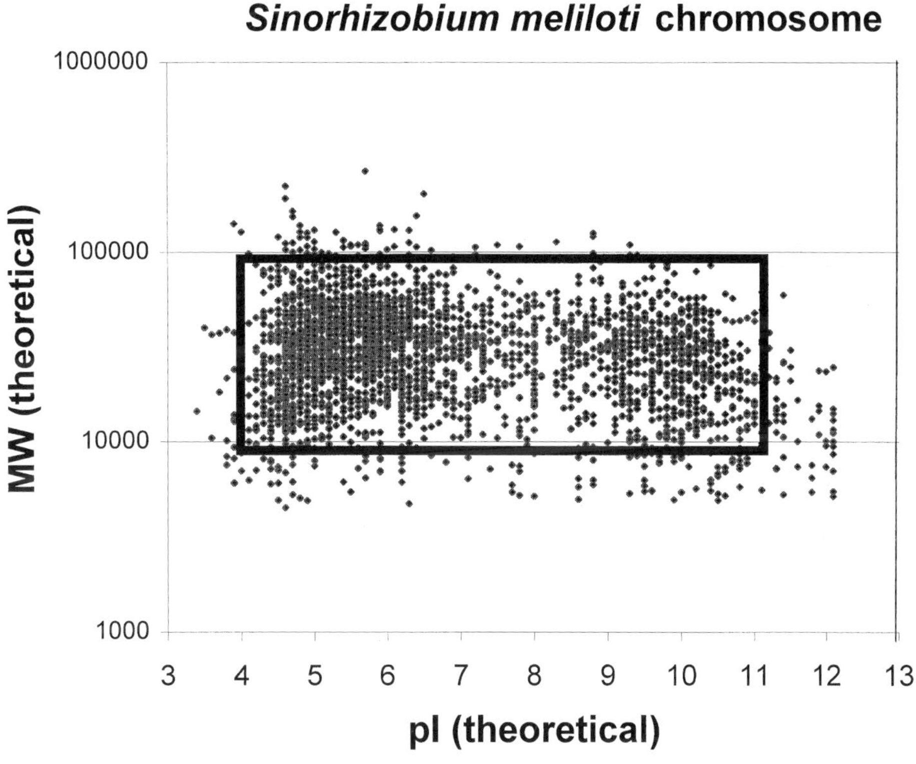

Figure 1. Diagramatic representation of *S. meliloti* proteins capable of being analyzed by proteome analysis. The dots represent the theoretical pI and molecular weight of proteins predicted from the *S. meliloti* chromosome. The box represents the pI and molecular mass window that is capable of being examined by 2-DGE analysis.

The procedures used to resolve, stain and quantify proteins in a gel are influenced by the experimental aims and by the type of post-gel analysis employed. Some of the major considerations that need to be understood prior to undertaking proteome analysis are briefly discussed here. *S. meliloti* strain 1021 encodes over 6200 open reading frames (Galibert *et al.* 2001) and if 50% of the genome is expressed at any one time, this would generate 3100 proteins of varying abundance. This protein number assumes that the proteins are not subjected to post-translational modifications that would lead to more than one protein product from one gene. Nevertheless, if one continues to work with this assumption, then there is a need to visualize at least 3000 individual proteins in order to obtain an overall perspective of the proteome of this organism and to conduct effective post-gel analysis of the proteins present. High loads of total protein are needed to achieve this aim, although gel resolution is compromised when loads are too high.

Another consideration is the limit of detection for colloidal Coomassie which is around 1 µg of protein. Since Coomassie stained protein spots are preferred for peptide mass fingerprint analysis (see below) this means that at least 3 mg of total protein needs to be separated over the entire pH range. We therefore load up to 1 mg of total protein for first dimensional separation for each pH range and use the sequential colloidal Coomassie staining method prior to gel analysis. Although MALDI-TOF mass spectrometry (Matrix Assisted Laser Desorption Ionization MS) is our preferred strategy for protein identification via peptide mass fingerprinting (Natera *et al.* 2000), we sometimes employ N-terminal sequencing (Chen *et al.* 2000 a, b; Guerreiro *et al.* 1997, 1998, 1999; Natera *et al.* 2000) or Western blotting to identify specific classes of proteins (see Figure 2).

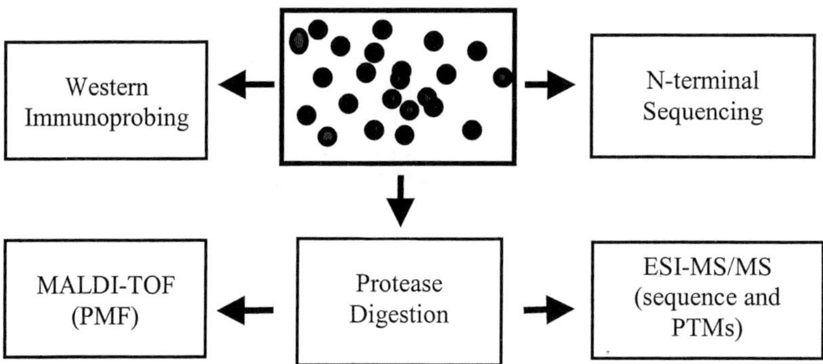

Figure 2. Different strategies for post-gel separation analyses. Proteins from 2-D gels (upper central panel) can be analyzed by Western immunoprobing, Edman sequencing or via mass spectrometry. MALDI-TOF generates peptide mass fingerprints of the peptides generated from protein digestion whereas ESI MS/MS can generate peptide sequence and identify post translational modifications (PTMs) of the peptides.

Cellular lysis of *S. meliloti* is achieved by disrupting the cells in reagents that are compatible with 2-DGE (Guerreiro *et al.* 1997). An overlapping series of IPG strips that can collectively separate proteins between pH 3 and pH 11 are used to establish a proteomic pI "contig". Proteins are then separated according to their size using either vertical or horizontal electrophoresis with the very thin pre-cast horizontal gels giving the best resolution. Protein visualization is best achieved by staining the gel directly using silver or fluorescent stains for analytical gels or sequential colloidal Coomassie staining (Chen *et al.* 2000 a, b). Image analysis of gels loaded with equal amounts of protein is achieved using Melanie 3 software and this enables differential protein expression to be quantified. Generally, we run at least three replicates to ensure that the changes in protein levels are due to the experimental regime.

Peptide mass fingerprinting is the cheapest and fastest method of post-gel protein identification. MALDI-TOF mass spectrometry, as stated above, can usually generate effective data when a minimum of 1 µg of protein is present in a stained protein spot. Proteins are first digested with a protease (usually trypsin) and the masses of the resulting peptides are determined with a high degree of accuracy to generate a peptide mass "fingerprint" from MALDI-TOF analysis. This information is then used to match theoretical databases of peptide fragments generated from analyzing the output of the genomic sequence information. Masslynx or similar software packages are used to determine the likelihood of the match and a diagramatic example of the matches generated is shown in Figure 3. Alternatively, N-terminal sequencing or Electrospray Ionization Mass Spectroscopy (ESI-MS^n) can be used to generate protein sequence from the peptides generated and determine many post-translational modifications.

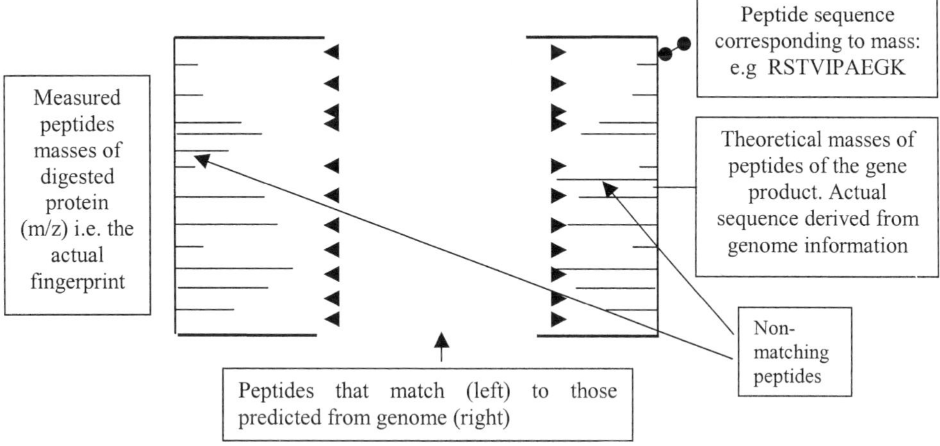

Figure 3. Diagramatic representation of the process of peptide mass fingerprinting.

3. Results and Discussion

Using these strategies we have examined the proteome of *S. meliloti*:

(i)   at different phases of growth (Guerreiro *et al.* 1999);
(ii)   after the deletion of the pSym (Chen *et al.* 2000b);
(iii)   in the endosymbiotic state compared to culture grown (Natera *et al.* 2000);
(iv)   to examine the effects of a mutation of *nolR* in strain Rm41 (Chen *et al.* 2000a).

We have also used proteome analysis to determine the response of *R. leguminosarum* to flavonoid exposure (Guerreiro *et al.* 1997), the effects of plasmid curing (Guerreiro *et al.* 1998) or to examine the unexpected pleiotropic effects of mutation (Guerreiro *et al.* 2000).

Several major points have resulted from the analysis of this data. First, some products of the nodulation genes can be detected in *R. leguminosarum* but thus far, not in *S. meliloti*. This may reflect the different extent of transcriptional activation of nodulation genes in these two organisms. Second, extensive pleiotropic effects in the proteome can be detected that result from either (a) the mutation of a regulatory gene (*nolR*) or (b) a structural gene involved in polysaccharide synthesis (*pssA*). The extent of these changes rival those seen when large areas of the pSym or other plasmids are deleted from the strains (Guerreiro *et al.* 2000) even though a large amount of genetic information is missing in the deleted strains. The high number of proteins that were induced by mutation of *pssA* in two species was unexpected especially if this gene product possesses the one function in polysaccharide synthesis. However, the increasing number of multifunctional proteins that are being discovered (Jeffery 1999) combined with the notion that multiple protein-protein interactions can occur in the cell, adds a further level of complexity to the analysis of biological systems and emphasizes that multiple approaches will be necessary to unravel these processes. It is not inconceivable that PssA may interact with other proteins in the cell or indeed possess another or other functions. Finally, extensive alterations occur in the expression profiles of proteins isolated from bacteroids when compared to culture grown cells. This result most likely reflects the alteration to metabolism that result from a low oxygen environment, the switch to the nitrogen-fixing state and the utilization of more specific sources of nutrients that are provided by the plant (Natera *et al.* 2000). We are currently undertaking a comprehensive analysis of over 2000 proteins isolated from either cultured cells or bacteroids. This endeavor will go a long way to define the major changes that occur as the bacteria make the transition from the cultured state to the bacteroid. We expect that over 85% of the proteins will be identified, far more than our previous analyses. This is to be expected since our initial analyses were made at a time when the full genome sequence was not available. Nevertheless, even with a 1x shotgun coverage of the genome (supplied by S. Long and M. Barnett, Stanford University), we were able to obtain a significant number of identities for our PMF queries (Natera *et al.* 2000).

4. References

Chen *et al.* (2000) Electrophoresis 21, 3833-3842
Chen *et al.* (2000) Electrophoresis 21, 3823-3832
Djordjevic *et al.* (1987) Ann. Rev. Phytopathol. 25, 145-168
Freiberg *et al.* (1997) Nature 387, 394-401
Galibert *et al.* (2001) Science 293, 668-672
Gottfert *et al.* (2001) J. Bacteriol. 183, 1405-1412
Guerreiro *et al.* (1997) Molec. Plant-Microbe Interact. 10, 506-516
Guerreiro *et al.* (1998) Electrophoresis 19, 1972-1979
Guerreiro *et al.* (1999) Electrophoresis 20, 818-825
Guerreiro *et al.* (2000) J. Bacteriol. 182, 4521-4532
Jeffery (1999) Trends Biochem. Sciences 24, 8-11
Kaneko *et al.* (2000) DNA Res. 7, 331-338
Natera *et al.* (2000) Molec. Plant-Microbe Interact. 13, 995-1009

5. Acknowledgements

We acknowledge access to the *S. meliloti* genome sequence prior to publication and access to the 1x shotgun genome sequence provided by the *S. meliloti* sequencing consortium. A grant from the Australian Government has facilitated the peptide mass fingerprinting analysis at the Australian Proteome Analysis Facility (Macquarie University, Australia).

COMPARISON OF CHROMOSOMAL GENES FROM *M. LOTI* AND *S. MELILOTI* SUGGEST AN ANCESTRAL GENOME

R.A. Morton

Dept of Biology, McMaster University, 1280 Main St. West, Hamilton, ON, L8S 4K1
Canada

1. Introduction

Although the origins of biological fixation of nitrogen are unclear, it is an ancient process, thought to have evolved more than 2 billion years ago, perhaps in response to a decline in abiotic nitrogen fixation as atmospheric CO_2 decreased (Navarro-González *et al.* 2001; Raven, Yin 1998). Biological nitrogen fixation is limited to prokaryotes, but in some cases microbes fix nitrogen in symbiosis with plants. It is not clear when such nitrogen-fixing associations may have originated (Raven, Yin 1998). The most recent are root-nodule symbioses. Nodule-forming plants are closely related, probably belonging to a single clade (Gualtieri, Bisseling 2000). Within this clade, however, only diverse subgroups of plants are nodulated and comparison of plant and bacterial phylogenies suggest multiple origins of symbiosis (Swensen 1996; Doyle 1998).

The availabilty of complete genomes of symbiotic, nodulating bacteria allows comprehensive comparison of their genetic systems and hypotheses about their evolutionary histories to be made. I have compared genes of *M. loti* and *S. meliloti* in order to identify those gene functions that were present in their ancestor and conserved in their present genomes. The results confirm the independent origin of current nitrogen-fixing bacteria and the key role played by horizontal gene transfer.

2. Material and Methods

Orthologous pairs of genes coalesce in the most recent common ancestor of their genomes. A pair of orthologs is a connection between two genes that can be visualized by a genome dot plot. In order to develop a set of orthologous connections, I used a principle of reciprocal similarity. Orthologous genes are more similar to each other than to any other gene in either genome. Each gene of one genome is BLASTed against all the genes of the second genome. This is repeated in the reciprocal direction. When the most similar pair of genes (the BLAST hit with the highest score) in reciprocal directions is the same, the pair is classified as orthologs. One difficulty with this method occurs when several target genes are similar to a single origin gene as, for example, is often the case with rRNA-coding genes. Therefore I made an adjustment when the BLAST hits a group of genes in the target genome with nearly the same score ("duplications"). Duplications were resolved by comparisons with neighboring genes. The "duplication" (if any) which maintained a continuous set of orthologous pairs was retained as the ortholog.

The *Mesorhizobium loti* MAFF303099 genome (http://www.kazusa.or.jp/rhizobase/) and a preliminary release of the *Sinorhizobium meliloti* 1021 genome (http://sequence.toulouse.inra.fr/rhime/Complete/doc/Complete.html) were obtained in February 2001. Orthologous ORFs were aligned with CLUSTALW and the expected substitution rate determined using the PROTDIST or DNADIST algorithms of Phylip (Felsenstein 1994). Distances are expected substitutions per site using the PAM 250 model of evolution for protein or the Kimura 2-parameter model with a transition/transversion ratio of 2 for DNA.

3. Results and Discussion

3.1. Ancestral genome. *M. loti* and *S. meliloti* are both nitrogen-fixing, symbiotic rhizobacteria which nodulate different host species. *M. loti* has a large chromosome (>6750 genes) and two small plasmids (pMLa and pMlb). The *S. meliloti* chromosome contains about 3400 genes and there are two large "megaplasmids" of 1.35 Mbp (pSymA) and 1.68 Mbp (pSymB) together containing more than 2800 genes. Genes of the *M. loti* chromosome were compared with those of each of the three *S. meliloti* replicons (Figure 1).

The results show that there is extensive orthology (2573 chromosomal gene pairs) between chromosomal genes of these two species. Orthologous pairs are scattered throughout their respective genomes. However, a pattern is seen of orthologs that form contiguous groups, sometime in inverted direction (Figure 1). These represent conserved chromosomal sequences of genes or syntenic groups. Genes in these syntenic groups can be identified by virtue of belonging to "runs" of orthologous pairs. When five or more pairs is used as a cut-off, 984 orthologs (approximately 40% of all chromosomal orthologs) were identified as belonging to a putative ancestral genome. The genome dot plots of *M. loti* chromosome genes against genes of the two *S. meliloti* megaplasmids show no evidence of extensive synteny although there are several regions of limited extent. In the case of pSymA, 620 potential orthologs were identified, but only 61 in runs of five or more while for pSymB there were 121 out of 825 orthologs. Thus, there is little support for the hypothesis that the smaller *S. meliloti* chromosome resulted from the transfer of ancestral genes into the megaplasmids. Rather, genes of the *S. meliloti* megaplasmids seem to be of divergent origin when compared to the *M. loti* chromosome (Figure 1).

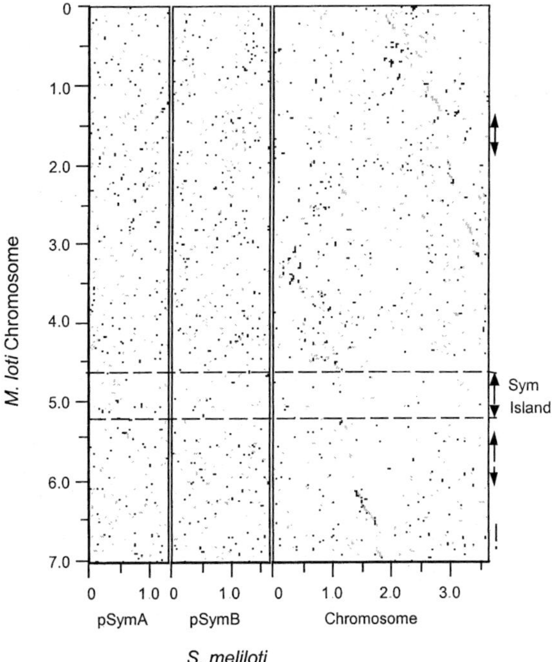

Figure 1. Genome dot plot comparing annotated genes of the *M. loti* MAFF303099 chromosome (vertical axis) with those of three replicons of *S. meliloti* 1021. Axes are in megabase pairs from an arbitrary origin. The *M. loti* "symbiotic island" is identified by "Sym Island" and the extent of several other regions of the *M. loti* chromosome that appear to be insertions relative to the *S. meliloti* chromosome are indicated to the right by double arrows.

3.2. Symbiotic insertions. Sullivan and Ronson (1998) showed that *M. loti* strain ICMP3153 could transfer a large symbiotic element to non-symbiotic strains which then allowed them to nodulate plants. This element was defined in *M. loti* MAFF303099 by Takakazu *et al.* (2000). It is shown on the *M. loti* chromosome right axis in Figure 1 as "Sym Island". Consistent with an origin by horizontal gene transfer, this *M. loti* symbiotic island was not identified as part of the ancestral chromosome. As well, there are a number of other regions of the *M. loti* chromosome

which appear to have received large insertions since divergence from the *S. meliloti*
Chromosomal rearrangements make difficult the determination of all such regions, bu
possible ones have been indicated by double arrows on the right side of Figure 1.

3.3. Gene divergence. Identification of genes descended vertically from an ancestral
allows study of rates of divergence. The 2526 putative protein-coding orthologs have an
divergence of 0.714 substitutions per amino acid site, while the average divergence of
protein-coding ORFs that were part of orthologous runs of 5 or more genes is 0.553 (Figur
fraction of those genes which are not part of extensive "runs" are clearly more d
(distances > 1). A similar analysis of orthologous, ancestral pairs between *S. typhi* and
gave an average distance of 0.14 expected amino acid substitutions per amino acid site (I
al. 2001). This is about four times the average distance (0.55) between the 967 proteir
genes conservatively identified as descended from an
ancestral genome.

Since *S. typhi* and *E. coli* are estimated to have
diverged approximately 100 million years ago, either
there was an ancient separation of the *M. loti/
S. meliloti* lineages (~400 million years ago), or
alternatively, there has been more rapid rate of
divergence of their proteins.

The greater average divergence of *M. loti/
S. meliloti* orthologs relative to *E. coli/S. typhi*
orthologs was confirmed for many individual,
homologous genes. Table 1 shows that equivalent
genes involved in ammonia assimilation are 5-10 times
more diverged in the rhizobial species than in *E. coli/S.
typhi.* On the other hand, the 16S rRNA genes are
only 1.5 times more diverged in rhizobia.

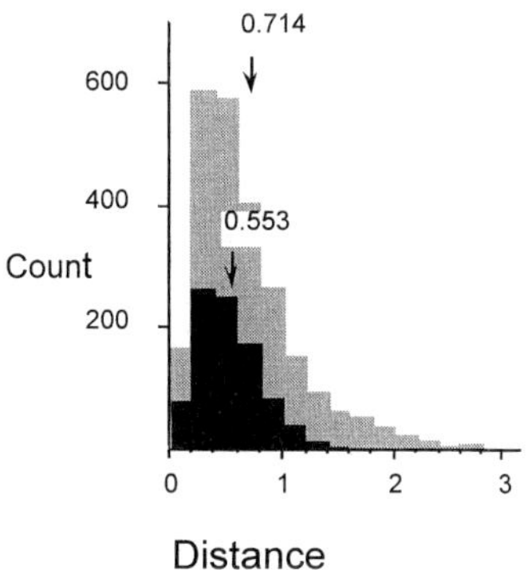

Figure 2.   Distribution of chro
protein-coding ortholog distances
M. loti and *S. meliloti.* Grey: all
(2526), black: orthologs in runs of
(967).

3.4. Origins of symbiotic nitrogen fixation. The
rhizobia have been divided into three distinct groups,
Rhizobium, Bradyrhizobium and Azorhizobium, on
the basis of 16S rRNA gene sequences (Young and Haukka 1996). These group:
monophyletic with regards to nitrogen fixation. Non-nodulating bacteria are inter-dispers
16S rRNA tree along with non-nitrogen-fixing species. Although plants that are ca
forming symbiotic relationships form a large monophyletic clade, nodulating groups are
among non-nodulating groups. There is little phylogenetic correlation between nitrog
bacteria and their legume hosts (Doyle, 1998). Related groups of bacteria can often
unrelated groups of plants.

Nodulation and nitrogen-fixing genes in *Mesorhizobium* have been located on ":
islands" and these have been demonstrated to be capable of transmission between bacter
(Sullivan and Ronson 1998). Taken together, these results suggest that symbiotic nitroge
has independently evolved and been lost many times.

Young and Haukka (1996) concluded from the 16S rRNA phylogeny that the
ancestor of rhizobia pre-dated the origin of higher plants. Among the rhizobia, *Meso*
and *Sinorhizobium* are distinct groups which cluster separately according to their 1
sequences. The extent of divergence of their 16S rRNA genes does not indicate as
separation of these two lineages as does their chromosomal orthologs (Table 1). Both,

Table 1. Protein distances between homologous gene pairs in *E. coli/S. typhi* compared to *M. loti/S. meliloti*

Chromosomal Gene	DNA or Protein Distance (sub./site)		
	E. coli *S. typhi*	*M. loti* *S. meliloti*	Ratio
glnA Glutamine Synthetase	0.018	0.106	5.9
gltD Glutamate Synthetase (β)	0.047	0.216	4.6
gltB Glutamate Synthetase (α)	0.058	0.263	4.5
amtB Ammonium Transport	0.088	0.789	9.0
glnK Nitrogen Regulation	0.013	0.148	11.4
glnB Nitrogen Regulation	0	0.118	NA
ntrB (glnL) Nitrogen Regulation	0.067	0.361	5.4
ntrC (glnG) Nitrogen Regulation	0.048	0.273	5.7
16S rRNA	0.031	0.046	1.5
pSymA (*S. meliloti*) to Chromosome (*M. loti*) Gene			
nifH Nitrogenase (Fe)	NA	0.076	NA
nifD Nitrogenase (FeMo α)	NA	0.154	NA
nifK Nitrogenase (FeMo β)	NA	0.123	NA

NA=not applicable.

are consistent with separation of lineages long before the evolution of symbiotic nitrogen fixation. Thus, nitrogen fixation and nodulation must have been acquired independently, apparently by horizontal transfer of alien genes into the chromosome of *M. loti* and into the pSymA plasmid of *S. meliloti*. Genes that have been transmitted vertically on the chromosomes of these two species appear to be those that are required for a free-living lifestyle, similar to that of their presumed progenitor.

4. References

Doyle JJ (1998) Trends Plant Sci. 3, 473-478
Felsenstein J (1994) PHYLIP Version 3.5, distributed by the author
Gualtieri G, Bisseling T (2000) Plant Mol. Biol. 42, 181-194
Koski LB, Morton RA, Golding GB (2001) Mol. Biol. Evol. 18, 404-412
Navarro-González R, McKay CP, Mvondo DN (2001) Nature 412, 61-64
Raven JA, Yip Z-H (1998) New Phytol. 139, 205-219
Sullivan JT, Ronson CW (1998) Proc. Natl. Acad. Sci. USA 95, 5145-5149
Swensen SM (1996) Amer. J. Bot. 83, 1503-1512
Takakazu K *et al.* (2000) DNA Res. 7, 331-338
Young JP, Haukka KE (1996) New Phytol. 133, 87-94

5. Acknowledgements

I would like to thank Dr Turlough Finan for suggesting this project and providing the opportunity to use the *S. meliloti* sequences before publication.

CHAIR'S COMMENTS: EVOLUTION, ECOLOGY AND ECOSYSTEMS

D. Prévost

Soils and Crops Research and Development Center, Agriculture and Agri-Food Canada,
2560 Hochelaga Blvd, Sainte-Foy, PQ, G1V 2J3 Canada

1. Introduction

This session brings significant new findings on the evolution of legume nodulating bacteria from a phylogenetic point of view as well as on the genome evolution in relation to ecological factors. During the last decade, a large number of reports on the diversity of nodulating bacteria revealed that they evolved and adapted in response to various environmental factors, host legume species and ecological niches. Evolutionary relationships among bacteria are usually estimated by sequence comparisons of the highly conserved 16S rRNA genes. However, phylogenetic analysis of genes involved in the establishment and the function of the symbiosis and of those associated in the adaptation to selective pressure increases our knowledge on their transfer and their origin. Furthermore, the use of elegant molecular techniques is leading to a more comprehensive understanding of the evolutionary changes that confer fitness to the environment and of the factors responsible for this evolution in the nodule forming bacteria.

2. From the Evolution of Nodulation Genes to That of 16S rRNA Genes

A very interesting finding presented by C. Boivin-Masson came out from an extensive phylogenetic analysis of the nodulation genes (*nodA*), for which a strong correlation was found between the NodA protein and the Nod factor (NF). The phylogenetic analysis of strains harboring novel NodA sequences showed that they belong to the genus *Burkholderia*. This genus (and a new *Ralstonia* sp. isolated from *Mimosa* spp. in Taiwan) is in the β-subclass of Proteobacteria. All other genera of nodulating bacteria ("*Rhizobium*" named species and *Methylobacterium*) reported up to now belong to the α-subclass. It has been already suggested that the nodulating capacity is distributed laterally among distant taxa. The present finding supports this hypothesis and suggests that *nod* genes from both α and β nodulating bacteria were acquired through horizontal gene transfer. Furthermore, a correlation was observed between *nodA* and 16S rRNA phylogenies of the strains studied. Other recent reports showed that the close relationship among symbiotic genes (*nodC* and *nifH*) of *Phaseolus* symbionts (different genera) was associated with their host range, but independent of their classification based on the 16S rRNA gene (Laguerre *et al.* 2000). However, with the different genera of rhizobia isolated from *Astragalus*, *Oxytropis* and *Onobrychis* spp., there was a phylogenetic congruence at the genus level between symbiotic (*nodC* and *nifH*) and 16S rRNA genes, but no relation was found with the host nodulation range, indicating a specific evolution pattern for these rhizobia (Prévost *et al.* 2000). Since bacterial symbionts of more than 90% of legumes genera have not been studied, it is evident that there are still exciting discoveries ahead.

While there have been major developments in our understanding of the evolution, we also appreciate that the methods used for phylogenetic analysis are in constant progress. The interpretation of the divergence estimated in the sequences of the 16S rRNA genes depends on the overall comprehension of bacterial evolution. In this session, P. van Berkum presented a critical examination of the use of 16S rRNA gene sequence for reconstruction of evolutionary histories. A major concern stems from the observation that there is genetic evidence that the evolution of 16S rRNA genes could be reticulate instead of hierarchical. The topologies of trees constructed from 16S rRNA genes differ depending on the taxa selected in the analysis, also the trees obtained with 16S rRNA gene, 23S rRNA gene and ITS region sequences are statistically not congruent. This calls for caution in the conclusions that can be drawn from the use of 16S rRNA gene alone, and

nstitutes a serious argument in favor of the standardization of the approaches. This could be
ιportant especially when the analysis is used for decisions in taxonomy. One example is the
stinction of *Sinorhizobium* from *Rhizobium*, based from the analysis of 16S rRNA gene, which
as not supported by the analysis of the 23S rRNA gene (Martinez-Romero *et al.* 2000).

The Acquisition of Genes for Adaptation and the Identification of Stress-induced Genes

ιe recent discovery of symbiosis islands (chromosomal elements) in *Mesorhizobium* strains from
ɔtus and the demonstration of their lateral transfer to non-symbiotic mesorhizobia present in soils is
rong evidence for the evolution of new symbiotic strains in the field. C. Ronson and his group
ιrsued their investigation with the aim to learn more about the contribution of these genomic
lands to the evolution and niche adaptation. In this session, C. Ronson reported on the presence of
enes downstream of the symbiosis island insertion sites (phe-tRNA locus). By grouping a
ɔpulation of 35 strains of mesorhizobia according to sequence similarity downstream of the
ɣmbiosis island, there was evidence for 15 different acquired DNA regions. The genomic structure
etermined in a few strains showed that some of these regions encode traits such as iron acquisition
ιd adhesins involved in seed colonization. These acquisitions may be advantageous for ecological
daptation, and competition and provide additional evidence of the adaptive value of horizontal gene
·ansfer.

The identification of genes that are expressed under specific environmental conditions has
onsiderably increased with the use of molecular tools, such as the marker-reporter genes.
'. de Bruijn and his team already got important data on genes induced in *S. meliloti*. They used a
'n5-*luxAB* reporter system to monitor gene expression in response to N, C, O_2 limitations and
essication. Many tagged genes were identified and studied for their regulation and their role in
ιersistence and competition. In this session, F. de Bruijn demonstrated the complementarity and the
ɔower of using comparative and functional genomics (micro-arrays) to study stress-induced
hizobial loci. Another promising approach, the IVET (*in vitro* expression technology) is in progress
o identify genes induced in the rhizosphere (Izallalen *et al.* 2000). These long-term studies will
ιelp to elucidate the processes involved in competition and persistence of introduced strains in soils
ιnd in the rhizosphere of legumes.

ł. Conclusions

Γhe progress in our understanding of the evolution and ecology will certainly be applicable to the
levelopment of new microbial technologies. For instance, the knowledge on lateral transfer of
iymbiotic genes and the identification of the genetic background for plant infection would allow the
levelopment of more efficient symbioses. Moreover, the understanding of the processes involved
ιnder environmental conditions for adaptation, persistence and competition will be useful in
noculant technology.

5. References

[zallalen *et al.* (2000) In Program and Abstracts, 17th NACSNF, pp. 69, Québec, Canada
Laguerre *et al.* (2001) Microbiol. 147, 981-993
Martinez-Romero *et al.* (2000) In Program and Abstracts, 17th NACSNF, pp. 33, Québec, Canada
Prévost D *et al.* (2000) In Pedrosa F *et al.* (ed) Nitrogen Fixation: From Molecules to Crop
    Productivity, pp. 205, Kluwer Academic Publishers, Dordrecht, The Netherlands

RHIZOBIA: THE FAMILY IS EXPANDING

L. Moulin[1], W.M. Chen[2], G. Béna[1], B. Dreyfus[1], C. Boivin-Masson[1]

[1]LSTM, IRD-INRA-CIRAD-ENSAM, TA 10/J, Baillarguet, 34 398 Montpellier Cedex 5, France
[2]Tajen Institute of Technology, No. 20, Wei-Shin Road, Shin-Erh Village, Yen-Pu, Ping-Tung 90703, Taiwan

1. Introduction

Whereas nitrogen fixation is widespread in bacteria and archae, nitrogen fixation in symbiosis with legumes is performed by a group of bacteria, known as rhizobia, that until recently belonged exclusively to the α-subclass of Proteobacteria (Young, Haukka 1996; Sy *et al.* 2001). The ability of rhizobia to nodulate legumes is determined by a set of genes, the nodulation genes, essential to trigger nodule induction (Perret *et al.* 2000). Nodulation genes are involved in the production of lipochito-oligosaccharides (Nod factors) that act as signaling molecules for nodulating specific legume hosts (Lerouge *et al.* 1990; Spaink *et al.* 1991). Rhizobia are distributed in four distinct branches of the α-Proteobacteria, each containing many bacterial species that are not rhizobia. It is now widely accepted that *nod* gene transfers have occurred among members of the four rhizobial clades and account for the polyphyletic origin of rhizobia (Suominene *et al.* 2001). Gene transfer leading to an efficient symbiosis was indeed demonstrated both in the field and in the laboratory (Sullivan *et al.* 1995, 1998). The clustering of rhizobia within a same bacterial phylum suggested that only α-Proteobacteria possessed the genetic background for legume symbiosis. Our recent results (Moulin *et al.* 2001) show that the ability to establish a symbiosis is more widespread in bacteria than anticipated to date since we found nodulating bacteria within the β-Proteobacteria.

We discuss how this finding may open the way for the discovery of new rhizobia and may contribute to a better understanding of the origin and evolution of rhizobium-legume symbioses.

2. Extension of Rhizobia from α- to β-Proteobacteria: Identification of Nodulating *Burkholderia*

The *nodA* gene is involved in the synthesis of the core Nod Factor by specifying the transfer of an acyl chain to the acceptor chitooligosaccharide. In the course of the phylogenetic analysis of the NodA protein from a collection of rhizobia (Moulin *et al.* this volume), we were intrigued by two sequences that did not group with other NodA sequences. One *nodA* was amplified from STM678, a strain isolated from *Aspalathus carnosa* in South Africa. The other *nodA* belonged to STM815, a strain isolated from *Machaerium lunatum* in French Guyana. By sequencing the 16S rDNA of these strains we found that they belonged to the *Burkholderia* genus within the β-subclass of Proteobacteria and are thus phylogenetically distant from known rhizobia (Moulin *et al.* 2001) (Figure 1). The phylogenetic position of STM678 was confirmed by 23S rDNA and *dnaK* partial sequencing. Recent investigations indicate that they correspond to two distinct species, close to *B. kururiensis* (P. Vandamme, unpublished results). The nodulation ability of strains STM678 and STM815 was confirmed by inoculation of *Macroptilium atropurpureum*, a broad host range legume, and by re-isolation and characterization of the bacteria isolated from the induced nodules. Nodules induced on *M. atropurpureum* were ineffective in terms of nitrogen fixation, probably because *M. atropurpureum* is not the original symbiotic partner. By screening among bacteria isolated from root nodules collected from various legumes in Senegal we found a third *Burkholderia* strain,

ORS1827 (Figure 1), isolated from *Alysicarpus glumaceüs* and fixing nitrogen in symbiosis with many tropical legumes.

3. Identification of a Second Rhizobial Genus, *Ralstonia*, within β-Proteobacteria

The taxonomic characterization of a collection of root isolates (about 180) from *Mimosa pudica* and *M. diplotricha* in Taiwan revealed that most of the strains (94% of the isolates) also belong to the β-Proteobacteria. 16S rDNA sequencing positioned all these strains in the *Ralstonia* genus, close to *R. eutropha* (Figure 1). These strains represent a novel *Ralstonia* species for which the name *R. taiwanensis* was proposed (Chen *et al.* in press). LMG19424 is the type strain. We checked the ability of 4 isolates to re-nodulate their host plants. All formed nitrogen-fixing nodules on *Mimosa pudica*. Characterization of the re-isolates confirmed the rhizobial status of these *Ralstonia* strains.

We propose to use the terms α-rhizobia and β-rhizobia to distinguish rhizobia belonging to α-Proteobacteria from rhizobia of β-Proteobacteria. This nomenclature will be used in the following text.

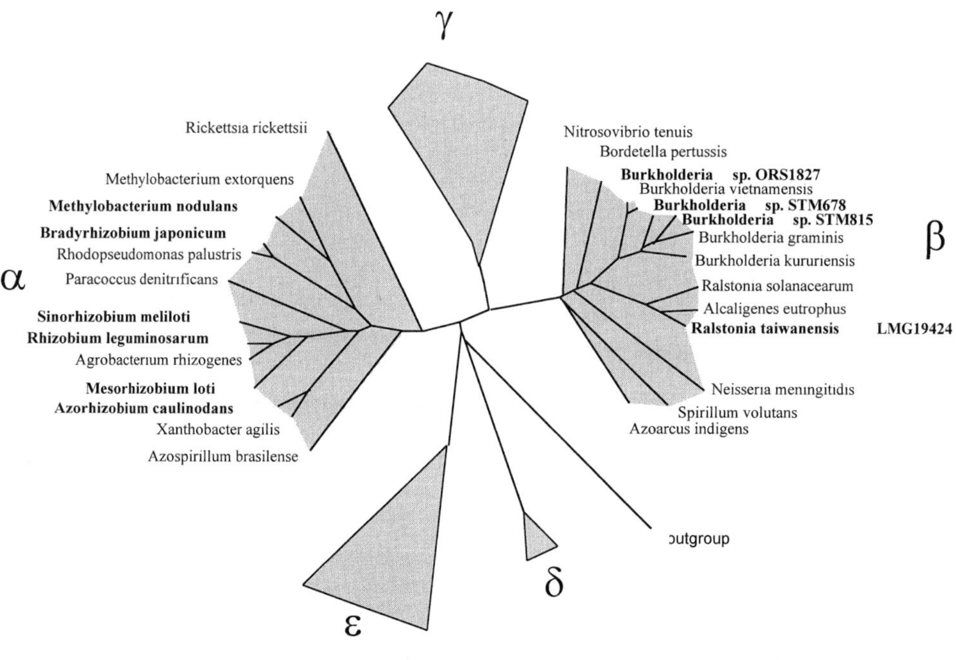

Figure 1. Unrooted 16SrDNA tree of Proteobacteria. The figure shows the phylogenetic relationships between the different rhizobial genera - as represented by type species in bold - including the new *Burkholderia* sp. and *Ralstonia* sp. strains. α, β, δ, γ and ε represent the different subdivisions of the Proteobacteria. The tree was constructed by using the neighbor-joining method and was adapted from van Berkum and Eardly, 1998.

4. α- and β-Rhizobia Use the Same Strategy to Nodulate Legumes

The *nodABC* genes are responsible for the synthesis of the core structure of the Nod factors and as such are present in all α-rhizobia. β-rhizobia are not an exception, since a *nodA* gene could be amplified and sequenced in *R. taiwanensis* LMG19424 and in the three *Burkholderia* sp. strains STM815, STM678 and ORS1827. Sequence similarity with the different complete rhizobial NodA protein sequences available in databases ranged from 67.5% (STM678/*A. caulinodans*) to 77.7% (STM678/*M. nodulans*). Further *nod* gene sequencing in *Burkholderia* sp. STM 678 and *R. taiwanensis* LMG14424 revealed a genetic organization of *nodABC* genes similar to that found in other rhizobia. In *Burkholderia* sp. STM678 *nodAB* are in the same orientation and overlapping and preceded by a *nodD*-dependent regulatory sequence (*nod* box). A *nodC* gene was found elsewhere in the genome. In *R. taiwanensis nodB* is in front of *nodC*, and preceded by a *nod* box whereas *nodA* was found elsewhere in the genome. Such genetic unlinkage of *nodABC* genes was already described (Zhang *et al.* 2000). A *nodA* mutant of *Burkholderia* sp. STM678, constructed by inserting a *lacZ*-kanamycin cassette into the *nodA* gene, did not nodulate *M. atropurpureum*, indicating that the *nodAB* genes are required for nodulation of this *Burkholderia* strain. Moreover this strain has been shown to produce Nod factors, that are N-methylated and 4,6-dicarbamoylated on the non-reducing end but not substituted on the reducing terminus (Boone *et al.* 1999).

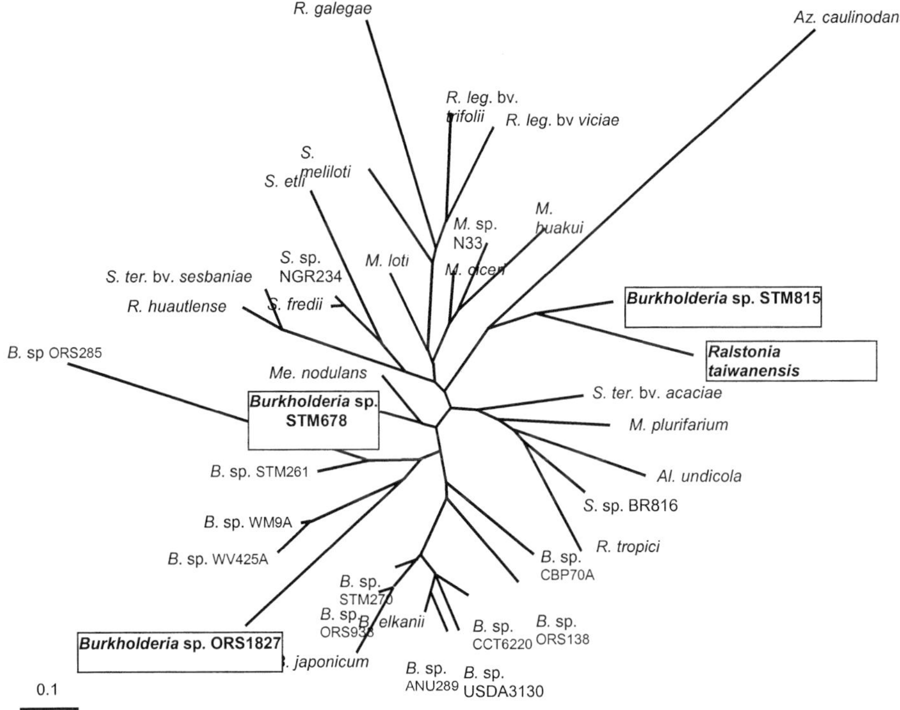

Figure 2. Unrooted *nodA* tree showing the close phylogenetic relationship between the *nodA* of α- and β-rhizobia (framed names). The maximum likelihood tree is based on full-length sequences. *Al., Allorhizobium, Az., Azorhizobium, B., Bradyrhizobium, Me., Methylobacterium, R., Rhizobium, S., Sinorhizobium, ter., teranga, leg., leguminosarum.*

5. Evidence of Multiple Lateral Transfers Between α and β-Rhizobia

The discovery of nodulating bacteria in the β-Proteobacteria phylum raises the question of the origin of nodulating genes, especially common *nodABC* genes for which hardly any homologous sequences in other organisms have been identified yet. Their occurrence in both α- and β-Proteobacteria cannot be explained in terms of descent through common ancestry, an hypothesis already rejected for the different rhizobial genera within α-Proteobacteria (Young and Haukka 1996). Phylogenetic analyzes indicate a much smaller phylogenetic distance between the *nodA* genes of β-rhizobia and other rhizobia than between the 16S rRNA genes of α- and β-Proteobacteria. This suggests again that the presence of *nod* genes in both α- and β-rhizobia occurred through horizontal gene transfer. It was however not clear whether a single transfer spread nodulation genes from one subclass to the other, or if recurrent transfers have occurred between the two subclasses.

To test these hypotheses we performed a maximum likelihood phylogenetic analysis of all available entire *nodA* sequences, including the four β-rhizobia, using PAUP* (Swofford 1998). The most likely tree obtained clustered the four β-rhizobial strains in 3 different clades (Figure 2). A single transfer would have given a tree with two main clades, discriminating the α- and β-groups. We tested whether such topology was statistically less likely than the obtained maximum likelihood tree. Constraining the four strains to be grouped in the same clade leads to a statistically less likely tree than the most likely tree (p=0.0372, Kishino-Hasegawa test). Constraint tree in which the three *Burkholderia* strains are grouped together was also rejected (p=0.0004).

These results favor recurrent transfers between the two subclasses of α- and β- over a single transfer between the two sub-classes.

6. Conclusion

The identification of rhizobia within Proteobacteria from the β-subclass shows that the ability to establish a symbiosis with legumes is more widespread in bacteria than anticipated to date. As a consequence, legumes are able to establish a symbiosis with phylogenetically distant bacteria. Such symbiosis is not a sporadic phenomenon, since *Ralstonia* appear to be the favorite partners of *Mimosa pudica* and *M. diplotricha* in Taiwan. We have identified nodulating *Methylobacterium* (Sy *et al.* 2001), *Burkholderia* and *Ralstonia*. Consequently the word rhizobium, originally a genus name, is now to be considered as a generic term grouping phylogenetically diverse bacteria sharing the ability to establish a legume symbiosis. Symbionts of less than 10% of the 750 legume genera being fully characterized, it is likely that further exploration of the rhizobial diversity may reveal the rhizobial nature of additional members of the β-Proteobacteria and possibly other taxonomic classes.

Our results show that the α- and β-rhizobia use the same strategy (*nod* genes and Nod factors) for establishing symbiosis with legumes. The spread of *nod* genes in α- and β-rhizobia probably originate from lateral gene transfer. This transfer may have occurred after the appearance of legumes on Earth, about 70 million years ago. Lateral transfer of symbiotic genes may occur between bacteria living in the same ecological niche, i.e. the rhizosphere. Although transfer is likely to be frequent within a bacterial population, only transfers occurring in bacteria that exhibit predisposition to the symbiosis (i.e. ability to overcome plant defenses, to infect the plant and to self-maintain in plant tissues) will be effective. The genome sequencing of phylogenetically different rhizobia should allow identifying the genetic background for plant infection and more generally the molecular basis of the preadaptation to legume symbiosis.

7. References

Boone C (1999) Carbohydr. Res. 317, 155-163
Chen *et al.* (2001) Inter. J. Syst. Evol. Microbiol.
Dénarié J *et al.* (1996) Annu. Rev. Biochem. 65, 503-535

Lerouge P *et al.* (1990) Nature 344, 781-784
Moulin L *et al.* (2001) Nature 411, 948-950
Perret X *et al.* (2000) Microbiol. Mol. Biol. Rev. 64, 180-201
Spaink HP *et al.* (1991) Nature 354, 125-130
Sullivan JT, Ronson CW (1998) Proc. Natl. Acad. Sci. USA 95, 5145-9
Sullivan JT *et al.* (1995) Proc. Natl. Acad. Sci. USA 92, 8985-9
Suominen L *et al.* (2001) Mol. Biol. Evol. 6, 907-16
Swofford DL (1998) PAUP Version 4, Sinauer Associates, Sunderland, MA
Sy A *et al.* (2001) J. Bacteriol. 183, 214-220
van Berkum P, Eardly BD (1998) In Spaink HP, Kondorosi A, Hooykaas PJJ (eds), The
        Rhizobiaceae, pp. 1-24, Kluwer Academic Publishers, Dordrecht
Young JPW, Hauk KE (1996) New Phytol. 133, 87-94
Zhang XX *et al.* (2000) Appl. Environ. Microbiol. 66, 2988-2995

GENOME DIVERSITY AT THE PHE-tRNA LOCUS IN A FIELD POPULATION OF MESORHIZOBIA

C.W. Ronson[1], J.T. Sullivan[1], G.S. Wijkstra[1], T. Carlton[1], K. Muirhead[1], J.R. Trzebiatowski[2], J. Gouzy[3], F.J. deBruijn[2,3]

[1]Dept of Microbiology, University of Otago, Dunedin, New Zealand
[2]Michigan State University, E. Lansing, MI USA
[3]LBMRPM, CNRS-INRA, Castanet-Tolosan, France

1. Introduction

In recent years, it has become apparent that many bacterial genomes have obtained a significant proportion of their genetic diversity through horizontal gene transfer, with acquisition presumably being counterbalanced by deletion of DNA in order to prevent excessive genome expansion (Ochman *et al.* 2000). Nevertheless genome size can vary widely between strains of the same species, as shown by the finding that the genome of *Escherichia coli* strain O157:H7 is 810 kb larger than the genome of *E. coli* strain K12. Comparison of the two strains indicates that their chromosomes comprise a conserved core interrupted by many lineage-specific "islands" (Hayashi *et al.* 2001; Perna *et al.* 2001). Many gram-negative bacterial pathogens are differentiated from benign relatives by the presence of horizontally-acquired "pathogenicity islands". These chromosomally encoded regions typically contain large clusters of genes required for virulence traits, such as cell invasion, iron uptake, adhesins and hemolysin. Several encode Type III secretion systems which deliver effector proteins directly into the host cell cytoplasm (Hacker, Kaper 2000). Many pathogenicity islands are situated at tRNA or tRNA-like loci, which are common sites for the integration of foreign sequences including bacteriophage, and some contain a gene encoding a phage-like integrase of the P4 family at one end. Most pathogenicity islands are no longer transferable, indicating that they have been fixed into the genome, probably through mutation of mobility genes and/or attachment site sequences. Thus acquisition of genomic islands has effected a stable change to the ecological and pathogenic character of many bacterial species (Ochman *et al.* 2000).

We have shown that non-symbiotic strains of mesorhizobia can evolve into *Lotus*-nodulating symbionts in a single quantum leap through the acquisition of a 500-kb chromosomal element. The element integrates into a phenylalanine tRNA gene, reconstructing the gene at one end (arbitrarily defined as the left end), and producing a 17-bp direct repeat of the 3' end of the tRNA gene at the right end. Within the left end of the element, a gene *intS* that encodes a product with similarity to members of the phage P4 integrase subfamily is located 198 bp downstream of the tRNA gene (Sullivan, Ronson 1998; Sullivan *et al.* 1995). This gene is required for excision of the element as a circle as well as its integration (Sullivan *et al.* 2000). The element was termed a symbiosis island on the basis of its similarities to pathogenicity islands of gram-negative bacteria. Like pathogenicity islands, the symbiosis island converts an environmental strain (a soil saprophyte) into a strain capable of forming a close association with an eukaryotic host. Examples of other genomic islands for which transfer has been demonstrated include the *Pseudomonas clc* element and the *Salmonella* conjugative transposon CTnscr94. The *clc* element is a 105-kb transferable element that contains chlorocatechol-degradative enzymes and integrates into a glycine tRNA gene using a P4-like integrase (Ravatn *et al.* 1998). CTnscr94 integrates into a phe-tRNA gene and contains genes for sucrose utilization (Hochhut *et al.* 1997). However, although it is clear that genomic islands play a key role in microbial evolution, the extent to which they contribute to the environmental adaptation of bacteria other than pathogens is unknown.

2. The Mesorhizobial Population Harboring the Symbiosis Island is Diverse

The symbiosis island was discovered during a study undertaken to examine generation of genetic diversity in a rhizobial population which developed under a stand of *Lotus corniculatus* established with a single inoculant strain in a region where there were no pre-existing rhizobia capable of nodulating the plant. Populations of indigenous rhizobia can rapidly supplant inoculant strains, presumably due to their superior environment-specific adaptive traits, even when the initial population is small or undetectable. Gaining an understanding of the ecology of indigenous rhizobial populations is a crucial step towards developing effective strategies to increase symbiotic nitrogen fixation through the addition of selected inoculant strains. The site examined was established using strain ICMP3153 as inoculant seven years prior to sampling. Differences in growth rate amongst strains isolated from nodules were noted, and RFLP profiling confirmed that considerable genetic diversity existed within the population. Only 20% of the nodule isolates were the same as ICMP3153, including strain R7A which has been used for subsequent studies. Subsequent molecular studies showed that the diverse strains were derivatives of indigenous non-symbiotic mesorhizobia that had acquired the symbiosis island from ICMP3153 (Sullivan *et al.* 1995).

The diverse symbiotic strains, together with seven non-symbiotic strains (strains CJ1–CJ7) that were isolated from the same site, were characterized by RFLP and multilocus enzyme electrophoresis (MLEE) analysis, full length 16S rRNA gene sequencing and total DNA:DNA hybridization analysis. The results showed that four non-symbiotic strains belonged to the same species as the diverse symbiotic strains, whereas the other three non-symbionts and the original inoculant strain represented further genomic species of mesorhizobia. Even within the same genomic species, the field isolates showed substantial genetic diversity (Sullivan *et al.* 1996). This diversity was further indicated by comparing the DNA sequence from six strains surrounding the phe-tRNA gene - the sequence immediately upstream from the tRNA gene was highly conserved, whereas the strains fell into three groups on the basis of sequence similarity downstream of the inserted symbiosis island (Sullivan, Ronson 1998). It was proposed that this diversity might represent further genomic islands integrated at the same tRNA locus that may adapt the indigenous strains to the local environment.

To learn more about the contribution of genomic islands to the evolution and niche adaptation of mesorhizobia, we have sequenced the symbiosis island from strain R7A and further characterized the DNA regions downstream of the phe-tRNA locus in several diverse strains. We also compared the sequences to the genome sequence of a Japanese isolate of *M. loti*, strain MAFF303099. The 7.6 Mb MAFF303099 genome consists of a chromosome and two plasmids, pMla and pMlb (Kaneko *et al.* 2000). Here we highlight the genetic diversity uncovered by a comparative analysis of the R7A and MAFF303099 symbiosis islands, and report on the identification of further genomic islands that may contribute to the diversity and adaptation of the bacteria.

3. Comparative Analysis the R7A and MAFF303099 Symbiosis Islands

Comparison with R7A indicates that the MAFF303099 chromosome contains a 610,975-bp symbiosis island integrated adjacent to the phe-tRNA gene (Kaneko *et al.* 2000). The R7A island at 501,801 bp in size is 109 kb smaller than the *M. loti* MAFF303099 island and encodes 416 potential genes. Comparisons of the two *M. loti* symbiosis islands indicate that they have similar metabolic and symbiotic potential. The two islands share a conserved backbone sequence of 248 kb with about 98% DNA sequence identity, indicating that the two islands evolved from a common ancestral source. The backbone contains the key symbiotic gene complement including all the genes required for Nod factor synthesis. It is interrupted by a series of strain-specific "islets" that represent DNA either lost or gained by each strain and range in size from a few base pairs up to 168 kb. The few non-syntenous regions that encode similar proteins show less than 90% nucleotide identity, suggesting that most were separately acquired by each island rather than arising through translocation. About 8% of the R7A island consists of insertion sequences (six identifiable intact genes) or fragments thereof, compared to 19% for MAFF303099 (Kaneko *et al.* 2000), which accounts for a significant portion of the size difference between the two islands. Analysis of the strain-specific segments of both islands reveals that in addition to IS genes, they contain mainly hypothetical genes, metabolic genes and ABC transporters. One significant difference is that the R7A island has a gene cluster with strong similarity to those *vir* genes from *Agrobacterium tumefaciens* that encode the Type IV pilus through which T-DNA is transferred to the plant. This cluster is missing from MAFF303099, which in turn has a gene cluster with strong similarity to the cluster encoding a type III secretion system in *Rhizobium* strain NGR234 (Viprey *et al.* 1998) that is missing from R7A. Another interesting feature is that of the 114 hypothetical genes detected in R7A that have no database matches in other bacteria, 102 are not present in *M. loti* MAFF303099 indicating that they are strain- rather than species-specific.

Overall the comparative analysis of the islands emphasizes that they are dynamic mosaics shaped by multiple recombination events and in particular acquisition and deletion of DNA segments. As well as strain-specific regions, variable G+C content and insertion sequences, there are several gene fragments or pseudogenes, some of which are in differing stages of decay in the two islands. Some have intact orthologs on the R7A island whereas others do not. In addition, a number of gene clusters found on the R7A island are also present on pMLa in MAFF303099 and some of these are absent from the MAFF303099 island.

4. Sequence Diversity Downstream of the phe-tRNA Locus or Symbiosis Island

DNA regions of about 1 kb downstream of the symbiosis island were amplified by inverse PCR and sequenced from 35 strains to determine whether the diverse mesorhizobial population contains additional acquired elements inserted adjacent to the phe-tRNA locus. In addition, cosmid clones of the DNA regions were isolated and partially sequenced for strains CJ3, CJ4, R7A and R88b. The sequences aligned into 15 similarity groups, with two groups found multiple times (Table 1).

Group A contains five strains with sequence similarity to a P4 integrase gene directly downstream of the 17-bp repeat of the phe-tRNA gene that demarcates the border of the symbiosis island. Strain MAFF303099 also contains a P4 integrase downstream of its symbiosis island (Kaneko *et al.* 2000). The gene products of that integrase, intR88B and intCJ4 have very high nucleotide sequence identity (94.5%) with each other. They do not share significant nucleotide sequence identity with the symbiosis island integrase, and show only 50% amino-acid identity with it. However DNA downstream of the integrase genes diverged sharply in each of the three strains, suggesting that the three genes are associated with different acquired DNA regions.

Table 1. Grouping of mesorhizobial strains according to sequence similarity directly downstream of the symbiosis island.

Group	Seq. Similarity	Nt ident. to MAFF303099
A (6 strains)	P4 integrase	Yes
B (9 strains)	FhuBD	500 bp only of 6 kb
C	Transposase	No
D	Aldehyde DH	No
E	Insertion seq	No
F, G, H, I, J, K, L	No similarities	No
M (R7A)	No similarities	Yes

The sequence of an 8-kb region for strain CJ3 was completed and revealed genes with strong similarity to fhuDB genes from a number of bacteria that are required for the transport of the ferric hydroxamate siderophore ferrichrome (Guerinot 1994). Hybridization analysis indicated that these genes were present in only a subset of strains, indicating that they are acquired. They are absent from MAFF303099. A gene *ggt* encoding γ-glutamyl transpeptidase was present directly upstream of the fhuBD genes. ggt is involved in glutathione metabolism and is expected to be a core chromosomal gene. Hybridization of the *ggt* gene to diverse strains showed that it was present in single copy in all strains. It is also present in MAFF303099. Hence the junction between the 6.6 kb acquired element in CJ3 and the core chromosome lies between the fhuD and *ggt* genes (Figure 1).

Strain R88B contains a P4 integrase and also the *fhuBD* genes. The *fhuDB* genes are adjacent to *ggt* and contiguity with MAFF303099 was found for *ggt* and several genes upstream of *ggt*, confirming that the junction between the core chromosome and acquired elements lies between *ggt* and *fhuD*. A cosmid containing the *int*[R88B] gene was also isolated and partially characterized. The region downstream of the *int* gene has a complex repeat structure and shows similarity to a gene encoding an outer membrane adhesin from *Pseudomonas putida* that has recently been identified as essential for seed colonization (Espinosa-Urgel *et al.* 2000). The region is absent from MAFF303099 and most other strains tested. The *fhuDB* and *mus-20* regions are yet to be linked (Figure 1).

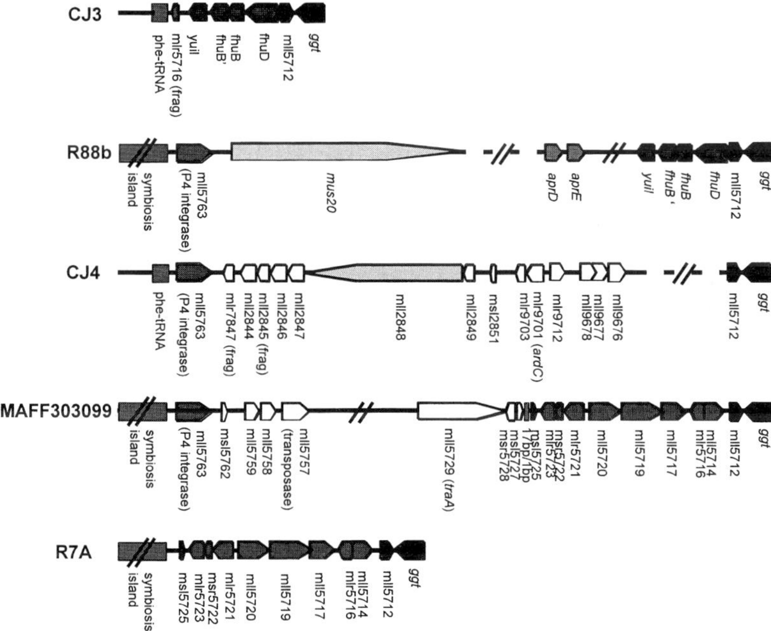

Figure 1. Genome structure downstream of the phe-tRNA gene or symbiosis island in strains CJ3, CJ4, R88B, MAFF303099 and R7A. The MAFF303099 sequence is from Kaneko *et al.* (2000). Genes designations are as for MAFF303099 where homologs are present. The R88B gene labeled *mus-20* is similar to *mus-20* of *P. putida* (Espinosa-Urgel *et al.* 2000).

Sequence downstream of the *int* gene in strain CJ4 (Figure 1) contained a cluster of genes found elsewhere in the genome of MAFF303099, including a gene *mll2848* whose product showed similarity to a family of outer membrane adhesins. This family includes *yadA* of *Yersinia* and *uspA* of *Moraxella* (Hoiczyk *et al.* 2000). Hence the *int* genes in R88B and CJ4 are both associated with genes encoding outer membrane adhesins that have been identified as essential colonization factors, strongly suggesting that the acquired DNA is of adaptive value.

The sequence immediately downstream of the symbiosis island in strain R7A showed identity to a region 35.4 kb downstream of the symbiosis island in MAFF303099 (Figure 1). The identity included the 3' 17-bp of the tRNA gene except that MAFF303099 showed a 1-bp deletion. These results clearly demarcate the island associated with the *int* gene in MAFF303099. The R7A DNA is not found in the other mesorhizobia analyzed, indicating that it represents another acquired region that R7A has in common with MAFF303099.

5. Concluding Remarks

In summary, the symbiosis island of R7A is a dynamically evolving element shaped by multiple recombination events that adapts a soil saprophyte to a symbiotic interaction with a plant host. In addition, we have found evidence for fifteen different acquired DNA regions adjacent to a phenylalanine tRNA gene in members of a population of soil mesorhizobia. Some of these acquired regions encode traits such as iron acquisition and seed adhesion that are likely to endow the strains carrying them with a competitive advantage. Overall, this work has shown remarkable genetic diversity in the soil mesorhizobial population that is due to the acquisition of chromosomal DNA and has provided insight into the extensive role of horizontal gene transfer in microbial evolution.

6. References

Espinosa-Urgel M *et al.* (2000) J. Bacteriol. 182, 2363-2369

Guerinot ML (1994) Annu. Rev. Microbiol. 48, 743-772

Hacker J, Kaper JB (2000) Annu. Rev. Microbiol. 54, 641-679

Hayashi T *et al.* (2001) DNA Res. 8, 11-22

Hochhut B *et al.* (1997) J. Bacteriol. 179, 2097-2102

Hoiczyk E *et al.* (2000) EMBO J. 19, 5989-5999

Kaneko T *et al.* (2000) DNA Res. 7, 331-338

Ochman H *et al.* (2000) Nature 405, 299-304

Perna NT *et al.* (2001) Nature 409, 529-533

Ravatn R *et al.* (1998) J. Bacteriol. 180, 5505-5514

Sullivan JT *et al.* (2000) In Triplett E (ed), Prokaryotic Nitrogen Fixation: A Model System for Analysis of a Biological Process, pp 693-704, Horizon Scientific Press, Wymondham, UK

Sullivan JT, Ronson CW (1998) Proc. Natl. Acad. Sci. USA 95, 5145-5149

Sullivan JT *et al.* (1996) Appl. Environ. Microbiol. 62, 2818-2825

Sullivan JT *et al.* (1995) Proc. Natl. Acad. Sci. USA 92, 8985-8989

Viprey VA *et al.* (1998) Mol. Microbiol. 28, 1381-1389

7. Acknowledgements

This work was supported by grants from the Marsden Fund administered by the Royal Society of New Zealand, the US DOE (DE FG02-91ER200021) and by Otago and Michigan State Universities.

HOW WELL DOES 16S rRNA GENE PHYLOGENY REPRESENT EVOLUTIONARY RELATIONSHIPS AMONG THE RHIZOBIA?

P. van Berkum[1], S. Reiner[2], B.D. Eardly[3]

[1]Soybean Genomics and Improvement Laboratory, USDA, ARS, Beltsville, MD 20705, USA
[2]Insect Biocontrol Laboratory, USDA, ARS, Beltsville, MD 20705, USA
[3]Penn State Berks-Lehigh Valley College, Berks Campus, P.O. Box 7009 Reading, PA 19610, USA

1. Introduction

Phylogeny is the analysis of character sets to reconstruct evolutionary paths of extant species. Often the character sets are based on morphological variation among extant and extinct organisms. However, rhizobia and other bacteria have few such informative characters and no fossil record is available. In such cases gene sequences are compared in an evolutionary context. This approach relies on the assumption that evolution throughout the genome progresses at an approximately constant rate in a hierarchical manner largely via the mechanism of mutation and Darwinian selection. Although many genes are potentially useful for the purpose of reconstructing evolutionary history, in bacteria it has become common practice to use the 16S rRNA genes. Use of the 16S rRNA gene further assumes that each genome contains a single copy or multiple identical copies and that evolution of the 16S rRNA gene approximates that of the entire genome. However, there are increasing concerns that these assumptions may not strictly apply. For instance, individual bacterial cells may harbor multiple divergent copies of the 16S rRNA gene (Amann *et al.* 2000; Carbon *et al.* 1979; Dreyden, Kaplan 1990; Rainy *et al.* 1996; van Berkum *et al.* 1999; Wang *et al.* 1997). Also, evolution of the 16S rRNA gene may be reticulate (Eardly *et al.* 1996; Sneath 1993; Wang, Zhang 2000; Yap *et al.* 1999). Because of these concerns we decided to examine the 16S rRNA gene to provide perspective on these concerns as they relate to rhizobia and closely related taxa. Our approach was to first evaluate how the addition of various taxa would influence the consistency of 16S rRNA gene tree topologies in a phylogenetic analysis of rhizobia and closely related non-rhizobial taxa; then to compare tree topologies reconstructed from the 16S rRNA gene, the 23S rRNA gene and the Internally Transcribed Space (ITS) region between them; and finally to investigate the possibility of gene conversion in the 16S rRNA gene of rhizobia and selected related bacteria.

2. Procedure (Material and Methods)

The sequences examined in this study (16S rRNA, 23S rRNA and ITS region) were generated either using an automated sequencing protocol described previously (van Berkum, Fuhrmann 2000) or in the case of the 16S rRNA genes were in part obtained from public databases. Sequences were aligned using PILEUP of the Wisconsin GCG package and alignments were checked manually by using Genedoc (Nicholas, Nicholas 1997). Neighbor-joining trees were constructed from Jukes-Cantor distances using MEGA (Kumar *et al.* 1993) or trees were assembled in a stepwise manner with Parsimony analysis using PAUP (Swofford 2001). Parsimony and distance trees also were generated with aligned sequences of the 16S rRNA genes constraining the topologies to resemble the 23S rRNA tree and were drawn with MacClade (Maddison, Maddison 1999). This was done to investigate differences in the number of steps required to construct trees and to compare tree topologies using the Shimodaira-Hasegawa test (Shimodaira, Hasegawa 1999). This was done by generating likelihood scores of tree files and then subtracting the score closest to zero from those of the other trees. The significance in differences among the likelihood scores was determined with a one-tailed bootstrap test using 1000 permutations of the data. Finally, the Geneconv program (Sawyer 1989) was used to test the possibility of a history of recombination among the 16S rRNA genes. With this method the distribution of polymorphic nucleotide positions along a gene is examined to estimate the likelihood that distinct segments have differing phylogenies.

3. Results and Discussion

We investigated consistency in 16S rRNA tree topology because topologies of published trees may vary, depending on differences in alignments, software used, or taxa sampled. To examine the effects of the latter, the 16S rRNA gene sequences for the three non-rhizobial taxa *Blastobacter aggregatus*, *Bl. capsulatus*, and *Ensifer adhaerans* were included in the analysis and effects on tree topology were noted. The resulting tree placed the two *Blastobacter* species within the genus *Agrobacterium*, and *E. adhaerens* was placed in close proximity to the genus *Sinorhizobium* (data not shown). The species *R. galegae* and *R. huautlense* also were moved from a position adjacent to *Ag. vitis* to one adjoining *R. gallicum*.

In the next phase of our analysis, we compared the tree topologies for three different loci within the rRNA operon. In order to determine whether the topologies of each were equally parsimonious, the 16S rRNA tree was constrained to the topology of the 23S rRNA tree (Figure 1). However, the constrained tree required more steps for construction than the unconstrained tree, indicating that the constrained tree was less parsimonious than the unconstrained tree. Subsequently we compared topologies of the constrained and the unconstrained 16S rRNA trees to determine whether the same or different phylogenetic information was obtained from analysis of the 16S rRNA

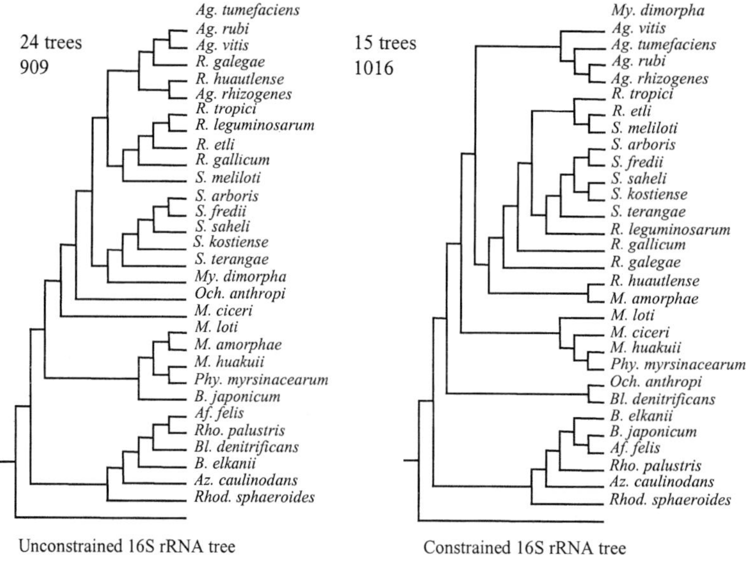

Figure 1. Comparison of the 16S rRNA tree to a 16S rRNA tree constrained to the topology of the 23S rRNA tree generated by Parsimony analysis and drawn in MacClade.

and 23S rRNA genes. Topologies of the 24 unconstrained trees (Figure 1) were similar as tested by the Shimodaira-Hasegawa test, while topologies of the unconstrained reference tree and the 15 constrained trees (Figure 1) were significantly different (P<0.001). A similar test was done comparing distance trees generated from the ITS region and the 16S rRNA gene (Figure 2). These two trees also had significantly different topologies as determined by the Shimodaira-Hasegawa test (P<0.001).

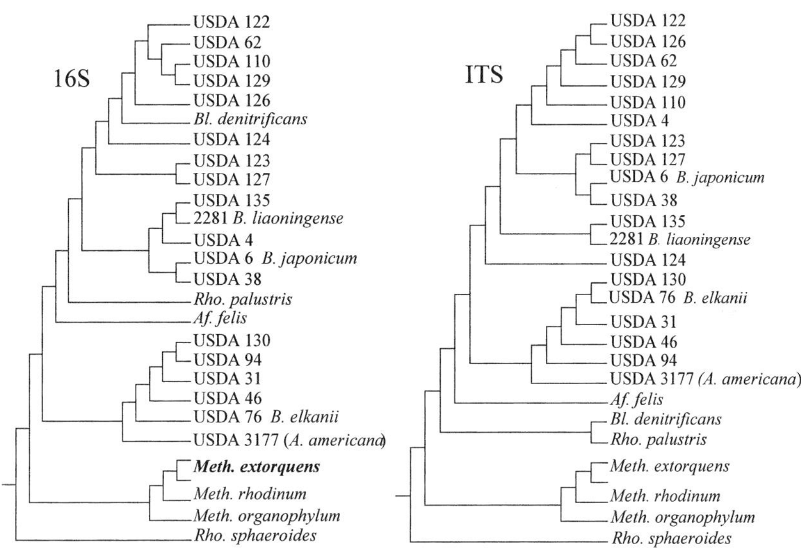

Figure 2. Comparison of trees reconstructed from the 16S rRNA gene and the ITS region.

We concluded from comparing tree topologies that the phylogenetic information obtained by analysis of the16S rRNA gene is incongruent with that of the 23S rRNA gene or with that of the ITS region. A comparison of the 23S rRNA gene and ITS region was not made.

Using tests for recombination we predicted that several different regions within the 16S rRNA gene potentially resulted from gene conversion (data not shown). We focused on a 229 bp segment identified from comparing the 16S rRNA gene sequences of *Mesorhizobium huakuii* and *S. fredii*. Topologies of trees reconstructed either from this small region or from the alignment after removing this short segment were significantly different from trees reconstructed from the entire length of the aligned 16S rRNA gene sequences. Therefore, it is possible that 16S rRNA gene sequences of rhizobia and related taxa contain short segments resulting from lateral gene transfer and genetic recombination with divergent alleles.

From our data we concluded that our concerns over the assumptions made when reconstructing bacterial evolutionary history from 16S rRNA gene sequence divergence are relevant to rhizobia and related taxa. Therefore, it may be inappropriate to use the 16S rRNA gene alone for reconstructing bacterial phylogenies and subsequently to use these results as primary evidence for deciding rhizobial nomenclature. Some examples where rhizobial classification decisions were made largely on the basis of analysis of the 16S rRNA gene include separation of *Sinorhizobium* from *Rhizobium* (DeLajudie *et al.* 1994), the proposal of the new genus *Allorhizobium* (DeLajudie *et al.* 1998), the suggestion for combining *Agrobacterium* and *Allorhizobium* into the genus *Rhizobium* (Young *et al.* 2001), and proposing separate genera for *Bradyrhizobium japonicum* and *B. elkanii* (Willems *et al.* 2001). Such changes in nomenclature may not be warranted since the evidence for phylogenetic placement of these genera is at best inconclusive. As an alternative we suggest that a more conservative approach be taken where taxonomic decisions are based on the analysis of a variety of loci, and that comparative analytical methods be used to estimate phylogenetic relationships among the species being considered.

4. References

Amann G *et al.* (2000) Extremophiles 4, 373-376

Carbon C *et al.* (1979) EMBO J. 11, 4175-4185

DeLajudie P *et al.* (1994) Int. J. Syst. Bacteriol. 44, 715-733

DeLajudie P *et al.* (1998) Int. J. Syst. Bacteriol. 48, 1277-1290

Dreyden SC, Kaplan S (1990) Nucleic Acids Res. 18, 7267-7277

Eardly BD *et al.* (1996) Plant Soil 186, 69-74

Kumar S *et al.* (1993) MEGA: Molecular Evolutionary Genetics Analysis, Version 1.01, The Pennsylvania State University, University Park, PA

Maddison WP, Maddison DR (1999) MacClade, Analysis of Phylogeny and Character Evolution, Version 3.08, Sinaur Associates, Sunderland, MA

Nicholas KB, Nicholas HB (1997) Alignment Editor and Shading Utility, Version 2.6.001

Rainey FA *et al.* (1996) Microbiol. 142, 2087-2095

Sawyer SA (1989) Mol. Biol. Evol. 6, 526-538

Shimodaira H, Hasegawa M (1999) Mol. Biol. Evol. 16, 1114-1116

Sneath PHA (1993) Int. J. Syst. Bacteriol. 43, 626-629

Swofford DL (2001) PAUP, Phylogenetic Analysis Using Parsimony, Version 4.0b8, Sinaur Associates, Sunderland, MA

van Berkum P, Fuhrmann JJ (2000) Int. J. Syst. Evol. Microbiol. 50, 2165-2172

van Berkum P *et al.* (1999) In Martinez E, Hernandez G (ed) Highlights of Nitrogen Fixation Research, pp. 267-269, Kluwer Academic/Plenum Publisher, New York

Wang Y, Zhang Z (2000) Microbiol. 146, 2845-2854

Wang E *et al.* (1997) J. Bacteriol. 179, 3270-3276

Willems A *et al.* (2001) Int. J. Syst. Evol. Microbiol. 51, 111-117

Yap WH *et al.* (1999) J. Bacteriol. 181, 5201-5209

Young JM *et al.* (2001) Int. J. Syst. Evol. Microbiol. 51, 89-103

GENETICS AND GENOMICS OF STRESS-INDUCED GENE EXPRESSION IN RHIZOBIA

F.J. de Bruijn [1,2,3], F. Ampe [3], J. Batut [3], H. Berges [3], M. Davalos [3], M.E. Davey [1,2], J. Gouzy [3], D. Kahn [3], E. Kiss [3], E. Lauber [3], C. Liebe [3], A. Milcamps [1], C. Ronson [4], P. Struffi [1], J. Sullivan [4], J. Trzebiatowski [1], C. Vriezen [1]

[1] MSU-DOE Plant Research Laboratory
[2] Dept Microbiol. Michigan State U., E. Lansing, MI 48824, USA
[3] INRA/CNRS Laboratoire de Biologie Moleculaire des Relations Plantes-Microorganismes, BP 27, 31326 Castanet-Tolosan Cedex, France
[4] Dept Microbiol., U. of Otago, Dunedin, New Zealand

1. Introduction

In nature, bacterial growth is restricted by a wide variety of environmental factors, including the lack of essential nutrients, osmotic stress, oxygen limitation and oxidative stress. In most natural settings at least some essential nutrients are limiting, and stress conditions are prevalent, therefore periods of negligible growth or dormancy are the more typical physiological state of bacteria. Understanding how bacteria are able to monitor, sense, and respond to their environment in nutrient-deprived, stressed physiological states or in eukaryotic host tissues is fundamental to our understanding of microbial biology and ecology, as well as symbiotic plant-microbe interactions.

While studies on the *E. coli* model system, and selected marine systems, have revealed a considerable number of basic genetic components of microbial stress responses, comparatively little is understood about how soil bacteria respond to nutrient limitation conditions. Soil is generally a harsh, oligotrophic environment. Nutrient limitation and oxygen limitation may represent the most prevailing stress conditions for soil bacteria. A paucity of organic matter is present in most soils, which is often insoluble or in a form inaccessible to bacteria (e.g. humus and lignin). In fact, non-growth or extremely limited growth, may be the rule, rather than the exception (Matin *et al.* 1989). Rhizosphere soil has been reported to be a somewhat more supportive environment for bacteria than bulk soil, due to the presence of plant root exudates as readily accessible nutrients. However, even in the rhizosphere, bacterial growth and activity are generally limited to short periods during which these exudates are available (Lynch, Whipps 1990).

We have focused our studies on stress mediated gene expression during nutrient deprivation and under other stress conditions in the indigenous soil bacterium, *Sinorhizobium meliloti*. This bacterium is a capable of establishing a symbiosis with the legume alfalfa (*Medicago sativa*), during which a new specialized organ is formed, the nitrogen-fixing root nodule. These nodules provide the proper physiological conditions for the bacteria to survive in the absence of competing microflora, and to reduce atmospheric dinitrogen to ammonia, which is then assimilated by the plant. Little is known about the manner in which these bacteria are able to persist in their free-living state, in the soil and rhizosphere or how they respond to specific physiological stress conditions *in planta*. *S. meliloti* has become one of the leading model organisms to study microbial persistence, competition and stress responses in soil, as well as plant-microbe interactions leading to indeterminate nodule formation, since its genetic analysis is highly developed, and its entire genomic DNA sequence has been determined and annotated (Galibert *et al.* 2001).

Moreover, one of its hosts, *Medicago truncatula*, has become a model organism to study the plant components of symbiotic nitrogen fixation, the formation of indeterminate nodules, as well as comparative legume genetics (Cook *et al.* 1997). Similar statements can be made for *Mesorhizobium loti* and its host *Lotus japonicus*, which have been developed as a model system for determinate nodule formation (Cook *et al.* 1997). The *M. loti* system is of particular interest, since the genomic sequence of an *M. loti* strain MAFF303099 (Kaneko *et al.* 2000) has been determined and most, if not all of the symbiotic genes of another strain, R7A, as well as several putative persistance- and competition-related loci have been found to be located on a transferable, well defined Symbiosis Island (Sullivan *et al.* 1995; Ronson *et al.* this volume).

Therefore, we have been interested in using molecular genetic, genomic and post-genomic approaches to study environmental control of gene expression in *S. meliloti*, and to employ comparative structural and functional genomics with other rhizobia, such as *M. loti* (and especially its transferable Symbiosis Island), in order to elucidate pathways involved in rhizosphere microbial persistence and competition, as well as other aspects of symbiotic plant-microbe interactions

We initially focused on a limited number of key issues in the area of *S. meliloti* gene regulation in response to environmental stress conditions that are relevant to molecular microbial ecology in general, and symbiotic plant-microbe interactions in particular, namely:

1. What is the nature of bacterial genes responding to environmental stress?
2. What is the role of stress-responsive bacterial genes in persistence or competition in bulk soil or the (endo)rhizosphere of plants?
3. Do novel, common global regulatory loci exist in (soil) bacteria that are responsible for persistence and competition in the bulk soil, in the rhizosphere of plants and/or bacterial differentiation (bacteroid formation)?

2. Isolation of Tn*5luxAB* Tagged *S. Meliloti* Loci

Using a promoterless Tn*5luxAB* (Wolk *et al.* 1991), transposon mutagenesis of *S. meliloti* strain 1021 was carried out and 5000 transcriptional fusions were analyzed for luminescence under N- or C- deprivation or O_2 limitation. Using this protocol, 22 *S. meliloti* strains carrying Tn*5luxAB* gene fusions induced by N-deprivation, 12 by C-deprivation and 24 by O_2-limitation were isolated. Cloning and DNA sequence analysis of the tagged loci revealed genes sharing similarities with previously identified bacterial loci involved in N- or C-metabolism and O_2-responses, as well as novel genes. The N- and/or C-deprivation induced loci include exopolysaccharide biosynthesis (*exoF* and *exoY*), nitrite and nitrate assimilation (*nasD,E, A*), amino acid transport (*braF* and *ilvH*), amino acid synthesis (*dapA*) or degradation (*speB, arcC*), as well as ribose transport (*rbsA*) genes. Among the loci carrying O_2-limitation induced gene fusions, similarities were discovered with exopolysaccharide (*exoO*), nitrogen fixation (*fixN*), cytochrome oxidase *(cyoC)*, heat shock (*htpG*), and novel genes.

Our genetic screen for N-, C-deprivation and O_2-limitation induced loci identified genes known to be induced under these conditions (e.g. *nas D, E, A* under N-deprivation; *rbsA* under C-deprivation; *fixN* under O_2-limitation), suggesting that our approach to isolate multiple environmentally controlled *S. meliloti* genes had been successful (Lim *et al.* 1993; Milcamps *et al.* 1998; Trzebiatowki *et al.* 2001). We do realize, however, that not all N- or C-deprivation or O_2-limitation induced genes would be picked up in a screen of 5000 Tn*5luxAB* insertions (Milcamps *et al.* 1998). Therefore our future studies will be focused on the use of micro- and macro-arrays to analyze gene expression patterns in a more global fashion (see below)

3. Characterization of Novel *S. Meliloti* Genes Induced Under Nutrient Deprivation Conditions

In order to identify genes involved in multiple stress responses and their regulators, we first focused on strains carrying gene fusions induced under N- and C-deprivation and/or O_2-limitation. Interestingly, we observed that only a limited number of Tn*5luxAB* tagged loci (13 out of a collection of 57) were induced by more than one stress. We chose strains N4 and C22 for further studies.

Strain N4 was found to carry a Tn*5luxAB* fusion inducible under N- and C-deprivation conditions, and in the presence of tyrosine. This strain was found to be unable to grow on tyrosine as sole C-source and produced a brown pigment, which was very pronounced in the presence of 0.2% tyrosine. The tagged locus was cloned from the *S. meliloti* genome by the plasmid rescue procedure described by Wolk *et al.* (1991), its DNA sequence was determined, and it was identified as the *hmgA* gene, encoding homogentisate dioxygenase. This enzyme is involved in the degradation of tyrosine. A very high similarity of the deduced protein with the corresponding eukaryotic enzymes of human, mouse and fungal origins was observed (50% identity, 57% similarity) for the human homolog. This was the first report of the presence of a homogentisate dioxygenase gene in bacteria, and a phenotype of the rhizobial mutant strain in culture closely resembling human phenylketonuria, namely the production of dark pigments in the urine of PKU patients. The Tn*5-luxAB* induced mutation in the *hmgA* locus does not affect the symbiotic properties (nodulation and nitrogen fixation) of *S. meliloti*, but does appear to be involved in stationary growth phase survival (Milcamps, de Bruijn 1999).

S. *meliloti* mutant strain C22 harbors a Tn*5luxAB* insertion in a gene that is induced by a variety of stresses, including carbon, nitrogen or iron deprivation, oxygen limitation, as well as high salt concentrations (400 mM NaCl). This Tn*5luxAB* tagged locus was cloned and its DNA sequence was determined. The tagged gene is part of an operon consisting of two ORFs, *ndiA* and *ndiB*, for nutrient deprivation induced genes A and B, which are unknown in other bacteria thus far. Comparison of the deduced amino acid sequences of both *ndiA* and *ndiB* to the protein databases at NCBI did not reveal significant similarity with any known gene products (Davey *et al.* 2000).

4. Isolation and Characterization of Regulatory Loci, Controlling the Nutrient Deprivation Responses in Strains N4 and C22

In order to identify the regulatory genes (circuits) responsible for the multiple nutrient and/or oxygen deprivation responses observed with strains N4 and C22, two types of studies were carried out. Since both N4 and C22 fusions are N-deprivation induced, we first examined whether the genes carrying the Tn*5luxAB* fusions were controlled by the well known *ntr* (nitrogen regulation) system, first described in enteric bacteria (Merrick, Edwards 1995). Therefore, we tested the expression of the tagged loci of strain N4 and C22 in *ntrA* and *ntrC* mutants of strain 1021 (Ronson *et al.* 1987; Szeto *et al.* 1987). Neither the expression of the N4 nor the C22 fusion were found to be *ntr* controlled, indicating that novel regulatory pathways are likely to be involved in their expression.

Our second approach to identify putative regulators, controlling the nutrient deprivation response of the tagged N4 and C22 loci, consisted of a secondary transposon mutagenesis of these strains. The aim of this approach was to inactivate gene(s) encoding trans-acting factor(s) which would alter the observed luminescence pattern of the Tn*5luxAB* reporter gene fusion. As secondary mutagen, the Tn*3* derivative Tn*1721* was chosen (Schoffl *et al.* 1981). With Tn*1721*, individual collections of 3600 double mutant derivatives of strains N4 and C22 were constructed

Out of 3600 strain N4 double mutant strains, two strains with a very reduced luminescence under N-deprivation conditions were selected, the Tn*1721* tagged loci were cloned and DNA sequence analysis revealed that both insertions were located in a single ORF. Database searches revealed significant similarity with a group of transcriptional regulators of the ArsR family. This gene was called *nitR* (nitrogen regulation; Millcamps *et al.* 2001), and will be one of the regulatory loci to be examined using macro- and micro-arrays (see below).

Out of 3600 strain C22 double mutant strains, one strain was isolated which no longer displayed luminescence under the conditions of N-, C-, Fe-deprivation or O_2-limitation. The Tn*1721* tagged locus was cloned and the Tn*1721* insertion was found to reside in an ORF encoding a protein with a high degree of similarity to the mitochondrial benzodiazepine receptor of rat, human, mouse and bovine (50% identity); as well as to the outer membrane oxygen sensor protein

TspO of *Rhodobacter sphaeroides* (42% identity; McEnery *et al.* 1992; Yeliseev, Kaplan 1995). This type of outer membrane receptor (pk18 /TspO) has been found in vertebrates and invertebrates, but has not been observed thus far in *Saccharomyces cerevisiae*, or in the majority of prokaryotes analyzed (Davey *et al.* 2000). Since a member of the alpha subdivision of purple bacteria is the likely source of the endosymbiont that gave rise to the mammalian mitochondrion (Yang *et al.* 1985), the finding of a pk18/TspO ortholog *in R. sphaeroides,* a member of this family, is intriguing. Its rhizobial equivalent will be the subject of future studies, including macro- and micro-arrays, to identify all the loci controlled by this interesting gene (see below).

5. Macro- and Micro-array Whole Genome Profiling

In addition to using genetic approaches, we have started to exploit the complete determination and annotation of the *S. meliloti* genome to enter the post-genomic era by using transcriptome analysis and proteomics to study the microbial genes involved in nutrient-deprivation and stress responses, as well as those involved in symbiotic plant-microbe interactions. The interest of large-scale gene expression analysis using DNA macro- and micro-array lies in the ability to simultaneously visualize the expression patterns of all the predicted ORFs of an organism under varying physiological conditions. Moreover, the analysis of expression patterns in particular mutant backgrounds allows the identification of unique or overlapping regulatory networks controlling the responses to particular conditions.

DNA sequence annotation performed by the Toulouse group has revealed that approximately 2500 *S. meliloti* genes (40% of the 6204 ORFs) cannot be defined functionally or via establishing similarities with gene/protein sequences in the available databases. Therefore the true post-sequencing challenge lies in the identification of the function of these unknown genes. One of the approaches available to achieve this goal is transcriptome analysis, since genes that are coordinately expressed/regulated are likely to be involved in particular functions, especially if they are clustered in the genome.

Towards this purpose, we are currently in the process of generating or using existing protocols to:

1. Use PCR to generate secondary and unique DNA probes corresponding to every ORF of *S. meliloti* identified;
2. Use the corresponding DNA products, or single oligonucleotides specific to each ORF, to generate macro-arrays;
3. Use the total genomic arrays to analyze *S. meliloti* gene expression under a variety of environmental stress conditions (including N-, C- and O_2-limitation; osmotic and oxidative stress), as well as during infection and *in planta*;
4. Use well characterized *S. meliloti* regulatory mutants generated previously in our laboratories or the regulatory loci described above, to study their genetic targets and the general genetic circuitries involved in environmental control of gene expression in this model organism.

We will investigate regulatory networks involved in several environmental challenges met by *S. meliloti* in the bulk soil, in the rhizosphere of plants and in symbiotic conditions in nitrogen-fixing nodules. An emphasis will be put on responses to oxygen, nitrogen, and carbon limitation and the impact of the corresponding genes and regulatory systems on the symbiotic interaction of *S. meliloti* with the model legume *Medicago truncatula*.

6. References

de Bruijn FJ *et al.* (1995) In Tikhonovich IA *et al.* (ed) Nitrogen Fixation: Fundamentals and
   Applications, pp. 195-200, Kluwer Academic Publishers, Dordrecht, The Netherlands
de Bruijn FJ (1998) In Elmerich C *et al.* (ed) Biological Nitrogen Fixation for the 21st century,
   pp. 195-200, Kluwer Academic Publishers, Dordrecht, The Netherlands
Cook D *et al.* (1997) Plant Cell 9, 275-281
Davey ME, de Bruijn FJ (2000) Appl. Environ. Microbiol. 66, 5353-5359
Galibert F *et al.* (2001) Science 293, 668-672
Kaneko T *et al.* (2001) DNA Res. 7, 331-338
Lim PO *et al.* (1993) In Guerrero R, Pedros-Alio C (ed) pp. 97-100, Spanish Society for
   Microbiology, Madrid, Spain
Lynch JJ, Whipps JM (1990) Plant Soil 129, 1-10
Matin A *et al.* (1989) Ann. Rev. Microbiol. 43, 293-316
McEnery MW *et al.* (1992) Proc. Natl. Acad. Sci. USA 89, 3170-3174
Merrick MJ, Edwards RA (1995) Microbiol. Rev. 59, 604-622
Milcamps AP *et al.* (1998) Microbiol. 144, 3205-3218
Milcamps AP, de Bruijn FJ (1999) Microbiol. 145, 935-947
Milcamps AP *et al.* (2001) Appl. Environ. Microbiol. 67, 2641-264
Ronson *et al.* (1987) Cell 4, 579-581
Schoffl R *et al.* (1981) Mol. Gen. Genet. 181, 87-94
Sullivan *et al.* (1995) Proc. Natl. Acad. Sci. USA 92, 8985-8989
Szeto *et al.* (1987) J. Bacteriol. 169, 1423-1432
Trzebiatowski J *et al.* (2001) Appl. Environ. Microbiol. 67
Wolk CP *et al.* (1991) Proc. Natl. Acad. Sci. USA 88, 5355-5359
Yang D *et al.* (1985) Proc. Natl. Acad. Sci. USA 82, 4443-4447
Yeliseev AA, Kaplan S (1995) J. Biol. Chem. 270, 21167-21175

THE GENOMES OF NITROGEN-FIXING ORGANISMS

R. Palacios

Nitrogen Fixation Research Center, National University of México,
 Ap. postal 565-A, Cuernavaca, Morelos, México

Genomic science is changing the focus of biology from the gene to a new paradigm centered in the genome. Genomics is rapidly developing through the integration of very powerful analytical techniques – DNA sequencing and transcript and protein detection from whole genomes – with the computational tools that allow the interpretation of the information obtained. The aim of genomics is to understand the biological significance of the information coded in a genome. This includes the acquisition of the sequence and its evolutionary correlation with that of other genomes; its integrated transcriptional patterns under different conditions; the complete set of proteins present at a given time and the structural characteristics and functional metabolic pathways that such sets of proteins confer to the organism.

Nitrogen fixation research has fully entered the era of genomics. The first landmark was sequencing the symbiotic plasmid of the broad host range *Rhizobium* strain NGR234 (Freiberg *et al.* 1997) that was reported on the 11th International Congress on Nitrogen Fixation. By the time of this 13th International Congress, we have the complete sequence of the genomes of different nitrogen-fixing organisms: the Archeae *Methanobacterium thermoautotrophicum* (Smith *et al.* 1997); and the symbiotic bacteria *Mesorhizobium loti* (Kaneko *et al.* 2000) and *Sinorhizobium meliloti* (electronic address: toulouse.inra.fr./meliloti.html; S. Long, this volume). In addition, sequences of the genomes of other nitrogen-fixing organisms has been obtained. These include the photosynthetic bacterium *Rhodobacter capsulatus* (R. Haselkorn, personal communication) and the filamentous cyanobacterium *Nostoc punctiforme* (T. Thiel, this volume).

A large amount of sequence information now exists in regard to the symbiotic regions of different Rhizobia. In addition to *Sinorhizobium meliloti, Mesorhizobium loti* and *Rhizobium* sp. NGR234, the nucleotide sequence of the symbiotic plasmid of *Rhizobium etli* (G. Dávila, this volume) and that of a region of 410 kb of the chromsome of *Bradyrhizobium japonicum* that contains most of the symbiotic genes (Göttfert *et al.* 2001) have been obtained. This information has led to the discovery and characterization of new genes that affect the symbiotic process.

From an integral viewpoint, the sequence information from symbiotic regions of different organisms is now being used for comparative genomic studies. Preliminary conclusions indicate that the overall order of symbiotic genes is not conserved and that the set of symbiotic genes present in a particular genome might be acquired from different evolutionary routes. This argues in favor of reviewing the taxonomy of *Rhizobium* and related bacteria in the light of the new integral genomic information.

Functional genomics of nitrogen-fixing organisms has also started both at the transcriptomic level (Perret *et al.* 1999) and the proteomic level (Natera *et al.* 2000; M. Djordjevic, this volume). Furthermore, comparative and functional genomics approaches are now being used to analyze complex ecological functions such as rhizobial soil persistence and competitivity for nodule formation (F. de Bruijn, this volume) as well as to characterize novel nitrogen-fixing endophytes (E. Triplett, this volume).

The knowledge of the DNA sequence of whole genomes or replicons has suggested new forms of genomic manipulation. In our laboratory we have developed an experimental strategy – natural genomic design – to obtain derivative rhizobial populations containing alternative genomic structures (Flores *et al.* 2000; P. Mavingui, unpublished).

It is clear that genomic science is introducing new horizons into nitrogen fixation research and that will be a key element to achieve the long term goals in our field.

References

Freiberg C *et al.* (1997) Nature 387, 394-401
Smith DR *et al.* (1997) J. Bacteriol. 179, 7135-7155
Kaneko T *et al.* (2000) DNA Res. 7, 331-338
Göttfert M *et al.* (2001) J. Bacteriol. 183, 1405-1412
Perret X *et al.* (1999) Mol. Microbiol. 32, 415-425
Natera S *et al.* (2000) MPMI 13, 995-1009
Flores M *et al.* (2000) Proc. Natl. Acad. Sci. USA 97, 9138-9143

Acknowledgements

This work was supported in part by grants L0013N and 0028 from CONACyT-México; and IN214498 from DGAPA-UNAM, México.

COMPLETE GENOME STRUCTURE OF *MESORHIZOBIUM LOTI* STRAIN MAFF303099

T. Kaneko, Y. Nakamura, S. Sato, E. Asamizu, T. Kato, S. Tabata

Kazusa DNA Research Institute, 1532-3 Yana, Kisarazu, Chiba 292-0812, Japan

1. Introduction

Mesorhizobium loti is a member of the rhizobia which performs nitrogen fixation on several *Lotus* species in determinant-type globular nodules. Nodule formation and nitrogen fixation result from interactions between symbiotic bacteria and host plants by the sequential expression of a series of genes from both bacteria and hosts. To understand the genetic systems required for the entire process of symbiotic nitrogen fixation, we have initiated the genome sequencing of *M. loti* strain MAFF303099. Here, we report the complete structure of the *M. loti* genome, which consists of a single chromosome and two large plasmids, and gene complements of both the chromosome and the plasmids. Characteristic features of the genes and the genome will also be presented.

2. Materials and Methods

2.1. Genome sequencing. *Mesorhizobium loti* strain MAFF303099 was obtained from the Genetic Resource Center, National Institute of Agrobiological Resources, Ministry of Agriculture, Forestry and Fisheries. The modified whole-genome shotgun strategy was adopted to determine the genome structure as previously described (Kaneko *et al.* 2000). A total of 90,706 sequence files corresponding to about 8-times genome equivalent were accumulated and subjected to assembly using the Phrap program (Phil Green, Univ. of Washington, Seattle, USA).

2.2. Gene assignment and annotation. Protein coding regions were assigned by a combination of computer prediction and similarity search as described previously (Kaneko *et al.* 2000). A computer program, Glimmer, was used for the prediction of protein-coding regions (Delcher *et al.* 1999). Genes for structural RNAs were assigned by similarity search against the structural RNA database in-house. Prediction by the tRNA scan-SE program was performed to assign tRNA-coding regions in combination with the similarity search (Lowe *et al.* 1997). Function of the predicted genes was assigned by the similarity to genes of known function. A BLAST score of less than e-20 was taken into account for genes encoding proteins of 100 amino acid residues or longer. Higher e-values were taken for genes encoding smaller proteins.

3. Sequence Determination of the Entire Genome

The genome sequence of *M. loti* strain MAFF303099 was deduced by assembly of 90,706 sequence files corresponding to approximately 8-times genome equivalent. Integrity of the final sequence was confirmed by comparing the distance of the end sequences of each of 361 cosmid clones on the generated sequence and its insert length for the entire genome. The length of the genome thus deduced was 7,596,297 bp long, which consists of three circular molecules, a chromosome of 7,036,071 bp and two plasmids, designated as pMLa and pMLb, of 351,911 bp and 208,315 bp, respectively. Overall GC contents of the chromosome and two plasmids, pMLa and pMLb, were 62.7%, 59.3% and 59.9%, respectively.

4. Assignment of Protein- and RNA-coding Genes

By taking the results of computer prediction and sequence similarity to known genes into account, the total number of the potential protein-coding genes finally assigned to the chromosome was 6,752 (Table 1). Two plasmids, pMLa and pMLb, had the capacity of coding for 320 and 209 proteins.

The putative protein-coding genes thus assigned to the genome starting with either ATG, GTG or TTG codon were denoted by serial number with three letters representing the species name (m), ORF longer than or less than 100 codons (l or s), and the reading direction on the circular map (r or l).

Two copies of rRNA gene clusters were found on the genome at coordinates of 2,745,482-2,751,894 and 2,752,970-2,759,407. The sequences of two clusters were identical except for two nucleotide residues downstream of *trnfM*. One gene for the RNA subunit of RNase P was identified. A total of 50 tRNA genes representing 47 tRNA species were assigned to the chromosome by sequence similarity to known bacterial tRNA genes and computer prediction using the tRNA scan-SE program. No RNA coding genes were found on the plasmid genomes. It should be remembered that the genes assigned merely represent the coding potentiality of proteins and RNAs under the defined assumptions.

Table 1. Features of the assigned protein-coding genes and the functional classification.

	Chromosome	%	pMLa	%	pMLb	%
Amino acid biosynthesis	177(2.6)		10	3.1	2	2
Biosynthesis of cofactors, prosthetic groups and carriers	145	2.1	14	4.4	1	1
Cell envelope	110	1.6	3	0.9	1	1
Cellular processes	176	2.6	14	4.4	16	16
Central intermediary metabolism	120	1.8	1	0.3	0	0
Energy metabolism	326	4.8	7	2.2	7	7
Fatty acid, phospholipid and sterol metabolism	163	2.4	4	1.3	1	1
Purines, pyrimidines, nucleosides and nucleotides	81	1.2	0	0	1	1
Regulatory functions	517	7.7	11	3.4	11	11
DNA replication, recombination and repair	85	1.3	7	2.2	11	11
Transcription	54	0.8	0	0	0	0
Translation	190	2.8	5	1.6	1	1
Transport and binding proteins	717	11	41	13	6	6
Other categories	814	12	43	13	17	17
Subtotal of genes similar to genes of known function	3675	54	160	50	75	36
Similar to hypothetical protein	1423	21	60	19	38	18
Subtotal of genes similar to registered genes	5098	76	220	69	113	54
No similarity	1654	25	100	31	96	46
Total	6752	100	320	100	209	100

5. Functional Assignment of the Protein-coding Genes

Of the 6752 potential protein-coding genes in the chromosome, 3673 (54%) were homologs to genes of known function, 1421 (21%) showed similarity to hypothetical genes, and the remaining 1678 (25%) showed no significant similarity to any registered genes (Table 1). Two plasmid genomes contained a larger number of genes of unknown function, 51% and 65% for pMLa and pMLb, respectively (Table 1). The number of genes in each functional category is summarized in Table 1.

6. Features of the Predicted Genes Characteristic of *M. loti*

6.1. Symbiotic island. A 610,975 bp DNA segment flanked by portions of phe-tRNA gene sequence on both sides was identified as a probable "symbiotic island" (Sullivan *et al.* 1998). P4 integrase

---

gene which was located near the end of the symbiotic island in *M. loti* strain ICMP3153 was also present (mll6432) in MAFF303099. A total of 580 protein-coding genes were assigned to the symbiotic island of MAFF303099 based on computer prediction and similarity search. The Glimmer program often failed to predict the genes of known function in the symbiotic island, suggesting exogenous origin of this DNA segment. As a result, the DNA segment contained 30 genes related to nitrogen fixation and 24 genes for nodulation (Figure 1).

Notable features of the genes in this segment are as follows:
(i)   Twelve genes for the conjugal transfer proteins were identified.
(ii)   Two gene clusters, each comprised of four genes for biotin synthesis (mll5828-mll5831,mll6003 and mll5004-mll6007), were assigned. Another gene cluster with the same gene set was found in plasmid pMLa. A cluster of genes for thiamine biosynthesis consisting of six genes was also identified (mll5788-mll5795), though the *thiG* gene seems to be split into two ORFs (mll5790 and msl5792) by a frameshift mutation.
(iii)   One hundred eleven out of 580 genes (19.6%) assigned in the symbiotic island were those for transposon-related function such as transposase, integrase, recombinase and resolvase.
(iv)   A cluster of genes for type III secretion system were found. Nine genes, *hrcN-y4yJ-hrcQ-hrcR-hrcS-hrcT-hrcU-y4yq-HrcV*, formed the cluster (mlr6342-mlr6348).
(v)   Two hundred and fifty genes showed a high degree of sequence similarity to those in the symbiotic plasmid, pNGR234a, of *Rhizobium* sp. NGR234 (536 kb) (Freiberg *et al.* 1997).

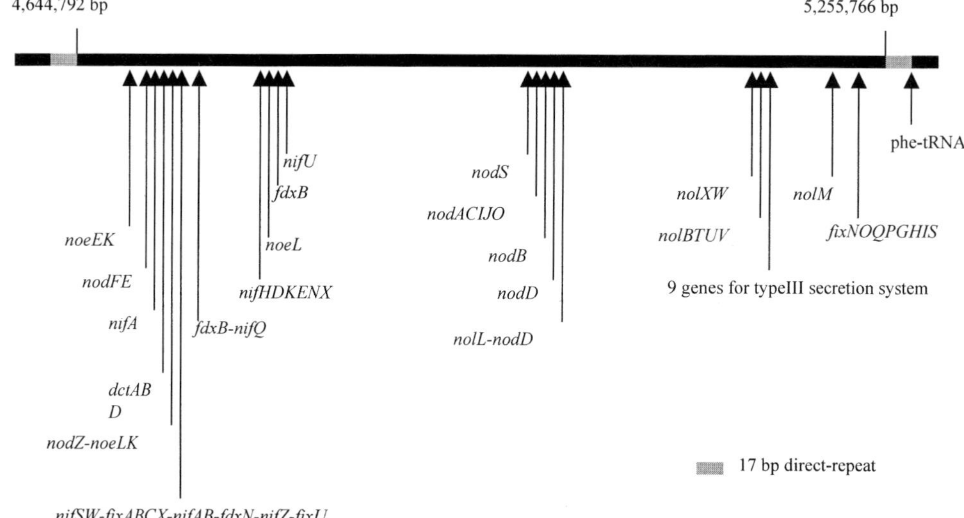

Figure 1. Genes for nodulation and nitrogen fixation in the symbiotic island.

6.2 Genes related to nodulation and nitrogen fixation. A total of 39 genes for nodulation were identified on the chromosome, and 24 of them were located in the symbiotic island. Forty-six genes were assigned as those for nitrogen fixation, of which 30 were found in the symbiotic island. Only one homolog of *noeC* gene for nodulation was present in the plasmid genome (pMLa). Nine genes and gene clusters contained the nod-box like sequences in the upstream. These include *nodZ-noeL-nolK* (mlr5848-5849-5850), *nodS* (mlr6161), *nolL* (mlr6181) and two-component response regulator y4xI (mlr6334).

The sequence and gene information are available in the Web database, RhizoBase, at http://www.kazusa.or.jp/rhizobase/.

7. References

Delcher A *et al.* (1999) Nucleic Acids Res. 27, 4636-4641
Freiberg C *et al.* (1997) Nature 387, 394-401
Kaneko T *et al.* (2000) DNA Res. 7, 331-338
Lowe T *et al.* (1997) Nucleic Acids Res. 25, 955-964
Sullivan J *et al.* (1998) Proc. Natl. Acad. Sci. USA 95, 5154-5149

8. Acknowledgements

We thank Drs K. Minamizawa and T. Uchiumi for valuable discussions. This work was supported by the Kazusa DNA Research Institute Foundation.

THE SYMBIOTIC PLASMID OF *RHIZOBIUM ETLI* CFN42

P. Bustos[1], M.A. Cevallos[1], J. Collado-Vides[2], V. González[1], A. Medrano[2], G. Moreno[2], M.A. Ramírez[1], D.R. Romero [3], J.G. Dávila[1]

[1]Programa de Evolución Molecular
[2]Programa de Biología Computacional
[3]Programa de Genética Molecular de Plásmidos Bacterianos
  Centro de Investigación sobre Fijación de Nitrógeno, UNAM., Av. Universidad S/N, Col.
  Chamilpa, Cuernavaca, Mor., México

The replicon p42d is the symbiotic plasmid (pSym) for the strain CFN42 of *Rhizobium etli*. This plasmid has all the required information to promote nodulation on beans when it is present in an *Agrobacterium tumefaciens* strain cured from its endogenous plasmids (S. Brom *et al.* 1988).

Phaseolus vulgaris the common bean, was originated and diversified in Mesoamerica, hence the genetic pool of *R. etli*, its symbiont, is also believed to be very large in Mexican fields. Accordingly the genetic diversity for the chromosome of a population of local isolates of *Rhizobium etli* is one of the largest found within a bacterial species (D. Piñero *et al.* 1988). Nevertheless at DNA level, the genetic sequences of p42d are highly conserved among the symbiotic plasmids of other strains of the species, independently of their geographical origin of isolation. Therefore, we propose that this plasmid was recently spread among soil bacteria where beans were cultivated. To further support this hypothesis, most of the strains isolated from bean rhizosphere, belong to the species but are devoid of the symbiotic plasmid (L. Segovia *et al.* 1991).

The genome of *Rhizobium etli* CFN42 is distributed among 7 replicons, one chromosome of approximately 5 Mb and six plasmids (p42a 0.2 Mb, p42b 0.15 Mb, p42c 0.27 Mb, p42d 0.37 Mb, p42e 0.5 Mb and p42f 0.7 Mb). The estimated number of reiterated sequence families for Rhizobium is 200, with an average of 2.5 elements per family (M. Flores *et al.* 1987). p42d contains 10 families of internally repetitive sequences, with a number of elements varying from 2 to 6. The sequences taken into account for this analysis are those that span at least 300 bp and had 80% of DNA identity (L. Girard *et al.* 1991). These reiterated sequences have been shown to participate in genome rearrangements mediated by homologous recombination (D. Romero *et al.* 1991).

Only with p42a, the symbiotic plasmid has an extensive conservation of reiterated sequences, many of them of the IS type are not present in the rest of the genome. The presence of this kind of sequences shared exclusively between these plasmids suggest that both plasmids arrived recently from a common genomic background. To further support this, p42a, a self-transmissible plasmid, conducts p42d during conjugation experiments (S. Brom, personal communication).

The sequence project for p42d was recently concluded. Its structure is a double stranded circular DNA with 371,256 bp. The plasmid has a RepABC replication system that confers a low copy number, 1 to 2 plasmids per cell, and a very high stability (M.A. Ramírez *et al.* 1997). The sequencing of the plasmid was carried out with a 373 ABI automatic sequencer with fluorescent ddNTP terminators. Templates were purified PCR products, and prepared from M13 clones of shot gun libraries from each one of the 13 cosmids that orderly cover the whole p42d plasmid. More than 500 reactions per cosmid were made to get the raw sequence, the sequences were initially assembled using the CONSED program. To fill the gaps between the CONTIGs, around 180 single strand DNA primers were prepared on the borders of each CONTIG, these primers were utilized with a p42d BamHI clone of the specific zone to get the sequence. A set of 1000 additional reactions was required for the refinement.

The final sequence of p42d, has less than one mismatch in every 10,000 bp. This sequence was tested by performing 72 PCR reactions with primers designed to successively cover the whole

---

pSym. As templates, total DNA from a CFN42 strain, recently isolated from a nitrogen-fixing nodule, were utilized. All the PCR products of the expected size were obtained.

In order to establish the open reading frames (ORFs) of the plasmid, GLIMMER software was employed in an iterative way, feeding the results of the first search as a seed of the second one, each time changing some of the parameters for the stringency of the analysis.

The amino acids sequence of the polypeptides, derived from the ORF sequences of the plasmid was then compared by BLASTX with those reported on the Data Bases. Specific comparisons were carried out with the genomes of *Mesorhizobium loti, Sinorhizobium meliloti, Methanobacterium thermoautotropicum*, the symbiotic island of *Bradyrhizobium japonicum* and the plasmids pNGR234a, from *Rhizobium* sp., pRi1724 from *Agrobacterium rhizogenes* and Ti from *Agrobacterium tumefaciens*. Codon adaptation index and G+C content were estimated for each ORF. IS elements were grouped according to reported families and compared with the reiterated sequences of the plasmid. Polypeptides derived from each ORF were organized in functional classes; 43% of the total are of unknown function.

References

Brom S *et al.* (1988) Appl. Environ. Microbiol. 54, 1280-1283
Flores M *et al.* (1987) J. Bacteriol. 169, 5782-5788
Girard ML *et al.* (1991) J. Bacteriol. 173, 2411-2419
Piñero D *et al.* (1988) Appl. Environ. Microbiol, 54, 2825-2832
Ramírez MA *et al.* (1997) Microbiol. 143, 2825-2831
Romero DR *et al.* (1991) J. Bacteriol. 173, 2435-2441
Segovia L *et al.* (1991) Appl. Environ. Microbiol. 57, 426-433

Acknowledgements

We are grateful to J.C. Hernández, V. Quintero, H. Salgado, R.E. Gómez, J.A. Gama, I. Hernández, J. Espíritu and E. Díaz for technical and computing assistance.

NITROGEN FIXATION: ANALYSIS OF THE GENOME OF THE CYANOBACTERIUM *NOSTOC PUNCTIFORME*

T. Thiel[1], J.C. Meeks[2], J. Elhai[3], M. Potts[4], F. Larimer[5], J. Lamerdin[6], P. Predki[7], R. Atlas[8]

[1]Department of Biology, University of Missouri, St. Louis, MO 63121, USA
[2]Section of Microbiology, University of California, Davis, CA 95616, USA
[3]Department of Biology, University of Richmond, Richmond, VA 23173, USA
[4]Department of Biochemistry, Virginia Polytechnic Institute and State University, Blacksburg,VA 24061, USA
[5]Computational Biology, Oak Ridge National Laboratory, Oak Ridge, TN 37831, USA
[6]Biology and Biotechnology Research Program, Lawrence Livermore National Laboratory, Livermore, CA 94551, USA
[7]DOE Joint Genome Institute, 2800 Mitchell Drive, Walnut Creek, CA 94598, USA
[8]Department of Biology, University of Louisville, Louisville, KY 40292, USA

1. Introduction

Nostoc punctiforme ATCC 29133 is a diazotrophic, filamentous cyanobacterium with oxygen-evolving photosynthesis. Although this strain is primarily an autotroph, it can grow heterotrophically in the dark using sucrose, glucose or fructose (Summers *et al.* 1995). Like all cyanobacteria it has chlorophyll *a* as well as light harvesting pigments called phycobiliproteins. This strain chromatically adapts to light by varying the amounts of one biliprotein, phycoerythrin, in response to the presence or absence of green light. Cyanobacteria, which form a monophyletic group, are classified on the basis of cell division (resulting in unicellular or filamentous growth) and the differentiation of specialized cells, such as heterocysts, akinetes, or hormogonia (Rippka *et al.* 1979). Heterocysts are specialized for nitrogen fixation in an oxic environment, hormogonia are motile filaments, and akinetes are resistant to many environmental stresses. *N. punctiforme* has one of largest bacterial genomes sequenced to date, which is not surprising since it encompasses virtually every characteristic of the cyanobacterial group. *N. punctiforme* has a complex life cycle that includes, at different times, the differentiation of vegetative cells into motile hormogonia, nitrogen-fixing heterocysts, or resistant akinetes. The patterning of spaced heterocysts in filaments as well as the interdependence between vegetative cells that supply fixed carbon and heterocysts that supply fixed nitrogen to the filament indicate cell-to-cell communication, suggesting that these and other heterocyst-forming cyanobacteria function as multicellular organisms (Thiel, Pratte 2001).

N. punctiforme forms symbiotic associations with a variety of species including the cycad *Macrozamia* sp., the angiosperm *Gunnera* sp., and the bryophyte hornwort *Anthoceros punctatus*. Cultured cells of this strain can establish a nitrogen-fixing symbiosis with *Anthoceros punctatus* (Enderlin, Meeks 1983) and *Gunnera* spp. (Johansson, Bergman 1994). In the association, photosynthesis in the symbiotic *Nostoc* species decreases and the rate of nitrogen fixation increases. In addition, heterotrophic metabolism, which supports nitrogen fixation, increases. The plant partner produces molecules that control the differentiation of hormogonia and heterocysts (Meeks 1998). The functional analysis of the *N. punctiforme* genome will help in understanding these microbe-plant interactions that promote a stable nitrogen-fixing association.

2. Procedures

2.1. Genome analysis. The genomes of *N. punctiforme* and *Rhodopseudomonas palustris* were sequenced using a whole genome shotgun strategy (Fleischmann *et al.* 1995). DNA preparation protocols are described on the web site: http://www.jgi.doe.gov/ under "Production protocols". Raw sequence data were assembled with PHRAP and through "auto-finishing" using software written by

Matt Nolan (JGI/Lawrence Livermore National Laboratory) and David Gordon (University of Washington). Open reading frames (ORFs) were identified using Critica, Glimmer and Generation. The set was searched against the KEGG GENES, Pfam, PROSITE, PRINTS, ProDom and COGS databases, in addition to BLASTP versus the non-redundant database. Putative genes were organized into functional categories based on KEGG categories and COGs hierarchies. Additional analyzes are based on sequence comparisons to Cyanobase, Genbank and Swissprot.

2.2. Phylogenic analysis. Phylogenetic analysis used the program Clustal W for amino acid sequence alignment and the Phylip programs Seqboot, Protdist, Neighbor, and Consense to produce distance trees. Trees were visualized in Treeview. Bootstrap values were based on 1000 replicates.

3. Results and Discussion

3.1. Genome analysis. The genome of *N. punctiforme* is significantly larger than any other sequenced cyanobacterial genome. In addition to the three genomes shown in Table 1, two additional cyanobacterial genomes are near completion: the 1.7 Mb genome of *Prochlorococcus marinus* MED4 and the 2.4 Mb genome of *Synechococcus* sp. strain WH 8102. The current *N. punctiforme* genome size is 9,757,495 bases (11X sequencing coverage); however, the annotation is based on only about 92% of the genome (8X coverage; 8,941,326 bases). Of the more than 5000 recognized ORFs only about 62% of those encode proteins with a known or probable known function, while the remainder encode conserved hypothetical proteins with no known function (Table 1). Interestingly, almost a quarter of the genome encodes ORFs that cannot be associated with a previously recognized ORF.

Table 1. Comparison of three cyanobacterial genomes

Strain	Genome size	ORFs	Recognized ORFs	Known or probable function	ORFs present in *Nostoc*	Unique to strain
Nostoc punctiforme	9.78 Mb	7,432	5,314	3,328	7,432	1,578 (23%)
Synechocystis 6803	3.57 Mb	3,215	3,215	1,521	2,547 (80%)	668 (20%)
Anabaena 7120	7.20 Mb	5,610	4,327		4,814 (86%)	797 (14%)

Comparison of the genomes of *N. punctiforme* and *Anabaena* sp. PCC 7120 reveal that they share about 80% of their genetic information, implying a close phylogenetic relationship between the two strains. *N. punctiforme* contains multiple copies of many genes in *Anabaena* 7120 and/or in *Synechocystis* 6803. *Anabaena* 7120 and *Synechocystis* 6803 each have about 700-800 ORFs that are unique, i.e. not present in any of the other strains, whereas *N. punctiforme* has about 1500 such ORFs. About 500 ORFs are shared by *N. punctiforme* and *Anabaena* sp. PCC 7120, but not with other cyanobacteria, suggesting that these ORFs may encode proteins involved in common phenotypic characteristics such as heterocyst differentiation.

The recognized genes encode proteins involved in all the aspects of cyanobacterial metabolism that are expected. The largest proportion of genes (about 5%) are associated with signal transduction mechanisms such as protein kinases and response regulators. Genes required for cell envelope synthesis, cell division, and chromosome segregation comprise almost 4% of all the ORFs as do genes involved in amino acid transport and metabolism. Almost 3% of the ORFs encode genes involved in organic carbon metabolism, presumably reflecting the heterotrophic capability of this strain. About 10% of the ORFs are probable enzymes or structural proteins; however, they cannot be assigned to a functional category. Clearly there is much to learn about *N. punctiforme* from a functional analysis of this large and complex genome.

In the *N. punctiforme* genome many of the gene categories contain a surprisingly large number of genes (Table 2). Among the largest categories are those representing the response regulator/sensor histidine kinase groups. There are also many copies of transcriptional regulators. This suggests a high level of regulation for the many processes in *N. punctiforme* that are required for cellular differentiation and for establishment and maintenance of symbiosis.

Table 2. Gene categories with multiple gene copies

Gene category based on COG assignments	Number of putative copies
Response regulator receiver domain	168
Transposases	148
Sensor histidine kinase	146
Transcriptional regulators (helix-turn-helix domain, AraC, ArsR, LuxR, LysR, TetR)	100
ATPase component of ABC Transporter	99
Tetratrico-peptide repeat protein	96

3.2. Nitrogenase genes.   Nitrogen fixation is mediated by an enzyme complex, containing dinitrogenase (encoded by *nifD* and *nifK*) and dinitrogenase reductase (encoded by *nifH*), whose assembly requires many *nif* gene products (Dean, Jacobson 1992). A large cluster of *nif* and *nif*-related genes from *nifB* to *fdxH* is highly conserved in cyanobacteria (Thiel *et al.* 1997, 1998) including *N. punctiforme* and *Anabaena* sp. PCC 7120 (Figure 1). A major difference between the latter two clusters is the excision elements that interrupt both *fdxN* and *nifD* in *Anabaena* sp. PCC

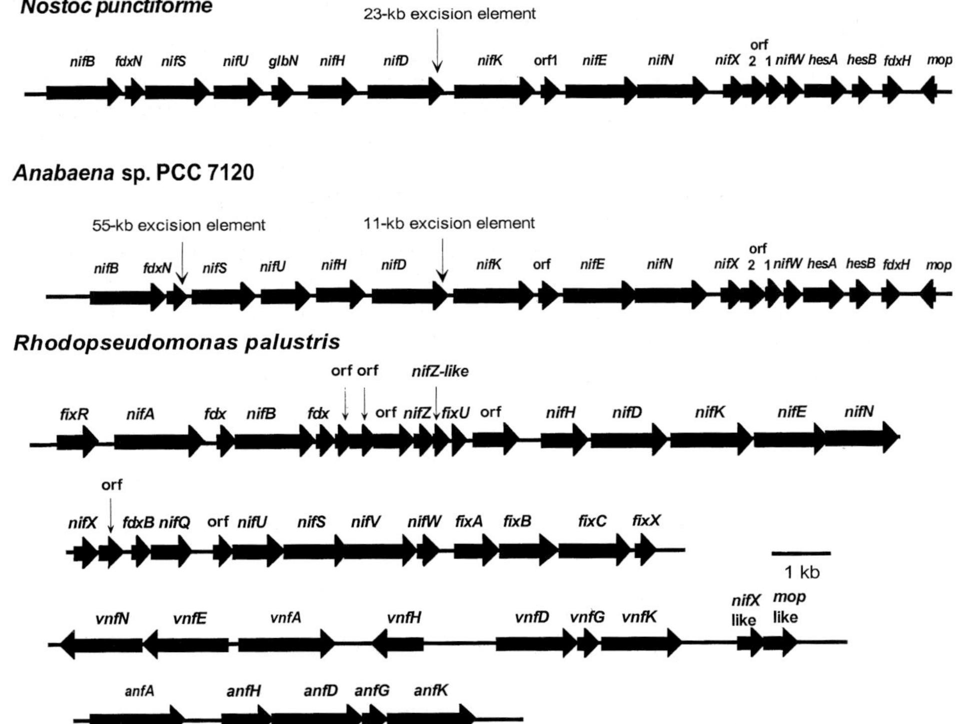

Figure 1. Organization of *nif* and *nif*-related genes in three diazotrophs. Note that genes shown on two lines representing the entire *nif* cluster of *R. palustris* are contiguous in the genome.

7120 (Golden *et al.* 1985, 1988). In *N. punctiforme* the *fdxN* excision element is missing and the 24-kb *nifD* element in *N. punctiforme* is almost completely different from the 11-kb *nifD* element in *Anabaena* sp. PCC 7120. The two *nifD* elements share only the *xisA* gene that is required for site-specific recombination that removes the element during heterocyst differentiation and a small open reading frame of unknown function. Within the *N. punctiforme* excision element there is a bacterial retron-like gene. *Anabaena* sp. PCC 7120 also has a gene that appears to be a retron; however, it is not located in the excision element and it is not closely related phylogenetically to the gene in *N. punctiforme*. Upstream of *nifH* in *N. punctiforme* and in *Nostoc commune* (Potts *et al.* 1992) is a hemoglobin-like gene called cyanoglobin (*glbN*) whose function is not known. The *nif* genes of *N. punctiforme* are most closely related to those of *Nostoc commune*, as determined by phylogentic analysis (Thiel, unpublished). Upstream of the major *nif* cluster in *N. punctiforme* are *nifP*, *nifZ* and *nifT* and downstream are genes for an uptake hydrogenase, including *hupS* and *hupL* which lack an excision element that is present in the *hupL* gene of *Anabaena* sp. PCC 7120 (Carrasco *et al.* 1995).

There is only one copy of most of the *nif* genes in *N. punctiforme*, supporting previous studies that indicated that this strain lacks alternative nitrogenases such as the V-nitrogenase in *Anabaena variabilis* ATCC 29413 (Thiel 1996). However, *N. punctiforme* has two additional copies of *nifH* and one additional copy of *nifE* and *nifN*. The organization of the multiple *nif* genes is shown in Figure 2 with multicopy *nif* genes of some other diazotrophic strains. Phylogenetic analysis indicates that the second copy of *nifH* in *N. punctiforme* (near the second copy of *nifEN*) clusters with the *N. punctiforme* and *N. commune* *nifH* genes that are part of the major *nif* cluster; however, it is less like those two genes than they are like each other (T. Thiel, unpublished). The third *nifH* in *N. punctiforme* is closely related to the *nifH* gene in *A. variabilis* that appears to encode the dinitrogenase reductase of the V-nitrogenase. None of the *nifH* genes in *N. punctiforme* is closely related to the second copy of *nifH* in *Anabaena* sp. PCC 7120

The genome of *Rhodopseudomonas palustris* was sequenced by the Joint Genome Institute at about the same time as the *N. punctiforme* genome. Analysis of the *R. palustris* genome indicates an unusually large number of *nif*-like genes. Like *Rhodobacter capsulatus*, *R. palustris* has genes that appear to encode a Mo-nitrogenase and an alternative Fe-nitrogenase. Unlike *R. capsulatus*, *R. palustris* also has genes that appear phylogenetically to be closely related to V-nitrogenase genes in *Azotobacter vinelandii* and *A. variabilis*. The organization of these three gene clusters is shown in Figure 1. In addition, *R. palustris* has additional copies of *nifB*, *nifH*, *nifE*, and *nifN* that are not associated with *nifDK* genes (Figure 2). Phylogenetic analysis indicates that these extra copies that are not associated with *nifDK* genes are only distantly related to homologs that have a known function. This is in contrast to the extra copies of *nifH* genes in cyanobacteria that cluster within the cyanobacterial *nifH* group. It will be of great interest to determine what role, if any, these multicopy *nif* gene families have in nitrogen fixation.

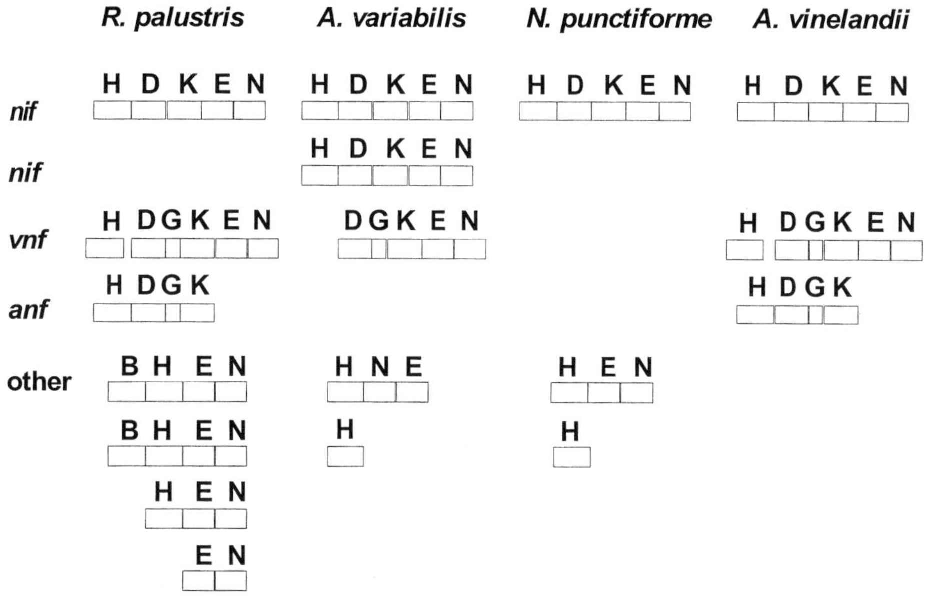

Figure 2. Schematic representation of the multicopy *nif* gene families of four diazotrophs.

4. References

Carrasco CD *et al.* (1995) Proc. Natl. Acad. Sci. USA 92, 791-795

Dean DR, Jacobson MR (1992) In Stacey G *et al.* (eds) Biological Nitrogen Fixation,
pp. 763-834, Chapman and Hall, Inc., New York

Enderlin CS, Meeks JC (1983) Planta 158, 157-165

Fleischmann RD *et al.* (1995) Science 269, 496-512

Golden JW *et al.* (1985) Nature 314, 419-423

Golden JW *et al.* (1988) J. Bacteriol. 170, 5034-5041

Johansson C, Bergman B (1994) New Phytol. 126, 643-652

Meeks JC (1998) BioScience 48, 266-276

Potts M *et al.* (1992) Science 256, 1690–1692

Rippka R *et al.* (1979) J. Gen. Microbiol. 111, 1-61

Summers ML *et al.* (1995) J. Bacteriol. 177, 6184-6194

Thiel T (1996) J. Bacteriol. 178, 4493-4499

Thiel T *et al.* (1997) J. Bacteriol. 179, 5222-5225

Thiel T *et al.* (1998) In Peschek GA *et al.* (eds), Phototrophic Prokaryotes, pp. 517-521,
Plenum Press, New York

Thiel T, Pratte B (2001) J. Bacteriol. 183, 280-286

5. Acknowledgements

The sequencing of *Nostoc punctiforme* and *Rhodopseudomonas palustris* was supported by the DOE through a contract with the Joint Genome Institute. Research in the laboratories of TT, JM, JE and MP is supported by grants from the NSF, USDA and DOE. We thank Elsie Campbell, Kari Hagen, John Ingraham and Francis Wong for sequence analyses.

Section 3:
Plant Genomics

INTEGRATED FUNCTIONAL GENOMICS TO DEFINE THE PLANT'S FUNCTION IN SYMBIOTIC NODULATION

P.M. Gresshoff[1], J. Stiller[1], T. Maguire[1], D. Lohar[5], S. Ayanru[1], D. Buzas[1], S. Habamunga[1], L. Smith[1], B. Carroll[2], I. Searle[2], K. Meksem[4], D. Lightfoot[4,] S. Grimmond[3], A.E. Men[1]

[1]Dept of Botany
[2]Dept of Biochemistry
[3]Institute for Molecular Bioscience
   The University of Queensland, Brisbane St. Lucia, QLD 4072, Australia
[4]Dept of Plant, Soil, and Agronomy, Southern Illinois University, Carbondale, IL, USA
[5]Center for Legume Research, University of Tennessee, Knoxville, TN, USA

1. Introduction

Recent advances of high throughput DNA sequencing, bioinformatics, robotics, BAC libraries, microarrays, insertional mutagenesis as well as promoter trapping open the opportunity for an integrated function and structure analysis of the genomes of soybean (*Glycine max*) and the model legume *Lotus japonicus*. We are specifically interested in the plant's role during the establishment of nodule morphogenesis, and the genes shared during seemingly related developmental programs leading either to nodule or lateral root formation. Plant mutations were induced using EMS, fast neutron deletion as well as insertion mutagenesis. Single recessive loci were mapped using molecular markers, which were used to isolate soybean BAC clones to generate contigs spanning mutant deletions. Special emphasis was given to the *Nts-1* locus of soybean that governs autoregulation of nodulation. Expression analysis of nodulation events using 4200 micro-arrayed root ESTs was initiated to detect gene products temporarily expressed during early nodulation. Insertion of a promoter-less gus-reporter gene into *Lotus japonicus* allowed the isolation of activated plant lines that showed development specific gus-gene expression. Isolation of flanking DNA sequences provided information of potential promoters and gene function as well as providing a link between structural and functional elements of nodulation-related genes. Evidence suggests that many nodule initiation functions evolved or are shared with lateral root related processes. The possibility exists that several non-legumes share such genes.

2. Positional Cloning of the Supernodulation (nts-1) Locus of Soybean

The original supernodulation mutants were isolated by EMS mutagenesis of wild-type cultivar Bragg (Carroll *et al.* 1985). Supernodulation in general is associated with: (a) an increase of nodule number and nodule mass, (b) decreased specific nitrogenase activity, (c) decreased lateral root growth in the inoculated state, (d) shoot control of supernodulation, and (e) nitrate tolerance of nodulation (but not nitrogen fixation) (Gresshoff 1993). These results lead to several important conclusions. First, nitrate regulation of nodulation and internal autoregulation share at least one functional step; second, lateral root initiation and growth are inversely coupled to nodule initiation and growth; and third, nodulation control and nitrogen fixation control are separate processes although they are epistatically related.

RFLP marker pUTG132a was mapped close to the nts-1 gene and used to isolate BAC clones from the Clemson soybean BAC library. End clones were isolated and used to generate a contig region spanning a region also defined by a fast neutron induced deletion of soybean, also exhibiting the supernodulation phenotype. The major BAC clone (138 kb in size; nearly 80% AT content) was sequenced and annotated (in collaboration with AGRF). About ten candidate ORFs with similarity to a variety of plant genes was revealed. Significant micro-synteny with the recently sequenced *Arabidopsis* genome revealed candidate genes. We are presently testing these for allelic variation among different soybean mutants. Several ESTs exist in the region including two loci for neutral amino acid transporters (pA381-1 and pA381-2), aldolase (Gm036), thymidylate synthase and a possible transcription factor.

Several conclusions can be taken from this chromosome walk: (a) it is feasible to walk in a complex genome like soybean (1100 Mb); (b) many endclone sequences reveal repeated DNA elements often related to retrotransposons; (c) molecular markers are preferential in the *G. max* genome supporting a *G. soja* to *G. max* genome expansion hypothesis; (d) a deletion of at least 150 kb leads to a minimal phenotype in nodule regulation but no other physiological alteration, suggesting that functional complementation occurs in a duplicated region of the soybean genome; (e) fast neutron induced deletions are valuable in gene isolation by defining breakage points, and could provide a useful tool if coupled with microarray analysis (see Figure 1). For example, 4200 root ESTs from soybean have been arrayed to investigate gene expression during early nodulation steps in both wild-type and mutant soybeans.

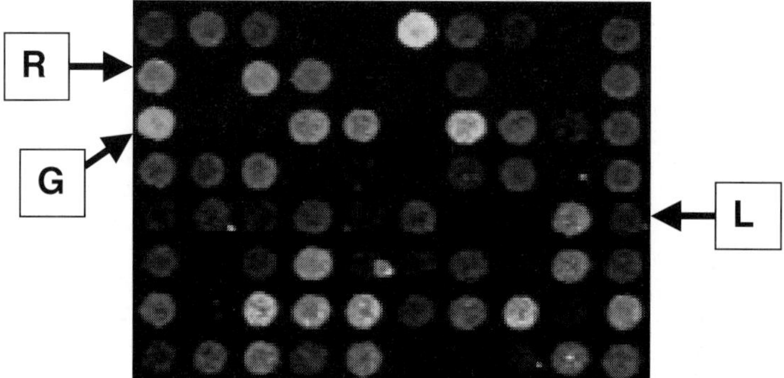

Figure 1. Partial exert of a soybean microarray of 4200 root EST clones hybridized with Cy3/Cy5 labeled RNA from soybean shoots and roots. R = red label from shoot; G = green label from root; L = low expression levels. Note: spot uniformity, low background as well as dynamic range.

3. Promoter Trapping in *Lotus japonicus*: a Gene Machine for Functional Analysis

Lotus japonicus is a model for its symbioses with both rhizobia and mycorrhizae (Handsberg, Stougaard 1992; Jiang, Gresshoff 1997). Both symbioses are not found in *Arabidopsis thaliana*. What distinguishes *Lotus* from most other legumes is its high transformation potential, small genome and true diploidy. Using *Agrobacterium tumefaciens* or *A. rhizogenes*, over 1000 transformed lines have been produced. Transformed plants were selected by kanamycin (geneticin, G418) or BASTA selection (Lohar *et al.* 2001).

Promoter trapping was achieved and generated a large range of developmentally tagged lines (Martirani *et al.* 1999). Expression patterns ranged from those entirely in the root tips, to broad constitutive expression in the entire root, to those in nodule primordia and those entirely within the nodule. The promoter trapping strategy for plant gene discovery has several advantages: (1) tagged gene expression is followed histologically, (2) the physiological regulation of the trapped promoter can be ascertained leading to prediction of possible function, (3) with a single insert line, inverse PCR gives flanking DNA regions defining the possible promoter and ORF (further substantiating functional predictions), (4) selfed progeny should contain 25% homozygous insertions, which may have a mutant phenotype provided physiological analysis is rigorous and the gene is essential. The approach is especially attractive as it allows discovery of genes intimately involved in nodule initiation but also expressed in lateral root development.

The utility of promoter trapping is best illustrated by line 'Cheetah'. This line contains a single promoter-less *gus* gene insert that is activated in root tips of the main and lateral roots, the basal root-vascular strand junction as well as nodule primordia and nodule basal regions (Figure 2). The left and right flanking regions of Cheetah were sequenced and allowed the detection of putative open reading frames. The putative promoter has been defined.

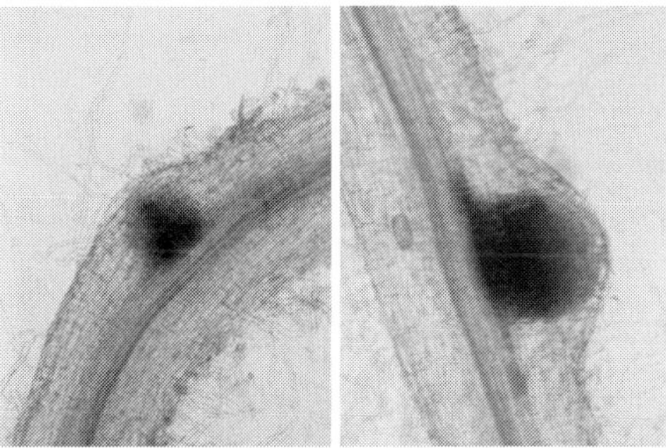

Figure 2. Cheetah *gus* gene expression in lateral root (left) and nodule (right) initials.

4. Molecular Physiology

High frequency transformation allows the testing of known plant genes or promoters in a large number of isolates. We transferred the *Arabidopsis* ethylene receptor mutant gene *etr1-1*, which confers ethylene insensitivity, to Gifu, resulting in plants with altered triple response when germinated in the dark. Plants also showed altered flower timing, fruit maturation and abscission of petals (see Figure 3). Insensitivity was directly correlated with the level of *etr1-1* expression and the degree of nodule initiation. Transgenic plants exhibited Mendelian segregation of the hypernodulation and ethylene insensitivity phenotypes. One concludes that ethylene controls the initiation of nodule primordia by inhibiting cell divisions off the phloem poles (Wopereis *et al.* 2000).

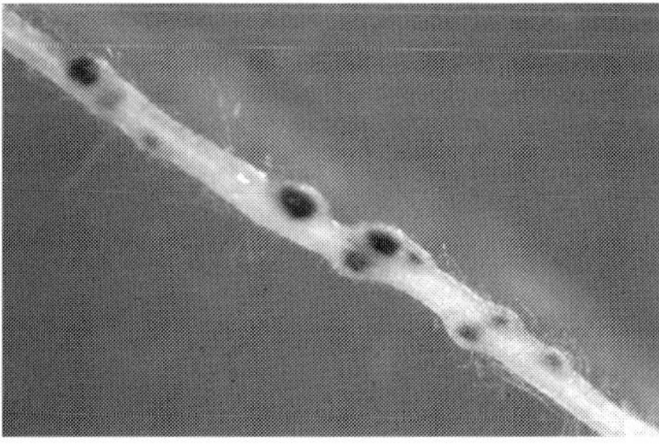

Figure 3. Altered nodule pattern (left) and flower maturation (right) in a transgenic *Lotus japonicus*, altered by the insertion of the ethylene insensitivity receptor gene *etr1-1*.

We constructed a BAC library of genotype 'Gifu' using *Hin*dIII digested DNA (Men *et al.* 2001). The library has 6.5-fold coverage and holds an average insert size of 94 kb. BACs carrying single genes have been detected using either hybridization of probes onto BAC clones arrayed on nylon membranes or by PCR screening of three-dimensionally pooled BAC pools. Discovery of plant genes from a legume is certain to provide extra insights into nitrogen and phosphorous acquisition and related plant-microbe interactions. As a by-product of this utilitarian goal is the recognition that seemingly organ-specific genes involved in legume nodule formation may have arisen from genes found important in basic plant processes such as lateral root proliferation, constraint of microbial invasion and alteration of phytohormone gradients.

5. References

Carroll BJ *et al.* (1985) Proc. Natl. Acad. Sci. USA 82, 4162-4166
Gresshoff PM (1993) Plant Breeding Rev. 11, 387-411
Handsberg K, Stougaard J (1992) Plant J. 2, 487-492
Jiang Q, Gresshoff PM (1997) Molec. Plant-Microbe Int. 10, 59-68
Lohar D *et al.* (2001) J. Exp. Bot.
Martirani L *et al.* (1999) Molec. Plant-Microbe Int. 12, 275-284
Men A *et al.* (2001) Molec. Plant-Microbe Int. 14, 375-379
Stiller J *et al.* (1997) J. Exp. Bot. 48, 1357-1365
Wopereis J *et al.* (2000) Plant J. 23, 97-114

GENOMICS TOOLS FOR DISSECTING NODULATION IN THE MODEL LEGUME *MEDICAGO TRUNCATULA*

K.A. VandenBosch[1], D.R. Cook[2]

[1]Department of Plant Biology, 220 BioSci Center, 1445 Gortner Ave, University of
  Minnesota, St. Paul, MN, 55108, USA
[2]Department of Plant Pathology, Room 254 Hutchinson Hall, University of California,
  Davis, CA 95616, USA

1. Introduction

Approximately ten years after the elucidation of Nod factor, the lipochito-oligosaccharide made by rhizobia in response to their host plants, investigators have identified many responses to this signal in plant roots. In addition, many details of cellular function in infected roots and nodules have been discovered, principally by identifying differentially regulated genes or by analyzing function of candidate genes. However, understanding of many aspects of nodulation remains sketchy. How is the Nod factor signal transduced in the host? Do the regulatory mechanisms that govern nodulation overlap with networks controlling other aspects of development? Are components of the nodulation machinery restricted to legumes, or perhaps to a larger clade of nodule-formers that includes actinorhizal plants? Do homologous rhizobia avoid triggering a defense response, or are such means used to optimize nodule number?

Realizing that answering such complex questions would require a facile genetic model organism, investigators set out more than a decade ago to identify legume species that could serve as a model for the legume family. *Medicago truncatula*, a self-compatible diploid species, has attracted many adherents because of its small genome, rapid life cycle and other attributes. This growing group of investigators has contributed to the development of tools for genetic analysis. Advancement of the system has been chronicled with a series of reviews (Barker *et al.* 1990; Cook 1999; Cook *et al.* 1997; Frugoli, Harris 2001; Harrison 2000). Because *M. truncatula* is congeneric with alfalfa, *M. sativa*, it is also host to the well-characterized *Sinorhizobium meliloti*, which is itself the topic of genomic and proteomic analysis. *M. truncatula* also hosts pathogens that attack alfalfa and related legumes. This enables the use of microbial partners on *M. truncatula* that have been well characterized on crop species, and indicates that this annual medic will be a tractable model for understanding resistance mechanisms in the legume family.

Motivated by a desire to dissect symbioses and to answer some of the questions posed above, a group of nodulation biologists has been at the core of several parallel efforts to develop genomic tools for *Medicago truncatula*. In the United States, the NSF Plant Genome Project and the Noble Foundation have sponsored these efforts, while in Europe, INRA has initiated genomic analysis in France, and efforts are continuing in several countries under the auspices of the EU. This overview will emphasize the accomplishments to date of the NSF-sponsored project, and will conclude by outlining the goals of the *Medicago* community for developing and applying genomic tools to additional problems in the biology of this important plant family.

2. Goals for the NSF-sponsored Project

The large number of host genes likely to be involved in legume/microbe interactions, and the genome complexity of the major crop legumes, necessitate a coordinated effort in a legume species with tractable molecular and genetic attributes. Such a species could be used as a node, or central point, for comparison with more complex related species. This multi-institutional project, entitled "*Medicago truncatula* as the Nodal Species for Comparative and Functional Legume Genomics," aims to develop *M. truncatula* to play a pivotal role, as a point of comparison with other members of the Fabaceae. The project is advancing detailed knowledge of the structural and functional genome elements that underlie aspects of plant biology unique to, or best-studied in, legumes. The biological emphases are symbiotic nitrogen fixation and nitrogen metabolism, mycorrhizal associates and phosphate metabolism, and key legume/pathogen interactions.

Research on the project encompasses the following approaches: (1) comparative genomics, which involve comparing the organization of genes between *M. truncatula* and pea, alfalfa and soybean; (2) functional genomics, emphasizing characterization of expressed genes and initiating large-scale analysis of expression patterns to study gene function; and (3) bioinformatics, which has emphasized development of database resources of the analysis and dissemination of *M. truncatula* genome information collected from the NSF project and other sources. The bioinformatics tools support both of the other goals, and relate the two to each other. The project web site, http://www.medicago.org, also serves as a discussion forum for the *Medicago* community, a catalog of published works, a directory of *M. truncatula* researchers, and points to other *Medicago* research sites. The discussion below highlights several advances in the project.

3. Advances in Comparative Genomics

The close relationship between *M. truncatula* and *M. sativa* is reflected in the extensive macro- and microsynteny between the two species. To date, only one locus, the nucleolar-organizing region, occupies a position on different linkage groups (LGs) in the two species. This enabled investigators to use the same nomenclature for LGs and chromosomes in both *Medicago* species. Even more important from a practical standpoint, the markers from the *M. sativa* map are being adapted for placement on the *M. truncatula* map as a means to swiftly populate the latter, and to enable comparative mapping projects in the two species. Gyorgy Kiss discusses one such successful comparative mapping project, and the contribution towards map-based cloning of an important symbiotic gene, in a chapter in this volume. Comparative mapping between *M. truncatula* and pea (*Pisum sativum*) indicates strongly conserved gene arrangement on five of the eight LGs in *M. truncatula*. This is expected to assist cloning efforts of loci identified in pea, where map-based cloning efforts are severely hampered by a large genome and ample repetitive sequences. One such project, directed by René Geurts, is underway in Wageningen Agricultural University (Gualtieri *et al.* submitted). The extent of microsynteny among these taxa is being further documented with an intensive study LG 5 in *Medicago* spp., which corresponds to LG 1 in pea.

The microsyntenic relationships among temperate legumes were expected because of their close phylogenetic relationships. Therefore, we are interested to see to what extent *M. truncatula* genome structure may be a good model for legumes that are less closely related, such as the tropical legume tribe Phaseolidae that contains important crops such as soybean (*Glycine max*) and beans (*Phaseolus* spp.). This work is being carried out collaboratively between the Cook lab and Nevin Young at the University of Minnesota. To date, macro- and microsynteny between *M. truncatula* and soybean has been demonstrated, although quantification of similarity is complicated by duplication within the soybean genome, and limited polymorphism in the soybean mapping populations. Preliminary work shows that synteny between *M. truncatula* and *G. max* is somewhat less than between duplicated regions of the *G. max* genome, but far exceeds synteny between the model dicot *Arabidopsis thaliana* and either of the two legumes (Yan *et al.* submitted). We predict that *M. truncatula* will have experimental value for at least some regions of the soybean genome, following continued development of a pan-legume map. A major, on-going objective that will facilitate this is the development of gene-based markers that will be used to coordinate the maps of *M. truncatula* and crop legumes.

Integration of the *M. truncatula* genetic and cytogenetic maps was achieved in collaboration with Ton Bisseling and colleagues in Wageningen. Working with pachytene chromosomes, the Dutch group has mapped at least two bacterial artificial chromosome (BAC) clones from the

M. truncatula genomic library onto each chromosome. Visualization indicates that the chromosome arms are composed of long, typically uninterrupted stretches of euchromatic DNA (Kulikova *et al.* 2001). Heterochromatin is largely clustered in the pericentromeric regions, and is estimated to comprise up to 80% of the *M. truncatula* genome. This simple genome arrangement, which predicts organization of the gene-rich regions in as little as 100 Mbp of mostly continuous euchromatin, will greatly aid map-based cloning efforts. In the longer term, this will also facilitate sequencing of the gene-rich portions of the genome.

4. Advances in Functional Genomics

Towards our goal of identifying legume genes involved in nutrient acquisition and microbial interactions, 17 cDNA libraries were made and subjected to high throughput sequencing. Five of these libraries are from *Sinorhizobium*-inoculated or infected tissues, contributing more than 19,000 ESTs from the approximately 55,000 ESTs from the project to date. All ESTs have been deposited in GenBank and used (with other public data) to construct a gene index that represents a minimally redundant set of sequences. The gene index (MtGI), constructed at The Institute for Genome Research, groups *M. truncatula* ESTs into tentative consensus sequences (TCs) according to sequence overlap (http://www.tigr.org//tdb/mtgi). The third release of MtGI (April, 2001) predicted ~13,000 TCs and ~19,000 singletons, totaling ~32,000 predicted unique sequences. The fourth edition of MtGI was due out in August, 2001. MtGI also assigns probable function to expressed genes, based on BLAST results.

The gene index can also indicate expression patterns of highly expressed genes, based on tissue of origin of the ESTs in the TCs. Analysis reveals many examples of tissue-specific patterns of expression. For example, about 3000 TCs, or nearly one fourth, are composed of ESTs exclusively from tissues responding to microbial infection or elicitation. Of these, approximately 1750 TCs appear to have a symbiosis-specific pattern of distribution, including about 900 TCs unique to rhizobium-inoculated tissues, and about 300 common between nodules and mycorrhizae. In addition to nodule-specific genes and putative markers of early responses to rhizobium, we have also identified root-specific markers and expressed sequences common to both symbiotic and pathogenic responses. This approach permits the identification of marker genes to be used as developmentally specific controls in planned transcriptional profiling studies.

Our long-term aims are to establish a unigene set of this EST resource and to investigate thoroughly the genome-wide patterns of gene expression by using hybridization of probes to DNA microarrays. As a first step, we have assembled a small-scale array of ~1000 cDNA clones (called the "kiloclone set"), that contains positive and negative controls and some developmental markers. The kiloclone set is also rich in clones encoding proteins with putative functions in signal transduction, transcriptional regulation, control of cell division and cell death, pathogen responses, secondary metabolism, and a number of genes of unknown function. Hybridization experiments to date have monitored differences in gene expression during the early steps of the symbiotic process or during pathogenic interactions, compared to uninoculated plants. These early experiments have instilled confidence in the assays by demonstrating both technical and biological repeatability. Novel genes expressed during nodulation and responses to the root rot pathogen *Phytophthora medicagenis* have been identified to date. Verification of the identity of the cDNA clones and their expression patterns is underway.

5. Concluding Remarks: The Long-term View

On-going work on our project will continue to emphasize structural, comparative and functional genomics approaches. The specific objectives of the structural genomics team are to create a comprehensive physical map for *M. truncatula*, and to determine the correspondence between 3000 ESTs and BAC addresses in the physical map. The combination of a physical map, BAC end sequencing, and the assignment of the location of several thousand ESTs are expected to delimit the gene-rich portions of the genome, and to lay the foundation for developing a low-pass gene inventory in these regions. The *M. truncatula* gene-based markers will also continue to be evaluated for their utility as markers for the genetic maps of alfalfa, pea and soybean. This comparative approach will be instrumental in extending the impact of *M. truncatula* genomics to other legumes. Fruits of the functional genomics activities will include assembly of a set of cDNA clones of minimal redundancy that represents the breadth of transcribed genes identified to date. The objective will be to include at least 24,000 representative clones in this "unigene set", which will be freely available to the public. The second major aim will be to use the unigene set in DNA microarray experiments that will monitor global patterns of gene expression. Of relevance here, we plan to identify genes that are differentially regulated from the earliest perception of Nod factor through nodule senescence, in wild type plants and developmental mutants

Genomic research on *Medicago truncatula* is having a significant impact on legume research, worldwide. Currently a database of *M. truncatula* researchers lists approximately 197 researchers, from 16 different countries (http://www.medicago.org). A strength of the *Medicago* community is its open communication and the development of public resources. A series of international workshops on this species has been key to involving a broad community of researchers in the development of resources, such as BAC and EST libraries, genetic markers and maps, and curated populations containing both natural and induced genetic variation. Beyond nodulation, these tools are being applied to many topics, including productivity and forage quality, seed development and nutrition, responses to abiotic and biotic stresses, and natural product biosynthesis. The long-term impact of these efforts will be to integrate genetic and functional information in *Medicago truncatula* specifically, and legumes generally. This knowledge will enable more efficient cloning and characterization of valuable genes and traits, and become the focus of crop improvement strategies throughout the world. Thus, the genetic and genomic system that started as a means to better understand nitrogen fixation has developed into a major resource for legume biology.

6. References

Barker DG *et al.* (1990) Plant Mol. Biol. 8, 40-49
Cook D (1999) Curr. Opin. Plant Biol. 2, 301-304
Cook D *et al.* (1997) Plant Cell. 9, 275-281
Frugoli J, Harris J (2001) Plant Cell. 13, 458-463
Harrrison M (2000) Trends Plant Sci. 5, 414-415
Kulikova O *et al.* (2001) Plant J. 26

7. Acknowledgements

We gratefully acknowledge support from the NSF Plant Genome Project, award number DBI-9872664, and the contributions of our co-principal investigators Steve Gantt, Michael Hahn, Maria Harrison, Dongjin Kim, Ernie Retzel, Debby Samac, Chris Town, Carroll Vance, and Nevin Young. A special acknowledgement is due to post-doctoral fellows and other associates on the project, including Maria Fedorova and Jinyuan Liu, who contributed to EST library construction, and Gabriella Endre and Silvia Penuela, who have pioneered the cDNA microarrays. We thank Jennifer Cho and Joe White for their roles in developing and utilizing MtGI. Lastly, we are grateful to collaborators Ton Bisseling, René Geurts and Gyorgy Kiss who have contributed to construction and utilization of comparative genomic tools outlined here.

A LARGE SCALE GENOME ANALYSIS OF *LOTUS JAPONICUS*, MG-20

S. Sato, T. Kaneko, Y. Nakamura, T. Kato, E. Asamizu, S. Tabata

Kazusa DNA Research Institute, 1532-3 Yana, Kisarazu, Chiba 292-0812, Japan

1. Introduction

The progress in DNA sequencing technology has allowed us to perform systematic and comprehensive analysis of genetic information in a variety of organisms. The entire genomes of several higher eukaryotes including a flowering plant, *Arabidopsis thaliana*, have already been sequenced, and the complete lists of potential gene complements in its genome have been compiled by computer-assisted analysis (The Arabidopsis Genome Initiative 2000). Furthermore, information and material resources obtained during the course of genome sequencing can be utilized to study functional aspect of genes in the genome.

Lotus japonicus has been proposed as a model system for molecular genetics in legume species, especially on interaction between the legumes and rhizobia. The genome was estimated to be 471 Mb long (Ito *et al.* 2000), which is comparable with that of a monocot model plant, *Oryza sativa*. Intensive genomic studies, including generation of DNA markers and a genetic linkage map, construction of genome and cDNA libraries, and collection of expressed sequence tags, are in progress. By taking advantage of the above circumstances, we initiated a large-scale analysis of the *L. japonicus* genome including EST analysis, genome sequencing and mapping.

2. Materials and Methods

2.1. EST analysis. Poly(A)$^+$ RNA was extracted from various organs of *Lotus japonicus* Miyakojima MG-20 and Gifu B-129 accessions, and cDNA libraries were prepared according to the standard method (Asamizu *et al.* 1999). ESTs from 5' and 3' ends of each cDNA insert were collected and analyzed as previously described (Asamizu *et al.* 2000).

2.2. Genome sequencing. Genomic DNA libraries of *L. japonicus* MG-20 for DNA sequencing and mapping were generated using TAC (transformation competent artificial chromosome) vector pYLTAC7 (Liu *et al.* 1999). For DNA sequencing, clones were selected by screening the genomic libraries by PCR using the primers designed on the basis of the EST sequences. Sequencing of the respective clones was performed by a bridging shotgun strategy. The average redundancy of the random data was 5 times. The raw sequence data produced were assembled using Phred-Phrap programs (Phil Green, Univ. Washington, Seattle). Additional sequencing was performed to maintain the secure phred score of 20 or higher for the entire region.

Assignment of the protein coding regions and gene modeling were performed as described previously (Sato *et al.* 2000). Briefly, similarity search against the non-redundant protein sequence database nr (compiled by NCBI) was carried out using the BLASTX program. In parallel, prediction of potential protein coding regions was performed with computer programs, Grail, GENSCAN and NetGene2. The transcribed regions were assigned by comparison of the nucleotide sequences with *Lotus* ESTs (Asamizu *et al.* 2000) in the public databases using the BLASTN program.

2.3. Linkage analysis. Simple sequence repeat (SSR) markers and dCAPS markers were generated using the sequence information of the clones. Linkage analysis was performed using the F2 population of Gifu B129 and Miyakojima MG-20 according to the standard method.

3. Results and Discussion

3.1. EST analysis. As the first step of EST analysis, 22,983 cDNA clones originated from whole plants were sequenced from their 5' ends. These ESTs were clustered into 7137 non-redundant groups, of which an overall GC content is approximately 49%. As cDNA clones were not always full-length, a possibility remains that different regions of a single gene are included as non-overlapping ESTs. To exclude this possibility, 3'-ESTs were analyzed. As of June 2001, a total of 38,685 3'-EST have been accumulated from the libraries of pods, roots nodules, nodule primordia of Gifu and Miyakojima accessions. These represented 16,896 non-redundant species. The latest status of the EST analysis is summarized in Table 1.

Table 1. The status of EST analysis

Organ	Accession	Library type	5' ESTs	3' ESTs
Whole plant	MG2	Normalized	1828	392
		Size-selected	470	146
Pod	MG2	Normalized		6532
		Size-selected		4811
Root	MG2	Normalized		7670
		Size-selected		2340
Nodule primordia	B12	Normalized		7307
		Size-selected		1158
Nodule	B12	Normalized		3478
Total			22983 *	3868
Non-redundant			7137	1689

* Data have been released

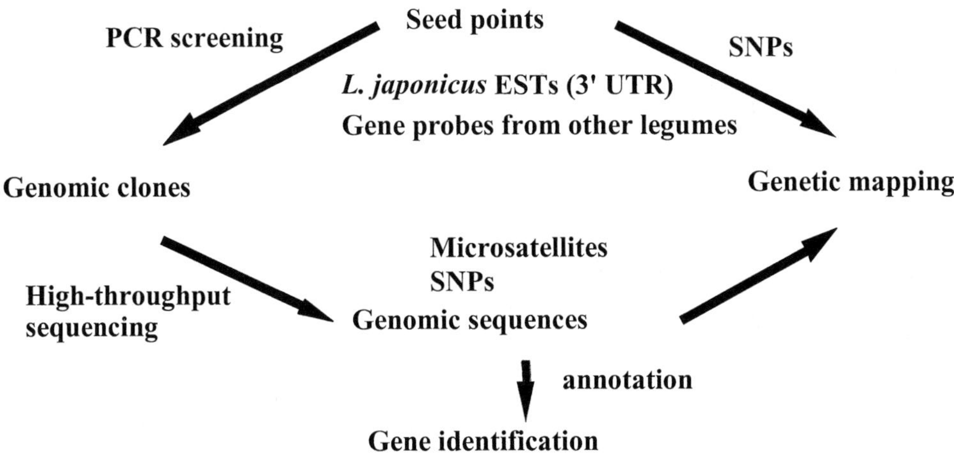

Figure 1. Strategy of the genome analysis of *Lotus japonicus* MG-20.

3.2. Genome sequencing. We have constructed a genomic library of *L. japonicus* in a transformation competent artificial chromosome (TAC) vector as a source of genome sequencing clones. High molecular weight DNA from *L. japonicus* accession MG-20 was partially digested with *Mbo*I or *Hin*dIII and ligated with pYLTAC7. The average insert sizes were 87 kb, 96 kb, 105 kb and 106 kb for four independent preparations, a total of which is 7.7 haploid genome equivalent. The TAC libraries thus generated were arrayed in 93 384-well microtiter plates, and 48 DNA pools each containing 384 clones were subjected to PCR screening.

Seed clones for sequencing were selected by high throughput PCR screening using primer sets designed on the basis of ESTs and cDNA markers of *L. japonicus* and other legumes (Figure 1). The nucleotide sequence of each selected clone was determined according to the bridging shotgun method with an approximate redundancy of 5. The finished sequences were subjected to gene assignment by similarity search and computer prediction. At present, a total of 688 seed clones have been selected. Ninety-four of them are ready for sequencing, 5 are being sequenced, 96 are in the finishing stage, and 44 are being annotated.

3.3. Linkage mapping. Mapping of the seed clones on the basis of genome sequence information was also performed. First of all, simple sequence repeats (SSR) were searched in the sequence of each sequenced clone (Figure 1). Primers were designed to amplify identified SSR regions and the polymorphism in the parents of the mapping family (accessions MG-20 and B-129) was tested. When the polymorphism was observed, linkage analysis was performed using the mapping population of 127 F2 plants.

If no SSR was found, single nucleotide polymorphism (SNP) was searched by comparing the sequences of the Miyakojima clones and those of the corresponding chromosomal regions of Gifu. dCAPS marker was generated at the position where the SNP was detected, then the linkage analysis was carried out using F2 mapping population. According to this procedure, a total of 57 clones have been mapped on the 6 linkage groups so far (Figure 2).

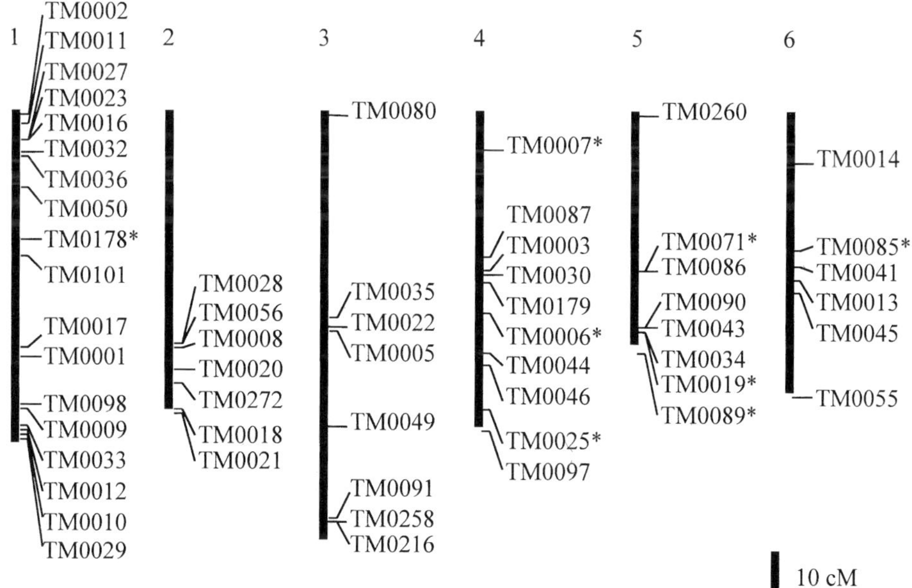

Figure 2. Positions of the sequenced clones on the linkage map.

4. References

The Arabidopsis Genome Initiative (2000) Nature 408, 796-815
Asamizu *et al.* (1999) DNA Res. 6, 369-373
Asamizu *et al.* (2000) DNA Res. 7, 127-130
Liu *et al.* (1999) Proc. Natl. Acad. Sci. USA 96, 6535-6540
Ito *et al.* (2000) J. Plant Res. 113, 435-442
Sato *et al.* (2000) DNA Res. 7, 131-135

5. Acknowledgements

We thank Drs T. Aoki, M. Kawaguchi, M. Parniske, J. Stougaard, and T. Uchiumi for providing information for selection of seed clones and for valuable discussions. This work was supported by the Kazusa DNA Research Institute Foundation.

GENETICS AND GENOMICS IN *LOTUS JAPONICUS*

N. Sandal[1], L. Krusell[1], S. Radutoiu[1], M. Olbryt[1], A. Pedrosa[2], S. Stracke[3], M. Parniske[3], A. Bachmair[2], T. Ketelsen[1], J. Stougaard[1]

[1]Laboratory of Gene Expression, Department of Molecular and Structural Biology, University of Aarhus, Gustav Wieds Vej 10, DK-8000 Aarhus C, Denmark
[2]Department of Cell Biology and Genetics, University of Vienna, Rennweg 14, A-1030 Vienna, Austria
[3]Sainsbury Laboratory, John Innes Centre, Colney, Norwich NR4 7UH, UK

1. Introduction

New possibilities for genetic studies in the *Lotus japonicus* model legume have recently opened as a result of the genome sequencing initiative and the EST sequencing programs started on *L. japonicus* (Cyranoski 2001; Asamizu *et al.* 2000). When combined with the diploid genetics of *L. japonicus,* a small genome size estimated to ~432 Mb and ample seed production from large self-fertile flowers, these genome initiatives will enable genetic linkage analysis and physical mapping to be done effectively. In addition several *L. japonicus* mutant populations have been generated and many mutant classes identified (Schauser *et al.* 1998; Szczyglowski *et al.* 1998; Imaizumi-Anraku *et al.* 1997).

Gross phenotypes divide the mutants into four classes:

1. Non-nodulating mutants that are impaired in early rhizobial interaction, Nod factor perception or downstream signaling;
2. Ineffective nodulating mutants that are either arrested during nodule organogenesis or impaired in nodule function;
3. Mutants with either increased or decreased nodule numbers. Here, inactivation of the normal autoregulatory mechanism leads to an excess of root nodules, whereas reduced nodule numbers may be caused by a variety of mutations including leaky mutations; and
4. Mutants with delayed nodulation. With these mutants, the infection and nodule-developmental process can be genetically dissected.

Although both transposon tagging (Schauser *et al.* 1999) and T-DNA tagging (Webb *et al.* 2000) have been accomplished, most of the mutant populations were produced by EMS mutagenesis and, in order to molecularly characterize the mutants from these collections, a map-based cloning procedure needs to be established. For this purpose, a genetic linkage map of *L. japonicus* Gifu has been developed.

2. Advantages of AFLP Technology

AFLP marker technology has proved to be reliable and effective for the generation of plant linkage maps. The AFLP technique combines restriction fragment analysis with PCR into a multi-locus DNA fingerprinting system that is independent of prior knowledge of genome sequence. DNA fragments amplified by PCR are resolved by electrophoresis in either gels or capillaries allowing the large numbers of fragments arising from complex genomes to be detected and analyzed. We have chosen to use AFLP for providing the backbone markers of the *L. japonicus* map and to supplement this analysis with markers generated with more time-consuming RFLP and gene specific PCR

---

technology. The resulting genetic linkage map was based on a highly polymorphic interspecific F2 mapping population established from a cross between *L. filicaulis* and *L. japonicus* ecotype Gifu.

3. Results and Discussion

A total of 518 anonymous AFLP markers, 3 anonymous RAPD markers, 38 gene specific markers and 6 recessive symbiotic mutant loci were mapped. This first generation map consists of six linkage groups corresponding to the six chromosomes in *L. japonicus.* Fluorescent *in situ* hybridization (FISH) with selected markers aligned the linkage groups to chromosomes. The length of the linkage map is 358 cM and the average marker distance is 0.6 cM. Distorted segregation of markers was found in sections of the map and linkage group I could only be assembled using color mapping with the assistance of localization of markers by FISH. In *Lotus*, two mapping populations are in use, one based on a highly polymorphic interspecific cross between *L. japonicus* ecotype Gifu and *L. filicaulis*, another originating from an ecotype cross between Gifu and Miyakojima (Kawaguchi *et al.* 2000). The possibilities for genetic analysis and map-based cloning using these two populations have been evaluated by mapping of three symbiotic loci, *Ljsym1*, *Ljsym5* and *Ljhar1-3*. A combination of marker analysis and BAC end sequencing has delimited *Ljsym1* within a 2 cM region, the *Ljsym5* locus within a 1 cM region and *Ljhar1* within 0.42 cM.

Although map-based cloning of these loci is progressing, cloning of tagged loci is still more effective. This technique will change in the future as map-based cloning will become faster and easier. Sequencing of the *Lotus* genome (ecotype Miyakojima) will provide a basis for identifying single nucleotide polymorphisms and short insertion/deletions between mapping partners. Lack of PCR markers is one of the main obstacles to fine mapping, so PCR markers based on nucleotide differences will greatly expedite map-based cloning. Effective use of this possibility requires additional sequencing of a mapping partner, but the potential resolution is very high. At present, the best partner for survey sequencing is the Gifu ecotype.

4. References

Asamizu E *et al.* (2000) DNA Res. 7, 127-130
Cyranoski D (2001) Nature 409, 272
Imaizumi-Anraku H *et al.* (1997) Plant Cell Physiol. 38, 871-881
Kawaguchi M (2000) J. Plant Res. 113, 507-509
Schauser L *et al.* (1999) Nature 402, 191-195
Schauser L *et al.* (1998) Mol. Gen. Genet. 259, 414-423
Szczyglowski K *et al.* (1998) Mol. Plant-Microbe Interact. 11, 684-697
Webb KJ *et al.* (2000) Mol. Plant-Microbe Interact. 13, 606-616

LOTUS JAPONICUS FUNCTIONAL GENOMICS: cDNA MICROARRAY ANALYSIS UNCOVERS NOVEL NODULINS

M. Udvardi[1], T. Altmann[1], B. Essigmann[1], G. Colebatch[1], S. Kloska[1], P. Smith[2], B. Trevaskis[1]

[1]Max Planck Institute of Molecular Plant Physiology, Am Mithlenberg 1, 14476 Golm, Germany
[2]Dept Botany, Univ. of West. Aust., Crawley, WA 6009, Australia

1. Introduction

Functional genomics brings together high-throughput genetics with multi-parallel analyses of gene transcripts, proteins, and metabolites to answer the ultimate question posed by all genome-sequencing projects: What is the biological function of each and every gene? Functional genomics is driving a shift in the research paradigm away from vertical analysis of single genes, proteins, or metabolites, towards horizontal analysis of full suites of genes, proteins, and metabolites. By identifying and measuring many of the molecular players that participate in a given biological process, functional genomics offers the prospect of obtaining a truly holistic picture of life.

Technologies for high-throughput analysis of transcripts, proteins, or metabolites have developed rapidly over the last decade. Several approaches are now available for quantifying transcript levels for thousands of genes simultaneously, including DNA microarray analysis (Gress *et al.* 1992), and sequencing-based methods such as serial analysis of gene expression (SAGE, Velculescu *et al.* 1995) and massively parallel signature sequencing (MPSS; Brenner *et al.* 2000). Approaches that utilize mass spectrometry enable hundreds of proteins or metabolites to be identified and quantified in a single experiment (Fiehn *et al.* 2000; Washburn *et al.* 2001). However, there may be hundreds of thousands of different proteins, and tens of thousands of different metabolites in a given organism. Thus, current methods can provide information on only a small fraction of an organism's entire proteome or metabolome. This is not the case for transcriptome analysis, where DNA microarrays representing entire genomes are already being used to provide a comprehensive profile of gene activity in some bacterial species as well as the yeast, *Saccharomyces cerevisiae*. It is expected that comprehensive analysis of the *Arabidopsis thaliana* transcriptome will become possible in the near future as a result of the completion of the *Arabidopsis* genome sequence late in 2000. However, useful transcriptome analysis of other plant species need not wait until their genomes are completely sequenced. In fact, libraries of expressed sequence tags (ESTs) derived from cDNA clones are already being put to good use in this regard, including those of several legume species.

Lotus japonicus is a legume with a number of attributes that make it useful for molecular genetics and functional genomics (Handberg, Stougaard 1992). It is a self-fertile, diploid species with a relatively small genome (approximately 450 Mbp), is easily transformed using *Agrobacterium tumefaciens*, and produces large amounts of seed in a relatively short time. *Lotus japonicus* has become a model species for studies on symbiotic nitrogen fixation. We are using Lotus to identify genes and proteins that play important roles in nodule primary metabolism and nutrient exchange between the plant and nitrogen-fixing bacteroids. To this end, we are producing an EST database from *Lotus japonicus* nodule cDNA clones, and are using it not only for gene discovery, but also for the production of DNA microarrays for transcriptome analysis. Here, we describe the status of our Lotus EST project and present the first results of transcriptome analysis in this species.

2. Procedure

2.1. Production of cDNA arrays. Two *Lotus japonicus* nodule cDNA libraries were used to generate clones for array analysis. The first library was provided by Dr. Jens Stougaard (Aarhus, Denmark), and constructed in the λZAP-XR vector (Stratagene). The second library was constructed in the pZL1 vector using the Superscript™ Lambda system for cDNA synthesis and λ cloning (Invitrogen™ Life Technologies). cDNA from 2307 bacterial plasmid clones was PCR amplified in 30 μl reactions in 384-well plates. For each clone, three independent reactions were performed and then pooled to reduce variation in PCR efficiency and to increase the concentration of product. PCR products were spotted in two positions onto Nytran® SuPerCharge nylon transfer membranes (Schleicher and Schuell, Germany) using the BioGrid robotic system (*Bio*Robotics Ltd, England). DNA on membranes was denatured with 0.5 M NaOH and baked at 80°C for two hours.

2.2. Reference hybridization. To estimate the amount of DNA spotted for each clone, filters were hybridized to a reference oligonucleotide that was complementary to vector sequence. For the λZAP-XR vector the oligonucleotide sequence was GCTGCAGGAATTC, and for the pZL1 vector the oligonucleotide sequence was ACGCGTGGGTCGA. Oligonucleotides were end labeled with γ^{33}P-ATP using T4 polynucleotide kinase (NEB). Filters were pre-incubated in SSARC (24% Sarcosyl NL3O, 4xSSC, 4 mM EDTA) for 2 hours at 5°C before addition of the probe and hybridization overnight at 5°C. Filters were washed for 30 min at 5°C and exposed to a phosphor screen (Fujifilm) for 4-6 hours, after which the imaging plate was scanned by the BAS-1800 II phosphor imager (Fujifilm) at a resolution of 50μm per pixel. Reference oligonucleotide probes were removed by stripping the filters at least twice in diluted SSARC (1/10 in 1 mM EDTA) for 30 min at 80°C, leaving the filters ready for use in complex hybridizations.

2.3. Complex hybridization. Total RNA was isolated from nodules of seven-week-old *L. japonicus* plants which had been inoculated at 7 days with *Rhizobium loti,* and with total RN55A from roots of seven-week-old uninoculated *L. japonicus* plants, as described by Jacobsen-Lyon *et al.* (1995). Five replicate complex hybridizations were performed for each tissue. Complex probes were prepared from 10 μg of total RNA by reverse transcription using SuperScript II reverse transcriptase (Invitrogen Life Technologies) in the presence of α^{33}P-CTP. Filters were pre-incubated for two hours at 65°C in Church buffer (0.5 M sodium phosphate, pH 7.2, 7% SDS, 1 mM EDTA) plus 100 μg/ml denatured salmon sperm DNA, after which complex probes were added in fresh Church buffer and hybridized for at least 24 hours at 65°C. Filters were then washed successively for 20 min in lxSSC containing 0.1% SDS and 4 mM sodium phosphate (pH 7.2), 20 min in 0.2xSSC containing 0.1% SDS and 4 mM sodium phosphate (pH 7.2), and 20 min in 0.lxSSC containing 0.1% SDS and 4 mM sodium phosphate (pH 7.2), all at 65°C. Air-dried filters were exposed to a phosphor screen overnight, which was then scanned as described above.

2.4. Data analysis. Detection and quantification of the signal intensities were performed using the AIS software package ArrayVision™ (4.0 Rev. 1.7; Imaging Research Inc), resulting in raw data representing the average pixel intensity for each spot. Raw data was normalized in a two-step procedure. Since we observed a linear dependency of the raw signal on the amount of PCR product as well as on the transcript abundance, we included the measurements from both the complex and the reference hybridization into our method. First, we normalized both hybridizations internally to compensate for varying amounts of total probe activity by dividing the background-subtracted signal by the average signal of all spots. In the second step, we divided the internally normalized complex signal by the internally normalized reference signal corresponding to the same spot in order to correct for different amounts of PCR product on the filter. Finally, the average of the two spots representing the same clone was calculated. Raw measurements that were not more than twice as high as empty local background spots were regarded as undetectable and replaced with an estimate of the missing value based on the local detection limit and the average and standard deviation of the replicate measurements of the same clone.

3. Results and Discussion

As of June 2001, we had sequenced 6000 Lotus nodule cDNA clones, which represented approximately 3000 different genes. Eighty percent of these encoded polypeptides that were homologous to known proteins. Twelve percent of ESTs appeared to encode enzymes, including an almost complete set of glycolytic enzymes and many enzymes involved in amino acid biosynthesis. Seventeen percent of encoded polypeptides were predicted to have at least one transmembrane domain. Amongst these were homologs to transporters for sugars, nucleotides, amino acids, peptides, and various anions or cations.

The first 2300 EST clones were used to produce DNA microarrays for transcriptome analysis. Microarrays were used to compare the transcriptome of nodules and roots from seven-week-old plants. While little technical and biological variation was apparent when replicate filters were probed with mRNA from the same organ (either root or nodule), there were marked and significant differences in gene expression between nodules and roots (Figure 1).

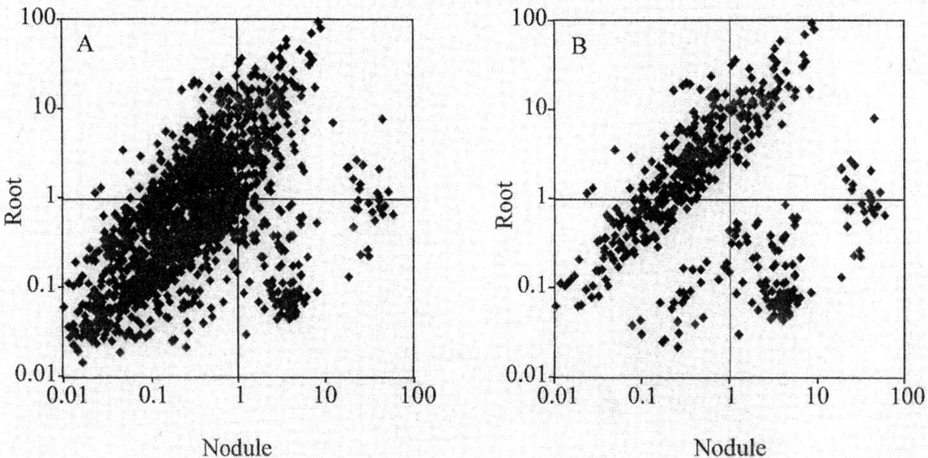

Figure 1. Scatter plot of gene activity (relative transcript level) in nodules vs. roots of seven-week-old plants. Panel A shows all genes, panel B shows genes that were expressed at a significantly higher or lower level in nodules than in roots.

Transcripts of 68 genes were significantly up-regulated in nodules compared to roots, and a larger number of genes appeared to be down-regulated during nodule development. Amongst the genes that were up-regulated during nodule development were a number encoding known nodulins, including 6 leghemoglobins, LjNOD21, and homologs of MtN24, pea early nodulin 12, and Enodl6. Several genes involved in C metabolism were up-regulated in nodules, including a sugar transporter (4-fold up) and two different carbonic anhydrases (CA, 3 to 4-fold up). Homologs of CA have been found to be up-regulated in the nodules of other legume species. The transporter may facilitate sugar uptake into nodule cells, which require large amounts of energy and carbon skeletons for nitrogen fixation and assimilation, respectively. The carbonic anhydrases may play a role in CO_2 recovery in nodules, a process in which another nodulin, phosphoenolpyruvate carboxylase, has been implicated in the past (Vance *et al.* 1994). Two genes involved in amino acid biosynthesis, aspartate amino transferase and aspartate kinase, were each up-regulated approximately two-fold in nodules. A number of nodulin genes involved in C and N metabolism in the nodules of other species were not present in the list of up-regulated genes in Lotus nodules. These included phosphoenolpyruvate carboxylase, glutamine synthetase, and glutamate synthase (Vance *et al.* 1994). Although homologs of these genes were present on the arrays, it is possible that the particular orthologs of those up-regulated in alfalfa were not represented on our filters. A less likely explanation for their absence is that there are major differences in the way primary metabolism is altered during nodule development in Lotus compared to other species.

Several transporter genes appeared to be up-regulated in Lotus nodules, including three different sulfate transporter genes (between 2- and 20-fold up) and a potassium transporter (2-fold up). It will be interesting to determine whether any of these transporters are involved in nutrient exchange between the plant and bacteroids.

Two genes involved in heme biosynthesis, succinate-CoA ligase and coproporphyrinogen oxidase, were found to be up-regulated 2.5-fold in nodules. The heme moiety is essential for the production of leghemoglobin, the most abundant of nodule proteins. Coproporphyrinogen oxidase was shown to be up-regulated in nodules of soybean and pea in the past (Santana *et al.* 1998).

Several genes encoding proteins that may have roles in signal transduction were also found to be up-regulated, including a homolog of the transcription factor tga, a MADS-box protein homolog, and a homolog of ER6, an ethylene-inducible serine/threonine receptor kinase. A homolog of the enzyme isoliquiritigenin 2'-O-methyltransferase, which is involved in the biosynthesis of an enhancer of rhizobial *nod* genes (Maxwell *et al.* 1993) was up-regulated in Lotus nodules (more than 10-fold). This raises the question of whether this signaling pathway remains active after rhizobia have taken up residence in nodule cells.

4. Summary

DNA microarray analysis can provide a wealth of new information about genes involved in different aspects of nodule development and function. In future, we plan to increase the number of genes represented on our microarrays, and to use these to profile changes in gene expression during normal nodule development and during development of nodules on Lotus mutants affected in nodule development or function. We will also use microarrays to monitor gene expression in nodules of wild-type plants infected with different mutants of the microsymbiont, *Mesorhizobium loti*. This should provide deeper insight into the interaction of the two genomes of these important symbionts.

5. References

Brenner S *et al.* (2000) Nat. Biotechnol. 18, 630-634
Fiehn O *et al.* (2000) Nat. Biotechnol. 18, 1157-1161
Gress TM *et al.* (1992) Mamm. Genome 3, 609-661
Handberg K, Stougaard J (1992) The Plant J. 2, 487-496
Jaconsen-Lyon K *et al.* (1995) The Plant Cell 7, 213-223
Maxwell CA *et al.* (1993) Plant J. 4, 971-981
Santana MA *et al.* (1998) Plant Physiol. 116, 1259-1269
Vance CP *et al.* (1994) Plant Science 101, 5 1-64
Velculescu VE *et al.* (1995) Science 270, 484-487
Washburn MP *et al.* (2001) Nat. Biotechnol. 19, 242-247

6. Acknowledgements

We would like to acknowledge the Max Planck Society and the Alexander von Humboldt Foundation for generous support.

Section 4:
Signal Transduction

CHAIR'S COMMENTS: SIGNAL TRANSDUCTION IN SYMBIOSIS: GETTING IN DEEPER AND DEEPER

M. Kahn

School of Molecular Biosciences, Institute of Biological Chemistry
Washington State University, Pullman, WA 99164-6340, USA

Nitrogen fixation that occurs in the context of a symbiotic relationship requires a multilevel exchange of signals between the two symbionts to elicit an appropriate developmental response. Establishing how this signal exchange takes place has been difficult, despite the fact that many clues have been available from the host-specific nature of the interaction. Landmarks along the way have included: (1) recognition that there was a correlation between surface antigens on the bacteria and lectins present in the host (Dazzo, Hubbell 1975); (2) realization that host defenses were activated even during successful nodulation (Bhuvaneswari *et al.* 1980); (3) that the defense response was systemic (Kosslak, Bohlool 1984); and (4) that it could be eliminated by mutation (Olsson *et al.* 1989); (5) identification of bacterial genes necessary for nodulation (Long *et al.* 1982); (6) that expression of these genes was induced by specific small molecules produced by the host (Peters *et al.* 1986); and (7) characterization of the Nod factor signals produced by these genes (Lerouge *et al.* 1990) and an appreciation that there were both specific and non-specific effects of applying these Nod factors to roots.

It has been over ten years since the first Nod factor was isolated and shown to initiate nodule formation on specific hosts in a way that correlated with the Nod factor's structure. Because Nod factors are diverse compounds based on modifying a lipochito-oligosaccharide backbone and are active at low concentrations they were hailed as a new type of plant signal molecule that behaved more like some of the animal hormones than did the recognized plant hormones, such as an auxin or cytokinin. The discovery of Nod factors implied that there should be a Nod factor receptor, which would presumably start a signal cascade leading to the cell division and other changes in root development characteristic of nodule formation. Finding this receptor has been very hard work. Nod factors have hydrophobic and hydrophilic domains and, like the classic detergents, they stick to many cellular components. The target tissue where they specifically act is not abundant and the observation that Nod factors will stimulate responses in non-host plant cells calls almost any assay other than nodulation into question. The report at this meeting by M. Etzler following on earlier work by her group (Etzler *et al.* 1999) strengthens the case that a lectin nucleotide phosphohydrolase (LNP) is a candidate for a Nod factor receptor. Information presented at the meeting showed that when substrates known to bind LNP are added to roots, they induce root deformation, that expression of LNP in *Arabidopsis* makes the transgenic roots sensitive to root hair curling by substrate or bacteria and that plants with antisense constructs of LNP do not show normal root hair curling. However, the affinity of isolated LNP for Nod factor is relatively low, suggesting that there is more to Nod factor binding than is currently appreciated.

LNP is not the only plant protein that binds Nod factor and it may be that there are several levels at which Nod factor acts. This was certainly suggested by results from A. Downie's group, who showed that some early responses to Nod factors do not require the fully decorated molecule and that the nodO protein, which is in no way related to Nod factor synthesis, is able to potentiate the activity of an incompletely decorated Nod factor. This suggests that there may be more than one pathway to some of Nod factor's effects. In that light, G. Stacey's description of the complex pattern of regulation of the *Bradyrhizobium* nodulation genes reinforces the idea that communication between bacteria and host can be subtle and is probably sensitive to many inputs, perhaps including some that we are not aware of yet.

One possible gene in the plant signal cascade was identified by G. Kiss and his group who identified the lesion in a classical alfalfa non-nodulating mutant, MN-108, as a change in the sequence of a potential receptor kinase gene. The approach taken was a technical *tour de force* of map-based cloning and sequencing which was capped by the finding that a similar lesion was present in a pea mutant also defective in nodulation. In addition to linking a point mutation to a function involved in nodule formation, the ability to cross species boundaries for confirmation gives hope that mutants in legume species with difficult molecular genetics will be able to contribute to the overall picture. While the evidence is fairly strong that a change in this receptor kinase is important in the mutant phenotype, the observation by Karlowski and Hirsch that expression of a zinc-finger binding protein is also altered in the mutant is intriguing. Changes in expression of this protein may also play a role in disconnecting the Nod signal from nodulation. Taken together with the isolation of a transposon-tagged *Lotus japonicus* nodulation gene (Schauser *et al.* 1999), the field appears to be moving a step at a time into the difficult developmental territory between infection and fixation.

At the other end of the signal cascade, various groups are examining the changes that occur in the plant cells within nodules. In particular, proteomics, EST analysis and mRNA profiling approaches are identifying large numbers of changes in expression that need to be integrated with models of how these changes relate to alterations in the cell cycle of nodule cells. Some changes are likely to result from blocking cell division, and would occur in any tissue where division was blocked but others are potentially nodule-specific. Working back from the mature nodules to see how the cells developed is potentially a very powerful approach to understanding nodule organogenesis. In these investigations, the ability to generate transgenic plants that have altered expression of particular proteins is turning into a powerful technique, though it is necessary to examine closely other potential phenotypes of the transgenics that might alter nodule maturation indirectly.

These advances, together with advances in mutant isolation in the two model legume species, *Lotus japonicus* and *Medicago truncatula*, indicate that a description of the plant response pathway downstream of Nod factor activation is on the verge of becoming a reality. With strong candidate genes in hand, techniques like two-hybrid analysis may be able to identify additional factors that are connected to proteins in the signal cascades. As patterns of gene expression in the pathway become more established, the genomic tools presented at the meeting will come to play a larger and larger role in discovering novel proteins that also fit the patterns.

While the last 20 years could be characterized as gene-by-gene discovery of elements involved in the legume-rhizobial conversation, we can expect that the future will be more concerned with assembling the flood of genetic information into a coherent picture of relationships. There is a strong temptation to view transduction as a linear device for attaching a signal to a response since it is often a single assay that establishes new connections. However, there are already many distinct mutants that affect nodulation, suggesting that there are many connections to be made. Signals are often transduced through a series of proteins as a way of bringing several inputs into the ultimate decision process. So, for example, legumes growing in sufficient nitrogen do not respond to bacteria in the same way as when they are nitrogen-stressed and bacteria arriving late at the root hair will not find the same welcoming reception as those that arrived earlier. How Nod factor signal transduction is connected to the perception of other signals, like nitrogen or infection status, are only two of the many questions that will hopefully be answered in the years ahead.

References

Bhuvaneswari TV, Turgeon BG, Bauer WD (1980) Plant Physiol. 66, 1027-1031
Dazzo FB, Hubbell DH (1975) Appl. Microbiol. 30, 1017-33

Etzler ME, Kalsi G, Ewing NN, Roberts NJ, Day RB, Murphy JB (1999) Proc. Natl. Acad. Sci. USA 96, 5856-61

Kosslak RM, Bohlool BB (1984) Plant Physiol. 75, 125-130

Lerouge P, Roche P, Faucher C, Maillet F, Truchet G, Prome J, Denarie J (1990) Nature 344, 781-4

Long SR, Buikema WE, Ausubel FM (1982) Nature 298, 485-488

Olsson JE, Nakao P, Bohlool BB, Gresshoff PM (1989) Plant Physiol. 90, 1347-1352

Peters NK, Frost JW, Long SR (1986) Science 233, 977-80

Schauser L, Roussis A, Stiller J, Stougaard J (1999) Nature 402, 191–195

SIGNAL EXCHANGE DURING THE EARLY EVENTS OF SOYBEAN NODULATION

G. Stacey, J. Loh, B. Zhang, Y.-H. Lee, C. Bickley, D. Lohar, G. Liao, G. Copley, M.G. Stacey

Center for Legume Research, Department of Microbiology, University of Tennessee, Knoxville, TN, 37996-0845, USA

1. Regulation of the *Bradyrhizobium japonicum* Nodulation Genes

The regulation of *nod* gene expression in several *(Sino)Rhizobium/ Bradyrhizobium/ Azorhizobium* species has been extensively studied. Current evidence shows that most *nod* genes are involved in the production of substituted lipo-chitin Nod signals that induce root hair curling and cortical cell division. The generic model for the regulation of these genes predicts that the *nodD* gene product is constitutively expressed and interacts with plant-produced flavonoid chemicals (e.g. isoflavones) leading to induction of other nodulation genes. This system of *nod* gene transcriptional control is essential for initiation of the infection process and for determining host specificity.

Continuing research in our laboratory has shown that the generic model of *nod* gene regulation defined in *(Sino)Rhizobium* species does not adequately account for the control of *nod* gene transcription in *B. japonicum*. Indeed, *nod* gene regulation in *B. japonicum* shows surprising complexity (Figure 1). Our data show that *B. japonicum* uses members of three different global regulatory protein families to control *nod* gene expression. Specifically, a LysR-type regulator,

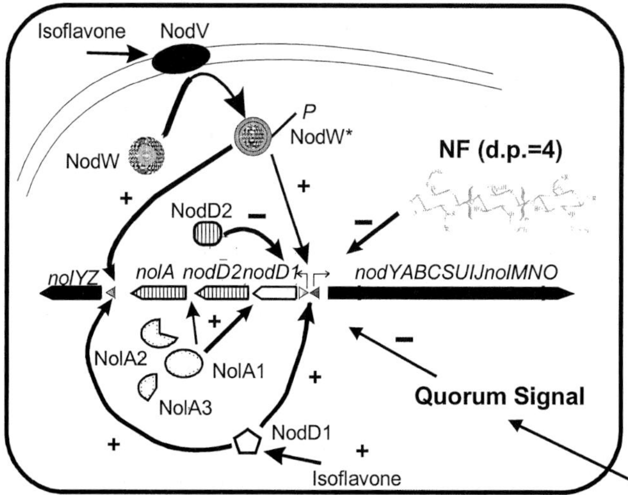

Figure 1. Regulation of *nod* gene expression in *Bradyrhizobium japonicum* is complex involving multiple positive (+) and negative (-) regulators. To this complexity, we can now add negative feedback regulation by tetrameric Nod signal (NF) and a novel quorum signal. These act through NolA1 and NodD2.

NodD1, and a two-component regulatory system, NodVW, positively regulate *nodYABC* gene expression in response to plant produced isoflavone signals (Loh *et al.* 1997, 1999, 2001; Loh, Stacey 2001). In addition, NolA, a member of the MerR-type regulatory family (see below), acts indirectly to repress *nod* gene expression (Dockendorff *et al.* 1994; Garcia *et al.* 1996). The *nolA* gene is a rare case in bacteria where a single gene encodes three, distinct polypeptides (Loh *et al.* 1999). The largest protein, NolA1, is a transcriptional regulatory protein that is required to induce *nolA2,3* as well as *nodD2*. The smallest of the NolA proteins, NolA3, appears to have an important, although undefined function, in nodulation. NodD2 is a repressor of *nod* gene expression. Thus, expression of NolA1 ultimately leads to a down-regulation of *nod* gene expression.

Recent data from our laboratory, described below, show that NolA is a central regulatory component that allows the cell to respond to a variety of signal molecules and finely regulate *nod* gene expression. Figure 1 presents our current working model for the regulation of the *nod* genes of *B. japonicum*. The net effect of this complex regulatory circuitry is to provide the organism with responsive mechanisms for fine regulation of *nod* gene expression and to integrate such regulation into the overall metabolism of the cell.

B. japonicum possesses two *nodD* genes arranged tandemly as *nodD₁nodD₂*. NodD2 is a repressor of *nod* gene expression in *B. japonicum* (Garcia *et al.* 1996), *Bradyrhizobium* sp. (*Arachis*) NC92 (Gillette and Elkan 1996) and *Sinorhizobium* sp. strain NGR234 (Fellay *et al.* 1998). Since NolA1 positively activates NodD2 expression, regulation of NolA expression effectively controls repression of the *nod* genes. Hence, NolA is a central, negative regulator of nodulation.

During the course of experiments designed to measure *nod-lacZ* expression, we made the observation that a *nodY-lacZ* fusion consistently showed higher activity when assayed in a NodC mutant. NodC is a chitin synthase involved in synthesizing the chitin backbone of the lipo-chitin Nod signal. These data, along with other observations, led us to postulate that Nod signals could act to feedback regulate Nod signal synthesis through the NolA-NodD2 repression pathway. As a first test of this hypothesis, we tested various chitin oligomers (d.p.-1-8) for their ability to induce *nolA-lacZ* and *nodD2-lacZ* expression. The data obtained indicated that chitin oligomers do induce *nolA* expression, with the chitin tetramer being the most active (Loh and Stacey 2001). The chitin tetramer was active as a *nolA* inducer at concentrations as low as 1 nM, with an optimal activity at 10 nM. Further experiments showed that the effect of chitin was specific for NolA1 expression (data not shown).

Different rhizobia produce NodC proteins that determine the chain length of the Nod signal. For example, the NodC from *S. meliloti* synthesizes primarily a chitin tetramer. In contrast, the NodC from *S.* sp. strain NGR234 produces primarily a chitin pentamer, while *B. japonicum* produces both chitin tetramer and pentamer. As a means to increase the level of chitin produced intracellularly, we expressed the *S. meliloti, S.* sp. NGR234, and the *B. japonicum nodC* genes in *B. japonicum* under the control of the constitutive *trp* promoter. Transconjugants expressing either the *S. meliloti* or *B. japonicum* NodC showed a 4-5 fold lower level of *nodC-lacZ* expression than a strain transformed with vector alone. In contrast, expression of the *S.* sp. NGR234 NodC protein did not affect *nodC-lacZ* induction by genistein (data not shown). These data are consistent with the previous results showing that a chitin tetramer is the most active inducer of NolA. Indeed, constitutive expression of the *S. meliloti* NodC had no effect on *nodC* expression when assayed in a NolA mutant background (data not shown). Addition of the Nod signal (an acylated, fucosylated pentamer) at concentrations as high as 1 μM to wild-type cultures did not induce *nolA* expression. However, addition of synthetic tetrameric LCOs (kindly provided by T. Ogawa, Riken Institute, Japan) activated NolA expression and repressed *nodC-lacZ* expression. This result was found regardless of chemical substitutions of the chitin backbone.

The above experiments suggest that NolA mediates feedback inhibition of *nod* gene expression in response to the level of intracellular chitin tetramer. Although the major Nod signal produced by *B. japonicum* is a substituted, chitin pentamer, our previous work documented the production of tetrameric Nod signals by this bacterium (Cohn *et al.* 1999).

A puzzling aspect of our studies on *nod* gene regulation was the requirement that cells be induced (with genistein or SSE) at low cell density in order to see maximal *nod-lacZ* expression. This phenomenon suggested that *nod* gene expression could be regulated in a population density-dependent manner, being repressed at high cell density. The discovery of the negative regulation system of NolA and NodD2 prompted us to look more closely at this question. As shown diagrammatically in Figure 1, the inducibility (fold-induction) of a *nodY-lacZ* fusion drops in response to the production of a quorum signal released from *B. japonicum* cells (Loh *et al.* 2001). Population density does not affect *nodY-lacZ* expression in a NolA mutant background. Thus, NolA appears to be directly involved in the repression of *nod* gene expression at high population density. Indeed, other experiments using strains with either a *nolA-lacZ* or *nodD2-lacZ* fusion showed that the level of these proteins increases with increasing cell density (data not shown)

Filtered, culture supernatants contain an inducer of *nolA-lacZ* which we have termed CDF, cell density factor (Loh *et al.* 2001). In most gram-negative bacteria such quorum sensing compounds are modified homoserine lactones. However, repeated attempts using a variety of methods failed to isolate a homoserine lactone from *B. japonicum* cultures. We have now succeeded in purifying small amounts of CDF, which exhibited the ability to induce a *nolA-lacZ* fusion. The purified CDF, as expected, also has the ability to repress *nod* gene expression. The chemical identity of this compound is currently under investigation.

A significant commercial industry exists for the production of *B. japonicum* inoculants for agricultural use. The companies that produce these inoculants routinely grow their cultures to high cell densities before eventually packaging the cells, together with a carrier (e.g. peat). A common problem with such commercial inoculants is that they compete very poorly for nodule occupancy against indigenous, soil *B. japonicum* strains. This is referred to as the "competition problem". Indeed, in some cases, the *B. japonicum* strain in the commercial inoculant may occupy only 0 to 1-2% of the nodules formed.

Upon the discovery of the CDF, its ability to induce NolA synthesis and, concomitantly, repress *nod* gene expression, we obtained several batches of commercial inoculant. The inoculants and an uninoculated peat (control) were extracted and the extract was tested for its ability to induce *nolA-lacZ* expression. Significant inducer activity was found in all of the inoculants tested (Loh *et al.* 2001). Controls of uninoculated peat showed little or no inducer activity. The conclusions from this study are that commercial soybean inoculant contains significant levels of a compound that represses *nod* gene expression and, therefore, likely reduces the efficacy of the inoculant strain. In order for the inoculant bacterium to nodulate the plant, it will have to initiate growth and overcome the action of the CDF. This delay could be a significant factor in the inability of the inoculant strain to compete against indigenous, soil rhizobia that are commonly present as low population densities (e.g. 10^{2-5} cfu g^{-1}).

It has been known for some time that, although *nod* gene expression is required for the earliest stages of nodulation, the expression of these genes is repressed within the nodule. Various authors have speculated about what may be shutting off *nod* gene expression *in planta*. Recently, we obtained convincing evidence that NolA1, likely through its role in quorum sensing, is required to repress *in planta* expression of *nodY*. We expressed a *nodY-GUS* fusion in wild-type *B. japonicum* and the NolA mutant, BjB3. Soybean plants were inoculated with both strains, nodules harvested 21 days later and stained for GUS activity. GUS activity was only detected in nodules formed by the NolA mutant strain. No *nodY* expression was seen in nodules formed by the wild type. Compared to broth cultures, *B. japonicum* does not attain a comparable cell density within the nodule. However, *in planta* each bacterium is enclosed in a symbiosome with little space between

---

the bacterium and symbiosome membrane. Therefore, even low synthesis of the quorum signal *in planta* could translate into a very high, localized concentration in the symbiosome.

2. Plant Response to Bacterial Nodulation Signals

The known lipo-chitin Nod signals produced by rhizobia are substituted chitin oligomers, usually of four to five *N*-acetylglucosamine (GlcNAc) residues, mono-*N*-acylated at the non-reducing end and carrying a variety of substitutions at both the reducing and non-reducing terminal GlcNAc residues. Each rhizobial species produces a variety of Nod signals with specific substitutions.

Our laboratory extensively studied the chemical specificity required for Nod signal action on soybean. In summary, these data showed that the presence of a 2-O-methylfucose residue on the terminal, reducing GlcNAc was critical for biological activity on soybean only when the LCO (lipo-chitooligosaccharide) was a pentamer (reviewed in Cohn *et al.* 1997). If the LCO was a tetramer, then fucosylation rendered the molecule inactive at eliciting root hair curling or cortical cell division in soybean. Hence, both the chemical substitutions and the LCO chain length were critical. Moreover, our results support the hypothesis that Nod signal perception in soybean involves at least two discrete recognition events with differing chemical specificity. For example, we found that a mixture of non-fucosylated Nod signals could surmount the requirement for a fucosylated Nod signal with regard to rice bean (*Vigna umbellata*) nodulation (Cohn *et al.* 1999). Other work examined nodulin gene expression in response to Nod signal addition. These studies showed that two, chemically distinct recognition events are involved in Nod signal action. One signaling step is somewhat non-specific in that even chitin oligomers (e.g. pentamer) would induce transient expression of the early nodulins. However, sustained expression of ENOD40 required a soybean-specific Nod signal. The existence of two recognition events was supported by finding that expression of the early nodulin ENOD2 could not be induced by the addition of any single LCO or Nod signal. ENOD2 expression was clearly induced only when a mixture of LCO was added. One member of this mixture could be a simple, chitin oligomer, if the other member was a soybean-specific Nod signal.

Currently, a bona fide Nod signal receptor has not been identified. Etzler *et al.* (1999) reported a unique lectin (DB46) isolated from the roots of the legume *Dolichos biflorus*. This lectin was found to bind to Nod signals from a variety of rhizobia. Etzler *et al.* (1999) demonstrated that the lectin possessed ATPase activity (i.e. apyrase activity), which was significantly increased upon addition of the Nod signal. For this reason, the lectin was termed a lectin-nucleotide phosphohydrolase (LNP). The *D. biflorus* LNP (i.e. DB46) was found on the surface of root hairs using fluorescent antibody labeling. Antibody directly against LNP blocked nodulation. Recently, we extended these studies by demonstrating the presence of orthologs of the *D. biflorus* apyrase (LNP) in other legumes [e.g. GS50, GS52 in soybean (Day *et al.* 2000) and Mtapy1-5 in *M. truncatula* (Cohn *et al.* 2001)]. We showed that GS52 in soybean and Mtapy1 and Mtapy4 from *M. truncatula* are early nodulins, induced within 3 h by rhizobial inoculation. Moreover, antibody against GS52 blocked soybean nodulation. *M. truncatula* mutants defective in very early nodulation events also were defective in Mtapy1 and Mtapy4 expression. Therefore, legume apyrases must be considered as candidates for a Nod signal receptor. The Cohn *et al.* (2001) and Day *et al.* (2000) studies showed the benefit of analyzing multiple legumes. Isolation of both the soybean and *M. truncatula* apyrases allowed the identification of a region of microsynteny between the two genomes containing a cluster of apyrase genes.

Chitin oligomers, which can be generated from fungal cell walls by endochitinase, can induce defense responses or related cellular responses in many monocots and some dicots. Our previous work on the specificity of Nod signal action (Cohn *et al.* 1997) suggested that a chitin binding protein could be involved in Nod signal recognition (i.e. as measured by ENOD40 induction). Therefore, Day *et al.* (2001) used ^{125}I-labeled APEA (aminophenyl ethylamine) conjugates of *N*-acetylchitooctaose and *N*-acetylchitopentaose as ligands to identify a chitin-binding site in microsomal membrane preparations from both soybean suspension cultured cells, as well as root preparations. Binding to this site was saturable with an apparently Kd of approximately 40 nM. Competition experiments using chitin oligomers (d.p.=2-8) demonstrated that this binding site preferred the higher molecular weight oligomers (d.p.=7-8). Affinity labeling using a ^{125}I-labeled *N*-acetylchitooctoase ligand identified an 85 kDa chitin-binding protein in the plasma membrane. The binding specificity of this 85 kDa protein for various chitin oligomers correlated with the ability of the same oligomers to induce an oxidative burst response and medium alkalinization in soybean suspension cultured cells. Treatment of soybean suspension cells with the pentameric, *B. japonicum* Nod signal also resulted in medium alkalinization. The response was similar to that shown by treatment with chitin pentamer or tetramer. Indeed, Nod signal binding to soybean plasma membrane preparations (as measured by competition against labeled ligand) also showed an affinity approximately that of the chitin tetramer. From these data, we concluded that the 85 kDa, chitin-binding protein was likely involved in the soybean defense response mediated by chitin fragments released from fungal pathogens. Although this protein appears to bind to the Nod signal, it does so at low affinity and is likely not involved in Nod signal action. However, Nod signal does act as a significant elicitor of defense-related responses in soybean. Therefore, the possibility arises that Nod signal elicitation of a defense response could affect nodulation.

To examine this possibility, we recently constructed transgenic *Lotus japonicus* plants expressing the bacterial NahG protein. This enzyme catalyzes the breakdown of salicylic acid (SA) to catechol. SA is a known mediator of induced defense responses in plants. Previous research showed that transgenic *nahG* plants were unable to accumulate SA and showed a reduced resistance to pathogen attack. Our preliminary data indicate that the *L. japonicus nahG* transgenic plants accumulated significantly less SA in their roots and showed an approximately 50% increase in nodulation. These results are consistent with a role for SA-mediated defense mechanisms in controlling legume infection by rhizobia.

3. References

Cohn J *et al.* (1997) Trends in Plant Sci. 3, 105-110
Cohn J *et al.* (1999) Mol. Plant-Microbe Int. 12, 1086-773
Cohn J *et al.* (2001) Plant Physiol. 125, 2104-2119
Day RB *et al.* (2000) Mol.-Plant Microbe Int. 13, 1053-1070
Day RB *et al.* (2001) Plant Physiol.
Dockendorff T *et al.* (1994) Mol. Plant-Microbe Int. 7, 596-602
Etzler ME *et al.* (1999) Proc. Natl. Acad. Sci. USA 96, 5856-5861
Garcia ML *et al.* (1996) Mol. Plant-Microbe Int. 9, 625-636
Gillette WK, Elkan GH (1996) J. Bacteriol. 178, 2757-21086
Fellay R *et al.* (1998) Mol. Microbiol. 27, 1039-1050
Loh J *et al.* (1997) J. Bacteriol. 179, 3013-3020
Loh J *et al.* (1999) J. Bacteriol. 181, 1544-1554
Loh J *et al.* (2001a) Mol. Microbiol.
Loh J, Stacey G (2001) Mol. Microbiol.

4. Acknowledgements

The work described was supported by grants from the US Department of Agriculture (# 99-35305-8323), National Science Foundation (# IBN-9728281, MCB-9808625), Department of Energy (#DE-FG02-97ER20260) and LiphaTech Corporation.

MAP BASED CLONING OF A RECEPTOR KINASE GENE (NORK) BY GENETIC MAPPING OF A MUTATION (nn_1) CONDITIONING NON-NODULATING PHENOTYPE IN THE TETRAPLOID ALFALFA MUTANT MN-1008

G.B. Kiss, S. Mihacea, Z. Kevei, P. Kaló, A. Perhald, A. Seres, A. Kereszt, G. Endre

Institute of Genetics, Biological Research Center of the Hungarian Academy of Sciences, Szeged, H-6701 Hungary

1. Introduction

MN-1008 is a non-nodulating (Nod⁻) mutant of alfalfa (*Medicago sativa* L.) isolated and described by Peterson and Barnes (1981). The mutant plants showed neither root hair deformation, cortical cell division nor calcium spiking after inoculation with *Sinorhizobium meliloti* or Nod factor (NF) treatment (Dudley, Long 1989; Ehrhardt *et al.* 1996; Endre *et al.* 1996). Variation in membrane depolarization activity was demonstrated by Felle *et al.* (1996) using individual plantlets from the MN-1008 seed population. In addition MN-1008 was resistant not only to *Rhizobium* infection but to vesicular-arbuscular mycorrhiza colonization as well (Brudbury *et al.* 1991). Spontaneous nodulation in the absence of *Rhizobium* (referred to as Nar phenotype, Truchet *et al.* 1989) is one of the unique features of alfalfa. Interestingly the non-nodulation alfalfa mutant, MN-1008, did not lose this character and was susceptible to spontaneous nodulation (Caetano-Annolés *et al.* 1993). In accordance with this phenomenon early nodulins (e.g. ENOD2 and ENOD40) could be induced by cytokinin (Hirsch *et al.* 1997) implying that in this mutant the extrinsic signal perception was abolished but the capacity of nodule initiation and development, as inner genetic program of the plant, was intact. As a consequence, the identification and isolation of the gene conditioning the non-nodulation phenotype in mutant MN1008 would shed light on a key function involved in the initiation of the signal transduction pathway leading to nodule formation. To achieve this goal the map-based cloning strategy was adapted for the tetraploid alfalfa.

2. Procedures

One mutant plant individual of *Medicago sativa* MN-1008 was crossed with a Nod⁺ alfalfa individual (*Medicago sativa* cv. Nagyszenasi) to generate the F1 plants which were grown in pots and self-mated to produce the F2 progenies. Individual plants from selected F2 families were used as mapping population to determine the map position of the non-nodulation trait and linked molecular markers. Plant nodulation test, DNA isolation, PCR and RFLP mapping were performed as described by Endre *et al.* (1996) and Kaló *et al.* (2000). Genetic analysis of the data was performed as described (Kaló *et al.* 2000; Kiss *et al.* 1998).

3. Results and Discussion

Genetic mapping of the non-nodulation trait was started with the identification of RAPD markers using the bulked segregant analysis method described by Michelmore *et al.* (1991). Using isolated DNA from five Nod⁻ and five Nod⁺ plants, respectively, four RAPD markers were identified and mapped in an extended tetraploid population in proportion to the nodulation phenotype. These Nod-linked RAPD markers were then mapped in the diploid alfalfa mapping population on linkage group (LG) 5. From this genetic region linked RFLP markers were selected and re-mapped in the tetraploid population. These RFLP markers mapped to the exact same region in the tetraploid map as the RAPD markers and the Nod⁻ phenotype. The summary of this mapping work is shown in Figure 1.

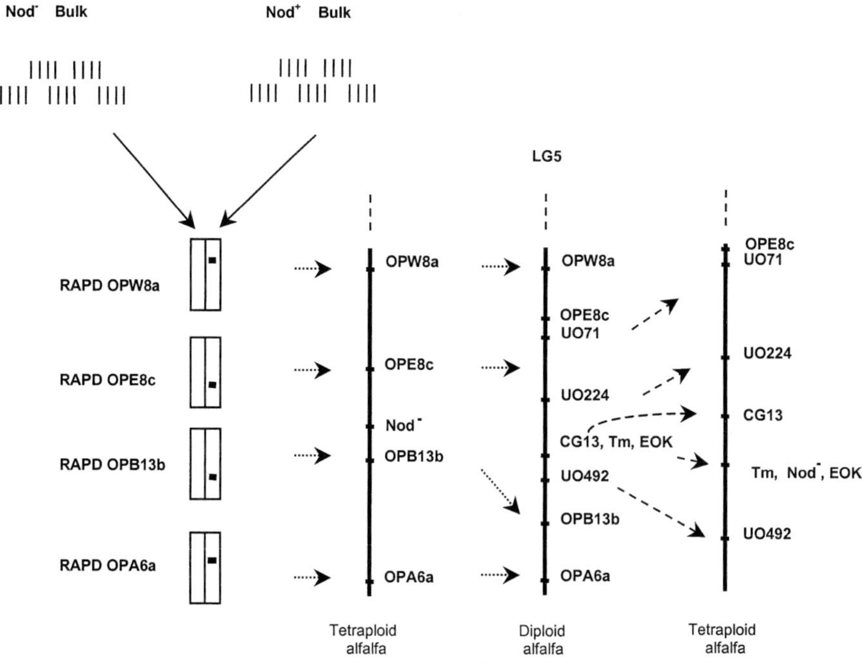

Figure 1. Genetic mapping of the Nod⁻ trait in tetraploid alfalfa (see text for explanation).

Two RFLP markers tightly linked to the non-nodulation trait were used to isolate primary BAC clones from the *Medicago truncatula* BAC library (Nam *et al.* 1999). The end sequences of these primary BACs were used to identify additional BAC clones. Restriction endonuclease mapping (BAC fingerprinting) was used to orient the isolated 10 BAC clones and construct the so-called Nod-contig, which overlaps more than 600 kb DNA sequence. Genetic markers generated from these BAC clones were used to map the Nod⁻ mutation more precisely within the contig.

The Nod-region (~200 kb) covering the Nod⁻ mutation (nn_1) was sequenced and the gene content of the region was determined. Twenty-six genes located in this region were used to search for similar gene content in the *Arabidopsis thaliana* genome. This analysis revealed partial synteny between the two species: one particular gene (see ORF8 in Figure 2) in this region was present as a one-copy gene in the genome of the *Medicago* species (in *M.t.* and *M.s.* as confirmed by DNA-DNA hybridization and RFLP mapping of this gene), while it was a two-copy gene in the *Arabidopsis* genome (see gene *A* and *B* in Figure 2). None of the genes in the flanking region of gene *A* was present in the Nod-region of *M. truncatula*. On the other hand, 9 out of the 26 genes were present in the vicinity of gene *B*. Some genes of the Nod-region were present elsewhere in the *Arabidopsis* genome with a configuration of same order and orientation. Interestingly, one gene in the Nod-contig (see ORF7 in Figure 2) was not present in the *Arabidopsis* genome according to any BLAST search available in the NCBI web site. From this finding it is concluded, that *Medicago* and *Arabidopsis* share limited microsysteny in certain locations of their genome (see Figure 2).

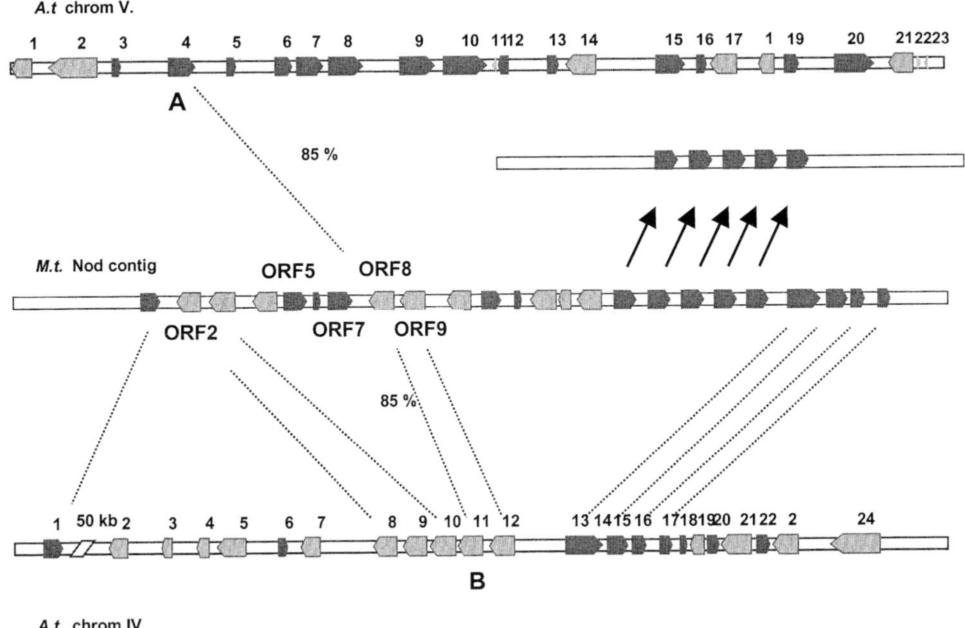

Figure 2. Comparison of the gene content of the Nod-region of *Medicago truncatula* with specific regions of the *Arabidopsis* genome (see text for explanation).

The non-nodulation phenotype of MN-1008 could be explained by potential mutations in the genes coded by ORF2, ORF5, ORF7 and ORF9 (see Figure 2). To search for possible genetic alterations in these ORFs, these genes were sequenced from the MN-1008 mutant plant material. Genomic DNA as well as RT-PCR-produced root cDNA template DNA was used in amplifications with exon-specific primer pairs designed according to the available sequence information. Amplified fragments were cloned into pUC19 vector and the inserts were sequenced in ABI377 automatic sequencer using the BigDye labeling procedure. Similarly, DNA templates originating from *M. truncatula* R38 mutant, a Nod⁻ but Myc⁺ *dmi2* allele (K. VandenBosch, personal communication), and *Pisum sativum* wild type (Frisson), and P4 and P55 mutants (Nod⁻, Myc⁻ *sym19* alleles, Schneider *et al.* 1999) were similarly sequenced. According to the obtained sequence information, mutation in ORF7 of each mutant plant could be identified. The nature of the mutations either generated a stop codon (*nn₁*) evoking premature translation termination or in two cases (R38 and P4) changed a highly conserved amino acid into a dissimilar residue.

The sequence of ORF7 is similar to receptor kinase genes identified in plants and therefore it was designated as NORK (NOdulation specific Receptor Kinase). The NORK gene of the above plant species contain 925 amino acid residues (as deduced from the cDNA sequence) with an N-terminal signal sequence. There is a transmembrane domain-like region in the middle of the molecule, therefore it is supposed that the N-terminal part of the protein is extracellular containing leucine rich repeats, while the C-terminal domain is intracellular coding for serine-threonine kinase sequences with conserved ATP binding sites and kinase active site.

Using the DNA region coding for the postulated extracellular part as the hybridization probe, distinct specific hybridizing bands could be detected in each legume species tested (*Desmodium, Glycine, Macroptilium, Medicago, Melilotus, Phaseolus, Trifolium, Sesbani, Vicia, Vigna*) even in *Cassia emerginata,* a legume in which symbiotic nodule formation capability had not evolved during evolution. On the other hand, representative members of non-legume plants (maize, tobacco, wheat, rice) did not dispose specific hybridization signals.

4. Concluding Remarks

The similar nodulation phenotypes of the alfalfa and pea mutants (Nod-, Hac-, no Ca^{2+} spiking, etc.), the similar map position of the mutations conditioning non-nodulation phenotype, and the mutation found in each NORK gene originating from the non-nodulating plants make NORK a promising candidate gene which must play a key role in the initiation of the signal transducing cascade leading to symbiotic nodule development by binding directly or indirectly the specific the Nod factor of *Sinorhizobium meliloti*. Transformation experiments are in progress to complement the mutations and to demonstrate the direct correlation between the non-nodulating phenotype and the mutation in the NORK gene.

5. References

Caetano-Annolés G *et al.* (1993) In Palacios R, Mora J, Newton WE (eds), New Horizons in
     Nitrogen Fixation, pp. 297-302, Kluwer Academic Publisher, Dordrecht, The Netherlands
Endre *et al.* (1996) Theor. Appl. Genet 93,1061-1065
Ehrhardt *et al.* (1996) Cell 85, 673-681
Felle *et al.* (1996) Plant J.10, 295-301
Dudley ME, Long SR (1989) Plant Cell 1, 65-72
Hirsch AM *et al.* (1997) Plant and Soil 194, 171-184
Kaló *et al.* (2000) Theor. Appl. Genet. 100, 641-657
Kiss GB *et al.* (1998) Acta Biol. Hung. 49, 125-142
Michelmore RW *et al.* (1991) Proc. Natl. Acad. Sci. USA 88, 9828-9832
Nam YW *et al.* (1999) Theor. Appl. Genet. 98, 638-646
Peterson, Barnes (1981) Crop Sci. 21, 611-616
Schneider *et al.* (1999) Mol. Gen. Genet. 262, 1-11
Truchet G *et al.* (1989) Mol. Gen. Genet. 219, 65-68

6. Acknowledgements

We are particularly indebted to K. VandenBosch and G. Duc for providing non-nodulating lines of *M. truncatula* R38 and *P. sativum* P4/P55, respectively, and to D. Cook for providing the *M. truncatula* BAC library. This work was supported by the BRC, Szeged, Hungary and by grants from the MTA (AKP 96 -360/62, and AKP 00-246/35), from the OTKA (F030408, T025467), from the OMFB (EU-97-D8-063), from the NKFP (Grant No. Medicago Genomics 4/023/2001, and from the European Union (EuDicotMap Grant No. BIO4 CT97 2170, and Medicago Grant No. QLTR-2000-30676).

PLANT AND BACTERIAL GENOTYPE INTERACTIONS IN EARLY STAGES OF INFECTION OF PEA BY *RHIZOBIUM LEGUMINOSARUM* BIOVAR *VICIAE*

J.A. Downie, S. Walker, B. Hogg, A.E. Davies, V. Viprey

John Innes Centre, Colney Lane, Norwich NR4 7UH, UK

1. Introduction

Many plant-host and bacterial mutants have been identified that block, or substantially reduce, the establishment of nitrogen-fixing symbioses between *Rhizobium leguminosarum* bv. *viciae* and pea. Here we focus on some bacterial and plant mutants that are affected in the initiation of nodules, particularly focusing on early stages of the interaction and plant-bacterial genotype interactions.

R. l. viciae strain A34 makes a mixture of four Nod factors which are oligomers of four or five N-acetyl glucosamine residues carrying either a C18:1 or a C18:4 acyl chain on the amino group of the terminal non-reducing residue, which also carries an O-linked acetyl group (Spaink *et al.* 1991; Firmin *et al.* 1993). Mutation of *nodE* results in the inability to produce the Nod factors carrying the C18:4 group, whereas the C18:1 Nod factors are still made (Spaink *et al.* 1991). *nodE* mutants are reduced for nodulation of pea, forming nodules which are delayed in emergence and are at around 30-40% of wild-type numbers. Nodulation is almost completely abolished if *nodO* is mutated in addition to the *nodE* (Economou *et al.* 1994). The *nodO* gene encodes a secreted calcium-binding protein that forms cation-selective pores in membranes (Sutton *et al.* 1994); it is thought that this pore-forming property in some way enhances the ability of the bacteria to infect peas and vetch.

Some strains of *R. l. viciae* contain an additional nodulation gene *nodX* (Davis *et al.* 1988) that mediates an acetylation of the terminal reducing glucosamine (Firmin *et al.* 1993). *nodX* is required for nodulation of cv. Afghanistan and cv. Iran peas, which carry an allele ($sym2^A$) that confers resistance to nodulation by strains of *R. l. viciae* lacking *nodX* (Kozik *et al.* 1995). The pea $sym2^A$ allele causes growth of infection threads to be arrested (Geurts *et al.* 1997), and this can be overcome if the bacteria carry *nodX*. *nodZ*, which causes a fucosylation of the Nod factor, can also overcome the $sym2^A$-mediated nodulation resistance (Ovtsyna *et al.* 1998) demonstrating that it is not the precise structure of the Nod factor that overcomes the nodulation resistance.

Another characteristic of cv. Afghanistan peas is that some strains of *R. l. viciae*, that are Nod⁻ because they lack *nodX*, can competitively inhibit nodulation by strains carrying *nodX* (Dowling *et al.* 1987). This phenomenon has been called competitive nodulation blocking (Cnb) and requires the fully decorated Nod factor structure, because mutations in *nodE* or *nodL* abolish the Cnb phenotype (Dowling *et al.* 1989; Firmin *et al.* 1993). It was predicted that the Cnb effect occurred prior to infection, because an exopolysaccharide-defective mutant that does not induce infections is Cnb⁺ (Firmin *et al.* 1993).

In addition to $sym2^A$ several pea mutants that are completely nodulation deficient have been identified (Sagan *et al.* 1994; Engvild *et al.* 1987; Kozik *et al.* 1996; Kneen *et al.* 1994; Tsyganov *et al.* 1998 and references therein). The study of these mutants has given an insight into aspects of the pea-*R. l. viciae* symbiosis (e.g. Albrecht *et al.* 1998; Schneider *et al.* 1999). Here we further that analysis and examine establishment of infections in mutant and wild-type plants inoculated with wild-type and mutant bacteria, that are compromised for nodulation due to mutations in *nod* genes, or blocked in competitive nodulation assays on cv. Afghanistan peas.

2. Nod Factor Induced Changes in Ca in Root Hairs of Pea Mutants

Ehrhardt *et al.* (1996) first demonstrated that Nod factors could induce periodic oscillations in calcium (calcium spiking) around the nuclear region of legume root hairs. On alfalfa, this required the appropriate (*Sinorhizobium meliloti*-made) Nod factor because the Nod factor from *R. l. viciae* had no effect. The host specific sulfation of the terminal reducing glucosamine was necessary for induction of Ca spiking. In *Phaseolus* bean, Nod factor induced a change in Ca at the root tip and subsequently an oscillation in intracytoplasmic Ca levels (Cardennas *et al.* 1999).

In pea, Nod factor-induced changes in intracellular Ca were analyzed using the calcium-sensitive fluorescent dye Oregon green-dextran, which was microinjected into young growing root hairs on lateral roots. The dye was excited at 488 nm and emitted fluorescent light was passed through a 515 nm-long pass filter. Sets of images were collected at 5 s intervals and the data obtained were calculated using the average pixel intensity for regions of the root hair. The Nod factor (at 10^{-9} M) clearly induced Ca spiking (Figure 1) and this occurred in all the cells that were imaged. This induction of Ca spiking occurred if the Nod factor lacked the *nodE*-determined C18:4 acyl group and the *nodL*-determined acetyl group. To determine the minimal Nod factor structure that would induce a response, we tested the effects of unsubstituted chitin backbone of the Nod factor, which consists of an oligomer of four or five *N*-acetyl glucosamine residues. At relatively high concentrations (10^{-6}-10^{-8} M), such chitin oligomers induced abnormal Ca spiking. Thus, although calcium spikes were observed, they were less frequent and more sporadic than those induced by Nod factor. Further details are described by Walker *et al.* (2000).

We analyzed the effect of Nod factor on intracellular calcium in the root hairs of several pea nodulation-defective mutants that do not show infections. Normal calcium spiking occurred in mutants carrying non-nodulating alleles of *sym2^A*, *sym7* or *sym9*. In contrast, no calcium spiking was induced in peas carrying mutant alleles of *sym8*, *sym10* or *sym19* (Walker *et al.* 2000). Representative traces from some mutants are shown in Figure 1.

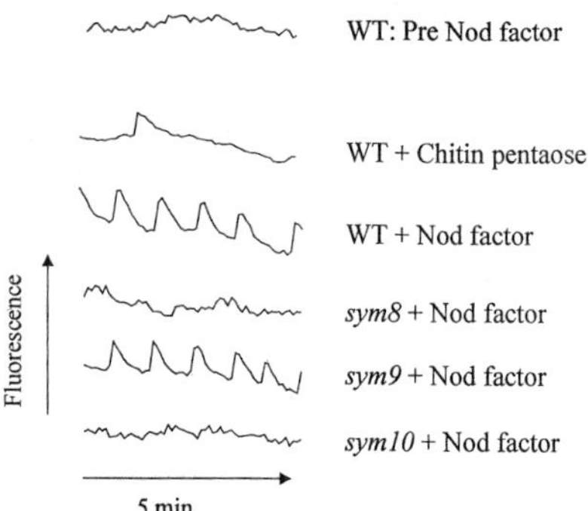

Figure 1. Calcium spiking induced by Nod factors. The traces show measurements of fluorescence in root hairs either before additions (top trace) or 20-25 min after the additions of chitin pentaose (10^{-6} M) or Nod factors (10-9 M). Representative traces obtained with only three of the mutants tested are shown. Further details are described in Walker et al (2000).

These results suggest that induction of calcium spiking occurs downstream of events that require the *sym8*, *sym10* and *sym19* gene products, but upstream of events that require the *sym2^A*, *sym7* and *sym9* gene products. If the other phenotypes of pea mutants carrying mutations at these loci are considered, it is possible to place the mutations into an order that could correspond to the relative order of function of the different gene products (Figure 2). Thus, plants carrying *sym2^A* can induce root hair curling and infection foci (Geurts *et al.* 1997, and see below), whereas *sym7* mutants induce root hair deformation but no infections (Markwei, LaRue 1992). Therefore $sym2^A$ is likely to influence events after $sym7$, and since the $sym9$ mutant induces little or no root hair deformation, this is likely to be upstream of $sym7$ and $sym2^A$. Therefore after calcium spiking, the events are likely to occur in the order determined by $sym9$, $sym7$ and $sym2^A$. Of the three loci that are required for calcium spiking, two ($sym8$ and $sym19$) have phenotypes that are, as yet, indistinguishable; pea lines carrying mutations at these loci do not induce root hair deformation and do not induce early nodulin gene expression (Albrechts *et al.* 1998; Schneider *et al.* 1999).

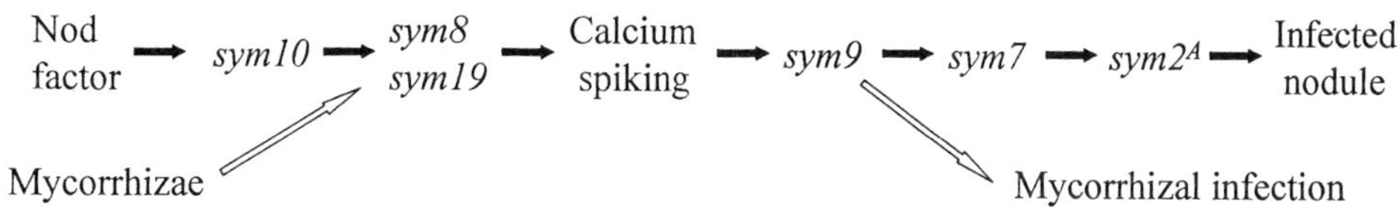

Figure 2. Model showing proposed sequence of early nodulation signaling events in relation to mutations blocking nodulating and infection thread growth in pea.

In previous work, Duc *et al.* (1989) and Albrecht *et al.* (1998) demonstrated that non-nodulating peas carrying alleles of $sym8$, $sym9$ and $sym19$ were also defective for mycorrhizal infections. In contrast, $sym10$ mutants established a normal mycorrhizal symbiosis. Taking these observations into account together with the data on calcium spiking, it seems likely that the $sym10$ gene product plays a role that is upstream of $sym8$ and $sym19$ (Figure 2). Therefore, of those pea mutant loci characterized, it is likely that $sym10$ is involved in the earliest step in nodulation signaling.

3. Analysis of Infection Events

Infection threads are absent from all of the pea mutants described above (Walker *et al.* 2000), and are greatly reduced in lines of pea homozygous for $sym2^A$ inoculated with a strain lacking $nodX$. However normal infections are induced if *R. l. viciae* carrying $nodX$ is inoculated onto $sym2^A$ lines of pea (Geurts *et al.* 1997). This observation, together with analysis of infection of *Medicago* spp. by mutants of *Sinorhizobium meliloti* (Ardourel *et al.* 1994), demonstrates a role for specific Nod factor structures in initiation of infections. We have examined infection events induced by *nod* mutants of *R. l. viciae* and the effect of Cnb$^+$ strains on infections in cv. Afghanistan, which is homozygous for $sym2^A$.

3.1. *nodE* or *nodO* required for infection thread growth but not for entry into root hairs. In the absence of *nodE*, the *nodO* gene is necessary for nodulation of pea and the vetch *Vicia hirsuta* (Economou *et al.* 1994). To assay where nodulation is blocked, *nodO*, *nodE* and *nodO-nodE* (double) mutants were marked with a constitutively-expressed *lacZ* gene and infection events were assayed by staining with X-gal. Normal infections were induced by the *nodO* mutant and the *nodE* mutant on both vetch and pea roots. These infection events could not be distinguished from those induced by a wild-type strain. However, with the *nodO-nodE* double mutant a strikingly different result was observed. Normally pea inoculated with a wild-type strain induces an average of 28 (± 6) infection threads per root section under the assay conditions used. These are found only very infrequently with the *nodO-nodE* double mutant (average 0.7 +/- 0.3 per root section). Instead, the mutant was found to have induced aberrant infection foci (average 24.3 +/- 3 per root section), that did not develop into infection threads. Such infection foci lacking infection threads were very rarely found on roots inoculated with the wild-type (average 0.6 +/- 0.3 per root section). This indicates that in the absence of *nodO* and *nodE,* entry can be made into root hairs. However infection thread growth can be stimulated by *nodO* or *nodE,* because if either gene is restored to the double mutant, normal infection threads are formed.

In vetch an analogous situation is seen (Walker, Downie 2000). Thus, normal infections are seen with *nodE* or *nodO* mutants but abnormal infection events are seen with the *nodO-nodE* double mutant. In vetch the number of infections induced by wild type averages about 130 and this is reduced by about two orders of magnitude in a *nodO-nodE* double mutant. Instead there are in excess of 700 aberrant infections in which bacteria are located within root hairs but normal infection threads are not seen in these infected root hairs (Walker, Downie 2000). Again, restoring either *nodO* or *nodE* restores normal infection. This implies that the pore-forming NodO protein may stimulate infection by forming an ion channel in the plant plasma membrane, where the plant cell wall has been degraded to enable the bacteria to make entry. Alternatively, if the host specific Nod factors (i.e. those carrying the *nodE*-determined C18:4 group) are made, infection threads can be established in the absence of NodO. This suggests that bacterial entry into root hairs requires minimal Nod factor structure, but that subsequent establishment of infection threads requires activation of some additional pathway.

3.2. Competitive nodulation blocking in cv. Afghanistan pea occurs at the infection stage. *R. l. viciae* strain TOM infects and nodulates cv. Afghanistan pea normally, but some strains of *R. l .viciae* (such as strain A34) competitively inhibit nodulation. Strain TOM was marked with a constitutively-expressed *lacZ* gene and infection events on cv. Afghanistan roots were assayed in the presence or absence of *R. l. viciae* A34 which is strongly Cnb[+]. Co-inoculation with A34 caused about a thirty-fold reduction in the number of infection threads formed by strain TOM. Previous work has demonstrated that the Cnb phenotype requires the host specific Nod factor structure, because *nodE* and *nodL* mutants of A34 are Cnb[-] (Dowling *et al.* 1989; Firmin *et al.* 1993).

One characteristic of the competitive nodulation blocking strain A34 is that it makes high levels of Nod factors, whereas strain TOM makes very low levels. These low levels are due in part to the presence of the *nolR* gene in strain TOM. NolR is predicted to be a repressor and it decreases expression of the *nodABC* operon and overall levels of Nod factors (Kiss *et al.* 1998). The cloned *nolR* gene was transferred to strain A34, and the resulting strain was greatly reduced for competitive nodulation blocking on cv. Afghanistan peas. This suggests that Cnb is due to high levels of Nod factor production.

Using a purified preparation of Nod factors from strain A34, it was possible to block nodulation of cv. Afghanistan by strain TOM, and this was correlated with a thirty-fold decrease in infection thread formation. Interestingly, a purified Nod factor preparation from a strain expressing the TOM nod genes was also found to inhibit nodulation of cv. Afghanistan by strain TOM. Therefore we conclude that if too much Nod factor is present, nodulation inhibition can occur with cv. Afghanistan, and it is the high levels of Nod factor, rather than the absence of the NodX-determined acetyl group, that determines nodulation blocking.

4. Conclusions

Nod factors with minimal structure (lacking host-specific decorations) can enable strains of *R. l. viciae* to enter root hairs, but host specific decorations (or alternatively the pore-forming protein NodO) are required for normal infection thread growth. Host-specific decorations are also required for competitive nodulation blocking, which occurs at the level of infection thread initiation. Taken together, these observations suggest that there are quantitative and qualitative differences in the recognition of undecorated or host-specifically decorated Nod factors. The mechanism by which the plant makes this discrimination has yet to be determined.

  © CAB *International* 2002 *Nitrogen Fixation. Global Perspectives* (eds. T Finan, M O'Brian, D Layzell, K Vessey, W Newton)

5. References

Albrecht C *et al.* (1998) Plant J. 15, 605-614
Ardourel M *et al.* (1994) Plant Cell 6, 1357-1374
Cardenas L *et al.* (1999) Plant J. 19, 347-352
Davis EO *et al.* (1988) Mol. Gen. Genet. 212, 531-535
Duc G *et al.* (1989) Plant Sci. 60, 215-222
Dowling DN *et al.* (1987) J. Bacteriol. 169, 1345-1348
Dowling DN *et al.* (1989) Mol. Gen. Genet. 216, 170-174
Economou A *et al.* (1994) Microbiol. 140, 2341-2347
Ehrhardt DW *et al.* (1996) Cell 85, 673-681
Engvild KC (1987) Theor. Appl. Genet. 74, 711-713
Firmin JL (1993) Mol. Microbiol. 10, 351-360
Geurts R *et al.* (1997) Plant Physiol. 115, 351-359
Kneen BE *et al.* (1994) J. Hered. 85, 129-133
Kiss E *et al.* (1998) Mol. Plant-Microbe Interact. 11, 1186-1195
Kozik A *et al.* (1995) Plant Sci. 108, 41-49
Kozik, A *et al.* (1996) Plant Mol. Biol. 31, 149-156
Markwei CM, LaRue TA (1992) Can. J. Microbiol. 38, 548-554
Ovtsyna AO *et al.* (1998) Mol. Plant-Microbe Interact. 11, 418-422
Sagan M *et al.* (1994) Plant Sci. 100, 59-70
Schneider A *et al.* (1999) Mol. Gen. Genet. 262, 1-11
Spaink HP (1991) Nature 354, 125-130
Sutton JM *et al.* (1994) Proc. Natl. Acad. Sci. USA 91, 9990-9994
Tsyganov VE *et al.* (1998) Mol. Gen. Genet. 259, 491-503
Walker SA *et al.* (2000) Proc. Natl. Acad. Sci. USA 97, 13413-13418
Walker SA, Downie JA (2000) Mol. Plant-Microbe Interact. 13, 754-762

6. Acknowledgements

We thank Karen Wilson for preparing Nod factors, and H. Spaink and E. Kondorosi for kindly giving us plasmids carrying *nodX* and *nolR*. This work was supported principally by the Biotechnology and Biological Sciences Research Council, with additional support from the European Union (Contract FMRX-CT98-0243).

Section 5:
Developmental Biology

LEGUME PLANT NITROGEN FIXATION SYMBIOTIC MUTANTS - A PROMISING TOOL FOR UNDERSTANDING FUNDAMENTAL PROCESSES AND MOLECULAR BASES OF PLANT DEVELOPMENT

I.A. Tikhonovich

All-Russia Research Institute for Agricultural Microbiology, St Petersburg,
Pushkin 8, Podbelsky chaussee, 3, 196608, Russia

Plant-microbe interactions provide a unique model based on the signaling process during nodule formation. The process of induction of nodule formation on the roots of legumes requires reorganization of a plant developmental program. Details of such reorganization remain unclear. The most exciting phenomenon is the action of a Nod factor in plant development. Among the problems to be solved in the connection with Nod factor action, the problem of receptor molecules should be treated as the most urgent. Also the intriguing data about the possible role of Nod factor during late stages of nodule development have been recently obtained (Timmers *et al.* 1998; V.A. Voroshilova *et al.* unpublished). We still do not understand what is the way of the regulation of the nodule formation genetic system of higher plants under the bacterial influences.

There are two sets of plant genes involved in symbiosis: nodulin-genes and sym-genes. Sym genes are thought to be good candidates for plant genes regulating the process of nodule development. To date the collections of legume mutants have been obtained and the terminus of sym-genes actions have been determined. Such data allow to describe the process of nodule formation as sequential functioning of the blocks of genes from both partners.

The highest number of legume plant symbiotic genes has been identified in pea (*Pisum sativum* L.) as a result of all efforts in the field of isolation of pea symbiotic mutants with abnormalities of nodule formation and function throughout the world. More than two hundred independently obtained symbiotic mutant lines are known to date. More than one hundred isolated symbiotic mutants have been isolated in various labs, and more than forty pea symbiotic genes have been identified to date (reviewed in Borisov *et al.* 2000). In parallel a majority of genetically characterized mutants have been characterized to identify nodule developmental stages (reviewed in Borisov *et al.* 2000). This characterization allowed subdivision of nodule morphogenesis into eight discrete developmental stages. These results have made it necessary to modify the previously used system of phenotypic codes describing the process of symbiotic nodule development (Vincent 1980; Caetano-Anollés, Gresshoff, Tsyganov *et al.* 1998). To date, the sequence of nodule developmental stages is defined as follows: (i) root hair curling (Hac), (ii) infection thread growth initiation (Iti), (iii) infection thread growth inside root hair (Ith), (iv) infection thread growth inside root tissue (Itr), (v) infection thread growth inside nodule tissue (Itn), (vi) infection droplet differentiation (Idd), (vii) bacteroid differentiation (Bad) and nodule persistence (Nop) (reviewed in Borisov *et al.* 2000).

Among the identified plant symbiotic genes there should be genes involved in the recognition and processing of a Nod factor. The molecular products of those genes and other sym genes, which are regulatory proteins, have been revealed with the use of the two-hybrid system (reviewed in Brent, Finley 1997) in proteomic analysis. The combination of this method with the mutational analysis of nodule formation provides a powerful tool for understanding the molecular bases of plant development as well as the process involving prokaryotes and eukaryotes into this highly integrated system, the nitrogen-fixing symbiotic nodule.

References
Borisov AY, Barmicheva EM, Jacobi LM, Tsyganov VE, Voroshilova VA, Tikhonovich IA (2000)
    Czech J. of Gen. Plant Breeding 36, 106-110
Brent R, Finley Jr. RL (1997) Annu. Rev. Genet. 31, 663-704
Caetano-Anollés G, Gresshoff PM (1991) Annu. Rev. Microbiol. 45, 345-382
Timmers ACJ, Auriac M-C, de Billy, Truchet G (1998) Develop. 125, 339-349
Tsyganov VE, Morzhina EV, Stefanov SY, Borisov AY, Lebsky VK, Tikhonovich IA (1998)
    Mol. Gen. Genet. 256, 491-503
Vincent JM (1980) In Newton WE, Orme-Johnson WH (ed) Nitrogen Fixation, Vol. 2,
    pp. 103-129, University Park Press, Baltimore, MD

Acknowledgements
This work was financially supported by the Russian Foundation for Basic Research (01-04-48580 and 01-04-49643) and the Netherlands Organisation for Scientific Research (NWO 047-007.017).

ROLE OF REACTIVE OXYGEN SPECIES AND ETHYLENE IN PROGRAMMED CELL DEATH DURING NODULE INITIATION ON *SESBANIA ROSTRATA*

W. D'Haeze, C. Chaparro, A. De Keyser, S. Deleu, R. De Rycke, S. Goormachtig, S. Lievens, R. Mathis, K. Schroeyers, W. Van de Velde, D. Vereecke, C. Verplancke, M. Holsters

Dept of Plant Genetics, V.I.B., K.L. Ledeganckstraat 35, B-9000 Ghent, Belgium

1. Introduction

Sesbania rostrata is a flooding-tolerant tropical legume with nodulation at lateral root bases and at the position of stem-located adventitious rootlets. One of the first visible features of *Azorhizobium caulinodans*-induced nodule initiation on *S. rostrata* is the formation of infection pockets. Infection pockets are big intercellular spaces occupied with proliferating bacteria, and are formed during both stem and root nodule formation. Infection pocket formation is accompanied by a phenomenon of local plant cell death. It was demonstrated that Nod factors purified from the wild-type strain complement ORS571-V44, abolished in the synthesis of Nod factors, to form infection pockets in the adventitious rootlets on *S. rostrata* stems. These observations suggested that Nod factors are required for infection pocket formation, since ORS571-V44 alone is unable to infect the outer cortical tissue (D'Haeze *et al.* 1998). This peculiar plant cell death-accompanied infection pocket formation and the fact that Nod factors are required for infection pocket formation, and thus may play a role in the induction of a plant cell to die, tempted us to further investigate this topic. Nod factor-induced root hair formation and nodulation were used as biological assays in a pharmacological and morphological approach, including the *in situ* localization of hydrogen peroxide.

2. Material and Methods

For bacterial growth conditions, strains, Nod factor preparations, plant growth and inoculations, and microscopy techniques, see the article by D'Haeze *et al.* (1998) and references therein.

Final concentrations of inhibitors used are the following: 7 μM L-α-(2-aminoethoxyvinyl)-glycine (AVG), 0.1 mM (aminooxy) acetic acid (AOA), 10 mM α-aminoisobutyric acid (AIB), 0.5 mM CoCl$_2$, 1 μM Ag$_2$SO$_4$, 0.05% (v/v) 2,5-norbornadiene, 1 μM diphenyleneiodonium chloride (DPI), 200 μM diethyldithio carbamic acid (DDC), 1 mM ascorbic acid, 1 mM LaCl$_3$, and 10^{-3}% (m/v) ruthenium red. All inhibitors delivered as powders were dissolved in water and filter sterilized, except DPI that was dissolved in dimethylsulfoxide (DMSO). In the latter case, the stock concentration was such that no more than 10 μl of DMSO solution was added per tube. No influence of this DMSO concentration was observed on plant growth or nodulation. Inhibitors were purchased from Sigma-Aldrich, except 2,5-norbornadiene (Avocado, La Tour du Pin, France). Roots were treated with inhibitors two days before Nod factors or the wild-type strain were added. For DDC, ascorbic acid, and ruthenium red, solutions with inhibitors were refreshed 4, 4, and 2 times, respectively, prior to observations. The effect of inhibitors on Nod factor-induced root hair formation and ORS571-induced nodulation was performed in at least four independent experiments using at least four plants.

Tubes were filled with Norris medium, leaving an air space of 1 cm to avoid leakage and contaminations. For scoring the effect of inhibitors on nodulation, the part of the root present in the upper airspace was not considered since it was not in contact with the medium and since some nodules were observed at the most upper part of the primary root. For the assay in which Ag$^+$ ions were added at different time points compared to addition of the wild-type strain, the solutions with inhibitor and wild-type strain were refreshed each time nodules were counted. The same was done when this experiment was carried out using AOA. The two latter assays were each done twice, independently, using 4 plants per time point at which Ag^+ or AOA was added.

Hydrogen peroxide and ethylene were applied in a final concentration of 1 mM and 21 µl/tube, respectively. When roots were treated with ethylene, a sticky substance was used to tighten the tube as well as possible from the air, in order to keep the ethylene inside the tube. These experiments were done at least three times, independently using 16 plants per treatment. For the assay in which ethylene was added at different time points compared to addition of ORS571, a second dose of ethylene was applied, five days after the first one.

3. Results and Discussion

3.1. Nod factor-induced root hair formation, a biologically relevant assay. Upon addition of 10^{-9} M PI Nod factors (Mergaert *et al.* 1997), bushes of axillary root hairs were observed, specifically at bases of lateral roots, at the sites where root nodules develop upon inoculation with *A. caulinodans* ORS571 (Ndoye *et al.* 1994). The efficiency of Nod factor-induced formation of axillary root hairs was high, since over 70% of the bases of lateral roots exhibited this phenomenon, when inspected five days after inoculation. Nod factors did not only induce the formation of axillary root hairs, but also root hair deformation. Most root hairs present in bushes of axillary root hairs showed moderate distortions, including tip swellings, branches, and corkscrews. Shepherd's crooks were not observed.

To determine whether root hair formation is also induced during nodule initiation, roots were inoculated with the ORS571 wild-type strain and scored for root hair formation at bases of lateral roots. As early as 15 hpi the presence of axillary root hair bushes was obvious. At 24 hpi, a similar phenotype was observed, and at 36 hpi, swelling of lateral root bases was apparent, indicative for nodule primordium formation. The root hair formation, occurring during nodule initiation, was Nod factor-dependent, since ORS571-V44, a *nodA* mutant that is unable to synthesize Nod factors, did not induce the formation of axillary root hairs. These observations suggested that Nod factor-induced root hair formation is a biologically relevant assay to study Nod factor-induced responses during nodule initiation by ORS571.

3.2. Ethylene synthesis and perception required for Nod factor-induced root hair formation and nodule initiation. The root hair formation assay was used to test the effect of inhibitors of ethylene synthesis (AVG, AOA, AIB, Co^{2+}) and perception (Ag^+, 2,5-norbornadiene) on Nod factor-induced root hair formation. Inhibitors were added prior to Nod factors. All inhibitors tested completely blocked induction of Nod factor-induced axillary root hair formation. Treatment of roots with exogenous ethylene induced the formation of root hairs. These appeared as bushes of axillary root hairs, but, in contrast to application of Nod factors, an enhanced number of root hairs was also detected on the primary root and lateral roots, only straight but not deformed root hairs could be observed, and the efficiency was ten-fold lower.

Some of the ethylene synthesis and perception inhibitors were tested for their influence on root nodule development on *S. rostrata*. When added two days prior to ORS571, all inhibitors tested blocked nodulation completely, scored approximately 10 days post inoculation. No swellings could be observed macroscopically. Controls for the effect of those inhibitors on flavonoid secretion, bacterial growth, and Nod synthesis and production were performed.

To investigate whether the Nod factor-induced ethylene production is only needed for nodule initiation or also during later stages of nodule development, the number of root nodules was counted during 14 days and compared between roots that were treated with Ag^+, added at different time points in respect to the wild type. When Ag^+ ions were added 1 or 2 days prior to the wild-type strain, or at the same moment, a complete block of nodule initiation was noticed. When the inhibitor was added 1 or 2 days after, approximately 3 and 7 nodules were formed, respectively, and the number did not increase in time. These were probably nodules that had been initiated during the time period at which the ethylene perception inhibitor was not yet present. This observation was in sharp contrast to the kinetics of nodule formation by ORS571 alone, that exhibited a continuous increase in nodule number within the period of inspection, suggesting a continuous process of nodule initiation. A similar experimental set up was achieved for AOA, an inhibitor of ethylene synthesis, and a so far identical graph was obtained. Thus, Nod factor-induced ethylene production is only required for nodule initiation.

3.3. Nod factor-induced root hair formation and nodulation require hydrogen peroxide. A proposed way for the synthesis of hydrogen peroxide in plants involves an NADPH oxidase complex and super oxide dismutase, the function of which is inhibited by DPI and DDC, respectively. These inhibitors, together with ascorbic acid, a scavenger of hydrogen peroxide, blocked Nod factor-induced root hair formation. Exogenous hydrogen peroxide induced the formation of axillary root hairs, that were, similar to ethylene-induced root hairs, not deformed. The efficiency of hydrogen peroxide-induced root hair formation was approximately five times lower than when ethylene was applied.

All compounds used above to inhibit the generation of reactive oxygen species also blocked nodulation when added two days prior to the wild-type strain. Thus both the ethylene production and the generation of reactive oxygen species are needed for Nod factor-induced root hair formation and nodulation. When ethylene was added to roots that were treated with DPI, bushes of axillary root hairs were observed. Although this observation is preliminary, it suggests that an oxidative burst may precede the ethylene production.

3.4. Morphological effects of Ag^+ and DPI on ORS571-induced root nodule initiation. To be able to evaluate the effect of Ag^+ and DPI on invasion and nodule development, a morphological approach was applied. Bases of lateral roots were imbedded in Technovit, semi-thin sections were made in a transversal direction, and stained with toluidine blue.

Initially, morphological aspects of lateral root bases of mock-inoculated roots, harvested 2, 6, and 10 dpi, were studied. Sections of material harvested at these three time points showed similar features. The vascular bundle is surrounded by cortical tissue that consists of typical, highly vacuolated cells, and big intercellular spaces. A few of these cortical cells stained strongly with toluidine blue. Since almost the entire cytoplasm and vacuole were greenish to blue colored, clearly distinguishable from neighboring cells, these cells will be further referred to as 'blue cells'. At positions in the lateral root bases, located close to the primary root, no clear epidermis was present, in contrast to locations further away from the primary root. It seems that, besides remainders of the epidermis, outer cortical tissue at bases of lateral roots is in direct contact with the environment. Most of these peripheral outer cortical cells, and occasionally also epidermal cells, have a swollen and/or enlarged appearance, which may explain the presence of bulge-like structures, observed under the binocular. In some cases, outer cortical cells seem to elongate thereby separating two neighboring covering peripheral cells.

As soon as 24 hpi with the wild-type strain, a nodule primordium, characterized by a focus of dividing cells and young cells with relatively small vacuoles, has been developed, opposite of a proto-xylem pole. Infection pockets were present in the outer cortical cell layers. Sections through developing root nodules, at 48 hpi, are characterized by the presence of nodule primordia with a shape of an open basket. Infection pockets are positioned deeper in the nodule tissues and infection threads, originating from the infection pockets, guide bacteria to the nodule primordia. Within the infection center, many intracellular infection threads were noticed.

The presence of either Ag^+ or DPI had a tremendous effect on nodule invasion and the initiation of nodule development. Apart from few bacteria present in the vicinity of outer cortical cells, neither invasion nor infection pocket formation could be observed, at six dpi. There were no signs of nodule primordium formation. These sections resembled those through mock-inoculated lateral root bases.

3.5. Morphology of Nod factor-induced pseudo-nodules. Inoculation of *S. rostrata* roots with 10^{-8} M Nod factors led to the formation of small swellings or pseudo-nodules at lateral root bases (Mergaert *et al.* 1993). Such pseudo-nodules were harvested 4, 6, 8, 9 and 10 dpi. Pseudo-nodules harvested at these time points showed similar features. At four dpi, the total number of cells present in a transversal section is higher than that of mock-inoculated controls, suggesting that cell division must have occurred. This is also suggested by the nicely organized appearance of cells in the inner cortex. Cell division was more clear in a pseudo-nodule harvested at nine dpi, exhibiting regions of small cells with relatively small vacuoles. A remarkable observation was the enhanced number of blue cells, compared to controls. A differential staining procedure was used to distinguish nuclei of dying cells from those of healthy cells (Kosslak *et al.* 1997). By this approach, the nucleolus of nuclei of healthy cells is generally dark blue stained, whereas the rest of the nucleus is lightly stained. Chromatin condensation, occurring within the nucleus of dying cells, caused that the demarcation between nucleolus and nucleus is less clear (Kosslak *et al.* 1997). The nucleus of healthy cells within Nod factor-induced pseudo-nodules was pink colored and the nucleolus dark blue. However, in blue cells, the whole nucleus became dark blue, suggesting chromatin condensation and/or DNA fragmentation. Strikingly, blue cells within Nod factor-induced pseudo-nodules were often present in the vicinity of large schizo-lysogenic holes, containing pieces of partially degraded cell walls. These schizo-lysogenic holes, resembled a sort of empty infection pocket-like structures and were never observed in sections through mock-inoculated lateral root bases. The observation of remainders of cell walls strongly suggested that a process of cell death must have occurred before.

When Ag^{+} ions were added, prior to 10^{-8} M Nod factors, none of the features characterizing Nod factor-induced pseudo-nodules could be observed, suggesting a complete block of root hair formation, cell division, and cell death processes leading to the formation of schizo-lysogenic holes.

3.6 Morphology of lateral root bases treated with exogenous ethylene. Lateral root bases treated with exogenous applied ethylene were harvested at 6 dpi and sectioned. Almost identical features were observed as those present within Nod factor-induced pseudo-nodules. Many blue cells were present, most often in the neighborhood of large schizo-lysogenic holes, that again contained remainders of plant cell walls. These schizo-lysogenic holes were so big that they were at some locations even in contact with the environment. This confirmed previous observations that ethylene may participate in the Nod factor signaling pathways.

3.7. *In situ* localization of hydrogen peroxide during initiation of stem nodule development. Because of the strong indications that hydrogen peroxide may be involved and required for nodule initiation on *S. rostrata*, hydrogen peroxide was localized during stem nodule initiation. We have opted to use the histochemical assay based on the reaction of hydrogen peroxide with $CeCl_3$ to produce electron-dense insoluble precipitates of cerium perhydroxides, $Ce[OH]2OOH$ and $Ce[OH]3OOH$ (Bestwick *et al.* 1997). Stems were inoculated with ORS571 and material was harvested at 50 and 70 hpi, immediately fixed or stained with $CeCl3$, and imbedded for electron microscopy.

At 50 hpi, it seems that the oxidative burst, represented by the generation of hydrogen peroxide, arises. Outer cortical cells showed dark regions, corresponding to the presence of hydrogen peroxide at the level of the cytoplasmic membrane and the cell wall. An outer cortical cell, located one cell layer beneath the peripheral outer cortical cells, is almost completely surrounded by hydrogen peroxide, localized in its cell wall and also in intercellular spaces. This particular cell showed signs of death including vacuole fragmentation and a loss of cell shape rigidity. No hydrogen peroxide could be observed in the nodule primordia.

At 70 hpi, large infection pockets were observed containing massive proliferating bacteria. Parts of cell walls of outer cortical cells in the direct vicinity of the infection pockets showed a strong staining, suggesting a massive production of hydrogen peroxide. The cytoplasmic membrane is strongly stained and hydrogen peroxide seems to diffuse from the cytoplasmic membrane toward the peripheral part of the cell wall. Bacteria that are close to these cell walls are imbedded in a matrix containing high concentrations of hydrogen peroxide. Remarkably, the bacterial body seemed to be protected against these high concentrations of hydrogen peroxide by the presence of a thick layer of low electron dense material, most probably EPSs, forming a clear barrier. Cells producing high amounts of hydrogen peroxide often show the onset of vacuole fragmentation and loosening of the cytoplasmic membrane from the cell wall, whereas the neighboring cells do not. No hydrogen peroxide was observed in nodule primordia. At 70 hpi, infection threads were present, and bacteria within infection thread were imbedded in an electron-dense matrix that contained high amounts of hydrogen peroxide. Hydrogen peroxide was also detected within cell walls of some intracellular infection threads. A longitudinal section through an intercellular infection thread revealed the presence of hydrogen peroxide within the matrix in which bacteria are imbedded, and illustrated again the possible role of the EPS layer to protect invading bacteria against those high concentrations of hydrogen peroxide.

3.8. A model for nodule initiation on *S. rostrata*. Nod factors probably induce an oxidative burst generating ROS, including hydrogen peroxide. ROS may subsequently induce a local ethylene production, which may eventually enhance ROS production. The preliminary observation that ethylene-induced axillary root hair formation was not inhibited in the presence of DPI suggests this sequence. This was also in agreement with our genetic approach by which it was demonstrated that the expression of both a peroxidase gene and an ACC synthase gene is induced very early in the symbiotic interaction. Interestingly, the peroxidase gene was expressed in cells surrounding infection pockets. Furthermore, a GA-20-oxidase, involved in giberellin synthesis, was expressed in cells that are going to form infection pockets, and nodulation and Nod factor-induced root hair formation was blocked in the presence of giberellin synthesis inhibitors. Possibly, giberellins may sensitize cells for hydrogen peroxide (Bethke, Jones 2001). Ethylene and/or ROS may be involved in the induction of a local plant cell death, required for infection pocket formation, a prerequisite for nodule initiation. Blocking the oxidative burst or ethylene production caused in addition abolishment of nodule primordium formation. Supposedly, ROS and/or other secondary signals may influence hormone landscapes or other factors.

4. References

Bestwick CS *et al.* (1997) Plant Cell 9, 209-221
Bethke PC, Jones RL (2001) Plant J. 25, 19-29
D'Haeze W *et al.* (1998) Mol. Plant-Microbe Interact. 11, 999-1008
Kosslak RM *et al.* (1997) Plant J. 11, 729-745
Mergaert P *et al.* (1993) Proc. Natl. Acad. Sci. USA 90, 1551-1555
Mergaert P *et al.* (1997) Mol. Plant-Microbe Interact. 10, 683-687
Ndoye I *et al.* (1994) J. Bacteriol. 176, 1060-1068

5. Acknowledgements

We would like to acknowledge G. Van De Sype, S. Pagnotta, and D. Hérouart (Université de Nice Sophia Antipolis, Nice, France) for help with hydrogen peroxide localization.

ENDOREDUPLICATION IS ESSENTIAL FOR SYMBIOTIC CELL DIFFERENTIATION IN *MEDICAGO TRUNCATULA*

E. Kondorosi, J.M. Vinardell, E. Fedorova, A. Cebolla, F. Roudier, S. Tarayre, G. Horvath, K. Fülöp, D. Vaubert, A. Kondorosi

Institut des Sciences du Vegetal, CNRS UPR 2355, 91198 Gif sur Yvette, France

1. Introduction

Plant organ development is a constant interplay between cell cycle and differentiation programs. Various stages of nodule development are controlled by differential regulation of the cell cycle. As a first step, the rhizobial Nod factors activate the cell cycle in the nodulation sensitive root zone. In *Medicago sativa* roots the G0-arrested cortical cells dedifferentiate and reenter the cell cycle in front of the protoxylem poles. These cells undergo the G1/S transition, however, the cell cycle is completed by cell division only in the inner cortex. Maintenance of mitotic activity and recruitment of the neighbouring cells lead to extensive but localized cell proliferation and formation of the nodule primordium.

As a second step, the nodule primordium differentiates. In the case of *Medicago*, an autonomous meristem is established in the apical region whereas submeristematic cells develop to various nodule cell types. Cell differentiation necessitates an irreversible cell cycle exit that is coordinated by the expression of unique genes to specify tissue identity. Terminal differentiation can occur by complete loss of cell cycle activity as it is in the peripheral nodule tissues. In contrast, the cell cycle remains partially active in the central region. This might be necessary for the differentiation of symbiotic nitrogen-fixing nodule cells which do not divide but replicate their genome. This form of the cell cycle, allowing genome duplication without mitosis is referred to as endoreduplication or endocycles. In contrast to the mitotic cycle, the endocycle consists of the DNA synthesis (S-phase) and a gap period, and inhibition of mitosis is due to the inactivation of the cyclin-dependent kinase (CDK)-mitotic cyclin complexes.

In *M. sativa* and *M. truncatula* nodules, the diploid cells remain uninfected and genome duplication seems to be a prerequisite for *Rhizobium* infection. Repeated rounds of endocycles result in the formation of cells from 8°C to 32°C nuclear DNA content (Truchet *et al.* 1978; Cebolla *et al.* 1999). The increased genome size and nuclear volume is accompanied by gradual and proportional enlargement of the cell size. This nodule zone (zone II) is characterized also by the expression of many early nodulins and cell cycle genes. The signals and the mechanisms which coordinate cell cycle and tissue-specific events remain unknown. *Rhizobium* infection or Nod factors are amongst the most likely candidates, however, many of these early events occur also in spontaneous nodules of *M. sativa* in the absence of rhizobia and Nod factors. This suggests the involvement of developmental signals in the coordination of cell cycle and nodule-specific events. One may speculate that the cell proliferation is limited and the size of the nodule primordium might be sensed in the plant. As a differentiation checkpoint, it could trigger further differentiation events during nodule development.

When rhizobia, as part of the symbiosomes, are released from the infection threads into the cytoplasm, they differentiate simultaneously with the host cell. Curiously, their development shares some common features with that of the host cells. During their maturation in nodule zone II, they grow and lose progressively their ability to divide. Measurement of DNA content indicated from 4- to 8-fold more DNA content in the bacteroids (for example in pea and *M. sativa* nodules) than in the free-living bacteria (Bisseling *et al.* 1977). Although this work was not followed up and the results were not confirmed with the recent techniques, the parallelism between bacteroid and host cell development might lead to the postulation of a common developmental switch/signal for both the prokaryote and eukaryote cell types.

Endoreduplication in plants is extremely widespread and genetically programmed. Multiplication of the genome is proposed to increase metabolic activity, rRNA synthesis and transcriptional activity. Moreover, the cell size for a given cell type is generally proportional to the amount of nuclear DNA, therefore endoreduplication constitutes an effective strategy for cell growth as well (Kondorosi *et al.* 2000). In *Medicago*, the highest ploidy levels were measured in the nodules. Below we give a short overview on studies focused on the better understanding of the regulation and mechanism of endocycles and on the significance of endoreduplication cycles in nitrogen-fixing nodule development.

2. Procedure

Transformation of *M. truncatula* was made according to Trinh *et al.* (1998). Molecular techniques, *in situ* hybridization and flow cytometry analysis of nuclear DNA content were according to Cebolla *et al.* (1999).

3. Results and Discussion

3.1. Characterization of nodule cell cycle activity. To reveal cell cycle activity we have been studying the expression pattern of several cell cycle marker genes during nodule development by *in situ* hybridization, RT-PCR analysis and by construction of transgenic *M. truncatula* plants carrying cell cycle gene promoter -GUS fusions. We identified a set of genes that were induced during dedifferentiation/reactivation of cortical cells (Figure 1). They included *Medsa;cycA2* whose cell cycle function was described recently (Roudier *et al.* 2000), *cycD3-1* (F. Foucher, unpublished), two *E2F* transcription factors required for S phase entry and function (J. Györgyey, unpublished), the S-phase-specific histone H3 and *Medsa;cycB2* (Savoure *et al.* 1995). All of these genes, except *cycD3-1* that was repressed when cell proliferation started, exhibited a constitutive expression during nodule development. Although the transcripts were present, their spatial distribution displayed differences. For example, expression of *cycA2* was restricted to proliferating cells (F. Roudier, unpublished) while expression of the other genes was detected both in the proliferating cells (nodule primordium, meristem) as well as in nodule zone II. Expression of histone H3 exhibited a spotty pattern in zone II and marked S-phase cells undergoing DNA replication. In zone II, a novel variant of *cycD3* (II) was switched on that likely contributes to G1-S progression during endocycles (F. Foucher, unpublished). In the symbiotic, nitrogen-fixing nodule zone III, all these genes were repressed reflecting the inactivation of the cell cycle. In *M. sativa* and *M. truncatula* nodules, expression of these cell cycle genes in nodule zone II coincided with the progressive maturation of symbiotic cells along 10-12 cell layers where cells more distal to the meristem became larger and contained larger nuclei and more symbiosomes. The cell and nuclear volumes in the terminally differentiated cells of nodule zone III were constant. These data on the expression pattern of cell cycle genes allowed to show that cells undergo endoreduplication in nodule zone II and inter-zone II-III.

3.2. Conversion of mitotic cycles to endocycles. Recently by screening a young nodule cDNA library for genes involved in nodule organogenesis in *M. sativa*, we identified a cell cycle switch gene, *ccs52* that controls transition from mitotic cycles to differentiation programs and conversion of mitotic cycles to endocycles (Cebolla *et al.* 1999). This cell cycle switch gene exhibited highly stimulated expression in the nodules compared to roots and encoded a 52 kDa protein. The protein, containing seven WD40-repeats, several CDK phosphorylation sites and conserved oligopeptide motifs, exhibited homology to regulatory proteins involved in ubiquitin-dependent proteolysis of mitotic cyclines. The effect of *ccs52* on cell cycle was first studied in fission yeast. Overexpression of the *ccs52* gene resulted in growth arrest, elongation and enlargement of cells. These cells contained large polyploid nuclei measured by flow cytometry as a consequence of repeated rounds of endoreduplication cycles. We have shown that production of the CCS52 protein resulted in the degradation of the fission yeast mitotic cyclin, CDC13 and thereby inhibition of the M-phase has led to the conversion of mitotic cycle to endoreduplication cycles

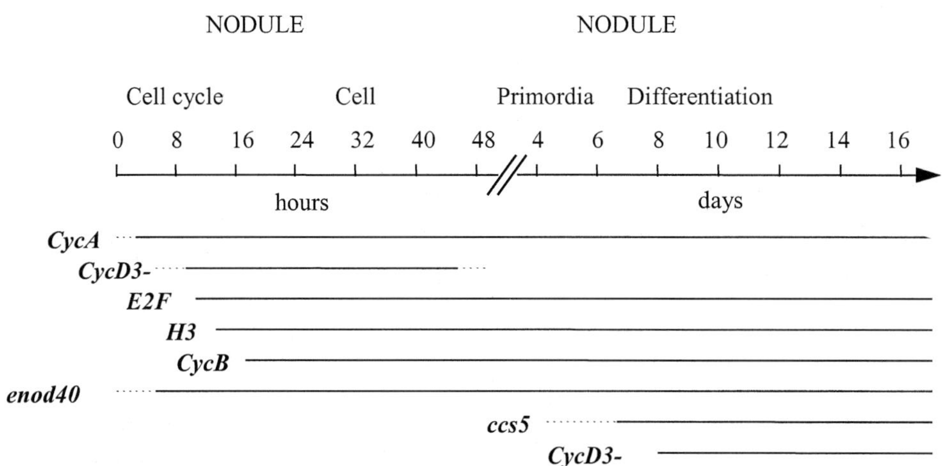

Figure 1. Expression of cell cycle genes and *enod40* during nodule development

In *Medicago* nodules, expression of *ccs52* was localized by *in situ* hybridization in zone I where cells exit from proliferation and in zone II which is the major site for endoreduplication and cell differentiation. RT-PCR analysis of the *ccs52* expression during nodule development revealed activation of the genes 4-5 days after *Rhizobium* inoculation. To study temporal and spatial expression patterns of *ccs52*, the genomic region was isolated and the promoter region was cloned in front of the *uidA* reporter gene (J. Vinardell *et al.*, unpublished). This construct was introduced in *M. truncatula* by *Agrobacterium tumefaciens*-mediated transformation and the *ccs52* promoter-driven GUS activity was tested in the transgenic roots as well as during different stages of nodule development. The GUS staining was absent in the growing primordia whereas gene expression was switched on prior to differentiation of primordia to the different the nodule zones (J. Vinardell *et al.*, unpublished). The *ccs52* promoter-driven GUS activity and expression pattern were consistent with the *in situ* hybridization experiments and with the predicted function of *ccs52* in mitosis arrest and endocycles.

M. truncatula displays a systemic endoploidy with the exception of leaves which are diploid. Northern analysis demonstrated that the *ccs52* transcripts were present also in other tissues. This indicated that the *ccs52* function might not be restricted to nodulation and could be involved in the development of other polyploid tissues and organs. To test this function in *M. truncatula*, transgenic plants were regenerated that carried the *ccs52* cDNA either in sense or antisense orientation downstream of the 35S promoter. By testing more than 30 transgenic plants in to independent transformations, no overexpression of the sense *ccs52* cDNA was found in the transgenic plants. This result indicates that the inhibitory effect of *ccs52* on cell proliferation likely interferes with callus formation and somatic embryogenesis. The antisense expression in three lines out of 75 transgenic plants resulted in slight downregulation of *ccs52*. The lack of knock-out phenotypes or significant reduction of *ccs52* transcript levels suggests that absence of gene function might be lethal and the level of gene expression might vary within a narrow window. However, even a 40% reduction in the mRNA level compared to the wild type was sufficient to reduce the degree of ploidy in the polyploid organs such as the petioles, roots and hypocotyls. Moreover, the reduced ploidy correlated with the formation of smaller cells demonstrating that expression level of *ccs52* affects directly both the degree of endoploidy and the size of the cell volume. Down-regulation of *ccs52* expression affected also nodule development. Though nodule primordia were formed with the same kinetics and number as in the control plants, later stages of nodule development halted. The ploidy level was significantly reduced, invasion of plant cells was less efficient and maturation of symbiotic cells were not completed that induced early senescence and cell death.

These results indicate that formation of highly polyploid cells is an integral part of nodule development. CCS52 appears to act as a key regulator of endocycles. Though its highest expression level correlates with the highest ploidy level in the nodule, *ccs52* is likely to have a general role in plant development, particularly in organs containing polyploid cells. CCS52, like its orthologs in yeast and animals, acts as substrate-specific activator of the Anaphase-Promoting Complex (APC), a cell cycle regulated ubiquitin ligase. These WD40-repeat proteins may target a great variety of proteins carrying D-box and KEN-box motifs in different stages of the cell cycle to APC for ubiquitination and then degradation by the 26S proteasome. Moreover, plants compared to other eukaryotes have evolved a multigenic family for these WD40-repeat proteins that likely regulate cell function and fate by fast and irreversible destruction of regulatory proteins.

4. References

Bisseling T *et al.* (1977) J. Gen. Microbiol. 101, 79-84
Cebolla A *et al.* (1999) EMBO J. 18, 4476-4484
Kondorosi E *et al.* (2000) Curr. Opin. Plant Biol. 3, 488-492
Roudier F *et al.* (2000) Plant J. 23, 73-83
Savouré A *et al.* (1995) Plant Mol. Biol. 27, 1059-1070
Trinh TH *et al.* (1998) Plant Cell Rep. 17, 345-355
Truchet G (1978) Ann. Sci. Nat. Bot. Paris 19, 3-38

IDENTIFICATION OF *TRANS*-ACTING FACTORS REGULATING NODULE DEVELOPMENT

A.C. Hansen, C. Johansson, H. Busk, A. Jepsen, D. Jensen, L. Fræmohs, E.O. Jensen

Laboratory of Gene Expression, Department of Molecular and Structural Biology, University of Aarhus, Denmark

1. Introduction

Functional studies of nodulin gene promoters in transgenic legumes have identified a number of *cis*-regulatory elements important for nodule specific expression.   One example is the pea *enod12* promoter which was investigated in details in transgenic *Vicia hirsuta* plants (Figure 1).   By analyzing 50 promoter deletions and hybrid promoters fused to the GUS reporter gene a tissue specific element was identified in the *Psenod12* promoter (Vijn *et al.* 1995; Christiansen *et al.* 1996; Hansen *et al.* 1999).

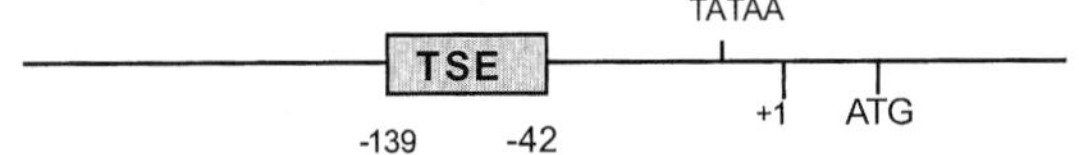

Figure 1. Schematic representation of the *cis*-regulatory element in the pea *enod12* promoter. TSE: tissue specific element

2. Identification of a putative transcription factor, ENBP1

Rather than searching for a single protein interacting with a specific sequence in the promoter, e.g. the TSE element, we decided to go for all proteins interacting with the proximal promoter region, since all the *cis*-regulatory elements required for nodule specific expression are located on this part. A set of 10 double-stranded overlapping oligonucleotides were synthesized covering the promoter region from -200 to +1 (Figure 2).   Each of the oligonucleotides were then concatenated before they were used as probes.

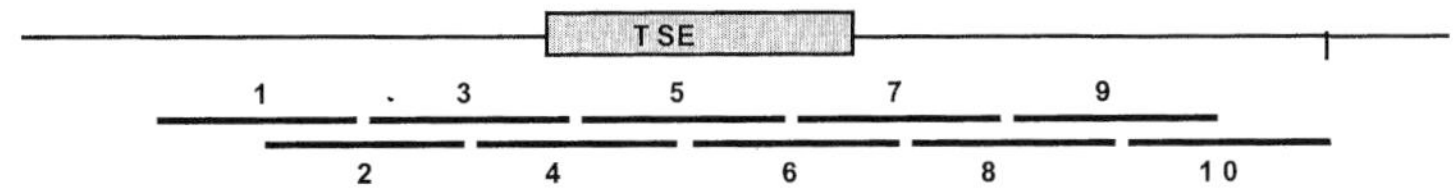

Figure 2. Positions of the oligonucleotides used as probes for the South-Western screening

Using a cocktail of all ten oligonucleotides 200,000 pfu were screened from an un-amplified *Vicia hirsuta* nodule λgt11 cDNA library.   The screening yielded 4 positive clones and the corresponding phages were purified. The clones were hybridized to each other and it turned out that 2 of the clones represented the same mRNA. The two clones encoded a putative transcription factor ENBP1 (Early Nodulin Binding Protein 1) that binds to the *Psenod12B* promoter *in vitro* (Figure 3). A bacterially expressed fusion protein, ATh, covering amino acids 188 to 728 of VsENBP1, bound to polypeptide includes the six AT-hooks present in VsENBP1 and constitutes the sequence-specific DNA binding domain of double-stranded oligonucleotides with VsENBP1.   DNaseI sequences, corresponding to *Psenod12B* promoter regions -45 to -139 and -160 to -206. The ATh footprinting revealed interactions at the sequences AATAA and TTATT present at position -78 to -82 and -92 to -96, respectively, in the *Psenod12B* promoter. A second domain in ENBP1 is a cysteine-rich region that binds to the *ENOD12* promoter in a sequence non-specific but metal dependent way.

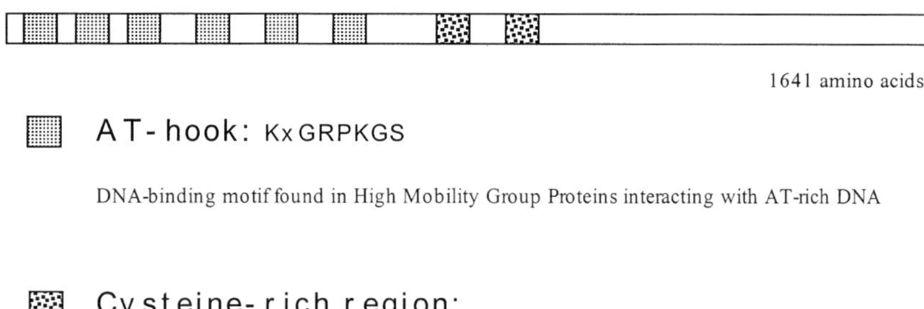

1641 amino acids

AT-hook: KxGRPKGS

DNA-binding motif found in High Mobility Group Proteins interacting with AT-rich DNA

Cysteine-rich region:

RNSRCHQCWKKSRTGLVVCSKCKKKKYCYECVAKWYQDKTREEIETACPFCLDYC
NCRMCLKKDDDRVYCDNCNTSIVNFHRSCSNPNCEYDLCLTCCTELRLGVH

The cysteine regions might form a DNA-binding zinc-finger found in many transcription factors

Figure 3. Schematic representation of ENBP1 from *Vicia hirsuta*

A region 139 bp immediately upstream of the transcription start site of *Psenod12B* is able to direct nodule-specific expression in transgenic *V. hirsuta* roots. Promoter fragments with introduced mutations in the AATAA-78 to -82 or TTATT-92 to -96 sequences either did not bind or had reduced binding capacity to the ATh polypeptide *in vitro*. Furthermore, promoters mutated in the AATAA-78 to -82 are inactive in transgenic vetch roots. Overexpression of *Vsenbp1* in transgenic vetch roots resulted in tissue-specific regulation of a leaky *Psenod12B* promoter, overriding the effect exerted by a *CaMV35S* promoter on *Psenod12B* promotergusA-int expression. The results presented here, therefore, indicate that VsENBP1 might be a *trans*-acting factor that through interaction with the AATAA-78 to -82 sequence regulates the tissue-specific expression of *Psenod12B*.

3. References

Christiansen H *et al.* (1996) Plant Mol. Biol. 32, 809-821
Hansen AC *et al.* (1999) Plant Mol. Biol. 40, 495-506
Vijn I *et al.* (1995) Plant Mol. Biol. 28, 1103-1110

RESPONSES OF *LOTUS JAPONICUS* TO NOD FACTORS

H.P. Spaink[1], C.P. Bras[1], M. Grønlund[1], P. van Spronsen[1], J. Kijne[1], H.R.M. Schlaman[1], A. Roussis[1], S. Wijting[1], J. Stougaard[2], E. Flemetakis[3], P. Katinakis[3]

[1]Leiden University, Institute of Molecular Plant Sciences, Wassenaarseweg 64, 2333 AL Leiden, The Netherlands
[2]Aarhus University, Laboratory of Gene Expression, Gustav WiedsVej 10, DK-8000 Aarhus C, Denmark
[3]Agricultural University of Athens, Department of Agricultural Biotechnology, Lera Odos 75,118 55 Athens, Greece

1. Introduction

Lotus japonicus has been shown to be a very suitable model system to study nodulation and nitrogen fixation (Handberg, Stougaard 1992). We are currently using this model system to study in detail the infection and nodulation process induced by *Mesorhizobium loti* (Pacios Bras *et al.* 2000). We have shown recently that *L. japonicus* responds to the addition of Nod factors by the formation of cytoplasmic bridges in the outer cortex (van Spronsen *et al.* 2001). In this respect it is identical in responsiveness to Nod factors as *Vicia sativa* and *Pisum sativum* (van Brussel *et al.* 1992). This response, which underlies the formation of the infection thread, is also dependent on the presence of ethylene, again similar to the situation in *V. sativa* and *P. sativum*. However, whereas the latter legumes form indeterminate nodules (of the *Galegae* form: in the inner cortex), *L. japonicus* forms determinate nodules. Interestingly, *Phaseolus vulgaris*, which also forms determinate nodules, does not respond to Nod factor by the formation of cytoplasmic bridges (van Spronsen *et al.* 2001). In conclusion, *L. japonicus* presents a sort of intermediate mechanism of nodulation in between that found in the *P. vulgaris* nodulation, where cells just below the epidermis form the primordia for nodulation, and the *Galegae* nodulation, where cells in the inner cortex form the nodule primordia. As a consequence, *L. japonicus* is a very useful model system to study the formation of cytoplasmic bridge since they are formed close to the outer surface of the root and therefore are more easily accessible for microscopic studies. Cytoplasmic bridge formation is interesting to study because it represents an early response to Nod factors which in many ways is similar to the responses of root hairs, e.g. both represent changes in cell polarity based on reorganization of the cytoskeleton.

2. Construction of Reporter Lines

In order to study the mechanism underlying cytoplasmic bridge formation and root nodulation, a number of marker lines of *L. japonicus* have been constructed. For the analysis of promoter activity it is often very useful to make use of a fusion construct consisting of the green fluorescent protein (GFP) and β-glucuronidase (GUS) genes (Quaedvlieg *et al.* 1998). Using this construct we have obtained various transgenic *L. japonicus* lines which will be used for analysis of responses to Nod factors. These promoters are either known promoters that were selected based on their known responsiveness to Nod factors or plant hormones. We have fused the *GH3* promoter to the *GUS-GFP* reporter construct in order to analyze the effect of Nod factors on auxin transport in *L. japonicus*. We have also fused the promoter of *ENOD40* of soybean and of *L. japonicus* (*ENOD40-1*: Flemetakis *et al.* 2000) to *GUS-GFP*. In addition to external addition of Nod factors and plant hormones we also use microtargeting following the strategy of Mathesius *et al.* (1998).

For the analysis of GFP various microscopic imaging techniques have been employed in our laboratory (Spaink *et al.* 2000). The most commonly used technique is confocal laser scanning microscopy (CLSM) which enables the four-dimensional (i.e. in space and time) visualization of GFP in living tissues. However, a pitfall that is often encountered is the high level of autofluorescence in various tissues (especially in leguminous plants). This problem can in several cases be solved with using spectral-resolved CLSM (e.g. using a Leica TCS SP system). This system makes a quantitative analysis of GFP in the case of the *GH3* promoter possible. However, for weak promoter activity, such as in the case of the *ENOD40* promoter, GFP fluorescence cannot be detected above the autofluorescence background using existing methods.

3. Responses of Reporter Lines

On basis of the preliminary results obtained with the above mentioned reporter lines we can conclude that *L. japonicus* responses to Nod factors display various differences with other legumes tested previously. For instance, the response of *GH3* promoter activity in the vascular bundle is different from the results obtained in the clover and pea system (Mathesius *et al.* 1998; Boot *et al.* 1999). Whereas clover and pea respond to the local application of Nod factor with a local decrease of *GH3* promoter activity in the vascular bundle, such a decrease was not observed in *L. japonicus*. The results indicate that in *L. japonicus* Nod factors do not redirect auxin to the nodule primordia by influencing auxin flow in the vascular tissue. Rather the results indicate that Nod factors might induce the local production of auxin in the nodule primordia. Also in the response of the *ENOD40* promoters clear differences are found when compared with the results for other plants, such as *Medicago sativa* (Fang *et al.* 1998). For instance, a very poor response could be detected towards cytokinin in transgenic *L. japonicus* plants containing the soybean or *L. japonicus* promoter fused to *GUS-GFP*. These lines responded specifically to *M. loti* Nod factors at nanomolar concentrations in the outer cortex and the root hairs. Interestingly, transgenic lines that contain multiple integrations of the soybean *ENOD40* reporter construct showed no constitutive GUS expression in the vascular tissue as was detected in the single integration lines. In none of the tested lines a response to the compound mastoporan was detected. In contrast, mastoporan is a strong inducer of the *ENOD12* promoter in *M. sativa* (Pingret *et al.* 1998). We are currently involved in a pharmacological approach to test the effect of many other compounds on the expression of the *ENOD40* promoter. For this purpose, the application of microtargeting techniques for direct introduction of the tested compounds in the plant tissue will be very valuable (Sautter *et al.* 1991).

4. References

Boot KJM *et al.* (1999) Mol. Plant-Microbe Interact. 12, 839-844

Fang Y *et al.* (1998) Plant Physiol. 116, 53-68

Flemetakis EN *et al.* (2000) Mol. Plant-Microbe Interact. 13, 987-994

Mathesius U *et al.* (1998) Plant J. 14, 23-34

Pacios Bras C *et al.* (2000) Mol. Plant Microbe Interact. 13, 475-479

Pingret JL *et al.* (1998) Plant Cell 10, 659-672

Quaedvlieg NEM *et al.* (1998) Plant Mol. Biol. 38, 861-873

Sautter C *et al.* (1991) Biotech. 9, 1080-1085

Spaink HP *et al.* (2000) In FO Pedrosa, M Hungria, MG Yates, WE Newton (ed) Nitrogen Fixation: From Molecules to Crop Productivity, pp. 219-222, Kluwer Acad. Pub, Dordrecht

van Brussel AAN *et al.* (1992) Science 257, 70-72

van Spronsen PC *et al.* (2001) Mol. Plant-Microbe Interact. 14, 839-847

5. Acknowledgements

This work was supported by the following network grants from the European community: ERB-FMR-XCT-960039, FMRX-CT98-0243 and HPRN-CT-2000-00086 (salaries of M. Grønlund and A. Roussis). We thank Drs D. Werner, T. Bisseling and M. Udvardi for coordination of these EU networks. This work was also supported by the Netherlands Organization of Scientific Research (NWO) (salary of H.R.M. Schlaman).

Section 6:
Signals
in the Soil

CHAIR'S COMMENTS: SIGNALS IN THE SOIL

L. Barran

Agriculture Canada, Soils and Crops Development Centre, Ste-Foy, QC, Canada

Microorganisms adapt to their environment by sensing signals and making necessary adjustments in their cellular processes. Much still remains to be learned about the nature of the signals, how they are sensed and transduced. The four presentations in this session dealt with chemical sensing in *Azospirillum brasilense*, rhizobial catabolic and chemoreceptor genes, biotin-dependent rhizobial carboxylases and the role of phosphate in the nodulation of actinorhizal plants.

Motile bacteria are able to detect and show taxis to agents involved in energy generation such as light, oxygen and oxidizable substrates. Energy taxis includes aerotaxis, phototaxis, redox taxis, taxis to alternative electron acceptors, and chemotaxis to oxidizable substrates (for review see Taylor *et al.* 1999). Bacteria utilize energy taxis to seek out environments most favorable for maintaining optimal cellular energy levels. *Azospirillum brasilense* colonizes the rhizosphere of agronomically important cereals and grasses, the roots of which exude significant amounts of organic acids, sugars and amino acids. Motility and chemotaxis are important factors for the bacterial colonization of the plant roots. *A. brasilense* shows a strong aerotaxis response which guides the bacteria to a low oxygen concentration that is optimal for energy generation and nitrogen fixation (Zhulin *et al.* 1996). The bacteria preferentially seek organic acids and sugars as carbon and energy sources thereby permitting the cells to achieve and maintain optimal energy levels in the plant rhizosphere. Zhulin and coworkers showed that energy taxis is the dominant behavior in *A. brasilense* with most chemoeffectors being processed by this method and that changes in the electron transport system govern most behavioral responses (Alexandre *et al.* 2000). The question as to whether the redox state of the electron transport system or an ion motive is the signal for chemotaxis remains to be established

Rhizobium and *Sinorhizobium* spp. contain plasmids on which symbiotic genes are located. These bacteria also contain a variable number of cryptic plasmids which can influence nitrogen fixation and nodulation as well as bacterial growth and survival. There is a need for data on the nature of the plasmid-encoded genes and their mechanism of action within the context of soil environment. The group of Michael Hynes has examined the function of many plasmid encoded genes of *S. meliloti* and *R. leguminosarum* with a view to elucidating their role in bacterial growth and survival as well as their impact on competition for nodulation. They have demonstrated that carbon catabolism genes are present on several plasmids of *R. leguminosarum* VF39. At least one of these genes, the malate synthase G gene, is required for the symbiotic nitrogen fixation of peas. Also of interest, was the finding of a number of genes that encode putative methyl-accepting chemotaxis proteins (MCPs) (Yost *et al.* 1998). Whether all of the MCP-like sequences encode genuine MCP proteins is not known. Mutation of *mcp*B resulted in impairment of chemotaxis towards a wide range of substrates. The *mcp*B and *mcp*C mutants were less competitive for the nodulation of peas than the wild type, while the *mcp*D mutant showed no decrease in competitiveness. The MCP-like receptors in *S. meliloti* differ from those of *R. leguminosarum* perhaps reflecting the observed differences in the chemotactic behavior of the two strains.

Biotin plays an important role in the growth and metabolism of diazotrophic bacteria via its participation in CO_2 fixation reactions. The Cuernavaca group headed by Michel Dunn has been studying the properties and metabolic role of biotin-dependent carboxylases (BDCs). Pyruvate carboxylase (PYC) catalyzes the formation of oxalacetate from pyruvate. This enzyme was shown to be important in determining growth defects of an *R. etli* poly-β-hydroxybutyrate synthetase mutant, however, the mechanism controlling the interaction of these enzyme activities remains to be elucidated (Cevallos *et al.* 1996; Dunn *et al.* 1996). A biotin independent phosphoenolpyruvate carboxylase (PCC) catalyzes the conversion of phosphoenolpyruvate to oxaloacetate. Dunn and colleagues have shown that some diazotrophs contain either PPC or PYC, while others have both enzymes, but it remains to be established what practical advantage accrues to the bacterium having a particular menu of these carboxylases. The Cuernavaca group has also characterized a propionyl CoA carboxylase (PCC) from *R. etli*. Future studies will be directed at integrating the roles of the BDCs in rhizobial metabolism and assessing the importance of these enzyme within the context of the bacterial environment.

The final presentation of the session related to the role of phosphate in the nodulation of actinorhizal plants by Huss-Danell's group. They showed that phosphate stimulated nodulation of *Alnus* by *Frankia* but that nitrogen inhibited both nodulation and nitrogen fixation (Ekblad, Huss-Danell 1995). Stimulation of nodulation by phosphate (1 mM) occurred even in the presence of relatively high nitrogen concentrations (7 mM N) and the stimulatory effect of phosphate was not due to a general stimulation of plant growth. High phosphate concentrations also stimulated nodule mass of clover roots suggesting a more generalized role for phosphate. Experiments using the split-root technique for *Alnus* and *Hippohaë* indicated that the effects of N and P on nodulation were systemic. The mechanism whereby P stimulates nodulation is unknown. There are a number of questions that need to be investigated, these include whether there is a requirement for P to be present during the entire nodulation phase. This can be approached by pulsing in P for different time periods. Other questions relate to the role of other ions. With regard to the stimulation of nodulation, to what extent if any can phosphate be replaced by other ions?

References
Alexandre G *et al.* (2000) J. Bacteriol. 182, 6042-6048
Cevallos M *et al.* (1996) J. Bacteriol. 178, 1646-1654
Dunn *et al.* (1996) J. Bacteriol. 178, 5960-5970
Ekblad A, Huss-Danell K (1995) New Phytol. 131, 453-459
Taylor BL *et al.* (1999) Annu. Rev. Microbiol. 53, 103-128
Yost CK *et al.* (1996) Microbiol. 144, 1945-1956
Zhulin *et al.* (1996) J. Bacteriol. 178, 5199-5202

CATABOLIC AND CHEMORECEPTOR GENES AND THEIR ROLE IN RHIZOBIAL ECOLOGY

M.F. Hynes[1], C.K. Yost[1], I.J. Oresnik[2], A. Garcia de los Santos[3], S.R.D. Clark[1], J.E. Macintosh[1]

[1]Dept. of Biological Sciences, University of Calgary, 2500 University Dr. NW, Calgary AB, T2N 1A9 Canada
[2]Dept. of Microbiology, University of Manitoba, Winnipeg MB, R3T 2N2 Canada
[3]CIFN, UNAM, Cuernavaca, Morelos, Mexico

1. Introduction

Although much effort has been devoted to the study of the interaction of rhizobia with their legume host plants, there are still many things about the life of rhizobia in the soil, and about the early interactions of the bacteria with the plant, which remain unclear. Specifically, we have little information about what carbon, nitrogen and other nutrient sources are important to the survival and proliferation of the bacteria during free-living growth, and during early stages of legume infection, and we also do not understand the role of motility and chemotaxis in the infection process and in the ability of the bacteria to thrive in legume and non-legume rhizospheres.

2. Plasmid-Encoded Catabolic Genes

Up to 40% of the genome of *Rhizobium* and *Sinorhizobium* species can consist of plasmids. The number and size of plasmids varies significantly from one strain to another, so we predicted that plasmid encoded properties would have a significant effect on interstrain differences in behavior. For this reason we have had a long-term interest in determining the function of cryptic plasmids in the rhizobia, and mapping plasmid encoded genes and investigating their importance in symbiosis and in inter-strain competition. The approach taken was to generate plasmid-cured derivatives using a Tn5 derivative carrying the *sacB* gene which is lethal to gram-negative bacteria in the presence of sucrose (Hynes *et al.* 1989). This strategy has allowed the isolation of plasmid-cured derivatives of strains of *R. leguminosarum* biovars *viciae* and *trifolii* (Hynes *et al.* 1989; Hynes, McGregor 1990; Baldani *et al.* 1992) and *Sinorhizobium meliloti* (Hynes *et al.* 1989; Oresnik *et al.* 2000). Examination of the properties of cured strains has resulted in the identification of a number of plasmid encoded phenotypes, the majority of which appear to involve catabolic genes. Initially, we screened for loss of catabolic activity in cured strains by growing the bacteria in minimal media supplemented with a variety of readily available and inexpensive carbon sources, but more recently, we have adopted more rapid screens based on BioLog[TM] plates, which simultaneously screen, through a colorimetric test, for use of a large number of carbon, nitrogen, sulfur and phosphorus sources. This latter approach has resulted in identification of a number of novel plasmid encoded catabolic phenotypes, some of which are summarized in the tables below. It should be mentioned that in some cases confirmation of a phenotype in a BioLog test is rendered difficult by the fact that the wild-type (wt) strain does not grow well enough on the C or N source as sole source in minimal media to distinguish it from the cured strain. Only results that seemed very clear, and which in many cases have been confirmed, are listed in the tables. Some of these results, using only BioLog GN plates, have previously been reported (Oresnik *et al.* 2000).

Identification of plasmid encoded catabolic phenotypes associated with plasmids is only one step in determining the importance of plasmids in soil ecology. Ideally, genes should be cloned, characterized, and mutated, and the mutants compared to wt for survival and proliferation in various environments as well as for competition for nodulation. We have initiated such experiments for a number of carbon catabolic genes, and are continuing to do so for new genes isolated from

R. leguminosarum and *S. meliloti* (Tables 1 and 2). We have already established a role for rhamnose catabolism and uptake genes in competition for clover nodulation (Oresnik *et al.* 1998) while at the same time showing that sorbitol and adonitol catabolism genes appear to be unimportant.

Table 1. Putative catabolic genes on plasmids in *R. leguminosarum* strain VF39.

Plasmid	Approximate size	Carbon catabolism genes on plasmid
pRleVF39b	220 kb	gluconate, malonate, glucuronate
pRleVF39c	300 kb	glycerol, adonitol*, melibiose
pRleVF39d	350 kb	alanine, adonitol*, trigonelline, hydroxyproline
pRleVF39e	500 kb	rhamnose, sorbitol, histidine, serine
pRleVF39f	600 kb	erythritol, ornithine, citrate, proline

* Only strains cured of both pRleVF39c and pRleVF39d are unable to grow on adonitol, suggesting there are genes for its use on both plasmids

Table 2. C and N utilization genes on plasmid pRme1021a (pSymA megaplasmid of *S. meliloti*), determined by comparison of wt with cured strain on BioLog GN and PM1, PM2 and PM3 plates.

Carbon utilization genes:	lyxose, psicose, acetate, acetoacetate, propionate, gluconate, arabinose, arbutin, sorbose, allose, lactate, xylitol, lactate, serine, alanine, gamma aminobutyrate, arginine, leucine, valine
Nitrogen utilization genes:	adenosine, xanthine, cytidine, guanine, thymidine alanine, serine, valine, arginine, ammonium, various dipeptides

An alternative approach to looking for catabolic genes important in the soil and rhizosphere is to look for genes induced by root exudates, or by specific compounds known to be abundant in the rhizosphere and plant root. We have generated several banks of mutants made with different promoter probe transposons in order to screen for inducible genes. One such mutant library was made with Tn5B22 (Simon *et al.* 1989), which generates *lacZ* fusions. This library was examined by plating on various media with X-gal, for transposons located in genes, which were induced by a variety of substrates, including arabinose. One arabinose inducible fusion turned out to be located in the malate synthase gene of *R. leguminosarum* VF39. This gene is homologous to the *E. coli* malate synthase G gene, and in addition to being strongly induced by arabinose, is inducible by glycolate and glyoxylate. It appears to be required for symbiotic nitrogen fixation on peas.

3. Chemotaxis Genes

It is logical to assume that rhizobia, in addition to being able to catabolize a wide variety of nutrient sources, would benefit from the ability to detect and move towards various nutrients, as well as toward plant roots. Several reports have indicated that chemotactic ability and/or motility confers a competitive advantage on rhizobia (Ames, Bergman 1981; Bauer, Caetano-Anolles 1990), and there is also evidence that rhizobia are chemotactic towards the same compounds which induce *nod* gene expression (Dharmatilake, Bauer 1992). Because rhizobia are able to use such a variety of different carbon and nitrogen sources, we set out to look for genes homologous to MCP type chemoreceptors, with the aim of finding receptors specific for compounds whose catabolism was plasmid-encoded. A family of at least 17 genes, and probably over 20, exhibiting homology to MCP genes was detected in *R. leguminosarum* VF39 (Yost *et al.* 1998). The complete sequences of three genes, *mcpB*, *mcpC* and *mcpD*, were reported by Yost *et al.* (1998) as was the fact that *mcpB* was required for a generalized chemotactic response, and that both *mcpC* (which is carried on plasmid pRleVF39f) and *mcpB* were important for competitive nodulation. Further studies were carried out on the *mcpG* gene, which is located on pRleVF39d, the pSym of VF39. The *mcpG* gene encodes a typical MCP; it is flanked by mono-oxygenase genes and is located next to an alanine transporter. Mutants in *mcpG* exhibit no obvious chemotactic defects. Homologs of *mcpG* are plasmid encoded in all strains of both *R. leguminosarum* and *R. etli* that we have examined, but are absent in *S. meliloti* and *R. tropici*. Gene fusions have been constructed to a number of our MCP genes and have been used to show that there is no obvious regulation during the free-living growth of VF39, but that MCP genes are down-regulated during nodulation.

While it is as yet unclear that all of the MCP gene homologs present in VF39 code for genuine MCPs, the four which we have sequenced completely, and the other three for which extensive sequencing has been done all seem to code for canonical MCPs possessing methylation sites and transmembrane domains. However, a search of the *S. meliloti* 1021 genome sequence reveals only nine genes potentially coding for MCP-like receptors, and none of these is particularly similar to any of the VF39 MCPs. This suggests that there may be important differences between the two genera in the way environmental signals are sensed. Further evidence for this comes from our observation that *R. leguminosarum* VF39 exhibits swimming patterns (i.e. tumbling) similar to those of *E. coli*, whereas *S. meliloti* has unidirectional flagella and does not tumble (Greck *et al.* 1995). We are interested in further elucidating these differences, and have isolated a number of chemotactic and motility mutants of VF39 which are currently being characterized.

4. References

Ames P, Bergman K (1981) J. Bacteriol. 148, 728-729

Baldani JI *et al.* (1992) Appl. Environ. Microbiol. 58, 2308-2314

Bauer WD, Caetano-Anolles (1990) Plant Soil 129, 45-52

Dharmatilake AJ, Bauer WD (1992) Appl. Environ. Microbiol. 58, 1153-1158

Greck M *et al.* (1995) Molec. Microbiol. 15, 989-1000

Hynes MF *et al.* (1989) Gene 78, 111-120

Hynes MF, McGregor NF (1990) Molec. Microbiol. 4, 567-574

Oresnik IJ *et al.* (1998) Molec. Plant-Microbe Interact. 11, 1175-1185

Oresnik IJ *et al.* (2000) J. Bacteriol. 182, 3582-3586

Simon R *et al.* (1989) Gene 80, 161-169

Yost CK *et al.* (1998) Microbiol. 144, 1945-1956

5. Acknowledgements

We thank K. Aarbo and C. Southward, two undergraduate students who contributed to some of the unpublished results reported here. The financial support of an NSERC operating grant to MFH is gratefully acknowledged.

CHARACTERISTICS AND METABOLIC ROLES OF BIOTIN-DEPENDENT CARBOXYLASES IN RHIZOBIA

M.F. Dunn[1], G. Araíza[1], S. Encarnación[1], T.M. Finan[2], J. Mora[1]

[1]Centro de Investigación sobre Fijación de Nitrógeno-UNAM, Cuernavaca, Morelos, Mexico
[2]McMaster University, Hamilton, ON, Canada L8S 4K1

1. Introduction

The rhizobia are gram-negative bacteria of the genera *Rhizobium*, *Bradyrhizobium*, *Mesorhizobium*, *Sinorhizobium* and *Azorhizobium* and are able to fix atmospheric nitrogen in symbiotic association with a compatible plant host. Carbon dioxide is essential for the growth of rhizobia (Lowe, Evans 1962), and the fact that biotin (vitamin H) participates in many CO_2 fixing reactions explains why this vitamin has such a profound effect on the growth and metabolism of these organisms (West, Wilson 1940; Allen, Allen 1950; Encarnación *et al.* 1995; Streit, Phillips 1996).

Biotin performs an essential metabolic role in all organisms as the prosthetic group of the biotin-dependent carboxylases (BDCs), in which it functions in the activation and transfer of CO_2. The generalized two-step reaction catalyzed by BDCs is:

$$\text{(i)} \qquad \text{ENZYME-biotin} + HCO_3^- + \text{Mg-ATP} \rightarrow \text{ENZYME-biotin-}CO_2^- + ADP + P_i + H^+$$
$$\text{(ii)} \qquad \text{ENZYME-biotin-}CO_2^- + \text{acceptor-H} \rightarrow \text{ENZYME-biotin} + \text{acceptor-}CO_2^-$$

BDCs contain three functional regions which participate in the carboxylation reaction. These functional regions are designated the biotin carboxylase (BC), biotin carboxyl carrier protein (BCCP) and carboxyltransferase (CT) domains or subunits. In some BDCs all of these functional regions are present as domains on a single protein, while in others they are distributed between two or more subunits of the enzyme.

The biotin prosthetic group is attached by the enzyme biotin protein ligase to a specific lysine residue in the BCCP. In the first-half reaction shown above, the BC region of the enzyme catalyzes the ATP-dependent carboxylation of the biotin attached to the BCCP. In the second step, the activated CO_2 is transferred from the biotin to the acceptor substrate in a reaction catalyzed by the CT region.

2. Materials and Methods

Complex and minimal media and culture growth conditions were as described previously (Dunn *et al.* 1996). Radiochemical assays to measure CO_2 fixation from $NaH^{14}CO_3$, and detection of BDCs on protein blots with streptavidin-horseradish peroxidase conjugate were performed as described previously (Dunn *et al.* 1996; M. Dunn, G. Araíza, T. Finan, submitted). Genome sequences putatively encoding BDCs in *Mesorhizobium loti* (Kaneko *et al.* 2000; http://www.kazusa.or.jp/rhizobase) were located by a keyword search followed by an open reading frame and BLASTP analyses of the sequences flanking the putative BDC. BDCs in the *Sinorhizobium meliloti* (http://sequence.toulouse.inra.fr/meliloti.html) genome were identified similarly, except that the identity of genes surrounding the BDCs were obtained using the Map function available at the site. Deduced genomic protein sequences of both genomes were also searched by BLASTP using various BDC sequences from other organisms as a query.

3. Results and Discussion

3.1. Biochemical and genetic studies on BDCs in rhizobia. Biotin-containing proteins are visualized with high specificity on protein blots developed with streptavidin-horseradish peroxidase conjugate. When protein extracts from various *Rhizobium*, *Sinorhizobium* and *Mesorhizobium* strains were analyzed by this method, three biotin-containing proteins with molecular masses of approximately 120, 78 and 14 kDa were detected (Araíza, Dunn 1998; Dunn *et al.* 1996, 2000). The 120 kDa protein is the subunit of the pyruvate carboxylase (PYC) in *Rhizobium etli*, *Rhizobium tropici* (Dunn *et al.* 1996), *Rhizobium leguminosarum* biovars viciae and trifolii (M. Dunn, unpublished) and *S. meliloti* (M. Dunn, G. Araíza, T. Finan, submitted). *pyc*::Tn*5* mutants of these rhizobia are unable to grow with pyruvate or sugars as sole carbon source but are symbiotically effective with their respective legume hosts (Arwas *et al.* 1986; Dunn *et al.* 1996; M. Dunn, G. Araíza, T. Finan, submitted; Ronson, Primrose 1979). PYC activity is present in *M. loti*, but not in *Bradyrhizobium* spp. and *Azorhizobium caulinodans* (M. Dunn, unpublished). The PYCs from *R. etli* CE3 and *S. meliloti* Rm1021 have been studied in the most detail and their characteristics are as follows:

(i)   Both PYCs function in the anaplerotic production of oxaloacetate from pyruvate, which is necessary for growth on pyruvate or sugars but not for symbiotic nitrogen fixation.

(ii)   Both the *R. etli* and *S. meliloti* PYCs are encoded by chromosomal genes whose deduced products (1154 and 1152 amino acids, respectively) are 87% identical and contain typical BCCP, BC and CT domains. The holo-PYCs produced by both species are homotetramers with molecular masses of about 500 kDa.

(iii)   PYC activity in both species is influenced primarily by the availability of biotin in the growth medium and to a lesser extent by the carbon source used for growth. Gene fusion studies have shown that *pyc* transcription is not significantly influenced under different growth conditions.

(iv)   The *R. etli* PYC exhibits maximum activity *in vitro* with a much lower concentration of the allosteric activator acetyl CoA than that required by the *S. meliloti* enzyme. Like all previously characterized prokaryotic homotetrameric PYCs, the *R. etli* enzyme is inhibited by the allosteric effector L-aspartate. The PYC from *S. meliloti* provides a surprising exception to this rule since its activity is unaffected by L-aspartate *in vitro*.

We have shown that PYC plays a major role in determining the growth defects of an *R. etli* poly-β-hydroxybutyrate (PHB) synthase (*phbC*) mutant. Because the growth defects of the *phbC* mutant are identical to those of the *R. etli pyc* mutant (Cevallos *et al.* 1996; Dunn *et al.* 1996), we measured PYC activity and protein levels in the *phbC* mutant and found that they were significantly lower than in the wild-type strain. Interestingly, the inactivation of a second gene, called *cfr* (carbon flux regulator) in the *phbC* mutant restores PYC activity and growth (Dunn *et al.* 2000; Dunn, Araíza, Encarnación, Vargas, Mora, submitted; Encarnación, Vargas, Dunn, Mendoza, Mora, Mora, submitted). We are currently working to unravel the molecular mechanism by which PYC activity is regulated in the *phbC* and *phbC cfr* mutants.

Phosphoenolpyruvate carboxylase (PPC) performs the same anaplerotic role as PYC by catalyzing the non-biotin-dependent conversion of phosphoenolpyruvate to oxaloacetate. Some bacteria contain solely PYC or PPC as an anaplerotic enzyme, while others contain both (Scrutton 1978). It is interesting to note that *Bradyrhizobium* species and *Azorhizobium caulinodans*, which are thought to have evolved before the other three genera (Martínez-Romero, Caballero-Mellado 1996), lack PYC and contain only PPC, while the opposite is true of *Rhizobium*, *Sinorhizobium* and *Mesorhizobium* species (Araíza, Dunn 1998; M. Dunn, G. Araíza, T. Finan, submitted).

The activity of propionyl CoA carboxylase (PCC) has been detected in extracts prepared from free-living cells and bacteroids of many rhizobia (DeHertogh *et al.* 1964; Dunn *et al.* 2000). Biochemical analysis of *R. etli* CE3 has shown that the 78 kDa biotinylated protein detected on western blots corresponds to the α subunit of PCC. By analogy to PCC α subunits characterized in other organisms, this protein would contain the BCCP and BC domains of the enzyme. The *R. etli*

PCC also contains a 47 kDa, non-biotinylated PCC β subunit, which would contain the CT domain. The partially purified PCC from *R. etli* showed a distinct kinetic preference for propionyl CoA as a substrate, but will also carboxylate acetyl- and butyryl-CoAs with a much lower efficiency (Dunn *et al.* 2000). The selection of *R. etli* PCC mutants has so far been unsuccessful, but an *S. meliloti* PCC mutant (*pccA*::Tn*5*) (Charles, Aneja 1999) is devoid of PCC activity and unable to utilize propionate as a carbon source (M. Dunn and G. Araíza, unpublished). Charles and Aneja (1999) noted the presence of a gene encoding methylmalonyl CoA mutase (*bhbA*) adjacent to *pccA* and suggested that Pcc and BhbA participate in the final steps of the propionate degradation. This is supported by earlier biochemical evidence indicating an anaplerotic role for these enzymes in the production of succinyl CoA from propionate in rhizobia (DeHertogh *et al.* 1964).

The 14 kDa biotinylated protein produced by rhizobia is similar in size to the BCCP subunits of many bacterial acetyl CoA carboxylases (ACCs). We have not succeeded in detecting ACC activity in *R. etli* preparations enriched in this protein, perhaps because the ACC complex is highly unstable (Fall 1976).

3.2. Analysis of rhizobial genome sequences for BDC-encoding genes. Computer analysis of the complete genome sequences of *S. meliloti* and *M. loti* provide valuable data on BDC-encoding genes (Figure 1) and allow us to predict the entire complement of BDCs produced by these organisms. Consistent with the biochemical and genetic analysis of the rhizobial PYCs, the genome sequences of both organisms encode a single, chromosomally-localized *pyc*. In *S. meliloti*, this gene is the same as that which we recently cloned and functionally characterized (M. Dunn, G. Araíza, T. Finan, submitted). Each genome contains single copies of the genes encoding the four subunits of ACC. The *accB* encodes the BCCP subunit of the ACC, and the putative *accB* products in *S. meliloti* and *M. loti* are 16.5 and 15.9 kDa, respectively, fairly similar to the 14 kDa biotinylated protein visualized on Western blots. These data suggest that rhizobia contain an ACC dedicated to fatty acid synthesis and that PCC, which can also carboxylate acetyl CoA, does not function in this capacity.

The genes encoding the subunits of PCC are located on pSymb in *S. meliloti* and the *pccA* subunit is the same as that described by Charles and Aneja (1999). A 3-methylcrotonyl CoA carboxylase (MCC) also appears to be encoded on the *S. meliloti* pSymb, and this gene product would be predicted to function in leucine degradation. The *mccA* and *mccB* genes are flanked by genes encoding other enzymes (*hmgL*; 3-hydroxy-3-methylglutaryl-CoA lyase and *ivdH*; isovaleryl-CoA dehydrogenase) involved in the degradation of leucine.

The prediction of genes encoding acyl CoA carboxylases in *M. loti* is complicated by the fact that the α and β subunit-encoding genes of PCCs and MCCs resemble one another at the primary sequence level. Two gene regions containing adjacent *pccB* and *mccA* genes are present on the *M. loti* chromosome, making it difficult to predict whether these encode the subunits of a MCC or a PCC. Non-adjacent, putative *pccA* and *pccB* genes are also present in *M. loti*, bringing the total number of BDCs in this organism to five.

Biochemistry, genetics and, more recently, genome analysis are providing insights into the characteristics and metabolic roles of BDCs in rhizobia. We should now be able to address whether any of the BDCs perform additional functions that are of unique importance to rhizobia, either as free-living populations in the soil or in symbiotic combination with a host plant. Because rhizobia produce multiple BDCs, they may provide a good model system to study how the biochemical and genetic regulation of this class of enzymes integrates with metabolism as a whole. Another interesting avenue of investigation involves determining how biotin protein ligase partitions biotin between the different BDCs.

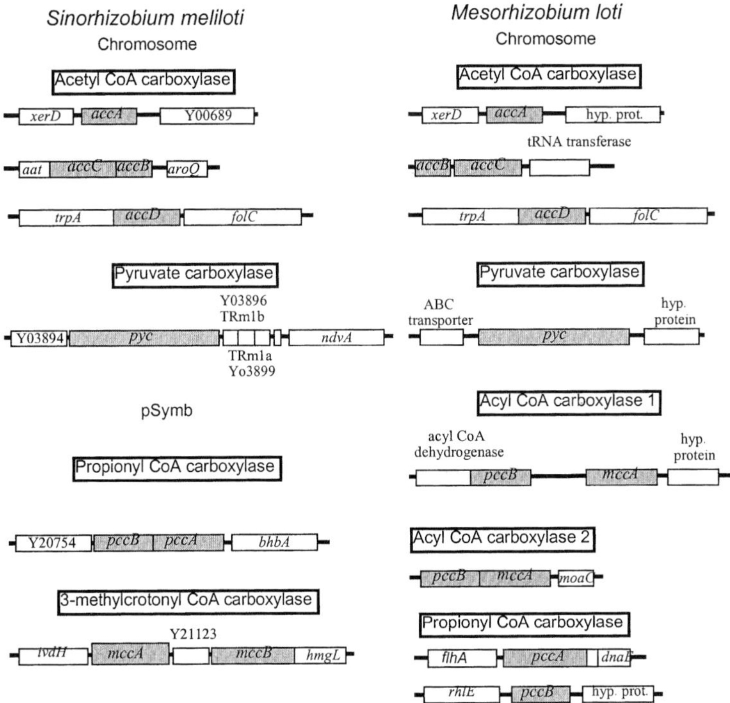

Figure 1. Genome sequence encoding BDCs

4. References

Allen EK, Allen GN (1950) Bacteriol. Rev. 14, 273-330

Araíza G, Dunn MF (1998) 16th North American Nitrogen Fixation Conference, abstract I.20

Arwas R *et al.* (1986) J. Gen. Microbiol. 132, 2743-2747

Cevallos MA *et al.* (1996) J. Bacteriol. 178, 1646-1654

Charles TC, Aneja P (1999) Gene 226, 121-127

DeHertogh AA *et al.* (1964) J. Biol. Chem. 239, 2446-2453

Dunn MF (1998) FEMS Microbiol. Rev. 22, 105-123

Dunn MF *et al.* (1996) J. Bacteriol. 178, 5960-5970

Dunn MF *et al.* (1997) FEMS Microbiol. Lett. 157, 301-306

Dunn MF *et al.* (2000) In Pedrosa FO, Hungria M, Yates G, Newton WE (eds), Nitrogen Fixation: From Molecules to Crop Productivity, pp. 379, Kluwer, Dordrecht, The Netherlands

Dunn MF *et al.* (2000a) 17th North American Conference on Symbiotic Nitrogen Fixation, pp. 27

Encarnación S *et al.* (1995) J. Bacteriol. 177, 3058-3066

Fall RR (1976) Biochim. Biophys. Acta 450, 475-480

Kaneko T *et al.* (2000) DNA Res. 7, 331-338

Lowe RH, Evans HJ (1962) Soil. Sci. 94, 351-356

Ronson CW, Primrose SB (1979) J. Gen. Microbiol. 112, 77-88

Scrutton MC (1978) FEBS Lett. 89, 1-9

Streit WR, Phillips DA (1996) Appl. Environ. Microbiol. 62, 3333-3338
West PM, Wilson PW (1940) Enzymol. 8, 152-162

5. Acknowledgements
This work was supported in part by Grant 3232-P from CONACyT (Mexico) and Grants IN209697 and IN227598 from DGAPA (Mexico).

PHOSPHORUS IS IMPORTANT IN NODULATION OF ACTINORHIZAL PLANTS AND LEGUMES

K. Huss-Danell[1], F. Gentili[1], C. Valverde[2], L. Wall[2], A. Wiklund[1]

[1]Dept of Agricultural Research for Northern Sweden, Crop Science Section, Swedish University of Agricultural Sciences, S-904 03 Umeå, Sweden
[2]Dept Ciencia y Tecnología, Universidad Nacional de Quilmes, Roque Saenz Peña 180, Bernal B1876BXD, Buenos Aires, Argentina

1. Introduction

Actinorhizal plants form N_2-fixing root nodules with the actinomycete *Frankia*. There are some 200 species of actinorhizal plants belonging to 25 genera in eight families (Huss-Danell 1997). Nearly all of them are woody plants. They are useful in soil reclamation, bioenergy production, as wind-breaks and as ornamentals. Some species serve as browse, and *Hippophaë rhamnoides* has edible and highly nutritious fruits and can provide raw material for phytochemical products. Representatives of actinorhizal plants are native in most regions of the world and several species, in particular *Casuarina* spp., are planted in many areas.

Strains of *Frankia* have been isolated from nodules of nearly all host genera (Benson, Silvester 1993). The majority of *Frankia* strains are from macerated nodules and only few originate from a single spore. *Frankia* grows by extension and branching of its septate, ca 0.5 to 1.5 μm wide filaments (sometimes called hyphae). In N-free culture and in nodules (except for genera *Casuarina, Allocasuarina*) tips of branched filaments differentiate into so-called vesicles, the site of nitrogenase. In symbiosis vesicles vary in size (up to 6 μm in diameter), shape and septation as determined by the host (Huss-Danell 1997). In culture and in some root nodules *Frankia* filaments also differentiate into sporangia with spores (Benson, Silvester 1993).

Actinorhizal nodules show a large variation in their morphology and anatomy. There is also a large variation in their physiological and biochemical solutions on how to cope with the dilemma of aerobic nodule metabolism coupled to oxygen sensitive nitrogenase (Huss-Danell 1997). All these characters are determined by the host. Actinorhizal nodules thus offer a range of fascinating interactions between plants and microbes for further investigation.

Depending on the host there are two different modes of infection by *Frankia* (Wall 2000). Intracellular penetration via root hairs occurs in the families *Betulaceae, Casuarinaceae* and *Myricaceae*. Cell divisions in root cortex give rise to a so-called prenodule but the actual root nodule originates in the pericycle and appears like a modified lateral root. Intercellular penetration occurs in the families *Elaeagnaceae, Rhamnaceae* and *Rosaceae*. Root hairs are not affected and prenodules are not formed. The root nodule originates in pericycle. In families *Coriariaceae* and *Datiscaceae*, the infection pathway is not yet described.

In intracellularly infected plants (Berry, Sunell 1990; Berg 1999), already from the very first ingrowth into a root hair, *Frankia* is encapsulated in a plant derived "capsule", a modified primary cell wall, and a membrane continuous with the plasmalemma. This encapsulation persists throughout the nodule and surrounds filaments as well as vesicles. In intercellularly infected plants (Miller, Baker 1985) *Frankia* grows between cells and becomes encapsulated once it grows into a cell of the emerging root nodule.

Actinorhizal nodules, or clusters of nodule lobes, are not evenly distributed on the root system. Rather, they occur singly or in groups with some spacing in between. This indicates that nodulation is regulated. Indeed, regulatory processes resembling "autoregulation" in legumes are seen in *Alnus* (Wall, Huss-Danell 1997) and in *Discaria* (Valverde, Wall 1999). Nodulation is also influenced by environmental factors such as nutrients. Like in legumes, N inhibits nodulation as well as N_2-fixation (Huss-Danell 1997). In *Alnus* a multivariate study comprising all six macronutrients showed that N_2 fixation per plant and the proportion of N in plants derived from N_2 fixation ($\%N_{dfa}$) was inhibited by N but stimulated by P. The highest $\%N_{dfa}$ was obtained at low N and high P (Ekblad, Huss-Danell 1995). Also in legumes a stimulating effect of P is seen and has been ascribed to either a generally stimulated growth (e.g. Robson *et al.* 1981) or a specific effect on nodules (Israel 1987). In our work on nodulation in actinorhizal plants we address the following questions. Can P stimulation counteract N inhibition? Does P act via stimulated plant growth or does P act specifically on nodulation? Does P act locally or systemically? Are there differences among host genera or infection pathways?

2. Procedure

We have used *Alnus incana* (L.) Moench (*Betulaceae*) as a representative of intracellularly infected actinorhizal plants and *Discaria trinervis* (Hooker et Arnot) Reiche (*Rhamnaceae*) and *Hippophaë rhamnoides* L. (*Elaeagnaceae*) as representatives of intercellularly infected actinorhizal plants. For comparison we have used the intracellularly infected legume *Trifolium pratense* L. Seedlings raised from surface sterilized seeds were grown in growth pouches (Mega, Minneapolis, MN, USA) or in pots with coarse sand or perlite. A nutrient solution (the modified Evans solution (Huss-Danell 1978) at 1/10 of full strength) had the P concentration according to recipe (medium level, 0.1 mM P) or was reduced 10 times (low P) or increased 10 times (high P). Ammonium nitrate was added to the solution to give low, medium or high N level (0.071, 0.71 or 7.1 mM N). Combinations of the N and P levels gave N/P ratios in the solution ranging from 0.071 to 710. Nodulation was evaluated as number and biomass of nodules. Growth of plants was recorded as their biomass at harvest. To distinguish between local and systemic effects plants were grown with split-root systems.

3. Results and Discussion

In *Alnus* (Wall, Hellsten, Huss-Danell 2000) both nodule number and nodule biomass per plant decreased when N concentration was increased, but only when P was at low or medium level. At high P level nodule number and biomass per plant were stimulated by increased N. At all N levels high P stimulated nodulation. To distinguish between a P effect via general growth stimulation and a P effect specifically on nodulation we analyzed nodule biomass in relation to plant biomass. Still, high P was stimulating at all N levels. The same pattern emerged when nodule biomass was related to root biomass, indicating that the P effect was not only a general growth effect on plants or below-ground plant parts but rather a more specific effect on nodulation.

An interaction between N and P was further seen when nodulation was plotted against N/P ratio in the nutrient solution. In particular the nodule biomass showed a negative correlation with increased N/P ratio. Individual nodule size and nitrogenase activity (ARA per plant) also showed a negative correlation with increased N/P ratio in the solution.

In *Trifolium* (Hellsten, Huss-Danell 2001) we used only a partial factorial design (low N at low and high P, medium N and P levels, high N at low or high P). Nodule number and nodule biomass per plant were inhibited by high N when P was low. High P was stimulating at low N and at high N, i.e. there was a counteracting effect of high P. Individual nodule size and ARA per nodule biomass was stimulated by low N. High P had no effect at low N but increased individual nodule size at high N.

Growth of the *Trifolium* plants responded strongly to the different nutrient treatments. Low N and low P gave the smallest plants and high N and high P gave the biggest plants, about a 10-fold difference in plant biomass. Plant given medium levels of N and P were the second biggest plants. In spite of these differences in plant growth, there were marked effects of N and P when nodulation was related to plant biomass. In general, nodulation decreased with increasing N, but there was a stimulation by high P at both low and high N level.

Effects of N and P on nodulation were systemic on both nodule number and biomass. This is concluded from split-root experiments where the two sides of *Alnus* and *Hippophaë* root systems were given different combinations of N and P in the nutrient solution (F. Gentili, K. Huss-Danell, unpublished). Inhibition by N was strong at the root side receiving high N and almost as strong in the second root side receiving medium N. When high P was given to one side of the root system a stimulation was seen mainly at the second side kept at medium P. There was an interaction between N and P with respect to how large N inhibition or P stimulation was obtained. Again, the effects were specific for nodulation rather than general growth effects.

In *Discaria* the nodule biomass in relation to plant biomass, but not the number of nodules per plant, was stimulated by P supply. N assimilation was stimulated by P, and the proportion of nodule biomass as part of plant biomass was negatively correlated to the leaf N/P ratio. This suggests that, in *Discaria*, P interacts with the feedback control of nodule growth that is associated with the plant demand for N, but not with the initiation of nodulation (C. Valverde, L.G. Wall, unpublished).

The importance of N/P ratio in the solution was thus evident in our experiments. We used several species and cultivation systems. On the other hand, we used the same nutrient solution in all our experiments. Moreover, the N/P ratio in solution does not necessarily reflect the internal N/P ratio of plants. Plants can adjust their N/P ratio in tissues by adjusting uptake rates and N_2-fixation rates. Further, when young seedlings are studied the seed reserves of N and P, and thus the N/P ratio, varies among species (C. Valverde, L.G. Wall, unpublished). In the literature there are only few nodulation studies where N and P in plant tissues were reported. Nodule biomass as proportion of plant biomass shows a negative correlation with increasing N/P ratio in leaves or in whole plants. This is concluded from data on the actinorhizal plants *Casuarina* (Yang 1995; Reddell, Yang, Shipton 1997), *Discaria* (C. Valverde, L.G. Wall, unpublished) and *Hippophaë rhamnoides* (F. Gentili, K. Huss-Danell, unpublished) as well as the legumes *Glycine* (Israel 1987; Drevon, Hartwig 1997), *Medicago* (Drevon, Hartwig 1997) and *Trifolium* (Almeida *et al.* 2000; Hellsten, Huss-Danell 2001). These studies comprise several species, growing conditions, nutrient solutions, plant ages, etc., and we conclude that N/P ratio in plant tissues has an important role in nodulation. Still, data are not fully conclusive because N and P in leaves or plants were measured at harvest several weeks after inoculation. It is not known how well these data reflect N and P concentrations in seedlings at time of nodule initiation.

Compared to other plant parts actinorhizal nodules are rich in P. A large need for ATP in nitrogen fixation and N assimilation is often proposed as an explanation. *Frankia* is a gram-positive bacterium with cell walls containing P. Further, *Frankia* is encapsulated by a membrane continuous with the plasmalemma and thus expected to be rich in P. However, the mechanisms and signals involved in P stimulation of nodulation and its possible interaction with N inhibition (Wall 2000) are not yet known.

4. Concluding Remarks

Nodulation responds to the interaction between N (inhibiting) and P (stimulating). Effects of P are specific on nodulation, not only acting via enhanced plant growth. N and P act systemically. Nodule biomass as part of plant biomass is negatively correlated with increased N/P ratio in nutrient solution. Nodule biomass as part of plant biomass is negatively correlated with increased foliar or plant N/P ratio in a number of actinorhizal as well as legume species and at a number of experimental conditions. The role of P in nodulation deserves further studies, and this applies to other nutrients as well.

5. References

Almeida JPF, Hartwig ÜA, Frehner M, Nösberger J, Lüscher A (2000) J. Exp. Bot. 51, 1289-1297
Benson DR, Silvester WB (1993) Microbiol. Rev. 57, 293-319
Berg H (1999) Can. J. Bot. 77, 1351-357
Berry AM, Sunell LA (1990) In Schwintzer CR, Tjepkema JD (eds), The Biology of *Frankia*
    and Actinorhizal Plants, pp. 61-81, Academic Press, San Diego, CA
Drevon JJ, Hartwig ÜA (1997) Planta 201, 463-469
Ekblad A, Huss-Danell K (1995) New Phytol. 131, 453-459
Hellsten A, Huss-Danell K (2001) Acta Agric. Scand. Sect. B
Huss-Danell K (1978) Physiol. Plant. 43, 372-376
Huss-Danell K (1997) New Phytol. 136, 375-405
Israel DW (1987) Plant Physiol. 84, 835-840
Miller IM, Baker DD (1985) Protoplasma 128, 107-119
Reddell P, Yang Y, Shipton WA (1997) Plant Soil 189, 213-219
Robson AD, O'Hara GW, Abbott LK (1981) Austr. J. Plant Physiol. 8, 427-436
Valverde C, Wall LG (1999) Can. J. Bot. 77, 1302-1310
Wall LG (2000) J. Plant Growth Regul. 19, 167-182
Wall LG, Huss-Danell K (1997) Physiol. Plant. 99, 594-600
Wall LG, Hellsten A, Huss-Danell K (2000) Symbiosis 29, 91-105
Yang Y (1995) Plant Soil 176, 161-169

6. Acknowledgements

Financially supported by the Swedish Natural Science Research Council, the Swedish Council for Forestry and Agricultural Research, the Swedish Foundation for International Cooperation in Research and Higher Education, Universidad Nacional de Quilmes, Agencia Nacional de Promoción Cientifica y Técnológia and CONICET (Argentina). Correspondence to Kerstin.Huss-Danell@njv.slu.se

Section 7:
Proteins in Regulation and Development

CHAIR'S COMMENTS: PROTEINS IN REGULATION AND DEVELOPMENT

P. Müller

Fachgebiet Zellbiologie und Angewandte Botanik, Fachbereich Biologie der Philipps-
Universität, D–35032 Marburg, Germany

The interaction between soil bacteria of the *Rhizobiaceae* and their leguminous host plants is a multistep process which finally leads to the formation of new plant organs, the nodules, a very efficient place of biological N_2 fixation. Two extremely different partners, the bacteria and a suitable plant, with a completely different organization of their genomes meet, and from the very beginning they will have a strong mutual influence on each other. The free-living bacteria will differentiate to intracellular bacteroids which fix N_2 to NH_4^+, and differentiated cortex cells of the plant regain meristematic activity to build up a shelter for the invading bacteria, where these are protected from competitors in the soil, fed by the plant-derived carbon sources, and exploited by the plant as a source of consumable N compounds. This process is beneficial for both partners, but – as demonstrated by plant or bacterial mutants – the symbiotic relationship might become a pathogenic one when the intimate biochemical dialogue between both representatives is disturbed by a single mutation in any gene related to this communication process, which is controlled by several cascades of transcriptional regulation systems.

The four papers of this session contribute to various aspects of the differentiation processes of the bacteria and the plants. Daniel Kahn and co-workers report on "The FixJ transcriptional activator: from structure to genome" (this volume). Yi-Ping Wang and co-workers study "CRP-cAMP-mediated repression effect on the *glnAp2* promoter in *E. coli*" (this volume). Furthermore, the "Functional analysis of regulatory genes involved in *M. truncatula* nodule organogenesis" is presented by Martin Crespi and co-workers (this volume). Finally Pawel Strózycki presents new data on "Iron proteins in legume plant development and nodulation – Ferritin" (this volume). The combination of these four contributions nicely represents the great diversity of aspects which play important roles in the formation and functional operation of nitrogen-fixing nodules.

With the establishment of the genomes of *Sinorhizobium meliloti* and *Mesorhizobium loti*, and the partial genomes of several others, e.g. the symbiotic plasmid of *Rhizobium* spp. NGR234 and the symbiotic region of *Bradyrhizobium japonicum*, the molecular analysis of the legume/Rhizobium symbiosis has now shifted from the specific level of single genes to whole sets of genes which are expressed under defined conditions (e.g. absence or presence of plant-derived inducers, availability of C compounds, oxygen concentration, etc.). The combination of transcriptomics and proteomics, based on the availability of bacterial genomes, is a powerful tool for elucidating the concerted events which are going on in the regulation and developmental processes in and between the symbiotic partners. Nevertheless, these experimental approaches cannot replace a detailed analysis of single genes in order to define the real functional importance of their gene products in the symbiotic differentiation process. It is necessary to keep in mind this particular aspect, because there are numerous examples for the existence of duplicate or even multiple allelic genes in the rhizobial genomes which might functionally substitute for each other.

In our laboratory, we have focused on the analysis of genes encoding extracytoplasmic proteins, because they might be required for the symbiotic relationships with the legume host plant. Surface components can play important roles in the exchange of metabolites and/or signals between the partners. As a rule, many of these proteins have N-terminal signal peptides and therefore are supposed to be translocated by the general secretory pathway (Pugsley 1993). To detect genes in *Bradyrhizobium japonicum* encoding extracytoplasmic proteins involved in symbiotic traits we previously used a delivery system for Tn*phoA* (Müller *et al.* 1995), because the promoterless reporter gene, '*phoA*, is a suitable means to identify putative secreted proteins, if a translational fusion has occurred by a proper insertion of the genetic element (right orientation and in frame fusion within a gene region encoding a periplasmic domain of the protein). Based on our experience with *B. japonicum* symbiotic mutants defective in two different signal peptidases, *sipS* (Müller *et al.* 1995) and *sipF* (Bairl *et al.* 1998), the construction of mutants, the subsequent screening of their symbiotic phenotypes and the detailed genetic analysis of the mutations turned out to be a successful, but very time consuming procedure. Thus, based on an *in vitro* transposon mutagenesis system (Reznikoff et Goryshin, 1999) and a '*phoA-aphII* cassette (Rodriguez-Quinones *et al.* 1994), Tn*KPK2* was constructed and used to establish an alternative and rapid strategy which enabled the identification of new *B. japonicum* loci required in symbiosis. The following experimental scheme should be generally applicable in many other systems:

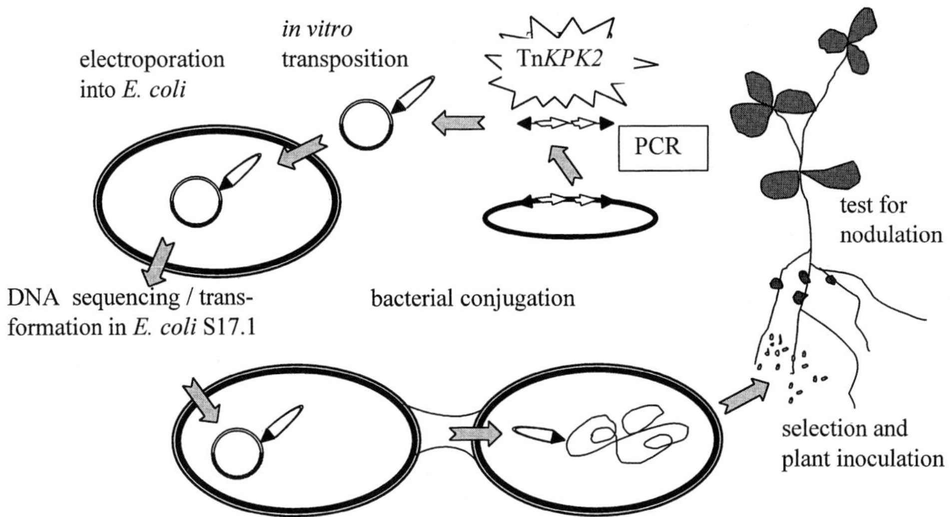

References

Bairl A, Müller P (1998) Mol. Gen. Genet. 260, 346-356
Manoil C, Beckwith J (1985) Proc. Natl. Acad. Sci. USA 8129-8133
Müller P *et al.* (1995) Mol. Microbiol. 18, 831-840
Müller P *et al.* (1995) Planta 197, 163-175
Pugsley AP (1993) Microbiol. Rev. 57, 50-108
Rezinkoff W, Goryshin I (1999) Epicentre Forum 6, 5-7
Rodríguez-Quinones *et al.* (1994) Gene 151, 125-130

THE FixJ TRANSCRIPTIONAL ACTIVATOR: FROM STRUCTURE TO GENOME

L. Ferrières, J. Schumacher, B. Ton-Hoang, J. Fourment, P. Roche, S. Rouillé, D. Kahn

Laboratoire de Biologie Moléculaire des Relations Plantes-Microorganismes
INRA / CNRS, BP 27, 31326 Castanet-Tolosan Cedex, France

1. Introduction

In the alfalfa symbiont *Sinorhizobium meliloti*, nitrogen fixation genes are controlled by the oxygen-regulated FixLJ 'two-component' regulatory system. Under microoxic conditions as provided by root nodules, the FixL histidine kinase phosphorylates FixJ, turning it into a transcriptional activator of the *nifA* and *fixK* promoters. As a consequence a genetic cascade is switched on, allowing expression of the nitrogen fixation apparatus during symbiosis (for review see Fischer 1994). Like other response regulators, FixJ is a modular protein, which has been dissected into an N-terminal phosphorylatable 'receiver' domain FixJN and a C-terminal transcriptional activator domain FixJC. In its non-phosphorylated form, the FixJN receiver domain inhibits the latent activity of FixJC at the *nifA* promoter (Kahn, Ditta 1991; Da Re *et al.* 1994). Phosphorylation simultaneously relieves this inhibition and triggers the dimerization of the protein via the FixJN receiver domain (Figure 1), resulting in a FixJ~P dimer which is the active form of FixJ (Da Re *et al.* 1999).

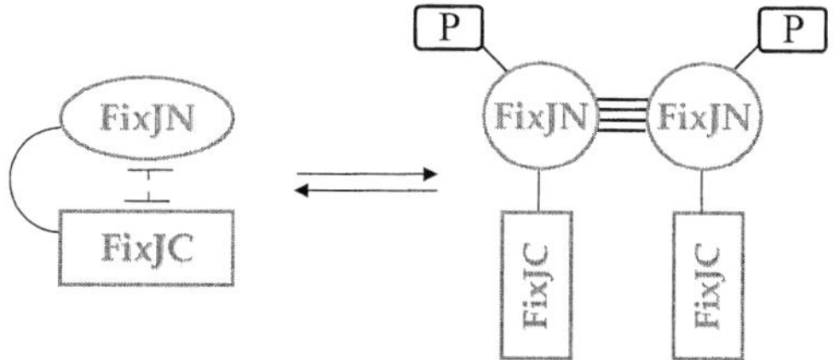

Figure 1. General scheme for FixJ activation.

The structure of FixJN (Gouet *et al.* 1999) shares basic features found in other receiver domains, with a parallel 2-1-3-4-5 β-sheet sandwiched between helices α1/α5 on one side and helices α2/α3/α4 on the other side. The phosphorylated Asp-54 residue participates in an acidic pocket located at the C-terminal edge of the central parallel β-sheet. The conformational change resulting from the phosphorylation of FixJN has recently been characterized at atomic resolution, showing how phosphorylation reshapes the α4–β5 face into a dimerization interface (Birck *et al.* 1999).

Here we used alanine-scanning mutagenesis in order to map functional interfaces of the FixJN receiver domain. Two interfaces were identified, one interacting with the FixJC output domain, the other required for transcriptional activation at the *fixK* promoter. In addition we present the outcome of a systematic search of FixJ targets in the *S. meliloti* genome.

2. The 'Aromatic Switch' Hypothesis for 2-Component Signal Transduction

A systematic alanine-scanning mutagenesis of the FixJN receiver domain was undertaken in order to map the interface with the FixJC transcriptional activator domain. Effects of mutations were tested *in vivo* in *nifA-lacZ* and *fixK-lacZ* reporter strains. Mutated proteins were purified from GST fusion proteins and assayed for acetyl-phosphate dependent phosphorylation, dimerization and DNA-binding ability. Mutations affecting the interaction between FixJN and FixJC were expected to confer a strong 'up' phenotype for *nifA* activation *in vivo*. Among 26 mutations tested, only one mutation (F101A) exhibited a strong 'up' effect both on *in vivo* activity and on DNA binding. This 'up' phenotype suggests that the mutated FixJ protein adopts an open conformation liberating the activity of the C-terminal domain. In addition, the F101A FixJ protein could not be phosphorylated with acetyl-phosphate, suggesting a propagating conformational change between Phe-101 and the phosphorylation site. This dual phenotype of the F101A mutation indicates that Phe-101 interacts both with the phosphorylation site and the C-terminal domain. Thus Phe-101 lies at the heart of signal transduction between the phosphorylation site and the output domain.

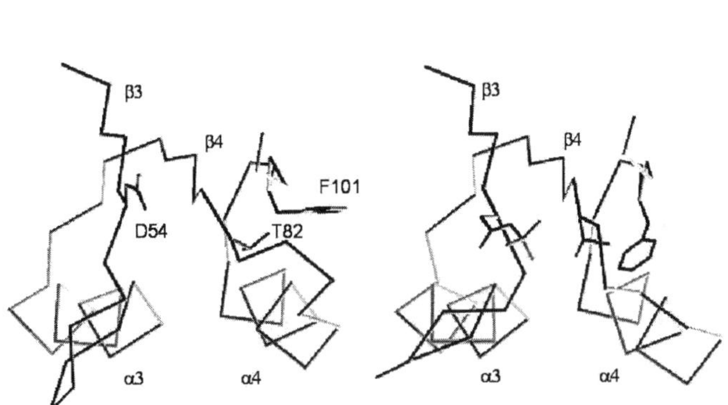

Figure 2. Conformational change induced by phosphorylation of the FixJ receiver domain.

Conversely crystallographic analysis of the phosphorylated receiver domain showed that phosphorylation of Asp-54 causes a marked change in the conformation of Phe-101, which switches from an outward to an inward oriented conformation (Birck *et al.* 1999; Figure 2). This conformational change is associated with a large displacement of the β4–α4 loop and the rotation of Thr-82 towards the phosphorylation site. We used a molecular dynamics approach to simulate this conformational change and found that the conformation of Phe-101 appears exquisitely sensitive to the conformation of the β4–α4 loop.

Because position 101 is usually conserved as an aromatic residue in receiver domains, we propose that the conformation of this aromatic residue plays a key role in 2-component signal transduction. This 'aromatic switch' hypothesis states that the inward or outward conformation of the aromatic ring determines the on/off state of the switch. The switch is triggered by a concerted displacement of the β4–α4 loop following phosphorylation of Asp-54, opening up a hydrophobic cavity filled by the aromatic residue. In the case of FixJ, this change results in two functionally important consequences: (i) liberating the FixJC output domain and (ii) reshaping the α4–β5 face into a dimerization interface.

3. *fixK*-Specific Involvement of the FixJ Receiver Domain in Transcriptional Activation

The FixJN receiver domain was understood until now as a molecular switch modulating both the activity of FixJC and the quaternary structure of FixJ, whereas the transcriptional activation function on *nifA* was assigned to the FixJC output domain. In the present study however, we identified yet another function for the FixJN receiver domain since it was found to be also specifically required for full activation of the *fixK* promoter. Indeed alanine scanning showed the involvement of FixJN strand β2 (Lys-31 and Gln-34) for activation of the *fixK* promoter. More specifically, mutation of these residues affected the recruitment of RNA polymerase at the *fixK* promoter without affecting FixJ~P DNA binding nor the recruitment of RNA polymerase at the *nifA* promoter. These results point to a specific contact between FixJ~P and RNA polymerase in the *fixK* preinitiation complex, involving strand β2 of the FixJ receiver domain. Therefore, the FixJ receiver domain appears to be much more than a simple molecular switch, since it acts as a transcriptional activator domain at the *fixK* promoter. This example illustrates the formidable functional plasticity of receiver domains.

4. The Quest for the FixJ Regulon

In the regulatory cascade controlling *nif* and *fix* genes, FixJ is a global regulator, while *nifA* and *fixK* are specialized regulators of *nif* and *fix* genes, respectively. This regulatory network architecture suggested that FixJ might control yet other sets of genes in response to microaerobiosis in the nodule environment.

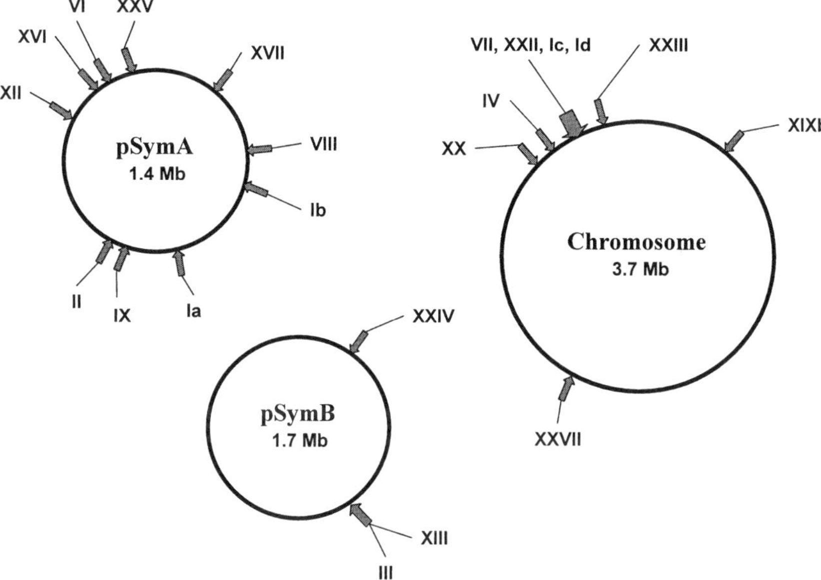

Figure 3. Distribution of FixJ binding sites on the *S. meliloti* genome.

In order to identify potential new FixJ targets in the genome of *S. meliloti* we developed an *in vitro* cyclic selection (SELEX) procedure using a fusion protein between GST and the DNA binding domain of FixJ. DNA fragments covering the entire *S. meliloti* genome were generated by random priming as described by Singer *et al.* (1997), protein-DNA complexes were adsorbed on glutathione-sepharose beads, bound DNA was eluted and amplified. Resulting DNA, enriched in FixJ binding fragments, was subjected to additional selection cycles until protein-DNA complexes could be detected by gel retardation analysis. Three selection rounds appeared to be sufficient. Selected fragments were cloned and sequenced, and fragments originating from the same region were clustered on the basis of sequence comparison. This analysis yielded 32 independent candidate FixJ targets. These candidate targets were further tested by gel retardation and DNase I foot-printing experiments in the presence of FixJ~P, which led to the identification of 22 FixJ~P binding sites on the *S. meliloti* genome. These include the known *fixK* and *fixK'* promoters (Waelkens *et al.* 1992) and 20 novel sites unequally distributed through the genome (Figure 3). pSymA contains 10 sites distributed throughout the replicon, while only 3 sites are located on pSymB. Similarly the chromosome contains very few FixJ~P binding sites, with the exception of a region which was previously recognized as resulting from horizontal transfer (Capela *et al.* 2001). Thus the *fixJ* gene and the majority of its targets appear to be carried by pSymA, which is consistent with the proposal that pSymA has been acquired relatively recently (Galibert *et al.* 2001), together with the *fixJ* regulon. In this scenario, the cluster of FixJ binding sites found on the chromosome would mostly derive from pSymA.

In agreement with this view, two of the new FixJ targets appear to result from a duplication of the *fixK* promoter, evident from the presence of a remnant truncated *fixK* ORF. This duplication of the *fixK* promoter confers *fixJ*-dependent microaerobic induction upon the downstream gene, as evidenced by RT-PCR and reporter gene experiments. Similar promoter duplications, including a truncated *nifH* ORF, were previously observed for the *nifH* promoter and were found to confer *nifA* regulation upon downstream genes (Better *et al.* 1983; Murphy *et al.* 1993). Such a 'promoter hijacking' may therefore be a more common phenomenon in the *S. meliloti* genome than originally thought, allowing for the recruitment of new genes into pre-existing regulatory networks.

5. References

Better M *et al.* (1983) Cell 35, 479-485
Birck C *et al.* (1999) Structure Fold Des. 7, 1505-1515
Capela D *et al.* (2001) Proc. Natl. Acad. Sci. USA
Da Re S *et al.* (1994) Nucleic Acids Res. 22, 1555-1561
Da Re S *et al.* (1999) Molec. Microbiol. 34, 504-511
Fischer HM (1994) Microbiol. Rev. 58, 352-386
Galibert F *et al.* (2001) Science 293, 668-672
Gouet P *et al.* (1999) Structure Fold Des. 7, 1517-1526
Kahn D, Ditta G (1991) Molec. Microbiol. 5, 987-997
Murphy P *et al.* (1993) J. Bacteriol. 175, 5193-5204
Singer B *et al.* (1997) Nucleic Acids Res. 25, 781-786
Waelkens F *et al.* (1992) Molec. Microbiol. 6, 1447-1456

6. Acknowledgements

This work was supported by grants from the European Union (BIO4 CT 97-2143), the Ministère de l'Education Nationale, de la Recherche et de la Technologie (Programme de Recherche Fondamentale en Microbiologie et Maladies Infectieuses et Parasitaires), INRA (Programme prioritaire "Microbiologie") and the Ministère des Affaires Etrangères (PROCOPE 97158).

A NOVEL REGULATORY LINKAGE BETWEEN CARBON METABOLISM AND NITROGEN ASSIMILATION IN *E. COLI* AND RELATED BACTERIA

Y.-P. Wang[1], Z.-X. Tian[1], G. Wu[1], M. Buck[2], A. Kolb[3]

[1]College of Life Sciences, Peking University, Beijing 100871, People's Republic of China
[2]Department of Biology and Biochemistry, Imperial College of Science, Technology and Medicine, London SW7 2AZ, UK
[3]Laboratoire des Regulations Transcriptionnelles, FRE 2364 CNRS, Institute Pasteur, 75724 Paris Cedex 15, France

1. Introduction

The quality and quantity of carbohydrate available to the cell is the major limiting factor for the capacity of energy consuming nitrogen assimilation and nitrogen fixation. In enteric-bacteria, in response to the PTS system, the cAMP receptor protein (CRP) is considered as a proximal activator interacting at the short distances with the major form σ^{70} RNA polymerase ($E\sigma^{70}$), at promoters of sugar catabolic operons (reviewed in Kolb *et al.* 1993; Busby, Ebright 1999). In contrast, the initiation of transcription at nitrogen-responsive, σ^{54} RNA polymerase ($E\sigma^{54}$)-dependent promoters is activated by an entirely different mechanism; it requires nitrogen regulator NtrC-phosphate usually bound at distal enhancer sequences and hydrolysis of NTP (reviewed in Magasanik 1996).

Analysis of σ^{54}-dependent *dctA* promoter reveals a novel negative regulatory function for CRP-cAMP in *E. coli* (Wang *et al.* 1993). CRP-cAMP is able to interact *in cis* from remote sites and *in trans* with the $E\sigma^{54}$-promoter closed complex. Moreover, such an interaction is kinetically linked to its repression effect (Wang *et al.* 1998). Due to the fact that CRP-cAMP can exert its effect on a 'core' promoter, which lacks a specific CRP-binding site, it is proposed that the effect might be general (Wang *et al.* 1998).

2. Results and Discussion

Among the nitrogen utilization regulons, the key promoter, *glnA*p2 is σ^{54}-dependent. It controls the expression of glutamine synthetase, the most important enzyme of nitrogen assimilation, and of NtrB and NtrC, the two regulatory proteins that controlling expression of the Ntr regulon (Reitzer, Magasanik 1986). The *glnA*p2- and its upstream mutated/deleted derivatives-*lacZ* fusions (pKU100, pKU101 and pKU102) were constructed and monitored under different conditions. Results show that in *E. coli* wild-type strain TP2101, the levels of the *glnA*p2 expression varied strongly when cells were grown on different carbohydrates. The levels are low when cells were grown on glycerol, and are high when grown on glucose (Table 1). It was previously reported that intracellular cAMP levels in cells growing on glucose are low (Dumay *et al.* 1996), while intracellular cAMP levels in cells growing on glycerol are high (Inada *et al.* 1996). It was decided, therefore, to investigate whether CRP-cAMP affected *glnA*p2 expression. The expression levels of *glnA*p2 (pKU101) were assayed in a *cya* mutant TP2006 and a *cya crp* double mutant TP2339, in the presence or absence of CRP-cAMP. When cells were grown on glycerol, the expression levels of *glnA*p2 were high in the absence of cAMP, and the expression levels were low in the presence of cAMP in the *cya* mutant TP2006 (Table 2). In contrast, the expression levels of *glnA*p2 were high in either presence or absence of cAMP, in *cya crp* double mutant TP2339, when cells were grown under similar conditions. Only when a plasmid carrying the *crp* gene (pLG339CRP) was used to complement TP2339's *crp* genotype, the repression effect of CRP-cAMP on *glnA*p2 was observed (Table 2). These results indicate that CRP represses *glnA*p2 in a cAMP-dependent manner.

Table 1. Carbon effect on *glnA*p2 expression.

Constructs	TP2101 (wt)		TP2006 (*cya*)	
	glucose	glycerol	glucose	glycerol
pKU100 CRP NtrC NtrC (*glnA*p1 *glnA*p2 *lacZ*)	5866±44	885±18	9105±201	9599±122
pKU101 CRP NtrC NtrC (*glnA*p1 *glnA*p2 *lacZ*)	5743±49	339±4	6573±219	7558±59

Table 2. CRP-cAMP-mediated repression on *glnA*p2.

Strain (genotype)	Plasmids	Exogenous cAMP	
		–	+
TP2006 (*cya*)	pKU101	7558±59	354±13
TP2339 (*cya crp*)	pKU101	6758±214	6626±67
TP2339 (*cya crp*)	pKU101&pLG339CRP	6498±64	142±5

The *glnA*p2 promoter has CRP-binding site (centered at -186.5 from *glnA*p2) located near, but not overlapping with the enhancer sequences for NtrC-P. It is required for CRP-cAMP-mediated activation of σ^{70}-dependent *glnA*p1. When this CRP-binding site is mutated (pKU101), or even the entire upstream sequence is deleted (up to –40 of *glnA*p2, pKU102), the CRP-cAMP-mediated repression effect remains (Table 2 for pKU101, data not shown for pKU102). These results indicated that specific CRP-binding site(s) are not essential for CRP-cAMP-mediated repression on *glnA*p2, a result similar to that observed on the heterologous *dctA* promoter (Wang *et al.* 1998). We rule out the possibility that CRP might interfere with NtrC mediated transcriptional activation via heterodimer/aggregate formation with NtrC, or somehow interfere with NtrC activity by disrupting the histidyl-aspartyl phosphorelay between NtrB and NtrC proteins. This is done by replacing the activator NtrC with an alternative transcriptional activator, NifA, in the *cya* mutant TP2006 harboring pKU101 or pKU102, and similar repression effects on *glnA*p2 and its derivatives were observed (data not shown).

In order to examine if CRP mediated repression on *glnA*p2 is at the transcriptional level, primer extension analysis was carried out. In *cya* mutant TP2006, when cells were grown without cAMP, it gave clear transcripts at +1 of *glnA*p2. In contrast, these clear transcripts were absent when cells were grown with cAMP (data not shown). To further examine the basis of CRP-cAMP-mediated inhibition of *glnA*p2, we assayed promoter DNA opening using *in vivo* $KMnO_4$ footprinting on the *glnA*p2 promoter. The results show that the presence of the CRP-cAMP complex prevents NtrC-dependent open complex formation on *glnA*p2 (data not shown). Similar results were obtained when NifA was used as the activator (data not shown). We conclude that the CRP-cAMP-mediated inhibition on *glnA*p2 expression is at the transcriptional level and may occur through limiting NtrC and NifA-dependent DNA opening.

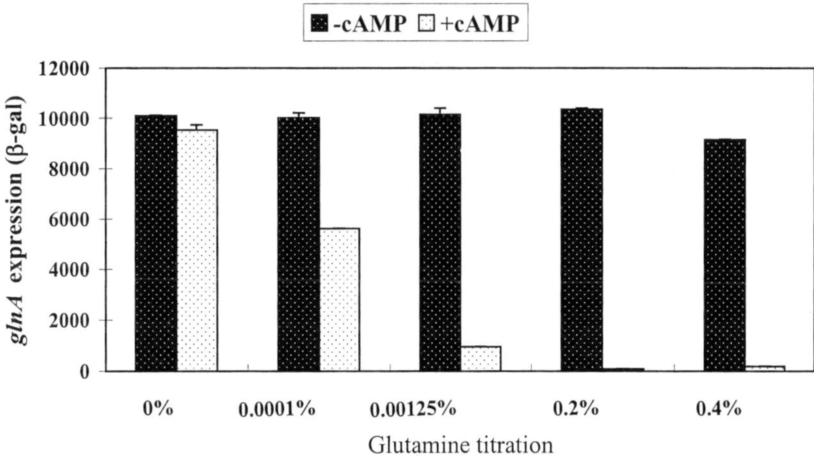

Figure 1. Alleviation of CRP-cAMP-mediated repression on *glnA*p2 by extremely low level of nitrogen source.

When the *cya* mutant TP2006 harboring pKU101 was grown on an extremely low level of nitrogen source, CRP-cAMP-mediated repression on *glnA*p2 was alleviated (Figure 1). This implies that over-production of NtrC-phosphate alleviates CRP-cAMP-mediated repression. Similarly, alleviation of CRP-cAMP-mediated repression was obtained when alternative activator NifA was over-expressed. Furthermore, CRP-cAMP-mediated repression effect on *glnA*p2 was abolished when the *rpoN* gene was over-expressed (data not shown). Taken together, the results indicate that over-expressions of either activator or *rpoN* could alleviate the repression effect on *glnA*p2 *in vivo*. Therefore, it is likely that CRP-cAMP exerts its repression effect by competing with activator for $E\sigma^{54}$ RNA polymerase on *glnA*p2.

Here we show that CRP represses transcription of *glnA*p2 in a binding site non-essential, activator independent, and cAMP-dependent manner. We propose that CRP-cAMP exerts its repression effect by competing with the activator(s) for $E\sigma^{54}$ RNA polymerase on *glnA*p2. In addition, we have observed that the *glnH*p2 and *glnK* promoters from *E. coli* and the *nifB*, *nifE*, *nifF*, *nifJ*, *nifLA* and *nifU* promoters from *K. pneumoniae* are also repressed by the CRP-cAMP complex. We predict that CRP-cAMP-mediated repression on σ^{54}-dependent promoters is quite common. Thus, dual roles of CRP (activation at σ^{70}-dependent sugar catabolic promoters and repression on σ^{54}-dependent nitrogen-responsive promoters) provide a novel regulatory linkage between carbon metabolism and nitrogen assimilation in *E. coli* and related bacteria.

3. References

Busby S, Ebright RH (1999) J. Mol. Biol. 293, 199-213
Dumay V *et al.* (1996) Microbiol. 142, 575-583
Inada T *et al.* (1996) Genes to Cells 1, 293-301
Kolb A *et al.* (1993) Ann. Rev. Biochem. 62, 749-795

Magasanik B (1996) In *Escherichia coli* and *Salmonella*: Cellular and Molecular Biology, 2nd edn, Neidhardt FC *et al.* (eds), pp 1344-1356, American Society for Microbiology, Washington, DC
Reitzer LJ, Magasanik B (1986) Cell 45, 785-792
Wang Y-P *et al.* (1993) Mol. Microbiol. 8, 253-259
Wang Y-P *et al.* (1998) EMBO J.17, 786-796

4. Acknowledgements

This work is supported in part by grants from National Natural Science Foundation of China (grants 39925017 and 39970168), the Ministry of Education of China and the 863 program from the Ministry of Science and Technology of China, the French Ministère de l'Education Nationale, de la Recherche et de la Technologie (Programme de Recherche Fondamentale de Microbiologie, Maladies Infectieuses et Parasitaires). We are grateful for the continuing support from Programme Franco-Chinois de Recherches Avancées (PRA), in association with the China National Center for Biotechnology Development of the Ministry of Science and Technology of China. Part of this work has been submitted to and accepted by Molecular Microbiology (2001).

FUNCTIONAL ANALYSIS OF REGULATORY GENES INVOLVED IN *MEDICAGO TRUNCATULA* NODULE ORGANOGENESIS

M. Crespi[1], A. Complainville[1], D. Chinchilla[1], F. Merchan[1,2], A. Campalans[1], C. Breda[1], C. Sousa[2], M. Megias[2], E. Dax[3], S. Wolf[3], A. Kondorosi[1]

[1]Inst. des Sci. du Végétal-CNRS, Gif sur Yvette, France
[2]Dept. Microbiol. and Parasitology, Univ. Sevilla, Sevilla, Spain
[3]Faculty Agric., The Hebrew Univ. Jerusalem, Israel

1. Introduction

The symbiotic interaction between leguminous plants and rhizobia results in the development of a novel organ on the plant roots, the nodule, where bacteria provide fixed nitrogen for the host. Briefly, the microsymbionts attach to and enter the root hair cells while the plant forms a tube-like infection thread through which the bacteria move into the root cortex. Simultaneously, cortical cells are induced to divide and the infection threads invade these dividing cells forming the nodule primordium. This primordium differentiates into a mature nodule where rhizobia are converted into bacteroids and start to fix nitrogen (Cohn *et al.* 1998). Rhizobia induce nodule morphogenesis on the plant through the production of nodulation signals (Nod factors; Schultze, Kondorosi 1998). These Nod signals trigger the earliest stages of nodule development, including root hair deformation and curling, cortical cell division and the expression of several early nodulin (*enod*) genes. Under starvation for combined nitrogen, plants of certain *Medicago* cultivars have the capacity to form root nodules even in the absence of rhizobia (NAR^+ phenotype), the so-called spontaneous nodules indicating that the developmental program of nodule formation preexists in legumes. These nodules show histological features similar to those of bacterium-induced nodules but the central zone cells contain only amyloplasts (Truchet *et al.* 1989). Hence, it has been suggested that spontaneous nodules might serve as carbon storage organs induced during nitrogen starvation and be the ancestors of the nitrogen-fixing nodules. Thus, Nod factors act as primary morphogenetic signals triggering a nodule developmental program although other plant factors are likely required for the regulation of nodule morphogenesis such as plant hormones (Fang, Hirsch 1998; Heidstra *et al.* 1997; Penmetsa, Cook 1997), the stele factor, uridine, (Smit *et al.* 1995) and the metabolic status of the plant (high carbon and low nitrogen) (Bauer *et al.* 1996).

The carbon metabolism of the plant host is adapted to altering demands during nodule development. Initially, amyloplast deposition is observed in the actively dividing cortical cells of the nodule primordia (Ardourel *et al.* 1994), demonstrating the accumulation of carbon translocated from the leaves. Later, the assimilation of symbiotically fixed nitrogen in the functional nodule requires a complex interplay with the carbon metabolism (Schultze, Kondorosi 1998) in order to satisfy (a) the demand of bacteroids for carbon and energy (required for nitrogen fixation), and (b) the provision of C4 carbon skeletons for the assimilation of fixed ammonia in the plant cells. The nodule primordium acts then as a sink and the plant tightly regulates nodule initiation through factors from the aerial parts of the plant ("shoot factor") (for review Caetano-Anollés, Gresshoff 1991), a phenomenon called autoregulation. The metabolic state of the root may be sensed by the plant, whereby intermediates may act directly as signals to modulate cellular responses or indirectly by influencing the activity of internal plant hormones required for nodule initiation or starch deposition.

Plasmodesmal (PD) function seems to be responsible for cell-to-cell communication and signaling in plant development and there is a growing body of evidence that a complex supracellular communication network acting through PD is present in plants (Lucas, Wolf 1999). This network was originally identified by studying viral movement between cells. Viral movement proteins (MPs) are localized to PDs and affect their function as first explored by expression of the MP gene in transgenic plants (Deom *et al.* 1990; Wolf *et al.* 1991). Based on the assumption that altering plasmodesmal size exclusion limit should affect the transport rate of small molecules, including sucrose (Tyree 1970), during the last few years, carbon transport and allocation in transgenic tobacco plants expressing the TMV-MP have been studied (Lucas *et al.* 1993; Olesinski *et al.* 1996). These data allowed the advancement of a hypothesis that trafficking of regulatory (information) molecules, through plasmodesmata may establish a special supracellular communication network regulating carbon partitioning. Plants may exploit this to create specialized physiological and developmental domains where signaling may occur during the formation of an organ primordium (e.g. the nodule initials) or between different cell layers of a meristem (Lucas, Wolf 1999 and references therein). Molecular mechanisms involved in the control of nodule organogenesis in the plant host are poorly understood. In our laboratory, several approaches are being attempted to understand how cell-to-cell communication processes may interact with key regulatory genes to understand the complex process of nodule development. In this paper, we describe the preparation of *M. truncatula* transgenic plants to characterize cell-to-cell communication processes and various functional approaches carried out on selected regulatory genes to understand their role in nodulation.

2. Procedures

Preparation of transgenic *Medicago truncatula* plants and RT-PCR experiments were done as described (Charon *et al.* 1999; Frugier *et al.* 2000). Western analysis was carried out according to standard techniques (Wolf *et al.* 1991). Bombardment of germinating alfalfa roots using "microtargeting" of DNA constructs has been described (Sousa *et al.* 2001).

3. Transgenic Plants Expressing Viral Movement Proteins and GFP Fusion Proteins Can Be Used to Study Cell-to-Cell Communication in Legumes

We have prepared diverse transgenic *M. truncatula* plants which may serve to analyze carbon partitioning during the symbiotic interaction. First, various independent transgenic lines expressing the movement protein of TMV (tobacco mosaic virus) were characterized. This MP has been shown to affect carbon partitioning in transgenic potato plants (Lucas, Wolf 1999). TMV-MP expression driven by the 35S promoter was confirmed using Western analysis of leaf extracts. No obvious visible phenotype was observed on these plants when growing under normal conditions or in symbiosis with *Sinorhizobium meliloti*. We could not find any significant differences in the number of nodules and further work is being carried out to analyze in detail the early steps of the interaction, such as primordium formation and amyloplast accumulation in dividing cortical cells. Second, we have expressed the CMV-MP (from cucumber mosaic virus) fused to GFP in *M. truncatula*. Selected plants with high levels of CMV-MP-GFP expression showed a dotted pattern of fluorescence corresponding to the labeling of PDs in various cell types. Third, plants containing phloem-specific expression of GFP have been characterized (using a Atsuc2-GFP construct, kindly provided by Dr N. Sauer). In these plants, phloem unloading of GFP into carbon "sinks" such as lateral roots and organ primordia was followed (see Imlau *et al.* 1999), by monitoring green fluorescence in tissues. These two latter transgenic lines are excellent tools to monitor *in vivo* the process of phloem unloading, carbon sink formation and nodule initiation.

4. The Role of *enod40* in Nodule Organogenesis

In addition to this physiological work, we have characterized several regulatory genes induced during nodule organogenesis in *Medicago* species using various approaches (Crespi *et al.* 1994; Frugier *et al.* 1998). One of the earliest nodulin genes associated to the nodule developmental program is *enod40*. In response to bacterial inoculation of the roots, *enod40* transcripts are detected first in the root pericycle opposite to the protoxylem pole, then in the dividing cortical cells and in all differentiating cells of the growing nodule primordia (Fang, Hirsch 1998 and refs therein). We have shown that, under nitrogen-limiting conditions, overexpression of this gene resulted in a significant increase of cortical cell divisions in *M. truncatula* roots (Charon *et al.* 1997). This was accompanied by a high level of accumulation of amyloplasts in the dividing cells. During nodulation, overexpression of *enod40* modified the early stages of development and induced proliferation of cortical cells all around the cortex in the infected region close to the root tip. This suggested that (a) *enod40* induction in the cortex may be a determinant of nodule initiation (Charon *et al.* 1999), and (b) its primary function is not exerted directly on triggering cell division *per se*. In contrast to auxin, *enod40* does not have the capacity to induce cell proliferation irrespective of the nitrogen status of the plants or the root position. In addition, a putative co-suppression phenomenon induced on selected lines yielded plants showing arrested nodule development, indicating that *enod40* expression is required for appropriate development of the primordium (Charon *et al.* 1999). Due to its expression pattern in vascular tissues and nodule initials, as well as to the phenotype of the transgenic plants, a role of *enod40* in cell-to-cell communication between the vascular tissue and specific cells of the cortex to allow proper organization of the nodule primordium seems very likely.

The *enod40* genes (Cohn *et al.* 1998; Schultze, Kondorosi 1998) are very peculiar because they code for about 0.7 kb RNAs containing only short ORFs. Modeling predicts that *enod40* RNA sequences have the tendency to form particularly stable secondary structures, a property shared with several biologically active RNAs (Crespi *et al.* 1994). On the other hand, a very small ORF corresponding to 10 to 13 amino acids in the 5' end of the transcripts is common among these genes and has been proposed to be the active gene product (van de Sande *et al.* 1996). Recently, we have demonstrated that several small ORF (sORFs) of *enod40* were translated when fused to a reporter gene. In addition, microtargeting of *Mtenod40* into *Medicago* roots induced a cell-specific growth response, division of cortical cells, that was used to test different gene derivatives containing specific point mutations and deletions (Sousa *et al.* 2001). These experiments indicated that translation of two sORFs present in the conserved 5' and 3' *enod40* regions was required for activity. Moreover, deletion of a *Mtenod40* region present in between the two sORFs and spanning the predicted RNA structure showed low activity in our assay, without affecting translation of the sORFs. Even though the encoded sORF-peptides maybe the functional gene products, the structured RNA region also participates in gene regulation. These data revealed that a complex cellular mechanism may be implicated in the translation and primary cellular function of *enod40*.

5. A Vascular Krüppel Transcription Factor Involved in the Formation of the Symbiotic Zone

Another regulatory gene identified was a Krüppel-like Zn-finger gene, *Mtzpt2-1*. This gene is strongly expressed in vascular bundles of roots and nodules, and antisense plants grew normally but developed Fix- nodules where differentiation of the nitrogen-fixing zone and bacterial invasion were arrested. These results indicate that a vascular bundle-associated Krüppel-like gene is required for the formation of the central nitrogen-fixing zone (Frugier *et al.* 2000). Interestingly, a homologous gene was shown to confer salt tolerance to yeast cells. Indeed, Mtzpt2-1 was also able to induce this response in yeast. Moreover, this transcription factor is strongly and rapidly induced after application of salt stress to root and nodules, suggesting that it may participate in osmotic stress responses in plant cells. Therefore, we think that *Mtzpt2-1* may be involved in the osmotic adaptation of the nodule vascular tissues to support nitrogen fixation.

6. Novel Kinase Associated to the Initial Steps of the Symbiotic Interaction

Very few regulatory kinases associated to nodulation are known. A gene, *Mtpk1*, encoding a novel protein kinase containing an ankyrin domain was identified as being induced during nodulation and in spontaneous nodules (Frugier *et al.* 1998). We have made translational fusions of the entire *Mspk1* gene to GFP in order to localize the protein in root tissues during symbiosis and in transfected onion cells. This kinase seems to co-localize with microtubules in the latter cells and may be associated to the microtubule rearrangements required during *Rhizobium* infection.

Expression of this gene was detected in different alfalfa organs and in the early stages of the symbiotic interaction.

The *Mtpk1* gene was isolated by screening of an *M. truncatula* BAC library and the sequence of its genomic region and of certain adjacent clones were determined. Several genes showing homologies to previously identified sequences in databanks were identified in the vicinity of the *Mtpk1* gene. The distribution of exons and introns was analyzed in detail for *Mtpk1* and compared with three homologous genes identified in *Arabidopsis thaliana*. These data suggest that *Mtpk1* may be involved in the early steps of the symbiotic interaction, though it is not exclusively associated with nodulation. The developed tools will serve to analyze the possible function and localization of *Mtpk1* during the initiation of nodule organogenesis.

7. Conclusion

Based on various functional and cell biological approaches, we have identified several regulatory genes which may be involved in different critical steps of the formation of the nodule. *Mtenod40* may participate in cell to cell communication processes influencing carbon partitioning and primordium formation in the cortical cells through interaction with root vascular tissues. *MtZpt2-1* seems to be required for osmotic adaptation of the nodule vascular tissues to support bacterial diferentiation and nitrogen fixation in the neighboring cells of the symbiotic zone. Finally, *Mtpk1* can be a determinant of the major microtubule rearrangements required for the penetration of the rhizobia in the roots and the morphogenesis of the primordium. Extending these results with other functional assays as well as coupling them with functional genomics and identification of insertional mutants may contribute to the identification of an integrative network of plant genes acting in nodulation.

8. References

Ardourel M *et al.* (1994) Plant Cell 6, 1357-1374

Bauer P *et al.* (1996) Plant J. 10, 91-105

Caetano-Anollés G, Gresshoff P (1991) Ann. Rev. Microbiol. 45, 345-382

Charon C *et al.* (1997) Proc. Natl. Acad. Sci. USA 94, 8901-8906

Charon C *et al.* (1999) Plant Cell 11, 1953-1965

Cohn J *et al.* (1998) Trends Plant Sci. 3, 105-110

Crespi M *et al.* (1994) EMBO J. 13, 5099-5112

Deom CM *et al.* (1990) Proc. Natl. Acad. Sci. USA 87, 3284-3288

Fang Y, Hirsch AM (1998) Plant Physiol. 116, 53-68

Frugier *et al.* (1998) Mol. Plant Microbe Int. 11, 358-366

Frugier *et al.* (2000) Genes Dev. 14, 475-482

Heidstra R *et al.* (1997) Development 124, 1781-1787

Imlau A *et al.* (1999) Plant Cell 11, 309-322

Lucas WJ *et al.* (1993) Planta 190, 88-96

Lucas WJ, Wolf S (1999) Curr. Opin. Plant Biol. 2, 129-197

Olesinski AA *et al.* (1996) Plant Physiol. 111, 541-550

Penmetsa RV, Cook DR (1997) Science 275, 527-530

Schultze M, Kondorosi A (1998) Ann. Rev. Genet. 32, 33-57

Smit G *et al.* (1995) Plant Mol. Biol. 29, 869-873

Sousa *et al.* (2001) Mol. Cell. Biol. 21, 354-366

Truchet G *et al.* (1989) Mol. Gen. Genet. 219, 65-68

Tyree MT (1970) J. Theor. Biol. 26, 181-214

Van de Sande K *et al.* (1996) Science 273, 370-373

9. Acknowledgements

This work was supported by a grant from AFIRST (French-Israeli Association for Research and Technology). FM was supported by a graduate scholarship from Fundacion "La Caixa", Spain.

FERRITIN AND IRON MANAGEMENT IN LEGUME PLANT DEVELOPMENT AND NODULATION

P.M. Stróżycki, A. Skapska, K. Kolaczkowska-Szcześniak, E. Sobieszczuk, A.B. Legocki

Institute of Bioorganic Chemistry Polish Academy of Sciences, Noskowskiego 12/14, Pozna
61-704, Poland

Iron is one of the most important nutrients for all eukaryotes. However, it is also one of the most dangerous elements. Because of its redox properties, iron is a critical element for such basic processes as DNA and hormone synthesis, respiration and photosynthesis (Briat *et al.* 1995). Although iron is the fourth most abundant element in the Earth crust, it is not easily available. This is because of low solubility of iron-containing minerals, especially in aerobic and neutral pH environments (Guerinot, Yi 1994). In order to cope with this problem, plants have developed several mechanisms of iron acquisition. Except for morphological changes leading to the extension of active root area, these mechanisms include proton pumping (acidification), secretion of organic acids and phenolics (chelation) and induction of the membrane bound reductase (dicots and nongraminaceous monocots). In addition, high-affinity chelators are used to dissolve Fe(III) oxides (Fox, Guerinot 1998; Hinsinger 1998; Jones 1998; Marschner, Romheld 1996; Schmidt 1999; Thoiron *et al.* 1997). However, in the presence of active oxygen species the same desirable iron may catalyze the generation of hydroxyl radicals (OH$^\bullet$) - the most powerful oxidizing agents known thus far (Fenton reaction) (Cadenas 1989; Meneghini *et al.* 1995; Nappi, Vass 2000). Attack of toxic oxygen species usually leads to severe results like lipid peroxidation, protein and DNA oxidation and eventually cell disintegration. Plant antioxidant defenses include such compounds like glutathione, ascorbate, carotenoids, tocopherols, etc. Antioxidant enzymes like catalase, peroxidases, dismutases and enzymes of the ascorbate-glutathione cycle are also activated (Becana *et al.* 1998; Larson 1995). It is obvious that there has to be a strict control of "free" iron in the cell, just to prevent generation of reactive oxygen species (ROS) in the first place. Limited generation of ROS, however, may be a part of plant defense systems against pathogens (Wojtaszek 1997). Predominant portion of cell iron is bound in proteins and enzyme cofactors, but it may easily be "set free" under stress situations.

Ferritins, iron storage proteins, are the major and the most effective elements necessary for iron homeostasis control. A ferritin molecule is a hollow protein shell, composed of 24 polypeptide subunits, capable of storing up to 4500 iron atoms in hydrous ferric oxide form (Harrison, Arosio 1996; Trikha *et al.* 1994). Thus iron biomineralized inside ferritin is safe for cell components. It may be released when needed for their functioning (Moore *et al.* 1992). Ferritins are widespread in all kingdoms of living organisms, animals, plants and microorganisms, showing high structural conservation (Theil 1987). Amino acid identities are lower except of residues required for a proper structure of the polypeptide (Andrews *et al.* 1992). Plant ferritin polypeptides consist of two additional (compared to animal ferritins) fragments. One, the so-called transit peptide (TP; 47-91 amino acids) is required for plastid targeting of ferritin precursor, the other, extension peptide (EP; 28-32 amino acids) is probably responsible for the stability of matured protein (Briat *et al.* 1999; Ragland *et al.* 1990).

As practically legume nitrogen fixation relies on iron (two of the key elements of nitrogen reduction, plant hemoglobin and bacterial nitrogenase, are iron-proteins), strict control of this metal during nodule development is particularly important. Moreover, leghemoglobin accounts for up to 30% of total soluble proteins in the nodule and has to be considered as a source of dangerous radicals (Moreau *et al.* 1996).

In our studies on iron management in plant tissues, we use yellow lupin (*Lupinus luteus*) as a model. Based on the plant hemoglobin sequence analysis we have assumed that lupins belong to one of the oldest species among legume plants (Kass, Wink 1995; Stróżycki *et al.* 2000; Stróżycki, Legocki 1995). Lupin plants infected with *Bradyrhizobium lupini* form nodules of a very characteristic type. As far as the general morphology is concerned, a lupinoid nodule is similar to that of undeterminate type – long-lasting meristems and clearly divided developmental zones. However, because of very early division of nodule meristem and its growth in all directions, it escapes from the cylindrical shape. During further development, meristematic activity remains only in nodule edges, and as a result of the activity of two lateral meristems, the initially spherical nodules grow laterally encircling the root (Golinowski *et al.* 1987). Because of the shape, this type of nodule is called a "collar type". Additionally, the first cell divisions that initiate nodule formation in lupin take place in the first layer of the primary root cortex, similar to the determinate type nodules, i.e. soybean or beans, and not in the deepest layers of the primary root cortex, which is typical of all other undeterminate-nodule type plants, i.e. pea, clover or alfalfa (Golinowski *et al.* 1992).

During lupin root nodule development, low "physiological" levels of ferritin polypeptide may be detected. There is a significant temporary increase in ferritin accumulation around the time when nitrogen fixation starts. This increase is correlated in time with massive synthesis of leghemoglobin (12-14 days after inoculation with *B. lupini*). In fact, strong accumulation of ferritin polypeptides starts with the first symptoms of nodule tissue decay and increases with nodule senescence. This picture of developmental regulation was also confirmed by RNA hybridization experiments and by *in situ* immuno detection. Apart from the clearly stronger hybridization signal on material from 14-day-old nodules, there is a distinct layer of cells just between the meristematic and matured nodule zones, characterized by a higher ferritin content. It is probably the equivalent of amyloplast containing interzone (or pre nitrogen-fixation zone) of alfalfa. It is also apparent in lupin senescing nodules that the main increase of ferritin signal is shifted to cortex and younger zones, close to meristems. It could be explained as protection of the still active and functional tissues against catalytic iron released from spreading "destruction". In addition, there is a strong increase in ferritin accumulation in root tissues surrounding vascular bundles and even after complete disintegration of a nodule, the adjacent root tissue seems to remain intact. It is likely that the main task of the antioxidant defenses of bacteroid tissues during nodule functioning and senescence repose on other systems (Becana *et al.* 2000). However, as it was reported before, a majority of the diverse antioxidant systems functioning in an active nodule decline with the aging process (Becana *et al.* 2000).

Observations presented above are generally in agreement with those of other authors (Lucas *et al.* 1998) describing ultra structural immuno-localization of ferritin in the cells of soybean, lupin and alfalfa nodules. The exception is the finding that meristematic activity was not observed at all stages of yellow lupin nodule development. According to our observations, meristems exist in *L. luteus* even in 96-day-old nodules (with clear mitotic figures).

Following the reports on ferritin accumulation in the cells of cortex and parenchyma in nodules of nitrate and dark stressed pea and bean plants, we checked a quantitative contribution of this protein to the general reaction in this type of induced senescence (Escuredo *et al.* 1996; Gogorcena *et al.* 1997). Surprisingly, there was no detectable change in ferritin accumulation levels in stems, roots or nodules of either nitrate or dark treated plants. Moreover, in contrast to the analysis performed for bean nodules (Matamoros *et al.* 1999), a slight decrease was noticed in the leaves of plants kept in the dark for 4-7 days.

To have a complete picture of lupin ferritin, we analyzed ferritin content in other than nodule parts of the unstressed plant. In lupin, we could detect ferritin polypeptide in all tested organs. The only exceptions were cotyledons during plant germination and young, forming seeds. The former is due to the fact that cotyledons are rather a source of iron (for a growing plant) and ferritin is degraded to release it. The latter is due to the fact that ferritin is synthesized at a later stage of seed development. The highest levels of ferritin were found in flowers and leaves. Flowers contain a whole range of iron containing proteins (and pigments), therefore, the presence of ferritin there is of no surprise. The case of lupin leaves is somewhat different from that reported for pea, where ferritins were detected only in the roots and leaves of young plantlets and remained undetectable in the corresponding organs of adult plants (Lobreaux, Briat 1991). In yellow lupin, the level of ferritin is much lower but significant in young leaves, and it increases with the plant age. Additionally, in all the cases tested, ferritin polypeptide levels rise during tissue senescence.

Nevertheless, ferritin may be considered to be a part of the oxidant stabilization system, which slows down the senescence processes by sequestration of released iron.

On the basis of the collected information it may be assumed that animal and plant ferritins are also coded by small gene families. Structures of these genes significantly differ between the families in the number (three vs. seven) and positions of introns (Proudhon *et al.* 1996).

In animal systems, ferritin synthesis is regulated on the translational level. Changes in conformation of iron regulatory proteins (IRP), caused by iron molecule binding, lead to removal of these proteins (iron sensing) from RNA structures (iron responsive elements; IRE). Freeing of 5' untranslated region of mRNA unblocks it for translation (Theil 1994). Regulation of synthesis of plant ferritins is much less understood. In general, it is believed that it is regulated mainly on the transcriptional level (Briat *et al.* 1999). Recently, some promoter sequence elements have been proposed to be important in the regulation of iron dependent soybean and maize ferritin gene expression (Petit *et al.* 2001; Wei, Theil 2000).

We have identified three classes of ferritin genes in yellow lupin tissues (cDNA and genomic). These genes code for polypeptides of 84-89% identity in the parts corresponding to matured protein and only 31-43% of identity of transit peptides. They reveal typical plant ferritin gene organization. We are currently analyzing promoter sequences of these genes in search for elements homologous to these reported.

The analysis of the expression patterns of ferritin genes in different organs on the RNA level generally has confirmed earlier results on ferritin polypeptides detection. The difference was that we detected ferritin transcripts in all analyzed tissues, even in germinating plant cotyledons and in young seeds. This fact indicates posttranscriptional regulation of ferritin synthesis in these organs. Using class specific probes we could also analyze patterns of expression of each class of lupin ferritin separately in different organs. Generally, one class RNA is almost undetectable, the other two show similar patterns in the above-ground organs and differential expression in nodules.

We also used hydroponically grown plants and tissue cultured cells to test inductivity of lupin ferritin genes. These experiments have revealed that transcription of each ferritin class gene is differently induced by iron and abscisic acid. These last results indicate a possibility of different contribution of each class of ferritin to oxidative stress and ABA mediated defenses. Additionally, ABA is a hormone known to take part in a whole range of stress responses in plants, i.e. water stress and pathogen attack (Giraudat *et al.* 1994). Apart from its scientific (cognitive) objectives, research into ferritins has also practical purposes.

It has been proposed that withholding iron is one of the mechanisms of plant defense against pathogen invasion. Compounds like phenolics, synthesized by plant tissues, can limit the pathogen growth by restricting vital iron. It is also known that some pathogenic organisms produce toxins which induce iron acquisition, thus promoting the growth of the invader. In experiments with transgenic tobacco, others have demonstrated that plants ecotopically expressing alfalfa ferritin are tolerant to oxidative damage and pathogens (Deak *et al.* 1999). In our experiments, however, we could not detect any increase in ferritin level after the infection of lupin plants with pathogens. We have analyzed proteins isolated from lupin leaf tissue, from and around necrotic spots after the infection with *Pseudomonas syringiae*, from crown gall induced on lupin root neck by *Agrobacterium tumefaciens* and from lupin stem infected by *Erwinia chrysantemi*. That could mean that ferritin synthesis in response to pathogen attack, at least in the cases and/or attack stages we examined, is not a part of the natural plant defense system.

As iron deficiency is the most common nutritional disorder, affecting more than 30% of the world's population (WHO, UNICEF, ASPP, USDA), iron fortified transgenic plants could help solve this problem. It has been demonstrated that plant ferritin can serve as a source of iron for animals, curing iron deficiency anemia (Beard *et al.* 1996). Further work is concentrated on improving iron content in plant tissues by overexpressing the ferritin gene. Iron fortification of rice seed by the overexpression of soybean ferritin has been demonstrated (Goto *et al.* 1999).

The increased iron storage capacity as a result of increased iron sequestration leads to certain changes in iron homeostasis in such transgenic plants. It has been shown that besides activation of iron transport systems (root ferric reductase), plants accumulating ferritin on the highest levels display "iron deficient", chlorotic phenotypes (Van *et al.* 1999). This result is only apparently contradicting previous observations of the linear relationship between the level of ferritin accumulation and increased iron content (Goto *et al.* 1901). Later experiments with transgenic tobacco plants overexpressing ferritin have shown high soil-dependent variability of leaf iron accumulation (Vansuyt *et al.* 2000).

Recently, the obtaining of transgenic plants, which overexpress the ferritin gene (lettuce) and show not only the ability to accumulate excess iron but also the ability to increase their own growth has also been reported (Goto *et al.* 2000).

In the course of our work we have also generated transgenic tobacco (lettuce) plants with S35-promoter driven lupin ferritin gene. We now have a collection of a whole range of ferritin accumulating plants. There are plants with almost undetectable (immuno-detection) amounts of lupin ferritin subunit as well as such in which the hybridization signal is very strong. Although we can observe a clear tendency toward the iron-starvation-like phenotype, which is strictly correlated with increased levels of detected lupin ferritin, we can easily manipulate the symptoms. In the tissue culture where iron availability becomes limited with time, we can observe chlorosis of diverse intensity, but general levels of iron accumulation in leaf tissues are similar. On the other hand, in artificial soil, where nutrient ratios may be supervised, these symptoms may be leveled by iron and phosphorous supplementation. However, even in these conditions the most severe effects of high ferritin accumulation, observed in flowers and fruits, are difficult to overcome. Additionally, some lines of plants with the highest accumulation of ferritin show strong physiological aberrations. In iron sufficient conditions, leaves of these plants, which are narrow and irregularly shaped, turn much deeper green than the leaves of other plants. Also the flowers are morphologically different, i.e. they have short stamens and long styles.

Information collected thus far proves that there is a natural correlation between metal storage and acquisition systems, and that in order to engineer a fully functional plant for biotechnological applications these systems cannot be evaluated separately.

Detailed data concerning our work on yellow lupin ferritins are currently being prepared for publication.

References

Andrews SC *et al.* (1992) J. Inorg. Biochem. 47, 161-174
Beard JL *et al.* (1996) J. Nutr. 126, 154-160
Becana M *et al.* (2000) Physiol. Plant 109, 372-381
Becana M *et al.* (1998) Plant and Soil 201, 137-147
Briat JF *et al.* (1995) Biol. Cell 84, 69-81
Briat JF *et al.* (1999) Cell. and Mole. Life Sci. 56, 155-166

Cadenas E (1989) Annu. Rev. Biochem. 58, 79-110

Deak M *et al.* (1999) Nature Biotech. 17, 192-196

Escuredo PR *et al.* (1996) Plant Physiol. 110, 1187-1195

Fox TC, Guerinot ML (1998) Ann. Rev. Plant Physiol. and Plant Mol. Biol. 49, 669-696

Giraudat J *et al.* (1994) Plant Mol. Biol. 26, 1557-1577

Gogorcena Y *et al.* (1997) Plant Physiol. 113, 1193-1201

Golinowski W *et al.* (1987) Acta Soc. Bot. Pol. 56, 687-703

Golinowski W *et al.* (1992) Acta Soc. Bot. Pol. 61, 307-318

Goto F *et al.* (1901) Transgenic Res. 7, 173-180

Goto F *et al.* (2000) Theor. and Appl. Gene (2000) 100, 658-664

Goto F *et al.* (1999) Nature Biotech. 17, 282-286

Guerinot ML, Yi Y (1994) Plant Physiol. 104, 815-820

Harrison PM, Arosio P (1996) Bba-Bioenergetics. 1275, 161-203

Hinsinger P (1998) Adv Agron, VOL 64 , 225-265

Jones DL (1998) Plant and Soil 205, 25-44

Kass E, Wink M (1995) Bot. Acta 108, 149-162

Larson RA (1995) Arch. Insect Biochem. Physiol. 29, 175-186

Lobreaux S, Briat J-F (1991) Biochem. J. 274, 601-606

Lucas MM *et al.* (1998) Protoplasma 204, 61-70

Matamoros MA *et al.* (1999) Plant Physiol. 121, 97-111

Meneghini R *et al.* (1995) Canc. J. 8, 109-113

Moore GR *et al.* (1992) J. Inorg. Biochem. 47, 175-181

Moreau S *et al.* (1996) J. Biol. Chem. 271, 32557-32562

Nappi AJ, Vass E (2000) Cell Mol. Biol. 46, 637-647

Petit JM *et al.* (2001) J. Biol. Chem. 276, 5584-5590

Proudhon D *et al.* (1996) J. Mol. Evol. 42, 325-336

Ragland M *et al.* (1990) J. Biol. Chem. 265(30), 18339-18344

Schmidt W (1999) New Phytologist 141, 1-26

Strózycki PM *et al.* (2000) Mol. Gen. Gene 263, 173-182

Strózycki PM, Legocki AB (1995) Plant Sci. 110, 83-93

Theil EC (1987) Annu. Rev. Biochem. 56, 289-315

Theil EC (1994) Biochem. J. 304, 1-11

Thoiron S *et al.* (1997) Plant Cell and Envir. 20, 1051-1060

Trikha J *et al.* (1994) Protein-Struct. Funct. Genet. 18, 107-118

Van WO *et al.* (1999) Plant J. 17, 93-97

Vansuyt G *et al.* (2000) Plant Physiol. and Biochem. 38, 499-506

Wei JZ, Theil EC (2000) J. Biol. Chem. 275, 17488-17493

Wojtaszek P (1997) Biochem. J. 322, 681-692

Section 8:
Stresses and Factors Limiting Nitrogen Fixation

ANTIOXIDANT PROTECTION OF LEGUME ROOT NODULES

I. Iturbe-Ormaetxe, M.A. Matamoros, J.F. Moran, J. Ramos, M.C. Rubio, M.R. Clemente, B. Heras, M. Becana

Estación Experimental de Aula Dei, CSIC, Apdo. 202, 50080 Zaragoza, Spain

1. Introduction

Legume nodules contain an impressive array of antioxidants to cope with reactive oxygen species (ROS), such as the superoxide radical and H_2O_2, that are generated in different subcellular compartments. ROS are involved in all stages of legume nodule development, from initiation to senescence (Becana *et al.* 2000; Santos *et al.* 2001). In the nodule cytosol, ROS are mainly formed in oxidative reactions involving leghemoglobin and are scavenged through the concerted action of CuZn-superoxide dismutase (SOD) and of the four enzymes of the Halliwell-Asada pathway (Dalton 1995). In this pathway, ascorbate peroxidase (APX) catalyzes the reduction of H_2O_2 to water by ascorbate, and the resulting monodehydroascorbate and dehydroascorbate are reduced back to ascorbate, respectively, by monodehydroascorbate reductase (MR) at the expense of NADH and by dehydroascorbate reductase (DR) plus glutathione reductase (GR) at the expense of NADPH. A key metabolite of the pathway for the detoxification of ROS in nodules is glutathione (GSH; γGlu-Cys-Gly), which is synthesized in both the bacteroids and plant fraction by two ATP-dependent enzymes, γ-glutamylcysteine synthetase and glutathione synthetase (GSHS), acting sequentially (Moran *et al.* 2000). However, some legume species and tissues can synthesize another thiol tripeptide, homoglutathione (hGSH; γGlu-Cys-βAla), in addition to or instead of GSH. It is generally assumed that hGSH is synthesized by a specific hGSH synthetase (hGSHS) and that it performs similar roles to GSH (Klapheck 1988; Matamoros *et al.* 1999).

The mitochondria are also a main site for generation of ROS in the nodules because of the abundance of these organelles in the cortex and infected region (Millar *et al.* 1995), their high rates of respiration required for active N_2 fixation (Dalton 1995; Millar *et al.* 1995), and their high content of heme and nonheme Fe, which can catalyze the formation of hydroxyl radicals from H_2O_2 (Becana *et al.* 1998). However, very little is known about the antioxidant composition of nodule mitochondria.

In this work we present data on three aspects on the antioxidant protection of nodules. First, we report on changes in transcript abundance for antioxidant enzymes during nodule senescence. Second, we have overexpressed in an heterologous system and characterized for the first time a functional hGSHS from nodules. Third, we analyze the antioxidant composition of nodule mitochondria and present a model for peroxide detoxification in these organelles.

2. Materials and Methods

2.1. Northern analysis. Total RNA was extracted from pea nodules by a phenol-LiCl procedure (Verwoerd *et al.* 1989), separated on agarose denaturing (formaldehyde) gels, and capillary transferred to Hybond-N+ nylon filters (Amersham). Blot hybridization with DNA probes prepared by random-primed ^{32}P-labeling and autoradiography were performed following standard protocols (Sambrook *et al.* 1989).

2.2. Overexpression, purification, and characterization of recombinant GSHS2. The open reading frame of *GSHS2* was PCR-amplified using gene-specific primers (Moran *et al.* 2000). The resulting 1.7-kb fragment was gel purified, subcloned into pCRII TOPO (Invitrogen), and transformed into DH5α competent cells. The inserted open reading frame of *GSHS2* was digested out with *Nco*I and

*Not*I, gel purified, and ligated into pFastBac HTb. Positive colonies of transformed DH5α cells were identified by PCR using pFastBac specific primers. Recombinant bacmid DNA was grown overnight in *E. coli*, isolated and used to transfect *Sf21* cells. Baculoviruses were harvested 72 h post-transfection and amplified by infecting monolayer cultures of insect cells. After optimizing infection conditions, protein production was scaled up by culturing insect cells at 27°C in 50 ml of TC-100 medium (Sigma) supplemented with 10% fetal calf serum (Sigma) and antibiotics. Cells were collected 48 h after virus infection and lysed by osmotic shock. Cell free extracts were loaded onto a cobalt affinity column (Clontech). Fractions were eluted with imidazole and subjected to Western analysis and protein staining to confirm the presence and purity of GSHS2.

2.3. Assay of antioxidant enzymes and metabolites. GPX activity was assayed with pyrogallol as substrate, including 0.5 mM p-chloromercuriphenylsulfonic acid (pCMS) in the reaction mixture (Amako *et al.* 1994). All the other enzyme activities were determined as described earlier (Gogorcena *et al.* 1997). Thiol tripeptides were quantified by HPLC with fluorescence detection (Matamoros *et al.* 1999).

2.4. Localization of enzymes in nodule mitochondria. Mitochondria purification was performed using modifications of published protocols and was monitored using organelle protein markers (Sandalio *et al.* 1987; Struglics *et al.* 1996). Solubilization and latency studies were carried out essentially as described by Jiménez *et al.* (1997).

3. Results and Discussion

3.1. Expression of antioxidants during nodule senescence. Northern hybridization of pea nodule RNA with specific probes for key antioxidant proteins revealed up-regulation of GSH peroxidase (reported for the first time in nodules) and catalase and down-regulation of the other antioxidants during natural (aging) senescence of nodules (Figure 1). Nitrate and dark treatments decreased the transcript abundance of plastidial CuZnSOD, mitochondrial MnSOD, cytosolic APX, and leghemoglobin (used as a control). Nitrate increased the transcript of cytosolic CuZnSOD whereas dark stress decreased it; the reverse occurred with ferritin. GSH peroxidase and catalase were up-regulated in both stress conditions. These results indicate that there is differential regulation of antioxidants at the transcript level during the various types of nodule senescence. General features of senescence, however, may be a lowering of important antioxidant defenses and a significant increase of GSH peroxidase and catalase transcripts. These increases may be a response to the enhanced production of lipid hydroperoxides and H_2O_2 at the later stages of senescence.

3.2. Functional characterization of GSHS2 from nodules. In a previous work (Moran *et al.* 2000) we reported the isolation from a pea nodule library of two cDNA clones that encoded enzymes with high homology to GSHS of other plants. One of the clones, *GSHS2,* encoded a cytosolic protein and was chosen for complete functional characterization. Repeated attempts to produce GSHS2 using *E. coli* expression systems were unsuccessful. In contrast, large amounts of virtually pure protein were produced efficiently in insect cells. This is important because GSHS enzymes are labile and of low abundance in plant tissues. The kinetic analysis of pure recombinant GSHS2 is shown in Table 1. The catalytic properties of the enzyme were determined using a fixed saturating concentration of γglutamylcysteine and a range of concentrations of Gly or βAla. The enzyme showed saturation kinetics and linear double-reciprocal plots with respect to both substrates. The much higher affinity of GSHS2 for βAla (K_m = 1.9 mM) than for Gly (K_m = 104 mM) as well as a specificity constant of 57 for the hGSHS/GSHS ratio permit us to conclude that the GSHS2 protein is a genuine hGSHS.

3.3. Antioxidants of nodule mitochondria. Bean plants were chosen to purify mitochondria because they produce large amounts of nodules and synthesize exclusively hGSH (Matamoros *et al.* 1999). A main thrust of this part of the work was to demonstrate the presence in nodule mitochondria of APX, guaiacol peroxidase (GPX), and other antioxidant enzymes, and to localize them inside the mitochondria. Peroxidase activity was assayed with ascorbate (APX activity) or pyrogallol (GPX activity) as substrates in the presence of inhibitors. Both activities were clearly detected and found to be drastically inhibited by KCN or azide, as would be expected for hemoproteins. The use of pCMS was critical to discriminate APX and GPX, as this compound markedly inhibits APX but has little effect on typical GPX enzymes (Amako *et al.* 1994). Previous work on the presence of GPX in

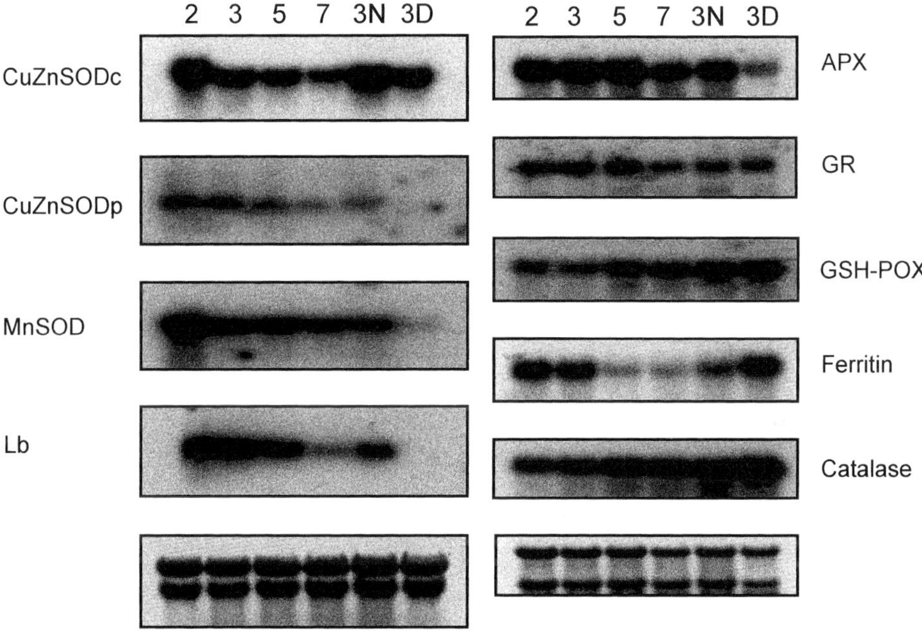

Figure 1. Expression of antioxidant enzymes during pea nodule senescence. *Lanes 2 to 7:* transcripts from 2- to 7-week-old nodules. *Lane 3N:* transcripts from 3-week-old nodules treated with nitrate for 4 days. *Lane 3D:* transcripts of 3-week-old nodules from plants exposed to continuous darkness for 4 days. Abbreviations: APX = ascorbate peroxidase; CuZnSODc = cytosolic CuZnSOD; CuZnSODp = plastidial CuZnSOD; GR = glutathione reductase; GSH-POX = glutathione peroxidase; Lb = leghemoglobin. Methylene blue was used to ensure uniform loading.

Table 1. Kinetic properties of the recombinant GSHS2 enzyme.

Constant[a]	GSHS activity	hGSHS activity	hGSHS/ GSHS
V^b (nmol min^{-1}mg^{-1} protein)	158 ± 22	3433 ± 137	21.7
V_m (nmol min^{-1}mg^{-1} protein)	4199 ± 108	4719 ± 430	1.12
K_m (mM)	104 ± 9	1.9 ± 0.3	0.018
Specificity constant (V_m/K_m)	41 ± 4	2350 ± 331	57.3

[a]Values are means $\pm$ SE (n=3).

[b]Activity rate using the standard concentration (5 mM) of Gly or βAla for the enzyme assay.

Table 2. Specific activities and intramitochondrial localization of antioxidant enzymes.

Enzyme[a]	Activity	Latency (%)	Solubilization (%)
MnSOD	42 ± 1	68 ± 5	85 ± 8
APX	552 ± 21	0	11 ± 6
GPX	2473 ± 205	46 ± 3	48 ± 6
MR	104 ± 9	20 ± 4	75 ± 1
DR	26 ± 1	45 ± 10	82 ± 1
hGR[b]	8 ± 1	53 ± 3	81 ± 5
Cytochrome c oxidase[c]	3984 ± 189	90 ± 1	3 ± 1
Malate dehydrogenase[c]	63 ± 3	86 ± 2	77 ± 1

[a] MnSOD activity is expressed in units mg^{-1} and malate dehydrogenase activity in μmol min^{-1} mg^{-1}. All other enzyme activities are expressed in nmol min^{-1} mg^{-1}. Latency was determined in hyposmotic and isosmotic media with or without 0.02% Triton X-100, respectively, and was calculated by the formula of Burgess *et al.* (1985). Values are means $\pm$ SE of 3-6 replicates, each corresponding to an independent nodule extract.

[b] hGR activity was assayed using oxidized GSH as substrate.

[c] Cytochrome c oxidase and malate dehydrogenase were used, respectively, as marker enzymes of the inner membrane and the matrix.

mitochondria has produced contradictory results. Prasad *et al.* (1995) detected membrane-bound GPX activity in maize leaf mitochondria, whereas Jiménez *et al.* (1997) did not find genuine GPX activity in pea leaf mitochondria. We have found GPX activity in nodule mitochondria. In fact, solubilization data indicate that 53% of the activity is in the matrix and that there is also significant activity associated with the membranes (Table 2). An interesting observation is that the mitochondrial APX is in the membrane and does not lose its activity in the absence of ascorbate. In this respect, the APX of nodule mitochondria behaves similarly to the cytosolic enzyme (Dalton 1995) but contrary to the enzymes of the chloroplasts (Amako *et al.* 1994), leaf peroxisomes (Jiménez *et al.* 1997), or potato tuber mitochondria (De Leonardis *et al.* 2000), which are inactivated by H_2O_2 in the absence of ascorbate. The zero latency of APX (Table 2) indicates that exogenous ascorbate is directly accessible to the enzyme and therefore that APX is located in the outer membrane or in the outside of the inner membrane. To localize APX more precisely, mitochondria were subjected to controlled osmotic lysis and separated into the outer membrane-enriched fraction, inner membrane-enriched fraction, and matrix. These fractionation studies revealed that most APX activity is associated with the inner membrane, with a small proportion of the activity being in the matrix. These data are consistent with the observation that only 11% of the APX activity remains in the soluble fraction (matrix) of mitochondria (Table 2).

The three other enzymes (DR, MR, GR) of the Halliwell-Asada pathway and MnSOD were found in the matrix of nodule mitochondria (Table 2). Mitochondria contained 13 ± 2 nmol of hGSH mg^{-1} protein, a level that is comparable to that of whole bean nodules (Matamoros *et al.* 1999). Therefore, the GR enzyme found in mitochondria should be functionally designated as a hGSH reductase (hGR). Because bean nodule mitochondria are devoid of the hGSH biosynthetic enzymes, γ-glutamylcysteine synthetase and hGSHS, whereas the latter activity is present in the nodule cytosol, it follows that the mitochondria obtain hGSH from the cytosol and utilize it as an antioxidant metabolite in place of GSH. Oxidized hGSH is recycled by hGR using NADPH.

We propose that the H_2O_2 formed in the inner membrane by the electron transport chain is scavenged by the APX located in the inner membrane. The ascorbate required for APX activity can be provided by the enzyme L-galactono-γ-lactone dehydrogenase, which is also localized in the inner membrane (Siendones *et al.* 1999). The resulting oxidized compounds, monodehydroascorbate and dehydroascorbate, can be recycled in the matrix or in the cytosol since both subcellular compartments contain MR, DR, and hGR. In the mitochondrial matrix, the H_2O_2 formed as a result of oxidative metabolism and of MnSOD activity could be scavenged directly by hGSH and antioxidant enzymes such as APX and GPX.

4. References

Amako K *et al.* (1994) Plant Cell Physiol. 35, 497-504

Becana M *et al.* (1998) Plant Soil 201, 137-147

Becana M *et al.* (2000) Physiol. Plant. 109, 372-381

Burgess N *et al.* (1985) Planta 166, 151-155

Dalton DA (1995) In Ahmad S (ed), Oxidative Stress and Antioxidant Defenses in Biology, pp. 298-355, Chapman and Hall, New York

De Leonardis S *et al.* (2000) Plant Physiol. Biochem. 38, 773-779

Gogorcena Y *et al.* (1997) Plant Physiol. 113, 1193-1201

Jiménez A *et al.* (1997) Plant Physiol. 114, 275-284

Klapheck S (1988) Physiol. Plant. 74, 727-732

Matamoros MA *et al.* (1999) Plant Physiol. 121, 879-888

Millar AH *et al.* (1995) Plant Cell Environ. 18, 715-726

Moran JF *et al.* (2000) Plant Physiol. 124, 1381-1392

Prasad TK *et al.* (1995) Plant Physiol. 108, 1597-1605

Sambrook J *et al.* (1989) Molecular Cloning, CSH Laboratory Press, New York

Sandalio LM *et al.* (1987) Plant Sci. 51, 1-8

Santos R *et al.* (2001) Mol. Plant-Microb. Interact. 14, 86-89

Siendones E *et al.* (1999) Plant Physiol. 120, 907-912

Struglics A *et al.* (1993) Physiol. Plant. 88, 19-28

Verwoerd TC *et al.* (1989) Nucleic Acid Res. 17, 2362

5. Acknowledgements

We thank David Dalton and Frank Minchin for helpful comments. This work was funded by grant PB98-0522 from the Dirección General de Enseñanza Superior e Investigación Científica (Spain).

ROLE OF REACTIVE OXYGEN SPECIES IN NODULE DEVELOPMENT

D. Hérouart[1], E. Baudouin[1], R. Santos[2], C. Mathieu[1], S. Sigaud[1], A. Jamet[1], P. Evans[3],
P. Frendo[1], G. Van de Sype[1], M.J. Davies[4], B. Halliwell[3], D. Touati[2], A. Puppo[1]

[1]Lab. Biol. Végétale. Microbiol. CNRS FRE 2294, U Nice-Sophia Antipolis, Nice, France
[2]Inst. Jacques Monod CNRS UMR 7592, U Paris 6-7, Paris, France
[3]Pharmacology Group, King's College, London, UK
[4]Heart Research Inst., Camperdown, Sydney, Australia

1. Introduction

Reduction of molecular oxygen proceeds through four steps, thus generating several active oxygen species (ROS) (Elstner 1982). The reaction chain requires initiation at the first step whereas subsequent steps are exothermic and can occur spontaneously, either catalyzed or not.

$$O_2 \rightarrow O_2^- \rightarrow H_2O_2 \rightarrow OH^\bullet + H_2O \rightarrow 2\,H_2O$$

The first step in oxygen reduction produces superoxide anion (O_2^-), which can cause lipid peroxidation and oxidize specific amino acids, such as histidine, methionine and tryptophan (Fridovich 1995). The second reduction step generates hydrogen peroxide (H_2O_2), a relatively long-lived molecule that can diffuse from its site of production (Levine *et al.* 1994). H_2O_2 is toxic through oxidation of SH groups. The toxicity of H_2O_2 can be enhanced in the presence of ferrous iron via the Fenton reaction.

$$H_2O_2 + Fe^{2+} \rightarrow Fe^{3+} + OH^- + OH^\bullet$$

The last species generated by this series of reductions is the hydroxyl radical ($OH^\bullet$). It reacts with biological molecules at its site of production with almost diffusion-controlled rates.

To protect against the damage caused by ROS, cells possess a number of antioxidant enzymes and low molecular weight antioxidants, such as superoxide dismutases (SOD), catalases and peroxidases, glutathione and ascorbate (Halliwell 1990). When an imbalance between production and removal of ROS occurs, an oxidative stress appears.

There are several known sources of ROS in plants. Some of the most common include the leakage of electrons to O_2 from electron transport chains in the chloroplasts (Foyer *et al.* 1994), mitochondria (Elstner, Osswald 1994) and peroxisomes (Corpas *et al.* 2001). Other sources include cell wall oxidases and peroxidases (Bolwell, Wojtaszek 1997). These ROS, generated by the reduction of molecular oxygen in plant systems, can have deleterious effects and lead to a situation of oxidative stress but are also, in contrast, interesting candidates for cell signaling (Dat *et al.* 2000). This dual role for ROS appears to be very important in legume nodule development.

Legume root nodules, which are characterized by an early senescence, are especially at risk from oxidative damage by ROS. Leghemoglobin autoxidation has been shown to produce O_2^-, which disproportionates to form H_2O_2 (Puppo *et al.* 1981). Additionally, the reaction of leghemoglobin with H_2O_2 generates oxidizing species such as ferryl heme proteins and protein radicals (Davies, Puppo 1992). H_2O_2 can also cause protein degradation to release "catalytic" iron, i.e. iron in a molecular form that can promote lipid peroxidation and $OH^\bullet$ generation (Puppo, Halliwell 1988). The formation of ROS may additionally occur as a result of the strong reducing conditions required for nitrogen fixation, and the action of several proteins including ferredoxin, uricase and hydrogenase (Dalton *et al.* 1991). Thus, a question arises: "Does oxidative stress occur during nodule senescence?"

2. Soybean Root Nodules and ROS

An increase in the concentration of lipid hydroperoxides was observed during soybean nodule aging. The concentration of H_2O_2 itself also increased in the same period of nodule development. The H_2O_2 increase paralleled that for peroxides (Evans *et al.* 1999). H_2O_2 production was detected in ultrathin sections of senescent soybean nodules as an electron-dense deposit stained with cerium chloride (Bestwick *et al.* 1997). A significant labeling was detected in the cell walls and around the peribacteroid membranes, which are amongst the first structures to be degraded during senescence. No H_2O_2 was detected in the non-infected cells. H_2O_2 was also evidenced in the senescing zone (zone IV) of alfalfa nodules, whereas it could not be detected in the nitrogen-fixing zone (zone III).

H_2O_2 is known to react with leghemoglobin *in vitro*, to generate modified hemoproteins which most probably occur from an intramolecular cross-link of a phenoxyl radical (involving a tyrosine residue) to the heme group (Moreau *et al.* 1996). One of these species is green in color and has a pI of 5.45. This species was detected in senescing nodules but not in extracts from four-week-old nodules (Mathieu *et al.* 1997). Electron paramagnetic resonance (EPR) spectroscopy has also been employed to examine the nature of radicals present in these organs. When EPR spectra of intact and largely senescent nodules were recorded at low temperature, typical signals were obtained. Exposure of young and mature nodules to oxidative stress, in the form of exogenous H_2O_2 addition, resulted in changes in the EPR spectra with the appearance of absorptions similar to those from untreated senescent nodules. In particular, one signal is believed to be due to an altered form of leghemoglobin (Mathieu *et al.* 1998).

The catalytic iron content greatly increased during nodule development and the increase appeared to be linear with age. When the peroxide level was plotted versus iron content, a linear correlation was obtained (Evans *et al.* 1999). Proteins are also damaged, including generation of carbonyls groups on certain amino acid side chains, by oxidative processes. There was also a rise in carbonyls with nodule age. Concentrations of xanthine and hypoxanthine, deamination products of the purines adenine and guanine, also increased during senescence, pointing to the existence of some DNA damage during nodule aging (Evans *et al.* 1999). All these results are consistent with the development of an oxidative stress during nodule senescence.

3. Alfalfa Root Nodules and ROS

On the other hand, alfalfa responds to infection with *Sinorhizobium meliloti* by production of O_2^- and H_2O_2. O_2^- was detected in infection threads and in infected cells in nodules up to nine days old. H_2O_2 production was seen in walls and infection threads of infected cells (zone II). H_2O_2 was not found in the meristematic zone I and fixation zone III (Santos *et al.* 2001). Thus, alfalfa produces a prolonged oxidative burst in response to *S. meliloti* infection. This is reminiscent of the oxidative burst occurring as an early response in plant defense reactions towards pathogens. To investigate the role of the microsymbiont superoxide dismutase (SOD) in protecting against oxidative stress in the symbiotic process, the *sodA* gene - encoding the sole cytoplasmic SOD of *S. meliloti* - was isolated and a null mutant was constructed. The resulting mutant, deficient in SOD activity, was able to grow normally and was only moderately sensitive to oxidative stress when free living. In contrast, its symbiotic properties in alfalfa were drastically affected (Santos *et al.* 2000). The SOD-deficient mutant nodulated poorly and displayed abnormal infection. Electron microscopy showed that bacteroid differentiation was blocked in most nodules at the level of infection zone II. In some cases, the bacteria were released into the cytoplasm without a peribacteroid membrane and did not differentiate. In other cases, the infection threads aborted and the bacteria degenerated without release (Santos *et al.* 2000).

As the SOD activity leads to the formation of H_2O_2, which has been detected in the infection threads, the symbiotic behavior of *S. meliloti* mutants lacking catalase activity was also investigated. *S. meliloti* contains three catalase genes, named *katA, katC* and *katB*, which encode two monofunctional catalases and a bifunctional catalase/peroxidase, respectively (Sigaud *et al.* 1999). Mutants lacking one of the catalases were not affected in their symbiotic capacities. However, a *katA/katC* double mutant exhibited a reduced efficiency of nodulation compared to the wild type. Other double mutants (*katA/katB* and *katB/katC*) and a triple mutant are currently being constructed and their nodulation capacities will be tested. Moreover, experiments using promoter-lacZ fusions showed that *sodA* was strongly expressed in infection threads, whereas a differential expression of the catalase genes was observed. This further confirms that the bacteria have to face an oxidative challenge during the infection process. In this framework, it must be noted that a bacterial gene, encoding a protein with a sequence similar to those of peroxiredoxins, has been isolated along with other genes induced during symbiosis (Oke, Long 1999).

4. Conclusions

All these results clearly show that O_2^- and H_2O_2 play an important role in the first step of nodule organogenesis. Furthermore, several experimental data suggest that another activated species - nitric oxide ($NO^•$) - should also be considered. Indeed, (i) the presence of a leghemoglobin-nitric oxide complex in young soybean nodules (Mathieu *et al.* 1998), (ii) the detection of $NO^•$ with a fluorescent probe in alfalfa nodules and (iii) the effect of a $NO^•$ scavenger on the nodulation process, strongly suggest that $NO^•$ is also involved in the oxidative burst described above and plays a role in the establishment of the symbiosis.

This oxidative burst in response to symbiotic infection can be consistent with rhizobia being initially perceived as invaders by the plant. However the plant reaction appears to limit only the infection and does not lead to the rejection of invading symbiont (Vasse *et al.* 1993). H_2O_2 and peroxidase may be involved in the regulation of infection (Cook *et al.* 1995), but a question arises: why does the oxidative burst not trigger plant defense reactions? It can be suggested that rhizobia inhibit signaling pathways, leading to the deleterious defense cascade. The prolonged maintenance of the oxidative burst might keep the plant's options open, allowing it, in the absence of the appropriate bacterial signal, to switch to a defense response. On the other hand, the oxidative burst can be considered as a process necessary for the establishment of the symbiotic interaction. In this framework, it could trigger the expression of plant and/or bacterial genes, which are essential for the nodulation process. It has been reported that an oxidative burst at the very early stage of infection is necessary for the induction of Rip1 peroxidase in the root of *Medicago truncatula* (Ramu *et al.* 1999). Moreover, Nod factors stimulate the oxidative burst and pharmacological effectors that inhibit the burst were shown to block induction of plant marker genes for nodulation (D.R. Cook, personal communication). In any case, it appears that ROS play a key role in the early steps of nodule development.

5. References

Bestwick *et al.* (1997) Plant Cell 9, 209-221
Bolwell GP, Wojtaszek P (1997) Physiol. Molec. Plant Pathol. 51, 347-366
Corpas FJ *et al.* (2001) Trends Plant Sci. 6, 145-150
Cook D *et al.* (1995) Plant Cell 7, 43-55
Dalton DA *et al.* (1991) Plant Physiol. 96, 812-818
Dat J *et al.* (2000) Cell. Mol. Life Sci. 57, 779-795
Davies MJ, Puppo A (1992) Biochem. J. 281, 197-201
Elstner EF (1982) Ann. Rev. Plant Physiol. 33, 73-96
Elstner EF, Osswald W (1994) Proc. R. Soc. Edinb. B 102, 131-154

Evans PJ *et al.* (1999) Planta 208, 73-79
Foyer CH (1994) Plant Cell Environ. 17, 507-523
Fridovich I (1995) Ann. Rev. Biochem. 64, 97-112
Halliwell B (1990) Free Radical Res. Comm. 9, 1-32
Levine A *et al.* (1994) Cell 79, 583-593
Mathieu C *et al.* (1997) Free Radical Res. 27, 165-171
Mathieu C *et al.* (1998) Free Radical Biol. Med. 24, 1242-1249
Moreau S *et al.* (1996) J. Biol. Chem. 271, 32557-32562
Oke V, Long SR (1999) Mol. Microbiol. 32, 837-849
Puppo A, Halliwell B (1988) Planta 173, 405-410
Puppo A *et al.* (1981) Plant Sci. Lett. 22, 353-360
Ramu SK *et al.* (1999) In Abstr. Inter. Soc. Molec. Plant-Microbe Interac., St Paul, MN
Santos R *et al.* (2000) Mol. Microbiol. 38, 750-759
Santos R *et al.* (2001) Molec. Plant-Microbe Interact. 14, 86-89
Sigaud *et al.* (1999) J. Bacteriol. 181, 2634-2639
Vasse *et al.* (1993) Plant J. 4, 555-566

NICKEL SEQUESTERING AND STORAGE BY *BRADYRHIZOBIUM JAPONICUM*

R.J. Maier, J.W. Olson

Dept of Microbiology, Univ. of Georgia, Athens, GA 30606, USA

1. Introduction

The major class of hydrogenases that facilitate energy input via H_2 into respiratory metabolism are the NiFe uptake-type hydrogenases. Although important in bacterial physiology in general, these uptake hydrogenases play a specific role in maintaining the energetic efficiency of symbiotic nitrogen fixation. This is because H_2 is an ample energy source readily available to diazotrophs while fixing N_2, as it is made available due to its inherent production by nitrogenase. Indeed, the energetic (ATP) and reductant input into nitrogenases can in some instances favor H_2 production over NH_3 production. This situation can result in over 50% of the reductant through nitrogenase being in H_2 rather than in ammonia. In cases where energy is the limiting factor for N_2 fixation, the ability to use H_2 would seem to be an important attribute for an efficient system.

The synthesis of NiFe hydrogenases is dependent on a supply of nickel and a mechanism to "sense" that the substrate H_2 is available and the proper redox environment exists (see Maier, Triplett 1996). In addition to the structural proteins (which contain the nickel-containing active center and iron sulfur clusters) a remarkably diverse array of accessory proteins are needed to assemble hydrogenases. These have putative roles in nickel or iron binding, or in maintaining the correct structure for insertion/delivery of these metals or other active site ligands into the apoproteins (Drapal, Bock 1998; Maier, Bock 1996). One of the most intriguing of accessory proteins needed for *B. japonicum* hydrogenase expression is HypB (Fu *et al.* 1995; Olson *et al.* 1997). The HypB protein contains a histidine-rich domain at its N-terminus (Fu *et al.* 1995; Rey *et al.* 1994). These his-rich areas are present (but to a lesser extent than for *B. japonicum*) in some other HypB proteins and in other nickel binding/storage proteins involved in urease and carbon monoxide dehydrogenase synthesis as well (see Figure 1). For references to the origin for each of the sequences shown in the figure, the reader is referred to Olson and Maier (2000).

```
Bj        15 HaHdHHHdHgHdHdHgHdgHHHHHHgHdqdHHHHHdHaH 55
Rl     15 HtHevgddgHgHHHHdgHHdHdHdHdHHrgdHeHddHHHaedgsvH 60
Ss        15 HsHHHHgdgnfaHsHddHdqqeHHHHH 41
Ac           18 HHHHgydHgHHHdHafvrrpapaeaaplvveglnlH 54
Av           18 HHHHgHdHHHHeHpfvrrpapaeaappaaggpnlH 53
Ms           26 HHHeHdHdHdHpHtHdH 42


Ka UreE    144 HgHHHaHHdHHaHsH 158
Rr CooJ     82 HspfHsHaHsHdHdHaHgHsHdHaHdHcHcHdH 114
```

Figure 1. Alignment of histidine (shown as H) rich regions in nickel accessory proteins.

Interestingly, the well-studied *Escherichia coli* protein lacks the histidine-rich domain; the protein is nevertheless very important for mobilizing nickel into the active center of the *E. coli* NiFe-hydrogenase (Maier, Bock 1996). Mutant strains of *B. japonicum* were obtained that synthesize an altered HypB lacking key components of the functional protein. Evidence for a role of HypB in nickel storage came from analysis of strain Δ23H (Olson *et al.* 1997; Olson, Maier 2000). When supplemented with nanomolar levels of nickel, hydrogenase activity of the mutant (strain Δ23H) was low in comparison to the wild type. However, this phenotype could be cured to wild type levels by adding μM levels of nickel to the medium. HypB also plays a role in transcription of hydrogenase, presumably because it is also likely to be involved in nickel donation to HupUV, the regulatory hydrogenase. The properties of HypB, along with the phenotypes associated with disruption of *hypB,* justify the conclusion that this protein (like that for *E. coli*) plays a key role in Ni donation to NiFe hydrogenases (Maier *et al.* 1995; Olson *et al.* 1997). We wished to use these mutant strains to study the symbiotic efficiency associated with HypB domains, and relate the function of these domains to nickel supplementation to the plants. Due to the nickel storage ability of HypB, it is sometimes referred to as nickelin.

2. Procedures

A sequence description list of the important areas within the mutant strains of *B. japonicum* is shown (Table 1), with histidine residues in bold. Note that the wild type contains 302 amino acids whereas the mutants Δ23H, ΔEg and K119T contain 264, 73, and 302 amino acid residues, respectively.

Table 1. Pertinent properties of mutants used.

Strain	Amino Acid Sequence at the N Terminus
Wild type	(X_{12})SIE**H**A**H**D**HHH**D**H**G**H**D**H**D**HGH**D**G**HHHHHH**G**H**D**Q**D**HHHHH**D**H**A**H**G(X_{247})
Δ23H	(X_{12})SIE--**H**G(X_{247})
ΔEg	First 67 a.a. like w.t.; #68-216 deleted in frame, last 6 a.a. (at C terminus) like w.t.
K119T	Lysine at position 119 changed to threonine

Importantly, strains Δ23H and K119T were shown to synthesize detectable HypB of the expected size (determined by immunoblotting, see Olson *et al.* 1997). In addition to mutant strains deficient in histidine residues, a mutation in the G-binding domain was desired (strain K119T), as the Ni donation/mobilization function has been attributed to a GTPase activity.

Growth of soybeans in nickel-deficient conditions has been shown to adversely affect the symbiosis, presumably by causing deficiencies in urease and hydrogenase activities (Dalton *et al.* 1988). Limitation of the nickel availability to the *Rhizobium leguminosarum*-pea symbiosis resulted in lower hydrogenase activities of the bacteroids (Brito *et al.* 1994). The *B. japonicum* mutant strains were inoculated onto soybeans, and the plants were grown in hydroponic nickel deficient conditions in cellophane pouches; three plants were grown per pouch and they were harvested at 38 days after planting. The trace elements were provided as ultrapure salts. The nodules were harvested and the bacteroids were assayed for (whole cell) H_2 oxidation activity. In addition, plant tissues, including nodules and bacteroids were assayed for nickel content by atomic absorption spectrophotometry. For most of the fractions, this was done after drying the harvested fraction, grinding it to a powder, and then acid-digestion of the powder.

3. Results and Discussion

The wild type bacteroid hydrogenase activities were affected only slightly by nickel supplementation to the plants (see Table 2), but the bacteroids of the Δ23H strain were deficient in hydrogenase activity when the plants were not supplemented with nickel.

Table 2. *hypB* mutations and symbiotic hydrogenase: effect of nickel supplementation.

Strain	Ni supplied[a]	Bacteroid activity[b]
JH (wild type)	None	1.07 ± 0.20
JH (wild type)	1 µM	1.62 ± 0.35
JH (wild type)	20 µM	1.16 ± 0.19
K119T	None	<0.01
K119T	1 µM	<0.01
K119T	20 µM	<0.01
Δ23H	None	0.24 ± 0.07
Δ23H	1 µM	0.31 ± 0.07
Δ23H	20 µM	0.99 ± 0.15

[a]Ni concentration supplied to soybeans; [b]Hydrogenase activity in µmol H_2 (hr.mg protein)$^{-1}$.

At the highest Ni-supplemented level, bacteroids of this mutant approached that of the wild type. This result indicates an important role for the histidine-rich portion of HypB in storing the limited supply of nickel for hydrogenase expression. The mutant strain with a single amino acid change in the lysine residue important for GTPase activity was completely devoid of activity. This mutant is probably unable to make hydrogenase even with Ni supplementation due to an inability to incorporate the metal into hydrogenase (Olson *et al.* 1997). Similarly, strain ΔEg, containing a large in-frame deletion within nickelin had undetectable bacteroid hydrogenase activity. It lacks both the storage and the GTPase functions (Olson, Maier 2000).

Table 3. Nickel content (ng/mg protein) of acid-solubilized fractions of nodules and bacteroids.

Location	Strain	Ni^{2+} supplied	Ni content
bacteroids	JH (wild type)	1 µM	3794 ± 211
	ΔEg	1 µM	1761 ± 136
	Δ23H	1 µM	1468 ± 136
nodules	JH (wild type)	1 µM	454 ± 46
	ΔEg	1 µM	260 ± 65
	Δ23H	1 µM	181 ± 78
bacteroids	JH (wild type)	20 µM	6301 ± 130
	ΔEg	20 µM	3863 ± 62
	Δ23H	20 µM	4407 ± 198
nodules	JH (wild type)	20 µM	1225 ± 169
	ΔEg	20 µM	1227 ± 311
	Δ23H	20 µM	1277 ± 512

Experiments to analyze the amount of nickel associated with plant tissues from soybeans nodulated by the various mutant strains (Table 3) showed that the histidine rich portion of *B. japonicum* HypB is helpful for retaining nickel. Both bacteroids and nodules of strain JH (the wild type) had significantly more nickel than the HypB mutant strains; this was especially evident in the lower nickel (1 µM nickel-supplied) condition. The nickel content of other plant tissues (stems, leaves, roots) increased with nickel supplementation to the nutrient solution, but we observed no differences in nickel in these tissues as a function of the different nodulating strains.

Legume nodulating *B. japonicum* and *R. leguminosarum* contain a domain very rich in histidine residues at the N-terminus of HypB. This is 24 out of 36 residues for *B. japonicum* and 20 out of 45 residues for *R. leguminosarum*. The need for a protein to store nickel for an organism that forms an intimate association with legumes could be related to the high rate of production of urease by soybeans. The latter could be considered to be a competing sink for available nickel within the same (root) tissue. The results we have obtained are consistent for a role played by nickelin in sequestering/storing (for the bacteroid) whatever pools of nickel the soybean had initially.

4. References

Brito B *et al.* (1994) J. Bacteriol. 176, 5297-5303
Dalton DA *et al.* (1998) BioFactors 1, 11-16
Drapal N, Bock A (1998) Biochem. 37, 2941-2948
Fu C *et al.* (1995) Proc. Natl. Acad. Sci. USA 92, 2333-2337
Maier T, Bock A (1996) In Hausinger R (ed), Mechanisms of Metallocenter Assembly,
        pp. 173-192,VCH Publ, New York
Maier T *et al.* (1995) Eur. J. Biochem. 230, 133-138
Maier RJ, Triplett EW (1996) Crit. Revs. Plant Sci. 15, 191-234
Olson JW, Maier RJ (2000) J. Bacteriol. 182, 702-706
Olson JW *et al.* (1997). Mol. Microbiol. 24, 119-128
Rey L *et al.* (1994) J. Bacteriol. 176, 6066-6073

5. Acknowledgements

We thank Susan Maier for technical assistance. This work was supported by the US Dept of Energy, Grant No. DE-FG02-99ER20321.

UNDERSTANDING ACID TOLERANCE IN ROOT NODULE BACTERIA

M.J. Dilworth[1], W.G. Reeve[1], R.P. Tiwari[1], A.I. Vivas-Marfisi[1], B.J. Fenner[1],
A.R. Glenn[2]

[1]Centre for *Rhizobium* Studies, School of Biological Sciences and Biotechnology, Murdoch
University, Murdoch, W.A. 6150, Australia
[2]Office of the Pro Vice-Chancellor (Research), University of Tasmania, Hobart, Australia

1. Introduction

The problems for agriculture on acidic soils are especially significant in relation to legume
nodulation on such soils. The low rainfall Mediterranean climate of the eastern wheat-belt of
Western Australia represents one particular type of problem. Here the nitrogen input for wheat
crops needs to come from N_2 fixation from pasture legumes, but the growing season is inadequate
for normal annual clovers and the acidity of the soils is too great for normal strains of *Sinorhizobium
meliloti* to persist and nodulate the annual species of *Medicago* with suitable short growing season
requirements. Introduction of strains of *S. meliloti* isolated from annual *Medicago* spp. growing on
acid soils in Mediterranean Europe (Howieson, Ewing 1986) have, however, helped establish wide
areas (ca. 400 000 ha) of medic pastures in this environment. Our laboratory is trying to understand
why these introduced strains of *S. meliloti* are able to colonize, persist and nodulate in such soils
when so many others cannot.

The initial question was whether strains were acid-tolerant because they carried genes, which
others did not, or whether they simply had more effective versions of genes which were of general
occurrence. Other questions concern how many genes are specifically required for acid tolerance,
how the acid environment is perceived, how the acid-tolerant cells respond to it, and what
physiological mechanisms are brought into play. Some of these questions are addressed in this brief
review.

2. Approaches

We have used three approaches to the problem of acid tolerance:
(1) the creation of acid-sensitive mutants from acid-tolerant strains of *S. meliloti* WSM419 and
Rhizobium leguminosarum bv. *viciae* WSM710, followed by the identification of the genes
essential to acid tolerance;
(2) the use of promoterless *gusA* fusion strains to identify genes which are acid-regulated; and
(3) the use of proteomics to identify proteins whose expression is affected by either transient or long-
term exposure to acid conditions.

3. Acid-sensitive Mutants

The frequency with which transposon Tn*5*-induced mutants of either WSM710 or WSM419 are
acid-sensitive suggests that there may be 20-50 *act* genes required for *ac*id *t*olerance. All but one
class of these mutants can be "rescued" (restored to acid tolerance) (Reeve *et al.* 1993) by higher
calcium concentrations (10 mM or more).

4. Essential Genes – *actA*

Mutations in this gene result in an inability to grow at pH 6 or to maintain intracellular pH (pH_i) if the
external pH (pH_e) is less than 6.5 (O'Hara *et al.* 1989). The mutant (TG2-6) is also abnormally
copper- and zinc-sensitive at pH 7.0, but removal of copper from the medium does not restore acid-
sensitivity. Complementation resulting in restoration of acid tolerance also restores copper- and zinc-
tolerance (Tiwari *et al.* 1996a). The deduced ActA protein sequence is suggestive of a membrane

protein; its closest similarity is to the CutE protein of *E. coli*, damage to which also results in Cu-sensitivity (Rogers *et al.* 1991). In *S. meliloti*, *actA* homologs occur in all strains tested, and expression in WSM419 is constitutive (Tiwari *et al.* 1996a). The mutant in *actA* is rescued by calcium.

5. Essential Genes – *actR* and *actS*

In *S. meliloti*, these two contiguous genes form a sensor-regulator pair, mutation of either of which leads to acid-, zinc- and cadmium-sensitivity (Tiwari *et al.* 1996b); acid sensitivity also occurs with *R. leguminosarum* (Boesten *et al.* 2000). Expression of promoterless *lacZ* fused to either *actS* or *actR* is constitutive (Tiwari *et al.* 1996b), consistent with a system necessary for a major cellular stress response. A multi-copy plasmid (pRT546-6) carrying *actS* restores acid and metal tolerance to an *actS⁻* mutant but not to an *actR⁻* mutant. However, a multi-copy plasmid carrying *actR* restores tolerance to both types of mutant, possibly implying "cross-talk" to other sensor(s) or sufficient phosphorylated ActR for normal function. Calcium allows mutants with lesions in *actS* or *actR* to recover acid-tolerance. The adaptive acid tolerance response (ATR) is abolished by mutation of either *actS* or *actR*.

The deduced ActS sequence shows five strongly hydrophobic domains likely to be membrane-located, and a typical sensor C-terminal domain (Tiwari *et al.* 1996b). The deduced ActR sequence has an H-T-H motif NVSETARRLNMNRRTLQRILAK suggestive of a DNA-binding role; it has aspartate and lysine residues typical of regulators.

The sequences of the ActS and ActR proteins show marked similarities to a major group of global regulator systems, particularly those involved in regulation of photosynthetic systems (Eraso, Kaplan 1994, 1995; Joshi, Tabita 1996; Bauer, Bird 1996; Masuda *et al.* 1999) and of N_2 fixation (RegSR) in *Bradyrhizobium japonicum* (Bauer *et al.* 1998; Fischer *et al.* 2000). However, in the N_2-fixing *S. meliloti* WSM419, mutations in *actS* or *actR* do not result in a non-N_2-fixing nodule phenotype as they do in *B. japonicum*, implying a marked difference in function despite the great overall similarity to the RegSR system.

5.1. Genes responsive to ActS-ActR. To address the question of which genes are controlled by ActS-ActR, we created random mutations in *S. meliloti* with a minitransposon carrying a protomoterless *gusA* (mTn5-GNm; Reeve *et al.* 1999) in a strain (RT295S) which was chromosomally *actS⁻* but complemented with a plasmid-borne *actS*, selecting for clones which expressed *gusA* differentially between pH 7.0 and 5.7. We then cured pH-selected mutants of the *actS*-carrying plasmid by introducing an incompatible plasmid, and then inserted plasmids carrying either *actR* alone or *actR* (Fenner *et al.* 2000) and *actS*. These strains were then examined to compare expression of GusA in broth cultures against that for strains carrying only the vector plasmid, the results leading to identification of a number of genes whose expression is modulated by the ActS-ActR system.

The genes so identified include two hypothetical *S. meliloti* proteins, ribulose-*bis*-phosphate carboxylase (small subunit), a nitrate reductase component, and *fixN2O2* (*cbb₃* cytochrome oxidase components) (Table 1), though only the last has any obvious connection to a possible mechanism for an acid protection mechanism.

6. Essential Genes – *actP*

Mutants with lesions in *actP* in *S. meliloti* (RT3-27) or *R. leguminosarum* bv. *viciae* (WR1-14) are acid-sensitive but not rescuable with calcium. They are both also specifically copper-sensitive (no change in sensitivity to cadmium, mercury, silver or zinc) compared to the wild type; removal of copper from the minimal medium also restores acid tolerance. The two species are markedly different in their copper tolerance. In a minimal medium at pH 5.7, *S. meliloti* WSM419 tolerates only 50 µM added $CuSO_4$ whereas *R. leguminosarum* bv *viciae* WSM710 tolerates 500 µM. This sensitivity is also markedly influenced by the medium; with glutamate as the N-source much higher concentrations of copper are tolerated than when NH_4Cl is used, because of chelation of copper by glutamate.

Table 1. *actS-actR*-responsive *gusA* fusions in *S. meliloti*.

Mutant	Gene	Putative function	Location
RTA6S	ORF4059	Hypothetical protein	Chromosome
RTA15S	*cbbS*	Ribulose-*bis*-phosphate carboxylase small subunit	pSymB
RTG47S	*narB*	Assimilatory nitrate reductase large subunit	pSymB
RTH48S	ORF2871	Binding-protein dependent transporter	Chromosome
RTL19S	ORF795	Hypothetical protein	Chromosome
RTM11S	*hyuA*	Hydantoin utilization protein	pSymB
RTN37S	*gst1*	Glutathione S-transferase	Chromosome
RTO33S	ORF 888	Two component receiver protein	Chromosome
BF1212S	*fixN2O2*	*cbb₃* cytochrome oxidase components	pSymA

The sequences of the inactivated genes are very similar and typical of P-type ATPases. They belong to the CPx sub-family of the ATPases that recognize and transport heavy metals. In *R. leguminosarum*WSM710, expression of an *actP-gusA* fusion is induced specifically by increasing copper concentrations at both pH 5.7 and 7.0. For the same concentration of added copper the induction is much greater at pH 5.7 than at 7.0, probably because decreased chelation at the lower pH results in a much greater copper concentration.

Downstream from *actP* in both organisms is *hmrR*, the sequence of which places it in the *merR* family of heavy metal regulators. Because of the copper induction of *actP*, we made sure that the inactivation of *actP* with Tn5 was not a polar effect on *hmrR* by showing complementation of the *actP* mutation with a DNA segment carrying *actP* with an active *hmrR*, or with one where the *hmrR* had been insertionally inactivated. When the chromosomal copy of *hmrR* is inactivated in WSM710, the basal level of expression of an *actP-gusA* fusion at both pH 7.0 and 5.7 is increased, but the response to copper concentration is abolished. HmrR therefore appears to be a second member of the MerR family responsive to copper and to belong to the CopR subgroup (Peterson, Moller 2000); unlike the *copR-copA* system in *E. coli*, HmrR appears to be both a positive and negative regulator.

7. Essential Genes – *exoH*

S. meliloti WRR1 is a Tn5-induced mutant of WSM419 isolated as being highly acid-sensitive; it also has a zinc-sensitive phenotype. As with the *exoH* mutant from *S. meliloti* 1021 (Leigh *et al.* 1987), its EPS1 is not succinylated, nor does the high molecular weight form normally get cleaved to low molecular weight forms. However, the symbiotic phenotype in WRR1 is not one of empty ineffective nodules as in strain 1021. Sub-cloning from an 18 kb fragment carrying *exoI* to *exoP* and contiguous genes from 1021 (Leigh *et al.* 1987) indicates that a fragment containing only *exoH* complements both acid- and zinc-sensitive phenotypes. The mechanism behind the acid- and zinc-sensitivity of WRR1 is still obscure, but may involve succinylation of a cellular component other than EPS1.

What is remarkable about these acid-sensitive mutants is that all show additional metal-sensitive phenotypes (Table 2). This connection is unclear except in the case of *actP*, where copper-associated induction of *actP* mediated by HmrR presumably allows the cell to export the increased amount of copper entering it as a result of low pH-induced increase in copper availability.

Table 2. Metal sensitivities of acid-sensitive mutants.

Organism	Mutant	Lesion in	Rescue by Ca^{2+}	Sensitivity response to			
				Low pH	Cu^{2+}	Cd^{2+}	Zn^{2+}
S. meliloti	TG2-6	*actA*	Yes	Yes	Yes		Yes
S. meliloti	TG5-46	*actR*	Yes	Yes		Yes	Yes
S. meliloti	RT295	*actS*	Yes	Yes		Yes	Yes
S. meliloti	RT3-27	*actP*	No	Yes	Yes		
R. leguminosarum	WR1-14	*actP*	No	Yes	Yes		
S. meliloti	WRR1	*exoH*	Yes	Yes			Yes

8. Acid-induced Genes Identified with Fusions

Mutants with genes transcriptionally activated at low pH have been obtained by using insertions carrying a promoterless *gusA* gene. The sequences surrounding the insertions have allowed identification of a number of them, including a putative regulator (*phrR*; Reeve *et al.* 1998), a membrane protein (*lpiA*), ABC transporters, *kdpBC* (for K^+ transport), a peptide synthetase, and *fixNO* (for cytochrome oxidase *cbb3*). These insertions do not destroy acid-tolerance, though the functional products may be required to mount an effective ATR.

8.1. *phrR* gene. This gene is adjacent to the essential *actA* gene. Its sequence suggests that it encodes a regulator protein; it has an H-T-H motif for DNA-binding, but thus far we do not know what it regulates. It is induced 5-fold by low pH, but its responses to other stresses (copper, zinc, ethanol, hydrogen peroxide) indicate that it functions in a range of stress response systems. However, not all stresses induce its expression – it does not respond to high temperature, high sucrose concentration, phosphate starvation or in stationary phase. Its expression is not dependent on an intact *actSR* system, and we do not know how it is controlled. Its inactivation results in no loss of symbiotic effectiveness in nodulation tests with *Medicago murex*, nor is there any impairment of acid tolerance (Reeve *et al.* 1998).

8.2. *lpiA* gene. This gene is the most strongly induced by low pH of any of those identified using *gusA* fusions. Unlike *phrR*, its induction is specific to low pH – none of the other stresses to which *phrR* responds results in induction. It nodulates *M. murex* normally and effectively, and staining with X-glc indicates that it is strongly expressed in the nodules. Increasing concentrations of calcium markedly affect the pH at which induction occurs, the calcium preventing induction until a lower pH is applied. Its induction, like that of *phrR*, does not depend on an intact *actSR* system.

9. The Proteomic Approach

This approach, a joint venture with the group of Dr Michael Djordjevic at the Australian National University, has examined the protein profiles for cells of WSM419 grown at pH 7.0 and exposed to pH 5.7 either transiently or long-term. The gels detected some 52 proteins whose expressions are changed by low pH exposure; of those identified by micro-sequencing, 9 were down-regulated and 7 up-regulated. There was no overlap between the group of proteins identified from proteome analysis and those predicted from the genes whose insertional fusions were induced by low pH.

10. Summary

The genes identified as essential for acid tolerance (Table 2) are probably only a small proportion of the total required, if the frequency of mutation to acid sensitivity correctly suggests 20-50 genes. The gene functions so far identified are quite diverse, and the lack of apparent linkage between them poses obvious problems for genetic transfer of acid tolerance from strain to strain. Thus far, all genes for acid tolerance appear to be present in all strains, raising the question of what subtleties of function or control are present in the acid-tolerant strains.

The connection between acid- and metal-tolerance may mean that for the root nodule bacteria the toxic effects of low pH may involve components of both proton excess and heavy metal toxicity.

While the *actS-actR* system is constitutively expressed and essential for acid tolerance, it is also clearly a global regulator akin to regulators in photosynthetic bacteria. What it perceives (pH, redox state, pO_2), where the signal is received, and how it passes from ActS toActR and onward to regulate gene expression, remain to be discovered.

In *S. meliloti* WSM419, it seems likely that there are two other systems responding to low pH – the *phrR* system and that controlling *lpiA*. The specificity of the low pH induction of *lpiA* makes it the more interesting in relation to low pH control, but that involving *phrR* is presumably part of the system integrating other cellular stress responses with that to low pH.

11. References

Bauer CE, Bird TH (1996) Cell 85, 5-8

Bauer E *et al.* (1998) J. Bacteriol. 180, 3853-3863

Boesten BA *et al.* (2000) In Olivares J, Palomares AJ (eds), Proc. 4th European Nitrogen Fixation Conf., p. 319, Sevilla, Spain

Eraso JM, Kaplan S (1994) J. Bacteriol. 176, 32-43

Eraso JM, Kaplan S (1995) J. Bacteriol. 177, 2695-2706

Fenner BJ *et al.* (2000) In Pedrosa, FO, Hungria M, Yates MG, Newton WE (eds), Proc. 12th Int. Conf. Nitrogen Fixation, p. 488, Kluwer, Dordrecht, The Netherlands

Fischer HM *et al.* (2000) In Olivares J, Palomares, AJ (eds), Proc. 4th European Nitrogen Fixation Conf., p. 108, Sevilla, Spain

Howieson JG, Ewing MA (1986) Austral. J. Agric. Res. 37, 55-64

Joshi MM, Tabita FR (1996) Proc. Natl. Acad. Sci. USA 93, 14515-14520

Lee JA *et al.* (1987) Cell 51, 579-587

Masuda S *et al.* (1999) J. Bacteriol. 181, 4205-4215

O'Hara GW *et al.* (1989) Appl. Environ. Microbiol. 55, 1870-1876

Petersen C, Moller LB (2000) Gene 127, 15-21

Reeve WG *et al.* (1993) Soil Biol. Biochem. 25, 581-586

Reeve WG *et al.* (1998) Microbiol. 144, 3335-3342

Reeve WG *et al.* (1999) Microbiol. 145, 1507-1516

Rogers SD *et al.* (1991) J. Bacteriol. 173, 6742-6748

Tiwari RP *et al.* (1996a) Microbiol. 142, 601-610

Tiwari RP *et al.* (1996b) Microbiol. 142, 1693-1704

Section 9:
Regulation of N_2 Fixation and Metabolism

CHAIR'S COMMENTS: RECENT PROGRESS IN THE REGULATION OF NITROGEN FIXATION AND NITROGEN ASSIMILATION GENES

C. Elmerich

Microbiologie et Environnement, URA CNRS 2172, Institut Pasteur, 25 rue du Dr. Roux, 75724 Paris Cedex 15, France

As new diazotrophic species are discovered (e.g. Moulin *et al.* 2001), there is continuous progress in the number of species for which we gather advanced molecular genetics of the *nif* and *fix* clusters (e.g. Lee *et al.* 2000; Desnoues *et al.* this volume). In addition, genomics projects for an increasing number of bacteria bring about new perspectives in our knowledge of the degree of conservation and evolution of nitrogen fixation and nitrogen assimilation genes.

1. Complexity of the Regulatory Cascades

Although several regulatory genes are common to most diazotrophs, in particular within the proteobacteria, the regulatory cascades differ from one species to the other. The case of *Bradyrhizobium japonicum* was reviewed by Hans-Martin Fischer. This bacterium contains two oxygen-responsive regulatory cascades: the FixLJ-K2 system controls the functions in anaerobic and microaerobic metabolism, whereas the RegSR-NifA system controls genes essential for symbiotic nitrogen fixation. The use of competitive DNA-RNA hybridization techniques identified new NifA regulated genes that appeared not to be essential for the symbiosis (Nienaber *et al.* 2000). Further analysis of the RegSR two-component system was presented including the identification of a RegR DNA-binding site and the description of RegSR homologs present in other bacteria. Cyanobacteria, which are phylogenetically distant from proteobacteria, possess a completely different network. Enrique Flores described the properties of NtcA, a transcriptional regulator protein from the cAMP receptor protein family, that binds DNA sequences in the promoter region of cyanobacterial genes governing nitrogen sources utilization and heterocyst differentiation (Herrero *et al.* 2000). A model of heterocyst development was then discussed. A substantial amount of information on the cascades in *R. etli*, *S. meliloti*, *A. vinelandii*, *Azospirillum*, *Herbaspirillum* and different photosynthetic bacteria was also reported during the Poster Presentation Session.

2. NifM is a Peptidylproline *cis-trans* Isomerase (PPIase)

Occurrence of a *nifM* (or *nifM*-like) gene seems to be limited to members of the gamma subgroup of proteobacteria, including non-nitrogen-fixing species, such as *Escherichia coli* and *Pseudomonas aeruginosa*. An analysis of the role of the NifM protein of *A. vinelandii* was reported by Nara Gavini. NifM was found to display PPIase activity. An interaction between NifM and the nitrogenase Fe protein was detected using the yeast two-hybrid technique.

3. PII and Ammonium Transporters

The involvement of PII and parologs in the regulatory cascades remain of major interest. PII proteins are small trimeric proteins playing an overall role in ammonia-sensing mechanisms (Arcondéguy *et al.* 2001). Most nitrogen fixers studied so far carry two copies of *glnB*-like genes, namely *glnB* and *glnK*. A single copy, *glnK*, was found in *A. vinelandii*, whereas three copies, *glnB*, *glnK* and *glnY* are present in *Azoarcus* (Martin *et al.* 2000) and in *Rhodospirillum rubrum* (Zhang *et al.* this volume). In general, bacterial genes encoding ammonium transporters (*amtB*) are located downtream of *glnK* genes. AmtB is a membrane-anchored protein that belongs to a family of transporters conserved in bacteria, fungi, plants and animals. The purification and properties of the

AmtB protein in *E. coli* was reported by Mike Merrick, who also showed how GlnK could associate with AmtB in the membrane and proposed a model of regulation of ammonium transport by PII.

4. Regulation of the Nitrogenase Activity

Two mechanisms for the regulation of the nitrogenase Fe-protein activity have been described in *Rhodobacter capsulatus*. One involves the classical ADP-ribosylation process, whereas the second one is independent of ADP-ribosylation (Yakunin, Hallenbeck 1998). *R. capsulatus* contains two *nifA* genes and one *anfA* (see Masepohl *et al.* this volume). In this bacterium, a double *glnB-glnK* mutant escapes ammonia control both at the level of the synthesis of nitrogenase and the regulation of its activity. A review of the current situation was presented by Patrick Hallenbeck, who discussed the possible involvement of the PII and Amt proteins in the regulatory processes.

5. Sensing of Redox and Nitrogen Status by the NifL Protein and Regulation of NifA Activity

It has been established for many years that NifL and NifA form an atypical two-component sensor-response regulator system. Ray Dixon reported refined information on the mechanisms of NifL modulation of NifA activity in *Azotobacter vinelandii* (Little *et al.* 2000; Reyes-Ramirez *et al.* 2001). NifLs N-terminal part contains a PAS domain typical of redox sensing proteins. Oxidation of the FAD moiety converts NifL into an inhibitory form. The C-terminal domain of NifL is responsible for sensing the nitrogen status through the binding of the PII (GlnK) protein in its non-uridylylated form. 2-Ketoglutarate was found to prevent the inhibition of NifA by NifL in response to ADP. The NifA N-terminal part itself carries a GAF-regulatory domain that is present in signaling molecules, which bind small cofactors. This domain is necessary to relieve inhibition when 2-ketoglutarate is present. Thus, the GAF domain of NifA can be considered as the sensor domain for the carbon status of the cell. This is also likely to be the case in different bacteria that do not contain a *nifL* gene. Indeed, it was previously reported that the N-terminal domain of NifA is inhibitory in *Azospirillum* and in *Herbaspirillum* and that PII plays a role in modulating NifA activity (Arsène 1996, 1999; Souza *et al.* 1999). Further documentation was reported at this meeting. In particular, using the yeast two-hybrid technique, PII was found to bind to the N-terminal domain of NifA in *Azospirillum* (Chen *et al.* this volume).

6. References

Arcondéguy T *et al.* (2001) Microbiol. Mol. Rev. 65, 80-105
Arsène F *et al.* (1996) J. Bacteriol. 178, 4830-4838
Arsène F *et al.* (1999) FEMS Microbiol Lett. 179, 339-343
Herrero A *et al.* (2001) J. Bacteriol. 183, 411-425
Lee S *et al.* (2000) J. Bacteriol. 182, 7088-7091
Little *et al.* (2000) EMBO J. 19, 6041-6050
Martin *et al.* (2000) Mol. Microbiol. 38, 276-288
Moulin *et al.* (2001) Nature 411, 948-950
Nienaber A *et al.* (2000) J. Bacteriol. 182, 1472-1480
Reyes-Ramirez *et al.* (2001) J. Bacteriol. 183, 3076-3082
Souza *et al.* (1999) J. Bacteriol. 181, 681-684
Yakunin AF, Hallenbeck PC (1998) J. Bacteriol. 180, 6392-6394

THE NETWORK CONTROLLING SYMBIOTIC NITROGEN FIXATION GENES IN *BRADYRHIZOBIUM JAPONICUM*

H.M. Fischer, M.A. Sciotti, H. Hennecke

Institute of Microbiology, Eidgenössische Technische Hochschule, Schmelzbergstrasse 7, CH-8092 Zürich, Switzerland

1. Introduction

Nitrogen fixation by diazotrophic bacteria is a strictly regulated process which is controlled at multiple levels to prevent futile synthesis and activity of nitrogenase and accessory functions. Aerobic diazotrophs face the "oxygen paradox" (Marchal, Vanderleyden 2000) in that they require oxygen to generate ATP via oxidative phosphorylation and, at the same time, oxygen should be kept at a very low level to prevent irreversible damage of nitrogenase. A number of regulatory principles is conserved among many diazotrophs while other control mechanisms are specific to individual species. In rhizobia, a largely common set of regulatory components have been identified which, however, are integrated in disparate regulatory networks with distinct features (for reviews, see Kaminski *et al.* 1998; Fischer 1994, 1996). Here, a summary is presented about the current knowledge of the regulatory circuits that control symbiotic nitrogen fixation genes in the soybean symbiont *Bradyrhizobium japonicum*. Special emphasis is given to a global regulatory system (RegSR) whose function is probably not restricted to nitrogen fixation.

2. Control of *B. japonicum* Symbiotic Genes by Two Linked Regulatory Cascades

On the basis of their regulation, symbiosis-related genes of *B. japonicum* can be divided into two groups which are controlled by two largely independent, oxygen-responsive regulatory cascades (Figure 1).

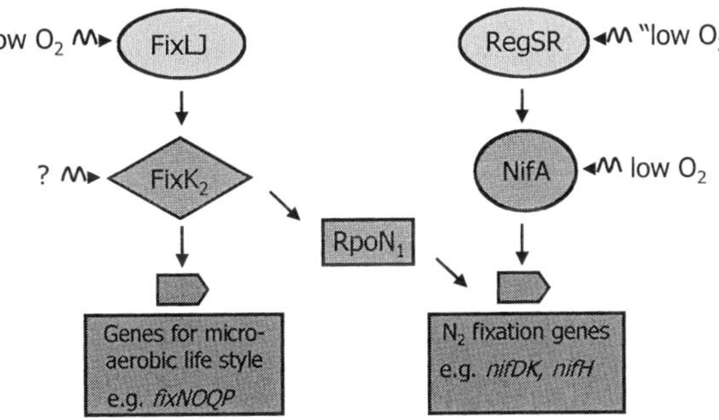

Figure 1. The regulatory cascades controlling symbiotic genes and accessory functions in *B. japonicum*. For details see text.

In the left cascade in Figure 1, oxygen sensing and transduction of the "low-oxygen" signal to the target genes is brought about by the FixLJ-FixK$_2$ cascade. FixLJ is a classical two-component regulatory system consisting of the heme-based sensor kinase FixL and the FixJ response regulator (see Tuckerman *et al.* 2001 and references therein). The only known target of FixJ in *B. japonicum* is *fixK$_2$* which encodes an FNR-type transcription activator that acts as a relay between FixLJ and the target genes (Nellen-Anthamatten *et al.* 1998). Currently, there is no evidence for an additional signal integrated by FixK$_2$. Many target genes belonging to this group are not directly involved in the process of nitrogen fixation but rather are related to the microaerobic lifestyle of bacteroids living in root nodules. Examples include genes required for synthesis and activity of a *cbb$_3$*-type terminal oxidase (*fixNOQP*, *fixGHIS*) which supports respiration in the microaerobic environment of root nodules (Preisig *et al.* 1993, 1996), heme biosynthetic genes (*hemA*; Page, Guerinot 1995); (*hemB*; Chauhan, O'Brian 1997); two *hemN* genes (Fischer *et al.* 2001) and genes for a symbiotic uptake hydrogenase (Durmowicz, Maier 1998). An additional target of the FixLJ-FixK$_2$ cascade is one of two *rpoN* genes found in *B. japonicum* (*rpoN$_1$*; Kullik *et al.* 1991). This gene encodes the specialized σ factor, σ^{54}, which is required for activation of –24/–12-type promoters associated with the *nif* and *fix* genes regulated by the second cascade.

Target genes of the right cascade in Figure 1 comprise those directly involved in nitrogen fixation (e.g. genes required for synthesis and functioning of nitrogenase) but also genes of other (e.g. *groESL* chaperonins; Fischer *et al.* 1993) or unknown functions (Nienaber *et al.* 2001). The major oxygen-responsive component of this cascade is NifA which is active only under low-oxygen conditions. Unfortunately, the sensing mechanism of NifA is only poorly understood. Indirect evidence suggests that it resembles that of FNR which harbors a redox-responsive iron-sulfur cofactor (for review see Kiley, Beinert 1999). *In vivo* activity of *B. japonicum* NifA is dependent on four conserved cysteine residues and is inhibited by the presence of metal chelators (see Fischer 1994 and references therein). Moreover, we recently obtained preliminary evidence that activity of *B. japonicum* NifA (but not that of *Klebsiella pneumoniae* NifA) is reduced in an *Escherichia coli* strain lacking the cysteine desulfurase IscS required for *in vivo* iron-sulfur cluster formation (H.M. Fischer, unpublished; Schwartz *et al.* 2000).

NifA itself is subject to a dual control mechanism in *B. japonicum*. It is encoded in the *fixR-nifA* operon which is preceded by two overlapping promoters, P1 and P2 (Barrios *et al.* 1995, 1998). The σ^{54}-dependent promoter P1 is autoregulated by NifA under low-oxygen conditions (Thöny *et al.* 1989). The activity of the P2 promoter is dependent on an upstream activator sequence centered at position –64 upstream of the *fixR-nifA* transcription start site (Bauer *et al.* 1998). It represents a binding site for the RegR response regulator of the superimposed RegSR two-component regulatory system. Under aerobic conditions, RegR mediates basal expression from the P2 promoter. Under low-oxygen conditions, *fixR-nifA* expression is increased approx. 5-fold. This is brought about not only by NifA-mediated activation of P1 but also by the elevated activity of P2 under decreased oxygen conditions as recently shown with the help of a *fixR-lacZ* reporter fusion in which P1 was inactivated by mutagenesis (M.A. Sciotti *et al.* this volume). It seems unlikely, however, that the RegSR system is responding directly to oxygen because of the lack of an obvious oxygen sensing moiety in RegS. Rather, we believe that the cellular redox-state is sensed indirectly via an alternative yet unidentified mechanism. Notably, evidence was recently presented that the RegSR-homologous system PrrBA of *Rhodobacter sphaeroides* is sensing the electron flow through the *cbb$_3$*-type cytochrome *c* oxidase (Oh, Kaplan 2000), and the redox state of the quinone pool was shown to act as a signal for the ArcBA and AppA-PpsR redox-responsive systems of *E. coli* and *R. sphaeroides*, respectively (Georgellis *et al.* 2001; Oh, Kaplan 2000).

3. Disparate Oxygen Responsiveness of the Regulatory Cascades

Using *lacZ* reporter fusions, we have compared the redox responsiveness of a FixLJ-dependent gene (*fixK$_2$*) with that of a NifA-dependent gene (*nifH*; see also M.A. Sciotti *et al.* this volume). As depicted in Figure 2, FixLJ-mediated activation was observed at relatively high oxygen concentrations (≤5%) whereas significant activation by NifA is observed only at oxygen concentrations of ≤0.5%. Induction factors were larger for the NifA- than for the FixLJ-controlled fusion because the latter was expressed at a significant level even under normal aerobic conditions. Interestingly, *fixK$_2$-lacZ* expression was rather low at ≤0.5% oxygen, which could be a consequence of the negative autoregulation of *fixK$_2$* (Nellen-Anthamatten *et al.* 1998). It is likely that these findings have physiological implications. During the early stages of symbiosis, i.e. infection and root nodule formation, the products of the FixLJ-FixK$_2$ regulon enable the invading bacteria to adapt their respiratory metabolism to the decreasing oxygen supply. Synthesis of the oxygen-labile nitrogen fixation apparatus under the control of the RegSR-NifA cascade is retarded until nodule formation and leghemoglobin synthesis have progressed to the point where free oxygen levels are sufficiently low to allow proper functioning of the nitrogenase complex.

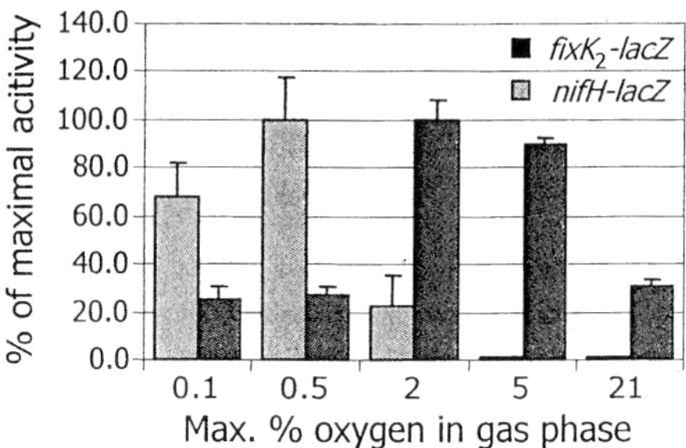

Figure 2. Disparate redox responsiveness of the *B. japonicum* regulatory cascades. Cultures of strains harboring the indicated translational *lacZ* reporter fusion-chromosomally integrated were grown at 30°C in Erlenmeyer flasks (21% oxygen) or in serum bottles that were flushed twice daily with oxygen-nitrogen gas mixtures containing the specified oxygen concentration. β-Galactosidase activity was assayed after 48 h of growth. Values are indicated as % of the maximum that each reporter strain reached.

4. RegSR: A Global Regulatory System

RegS and RegR exhibit the typical structural features of two-component regulatory proteins (Bauer *et al.* 1998). Biochemical studies with purified proteins demonstrated that a soluble variant of RegS autophosphorylates and can donate the phosphoryl group to RegR (Emmerich *et al.* 1999). In addition, RegS is able to dephosphorylate RegR-phosphate. Binding of RegR to the *fixR-nifA* upstream element is strongly stimulated by phosphorylation. A minimal RegR-binding site was defined by performing DNA binding studies with mutated derivatives of the *fixR-nifA* upstream activator sequence and an *in vitro* binding-site selection assay (SELEX; Emmerich *et al.* 2000a). The 'RegR box' comprises 11 critical nucleotides within a 15-bp imperfect inverted repeat.

RegSR-homologous systems are present in *Rhodobacter capsulatus* (RegBA), *R. sphaeroides* (PrrBA), *Synechocystis* sp. strain PCC 6803 (RppBA) and *Sinorhizobium meliloti* (ActSR) where they are involved in the control of such diverse processes as photosynthesis, aerobic respiration, fixation of CO_2 and N_2, H_2 oxidation, regulation of C_1 metabolism and acid tolerance (for references, see Swem *et al.* 2001; Emmerich *et al.* 2000b; Fenner *et al.* 2000). The functional similarity of RegR and RcgA was substantiated by heterologous complementation studies. Specifically, *regA* was able to restore symbiotic nitrogen fixation of a *B. japonicum regR* mutant, and RegR activated in *R. capsulatus* the expression of the photosynthesis operon *puc*, normally a target for RegA (Emmerich *et al.* 2000b). These results are in good agreement with the striking similarity of the RegR box with a consensus DNA-binding site for RegA that was derived from footprinting studies with a number of RegA target promoters (Swem *et al.* 2001). Thus, the *B. japonicum* RegSR systems belong to a growing class of global regulatory systems that control diverse processes involved in redox metabolism.

5. Which Other Functions Has the RegSR System in *B. japonicum*?

On the basis of its diverse regulatory functions the RegBA regulatory system of *R. capsulatus* is believed to play a key role in the maintenance of a balanced cellular redox state (see above; Swem *et al.* 2001). For example, the reductive pentose phosphate pathway (Calvin-Benson-Bassham pathway) which can serve as an electron sink during photoheterotrophic growth of *R. capsulatus* belongs to the RegBA regulon (Vichivanives *et al.* 2000; Tichi, Tabita 2000). With this in mind, we have investigated whether the *cbb* operon of *B. japonicum* encoding the enzymes of the Calvin-Benson-Bassham pathway belongs to the RegSR regulon. Unlike the results reported for

R. capsulatus, we found no effect of a *regR* null mutation on the expression of the *B. japonicum cbbFPTALSXE* operon when assayed in a strain with an intact *cbb* operon (H.M. Fischer *et al.*, unpublished). Interestingly, we observed a 5–10-fold increase of *cbb* expression in a strain carrying a chromosomal *cbbP-lacZ* fusion which reduces transcription of the distal genes, and this increase was dependent on RegR. We conclude that *cbb* expression is negatively autoregulated by a mechanism that involves RegR. It is likely that RegR exerts its effect through CbbR, the LysR-type activator required for *cbb* expression.

   B. japonicum regR null mutants are symbiotically defect (Bauer *et al.* 1998). We were interested to find out whether this phenotype is simply a consequence of the altered *fixR-nifA* expression level or whether *regR* has additional target genes required for symbiotic nitrogen fixation. To this end, we constructed a strain lacking *regR* and expressing *nifA* from the promoter of a kanamycin resistance cassette (*aphII*) inserted upstream of *nifA* in the *fixR* gene (*fixR*, whose function is not known, was shown previously to be dispensable for symbiotic nitrogen fixation; Fischer *et al.* 1986). To monitor NifA activity, a *nifH-lacZ* fusion was integrated into the same background. The symbiotic properties of these strains were determined in a plant infection test and *nifH-lacZ* expression levels were assayed in microaerobically grown cultures (Table 1).

Table 1. Symbiotic properties and microaerobic NifA activity (monitored as *nifH-lacZ* expression) of a *B. japonicum regR* mutant harboring a *nifA* expression cassette (*aphII::nifA*).

Strain	Relevant genotype	Fix phenotype	*nifH-lacZ* expression
110-48	*nifH-lacZ*	+	100 %
A11-48	*nifH-lacZ, aphII::nifA*	+	184 %
2426-48	*nifH-lacZ, ΔregR*	–	3 %
2426A11-48	*nifH-lacZ, ΔregR, aphII::nifA*	–	<1 %

   Forced expression of *nifA* was not sufficient to correct the symbiotic defect of the *regR* deletion mutant, and, most notably, no NifA activity could be detected in this strain. This result suggests that RegR is not only required for proper expression levels of *nifA* but also for NifA activity. We speculate that *regR* mutants have altered cellular redox conditions which may interfere with NifA activity. The molecular basis for the link between RegR and the formation of active NifA is currently not known. Also, it remains open whether RegR controls additional functions which are essential for symbiosis but independent of NifA.

6. References

Barrios H *et al.* (1995) J. Bacteriol. 177, 1760-1765

Barrios H *et al.* (1998) Proc. Natl. Acad. Sci. USA 95, 1014-1019

Bauer E *et al.* (1998) J. Bacteriol. 180, 3853-3863

Chauhan S, O'Brian MR (1997) J. Bacteriol. 179, 3706-3710

Durmowicz MC, Maier RJ (1998) J. Bacteriol. 180, 3253-3256

Emmerich R *et al.* (1999) Eur. J. Biochem. 263, 455-463

Emmerich R *et al.* (2000a) Nucleic Acids Res. 28, 4166-4171

Emmerich R *et al.* (2000b) Arch. Microbiol. 174, 307-313

Fenner BJ *et al.* (2000) In Pedrosa FO *et al.* (eds), Nitrogen Fixation: From Molecules to Crop
   Productivity, pp. 89-90, Kluwer Academic Publishers, Dordrecht, The Netherlands

Fischer HM (1994) Microbiol. Rev. 58, 352-386

Fischer HM (1996) Trends Microbiol. 4, 317-320

Fischer HM *et al.* (1986) EMBO J. 5, 1165-1173

Fischer HM *et al.* (1993) EMBO J. 12, 2901-2912
Fischer HM *et al.* (2001) J. Bacteriol. 183, 1300-1311
Georgellis D, Lin ECC (2001) Science 292, 2314-2316
Kaminski P *et al.* (1998) In Spaink HP *et al.* (eds), The Rhizobiaceae, pp. 431-460, Kluwer
    Academic Publishers, Dordrecht, The Netherlands
Kiley PJ, Beinert H (1999) FEMS Microbiol. Rev. 22, 341-352
Kullik I *et al.* (1991) J. Bacteriol. 173, 1125-1138
Marchal K, Vanderleyden J (2000) Biol. Fertil. Soils 30, 363-373
Nellen-Anthamatten D *et al.* (1998) J. Bacteriol. 180, 5251-5255
Nienaber A *et al.* (2000) J. Bacteriol. 182, 1472-1480
Oh JI, Kaplan S (2000) EMBO J. 19, 4237-4247
Page KM, Guerinot ML (1995) J. Bacteriol. 177, 3979-3984
Preisig *et al.* (1993) Proc. Natl. Acad. Sci. USA 90, 3309-3313
Preisig *et al.* (1996) Arch. Microbiol. 165, 297-305
Schwartz CJ *et al.* (2000) Proc. Natl. Acad. Sci. USA 97, 9009-9014
Swem LR *et al.* (2001) J. Mol. Biol. 309, 121-138
Thöny B *et al.* (1989) J. Bacteriol. 171, 4162-4169
Tichi MA, Tabita FR (2000) Arch. Micobiol. 174, 322-333
Tuckerman JR *et al.* (2001) J. Mol. Biol. 308, 449-455
Vichivanives P *et al.* (2000) J. Mol. Biol. 300, 1079-1099

7. Acknowledgements

We thank Astrid Chanfon for excellent technical assistance. Work in our laboratory is supported by grants from the Swiss Federal Institute of Technology, Zürich and the Swiss National Foundation for Scientific Research.

ROLE OF NtcA IN NITROGEN CONTROL AND HETEROCYST FUNCTION IN CYANOBACTERIA

E. Flores, A. Valladares, M.F. Vázquez-Bermúdez, J. Paz-Yepes, P. Barraillé, S. Picossi, J.E. Frías, A.M. Muro-Pastor, A. Herrero

Instituto de Bioquímica Vegetal y Fotosíntesis, C.S.I.C.-Universidad de Sevilla, Centro de Investigaciones Científicas Isla de la Cartuja, c/ Américo Vespucio s/n, E-41092 Seville, Spain

1. Introduction

The cyanobacteria are prokaryotes that belong to the domain of Bacteria and are characterized by their capability to perform oxygenic photosynthesis. The eubacterial character of cyanobacteria is reflected in the structure of their transcriptional apparatus which is similar to that of well-characterized eubacteria like *Escherichia coli* (Houmard 1994) and in their ability to act as recipients of DNA in conjugation with *E. coli* promoted by wide host-range plasmids like RP-4 (Wolk *et al.* 1984). The dominant mode of growth of cyanobacteria is photoautotrophy, fixing CO_2 through the reductive pentose phosphate pathway (Stanier, Cohen-Bazire 1977). Other metabolic points contributing to CO_2 fixation in cyanobacteria are those of carbamoyl phosphate synthetase and phosphoenolpyruvate carboxylase. Because the cyanobacteria lack 2-oxoglutarate dehydrogenase, they have an incomplete Krebs cycle that essentially functions anabolically rendering 2-oxoglutarate, which is used for the incorporation of N into the cellular organic materials. The main route for N incorporation in cyanobacteria is the glutamine synthetase-glutamate synthase pathway, although some cyanobacteria also express a glutamate dehydrogenase activity (Flores, Herrero 1994). The inorganic nitrogenous compound that is incorporated into C skeletons is ammonium, which can be taken up from the extracellular medium or produced intracellularly from other N sources.

2. Nitrogen Sources

The cyanobacteria have Amt proteins that can mediate the uptake of ammonium (as well as of its structural analog methylammonium) when it is present at very low concentrations in the extracellular medium. We refer to these proteins as Amt1 (Montesinos *et al.* 1998). Analysis of sequenced genomes has shown however that some cyanobacteria posses additional *amt* genes, like the *amt2* and *amt3* genes of the unicellular strain *Synechocystis* sp. PCC 6803 whose function remains unknown. Nonetheless, mutation of the three *amt* genes of strain PCC 6803 renders a strain that can still incorporate some ammonium from the extracellular medium (J. Paz-Yepes, unpublished), probably through a mechanism which involves diffusion of unprotonated ammonia followed by trapping of ammonium by glutamine synthetase.

Sources of N other than ammonium which are widely used by cyanobacteria include nitrate (and nitrite), urea, and dinitrogen. Nitrate and nitrite are transported into the cell in many cyanobacteria through an ABC-type transporter known as NrtABCD (Flores, Herrero 1994) or, in some other cyanobacteria, through a major facilitator superfamily permease, NrtP (Sakamoto *et al.* 1999). Nitrate is intracellularly reduced to nitrite by the ferredoxin-dependent nitrate reductase, and nitrite is reduced to ammonium by the ferredoxin-dependent nitrite reductase (Flores, Herrero 1994). A genetic structure commonly (although not universally) found for the nitrate assimilation genes is: *nir* (nitrite reductase) – gene(s) encoding the permease – *narB* (structural gene for nitrate reductase). This gene cluster constitutes an operon (Flores, Herrero 1994) with a clear polarity, i. e. transcripts corresponding to the 5' end of the operon are more abundant than those corresponding to the 3' end (Frías *et al.* 1997).

As is the case for ammonia, urea can readily diffuse through biological membranes. However, in *Synechocystis* sp. PCC 6803 and in the N_2-fixing strain *Anabaena* sp. PCC 7120 an ABC-type transporter which exhibits a high affinity for urea ($K_s \sim 100$ nM) has been identified (A. Valladares, unpublished). This transporter, encoded in *Anabaena* sp. by the *urtABCDE* operon, may have an important role in urea assimilation when urea is available at concentrations too low to allow a significant rate of diffusion into the cell. In cyanobacteria, intracellular urea is degraded to ammonium and CO_2 by a conventional bacterial-type, Ni^{2+}-containing urease.

Some cyanobacteria can also assimilate N_2 while growing phototrophically, and they have developed different strategies to protect the oxygen-sensitive nitrogenase from both atmospheric and photosynthetically generated oxygen. Separation, either in space or in time, of the nitrogenase and photosynthetic activities is commonly observed in cyanobacteria (Fay 1992). In some filamentous cyanobacteria, nitrogenase is confined to heterocysts, differentiated cells that lack oxygen-evolving photosystem II activity and exhibit a metabolism directed to provide an environment adequate for the nitrogenase activity (Wolk *et al.* 1994). The nitrogenase structural genes, *nifHDK*, are expressed in the heterocysts of aerobically grown filaments of *Anabaena* sp. PCC 7120. Heterocysts and vegetative cells exhibit an intense exchange of metabolites in the N_2-fixing cyanobacterial filament (Wolk *et al.* 1994).

3. Nitrogen Control

Growth of cyanobacteria in the presence of ammonium results in repression of most of the proteins mentioned above involved in the assimilation of sources of N alternative to ammonium or of ammonium itself. Levels of the nitrate/nitrite transporter, nitrate and nitrite reductases, ammonium/methylammonium permease, urea transporter, and glutamine synthetase are significantly lower in cells grown with ammonium than in cells grown with nitrate or incubated in the absence of combined N (Herrero *et al.* 2001). In the N_2-fixing cyanobacteria, the presence of nitrate is additionally required to attain significant levels of the nitrate and nitrite reductases. Interestingly, as has been shown for the ammonium/methylammonium transport activity in *Synechocystis* sp. PCC 6803 (Montesinos *et al.* 1998), nitrate can also determine repression when the cells are incubated with a low supply of CO_2. This observation suggests that it is a low C/N ratio, rather than simply the presence of an easily assimilable N source, that determines N repression. In N_2-fixing cyanobacteria, heterocyst development and nitrogenase synthesis are repressed not only by ammonium but also by other sources of combined N like nitrate or urea. Nonetheless, these N sources have to render ammonium in order to cause repression, and ammonium metabolism through glutamine synthetase is required for the repressive effect to be manifest (Flores, Herrero 1994).

4. The NtcA Transcriptional Regulator

ntcA was first identified in the unicellular cyanobacterium *Synechococcus* sp. PCC 7942 as a gene whose mutation caused the inability to derepress N assimilation proteins in response to ammonium withdrawal; the *Synechococcus ntcA* mutant can grow with ammonium but not with nitrate as the N source (Vega-Palas *et al.* 1990). Northern analysis has confirmed that a number of genes or operons in *Synechococcus* sp. PCC 7942 (e.g. *amt1*, the *nir* operon) are not expressed in an *ntcA* mutant, and primer extension analysis permitted to identify transcription start points whose use requires an intact NtcA protein (reviewed in Herrero *et al.* 2001). This observation, together with the regulatory behavior of revertants of the original *ntcA* mutant (Vega-Palas *et al.* 1990), allowed identification of NtcA as a positive-acting transcription factor for those genes or operons. The *ntcA* gene is widespread in cyanobacteria (Frías *et al.* 1993), and *ntcA* mutants have also been reported for the heterocyst-forming cyanobacteria *Anabaena* sp. PCC 7120 (Frías *et al.* 1994; Wei *et al.* 1994), *Anabaena variabilis* (Thiel, Pratte 2001), and *Nostoc punctiforme* (Wong, Meeks 2001). None of these mutants can develop heterocysts, which suggests that NtcA also links heterocyst development to the environmental cue of N limitation.

The deduced amino acid sequence of the NtcA protein shows a high similarity in all the cyanobacteria for which the *ntcA* gene has been identified and sequenced. NtcA consists of ~220 amino acid residues and is homologous to proteins of the CAP family of bacterial transcriptional regulators (Herrero *et al.* 2001). NtcA bears three strongly conserved regions: a 73-amino acid N-terminal region which corresponds to the CAP region of β-roll structure, a 39-amino acid central region that may correspond to the region of CAP involved in protein dimerization, and a 24-amino acid C-terminal region exhibiting features of a helix-turn-helix structure characteristic of DNA binding domains (Herrero *et al.* 2001). Band-shift assays have enabled to identify DNA fragments carrying NtcA-binding sites, and DNase I footprinting analysis has identified more precisely the binding site for NtcA upstream from a number of NtcA-regulated genes (Luque *et al.* 1994; Frías *et al.* 2000). Scrutiny of the NtcA binding sites in the promoters of twenty-one NtcA-activated genes (reviewed in Herrero *et al.* 2001) has shown that their sequences correspond to the following consensus sequence.

The strongly conserved $GTAN_8TAC$ sequence is located in these promoters ~22 nucleotides upstream from a –10 box with the sequence TAN_3T, and the spacing between the –10 box and the transcription start point is ~6 nucleotides. Thus, the characterized NtcA-dependent promoters are similar to the class II CAP-activated promoters, where the binding site for the regulator is centered at about –41.5 with respect to the transcription start point.

The sequence of an NtcA-binding site determines the affinity of NtcA for such binding site. Thus, different dissociation constants (K_d) have been found for the binding of NtcA to different NtcA-binding sites, ranging from about 33 nM to about 1.4 µM for the NtcA-binding sites of the *Synechococcus glnA* and *glnB* gene promoters, respectively (M.F. Vázquez-Bermúdez, unpublished). It is possible that the affinity of NtcA towards its binding sites in different regulated promoters helps to establish a hierarchy of gene expression under N limitation. Additionally, we have used the determination of the K_d of NtcA towards some mutated NtcA-binding sites to test the importance of certain nucleotides in the NtcA-binding site. The data obtained have corroborated the essential role of the most conserved nucleotides in establishing the binding affinity of NtcA.

5. How is NtcA Regulated?

In *Synechococcus* sp. PCC 7942, *ntcA* is a positively autoregulated gene. While in the presence of ammonium it is expressed at a low level from a constitutive promoter, after ammonium withdrawal it is induced being transcribed from a canonical NtcA-dependent promoter (Luque *et al.* 1994). It appears that the low level of NtcA protein present in ammonium-grown cells is able to activate transcription upon removal of ammonium and, therefore, that the NtcA protein is subjected to N-regulation. Consistent with this hypothesis, overexpression of the *ntcA* gene in *Synechococcus* sp. PCC 7942 and *Anabaena* sp. PCC 7120 does not result in the constitutive expression of genes activated by NtcA (I. Luque, M.F. Vázquez-Bermúdez, J.E. Frías, unpublished).

We have observed that 2-oxoglutarate stimulates the binding *in vitro* of NtcA to the *Synechococcus glnA* promoter, but not to other *Synechococcus* promoters like that of the *nir* operon (M.F. Vázquez-Bermúdez, unpublished). The K_d for NtcA binding to the *glnA* promoter in the presence of 2-oxoglutarate is about four-fold lower than in its absence. P_{II}, the *glnB* gene product, is a C/N balance signal transduction protein that is widespread in bacteria and has also been characterized in cyanobacteria (Arcondéguy *et al.* 2001). Although previous studies indicated that N-regulation of expression of the *Synechococcus nir* operon proceeds normally in a *glnB* mutant (Lee *et al.* 1998), we have now observed that induction of *amt1*, another NtcA-dependent gene, is impaired in a *Synechococcus glnB* mutant (J. Paz-Yepes, unpublished). We are currently investigating the possible role of 2-oxoglutarate and the P_{II} protein as additional elements for transcriptional N-control in cyanobacteria. The *ntcA* gene has been expressed in *E. coli* and, using constructs designed to report repression or activation of gene expression, we have observed repression provoked by NtcA as well as NtcA-dependent gene expression in this heterologous system (P. Barraillé, unpublished). These results open the possibility of studying aspects of the NtcA function in a well-characterized system like *E. coli*.

6. Role of NtcA in Heterocyst Development and Function

Heterocyst development takes place in response to N deprivation and, as mentioned above, requires NtcA. After NtcA, a number of genes whose products appear to act as early regulators of the process of development, such as *hetR*, *hetC* and *hetP*, are induced (Wolk 2000). Numerous genes that are required for the morphological and biochemical differentiation of the heterocyst are then expressed. These include genes encoding proteins involved in the biosynthesis of the heterocyst envelope polysaccharide (*hep* genes) and glycolipid (*devBCA*, *hgl* genes) layers, as well as genes encoding enzymes like nitrogenase, glutamine synthetase (*glnA*) or ferredoxin-NADP$^+$ reductase (*petH*) that function in the mature heterocyst (Wolk 2000).

The expressions of the *hetC* gene and of the *devBCA* operon are dependent on NtcA and take place from NtcA-type canonical promoters, and the binding of NtcA to DNA fragments containing these promoters has been shown *in vitro* (Muro-Pastor *et al.* 1999; Fiedler *et al.* 2001). In the mature heterocyst, the *glnA* gene is transcribed mainly from the so-called P$_1$ promoter (A. Valladares, A. M. Muro-Pastor, unpublished), a well-characterized NtcA-dependent promoter (Frías *et al.* 1994). These results indicate that in addition to being required for the process of heterocyst development to start, NtcA also has an active role in gene expression during the differentiation of the heterocyst and in the mature heterocyst. It appears therefore that N-control may be operative at different steps of heterocyst development and even in the mature heterocyst.

7. Novel, Non-Canonical NtcA-Dependent Promoters

The NtcA-dependent promoters discussed above all conform to the canonical "class II" structure on which NtcA likely interacts directly with the RNA polymerase. However, a number of cyanobacterial genes have recently been characterized whose expression is NtcA-dependent but for which a canonical NtcA-type promoter is not evident. Among these genes, there are some that may be regulated only indirectly by NtcA (e.g. the *hetR* gene), but there are others whose promoters contain some kind of NtcA-binding site. Thus, the promoters of the *cox2* (encoding a cytochrome *c* oxidase) and *cph* (encoding enzymes of cyanophycin metabolism) operons of *Anabaena* sp. PCC 7120 contain standard NtcA-binding sites which are situated farther upstream from the −41.5 position (A. Valladares, S. Picossi, unpublished). On the other hand, the promoters of the *petH* gene and of the *nifHDK* operon show, at about position −41.5, sequences that resemble but do not match that of the standard NtcA-binding site (Valladares *et al.* 1999). In these promoters, the participation of other regulators or transcription factors in NtcA-dependent activation of gene expression is an interesting possibility to be addressed in the future.

8. References

Arcondéguy T *et al.* (2001) Microbiol. Mol. Biol. Rev. 65, 80-105
Fay P (1992) Microbiol. Rev. 56, 340-373
Fiedler G *et al.* (2001) J. Bacteriol. 183, 3795-3799
Flores E, Herrero A (1994) In Bryant DA (ed), The Molecular Biology of Cyanobacteria, pp. 487-517, Kluwer Academic Publishers, Dordrecht, The Netherlands
Frías JE *et al.* (1993) J. Bacteriol. 175, 5710-5713
Frías JE *et al.* (1994) Mol. Microbiol. 14, 823-832
Frías JE *et al.* (1997) J. Bacteriol. 179, 477-486

Frías JE *et al.* (2000) Mol. Microbiol. 38, 613-625
Herrero A *et al.* (2001) J. Bacteriol. 183, 411-425
Houmard J (1994) Microbiol. 140, 433-441
Luque I *et al.* (1994) EMBO J. 13, 2862-2869
Montesinos ML *et al.* (1998) J. Biol. Chem. 273, 31463-31470
Muro-Pastor AM *et al.* (1999) J. Bacteriol. 181, 6664-6669
Sakamoto T *et al.* (1999) J. Bacteriol. 181, 7363-7372
Stanier RY, Cohen-Bazire G (1977) Annu. Rev. Microbiol. 31, 225-274
Thiel T, Pratte B (2001) J. Bacteriol. 183, 280-286
Valladares A *et al.* (1999) FEBS Lett. 449, 159-164
Vega-Palas MA *et al.* (1990) J. Bacteriol. 172, 643-647
Wei T-F *et al.* (1994) J. Bacteriol. 176, 4473-4482
Wolk CP *et al.* (1984) Proc. Natl. Acad. Sci. USA 81, 1561-1565
Wolk CP *et al.* (1994) In Bryant DA (ed), The Molecular Biology of Cyanobacteria, pp. 769-823,
    Kluwer Academic Publishers, Dordrecht, The Netherlands
Wolk CP (2000) In Brun YV, Shimkets LJ (eds), Prokaryotic Development, pp. 83-104,
    American Society for Microbiology, Washington, DC
Wong FCY, Meeks JC (2001) J. Bacteriol. 183, 2654-2661

9. Acknowledgements

Work supported by grants PB97-1137 and PB98-0481 from the Ministerio de Ciencia y Tecnología and by Plan Andaluz de Investigación (research group CVI 129), Spain. S.P. and J.P.Y. were the recipients of fellowships from the Plan de Formación de Personal Investigador, Spain. P.B. was supported by a fellowship from the European Science Foundation programme on Cyanobacterial Nitrogen Fixation (CYANOFIX).

REGULATION OF NITROGENASE IN THE PHOTOSYNTHETIC BACTERIUM, *RHODOBACTER CAPSULATUS*

P.C. Hallenbeck[1], A.F. Yakunin[1], T. Drepper[2], S. Gross[2], B. Masepohl[2], W. Klipp[2]

[1]Dept. Micro. et Immuno., U de Montréal, Montréal, Canada
[2]Lehrstuhl Biol. Mikro., Ruhr-Univ., Bochum, Germany

1. Introduction

Previous studies have shown that the regulation of nitrogenase synthesis and activity in *Rhodobacter capsulatus* is complex (Masepohl, Klipp 1996). Transcription of Fe-protein requires active NifA protein and *R. capsulatus* contains two copies of NifA, NifA1 and NifA2, which can function in this regard. Transcription of NifA requires the *R. capsulatus* NtrC and NtrB homologs whose activity is modulated in response to the nitrogen status of the cell. Once synthesized, the activity of NifA is controlled by fixed nitrogen in a manner that has been unclear. After its synthesis, nitrogenase is subject to several controls. Reversible ADP-ribosylation of Fe-protein, carried out by DraT and DraG, has the potential for modulating nitrogenase activity in response to the addition of NH_3 or sudden changes in light intensity. As well, there appears to be an additional mechanism that is capable of regulating nitrogenase activity in *R. capsulatus*, a control over nitrogenase activity that is independent of the ADP-ribosylation system since it can be observed in strains that lack DraT, DraG. These systems for the regulation of nitrogenase activity are in turn sensitive to the cellular nitrogen status. Not only are these responses seen in response to sudden changes in environmental conditions, e.g. addition of NH_3, light deprivation, but they are also observed with cells that are under steady state conditions. Thus cells that are actively fixing N_2, cells growing on glutamate, or cells grown on moderate amounts of limiting NH_3 all show some degree of ADP-ribosylation of Fe-protein (Yakunin *et al.* 1999). The degree of modification observed has been shown to be correlated with the intracellular pool of fixed nitrogen.

In terms of fast, short-term responses, highly nitrogen-limited cultures show a classical nitrogenase switch-off effect, with complete inhibition of nitrogenase activity following ammonia addition, a period of no nitrogenase activity that is proportional to the amount of NH_3 added, and then complete recovery of nitrogenase activity upon exhaustion of the added NH_3. Moderately nitrogen-limited (MNL) cultures show a different response, a "magnitude" response, with a decrease in *in vivo* nitrogenase activity that is proportional to the amount of added ammonia (Yakunin, Hallenbeck 1998). Since we found that MNL cultures appeared to have a decreased level of high-affinity ammonium transport compared with HNL cultures, we hypothesized that the switch-off process responded to a signal generated by transport of ammonium by the high-affinity system. Here we describe the presence of two putative ammonium transporters in *R. capsulatus*, AmtB and AmtY and describe the effects of mutations in the corresponding genes on these processes. As well, *R. capsulatus* possesses two P_{II} homologs, GlnB and GlnK, and we have carried out mutagenesis of the corresponding genes. GlnK and GlnB appear to be involved at all levels of nitrogenase regulation.

2. Procedures

Standard cloning procedures were used to produce a strain constitutively expressing NifA1 and the following constructs on suicide plasmids: $amtB::Km^R$, $amtY::Gm^R$, $glnK::Gm^R$. Plasmids were introduced into *R. capsulatus* using conjugation and mutant strains selected and verified. Nitrogenase activity was measured by the acetylene reduction method and ADP-ribosylation of Fe-protein was detected using Western blots as previously described (Yakunin *et al.* 1999).

3. Results and Discussion

GlnB and GlnK are involved in the regulation of NifA1 synthesis and activity. The influence of glnB and glnK single and double mutations on the nif-encoded nitrogenase system was analyzed at different levels of regulation. First, we examined the accumulation of the transcriptional activator NifA1 in *R. capsulatus* wild-type and mutant strains using an antiserum specific for this protein (results not shown). In the wild-type strain, NifA1 accumulated only in N-limited cells, which is consistent with NtrC-mediated transcriptional control of the *nifA1* gene. Inactivation of *glnB* resulted in the partial escape from NH_3 control of the regulation of *nifA1* expression, and the NifA1 protein accumulated in both N-limited, and to a lesser extent in ammonium-grown cells. Similar results were found for a *glnK* mutant that was constructed to allow the constitutive expression of *amtB*. In a *glnB-glnK* double mutant, synthesis of NifA1 was greatly enhanced both under N-limiting and N-sufficient conditions.

In order to ascertain the effects of mutations in *glnB* or *glnK* on the post-translational control of NifA activity by NH_3, a strain that constitutively expresses *nifA1* was constructed. NifA1 activity was indirectly assayed by measuring levels of Fe-protein using Western blotting with Fe-protein specific antiserum (results not shown). As described elsewhere, overexpression of NifA1 in the wild type did not lead to the significant synthesis of nitrogenase under nitrogen-replete conditions (Paschen *et al.*, submitted). These data further corroborate the hypothesis that NifA1 activity is posttranslationally controlled by the nitrogen status of the cells. Ammonium suppression of NifA1 activity is mainly mediated by GlnK as demonstrated by Fe-protein synthesis in the *glnK* mutant strain. The level of Fe-protein, but not of NifA1, varied drastically in strains lacking *glnK* depending on slight variations in culturing conditions including culture volume, growth phase, temperature and light intensity, indicating that GlnK might act in the "fine-tuning" of NifA1 activity in response to subtle changes in the environmental conditions. While a single *glnB* mutation had no effect on the regulation of NifA1 activity, extremely high amounts of Fe-protein accumulated in the *glnB-glnK* double mutant independently of whether or not fixed nitrogen was available. Thus GlnB as well as GlnK seems to be involved in the ammonium-dependent control of NifA activity.

These results were confirmed and amplified by examining the effects of these mutations on the concentrations of NH_3 needed to repress the synthesis of Fe-protein in the various strains. Batch cultures were grown on various initial concentrations of NH_3 and the quantities of Fe-protein produced were visualized by Western analysis. In the wild-type strain (Figure 1A), nitrogenase synthesis was repressed by 18 mM NH_3. The *glnB* strain was less sensitive to fixed nitrogen and 22 mM NH_3 was required to repress nitrogenase synthesis (Figure 1B). However, nitrogenase synthesis was completely derepressed in the *glnK* and *glnB-glnK* mutants where nitrogenase synthesis was observed at the highest levels of NH_3 examined (Figure 1C, D).

A : wild-type

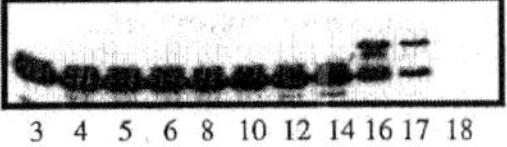

B : *glnB*

C : *glnK*

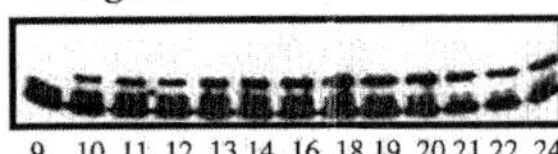

D : *glnB,glnK*

Figure 1. Batch cultures were grown on the indicated concentrations of NH_3 (mM).

3.1. The role of GlnB and GlnK in regulating nitrogenase activity. In addition to the effects observed on the ammonium control of nitrogenase synthesis, the immunoblot shown in Figure 1 also demonstrates that there are effects of mutations in *glnB* and *glnK* on nitrogenase modification by ADP-ribosylation. This long-term ADP-ribosylation previously observed with wild-type cultures under a variety of growth conditions is responsive to the cellular nitrogen status (Yakunin *et al.* 1999). In the wild-type cultures, modification is not seen until the initial NH_3 concentration used for growth is raised to 16 mM (Figure 1A). In the *glnB* strain, even higher initial NH_3 concentrations (19 mM) are required before modification is observed (Figure 1B). This suggests that GlnB normally communicates, possibly through direct interactions, the nitrogen status to DraT, the enzyme responsible for adding the ADP-ribose to the Fe-protein, and that higher levels of fixed nitrogen are needed to overcome its absence. Contrary to what is observed with the *glnB* strain, in the *glnK* strain ADP-ribosylation is seen at much lower initial NH_3 concentrations, 10 mM and higher (Figure 1C). This suggests that GlnK communicates primarily with DraG, the enzyme responsible for removing the modifying group.

Thus the accompanying model can be proposed for the various GlnB, GlnK functions. At one level (I), GlnB and GlnK interact with NtrB to control the synthesis of the NifA proteins with the respect to the nitrogen status. Both proteins appear to be necessary for the complete repression of the *nifA*s in nitrogen replete media. At a second level (II), it is primarily GlnK that controls the activity of NifA, probably by directly interacting with it. At a third level (III), GlnB and GlnK communicate with the enzymes that carry out the Fe-protein modification/demodification cycle, DraT and DraG, respectively. It is likely that many of these relationships represent direct protein-protein interactions and this is being tested using a yeast two-hybrid system. So far protein-protein interactions between GlnB and NtrB, NifA1, NifA2, GlnK, and DraT as well as between GlnK and NifA1 and NifA2 have been detected (Masepohl *et al.* 2001).

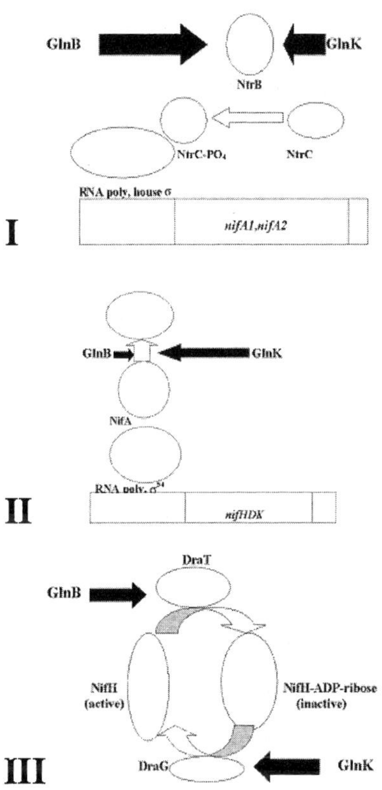

Figure 2. Model for GlnB, GlnK actions.

3.2. The role of AmtB and AmtY in regulating nitrogenase activity. What factors might control the activities of GlnK and GlnB in mediating the nitrogenase responses to NH_3 addition? Previously we had suggested that the high-affinity ammonium transport system, which could be encoded by *amtB*, might be involved since we found that its activity varied under the same culture conditions that caused a variation in the nitrogenase responses (Yakunin, Hallenbeck 1999).

The AmtBs (ammonium transport B, TC no 2.A.49.3.2) are ubiquitous membrane proteins found in all three domains of life. They are thought to either transport ammonium or to facilitate the equilibration of ammonia across the cytoplasmic membrane. It has been proposed that the yeast high affinity ammonium permease Mep2p, an AmtB homolog, functions as an ammonium sensor, generating a signal which activates signal transduction cascades that regulate filamentous growth in response to ammonium starvation (Lorenz, Heitman 1998).

R. capsulatus contains two AmtB homologs: AmtB, which is organized in a *glnKamtB* operon as in many other bacteria (Thomas *et al.* 2000), and AmtY, which is monocistronic and found a short distance, 2.5kb, from *ntrBC*. The predicted AmtB protein is highly similar to other putative bacterial and archaeal AmtBs. However, the *R. capsulatus* AmtB and AmtY proteins are only distantly related. Both *glnKamtB* and *amtY* are Ntr regulated (Gross *et al.* 2001). We created knockout insertions in these genes and studied their effects. Strains carrying *amt* insertions showed wild-type growth on all media tested, including RCV-NH₃. When their ammonium uptake activities were measured, nitrogen-limited cultures of both the *amtB* and *amtY* mutants as well as the *amtB-amtY* double mutant demonstrated very nearly wild-type rates of ammonium uptake, which is consistent with the observed growth characteristics and shows that AmtB and AmtY are not required for this activity in *R. capsulatus*. We then measured methylammonium uptake, which is diagnostic for the activity of the high-affinity ammonium transport system. The *amtB*, but not the *amtY* strain, was completely deficient in this activity.

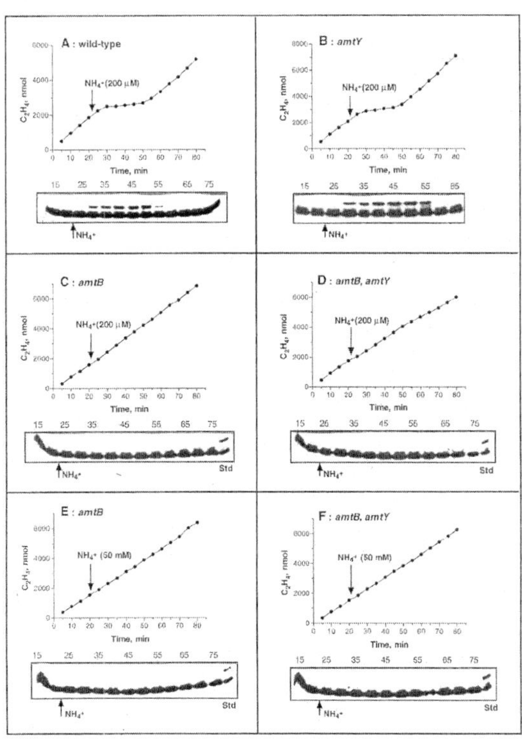

Figure 3. Nitrogenase modification in *glnB, glnK* and *glnB-glnK* mutant strains.

The ability of *amt* strains to modulate nitrogenase activity and to ADP-ribosylate the Fe-protein of nitrogenase was examined. Both the wild type and *amtY* strains demonstrated a classical *in vivo* nitrogenase switch-off response to the addition of ammonium (Figure 3). Almost complete inhibition of acetylene reduction was evident within 5 min of ammonium addition, and full recovery of the initial rate of nitrogenase activity was obtained within a relatively short period of time. In both strains the addition of ammonium also induced fast, short-term, ADP-ribosylation of Fe-protein (Figure 3). The *amtY* mutant showed higher levels of Fe-protein ADP-ribosylation than the wild-type strain in response to the addition of the same amount (200 µM) of ammonium. Moreover, in this mutant ADP-ribosylation response was faster than in wild-type strain, and the modified Fe-protein can be detected within 5 min after ammonium addition, while it takes a longer time for the wild-type strain (Figure 3). In striking contrast, the addition of the same amount of ammonium, or higher (50 mM), had no effect on the *in vivo* nitrogenase activity and Fe-protein modification state in both *amtB* and *amtB-amtY* double mutants (Figure 3). To directly demonstrate that this effect was due to the introduced mutation, we studied the effects of complementation with a plasmid-borne wild-type copy of *amtB*. Analysis of the effects of ammonium addition on nitrogenase activity and Fe-protein modification showed that the wild-type response had been restored in the plasmid-carrying strains (results not shown). This shows that the defect is caused by the mutation in *amtB*.

We have shown that the DraT/DraG system is functionally intact by examining cultures that had gone through a period of light deprivation followed by reillumination. Previous studies have shown that this treatment results in a rapid ADP-ribosylation of the Fe-protein upon removal of the light followed by fast demodification upon light shift-up. All the mutant strains gave a wild-type response in this analysis. Thus the modification system is intact in the *amtB* mutants but is apparently not receiving the proper signal upon the addition of NH_3. Several lines of evidence indicate that NH_3 is entering the cell under these conditions and that therefore the lack of response is not merely due to a defect in transport. First, gross ammonium uptake activity was unaffected. Second, we examined the regulation of glutamine synthetase. The activity of glutamine synthetase is subject to two forms of fast short-term inhibition that respond to ammonium additions, covalent modification (adenylylation) and feedback inhibition. The activity of glutamine synthetase in *amt* mutants showed the same fold decrease as the wild type upon ammonium addition. Thus, AmtB appears to specifically signal the presence of NH_3 in the external medium to the systems that control nitrogenase activity.

In conclusion, we have shown that: (1) AmtB is necessary for sensing external ammonium and initiating the switch-off and ADP-ribosylation responses; (2) GlnB and GlnK both have a role in the control of NifA synthesis; (3) it is primarily GlnK that controls NifA activity; and (4) GlnB and GlnK are involved in modulating the switch-off and ADP-ribosylation responses.

4. References

Lorenz MC, Heitman J (1998) EMBO J. 17, 1236-1247

Masepohl B, Klipp W (1996) Arch. Microbiol. 165, 80-90

Thomas *et al.* (2000) Trends Genet. 16, 11-14

Yakunin AF, Hallenbeck PC (1998) J. Bacteriol. 180, 6392-6395

Yakunin AF *et al.* (1999) J. Bacteriol. 181, 1994-2000

5. Acknowledgements

This research was supported by a grant from the Natural Sciences and Engineering Council of Canada.

ROLE OF THE NifM IN MATURATION OF THE Fe-PROTEIN OF NITROGENASE

N. Gavini, L. Pulakat

Dept Biological Sciences, Bowling Green State University, Bowling Green,
OH 43403, USA

1. The Fe-protein

Nitrogenase is composed of two separate proteins designated the Fe-protein and the MoFe-protein. The Fe-protein of nitrogenase has been reported to have four distinct functions. In order for N_2 reduction to occur, the Fe-protein must transfer electrons to the MoFe-protein (Howard, Rees 2000; Burgess, Lowe 1996). The Fe-protein is a dimer that couples the hydrolysis of two MgATPs to the transfer of a single electron; and in addition, it is also associated with the precursor of the MoFe-protein, apodinitrogenase, to facilitate its activation with the FeMo-cofactor. Our comparison of the polypeptide sequences of different Fe-proteins has revealed the existence of a remarkable degree of conservation of the amino acid sequence. A consensus sequence for the Fe-protein derived from this analysis is shown in Figure 1.

```
m-m---lrqiafYGKGGIgKSTtsqntlaalae-mgqkilivGCDPkaDsTrlilhskaqd
           DOMAIN#I                              DOMAIN#II

                                                  *
tvldaae-gsvedledledvlkeGyggikCvEsGGPePGvGCAGRGvItsinfleengay--
                             DOMAIN#III

              *
ddldyvsyDvLGDVVCGGFAmPirenkAqeiYiVmsgemMAmYAANNIskGilk
          DOMAIN#IV                       DOMAIN#V

yansggvrLGGlicNsRtdrelelieala-klgtqlihfvPrdnivqhaElrrmTviey-

pdskqadeyr-LArkihnn.gkgviPtPitmdelee.lmefgim..ede
```

Figure 1. NifH consensus sequence. Upper case letters indicate residues that are invariant. Star above the cysteins (C) indicates cluster ligands. The conserved prolines are shown in bold. Underlined regions marked with domain numbers are arbitrary.

Apart from the structural genes *nifHDK* that encode the Fe- and MoFe-proteins, a number of *nif* accessory genes are involved in biological nitrogen fixation (Jacobson *et al.* 1989). The ability of the Fe-protein to accomplish these different tasks in the N_2-fixation process, depends upon certain specific post-translational events that convert the newly synthesized inactive protein into its matured, functional form (Rangaraj *et al.* 2000 and references therein). It was reported that the products of three *nif* genes - *nifM*, *nifS* and *nifU* - have been implicated in playing a role in the maturation and assembly of the Fe-protein. The role of *nifS* and *nifU* in the activity of nitrogenase is discussed in detail by D.R. Dean in this volume.

2. The NifM Protein

The role of the *nifM* product in the maturation of the Fe-protein component of nitrogenase is obscure. Thus far, the *nifM* gene has been cloned from *Klebsiella pneumoniae*, *A. vinelandii* and *A. chroococcum*. A comparison of the amino acid sequences deduced from the nucleotide sequences of these genes clearly demonstrates the existence of a very low level of interspecies sequence identity. This lack of sequence identity could be the reason for the inability to identify *nifM* homologs in various nitrogen-fixing bacteria by using *nifM*-specific DNA probes. In fact, the confirmation that the *nifM* DNA cloned from *A. vinelandii* or *A. chroococcum* has the same function as that of the *nifM* gene of *K. pneumoniae* was established only after demonstrating that the *nifM* gene from *A. vinelandii* and *A. chroococcum* could compliment a point mutation in the *nifM* DNA of *K. pneumoniae*. However, the carboxyl terminal regions of these *nifM* products do show significant homology suggesting that the active portion of the *nifM* product is likely to be located near the C-terminal region of the polypeptide. Characterization of the activities of the nitrogenase components in *nifM* mutants of *K. pneumoniae* revealed that such strains are primarily deficient in Fe-protein activity (Roberts, Brill 1981).

The role of different *nif* genes in the maturation of nitrogenase component proteins was examined by expressing the *nifH* gene of *K. pneumoniae* in an *E. coli* background in combination with these different *nif* genes. This was accomplished by using a binary plasmid system (Howard *et al.* 1986). The *nifH* gene's expression was conducted under anaerobic conditions; and it was observed that its expression in *E. coli* in the absence of the *nifM* gene resulted in the accumulation of only very low levels of the Fe-protein polypeptide. This Fe-protein had no detectable activity as determined by an *in vitro* C_2H_2-reduction assay. These results suggested that the *nifM* gene product plays a role in conferring activity and some stability to the Fe-protein. Since isolated Fe-protein does not contain any NifM protein, it is unlikely that NifM is a subunit of the Fe-protein complex. Moreover, it was demonstrated that when the *nifH* gene was expressed in foreign hosts (*E. coli* and yeast) in the absence of the *nifM* gene product, the homodimers of the Fe-protein were still detectable. Therefore, the role of the NifM protein could be to impart activity and stability to the Fe-protein through some sort of catalytic event.

3. A Possible Function for NifM

As mentioned above, the *K. pneumoniae nifM* gene product shares very low homology with the products of the *nifM* genes from *Azotobacter*. However, some significant homology is present in the carboxyl terminal region of these proteins suggesting that the observed functional similarity (interspecies complementation) is probably confined to this region. We compared the carboxyl terminal region of the *nifM* genes and generated a consensus sequence which was then compared to the conceptually translated nucleotide sequence databases. This comparison showed that the carboxyl terminal region of the NifM proteins shares significant homology with the family of proteins called peptidyl-prolyl *cis/trans* isomerases (Figure 2).

Peptidyl-prolyl *cis/trans* isomerases (PPIases) are thought to be involved in assisting protein folding (Fisher, 1994). These proteins catalyze the *cis/trans* isomerization of the peptidyl-prolyl peptide bond in oligopeptides and proteins – a rate-limiting step in the process of protein folding essential for generating functional proteins. Some denatured proteins regain their native conformation within milliseconds to seconds; whereas, others refold very slowly and the time range varies from minutes to hours. The slow conformational changes arise from the well-known delocalization of electrons in the amide bond and are even more pronounced if additional steric constraints are imposed by the proline ring. In contrast to C(O)NH bonds, the prolyl-peptide bonds exist in the two lowest energy conformational forms – *cis* (w = 0°) and *trans* (w = 180°) – which are thermodynamically comparable. Therefore, polypeptides with *n* proline residues can exist in 2^n isomeric forms (Figure 3). When the proteins form three-dimensional structures, the *cis*- and *trans*-

isomers in these structures are no longer of equal energy and the less preferred isomer is not likely to exist in the folded chain. According to the proline hypothesis, it is assumed that the slow folding forms of the unfolded protein possess non-native isomers of the peptide bonds between proline and another residue and that the crucial steps in the refolding of the slow folding molecules are limited in rate by the slow isomerization of such incorrect proline-peptide bonds. PPIases have been demonstrated to catalyze the *cis/trans* isomerization of these peptidyl-prolyl bonds and, thus, to enhance the rate of refolding of the slow folding forms of denatured proteins.

```
                            Domain 1                      Domain 2

                           ___________            ________________________________
NifM Consensus             --h-l-t---d--el-r---------fa--a--hs-cp--a---q-lg----
A.vinelandii NifM          KAHILVTINEDFPELKRLRGK PERFAEQAMKHSECPT-AMQGGLLGEVVP
A.chroococcum NifM         ARHILVTINEDFPELKRLRGK PERFAEQAMKHSECPT-AMQGGLLGEVVP
K.pneumoniae NifM          TRHLLLTVDNDR-ELYRQINASRDAFAPLAQRHSHCPS-ALEEGRLGWISR
E.coli PpiC                ALHILVKEEKLALDLLEQIKNGAD-FGKLAKKHSICPS-GKRGGDLGEFRQ
B.subtillis PrsA           ASHILVADKKTAEEVEKKLKKGEK-FEDLAKEYSTDSS-ASKGGDLGWFAK
E.coli SurAn               LSHILIPLPENPTSIVDQARNGAD-FGKLAIAHSADQQ-ALNGGQMGWGRI
E.coli SurAc               ARHILLKPSPIMTDAADIESGKTT-FAAATKEFSQDPVSANQGGDLGWATP
B.subtillis Orf            IRHIVVKDEEEAREVLKELKGGSS-FEAVAAERSTDRYTSPYGGDLGFVTE
L.la PrtM                  VQHILTSDEDTAKQVISDLAAGKD-FAMLAKTDSIDTATKDNGGKISFELN
EST human                                          FQSPASQFSDC-SSAKARGDLGAFSK
D.disc. ORF                                              GDP--RQRGGDLGWAPA
A.thaliana PPIASE                                   FEEVATRVSDC-SSAKRGGDLGSFGR


                            Domain3

                           ___________________________
Consensus      --q-lyp-l---lf--a---ls-----s--g-h-l-ce---
NifM-A.v.      --GTLYPELDACLFQMARGELSP-VLESPIGFHVLYCESVS      [P14890]
NifM-A.c.      --GTLYPELDACLFQMARGELSP-VLESPIGFHVLFCESVS      [P23119]
NifM-K.p.      --GLLYPQLETALFSLAENALSL-PIASELGWHLLWCEAIR      [P08534]
PpiC-E.c.      --GQMVPAFDKVVFSCPVLEPTG-PLHTQFGYHIIKVLYRN      [M87049]
PrsA-B.s.      -EGQMDETFSKAAFKLKTGEVSD-PVKTQYGYHIIKKTEER      [P24327]
SurAn-E.c.     --QELPGIFAQALSTAKKGDIVG-PIRSGVGFHILKVNDLR      [S40574p]
SurAc-E.c.     --DIFDPAFRDALTRLNKGQMSA-PVHSSFGWHLIELLDTR      [S40574p]
Orf-B.s.       ASDNIPSAYIEEAKTLKEDEWSQEPIKVSNGYAIIQLKEKL      [D26185_135g]
PrtM-L.l.      -NKTLDATFKDAAYKLKNGDYTQTPVKVTDGYEVIKMINHP      [Q02473]
EST human      --GQMQKPFEDPWFARRTGEMSG-TVFTDSGIHVIVRTE        [M86110g]
Orf-D.d.       -TNYVQPFAEAVTKLKKGQLVDK-PVQTQFGWHVIQVDDTR      [X70280]
PPIASE-A.t.     GQMQKPFEEATYALKVGDISD IVDTDSGVHII             [F13919]
```

Figure 2. Comparison of the consensus sequence of the NifM protein to the amino acid sequences of the peptidyl-prolyl *cis/trans* isomerases. The three domains that share high homology are identified by overlining and are numbered arbitrarily. The fully conserved residues in each peptide are shown in bold letters.

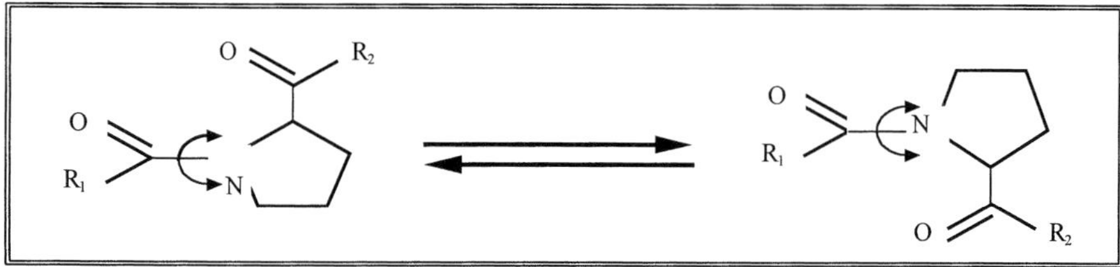

Figure 3. Differences in the structure between the *cis* and *trans* isomers of a peptide bond N-terminal to proline.

PPIases are expressed ubiquitously and the genes encoding these proteins have been identified in plants and animals as well as in lower eukaryotes and bacteria (Fisher 1994 and references therein). These proteins are also called immunophilins since they bind immunosuppressive drugs such as cyclosporin, FK506 and rapamycin. The two major families of PPIases are called the cyclophilins (cyclosporin-binding proteins) and FKBPs (FK506-binding proteins that also bind rapamycin). The two families have little sequence similarity to each other. However, within each family there are regions containing amino acids that have been highly conserved throughout evolution. These highly conserved regions span about 100 residues for cyclophilins and about 80 residues for FKBPs. In *E. coli*, two genes encoding cyclophilins and two genes encoding FKBP-related proteins have been identified so far. Apart from these genes, a new gene was recently isolated from *E. coli* that encodes a protein with PPIase activity that is not inhibited by either cyclosporin or FK506. This protein, which does not show any significant homology to the PPIases belonging to either the cyclophilin or the FKBP family, has a very low molecular mass and was called PpiC or Parvulin (Rudd *et al.* 1995). However, comparison of this protein with the predicted amino acid sequences in the database showed three regions of significant amino acid similarity between the PpiC protein and the *Bacillus subtilis* lipoprotein PrsA, the *E. coli* protein SurA and the NifM proteins (Figure 2). Based on this comparison, it has been suggested that the PpiC protein and the proteins that share amino acid similarity with PpiC define a new family of PPIases.

Thus, our comparison studies as well as others (Rudd *et al.* 1995) indicate that one of the possible functions of the NifM protein is to assist in the proper folding of the Fe-protein by catalyzing the conformational interconversions by the *cis/trans* isomerization of the peptide bond N-terminal to the proline residues present in this peptide. An examination of the consensus peptide sequence of the Fe-protein derived by comparing different Fe-proteins indeed shows the existence of seven fully conserved proline residues (Figure 1). Therefore, it is reasonable to assume that the stability and activity of the protein will largely depend upon the appropriate conformation of the peptidyl-prolyl bonds present in this protein. To test this idea, we proceeded to overexpress and purify the wild type *K. pneumoniae* and *A. vinelandii* NifM proteins and analyze these proteins for PPIase activity. Our analysis made use of the standard protease-coupled assay that is used to determine PPIase activity (Fisher 1994) and allowed us to assess the ability of the NifM protein to catalyze the *cis/trans* isomerization of the peptidyl-prolyl bond in an oligopeptide. The results suggested that the extent of PPIase activity that the NifM exhibited was comparable to that of PpiC of *E. coli*, a PPIase that shares homology with NifM.

In order to understand the NifM protein's role in modifying the stability and activity of the Fe-protein, we have also investigated existence of direct interaction between the NifM protein and the Fe-protein using the 'MATCHMAKER Two-Hybrid' yeast-based genetic assay as shown in Figure 4. Our preliminary results indicated existence of direct interaction between the Fe-protein (translationally fused to GAL4 DNA binding domain [GAL4BD]) and the NifM protein (translationally fused to GAL4 activation domain [GAL4AD]) in yeast two-hybrid protein-protein interaction assay.

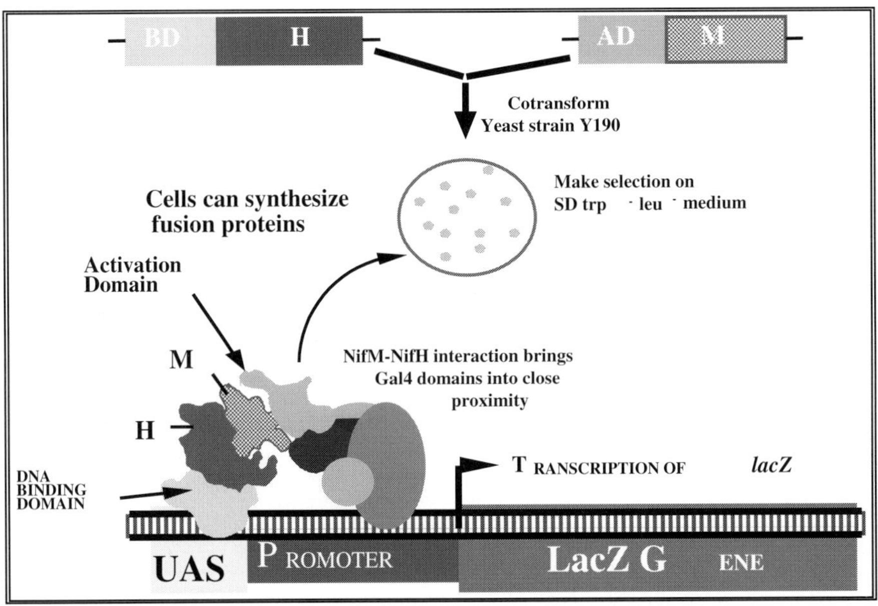

Figure 4. The general strategy used for analyzing the interaction between the Fe-protein and the NifM protein by using the 'MATCHMAKER Two-Hybrid' yeast-based genetic assay.

4. References

Burgess BK, Lowe DJ (1996) Chem. Rev. 96, 2983-3011

Fisher G (1994) Angew. Chem. Int. Ed. Engl. 33, 1415-1436

Howard KS *et al.* (1986) J. Biol. Chem. 261, 772-778

Howard JB, Rees DC (2000) In Tiplett EW (ed), Prokaryotic Nitrogen Fixation: A Model System for Analysis of a Biological Process, pp. 43-53, Horizon Scientific Press, Wymondham, UK

Jacobson MR *et al.* (1989) J. Bacteriol. 17, 1017-1027

Rangaraj P *et al.* (2000) In Tiplett EW (ed), Prokaryotic Nitrogen Fixation: A Model System for Analysis of a Biological Process, pp. 55-79, Horizon Scientific Press, Wymondham, UK

Roberts GP, Brill WJ (1981) Ann. Rev. Microbiol. 35, 207-235

Rudd *et al.* (1995) TIBS 20, 12-14

5. Acknowledgements

The research in PI's laboratory during the past several years is supported by funding from the USDA, NSF and NIH. We also thank past and present undergraduate and graduate students of Gavini/Pulakat laboratory at BGSU for their participation.

BACTERIAL AMMONIUM TRANSPORT PROTEINS: STRUCTURE AND FUNCTION

M. Merrick[1], D. Blakey[1], G. Coutts[1], G. Thomas[2]

[1]Dept of Molecular Microbiology, John Innes Centre, Norwich, NR4 7UH, UK
[2]Dept of Molecular Biology and Biotechnology, Univ of Sheffield, Sheffield, S10 2TN, UK

1. Introduction

The transport of ammonium across biological membranes is an important physiological process in all domains of life (Kleiner 1993; Knepper *et al.* 1989). Genes encoding high affinity ammonium transporters were first described in *Saccharomyces cerevisiae* and *Arabidopsis thaliana* (Marini *et al.* 1994; Ninnemann *et al.* 1994) since when more than 50 homologs have been identified. The genes encode highly hydrophobic proteins generally of between 400 and 450 amino acids that represent a unique family (Amt) of transporters found in archaea, bacteria, fungi, plants and animals (Saier Jr. *et al.* 1999). The Rhesus (Rh) family of blood antigens also show significant similarity to the Amt proteins (Marini *et al.* 1997b) and by complementation of a *S. cerevisiae* Δ*mep1,2,3* mutant it has recently been demonstrated that the Rhesus-associated glycoprotein (RhAG) and the non-erythroid Rhesus-related glycoprotein (RhCG) can indeed transport ammonium (Marini *et al.* 2000b).

With the recent explosion in bacterial genome sequences it has become apparent that nearly all bacteria and archaea encode members of the Amt family. In many cases organisms encode multiple Amt proteins, with as many as three copies being found in *Synechocystis* sp. PCC6803 and *Archaeoglobus fulgidus*. There are also three copies in *Saccharomyces cerevisiae* and six copies in *Arabidopsis thaliana*. The functions of the different paralogs are not understood but they often have different K_m values and in plants they exhibit different tissue specificities and expression profiles. In both the bacteria and the archaea the *amt* genes are almost invariably linked to a gene, designated *glnK*, encoding a small signal transduction protein that is a member of the P_{II} protein family. Conservation of gene linkage in distantly related bacteria often reflects a functional relationship between the gene products, and the linkage of *amtB* and *glnK* led us to propose that the GlnK and AmtB proteins may interact physically (Thomas *et al.* 2000a).

While there is a considerable body of data to support the proposal that the Amt proteins facilitate the entry of ammonium into the cell, the mode of action of these proteins is still a matter of debate. In its uncharged form ammonia (NH_3) is highly membrane-permeable, but energy-dependent uptake systems for ammonium have been reported for many organisms (Kleiner 1993) and the prevailing view has been that these systems function in active transport of the charged species, NH_4^+ (Kleiner 1993; Ninnemann *et al.* 1994; Marini *et al.* 1997a). However an alternative model for ammonium acquisition has been proposed by Soupene *et al.* (Soupene *et al.* 1998) who investigated the physiological role of the *E. coli* AmtB protein. The authors concluded that the AmtB protein recognizes the ammonia molecule as its substrate and that it uses a facilitated diffusion mechanism, similar to the glycerol facilitator (Heller *et al.* 1980), to catalyze equilibration of ammonia across the cytoplasmic membrane.

Secondary structural predictions for members of the Amt family suggest that they encode proteins with 10-12 trans-membrane (TM) helices with a C-terminal cytoplasmic extension (Ninnemann *et al.* 1994; Marini *et al.* 1994; Marini *et al.* 2000a; Taté *et al.* 1998). In a detailed *in silico* and *in vivo* empirical topological analysis of the *E. coli* AmtB protein we concluded that this protein has 12 TM helices with both the N-terminus and C-terminus in the cytoplasm (Thomas *et al.* 2000b). A detailed understanding of both the mode of action and the structure of Amt proteins can only be obtained through purification and biochemical analysis for which we believe the *E. coli* AmtB protein provides an excellent model system. In our recent work we have used this system to address a number of these questions.

2. Results

2.1. The mode of action of AmtB.

Since the discovery of the Amt proteins many members of this family have been demonstrated to transport the ammonium analog [^{14}C]-methylammonium ([^{14}C]-MA) (Marini *et al.* 1994; Ninnemann *et al.* 1994; Taté *et al.* 1998; Siewe *et al.* 1996). However, it is likely that most, if not all, of these assays do not directly measure accumulation of [^{14}C]-MA. This is due to the metabolic conversion of [^{14}C]-MA to impermeable species which, due to a washing step during sampling, are all that remain inside the cell after the assay. In many bacteria, including *E. coli*, this assimilation is through the action of glutamine synthetase (GS), which converts [^{14}C]-MA to [^{14}C]-methylglutamine (Barnes *et al.* 1983; Boussiba *et al.* 1991; Meier-Wagner *et al.* 2001). Soupene *et al.* concluded that AmtB does not concentrate NH_4^+ but rather that it facilitates the equilibration of NH_3 across the cytoplasmic membrane and that any apparent concentration of [^{14}C]-MA is due to metabolism by GS (Soupene *et al.* 1998). The problem of the [^{14}C]-MA transport assay was recognized by Jayakumar and Barnes (Jayakumar *et al.* 1983) who developed a novel filtration method so that unassimilated [^{14}C]-MA as well as assimilated [^{14}C]-MA could be detected. In this way they were able to show that the initial uptake phase for [^{14}C]-MA lasts no more than 5 seconds (Jayakumar *et al.* 1985). However these crucial aspects of the assay procedure were not taken into account by Soupene *et al.*

We have revisited this question and compared wild-type and Δ*amtB* strains using both assay methods. In the wild-type strain there is a clear difference between the two assays. There is rapid phase of uptake in the absence of assimilation (non-washed assay) and after 30 seconds the rates of uptake become similar in both conditions (Figure 1). By contrast the *amtB* mutant shows a negligible rapid uptake in the absence of washing and, importantly, uptake tails off rapidly as there is no assimilation.

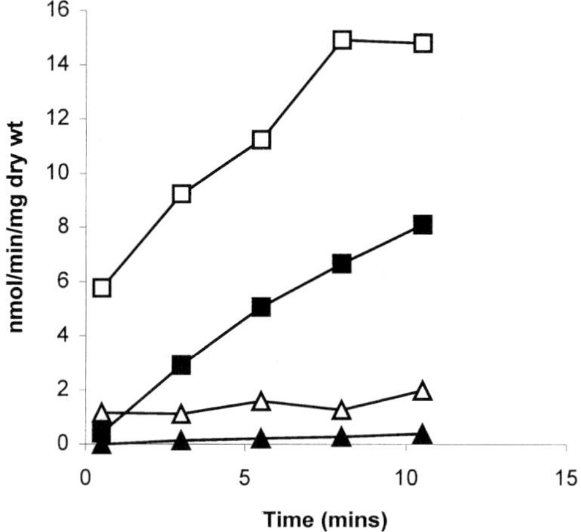

Figure 1. Effect of washing filters in [^{14}C]-MA uptake assays. Standard uptake assays were modified by using a double filter arrangement as described by Jayakumar and Barnes (Jayakumar et al 1983). Samples were either washed (filled symbols) or left unwashed (unfilled symbols). Data using the wild-type strain ET8000 (□) and the *amtB* mutant GT1001 (Δ), are shown.

The apparent low level of rapid uptake in the *amtB* mutant is likely to be due to a combination of the small amount of extracellular radiolabel remaining on the filters and non-specific association with the biomass. However, it is clear that in the presence of AmtB there is a much greater initial uptake of methylammonium, indicating that, in contrast to the report of Soupene *et al.* (Soupene *et al.* 1998), there is AmtB-dependent concentration of unmodified methylammonium by *E. coli*.

Given these results we have used the washed assay method to calculate the apparent K_m of AmtB for methylammonium. We determined the rates of [14C]-MA assimilation for the wild-type strain over a substrate concentration range of 1 μM to 50 mM. The process is saturable and follows Michaelis-Menton kinetics giving a K_m of 200 μM. This compares with two independent estimates in the literature of 78-79 mM for the K_m of GS for methylammonium. From these data we conclude that the kinetic measurements we have made reflect the transport step of the reaction. The data also support our thesis that AmtB functions as a secondary transporter that concentrates ammonium. If AmtB functioned in an analogous manner to facilitator proteins such as GlpF, one would expect the two K_m values to be very similar.

2.2. Purification of *E. coli* AmtB

To facilitate the purification of *E. coli* AmtB we chose to add a C-terminal hexahistidine (His) tag. As there is considerable conservation of primary sequence within the C-terminus we separated the His tag from the native sequence by a linker of ten residues. The intention was that this might improve accessibility of the His tag in the protein and thereby increase the efficiency of Ni^{2+} affinity chromatography. Addition of the linker sequence and of the His tag did not impair AmtB activity and the engineered *amtB* gene was expressed from an inducible T_7 polymerase-dependent promoter in a derivative of *E. coli* strain BL21(DE3). After induction the [14C]-MA transport activity was around 40 times greater than that of the host strain alone. The AmtB6H protein was solubilized from isolated cell membranes using dodecyl-β-D-maltoside (DDM) and then subjected to Ni^{2+} affinity chromatography yielding a preparation of >95% purity (Figure 2).

The estimated monomeric mass of AmtB from primary sequence is 44.5 kDa but the purified protein runs on SDS PAGE with an apparent molecular mass of around 90 kDa, with a minor species at around 33 kDa. However membrane proteins often run with an aberrantly low apparent molecular mass on SDS-PAGE and we assume that the lowest (33 kDa) band is the monomeric species. The fact that the majority of the protein ran at a significantly higher molecular weight suggested that this could be an oligomeric species. Analysis of the purified material by both size exclusion chromatography and dynamic light scattering showed the protein to behave as a homogeneous species. An accurate estimate of the molecular mass of the AmtB:DDM complex was obtained by sedimentation equilibrium analytical ultracentrifugation. This analysis requires an independent estimate of the molar ratio of DDM:AmtB in the complex which we determined using equilibrium column desorption with [14C]-DDM (Friesen *et al.* 2000).

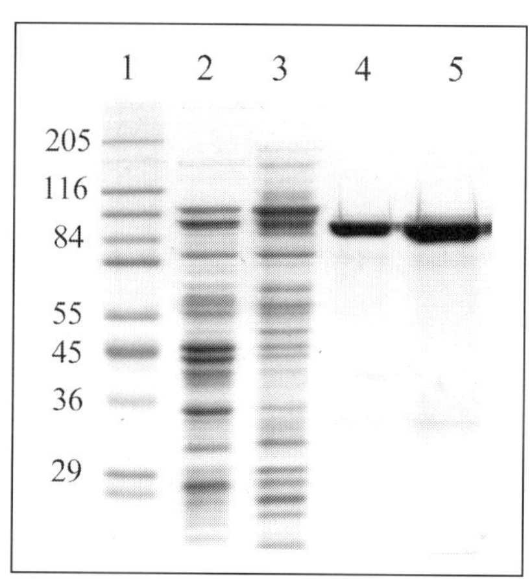

Figure 2. SDS PAGE showing steps in purification of His-tagged AmtB from *E. coli.* Lane 1 – mol. wt. markers, Lane 2 – whole cell lysate, Lane 3 – membrane extract, Lane 4 – after Ni^{2+} affinity purification, Lane 5 – after ion exchange chromatography.

The resultant data gave a molecular weight of 127 ± 17 kDa which implies that the protein in the particles is present as a trimer (predicted as 133.5 kDa). There is evidence that other members of the Amt family, namely the *S. cerevisiae* Mep proteins and the human Rhesus proteins form oligomeric complexes and it is therefore possible that the trimeric conformation of *E. coli* AmtB is representative of all members of this protein family.

2.3. The AmtB and GlnK proteins interact.

In *E. coli* there are two members of the P_{II} signal transduction protein family, GlnB and GlnK. GlnB facilitates the sensing of the intracellular nitrogen status and regulates the activities of the NtrBC system and adenylyltransferase. While GlnK can partially substitute for some of the functions of GlnB, the primary function of GlnK is unknown (Atkinson, Ninfa 1999). In order to test our hypothesis that AmtB interacts with the signal transduction protein GlnK we have employed cell fractionation together with Western blotting to investigate the cellular localization of the P_{II} proteins of *E. coli* in both the wild-type and a range of mutant strains. For all of these experiments cells were grown in a nitrogen-limited minimal medium. The cells were harvested and broken by sonication. Samples of the whole cell lysate were retained and the membrane fraction was then separated by ultracentrifugation at $250,000g$. To distinguish between proteins that were weakly or tightly membrane-associated, the membrane fractions were then subjected to extensive washing with 50 mM sodium phosphate buffer supplemented with 600 mM sodium chloride.

In intial experiments we found that in cells grown under nitrogen limitation a significant percentage of the P_{II} protein in the cell is membrane associated. Studies of *glnB* and *glnK* mutants suggested that this was the case for both P_{II} paralogs although it was much more pronounced for GlnK. Most significantly this association is entirely absent in a strain lacking AmtB. These observations fully support our hypothesis that the GlnK and AmtB proteins are capable of forming a complex and raise the question of the likely function of this association. We therefore examined the effects of ammonia-shock, in which nitrogen-limited cells are subjected to a rapid increase in their nitrogen status, typically by the addition of 30 mM ammonium to the culture. Under these conditions glutamine synthetase is known to be rapidly adenylylated and it would make physiological sense to inactivate transport activity of AmtB. Analysis of fractions from cells harvested 15 minutes after ammonia shock showed the expected rapid deuridylylation of P_{II} and a concomitant dramatic increase in the amount of membrane-associated P_{II} (Figure 3). Our current hypothesis is that the membrane-sequestration of GlnK by AmtB in response to a rapid increase in the cellular nitrogen status serves to inactivate AmtB. However it is also possible that this change in the localization of GlnK may have important consequences for other functions of GlnK by rapidly changing the intracellular pool of the protein.

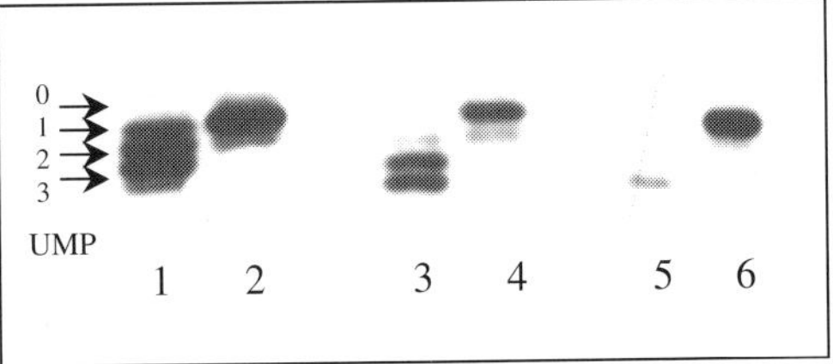

Figure 3. Effect of ammonia shock on GlnK association to the membrane, in a $\Delta glnB$ strain. Lane 1, whole cells pre ammonia shock; Lane 2, whole cells post ammonia shock; Lane 3, cytoplasm pre ammonia shock; Lane 4, cytoplasm post ammonia shock; Lane 5, membrane pre ammonia; Lane 6, membrane post ammonia shock.

3. References

Atkinson MR, Ninfa AJ (1999) Mol. Microbiol. 32, 301-313

Barnes EM *et al.* (1983) J. Bacteriol. 156, 752-757

Boussiba S *et al.* (1991) FEMS Microbiol. Rev. 88, 1-14

Friesen RH *et al.* (2000) J. Biol. Chem. 275, 33527-33535

Heller KB *et al.* (1980) J. Bacteriol. 144, 274-278

Jayakumar A *et al.* (1983) Anal. Biochem. 135, 475-478

Jayakumar A *et al.* (1985) J. Biol. Chem. 260, 7528-7532

Kleiner D (1993) NH_4^+ Transport Systems, In Bakker EP (ed), Alkali Cation Transport Systems in Prokaryotes, pp. 379-396, CRC Press, Boca Raton, Florida

Knepper MA *et al.* (1989) Physiol. Rev. 69, 179-249

Marini A-M *et al.* (1994) EMBO J. 13, 3456-3463

Marini A-M *et al.* (1997a) Mol. Cell Biol. 17, 4282-4293

Marini A-M *et al.* (1997b) Trends Biochem. Sci. 22, 460-461

Marini A-M *et al.* (2000a) Mol. Microbiol. 38, 552-564

Marini A-M *et al.* (2000b) Nat. Genet. 26, 341-344

Meier-Wagner J *et al.* (2001) Microbiol. 147, 135-143

Ninnemann O *et al.* (1994) EMBO J. 13, 3464-3471

Saier Jr. MH *et al.* (1999) Biochem. Biophys. Acta. 1422, 1-56

Siewe RM *et al.* (1996) J. Biol. Chem. 271, 5398-5403

Soupene E *et al.* (1998) Proc. Natl. Acad. Sci. USA 95, 7030-7034

Taté R *et al.* (1998) Mol. Plant-Microbe Interac. 11, 188-198

Thomas G *et al.* (2000a) Trends Genet. 16, 11-14

Thomas G *et al.* (2000b) Mol. Microbiol. 37, 331-344

4. Acknowledgements

D. Blakey and M. Merrick were supported by a grant-in-aid from the BBSRC to the John Innes Centre. Graham Coutts acknowledges a BBSRC studentship. G. Thomas was supported by a grant from CEE (EURATINE:BIOTECH94-2310).

INTEGRATION OF NITROGEN, CARBON AND REDOX STATUS BY THE *AZOTOBACTER VINELANDII* NifL-NifA REGULATORY COMPLEX

R. Little[1], S. Perry[1], V. Colombo[1], F. Reyes-Ramirez[2], R. Dixon[1]

[1]Dept of Molecular Microbiology, John Innes Centre, Norwich, NR4 7UH, UK
[2]School of Biological Sciences, University of East Anglia, Norwich, NR4 7TJ, UK

1. Introduction

The *Azotobacter vinelandii* NifL-NifA two-component regulatory system integrates metabolic signals for redox, carbon and nitrogen status to fine tune regulation of nitrogenase synthesis (Dixon 1998). The NifL protein utilizes discrete mechanisms to perceive these signals leading to the formation of a protein-protein complex which inhibits NifA activity. The binding of adenosine nucleotides to NifL plays a key role in transducing environmental signals to form the inhibitory protein complex (Eydmann *et al.* 1995; Money *et al.* 1999). We have recently demonstrated that an additional ligand, 2-oxoglutarate, allosterically modulates the activity of the complex to antagonize the influence of adenosine nucleotides on NifL activity (Little *et al.* 2000). Redox signaling is mediated by the N-terminal FAD-containing PAS domain in NifL (Söderbäck *et al.* 1998) and the nitrogen status is sensed via interaction with the non-modified form of the PII-like signal transduction protein (Av GlnK) (Little *et al.* 2000) encoded by *A. vinelandii glnK* gene (Meletzus *et al.* 1998).

The mechanism of nitrogen sensing by the *A. vinelandii* system (Figure 3) is clearly different from that in *K. pneumoniae* since in the latter system, GlnK is required to prevent NifL from inhibiting NifA under nitrogen-limiting conditions (Jack *et al.* 1999; He *et al.* 1998), whereas in the former system both *in vitro* and *in vivo* experiments suggest that Av GlnK increases the inhibitory activity of NifL under conditions of nitrogen excess (Little *et al.* 2000; Reyes-Ramirez *et al.* 2001). Moreover, whereas the uridylylation status of GlnK has no apparent influence on nitrogen regulation of the *K. pneumoniae* NifL-NifA system, *glnD* mutants of *A. vinelandii* are unable to fix nitrogen (Contreras *et al.* 1991; Colnaghi *et al.* 2001) and uridylylation of Av GlnK abrogates its ability to enhance the inhibitory activity of NifL (Little *et al.* 2000). In order to further dissect the molecular mechanisms of signal transduction in this system, we have analyzed the interaction of the PII protein and 2-ketoglutarate with the individual component proteins and have isolated mutant forms of NifL and NifA which discriminate between redox and fixed nitrogen sensing, suggesting that these signals generate different inhibitory conformers.

2. Ligand Binding: Interaction with Nucleotides and 2-Oxoglutarate

A. vinelandii NifL contains a C-terminal HATPase_c domain (Figure 1) which is found in histidine protein kinases and phytochrome-like ATPase. Although NifL does not apparently exhibit either histidine protein kinase or ATPase activity, the C-terminal domain binds adenosine nucleotides (Söderbäck *et al.* 1998). The presence of Mg ADP increases the stability of the NifL-NifA complex (Money *et al.* 1999), leading to inhibition of transcriptional activation by NifA *in vitro* even when the NifL protein is in its reduced form (Hill *et al.* 1996). Limited proteolysis experiments suggest that adenosine nucleotide binding promotes conformational changes in the C-terminal domain of NifL (Money *et al.* 2001; Söderbäck *et al.* 1998). Mutagenesis of conserved residues in the HATPase_c domain of NifL which influence nucleotide binding in other members of this family, inactivate the signal transduction function of NifL, so that it is unable to inhibit NifA activity in response to environmental effectors (Perry *et al.* this volume). Hence nucleotide binding is a major determinant of NifL activity.

We have previously suggested that nucleotide binding by NifL might provide a mechanism for sensing the ATP/ADP ratio, but this seems unlikely, since the binding site would be anticipated to be saturated *in vivo*. Moreover, the ability of MgADP to override the redox response of NifL *in vitro* presents a physiological quandary, since NifL would be expected to constitutively inhibit NifA activity, independent of the redox status. However, we have recently observed that the NifL-NifA system is directly responsive to an additional ligand, 2-oxoglutarate, which antagonizes the influence of adenosine nucleotides on NifL activity, thus relieving inhibition by NifL in the presence of ADP (Little *et al.* 2000). Since 2-oxoglutarate is a key intracellular signal of the carbon status, the response of the NifL-NifA system to this metabolite may reflect an additional physiological switch which deactivates NifL in response to carbon availability. The response of the NifL and NifA proteins to 2-oxoglutarate is within the physiological range and the *A. vinelandii* NifL-NifA system is apparently responsive to the intracellular level of 2-oxoglutarate *in vivo* when introduced into *E. coli* (Reyes-Ramirez *et al.* 2001).

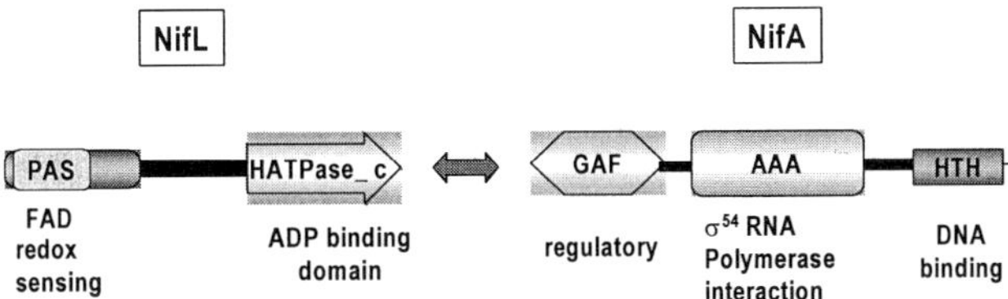

Figure 1. Domain structure of NifL and NifA. Domains were assigned using SMART (http://SMART.embl-heidelberg.de) with the exception of the DNA binding domain of NifA.

The amino terminal region of *A. vinelandii* NifA comprises a GAF domain (Figure 1), a ubiquitous motif found in signaling proteins from all kingdoms of life, some of which bind cyclic GMP (Aravind, Ponting 1997). Recent structural determination of *Saccharomyces cerevisiae* YKG9, a member of the GAF family, reveals that the fold of this motif resembles that of the PAS domain and similarities in the binding pockets of the two motifs have led to the suggestion that the GAF fold may also bind a variety of different co-factors (Ho *et al.* 2000). The GAF domain of NifA is predicted to have a regulatory function and when this domain is removed, producing a truncated variant of NifA comprising the central and C-terminal domains, the truncated protein is no longer susceptible to inhibition by the oxidized form of NifL *in vitro* (Barrett *et al.* 2001). However, the truncated protein is still inhibited by NifL when Mg ADP is present, although co-chromatography experiments suggest that the complex formed under these conditions is less stable than with native NifA (Money *et al.* 1999). The GAF domain apparently has a role in regulating the nucleoside triphosphatase activity of the central domain of NifA in response to NifL, since the ATPase activity of NifA is no longer inhibited by NifL when the GAF domain is absent (Barrett *et al.* 2001).

We have observed that the GAF domain is necessary for the response of the NifL-NifA system to 2-oxoglutarate, since when this domain is absent, NifL inhibits NifA activity even at high 2-oxoglutarate concentrations (Figure 2A). Moreover, limited proteolysis experiments suggest that the conformation of NifA alters in the presence of 2-oxoglutarate such that the linker region between the GAF and central domains of NifA is more susceptible to trypsin cleavage (Figure 2B). Since this effect is observed in the absence of NifL, it is likely that 2-oxoglutarate binds to NifA, commensurate with the proposed role of the GAF domain in binding small molecules. The proposed conformational change in NifA elicited by 2-oxoglutarate may thus prevent NifL from inhibiting NifA in the presence of adenosine nucleotides, thus favouring *nif* gene transcription when both the carbon and nitrogen status are appropriate.

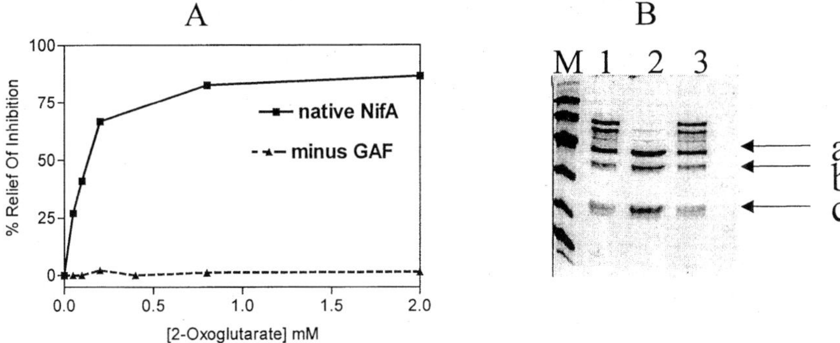

Figure 2. Influence of 2-oxoglutarate on NifA. (A) Role of the GAF domain in the response to 2-oxoglutarate. Inhibition by NifL is relieved in the presence of 2-oxoglutarate with native NifA (squares) but in the absence of the GAF domain no response is observed (triangles). (B) Influence of 2-oxoglutarate on limited digestion of NifA by trypsin. Lane M, molecular weight markers; lane 1, NifA with no addition; lane 2 NifA plus 2-oxoglutarate (2 mM); lane 3, NifA plus 3-oxoglutarate (2 mM). Bands indicated as a, b and c represent previously identified trypsin fragments (Money *et al.* 2001).

3. Nitrogen Sensing: Av GlnK Interacts with NifL

PII-like signal transduction proteins have been implicated in the regulation of nitrogen fixation in several diazotrophs. The activity of these trimeric proteins is regulated by reversible uridylylation catalyzed by uridylyltransferase (UTase) (Arcondeguy *et al.* 2001; Ninfa and Atkinson 2000). Under nitrogen-limiting conditions PII-like proteins are uridylylated by the UTase and de-uridylylated under conditions of nitrogen excess. A direct role for PII-like proteins in regulating nitrogen fixation has only recently been established. We have shown that the inhibitory activity of NifL:NifA complex is stimulated by interaction with the non-modified forms of *E. coli* PII and *A. vinelandii* GlnK in the presence of adenosine nucleotides (Little *et al.* 2000). The interaction with PII-like proteins overrides the relief of inhibition observed in the presence of high concentrations of 2-oxoglutarate. We propose that Av GlnK signals the nitrogen status by interaction with the NifL-NifA system under conditions of nitrogen excess and that the inhibitory activity of NifL is relieved by elevated levels of 2-oxoglutarate when Av GlnK is uridylylated under conditions of nitrogen limitation.

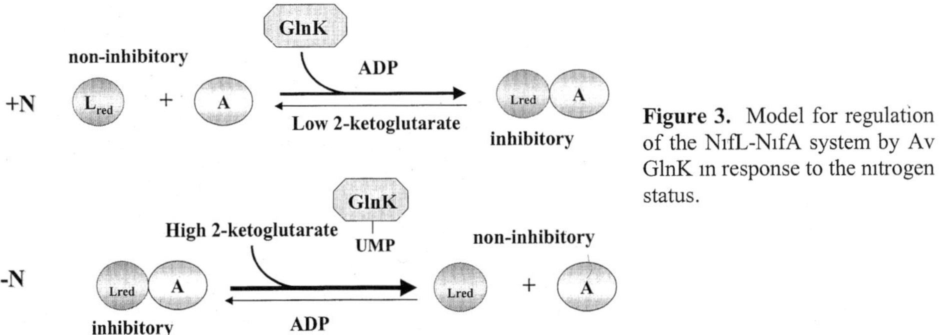

Figure 3. Model for regulation of the NifL-NifA system by Av GlnK in response to the nitrogen status.

In order to elucidate the mechanism of nitrogen sensing by the NifL- NifA system it is necessary to determine which protein component(s) interact with PII-like proteins and to analyze the effectors required for this interaction. Using co-chromatography assays with histidine tagged proteins, we have observed that Av GlnK interacts with NifL in the presence of 2-oxoglutarate and ATP (Figure 4). In contrast, no interaction was observed with NifA (data not shown).

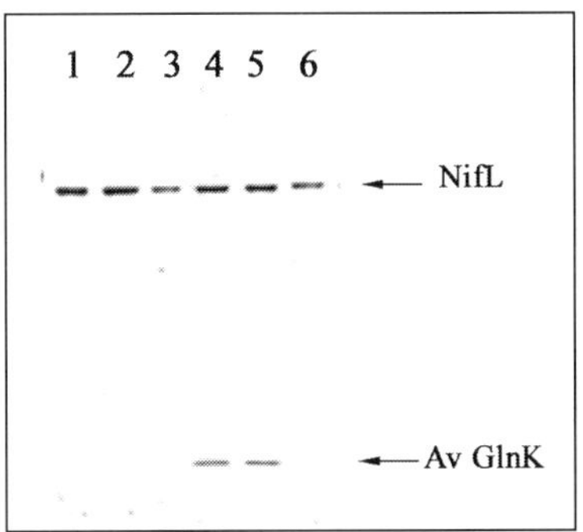

Figure 4. Co-chromatography of histidine-tagged NifL with Av GlnK. Final ligand concentrations were: 2-oxoglutarate (2mM); ATP (3.5 mM)
Lane 1, NifL + Av GlnK
Lane 2, NifL + Av GlnK + ATP
Lane 3, NifL + Av GlnK + 2-oxoglutarate
Lane 4, NifL + Av GlnK + 2-oxoglutarate + ATP
Lane 5, NifL + Av GlnK + 2-oxoglutarate + ATP
Lane 6, NifL + Av GlnK E44C + 2-oxoglutarate + ATP

The role of 2-oxoglutarate in this case probably reflects the requirement for the binding of this ligand to Av GlnK, since no interaction of 2-oxoglutarate was observed with NifL alone in ligand binding assays. As controls for these experiments we have used a mutant form of Av GlnK with a substitution, E44C, in the surface exposed T-loop which has a major role in interaction with receptors. In contrast to wild-type Av GlnK, the purified E44C protein does not stimulate the inhibitory action of NifL *in vitro* and does not interact with NifL in the co-chromatography assay (Figure 4). Surface plasmon resonance (SPR) experiments using BIAcore instrumentation also indicate that NifL interacts with the Av GlnK protein. In these experiments a histidine-tagged derivative of NifL lacking the redox-responsive PAS domain was coupled to a Ni-NTA chip and Av GlnK was added in the presence of 2-oxoglutarate and Mg ATP. The interaction between non-uridylylated Av GlnK and NifL was Mg and adenosine nucleotide-dependent and was not seen with the Av GlnK E44C mutant, confirming the results of the co-chromatography experiments. Since the interaction was not detectable with a truncated fragment of NifL, which lacks the C-terminal nucleotide binding domain, it apparently requires the C-terminal domain of NifL.

4. Mutations which Influence Signal Transduction

We have isolated a series of mutations in both the GAF domain and the central domain of NifA which prevent inhibition by NifL. Most of these mutations block inhibition by NifL in response to both fixed nitrogen and oxygen, but some mutant NifA proteins can apparently discriminate between the inhibitory forms of NifL present under different environmental conditions. Notably, mutant NifA Y254N shows complete resistance to inhibition by NifL *in vivo* when cultures are grown anaerobically under conditions of nitrogen excess, but is sensitive to NifL under aerobic growth conditions. The Y254N mutation is located in the central domain of NifA in a predicted α-helix that forms part of the Walker A motif in the AAA+ family of ATPases (Neuwald *et al.* 1999). Purified NifA Y254N is resistant to inhibition by the ADP bound form of NifL *in vitro* and is also not responsive when Av GlnK, and 2-ketoglutarate are added to the reaction. However, the mutant protein is susceptible to inhibition by the oxidized form of NifL in the presence of ADP, albeit at higher NifLox concentrations than observed with the wild-type NifA protein. The biochemical results thus concur with the *in vivo* phenotype of the mutation and indicate that the mutant NifA protein is unable to respond to the nitrogen signal generated via the interaction of NifL with Av GlnK and adenosine nucleotides, but is responsive to the oxidized form of NifL when ADP is present. These results suggest that the nature of the complex formed between reduced NifL and NifA in the presence of Av GlnK is different from that formed between oxidized NifL and NifA. Potentially, different conformers of NifL may be generated in response to discrete signal transduction events. The discrete nature of these events is also notable from the properties of mutations in the central region of NifL which inactivate the redox response but do not influence the response to fixed nitrogen. These mutations presumably do not disable perception of the redox signal *per se* but interfere with communication of this signal to NifA without apparently affecting transduction of the nitrogen signal (Perry *et al.* this volume).

5. Conclusions

Our results suggest that while NifL senses both the redox and fixed nitrogen status, the NifA protein also plays a major role in signal perception through the response to 2-oxoglutarate. Although 2-oxoglutarate can be considered as a key physiological signal of the carbon status, the level of this metabolite is also influenced by the nitrogen status. In terms of signal integration, under oxidizing conditions we propose that signal transduction is dominated by the "ON" status of the flavin in the PAS domain of NifL, which promotes formation of the inhibitory binary complex, whereas under reducing conditions, signaling is dominated by the uridylylation state of Av GlnK and the concentration of 2-oxoglutarate. When Av GlnK is in its non-modified form our model predicts that the interaction with NifL will favor the formation of a GlnK-NifL-NifA ternary complex. Several years ago, it was predicted from studies with *K. pneumoniae* that "NifL may sense one signal and the N-terminal domain of NifA the other, both proteins having to be in their derepressing state to prevent inhibitory binding" (Drummond *et al.* 1990). In the light of our finding that the GAF domain of NifA is required for the response to 2-oxoglutarate, this was a very visionary hypothesis.

6. References

Aravind L, Ponting CP (1997) Trends Biochem. Sci. 22, 458-459
Arcondeguy T *et al.* (2001) Microbiol. Mol. Biol. Rev. 65, 80-105
Barrett J *et al.* (2001) Mol. Microbiol. 39, 480-494
Colnaghi R *et al.* (2001) Microbiol. 147, 1267-1276
Contreras A *et al.* (1991) J. Bacteriol. 173, 7741-7749
Dixon R (1998) Arch. Microbiol. 169, 371-380
Drummond *et al.* (1990) Mol. Microbiol. 4, 29-37
Eydmann T *et al.* (1995) J. Bacteriol. 177, 1186-1195
He L *et al.* (1998) J. Bacteriol. 180, 6661-6667
Hill S *et al.* (1996) Proc. Natl. Acad. Sci. USA. 93, 2143-2148
Ho YS *et al.* (2000) EMBO J 19, 5288-5299
Jack R *et al.* (1999) J. Bacteriol. 181, 1156-1162
Kamberov E *et al.* (1995) J. Biol.Chem. 270, 17797-17807
Little R *et al.* (2000) EMBO J 19, 6041-6050
Meletzus D *et al.* (1998) J. Bacteriol. 180, 3260-3264
Money T *et al.* (1999) J. Bacteriol. 181, 4461-4468
Money T *et al.* (2001) J. Bacteriol. 183, 1359-1368
Neuwald AF *et al.* (1999) Genome Res 9, 27-43
Ninfa A, Atkinson M (2000) Trends in Microbiol. 8, 172-179
Reyes-Ramirez F *et al.* (2001) J. Bacteriol. 183, 3076-3082
Söderbäck E *et al.* (1998) Mol. Microbiol. 28, 179-192

Section 10:
Nodule
Metabolism

CHAIR'S COMMENTS: NODULE METABOLISM: OVERVIEW

B.T. Driscoll

Department of Natural Resource Sciences, McGill University
Ste-Anne-de-Bellevue, Québec, Canada

Nutrient exchange is at the heart of the symbiosis between legumes and rhizobia. The basics of this exchange are fairly well understood: in exchange for an energy source from the plant, rhizobia fix nitrogen within nodules and supply the host with nitrogenous compounds that are readily assimilated. As with signaling and nodulation however, there appear to be differences in nutrient metabolism and exchange between the different rhizobium-legume symbioses. There are still many unanswered questions such as; which compounds are being exchanged, how are the metabolic pathways regulated, what is the role of storage compounds and nutrient supply during infection, and what are the metabolic changes that occur during the transition from infecting rhizobia to bacteroids. In addition, there are questions regarding the role of metabolism in bacterial survival and competition, and how legume yield may be improved through engineered changes to rhizobial and plant metabolic processes. The metabolism and transport of carbon and nitrogen in root nodules has been the subject of many reviews over the years (for example: Dunn 1998; McDermott *et al.* 1989; Poole, Allaway 2000; Tajima, Kouchi 1996; Udvardi, Day 1997; Vance, Heichel 1991). This overview is very brief, and focused mainly around the topics covered by the session speakers (CP Vance, TC Charles, PS Poole, C Atkins).

C_4-dicarboxylic acids (C_4-DCA) such as malate are found in high concentrations in root nodule cells, and are the principal source of energy supplied by the plant to N_2-fixing bacteroids. The role of malate in symbiosis has been reviewed by Vance and Heichel (1991). There is evidence that some legume enzymes in the pathway which convert photosynthate to malate are highly expressed in nodule tissue, such as the nodule-enhanced malate dehydrogenase (MDH) of alfalfa (Miller *et al.* 1998).

Bacteroids rely on the oxidation of C_4-DCA via the tricarboxylic acid (TCA) cycle to support the energy requirements of the nitrogenase enzyme. In *S. meliloti* bacteroids, malate appears to be channeled through two pathways (i) MDH, and (ii) NAD^+ malic enzyme and pyruvate dehydrogenase, to produce the oxaloacetate and acetyl-CoA, respectively, required for synthesis of citrate (Cabanes *et al.* 2000; Driscoll, Finan 1993). There is a large body of evidence indicating that C_4-DCA are the principal source of carbon supplied by the plant to N_2-fixing bacteroids, and that they are metabolized by bacteroids to generate the energy necessary to support N_2-fixation. Nitrogenase activity in isolated bacteroids is highly stimulated by C_4-DCA, but not by sugars (Bergersen, Turner 1967; Miller *et al.* 1988). Transport of C_4-DCA via the bacterial C_4-DCA transport (*dct*) system is essential in N_2-fixing bacteroids, as Dct$^-$ mutants induce root nodules, but are unable to fix N_2 (Ronson *et al.* 1981, and many others). Transport of C_4-DCA across the peribacteroid membrane has also been demonstrated (see Udvardi, Day 1997 for review).

C_4-DCA are metabolized directly via the TCA cycle. All of the TCA cycle enzymes have been detected in *Rhizobium* species, and they appear to be highly active in bacteroids (see reviews by Dunn 1998; Poole, Allaway 2000). While most mutants that completely lack a TCA cycle enzyme form nodules that fail to fix nitrogen, *B. japonicum sucA* mutants are not completely Fix$^-$, possibly due to a metabolic bypass of 2-oxoglutarate dehydrogenase via succinic semialdehyde (Green *et al.* 2000). Poole *et al.* (1999) showed that the *R. leguminosarum mdh* is in an operon with *sucCD*, and possibly with *sucAB*. Their results indicate that arabinose stimulates expression of an *mdh::lacZ* gene fusion, and that the very high expression of this fusion in a *sucD* mutant background may be the result of the accumulation of a metabolic intermediate, perhaps arabinose itself. In

S. meliloti, the *mdh* gene is in an operon with *sucCD*, while *sucAB* and *sdhCDAB* appear to constitute separate operons even though the former are directly downstream of *mdh sucCD* (S.I. Dymov *et al.*, unpublished).

While it has long been recognized that legumes assimilate nitrogen either via the amide or the ureide pathway, it has been recently shown that bacteroids of different rhizobia may export different forms of nitrogen. *B. japonicum* bacteroids export alanine rather than ammonium, as previously believed (Waters *et al.* 1998). *R. leguminosarum* bacteroids appear to supply both ammonium and alanine, and while pea plants inoculated with alanine dehydrogenase mutants are Fix[+], they show reduced dry weight values, indicating that both compounds contribute to plant growth (Allaway *et al.* 2000). In tropical legumes such as cowpea and soybean, enzymatic (Shelp *et al.* 1983) and gene expression studies (Smith *et al.* 1998) indicate that nitrogen assimilation is via purine synthesis. The mode by which amino groups are transferred from alanine to this pathway is not yet clear, however.

The emergence of complete genome sequences, such as that of *S. meliloti* (Galibert *et al.* 2001), will enhance the ongoing goal to understand nodule metabolism, making the tedious process of completely sequencing metabolic genes an unnecessary step. The post genomics era will return the focus to the genetic and biochemical characterization of biochemical processes, and allow a better comparison of the differences observed between the various symbioses. There are many metabolic processes that occur in nodules that we do not understand well enough yet, such as the processes that allow legumes to balance the supply of energy to bacteroids with the supply of carbon to the rest of the plant, bacterial growth in the infection thread, the transition from bacteria to bacteroids, how bacteroids generate energy under nodule conditions (with a highly controlled O_2 supply), and the assimilation of alanine and/or ammonium. Interesting questions outside of the nodule also exist, such as the role of bacterial storage compounds like polyhydroxybutyrate (Aneja, Charles 1999) in the soil rhizosphere, and how these compounds affect survival and competition for nodulation.

References

Allaway D *et al.* (2000) Mol. Microbiol. 36, 508-515

Aneja P, Charles TC (1999) J. Bacteriol. 181, 849-857

Bergersen FJ, Turner GL (1967) Biochim. Biophys. Acta 141, 507-515

Cabanes D *et al.* (2000) Molec. Plant-Microbe Interac. 13, 483-493

Driscoll BT, Finan TM (1993) Molec Microbiol 7, 865-873

Dunn MF (1998) FEMS Microbiol. Rev. 22, 105-123

Galibert F *et al.* (2001) http://sequence.toulouse.inra.fr/meliloti.html

Green LS *et al.* (2000) J. Bacteriol. 182, 2838-2844

McDermott TR *et al.* (1989) FEMS Microbiol. Rev. 63, 327-340

Miller SS *et al.* (1998) Plant J. 15, 173-184

Miller RW *et al.* (1988) J. Cellul. Biochem. 38, 35-49

Poole PS *et al.* (1999) FEMS Microbiol. Lett. 176, 247-255

Poole PS, Allaway D (2000) Adv. Microbial Physiol. 43,117-163

Ronson CW *et al.* (1981) Proc. Natl. Acad. Sci. USA 78, 4284-4288

Shelp BJ *et al.* (1983) Arch. Biochem. Biophys. 224, 429-441

Smith PM *et al.* (1998) Plant Mol. Biol. 36, 811-820

Tajima S, Kouchi H (1996) In Stacey G, Keen NT (eds), Plant-Microbe Interactions, Volume 2, pp. 27-60, Kluwer Academic Publishers, Boston, MA

Udvardi MK, Day DA (1997) Annu. Rev. Plant Physiol. Plant Molec. Biol. 48, 493-523

Vance CP, Heichel GH (1991) Annu. Rev. Plant Physiol. Plant Molec. Biol. 42, 373-392

Waters *et al.* (1998) Proc. Natl. Acad. Sci. USA 95, 12038-12042

THE PIVOTAL ROLE OF MALATE IN ROOT NODULE METABOLISM AND LEGUME GROWTH

C.P. Vance[1,2], B. Bucciarelli[2], J. Schulze[3], R.H. Litjens[2], G.B. Trepp[2], D.A. Samac[1,4], D.L. Allan[5], M. Tesfaye[4]

[1]USDA, Agricultural Research Service, Plant Science Research
[2]Department of Agronomy and Plant Genetics, Univ. Minnesota, St. Paul, MN 55108, USA
[3]University of Halle Wittenberg, Halle/Saale, Germany
[4]Department of Plant Pathology [5]Dept Soil, Water, and Climate, Univ. Minnesota, St. Paul, MN 55108, USA

1. Introduction

In plants, malate is a key product of metabolism. It is thought by many (see Lance, Rustin 1984) to be the ultimate product of glycolysis, rather than pyruvate. As such it plays important roles in photosynthesis (both C_3 and C_4), energy generation, fatty acid oxidation, ion balance, energy generation, pulvinal and stomatal function, amino acid synthesis, and nitrogen (N_2) fixation (Gietl 1992; Martinoia, Rentsch 1994).

In N_2-fixing nodules malate is the predominant source of energy for bacteroid respiration (Driscoll, Finan 1993) and provides a significant portion of the carbon skeletons for assimilation of fixed N_2 (Rosendahl *et al.* 1990). Malate may also be involved in regulation of the nodule oxygen diffusion barrier through an osmoelectrical mechanism (Vance, Heichel 1991; Denison 1998; Galvez *et al.* 2000). The critical role that malate plays in root nodules is evidenced by the fact that ineffective nodules, whether induced by changes in either the bacterial or plant genotype, have strikingly reduced levels of malate as compared to effective nodules. Moreover, mutations in rhizobia that block organic acid use result in ineffective nodules while those that block amino acid and carbohydrate use generally have no effect on N_2 fixation (Ronson *et al.* 1981; Driscoll, Finan 1993).

The enzyme malate dehydrogenase (MDH; EC 1.1.1.82) catalyzes the reversible reduction of oxaloacetate to malate. Because malate is important in many metabolic pathways in higher plants, it occurs in multiple forms that differ in co-enzyme specificity and subcellular localization (Gietl 1992). We have identified five different forms of MDH in alfalfa (Miller *et al.* 1998): glyoxysomal, gMDII; mitochondrial, mMDH; chloroplast, chMDH; cytosolic, cMDH; and nodule enhanced, neMDH. Kinetic analysis along with mRNA and protein expression data indicated that neMDH and cMDH may have important functions in effective root nodules. To further understand the role of malate in root nodule function we thought it important to: (1) isolate and characterize the plant genes encoding ne- and cMDH; (2) identify cellular expression patterns for ne- and cMDH protein and mRNA in alfalfa root nodules; and (3) assess whether constitutive overexpression of neMDH had an effect on symbiotic N_2 fixation.

2. Materials and Methods

Alfalfa plants were grown in the glasshouse as previously described by Vance *et al.* (1979). Procedures described by Gregerson *et al.* (1994) were used to isolate the alfalfa ne- and cMDH genes. Comparable *Medicago truncatula* homologs were isolated by screening a *M. truncatula* BAC library (courtesy of Nevin Young) with radiolabeled *M. sativa* ne- and cMDH cDNAs (Miller *et al.* 1998). *In situ* hybridization and immunolocalizations were performed as described (Trepp *et al.* 1999a, 1999b). Chimeric c- and neMDH promoter::GUS reporter constructs were prepared and transformed into alfalfa according to Yoshioka *et al.* (1999). Plants overexpressing neMDH were developed essentially as described by Schulze *et al.* (1998) and Schoenbeck *et al.* (2000), except constructs were in the sense orientation.

3. Results

Protein immunoblots and RNA blots were used to evaluate expression of c- and neMDH. Similar to our previous report (Miller *et al.* 1998), cMDH protein and mRNA were expressed rather uniformly in all tissues and were unrelated to root nodule formation. By contrast, neMDH protein and mRNA were most highly expressed in effective root nodules, with maximum expression associated with effective nodule development.

An alfalfa genomic library was screened with ^{32}P-labeled c- and neMDH cDNAs and two full-length genomic clones were isolated which encoded *c*- and *neMDH*, respectively. The alfalfa *cMDH* and *neMDH* were sequenced and 2 kb of the 5' upstream putative promoter regions defined for each. *Cytosolic MDH* was comprised of seven exons interrupted by six introns. This gene structure was conserved in the *Arabidopsis* genome, however, introns in *Arabidopsis* were much smaller than those in alfalfa. Preliminary sequencing of a *Medicago truncatula cMDH* showed that intron-exon structure and sequence similarity was very conserved between alfalfa and *M. truncatula*. By comparison, alfalfa *neMDH* contained only one intron and it was located in the 5'-untranslated region of the gene. Again intron structure was conserved between alfalfa, *M. truncatula*, and *Arabidopsis*.

The promoter region of *cMDH* and *neMDH* were translationally fused to β-glucuronidase (cMDH::GUS; neMDH::GUS) to form a reporter construct and transformed into alfalfa. Both reporter genes directed GUS activity to root nodule but the pattern of staining varied. While neMDH::GUS plants showed staining throughout the nodule, cMDH::GUS nodules stained primarily in the nodule meristem and invasion zone. Interestingly, leaf stomata of cMDH::GUS plants showed intense staining while leaf stomata of neMDH::GUS plants did not stain.

Protein immunolocalization and RNA *in situ* hybridization were used to evaluate protein and transcript location at the cellular level. Protein immunolocalization suggested that neMDH was localized in both infected and uninfected cells of root nodules. The most intense staining was at the periphery of infected cells and in amyloplasts of uninfected cells. Whereas, the most intense staining for cMDH appeared to be in uninfected cells of the nodule interior and inner cortex. Similar to protein localization, transcripts for neMDH were found primarily in the infected cells of the N_2-fixing zone in nodules while those for cMDH were most apparent in the nodule meristem and inner cortex.

Transgenic alfalfa plants were produced that overexpress neMDH by fusing the CaMV^{35}S promoter to the neMDH cDNA and transforming plants with *Agrobacterium tumefaciens*. Several independent transformants were evaluated for overexpression of neMDH through activity analysis and protein immunoblots. Two plants with greatest neMDH expression in roots were evaluated for N_2 fixation using $^{15}N_2$. Both plants (neMDH 10-20 and neMDH 16-27) had increased nodule efficiency as measured by ^{15}N incorporation mg nodule^{-1} h^{-1}. Nodule efficiency was increased more than 10%. However, little effect was seen on plant dry matter accumulation.

4. Discussion

We have extended the understanding of malate in root nodule symbiosis by showing c- and neMDH are encoded by separate distinct genes, and the promoters for these genes target expression to different cell types. The localization or targeting to different cell types was confirmed and strengthened by immunolocalization and *in situ* hybridization. The fact that neMDH appears to be localized to the symbiotic zone, particularly infected cells, suggests that this isoform is involved with providing malate to bacteroids and malate for amino acid synthesis. Initial subcellular localization of neMDH to plastids in the infected cell zone, along with the localization of aspartate aminotransferase and NADH-glutamate synthase to plastids (Robinson *et al.* 1996; Trepp *et al.* 1999a) provides good evidence for neMDH's involvement in nodule amino acid synthesis. This would be consistent with earlier radiolabeling studies showing incorporation of nodule fixed $^{14}CO_2$

and ^{14}C-glutamate into aspartate (Rosendahl *et al.* 1990). Likewise, rapid labeling of bacteroids by nodule fixed [14]CO$_2$ and ^{14}C-malate support a role for ne-MDH in providing bacteroids with energy.

The localization of cMDH protein to uninfected cells of the nodule inner cortex and N$_2$-fixing zone along with the similar distribution pattern seen for nodule forms of carbonic anhydrase (CA) (Galvez *et al.* 2000) and phosphoenolpyruvate carboxylase (PEPC) (Pathirana *et al.* 1997) place the requisite carbon metabolism enzyme involved in osmoelectrical contraction of cells in the variable oxygen diffusion barrier. Moreover, the role of MDH, PEPC and CA in stomatal aperture and pulvinal function is well documented (Assman 1999). Consistent with an osmoelectrical function, the promoter of cMDH targets reporter gene expression to both leaf and stem stomata as well as the nodule meristem and inner cortex. Taken inclusively, our data demonstrate exquisite, independent cellular functions for neMDH and cMDH.

The unusual kinetics and high turnover rate for neMDH (Miller *et al.* 1998) made overexpression of this gene a potential target for improving N$_2$ fixation. With plants overexpressing neMDH we did find an increase in nodule N$_2$ fixation efficiency based upon ^{15}N incorporation. However, there appeared to be little effect on total plant growth and N accumulation because transgenic plants produced fewer nodules. Similar types of molecular compensation have been noted when nodule aspartate aminotransferase was altered in alfalfa (Farnham *et al.* 1992).

Accompanying enhanced nodule efficiency in transgenic MDH plants was improved resistance to aluminum (Al) toxicity (Tesfaye *et al.* 2001). This increased Al resistance was attributed to enhanced synthesis of malate. This finding is consistent with enhanced resistance to Al in wheat and tobacco being related to higher amounts of malate and citrate, respectively, being released from roots.

From the studies reported here it is apparent that malate plays a pivotal role in legume N$_2$ fixation and growth.

5. References

Assman SM (1999) Plant. Cell. and Envir. 22, 629-637

Denison RF (1998) Bot. Acta 111, 191-192

Driscoll BT, Finan TM (1993) Mol. Microbiol. 7, 8855-8873

Farnham M *et al.* (1992) Theor. Appl. Genet. 241, 124-128

Galvez S *et al.* (2000) Plant Physiol. 124, 1059-1068

Gietl C (1992) Biochem. Biophys. Acta 1100, 217-234

Gregerson RG *et al.* (1994) Plant Mol. Biol. 25, 387-399

Lance C, Rustin P (1984) Physiol. Veg. 22, 625-641

Martinoia E, Rentsch D (1994) Annu. Rev. Plant Physiol. and Plant Mol. Biol. 45, 447-467

Miller SS *et al.* (1998) Plant J. 15, 173-184

Pathirana S *et al.* (1997) Plant J. 12, 293-304

Robinson DL *et al.* (1996) Plant. Cell. and Envir. 19, 602-608

Ronson C *et al.* (1981) Proc. Natl. Acad. Sci. 78, 4284-4288

Rosendahl L *et al.* (1990) Plant Physiol. 93, 12-19

Schoenbeck M *et al.* (2000) J. Exp. Bot. 51, 29-39

Schulze J *et al.* (1998) Phytochem. 49, 341-346

Tesfaye M *et al.* (2001) Plant Physiol. 127

Trepp GB *et al.* (1999a) Plant Physiol. 119, 817-828

Trepp GB *et al.* (1999b) Plant Physiol. 119, 829-837

Vance CP *et al.* (1979) Plant Physiol. 64, 1-8

Vance CP, Heichel GH (1991) Annu. Rev. Plant Physiol. and Plant Mol. Biol. 42, 373-392

Yoshioka H *et al.* (1999) Mol. Plant-Microb. Interact. 12, 263-274

SINORHIZOBIUM MELILOTI PHB CYCLE GENETICS

T.C. Charles

Dept of Biology, University of Waterloo, 200 University Ave. West, Waterloo, ON, N2L 3G1 Canada

1. Introduction

Bulk soils are characteristically oligotrophic, carbon-limited environments (van Elsas, van Overbeek 1993; van Veen, van Overbeek *et al.* 1997), due to recalcitrance of carbon substrate to degradation, and sequestration in soil sites beyond the reach of microorganisms. Microbial cell growth rates in soil environments are therefore extremely low, on the order of a few cell divisions per year (van Elsas, van Overbeek 1993), and most cells exist in a state of carbon starvation-induced stationary phase. In contrast, rhizosphere environments are rich in carbon nutrients, due to exudation of carbon compounds from plant roots. Of particular importance to local microbial ecology is the organisms' ability to traverse between bulk soil and rhizosphere, and to take advantage of nutrients as they become available. In the absence of cell growth, the organisms are unable to compete for a particular nutrient unless they are able to take up and store that compound and then use the stored material for cell growth when conditions are more favorable. Intracellular storage compounds thus have crucial roles to play in the proliferation and persistence of bacterial cells in the environment, and in their ability to colonize plant roots. One of the best-known carbon storage compounds is poly-3-hydroxybutyrate (PHB), commonly found as deposits in bacterial cells (Anderson, Dawes 1990). These PHB deposits typically accumulate intracellularly under conditions where growth is limited by a factor other than carbon nutrient availability. If later exposed to carbon starvation conditions, these cells may catabolize these deposits as a source of carbon and energy.

Root nodule bacteria such as *Sinorhizobium meliloti* have complex and specific nutrition requirements before and during their interactions with the plant host (Finan, McWhinnie *et al.* 1991). *S. meliloti* is also very good at persisting in the soil as a saprophyte in the absence of the plant host. Interest in PHB metabolism in root nodule bacteria is in part due to the prominence of PHB as the predominant carbon storage compound in some bacteroids (Forsyth, Hayward *et al.* 1958; McDermott, Griffith *et al.* 1989). Possible roles for PHB are (i) fueling the extensive cell division (Gage, Bobo *et al.* 1996) that takes place within the infection thread, (ii) protection of nitrogenase enzyme from oxygen during periods of darkness by providing reducing power for the maintenance of the O_2 diffusion barrier in the absence of photosynthesis (Bergersen, Peoples *et al.* 1991), (iii) as an aid in the recovery of bacteroids on their release into the rhizosphere following nodule senescence (Klucas 1975), and (iv) to increase the survival of bacteria in the soil and rhizosphere. Although *S. meliloti* cells within the infection thread are often observed to have PHB deposits, these deposits are not present in mature bacteroids (Paau, Bloch *et al.* 1980; Hirsch, Long *et al.* 1982; Hirsch, Bang *et al.* 1983). If PHB is important in the interaction of *S. meliloti* with the host plant, it is more likely to be at the early stages leading up to infection.

Most of the early studies of PHB metabolism were targeted at the possible direct role of PHB in the energetics of N_2 fixation, and did not address the contribution of PHB to nodulation or saprophytic growth. Recent years have brought the application of genetic and molecular biological techniques to the study of PHB metabolism (Povolo, Tombolini *et al.* 1994; Sikora, Kuykendall *et al.* 1994; Tombolini, Povolo *et al.* 1995; Cevallos, Encarnación *et al.* 1996; Mandon, Michel-Reydellet *et al.* 1998; Willis, Walker 1998), but until recently (Charles, Cai *et al.* 1997; Aneja, Charles 1999; Cai, Driscoll *et al.* 2000), few studies have considered the degradation part of the PHB cycle. The physiological effects of PHB metabolism mutations are various. *Rhizobium leguminosarum phbC* mutants exhibit reduced amino acid uptake activity (Walshaw, Wilkinson *et al.* 1997), and *S. meliloti phbC* mutants have reduced growth rates on PHB cycle intermediates (Cai, Driscoll *et al.* 2000). While *Rhizobium etli phbC* mutants exhibit enhanced N₂-fixation activities (Cevallos, Encarnación *et al.* 1996), *Azorhizobium caulinodans phbC* mutants are deficient in N₂-fixation due to inhibition of *nifA* transcription (Mandon, Michel-Reydellet *et al.* 1998). *S. meliloti phbC* mutants exhibit slight deficiencies in N₂-fixation, and are defective in competition for nodulation (Willis,Walker 1998).

2. Isolation of PHB Degradation Pathway Mutants

The genetic approach that was used in the initial stages of the study of the PHB degradation pathway was to screen Tn*5*-generated mutants for those that were unable to utilize the PHB cycle intermediates 3-hydroxybutyrate and acetoacetate as sole carbon source (Charles, Cai *et al.* 1997). This screen resulted in a number of mutant classes, and subsequent characterization contributed to the current understanding of the PHB cycle (see Figure 1). Members of the first class were unable to utilize 3-hydroxybutyrate but retained the ability to utilize acetoacetate and acetate as sole carbon source. They were deficient in 3-hydroxybutyrate dehydrogenase activity and the mutation mapped to megaplasmid pRmeSU47b. Members of another major class, which were not able to use acetoacetate or 3-hydroxybutyrate, were mutated in a chromosomal gene that exhibited predicted amino acid similarity to the acetyl-CoA synthetase enzymes. Another major class, with a similar growth phenotype as the second class, was determined by complementation analysis to be in a locus comprised of at least four transcriptional units. This locus mapped to megaplasmid pRmeSU47b, and corresponds to the locus first identified by deletion analysis as *bhb* (Charles, Finan 1991). One of these transcripts, *bhbA*, was characterized, and was shown to encode the enzyme methylmalonyl-CoA mutase (Charles, Aneja 1999). Whether methylmalonyl-CoA mutase and the other gene products in this locus are really involved in the PHB degradation pathway remains to be determined. An unexpected outcome of the screen for altered utilization of PHB cycle intermediates was the isolation of a mutation in *phbC*, encoding PHB synthase (Cai, Driscoll *et al.* 2000).

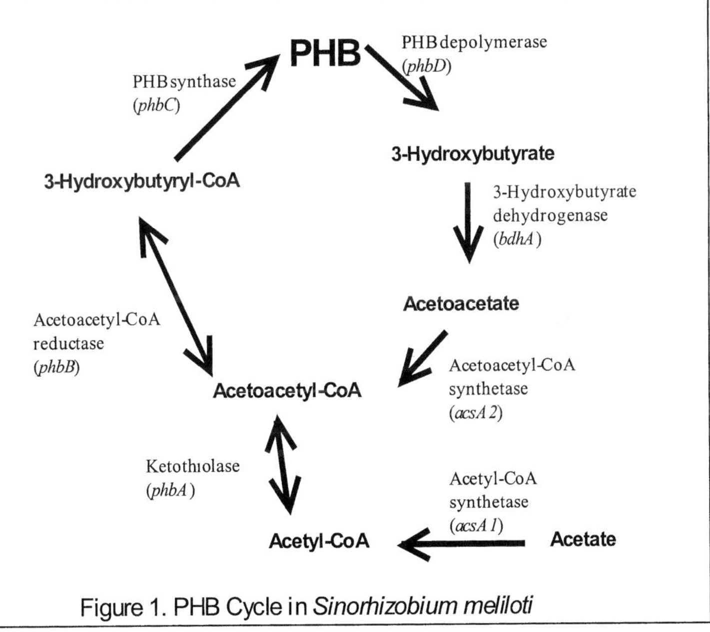

Figure 1. PHB Cycle in *Sinorhizobium meliloti*

3. Acetoacetyl-CoA Synthetase

The gene encoding a protein with predicted amino acid similarity to the acetyl-CoA synthetase enzymes in fact encodes the enzyme acetoacetyl-CoA synthetase (Cai, Driscoll *et al.* 2000). This enzyme activates acetoacetate to acetoacetyl-CoA, and is not able to use acetate as substrate. A true acetyl-CoA synthetase, with greater similarity to other bacterial acetyl-CoA synthetase enzymes, is encoded elsewhere on the chromosome, and this enzyme is not able to use acetoacetate as substrate (Thaha 1999). The genes have been designated *acsA2* (for the acetoacetate-specific) and *acsA1* (for the acetate-specific) respectively. The *acsA1* mutants are able to grow on acetoacetate or 3-hydroxybutyrate, but not acetate. This is consistent with the generation of acetyl-CoA via ketothiolase activity on acetoacetyl-CoA during growth on 3-hydroxybutyrate or acetoacetate. A key outcome of this work was the realization that acetoacetate is activated by acetoacetyl-CoA synthetase, rather than by a CoA transferase, as in *E. coli* (Jenkins, Nunn 1987).

4. 3-Hydroxybutyrate Dehydrogenase

The depolymerization product of the action of intracellular PHB depolymerase on PHB is 3-hydroxybutyrate, which is further oxidized to acetoacetate by the enzyme 3-hydroxybutyrate dehydrogenase. This enzyme is encoded by the gene *bdhA*, and sequence analysis indicated that it is a member of the short chain alcohol dehydrogenase superfamily (Aneja, Charles 1999). Although the enzyme had been purified and biochemically characterized from a number of bacteria, this was the first description of a bacterial gene encoding 3-hydroxybutyrate dehydrogenase.

The *bdhA* gene is the first gene in an operon also containing *xdhA2* and *xdhB2*, which encodes the two subunits of xanthine dehydrogenase/oxidase. Examination of cell extracts for xanthine oxidase activity on native gels confirmed that Tn*5* insertion in *bdhA* abolishes downstream-encoded xanthine oxidase activity (Aneja, Charles 1999). This analysis also revealed the presence of a second xanthine oxidase activity that remains in the mutant strain. The recently completed genome sequence (Galibert *et al.* 2001) made it clear that this second activity is also encoded on megaplasmid pRmeSU47b, but in a different region of this replicon. The genes encoding this second activity have been designated *xdhA1* and *xdhB1*. These genes are found in an operon along with the downstream gene *xdhC*. The gene product of *xdhC* is probably involved in insertion of the molybdenum co-factor (Leimkühler, Kern *et al.* 1998).

5. Regulation of the *bdhA-xdhA2-xdhB2* operon.

The accumulation of PHB in biotin-limited cultures of *S. meliloti*, but not in biotin-sufficient cultures, suggested that the regulation of a controlling step in the PHB pathway might be influenced by biotin (Hofmann, Heinz *et al.* 2000). A similar accumulation of PHB by *R. etli* under biotin-limited conditions had been reported earlier (Encarnación, Dunn *et al.* 1995). The presence of 4 nM biotin resulted in more than four-fold increase in *bdhA-lacZ* expression over the level of expression in the absence of biotin (Hofmann, Heinz *et al.* 2000). This finding is consistent with an enhanced rate of PHB degradation under biotin-sufficient conditions. The location of *xdhA2-xdhB2* in the same operon as *bdhA* means that xanthine dehydrogenase/oxidase activity also probably increases in the presence of biotin. This mixed-function operon arrangement thus links the degradation of intracellular carbon and nitrogen stores to the same biotin-regulated expression.

6. Concluding Remarks

The strategy that we have used, the isolation of mutants unable to grow using PHB cycle intermediates as sole carbon source, has allowed us to identify the genes involved in the degradation part of the PHB cycle. The genetics of the system is leading us towards a more complete understanding of the role of the PHB cycle in the soil, rhizosphere and during nodulation. The linkage of the metabolism of PHB carbon stores and intracellular purine stores by the *bdhA-xdhA2-xdhB2* mixed function operon suggests a model for the colonization of the alfalfa root. In the rhizosphere, high levels of available carbon source coupled with the relatively lower levels of nitrogen would result in PHB accumulation. The nitrogen limitation limits growth, leading to rRNA degradation, and some of the purines from the degraded rRNA would be accumulated intracellularly as hypoxanthine. Thus, cells in the rhizosphere would contain PHB and hypoxanthine stores. Biotin in the alfalfa rhizosphere would then induce transcription of *bdhA-xdhA2-xdhB2* operon, resulting in degradation of both PHB and hypoxanthine stores and concomitantly increased cell growth and colonization of the alfalfa root. This model remains to be tested experimentally.

7. References

Anderson AJ, Dawes EA (1990) Microbiol. Rev. 54, 450-472

Aneja P, Charles TC (1999) J. Bacteriol. 181, 849-857

Bergersen FJ, Peoples MB *et al.* (1991) Proc. Royal Soc. London Series B Biol. Sci. 245, 59-64

Cai G-Q, Driscoll BT *et al.* (2000) J. Bacteriol. 182, 2113-2118

Cevallos MA, Encarnación S *et al.* (1996) J. Bacteriol. 178, 1646-1654

Charles TC, Aneja PA (1999) Gene 226, 121-127

Charles TC, Cai G-Q *et al.* (1997) Genetics 146, 1211-1220

Charles TC, Finan TM (1991) Genetics 127, 5-20

Encarnación S, Dunn M *et al.* (1995) J. Bacteriol. 177, 3058-3066

Finan TM, McWhinnie E *et al.* (1991) Mol. Plant-Microbe Interact. 4, 386-392

Forsyth WGC, Hayward AC *et al.* (1958) Nature 182, 800-801

Gage DJ, Bobo T *et al.* (1996) J. Bacteriol. 178, 7159-7166

Galibert F *et al.* (2001) *Sinorhizobium meliloti* strain 1021 genome project,
    http://sequence.toulouse.inra.fr/meliloti.html

Hirsch AM, Bang M *et al.* (1983) J. Bacteriol. 155, 367-380

Hirsch AM, Long SR *et al.* (1982) J. Bacteriol. 151, 411-419

Hofmann K, Heinz EB *et al.* (2000) FEMS Microbiol. Lett. 182, 41-44

Jenkins LS, Nunn WD (1987) J. Bacteriol. 169, 42-52

Klucas RV (1975) Plant Physiol. 54, 612-616

Leimkühler S, Kern M *et al.* (1998) Mol. Microbiol. 27, 853-869

Mandon K, Michel-Reydellet N *et al.* (1998) J. Bacteriol. 180, 5070-5076

McDermott TR, Griffith SM *et al.* (1989) FEMS Microbiol. Rev. 63, 327-340

Paau AS, Bloch CB *et al.* (1980) J. Bacteriol. 143, 1480-1490

Povolo S, Tombolini R *et al.* (1994) Can. J. Microbiol. 40, 823-829

Sikora LJ, Kuykendall LD *et al.* (1994) Microbiology 140, 2761-2767

Thaha FZ (1999) M.Sc. Thesis, McGill University

Tombolini R, Povolo S *et al.* (1995) Microbiology 141, 2553-2559

van Elsas JD ,van Overbeek LS (1993) In S. Kjelleberg (ed), Starvation in Bacteria, pp.55-79,
    Plenum Press, New York.

van Veen JA, van Overbeek LS *et al.* (1997) Microbiol. Mol. Biol. Rev. 61, 121-135

Walshaw DL, Wilkinson A *et al.* (1997) Microbiol. 143, 2209-2221

Willis LB, Walker GC (1998) Can. J. Microbiol. 44, 554-564

8. Acknowledgements

Rhizobium PHB research in the Charles laboratory is supported by Natural Sciences and Engineering Research Council of Canada. The author wishes to thank Brian Driscoll and Wolfgang Streit for fruitful collaboration, and members of the Charles laboratory for contributions to this work. Public access to the *S. meliloti* genome sequence prior to publication is also very much appreciated.

ALANINE SYNTHESIS AND SECRETION BY *RHIZOBIUM LEGUMINOSARUM*

P.S. Poole, E.M. Lodwig, D.A. Allaway, A. Bordes

Division of Microbiology, School of AMS, University of Reading, Whiteknights, Reading, RG6 6AJ, UK

On the basis of ^{15}N-labeling studies it had been generally accepted that ammonium is the sole secretion product of N_2-fixation by the bacteroid and that the plant is responsible for assimilating it into amino acids. However, ^{15}N-labeling studies indicated that soybean bacteroids only secrete alanine and not ammonia (Waters *et al.* 1998). We investigated this in pea bacteroids and found that both ammonium and alanine are secreted.

Alanine secretion could account for up to 20% of the combined secretion of alanine and ammonium (Allaway *et al.* 2000). The *in vitro* partitioning between them will depend on whether the system is open or closed, as well as the ammonium concentration and bacteroid density. This may be explained by the low affinity of alanine dehydrogenase for ammonium, which was determined as 5.1 mM. Similar to this whole cells excrete alanine with a K_m for ammonium of 3.2 mM. Thus the ability to detect alanine synthesis *in vitro* depends on the accumulation of ammonium in the assay system.

The activity of alanine dehydrogenase is higher in soybean compared to pea bacteroids, potentially favoring alanine synthesis in soybean. However, since the assay conditions can alter the amount of alanine formed by isolated bacteroids we identified and mutated the gene for alanine dehydrogenase (*aldA*). This confirmed that AldA is the primary route for alanine synthesis in isolated bacteroids. Bacteroids of the *aldA* mutant fix nitrogen but only secrete ammonium at a significant rate, resulting in lower total nitrogen secretion. Peas inoculated with the *aldA* mutant are green and healthy, demonstrating that ammonium secretion by bacteroids can provide sufficient nitrogen for plant growth. Plants inoculated with the mutant were reduced in biomass compared to those inoculated with the wild type. The labeling and plant growth studies suggest that alanine synthesis and secretion contributes to the efficiency of N_2-fixation and therefore biomass accumulation.

We are currently over-expressing AldA to test the hypothesis that the level of AldA will be crucial to the amount of alanine formed by bacteroids. Full expression of AldA appears to require that both *aldR* and *aldA* are present on a plasmid, presumably due to autoregulation of *aldR* (see below).

Divergently transcribed from *aldA* is a leucine regulator protein (LRP) homolog, *aldR*. Insertion of an interposon in *aldR* prevented transcription of *aldA,* as measured by *gusA* fusions and Northern blotting, as well as AldA enzyme activity. AldR is autoregulatory, since *aldR* insertions also prevent transcription of *aldR::gusA* fusions. AldA activity is induced by growth on carboxylic acids, including pyruvate, malate and succinate as well as alanine. We are currently investigating whether alanine is excreted by free-living cultures and how factors such as O_2 tension and growth phase may affect this. Alanine does not appear to be secreted by free-living cultures in exponential phase but a burst of alanine synthesis can be induced in stationary phase cultures by the addition of excess L-malate and ammonium.

Alanine synthesis can also be considered as a possible overflow pathway in the same sense as synthesis of polyhydroxybutyrate (PHB) and glycogen. We have therefore been investigating the effects of mutating each of these pathways individually as well as together. This is also being done for both indeterminate and determinate plants (pea and bean respectively) using isogenic parent strains that differ only in the *sym* plasmid. Glycogen synthase mutants (*glgA*) show a large increase in plant starch in the nodule but remain able to fix nitrogen. This suggests that glycogen synthesis in pea bacteroids may be a major carbon storage compound able to profoundly affect carbon metabolism in the plant. The construction of double and triple mutants should help us understand how plastic these pathways are. We are particularly interested to know whether blocking one or more pathways causes a diversion of carbon and reductant to other overflow pathways.

References

Allaway D *et al.* (2000) Mol. Microbiol. 36, 508-515
Waters JK *et al.* (1998) Proc. Natl. Acad. Sci. USA 95, 12038-12042

ASSIMILATION OF FIXED-N IN A UREIDE-FORMING SYMBIOSIS

C.A. Atkins [1], P.P. Thumfort [2]

[1]Botany Dept, University of Western Australia, Crawley WA 6009, Australia
[2]Dept of Biology, MIT, Cambridge, MA 02139, USA

Recent data have challenged the long held view that the product of nitrogenase activity, ammonia, is transferred from the symbiosome to the host cell cytosol where it is assimilated (reviewed in Day *et al.* 2001). Waters *et al.* (1998) have presented evidence for the excretion of fixed N as alanine by bacteroids isolated from soybean (*Glycine max* [L.] Merr.) and purified by sucrose density gradient fractionation. However, excretion and accumulation of alanine by bacteroids was highest at pO_2 less than 0.01 while nitrogenase activity was maximal at pO_2 of 0.06. On the other hand, flow-chamber experiments under conditions where nitrogenase activity was optimized, also with isolated bacteroids from soybean, identified ammonia as the major excreted product of fixation (Li *et al.* 2001). A significant difference between the two types of experimental approach was the removal of the excreted products of fixation in the flow-through system while they accumulated in the closed system used by Waters *et al.* (1998) and Allaway *et al.* (2002) have suggested that this is the explanation for the conflicting results.

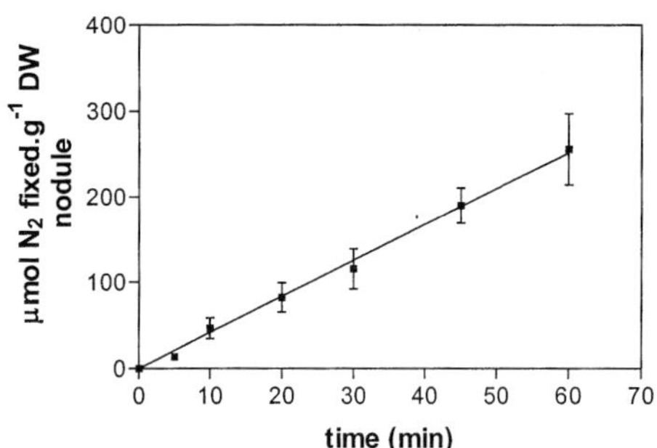

Figure 1. Fixation of N_2 assayed by $^{15}N_2$ labeling of nodulated roots of cowpea. Data are means ± SE (n = 5) and the line is fitted by non-linear regression. The atmosphere in each incubation vessel was assayed for isotopic enrichment just prior to collecting the root system and found to be 74.7 ± 0.5 at %xs ^{15}N (n = 30).

The present study reports results of steady state $^{15}N_2$ labeling of nodulated roots of intact cowpea plants (*Vigna unguiculata* [L.] Walp. cv Vita 3: *Bradyrhizobium* strain CB756) under conditions that maintained high rates of nitrogenase activity (250 µmol N_2 h^{-1} g^{-1} DW nodule; Figure 1). Root systems were frozen in liquid N_2 after 5, 10, 20, 30, 45 or 60 min and nodules removed. The experiment was replicated five times and extracted solutes analyzed for isotopic enrichment by GC-MS of their TBDMS derivatives. Figure 2 indicates rapid labeling of the amide group of glutamine (single-labeled molecules) reaching equilibrium at ca. 50% of the isotopic enrichment of the $^{15}N_2$ supplied after 5 min. Double-labeled molecules (amide and amino groups) accumulated more slowly, consistent with the slower transfer of label to the amino group of glutamine. In cowpea nodules, as in those of soybean, the amide group of glutamine is used by two amidotransferases of the *de novo* purine pathway to form IMP that is oxidized to xanthine and ureides (reviewed by Atkins, Smith 2000). The accumulation of double-labeled xanthine molecules (Figure 3) at a rate greater than either single-labeled molecules or those with 3 or 4 ^{15}N atoms is consistent with this pathway. Furthermore, the slightly slower labeling kinetics of xanthine would be expected for an intermediate of the pathway that is some 16 enzymic steps removed from the second of the glutamine-dependant steps (FGAR amidotransferase). The labeling of the 4 amino acids shown in Figure 4 is consistent with their acquisition of N from the amido group of glutamine through GOGAT and the activity of aminotransferases.

While this study is unable to take account of the influence that multiple pools of these compounds might have on the labeling kinetics shown there is no evidence that supports the idea that alanine is a precursor of the amide group of glutamine or of the purine ring.

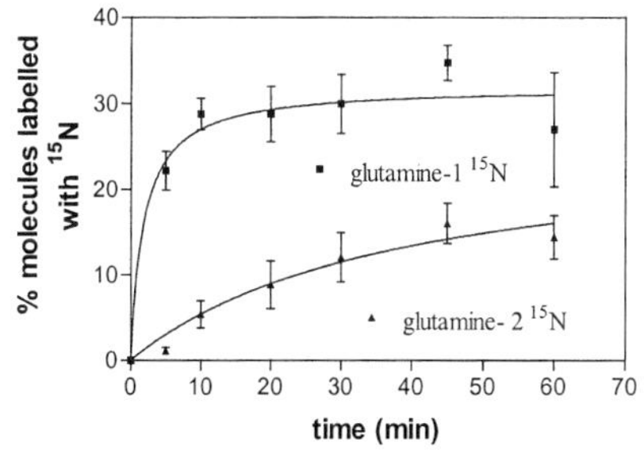

Figure 2. ^{15}N labeling of glutamine in nodules of cowpea during exposure to ^{15}N$_2$. Data are means ± SE (n = 5) and the lines are fitted by non-linear regression.

Figure 3. ^{15}N labeling of xanthine in nodules of cowpea during exposure to ^{15}N$_2$. Data are means ± SE (n = 5) and the lines are fitted by non-linear regression.

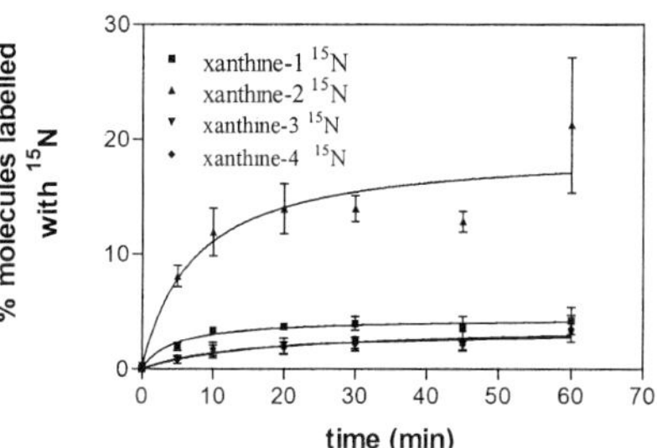

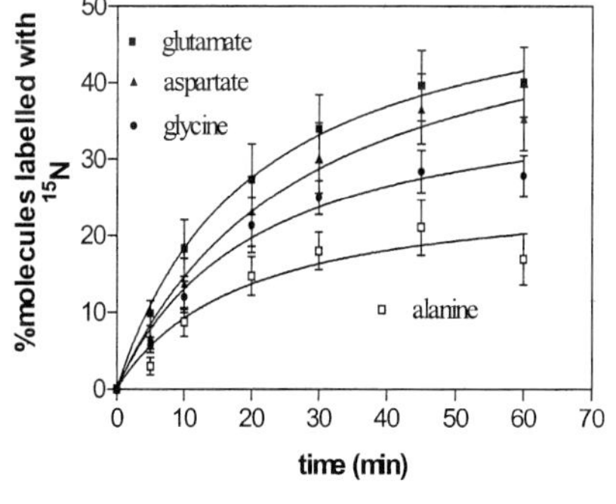

Figure 4. ^{15}N labeling of amino acids in nodules of cowpea during exposure to ^{15}N$_2$. Data are means ± SE (n = 5) and the lines are fitted by non-linear regression.

References

Allaway D *et al.* (2000) Mol. Microbiol. 36, 508-515
Atkins CA, Smith PMC (2000) In Triplett E (ed), Prokaryotic Nitrogen Fixation. pp. 559-587, Horizon Scientific Press, Wymondham, UK
Day DA *et al.* (2001) Cell. Mol. Life Sci. 58, 61-71
Li Y *et al.* (2001) Microbiol. 147, 663-670
Waters JK *et al.* (1998) Proc. Nat. Acad. Sci. USA 95, 12038-12042

Section 11:
Endophytic /
Associative
Plant–Microbe
Interactions

CHAIR'S COMMENTS: ENDOPHYTIC AND ASSOCIATIVE PLANT-MICROBE INTERACTIONS

C. Kennedy

Dept Plant Pathology, University of Arizona, Tucson, AZ 85721, USA

Associative nitrogen-fixing bacteria colonize the rhizosphere and sometimes the interior surfaces of outer root cortical cells of host plants. Those identified include species of several genera, mainly members of Proteobacteria groups. The best studied of these is *Azospirillum brasilense* and other *Azospirillum* species. Plant hosts for these root-colonizing bacteria include wheat, maize and other monocots and a few dicots.

Diazotrophs colonizing the interior of plants, mainly in intercellular spaces, are endophytes, the best studied being *Gluconacetobacter diazotrophicus* (formerly *Acetobacter diazotrophicus*), and *Herbaspirillum* species. These have been mainly isolated from sugarcane grown agriculturally throughout the world, and also from a few other plants, both monocots and dicots, including coffee, pineapple, sorghum and others. Another endophytic relationship occurs between a nitrogen-fixing species of *Azoarcus*, originally isolated from Kallar grass in Pakistan and now studied as a colonizer of rice.

Results from several laboratories have shown that many associative and endophytic diazotrophs do enhance the growth of their plant partners in plant inoculation experiments. A challenge to this field of research is to establish, definitively, whether the transfer of fixed nitrogen from the identified associative or endophytic colonizers is significant. N balance and isotope dilution experiments would suggest this is true. It is also true that other factors such as indole acetic acid (IAA) production can be beneficial to plant growth, possibly by increasing efficiency of fixed N source uptake by plant roots.

This field of research is growing rapidly – about 20% of the posters at this meeting concern associative and endophytic diazotrophs, studied not only for their benefit to plant growth but also to address basic questions of bacterial physiology, genetics, and gene regulation in these organisms. The next few years will be important in establishing the exact nature of the contribution of individual endophytes to plant health and nutrition. The papers presented in this session address some of these and other issues. An overview follows.

Eric Triplett described diazotrophic endophytes isolated from maize and switchgrass. A strain of *K. pneumoniae* isolated from maize, named 342, and its *nifH* mutant derivative, along with switchgrass isolate *Pantoea* sp. P102 were used to inoculate rice and wheat varieties. These plants were tested because in previous experiments, *K. pneumoniae* isolates failed to relieve N-deficiency in maize although inoculated plants showed growth responses to inoculation in N-sufficient conditions. The smaller size of rice and wheat might allow the benefits of bacterial nitrogen fixation to be more easily evident. Gfp-tagged bacteria were observed to colonize intercellular spaces of the root cortex in both host plants. Growth benefit to both plants grown under N deficiency was observed with wild-type strains but not with the *nifH* mutant of strain 342, consistent with but not proving that bacterially fixed nitrogen was supplied to host plants. This work adds additional members to the growing family of diazotrophic endophytes and widens the potential for beneficial plant-microbe associations.

Paula Bonfante and colleagues have discovered diazotrophic bacteria to be endosymbionts of arbuscular mychorrhizal fungi: non-culturable species of *Burkholderia* appear in the cytoplasm of *Gigaspora margarita* BEG34 and other species of Gigasporaceae. These bacteria were found to contain DNA encoding the three nitrogenase subunits, the *nifHDK*, genes. The fungi apparently transmit the bacteria vertically from one generation to the next. This novel association could represent a mechanism for extending the known benefit of AM fungi on plant growth to conditions where nitrogen is otherwise limiting.

Species of *Azospirillum* were the first associative diazotrophs to be isolated from the rhizosphere of monocot host plants and remain the best characterized of this category of nitrogen-fixing bacterial types. It is now widely considered that while little bacterially fixed nitrogen is provided for host plant growth, *Azospirillum* is a beneficial organism by its ability to produce the auxin, indole acetic acid (IAA). Jos Vanderleyden and co-workers extended these studies and report that wheat growth promotion by *A. brasilense* occurs under sub-optimal levels of fixed N supply but not in the absence of provided fixed N. IAA production by *Azospirillum brasilense* is growth phase dependent, highest at pH 5.5 and is stimulated by external supply of IAA. The *ipdC* gene encodes a key enzyme in the IAA biosynthetic pathway, indole pyruvate decarboxylase. Two promoter-swapping strategies were described for constitutitive IPD expression and for plant-induced IPD expression. Plants inoculated with *A. brasilense* carrying these elements had shorter, more highly developed roots with greater root hair proliferation than plants inoculated with wild-type. Also, the total mass of plants was greater in those inoculated with the *ipdC* expression construct strains than in those inoculated with wild-type. These studies confirm and extend the hypothesis that IAA is a plant-growth promoting substance produced by *A. brasilense* and further indicate that genetic manipulation of levels/patterns of genes involved in IAA biosynthesis can be used to improve the degree of benefit that associative rhizosphere bacteria have on plant growth.

Barbara Reinhold-Hurek and colleagues originally isolated *Azoarcus* sp. strain BH72 from Kallar grass in Pakistan. Here she reported that further attempts to isolate the strain again from Kallar grass were not successful. However, *nifH* mRNA was abundant in plant tissues; preparation of cDNA from RNA and its sequencing showed that the *nifH* transcript was from *Azoarcus* sp. strain BH72. Thus this organism apparently has different phases during which it can be culturable or unculturable, depending on factors not known at this stage. One speculation is that the cells in plants are in a state of hyperderepression, characterized by nitrogenase being bound to a membrane array in a diazosome structure; such cells could possibly be unable to form colonies. Turning to the use of rice as a host for *Azoarcus* sp. strain BH72, results were shown suggesting that the degree of interaction between the bacteria and rice plants is cultivar-specific, with the suggestion that the degree of expression of *nifHDK* correlates with the degree to which the host plant carries out a defense response, typified by lignification and browning. Finally, the three PII-like proteins in *Azoarcus* sp. strain BH72, GlnB, GlnK and GlnY, were further characterized, along with the ammonium transporter encoded by *amtB* adjacent to *glnK*. Mutations in these genes variably influence the capacity for switch-off of nitrogenase activity that occurs after ammonium addition to nitrogen-fixing cultures. In addition both GlnK and GlnY have an ability to associate with cell membranes according to N status, apparently by binding to the membrane spanning AmtB protein.

Future work on these and other associative and endophytic/endosymbiotic diazotrophs will focus on identifying specific factors in both the bacteria and plant partners that lead to beneficial effects and effective colonization.

ISOLATION AND CHARACTERIZATION OF DIAZOTROPHIC ENDOPHYTES FROM GRASSES AND THEIR EFFECTS ON PLANT GROWTH

P.J. Riggs, R.L. Moritz, M.K. Chelius, Y. Dong, A.L. Iniguez, S.M. Kaeppler, M.D. Casler, E.W. Triplett

Dept of Agronomy, University of Wisconsin-Madison, Madison, WI 53706, USA

1. Introduction

The discovery of nitrogen fixation by diazotrophic endophytes in sugarcane by Döbereiner, Kennedy and coworkers (for reviews see Boddey *et al.* 2000; Sevilla, Kennedy 2000) has encouraged several laboratories around the world to identify nitrogen-fixing associations between other grasses and bacteria. Here, we review our efforts to discover such associations, present some new data on the characterization of some of these associations, and describe our plans for the future.

We first isolated a diazotrophic endophyte from a field-grown maize plant collected from the University of Wisconsin Agricultural Experimental Farm late in the growing season (Palus *et al.* 1996). Using the procedure of Dong *et al.* (1994), apoplastic fluid was collected from the stem and was plated on a modified LGI medium of Cavalcante and Dobereiner (1988). Three isolates were capable of growth on an N-free medium and could reduce acetylene as well as incorporate $^{15}N_2$ into cells *in vitro*. These isolates were identified as members of the genus *Klebsiella*. We estimated that the number of cells in these stems was approximately 2000 per gram fresh weight of tissue. This was about 2–3 orders of magnitude lower than the number of *Gluconacetobacter diazotrophicus* in sugarcane. Since that time, we have made four other independent isolations of *Klebsiella* strains from the interior of maize (Chelius, Triplett 2000a, 2000b, 2001). *Klebsiella* appear to be common endophytes of maize.

We next wanted to show that these *Klebsiella* endophytes were inhabiting the interior of maize and that they could re-enter the plant upon inoculation (Chelius, Triplett 2000a). After labeling the *Klebsiella* cells with a constitutively expressed green fluorescent protein gene, the cells were found to re-enter maize upon inoculation and inhabit the intercellular spaces of the root cortex in large numbers. They are present in stems but in much lower numbers than in roots. In addition, these *Klebsiella* would produce NifH protein *in planta* provided that sucrose was added to the nutrient solution for the plants (Chelius, Triplett 2000b). Egener *et al.* (1998, 1999) previously showed that the expression of *nifH* by *Azoarcus* sp. in rice seedlings required the addition of a carbon source.

We have also shown by culture-dependent and culture-independent methods that the phylogenetic diversity of prokaryotes within maize is very extensive (Chelius, Triplett 2000c, 2001). Several divisions of bacteria are present in maize with the largest proportion of the bacteria belonging to the proteobacteria. Two divisions of archaea, the euryarchaea and crenarchaea, are also present in maize but given that nested PCR was required to detect the presence of archaea in maize, we suspect that their numbers *in planta* are very low compared to the bacteria. Very novel bacteria have also been cultured from the interior of maize. This includes a new genus and species, *Dyadobacter fermentens* (Chelius, Triplett 2000c).

The frequency with which *Klebsiella* have been found in maize has encouraged us to characterize these endophytic diazotrophs more thoroughly. We are particularly interested in comparing these endophytes to clinical isolates of *K. pneumoniae*. To begin this work, we used genomic interspecies microarray hybridization to identify 3000 genes in common with *E. coli* that are present in one of our endophytic *Klebsiella* (Dong *et al.* 2001). Genes present in one of our endophytes but absent in a clinical isolate of *K. pneumoniae* are being isolated by subtraction hybridization.

In addition to isolating endophytes from maize, we have also isolated many endophytes from switchgrass (*Panicum virgatum* L.). Switchgrass was collected from remnant prairies in the central sands region of Wisconsin. These native plants have never seen nitrogen fertilizer. The isolation and characterization of several diazotrophs collected from these plants will be described here as well as their effect on the growth of wheat and rice.

We have discovered that several of these diazotrophic endophytes enhance maize growth in the presence of nitrogen fertilizer both in greenhouse and field experiments (Riggs *et al.* 2001). A novel approach to discover the molecular basis for these growth increases will be presented here.

2. Procedure

2.1. Collection and culture of switchgrass. Switchgrass plants were collected (WS98.1) from Buena Vista Quarry Prairie, a native prairie remnant near Plover, Wisconsin on an extremely sandy soil. From each of 10 random plants, we collected 10 tillers, a sample of soil surrounding the plant, and a sample of open-pollinated seed from the plant. Tillers were planted in about 200 ml of prairie soil surrounded by sand in the greenhouse. Plants were watered with a minus-N nutrient solution.

2.2. Isolation of endophytes from switchgrass and maize. Endophytes were isolated from maize and switchgrass as described previously (Chelius, Triplett 2000a, 2000b, 2001). *Klebsiella pneumoniae* 342 was isolated from an N-efficient line of maize obtained from CIMMYT, the International Maize and Wheat Improvement Center, in Mexico. *Pantoea* sp. P101 and P102 were isolated from switchgrass lines collected near Plover, Wisconsin.

2.3. Inoculation, presence, and NifH expression of *K. pneumoniae* in wheat. These experiments were done as described previously for our work on maize (Chelius, Triplett 2000a, 2000b).

2.4. Culture and harvest of rice, wheat, maize and *Arabidopsis thaliana* in the greenhouse. The rice and wheat lines used in this work were *Oryza sativa* spp. *japonica* cv. Nipponbare and *Triticum aestivum* L. cv. Trenton. Various ecotypes of *Arabidopsis thaliana* were obtained from the Arabidopsis Biological Resource Center (Ohio State University, Columbus, OH, USA). The seeds were germinated and plants cultured using the protocols of the Arabidopsis Biological Resource Center (http://aims.cse.msu.edu/~aimsweb/catalog97/IV-A.html). Grasses were cultured in a 1:1 sand:vermiculite mix in two liter plastic pots. Plants were given a nutrient solution that contains the following ingredients: 5 µM $CaCl_2$, 1.25 µM $MgSO_4$, 5 µM KCl, 1 µM KH_2PO_4, 0.162 µM $FeSO_4$, 2.91 nM H_3BO_3, 1.14 nM $MnSO_4$, 0.76 nM $ZnSO_4$, 0.13 nM $NaMoO_4$, 0.14 nM $NiCl_2$, 0.013 nM $CoCl_2$ and 0.19 nM $CuSO_4$. Plants were given supplemental lighting. To obtain the dry weight data presented for all of these plant species, the above-ground portions of the plants were placed in paper bags and dried at 65°C for at least 48 h prior to weighing. Nitrogen content of wheat tissue was determined after grinding the tissue in a Wiley mill. Nitrogen concentration was measured by rapid combustion at 850°C followed by conversion of all N-combustion products to N_2 and subsequent measurement by a thermoconductivity cell (LECO Model FP-528; LECO Corp., St Joseph, MO).

2.5. 16S rDNA amplification and sequencing. PCR amplification and sequencing of the 16S rRNA gene from grass endophytes was done as described previously (Chelius, Triplett 2000c). The accession numbers for the 16S rRNA genes of *K. pneumoniae* 342, *Pantoea* sp. P101 and *Pantoea* sp. P102 are AF394537, AF394538 and AF394539, respectively.

2.6. Construction of *nifH* insertion mutant of *Klebsiella peumoniae* 342. Amplification and sequence analysis showed that *nifH* from strain 342 has 88% sequence identity with *nifH* from *K. pneumoniae* M5a1. Primers nifH1f (5'–GCCTGCAGATGACCATGCGTCAATGCGCC–3') and nifH876r (5'–GCGAATTCCGCGTTTTCTTCGGCGGCGGT–3') were designed based on the *nifH* sequence of *K. pneumoniae* M5a1 (GenBank accession number X13303). One hundred ng of strain 342 DNA was used as the template in a 25 µl PCR mixture containing 1X PCR Buffer (Promega), 2.5 mM MgCl₂, 0.2 mM dNTPs and 0.5 U of *Taq* polymerase. Thermal cycle conditions were 4 min of denaturation at 95°C; 30 cycles of 94°C for 30 s, 54°C for 30 s, and 72°C for 1 min; and then an extension at 72°C for 7 min. The PCR product was purified using a Qiagen PCR purification kit and then ligated to pGEM-T Easy vector (Qiagen). White colonies were selected and plasmids were isolated. ABI's cycle sequencing kit was used by the University of Wisconsin-Madison Biotechnology Center to sequence two of those plasmids using primers T7 and Sp6. As a result of this high identity, plasmid pSA30 containing *nifHDKYE* from *K. peumoniae* M5a1 (Brown, Ausubel 1984) was used to make the insertion and subsequent marker exchange. A 1.7 kb fragment containing *nifH* gene (882 bp) and part of *nifD* (632 bp) was excised from pSA30 by double digestion with *Eco*RI and *Bam*HI. This fragment was inserted into *Eco*RI/*Bam*HI doubly digested vector pUC18, resulting in plasmid pH1. A 1.4 kb fragment from pKRP11 (Reece, Phillips 1995) containing a kanamycin resistance gene downstream of a constitutive promoter was excised with *Hin*dIII and then blunted with Klenow. Following *Bgl*II digestion of pH1 and subsequent blunting with Klenow, the fragment from pKRP11 was inserted into the *Bgl*II site of pH1. The *Bgl*II site in pH1 was located in the middle of *nifH* gene. This new plasmid was named as pI1. To exchange the inserted allele of *nifH* for the wild type allele on the *K. pneumoniae* 342 chromosome, the 3.1 kb fragment containing nifHD'-Km was excised from pI1 by double digestion with *Eco*RI and *Pst*I, and *Eco*RI site was blunted. This fragment was ligated into the *Pst*I/*Sma*I doubly digested suicide plasmid pJQ200KS+ and marker exchanged as described previously (Scupham, Triplett 1997). Nif isolates were confirmed by Southern hybridization with an *npt*II probe

3. Results and Discussion

The effects on plant growth of three strains isolated in this laboratory from N-efficient lines of maize or switchgrass are described here. These strains are compared with two strains that are nitrogen-fixing endophytes of sugarcane. These strains often increase the productivity of maize when N fertilizer is supplied but they do not relieve N-deficiency conditions of those plants not given N fertilizer (Riggs *et al.* 2001). Here we tested the growth responses of these strains on rice and wheat in the absence of N and the growth effect of these strains on *A. thaliana* with the addition of fixed N. As wheat and rice plants are much smaller than maize plants, our hope was that these strains might provide enough fixed N to relieve the nitrogen deficiency of smaller grasses that could not be observed on maize (Figures 1 and 2).

None of the strains used here relieved nitrogen deficiency symptoms in wheat since N-fertilized plants grown simultaneously were much larger and more vigorous. However, without addition of fixed N, the growth, nitrogen concentration, and total N per plant of

Figure 1. Growth of Trenton spring wheat six weeks after planting. Seeds were inoculated with *K. pneumoniae* 342 or a *nifH* mutant of strain 342. An uninoculated control is also shown. Plants were cultured in the absence of added N.

strain 342-inoculated plants was significantly higher (4.22 mg N/plant) than in the uninoculated plants (3.54 mg N/plant) as well as those plants inoculated with the *nifH* mutant of strain 342 (3.53 mg N/plant). Thus, this growth increase in N-starved wheat upon inoculation with strain 342 can be attributed to nitrogen fixation by this endophyte (Figure 1). Strain 342 is an endophyte of wheat (Dong, Triplett, unpublished) in addition to its ability to survive endophytically in maize (Chelius, Triplett 2000a, 2000b).

As these strains can enhance maize growth with added N (Riggs *et al.* 2001), we are interested in the mechanism of these responses. We have decided to determine whether any of these strains can enhance the growth of *Arabidopsis* in the presence of fixed N. If *Arabidopsis* is increased in size, we can then screen mutants of *Arabidopsis* for those that do not respond to the inoculum. The non-responsive mutants will then be used to identify host genes involved in this response whose identity should give us clues as to the role of the bacteria in the response. We have found several ecotypes whose growth is dramatically increased upon inoculation with one or more of the endophytes used here (Figure 3) so we can now begin to screen EMS mutants of Ws-2. Hormone-insensitive mutants of *Arabidopsis* will also be tested.

Figure 2. Growth of rice 13 weeks following inoculation with *Pantoea* sp. P101. An uninoculated control is also shown. Plants were cultured in the absence of added N.

We continue to maintain in the greenhouse the switchgrass plants that were collected in the fall of 1998 from remnant prairies near Polver, Wisconsin. Some of these plants continue to grow, although very slowly, despite having no fixed nitrogen source since November, 1998. Native grasses growing under low nitrogen conditions appear to be a rich source of diazotrophic bacteria that may provide fixed N to plants as well as increase plant growth by mechanism(s) independent of nitrogen fixation. In future work, we intend to test the ability of the diazotrophs collected from maize and switchgrass to provide fixed N to wheat and rice by $^{15}N_2$ reduction assays.

Figure 3. Growth of *Arabidopsis thaliana* Ws-2 following inoculation with the switchgrass isolate *Pantoea* sp. P102 or the maize isolate *K. pneumoniae* 342. An uninoculated control is also shown. Plants were cultured with added N.

References
Boddey RM *et al.* (2000) In Triplett EW (ed) Prokaryotic Nitrogen Fixation: A Model System for the Analysis of a Biological Process, pp. 705-726, Horizon Scientific Press, Norfolk, UK
Brown SE, Ausubel FM (1984) J. Bacteriol. 157, 143-147
Cavalcante VA, Dobereiner J (1988) Plant and Soil 108, 23-31

INTERACTIONS BETWEEN ENDOBACTERIA AND ARBUSCULAR MYCORRHIZAL FUNGI

P. Bonfante, V. Bianciotto, D. Minerdi

Dept of Plant Biology, University of Torino, CSMT-CNR, Viale Mattioli 25, 10125 Torino, Italy

1. Introduction

Arbuscular mycorrhizal (AM) fungi constitute one of the most widespread microbial communities of the rhizosphere, since they establish symbiotic associations with the roots of about 80% of plant species (Smith, Read 1997). Fossil and molecular data suggest that roots and arbuscular mycorrhizal (AM) fungi have shared a cooperative life since Devonian times (Simon *et al.* 1993). The success of mycorrhizas in evolution is mainly due to the central role that AM fungi play in the capture of nutrients from the soil in almost all ecosystems (Smith, Read 1997). As a consequence, they are crucial determinants of plant biodiversity, ecosystem variability and productivity of plant communities (van der Heijden *et al.* 1998). AM fungi are not only an essential feature of the biology and ecology of most terrestrial plants, they also interact with different classes of bacteria during their life cycles. AM fungi establish in fact interactions both with bacteria living in the rhizosphere during their extraradical phase and with endosymbiotic bacteria which live in the cytoplasm of some fungal isolates (Perotto, Bonfante 1997; Bianciotto *et al.* 2001).

The understanding of these multiple interactions is one of the most exciting challenges of current research in the field of molecular microbe-plant interactions and of their application in low chemical input agricultural systems.

In recent years a wealth of experimental investigations, together with the development of new technologies, has led to substantial advances in the knowledge of arbuscular mycorrhizal functioning, mostly in the field of molecular biology (Harrison 1999). Despite these recent achievements in our knowledge of the molecular basis of plant-fungal interactions, many aspects of the biology of AM fungi are still obscure and hamper a full exploitation of their potentialities as biotechnological tools. This is mostly due to their obligate biotrophic status, their multinuclear condition, and an unexpected level of genetic variability (Bonfante, Perotto 1995; Gianinazzi-Pearson 2000; Hosny *et al.* 1999; Lanfranco *et al.* 1999).

The aim of this short review is to analyze a feature which is peculiar of some AM fungi and increases the level of their genetic complexity: the presence of endobacteria living in their cytoplasm. We will provide information on their identification, phylogeny and possible functions.

2. The Identification of Bacteria-like Organisms in AM Fungi

Intracellular structures very similar to bacteria and called Bacteria-like Organisms (BLOs) were first described in the 1970s (Mosse, 1970; Scannerini, Bonfante 1991 for a review). Ultrastructural observations clearly revealed their presence in many field-collected fungal isolates. Further investigation on these BLOs, including the demonstration of their prokaryotic nature, was long hampered because of their inability to grow on plate. Only a combination of morphological observations (electron and confocal microscopy) and molecular analyzes allowed us to identify BLOs as true bacteria and to start unraveling their symbiotic relationship with AM fungi (Bianciotto *et al.* 1996).

Isolate BEG 34 of *Gigaspora margarita* contains a large number of BLOs which can be easily detected by staining with fluorescent dyes specific for bacteria and capable of distinguishing between live and dead bacteria. About 250,000 live bacteria were counted in a single spore, which is a large structure of about 260-400 μm. Ultrastructural observations performed on high-pressure freezing/freeze-substituted samples revealed a large number of rod-shaped BLOs in the vacuoles of germinating spores, often associated to the abundant protein bodies. On the basis of the 16S rDNA sequences the bacterial endosymbionts living in the fungus *Gigaspora margarita* (BEG 34) were identified as belonging to the genus *Burkholderia*.

3. A Phylogenetic Tree for the Endosymbiotic *Burkholderia*

To determine whether bacteria are also harbored by other members of *Gigasporaceae*, eleven fungal isolates collected from different geographic areas and belonging to six different species were analyzed by morphological and molecular approaches. A fluorescent dye was used to visualize the bacteria inside the spores, while 16S ribosomal genes were amplified by PCR using universal eubacterial primers and primers specific for the endobacteria identified in *Gi. margarita* BEG 34. With the exception of *Gigaspora rosea*, isolates from all other species harbored bacteria in their cytoplasm and gave an amplified fragment of the expected size with the universal primers (Bianciotto *et al.* 2000). Six out of seven isolates (belonging to five different species) could be also amplified with the specific primers. These specific primers were used as probes for *in situ* hybridization on *Gi. margarita* spores, where they successfully identified bacteria. The 16S rDNA amplified from isolates of *Scutellospora persica*, *S.castanea* and *Gi. margarita* was sequenced and aligned with the closer bacterial sequences available in databases. With neighbor-joining analysis a strongly supported branch containing all endosymbiotic bacteria so far sequenced in *Gigasporaceae* could be identified nested in the genus *Burkholderia*. These results demonstrate therefore that endobacteria are widespread in *Gigasporaceae*. In addition, preliminary experiments showing that they pass through one fungal generation to another with a vertical transmission mechanism (V. Bianciotto and G. Becard, in preparation) suggest that they represent a stable cytoplasmic component.

4. How to Study the Genome of the Endosymbiotic *Burkholderia*

The genus *Burkholderia* is an extremely heterogeneous group which includes soil bacteria, plant growth promoting rhizobacteria and human and plant pathogens (Bevivino *et al.* 1994). Strains of *Burkholderia cepacia* complex can survive within vacuoles in different isolates of the genus *Acanthamoeba* (Marolda *et al.* 1999). In addition, rhizospheric *Burkholderia* isolates have been found to fix nitrogen (Gillis *et al.* 1995). In order to understand the role of the intracellular *Burkholderia*, we have therefore set out to determine whether they have the molecular machinery for the the colonization of eukaryotic cells and for the uptake of mineral nutrients. Since intracellular *Burkholderia* cannot be grown on a cell-free medium, we took advantage of a genomic library constructed with DNA extracted from *Gi. margarita* spores (van Buuren *et al.* 1999): since it was also representative of the bacterial genome, it was used to investigate the prokaryotic genome by screenings with bacterial heterologous probes. Following this approach, a putative phosphate transporter operon (*pst*) was discovered as well as a gene involved in colonization events by bacterial cells (*vacB*) (Ruiz Lozano, Bonfante 1999, 2000).

5. The Organization of the *nif* Operon

A DNA region containing putative *nif* genes and belonging to the endosymbiont *Burkholderia* has been identified and characterized (Minerdi *et al.* 2001). Screening of the library with *Azospirillum brasilense nif*HDK genes as the prokaryotic probes led to the identification of a 6413 bp region. Analysis revealed three open reading frames (ORFs) encoding putative proteins with a very high degree of sequence similarity with the two subunits (NifD and NifK) of the component I and with component II (NifH) of nitrogenase from different diazotrophs. The three genes were arranged in an operon similar to that shown by most archaeal and bacterial diazotrophs.

PCR experiments with primers designed on the *Burkholderia nif*HDK genes and Southern blot analysis demonstrate that they actually belong to the genome of the *Gi. margarita* endosymbiont. RT-PCR experiments with primers designed on the *Burkholderia nifD* and *nifK*

---

genes and performed on total RNA extracted from germinated spores demonstrate the genes expression during this step of the fungal life cycle.

6. Concluding Remarks

The constant presence of endobacteria in some species of *Gigasporaceae* through the fungal life cycle and their pattern of distribution in AM species, the origin of which is geographically very far, open intriguing questions on the biology of this association and on the possibility of co-evolution events which have linked the two partners, as in other endosymbioses (Clark *et al.* 2000). The discovery of relevant genes in the genome of the endosymbiotic *Burkholderia* (i.e. nitrogen fixation, P-transporter genes) demonstrate that these bacteria possess the molecular mechanisms for playing an active role in the nutrient metabolism, adding a further level of complexity to the nutritional exchanges between plants and mycorrhizal fungi.

7. References

Bevivino *et al.* (1994) Microbiol. 140, 1069-1077

Bianciotto V *et al.* (1996) Appl. Environ. Microbiol. 62, 3005-3010

Bianciotto V *et al.* (2000) Appl. Environ. Microbiol. 66, 4503-4509

Bianciotto V *et al.* (2001) Mol. Plant Microbe Interactions 14 (2), 255-260

Bonfante P, Perotto S (1995) New Phytol. 130, 3-21

Clark MA *et al.* (1997) Evolution Int. J. Org. Evolution 54, 517-25

Gianinazzi-Pearson V (2000) In Hoch B (ed), The Mycota ix. pp. 45-61, Springer-Verlag, Berlin

Gillis *et al.* (1995) Int. J. Syst. Bacteriol. 45, 508-516

Harrison MJ (1999) Annu. Rev. Plant Physiol. Plant Mol. Biol. 50, 361-389

Hosny M *et al.* (1999) Gene 226, 61-71

Lanfranco L *et al.* (1999) Mol. Ecol. 8, 37-45

Marolda CL (1999) Microbiol. 145, 1509-1517

Minerdi D *et al.* (2001) Appl. Environ. Microbiol. 67, 725-732

Mosse B (1970) Arch. Mikrobiol. 74, 146-159

Perotto S, Bonfante P (1997) Trends Microbiol. 5, 496-501

Ruiz-Lozano M, Bonfante P (1999) J. Bacteriol. 181, 4106-4109

Ruiz-Lozano M, Bonfante P (2000) Microb. Ecol. 39, 137-144

Scannerini S, Bonfante P (1991) In Margulis L, Fester R (eds), Simbiosis as Source of Evolutionary Innovation: Speciation and Morphogenesis, pp. 273-287, The MIT Press, Cambridge, MA

Simon L *et al.* (1993) Nature 363, 67-68

Smith SE, Read DJ (1997) Mycorrhizal Symbiosis, Academic Press, London, UK

van Buuren, ML *et al.* (1999) Mycol. Res. 103, 955-970

van der Heijden MGA *et al.* (1998) Nature 396, 69-72

7. Acknowledgements

The research illustrated in this review has been funded by the EU GENOMYCA project, QLK5-CT-2000-01319, the Italian National Council of Research and the National Project Produzione Agricola nella Difesa dell'Ambiente (PANDA).

NITROGEN FIXATION AND INTERACTIONS WITH RICE BY *AZOARCUS* SP. STRAIN BH72

B. Reinhold-Hurek[1], T. Hurek[2], D.E. Martin[1], A. Sarkar[1], S. Wiese[1]

[1]Dept of General Microbiology, University Bremen, FB 2, PO Box 33 04 40, D-28334 Bremen, Germany
[2]Cooperation Laboratory Max-Planck-Institute for Marine Microbiology/ University Bremen, FB 2, PO Box 33 04 40, D-28334 Bremen, Germany

1. Introduction

Plant roots offer a variety of microhabitats for microbial colonization including the rhizosphere soil, the plant surface (rhizoplane) and inner root tissues. Plant-bacteria systems which have recently gained attention are associations between endophytic, nitrogen-fixing bacteria and grasses or cereals (Reinhold-Hurek, Hurek 1998). Well-studied examples are members of the beta-subclass of the *Proteobacteria, Azoarcus* spp. and *Herbaspirillum seropedicae*, or a member of the alpha-subclass, *Gluconacetobacter diazotrophicus* (formerly *Acetobacter diazotrophicus*). They infect Kallar grass and rice (Hurek *et al.* 1994), a wide range of grasses and cereals (James *et al.* 1997, 1998) or sugar cane and coffee plants (James *et al.* 1994; Jiminez-Salgado *et al.* 1997), respectively. Surprisingly, rhizobia have also been found to be natural endophytes of non-legumes, e.g. in rice (Yanni *et al.* 1997; Engelhard *et al.* 2000); other endophytes were detected in maize (Palus *et al.* 1996).

2. Plant Colonization

These diazotrophic endophytes have several features in common. Unlike rhizobia, they do not survive well in soil. In most cases they cannot be isolated from soil but only from plant material and are thus ecologically dependent on plants (James, Olivares 1998; Reinhold-Hurek, Hurek 1998). In gnotobiotic culture, it was demonstrated that pure cultures of these bacteria are capable of invading plant roots. Points of invasion are the zone of elongation and differentiation close to the root tip and the emergence points of lateral roots (Hurek *et al.* 1994; James *et al.* 1994, 1997). The colonization is mostly intercellular, only rarely intracellular plant cell colonization is observed. However, these plant cells are not viable (Hurek *et al.* 1994), and there is no evidence for an endosymbiosis in living plant cells (Reinhold-Hurek, Hurek 1998) in contrast to the rhizobium-legume symbiosis.

The major site of endophytic colonization appears to be the root cortex, especially in the aerenchyma of flood-tolerant plants such as Kallar grass (*Leptochloa fusca* L. (Kunth)) and rice, where large microcolonies of *Azoarcus* sp. BH72 can be found (Hurek *et al.* 1994; Egener *et al.* 1998). However, rarely these endophytes do also invade the stele and penetrate xylem vessels, which were previously thought to be sterile in healthy plants. Xylem colonization may facilitate the systemic infection of plants, the bacteria spreading from roots to shoots of young rice plants (Hurek *et al.* 1994; Gyaneshwar *et al.* 2000). For other diazotrophic endophytes (*Heraspirillum* spp., *Gluconacetobacter diazotrophicus*), xylem cells appear to be a more frequent colonization site (James, Olivares 1998; Olivares *et al.* 1997).

Despite the relatively dense colonization (up to almost 10^8 bacteria per g root dry weight for *Azoarcus* sp. BH72 in field-grown Kallar grass (Reinhold *et al.* 1986)), the endophytes do not cause symptoms of plant disease in their host plants. In contrast, in several plants they have been shown to promote plant growth (Hurek *et al.* 1994; Sevilla *et al.* 2001). Whether these growth responses are due to bacterial phytohormone production, nitrogen fixation or other mechanisms may vary with the system and is still under investigation.

3. Rice Cultivars Show Differences in Endophytic *nif*-gene Expression by *Azoarcus* spp.

In order to visualize bacterial gene expression in the plant, we used transcriptional fusions of target genes such as *nifH*, one of the structural genes for the nitrogenase complex reducing N_2 to ammonia, with the reporter gene *gfp* (encoding the jellyfish green fluorescent protein) or *gus* (encoding β-glucuronidase). Infection studies using *Azoarcus* sp. BH72 on rice seedlings in gnotobiotic culture revealed, that bacterial *nifH::gfp* is expressed at high levels inside the roots (Reinhold-Hurek, Hurek 1998). The sites of nitrogenase gene expression are the infection sites at emerging lateral roots and root tips as well as microcolonies inside the cortex in the aerenchymatic air spaces (Egener *et al.* 1999). It is remarkable that this endophytic gene expression occurs in the apoplast, apparently depending on apoplastic nutrient flow. Only minor amounts of carbon source (5 mg of malate per l) were added as "starter" to the plant medium, thus the endophytic *nifH* gene expression appears to be driven by carbon- and energy sources supplied by the plant. In contrast, in other systems studied such as *Gluconacetobacter diazotrophicus* in sugar cane (James, Olivares 1998), significant *nif* gene expression was only detected in gnotobiotic systems when high amounts of carbon source were added to the plant medium.

Since *nifH* gene expression in *Azoarcus* sp. BH72 is similarly regulated in response to O_2 and ammonium as the nitrogenase activity is (Hurek *et al.* 1994), the reporter gene studies demonstrate that the rice root apoplast may provide a suitable microhabitat for endophytic nitrogen fixation of *Azoarcus* sp. BH72. This is also the case for the host plant from which this strain was originally isolated, Kallar grass (Reinhold *et al.* 1986). It was shown by *in situ* hybridization studies that field-grown plants harbor *Azoarcus* sp. *nifH* mRNA in the aerenchyma (Hurek *et al.* 1997).

Recent findings suggest that the assumption that diazotrophic endophytes do not cause damage of plants cannot be generalized. Previously, a phytopathogenic species closely related to *Herbaspirillum seropedicae, H. rubrisubalbicans*, was detected to cause mild symptoms of disease in leafs of certain sugar cane cultivars, in others it grew as a non-pathogenic endophyte (Olivares *et al.* 1997). In our laboratory, a recent screening of rice cultivars with respect to endophytic *nifH::gus* expression revealed significant, cultivar-specific differences in response to *Azoarcus* sp. strain BH72. Some cultivars supported high levels of root-associated *nifH::gus* expression, while no significant expression could be detected in others. All of the latter cultivars showed browning of roots upon inoculation, probably due to lignification and accumulation of phenolic substances. This plant reaction, which resembled a plant defense response, required bacterial colonization: a mutant of *Azoarcus* sp. strain BH72 lacking type IV pili was previously shown to be deficient in effective colonization of rice roots (Dörr *et al.* 1998). Type IV pili are also known in animal and human pathogens as virulence factors mediating attachment to the host tissue. This mutant did not cause the plant response upon inculation, indicating physical contact of roots and bacteria is necessary to elicit that plant response. Further studies on plant defense genes which might be induced are on the way.

4. Regulation of Nitrogen Fixation in Azoarcus sp. Strain BH72

P_{II} proteins are small signal transduction molecules involved in nitrogen fixation and assimilation. In many *Proteobacteria*, two different P_{II}-like proteins occur (Arcondéguy *et al.* 2001). In contrast, *Azoarcus* sp. strain BH72 harbors three paralogs of P_{II}-like proteins (Martin *et al.* 2000). All three proteins can be covalently modified by uridylylation in response to ammonium starvation (Martin *et al.* 2000). GlnK and GlnY both occur in wild type cells, whereas GlnY could only be detected in a *glnB⁻-glnK⁻* double mutant. As most other P_{II}-like proteins, GlnK and GlnB occur in their native form under conditions of ammonium excess, or become deuridylylated in when ammonium is added to a N-starved culture. However GlnY is unusual in that the uridylyl residue is not removed under ammonium excess, as demonstrated in the *glnK⁻-glnB⁻* double mutant (Martin *et al.* 2000). But what is the differential role of these proteins? Recently it was demonstrated that active nitrogenase of strain BH72 is rapidly inactivated by addition of ammonium (Egener *et al.* 2001). This ammonium "switch-off" requires a ferredoxin contranscribed with the structural genes of nitrogenase, *nifHDK* (Egener *et al.* 2001). In addition, the P_{II}-like proteins are differentially involved in this process. Mutational analysis has shown that GlnK but not GlnB or GlnY is required for the "switch-off". The modification of the iron protein of nitrogenase, which occurs concomitantly with the ammonium "switch-off", is in contrast dependent on the presence of both proteins, GlnK and GlnB. Interestingly, a *glnB⁻* mutant did still show ammonium "switch-off", however it did not show modification of the iron protein of nitrogenase any more. This clearly demonstrates that inhibition of nitrogenase activity and its covalent modification are distinct processes.

Moreover, the *amtB* gene encoding a putative high-affinity ammonium transporter is also involved in the "switch-off" process. In an *amtB⁻* mutant, rapid inactivation of nitrogenase activity was not observed any more. Since in this mutant, ammonium was removed from the medium at the same rate as in wild type cells, it is not likely that this effect is due to the lack of ammonium uptake. Western blot analyses of cell fractions of strain BH72 have shown that GlnK occurs not only in the cytoplasm, but also membrane-associated. This is not the case for GlnB. This indicates that GlnK and AmtB might physically interact, as recently hypothesized by others (Thomas *et al.* 2001). Therefore, AmtB might act as a sensor involved in the signal transduction process of sudden ammonium access. Interestingly, even in an *amtB⁻* mutant, GlnK was found to be associated with bacterial membranes, indicating that other membrane proteins in *Azoarcus* sp. strain BH72 might interact with GlnK, as well.

5. References

Arcondéguy T *et al.* (2001) Microbiol. Mol. Biol. Rev. 65, 80-105
Dörr J *et al.* (1998) Mol. Microbiol. 30, 7-17
Egener T *et al.* (2001) J. Bacteriol. 183, 3752-3760
Egener T *et al.* (1999) Mol. Plant-Microb. Interact. 12, 813-819
Egener T *et al.* (1998) Mol. Plant-Microb. Interact. 11, 71-75
Engelhard M *et al.* (2000) Environ. Microbiol. 2, 131-141
Gyaneshwar P *et al.* (2000) J. Bacteriol.
Hurek T *et al.* (1997) J. Bacteriol. 179, 4172-4178
Hurek T *et al.* (1994) J. Bacteriol. 176, 1913-1923
James EK, Olivares FL (1998) Crit. Rev. Plant Sci. 17, 77-119
James EK *et al.* (1997) J. Exp. Bot. 48, 785-797
James EK *et al.* (1994) J. Exp. Bot. 45, 757-766
Jiminez-Salgado T *et al.* (1997) Appl. Environ. Microbiol. 63, 3676-3683
Martin D *et al.* (2000) Mol. Microbiol. 38, 276-288
Palus JA *et al.* (1996) Plant Soil 186, 135-142
Olivares FL *et al.* (1997) New Phytol. 135, 723-737
Reinhold-Hurek B, Hurek T (1998) Trends Microbiol. 6, 139-144
Reinhold-Hurek B, Hurek T (1998) Crit. Rev. Plant Sci. 17, 29-54
Reinhold B *et al.* (1986) Appl. Environ. Microbiol. 52, 520-526
Sevilla M *et al.* (2001) Mol. Plant-Microb. Interact. 14, 358-366
Thomas G *et al.* (2000) Trends in Genetics 16, 11-14
Yanni YG *et al.* (1997) Plant Soil 194, 99-114

Section 12:
Common Themes in Symbiosis and Pathogenesis

CHAIR'S COMMENTS: EXTENSIN-LIKE GLYCOPROTEINS IN THE LUMEN OF INFECTION THREADS IN PEA ROOT NODULES

N.J. Brewin, E.A. Rathbun

John Innes Centre, Norwich NR4 7UH, UK

1. Introduction

Rhizobium leguminosarum bv *viciae* gains entry to pea root tissues through tip-growing infection threads (Brewin 1998). Growth of bacteria in close association with a deformed root hair cell wall results in the initiation of an infection thread as a tubular ingrowth of the plant cell wall. Rhizobia are propagated from cell to cell because transcellular threads are reinitiated in adjacent cells in the root cortex and subsequently in the invasion zone of the developing nodule (Rae *et al.* 1992). The lumen of the infection thread is topologically equivalent to an intercellular space and the luminal matrix apparently shares many components with the extracellular matrix. Here, we describe the isolation and characterization of a plant matrix glycoprotein, MGP. This 110 kDa component was originally identified in the lumen of the infection thread using monoclonal antibody, MAC265, which recognizes a carbohydrate epitope.

2. Procedure

MAC265 was used for immunoaffinity purification of the glycoprotein from pea roots exudates. Subsequently, a peptide fragment was obtained and subjected to N-terminal sequencing. Using PCR-primers based on this sequence, clones were isolated from the cDNA of inoculated pea roots and nodules. DNA sequencing revealed a family of closely related and repetitive polypeptides with proline-rich motifs, characteristic of extensins.

3. Results and Discussion

```
MRSLMASAALILALAMLFLSFPSEISANQYSYSSPPPPVHSPPPPKDPYHYSSPPPPP

                      VHTYPHPHPVYHSPPPP

  VHTYPHPHPVYHSPPPPTPHKKPYKYPSPPPPPVHTYPHPHPVYHSPPPP

            HKKPYKYSSPPPPPVHTYPHPHPVYHSPPPP

                  VHTYPHPHPVYHSPPPPP

TPHKKPYKYPSPPPPPAHTYPPHVPTPVYHSPPPPAYSPPPPAYYYKSPPPPPYHH*
```

Figure 1. Consensus of translated MGP sequences deduced from 30 cDNA clones.

Figure 1 shows the consensus of amino acid sequences from N-terminal methionine to stop codon (*) deduced from 30 cDNA clones. The conserved N-terminal, "central" and C-terminal regions are shown in bold. The central conserved region can be divided into two motifs, each of which is reiterated in other parts of the sequence. Motif I, Thr-Pro-His-Lys-Lys-Pro-Tyr-Lys-Tyr-Pro-Ser-(Pro)$_5$, is also present within the C-terminal domain. Motif II, Val-His-Thr-Tyr-Pro-His-Pro-His-Pro-Val-Tyr-His-Ser-(Pro)$_4$, is repeated twice in the central domain and is also repeated a variable number of times elsewhere. Variant residues are underlined and blocks of SP_4 or SP_5 are shaded. Note that Tyr residues generally occur in pairs, except at the C-terminus. This might be a significant feature in relation to peroxide-driven cross-linking (see below).

MGP as an extensin corresponds most closely to the deduced amino-acid sequence of an extensin from *V. faba*, VfNDS-E, and from *Medicago truncatula* (Gamas *et al.* 1996), both of which are apparently up-regulated in nodules relative to root tissues. We propose to term this sub-class of extensins "nodule extensins". Because nodule extensin is a key component of the extracellular matrix, we hypothesize that its physical and biochemical properties may have an important influence on the process of tissue and cell colonization by Rhizobium bacteria (Peters *et al.* 2000).

The solubility of MGP (i.e. its extractability from tissues) is apparently reduced by hydrogen peroxide (Wisniewski *et al.* 1999). Protein cross-linking is also suggested by the fact that these antigens often have different apparent molecular masses, following isolation from different tissues of pea and lupin (de Lorenzo *et al.* 1998). These preliminary observations suggest that protein cross-linking may be an important factor controlling the fluidity of the matrix that surrounds rhizobia in the infection thread lumen. Propagation of the infection thread may depend on the dynamic equilibrium between secretion of nodule extensin into the lumen and insolubilization of this matrix by locally generated hydrogen peroxide (Wisniewski *et al.* 1999). During incompatible interactions, overproduction of peroxide might be expected to result in abortion of infection threads (Vasse *et al.* 1993).

4. References

Brewin NJ (1998) In HP Spaink *et al.* (eds.) The *Rhizobiaceae*, pp. 417-429, Kluwer, Dordrecht
de Lorenzo CA *et al.* (1998) Protoplasma 201, 71-84
Gamas P *et al.* (1996) Mol. Plant Microbe Interact. 9, 233-242
PetersWS *et al.* (2000) Comp. Biochem. and Physiol. A-Mol. and Integ. Physiol. 125, 151-167
Rae AL (1992) Mol. Plant Microbe Interact. 4, 563-570
Vasse J *et al.* (1993) Plant J. 4, 555-566
Wisniewski JP (2000) Mol. Plant Microbe Interact. 13, 413-420

5. Acknowledgements

We thank Mike Naldrett for help with peptide sequencing and BBSRC for financial support.

NODULE INVASION AND INTRACELLULAR SURVIVAL BY *SINORHIZOBIUM MELILOTI*

J. Lloret[1], L. Barra[1,2], G.R. Campbell[1], G.P. Ferguson[1], K. LeVier[1], B.J. Pellock[1], T.Y. Shcherban, C. Blanco[2], G.C. Walker[1]

[1]Dept. of Biology, Massachussetts Institute of Technology, Cambridge, MA, USA
[2]UMR-CNRS 6026, Universite de Rennes-1, 35042 Rennes Cedex, France

1. Introduction

The *Rhizobium*-legume symbiosis is a complex process that involves an exchange of different signals between the bacterium and its host. Plant flavonoid compounds secreted by the root function both to attract the bacteria and to induce production of rhizobial Nod factor, a lipochito-oligosaccharide that triggers hair root curling and nodule formation (Ehrhardt *et al.* 1995; Ardourel *et al.* 1994; Roche *et al.* 1992). Once trapped inside the curled root hair, the bacteria elicit an extended invagination of the root hair cell membrane called an infection thread. The rhizobia travel down the infection thread into the developing nodule and are released into nodule cells where they differentiate into the nitrogen-fixing bacteroid form. This paper summarizes our laboratory's efforts to understand the rhizobial functions required for nodule invasion and for the establishment of a chronic infection of the plant cells within the nodule.

2. Role of Polysaccharides in the Infection Process

To invade nodules through infection threads, *Sinorhizobium meliloti* must synthesize at least one of the following polysaccharides: succinoglycan, EPS II or a symbiotically active form of K-antigen (Figure 1). Mutants defective in the production of these polysaccharides are blocked at the nodule invasion step and induce the formation of small white root nodules devoid of bacteria (Leigh 1985; Reuber *et al.* 1991; Gonzalez *et al.* 1996). Many studies have utilized strains, Rm1021 and Rm2011, which under non-limiting nutrient conditions, produce only succinoglycan. Rm1021 and Rm2011 have a cryptic capacity to synthesize symbiotically active forms of EPS II (Glazebrook *et al.* 1989).

Succinoglycan is a polymer of an octasaccharide repeated unit, composed of one galactose and seven glucoses residues. Each repeated unit is modified with approximately one acetyl, one succinyl, and one pyruvyl substituent (Reuber *et al.* 1991; Gonzalez *et al.* 1998; Wang *et al.* 1999). Because Calcofluor can bind succinoglycan and then fluoresce under UV light, genetic screens for non-fluorescent mutants were used to identify the *exo* gene cluster involved in succinoglycan biosynthesis (Leigh *et al.* 1985; Doherty *et al.* 1988; Reuber *et al.* 1993). Many *exo* mutants were blocked in succinoglycan production and in nodule invasion thereby indicating the importance of extracellular polysaccharides in the nodulation process (Leigh *et al.* 1985; Reuber *et al.* 1993). A biochemical approach using radiolabeled sugar precursors was used to construct a detailed model for succinoglycan biosynthesis. Briefly, UDP-glucose and ^{14}C-UDP-galactose were added to different permeabilizing *exo* mutant backgrounds. Analysis of the lipid-linked intermediates accumulated by various *exo* mutants showed that galactose is the first sugar added to an undecaprenol carrier, followed by the seven glucoses. The acetyl, succinyl, and pyruvyl modifications are added during the synthesis of the octasaccharide (Reuber *et al.* 1993).

S. meliloti synthesizes high molecular weight (HMW) forms of succinoglycan, consisting of several hundred or greater octasaccharide repeating units, and low molecular weight (LMW) forms, consisting of monomers, dimers and trimers of the octasaccharide repeating unit (Gonzalez *et al.* 1998; Wang *et al.* 1999). The LMW form of succinoglycan, in particular the trimer form, is the symbiotically active species as small added amounts of trimer molecules can partially rescue the symbiotic defect of a non-succinoglycan producing mutant (Wang *et al.* 1999). Two non-exclusive pathways have been proposed for how *S. meliloti* produces LMW succinoglycan: (i) a direct synthesis pathway mediated by *exoP* and *exoT* in which LMW succinoglycan is produced by limited polymerization of the octasaccharide and (ii) a degradative pathway in which glycanases cleave HMW chains to LMW forms (Gonzalez *et al.* 1998; York *et al.* 1998).

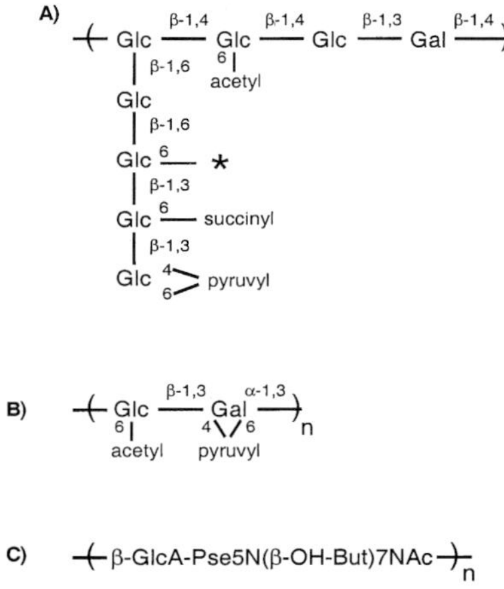

Figure 1. Succinoglycan (A), EPS II (B) and K-antigen (C) repeating units.

Succinoglycan production is needed for the nodule invasion step of symbiosis, but once the rhizobia are inside the plant cell and have differentiated into bacteroids, the *exo* genes are no longer expressed (Reuber *et al.* 1991). This suggests that the regulation of succinoglycan production may also be an important factor in nodule development. Several genes have been implicated in the regulation of succinoglycan production: *exoR*, *exoS*, *exsB*, *mucR*, and *exoX* (Doherty *et al.* 1988; Reed *et al.* 1991a, 1991b; Becker *et al.* 1993; Bertram-Drogatz *et al.* 1998). To date, the ways that these genes affect succinoglycan production are unclear. We have shown that ExoS is the sensor of the ExoS/ChvI two-component regulatory system (Cheng *et al.* 1998a), similar to the ChvG/ChvI system of *A. tumefaciens* (Charles *et al.* 1993), which in turn is part of the two-component regulatory system family that includes EnvZ/OmpR. This system is able to stimulate the transcription of the *exo* genes, but the environmental signal sensed by ExoS that triggers this response remains obscure.

S. meliloti Rm1021 is able to produce the EPS II under phosphate-limiting condition or when it carries either the *expR101* allele of the *expR* locus or a *mucR::Tn5* mutation. EPS II is a polymer of a glucose-galactose repeating unit and is modified with acetyl and pyruvyl substituents. Like succinoglycan, it is found in two forms, HMW and LMW EPS II. Low phosphate (Zhan *et al.* 1991) and *mucR* mutant (Ruberg *et al.* 1999; Keller *et al.* 1995) induce HMW EPSII product, but the *expR101* allele produces also a LMW form, which is symbiotically active and can substitute for succinoglycan during the invasion process (Gonzalez *et al.* 1996; Glazebrook *et al.* 1989). We find that the *expR101* mutation results in the functional restoration of the *expR* gene, which in the parental strain Rm1021 is interrupted by an insertion sequence (Pellock *et al.*, unpublished). The cluster of *exp* genes involved in the synthesis of EPS II has also been identified (Becker *et al.* 1997), but a model for EPS II synthesis has not yet been proposed. A number of the genes found in the *exp* cluster have putative functions that are not easily explained in an EPS II biosynthetic model. Some strains of *S. meliloti* (AK631) produce a capsular polysaccharide, structurally analogous to group II K antigens found in *Escherichia coli* known as the K-antigen (Reuhs *et al.* 1993). The K-antigen is a polymer of a disaccharide composed of an aminohexulosonic acid and α-keto-3,5,7,9-tetradeoxy-5,7-diaminononulosonic acid, a variant of 3-deoxy-D-manno-2-octulosonic acid (KDO). Some strains produce a strain-specific K-antigen that can substitute exopolysaccharide function in symbiosis in *S. meliloti* mutants unable to synthesize succinoglycan and/or EPS II. Certain functions are required in common for the synthesis of K antigen and LPS (Campbell *et al.* 1998).

Interestingly, the ability of these three polysaccharides to promote infection thread initiation and extension shows differences, which can be monitored through strains that constitutively express green fluorescent protein (Cheng *et al.* 1998b; Pellock *et al.* 2000). Succinoglycan is the most efficient of all three in promoting infection thread growth. EPS II and K-antigen mediated infection threads frequently exhibit aberrant morphologies. Furthermore, K-antigen is less efficient than succinoglycan in infection thread extension and EPS II mediated infection is more likely to be aborted in the infection thread initiation or extension steps. The latter may explain the stunted host plant growth observed in EPS II mediated symbiosis.

Recent studies in our lab have shown that the proper *S. meliloti* LPS structure is necessary for the establishment of chronic infection within alfalfa. Different mutants exhibiting anomalies upon entry into plant cells have been identified. One of these mutants, originally termed *fix-389*, has a disrupted *lpsB* gene, which encodes a putative glycosyl transferase one family protein (Lagares *et al.* 2001; Campbell *et al.* submitted). The loss of LpsB causes striking changes in the carbohydrate core of lipopolysaccharide, including the absence of uronic acids and a 40-fold relative increase in xylose. We have also shown that the *lpsB389* mutant exhibits sensitivities to the cationic peptides melittin, polymixin B and poly-L-lysine, in a manner that parallels of *Brucella abortus lps* mutants (Freer *et al.* 1999; Ulgade *et al.* 2000). Sensitivity to these components of the plant's innate immune system may be part of the reason that this mutant is unable to properly sustain a chronic infection within the cells of its host plant alfalfa (Campbell *et al.* submitted).

3. Involvement of BacA in Intracellular Survival

Upon release from the infection thread into acidic plant membrane-bound compartments, wild-type rhizobia differentiate into nitrogen-fixing bacteroids. In contrast, *S. meliloti* mutants lacking a functional BacA protein are unable to survive intracellularly, senescing shortly after their release from the infection thread (Le Vier *et al.* 2000; Glazebrook *et al.* 1993, 1996; Ichige *et al.* 1997). The *bacA* gene was first isolated in a screen that identified *S. meliloti* mutants with symbiotic deficiencies (Long *et al.* 1988). This ability to establish a chronic infection into eukaryotic cells is shared with a number of pathogen bacteria including *Brucella abortus,* a mammalian pathogen, which causes bovine abortions and brucellosis in humans. Intriguingly a *bacA* mutant of *Brucella abortus*, a close phylogenetic relative of *S. meliloti*, is unable to establish a chronic infection in host cells in experimentally infected mice (Le Vier *et al.* 2000). Replication and survival into murine macrophages is also affected in *B. abortus* lacking active BacA protein. Moreover, as with *S. meliloti*, *bacA* mutants of *B. abortus* behave like the wild-type bacteria during the initial stages of host infection. Therefore, a parallel can be established between the survival pattern of a plant symbiont, *S. meliloti* and a mammalian pathogen *B. abortus*.

This is consistent with common environmental stresses both pathogen and symbiont have to face while invading their host cells. For example, oxidative stress is one of the host defense responses produced both by infected macrophages as well as by plant cells (Lamb *et al.* 1997; Ugalde 1999). The production of reactive oxygen species by the infected plant host, is an early event of the hypersensitive response against pathogenic bacteria. A prolonged oxidative burst has previously been shown to occur subsequent to the entry of *S. meliloti* into alfalfa roots (Santos *et al.*

2001). Non-opsonized *B. abortus* do not induce this oxidative defense response, while opsonized bacteria lead to a significant production of oxygen reactive species. In this last case, reactive nitrogen is involved in clearing the macrophage of parasites (Ugalde 1999).

It is possible that BacA may be involved in the resistance to at least one type of intracellular stress. The BacA protein is predicted to be an internal cytoplasmic membrane transporter (Glazebrook *et al.* 1993). Because a *bacA* mutant has increased resistance to bleomycin (Ichige *et al.* 1997; Ferguson personal communication; LeVier submitted) it was also suggested that BacA could act as a bleomycin uptake system. However, recent physiological studies (G.P. Ferguson, unpublished) and genetic analyses (LeVier submitted) have highlighted new *bacA* phenotypes and have led to the hypothesis that BacA can carry out more than one function. BacA could act as a signal receptor, transducing environmental signals or stresses; these signals could be essential to the bacteria adaptation to an intracellular lifestyle. Another possible role for BacA, suggested by the involvement of BacA in deoxycholate, SDS and ethanol resistance, is that BacA could also be involved in the homeostasis of the cell envelope. BacA could also be an efflux pump. We are now attempting to determine the role of BacA and to model its behavior during the cell colonization.

4. References

Ardourel DN *et al.* (1994) Plant Cell. 6, 1357-1374

Bertram-Drogatz P *et al.* (1998) Mol. Gen. Genet. 257, 433-441

Becker A *et al.* (1997) J. Bacteriol. 179, 1375-1384

Campbell GR *et al.* (1998) J. Bacteriol. 180, 5432-5436

Charles *et al.* (1993) J. Bacteriol. 175, 6614-6625

Cheng HP *et al.* (1998a) J. Bacteriol. 180, 20-26

Cheng HP *et al.* (1998b) J. Bacteriol. 180, 5183-5191

Doherty D *et al.* (1988) J. Bacteriol. 170, 4249-4256

Ehrhardt D *et al.* (1995) J. Bacteriol. 177, 6237-6245

Freer E *et al.* (1999) Infect. Immun. 67, 6181-6186

Glazebrook JIA *et al.* (1989) Cell. 56, 661-672

Glazebrook JIA *et al.* (1996) J. Bacteriol. 178, 745-752

Glazebrook JIA *et al.* (1993) Genes Dev. 7, 1485-1497

Gonzalez JE *et al.* (1998) Proc. Natl. Acad. Sci. USA 95, 13477-13482

Gonzalez JE *et al.* (1996) Proc. Natl. Acad. Sci. USA 93, 8636-8641

Gonzalez JE *et al.* (1996) Gene. 179, 141-146

Ichige A *et al.* (1997) J. Bacteriol. 179, 209-216

Keller M *et al.* (1995) Mol. Plant Microbe Interact. 8, 267-277

Lagares A *et al.* (2001) J. Bacteriol. 183, 1248-1258

Lamb C *et al.* (1997) Annu. Rev. Plant Mol. Biol. 48, 251-275

Leigh JA *et al.* (1985) Proc. Natl. Acad. Sci. USA 82, 6231-6235

LeVier K *et al.* (2000) Science 287, 2492-2493

Liu MGJ *et al.* (1998) Appl. Environ. Microbiol. 64, 4600-4602

Long S *et al.* (1988) J. Bacteriol. 170, 4257-4265

Long S (2001) Plant Physiol. 125, 69-72

Osteras M *et al.* (1995) J. Bacteriol. 177, 5485-5494

Pellock BJ *et al.* (2000) J. Bacteriol. 182, 4310-4318

Reed *et al.* (1991a) J. Bacteriol. 173, 3776-3788

Reed *et al.* (1991b) J. Bacteriol. 173, 3789-3794

Reuber TL *et al.* (1991) Ann. NY Acad. Sci. 646, 61-68

Reuber TL *et al.* (1993) Cell 74, 269-280

Reuhs BL *et al.* (1993) J. Bacteriol. 175, 3570-3580

Roche P *et al.* (1992) Biochem. Soc. Trans. 20, 288-291
Ruberg S *et al.* (1999) Microbiol. 145, 603-611
Santos R *et al.* (2001) Mol. Plant-Microbe Interact. 14, 86-89
Ugalde JE (2000) Infect. Immun. 68, 5716-5723
Wang LXY *et al.* (1999) J. Bacteriol. 181, 6788-6796
York GM *et al.* (1998) Proc. Natl. Acad. Sci. USA 95, 4912-4917
Zhan HJ *et al.* (1991) J. Bacteriol. 173, 7391-7394

5. Acknowledgements

Supported by Grant GM31030 from the NIH (to GCW), a Fulbright-MECD postdoctoral fellowship from Spain (to JL), and a doctoral fellowship from "Region Bretagne - Bourse de These en cotutelle" from "Le conseil Regional de Bretagne", France (to LB).

FINE TUNING OF NODULATION BY RHIZOBIA

N.M. Boukli, W.J. Broughton, W.J. Deakin, H. Kobayashi, C. Marie, X. Perret, M. Saad,
P. Skorpil

L.B.M.P.S., Université de Genève, 1 ch. de l'Impératrice, 1292 Chambésy/Genève,
Switzerland

1. Introduction

Nitrogen-fixing symbioses between legumes and the soil bacteria *Azorhizobium*, *Bradyrhizobium*,
Mesorhizobium and *Rhizobium* (collectively rhizobia) contribute substantially to plant productivity.
Legumes and rhizobia have vastly different genomes. All rhizobia have a chromosome along with
none to many plasmids which, combined, total six to nine mega-base pairs (Mbp) (Perret *et al.*
2000). In contrast, genomes of legumes are much larger with many chromosomes and total DNA
contents that range from about 450 Mbp/1C to 4500 Mbp/1C. As examples, the model legumes
Lotus japonicus, *Medicago truncatula* and *Phaseolus vulgaris* have haploid genomes made up of
six, eight or eleven chromosomes with DNA contents of ± 450 to 550 Mbp/1C. Legume genomes
are thus at least 50 times larger than those of their micro-symbionts. Despite these differences,
available information suggests that the contributions of both symbiotic partners are roughly equal.
Like most plants, legumes release a variety of compounds into the soil surrounding their roots (the
rhizosphere). Amongst these substances, phenolic compounds and especially flavonoids are
perceived by rhizobia as inducers of <u>nod</u>ulation (*nod*)-genes. In turn, rhizobia secrete a variety of
lipochito-oligosaccharides called Nod factors. By themselves Nod factors are fully able to mimic
the initial effects of rhizobia on legumes. Once rhizobia enter the infection thread, Nod factors seem
to turn their guiding role over to other molecules however. This second set of rhizobial "factors"
include extra-cellular polysaccharides as well as proteins and are perceived by different plants in
various ways (see Broughton *et al.* 2000). Their possible roles are discussed below.

2. Results and Discussion

2.1. Polysaccharides/surface components. In *R. meliloti* Rm1021, a class of exopolysaccharides
(succinoglycans) is necessary for the invasion of *Medicago sativa* nodules. Although mutants
incapable of synthesizing succinoglycan (SG) produce normal root-hair curling, the nodules are
devoid of bacteria and bacteroids, and thus ineffective. Fluorescence-microscope analyses show that
nodule invasion is aborted at various stages in different *exo*-mutants (Cheng, Walker 1998). Cells
which lack ExoR, a negative regulator of *exo*-gene expression (Reed *et al.* 1991), vastly overproduce
succinoglycan and are unable to colonize curled root-hairs or form infection threads. In contrast, a
mutant in *exoY* that is incapable of synthesizing succinoglycan since it lacks the first enzyme in the
biosynthetic pathway (Reuber, Walker 1993) colonizes curled root-hairs, but forms few, very short
infection threads. Although the *exoH*$^-$ mutant that produces symbiotically dysfunctional
succinoglycan forms infection threads longer than those of the *exoY* mutant, they never extend as far
as the base of the root-hair cell. Thus in *R. meliloti*, succinoglycan is a symbiotic signal. Mutants
that fail to produce the symbiotically active forms of these exopolysaccharides in adequate
quantities, are incapable of penetrating the adjacent cell and thus remain blocked or trapped in the
infected root-hair.

A number of mutations giving similar phenotypes, but in a number of other genes have been
found (see Perret *et al.* 2000). Where production of cyclic-β-(1→2)glucans, rhamnose-rich SPS,
LPS, SG, β-(1→2)glucans, and KPS is impaired, Nod$^+$ but Fix$^-$ phenotypes are observed. Different
regulators appear to control the expression of these diverse genes, and the resultant phenotype
depends on the plant. Quite often an Inf$^+$ Fix$^-$ phenotype is observed, suggesting that the genes play

similar roles. One possibility is that such surface components are necessary for or form part of the developing infection thread. In this scenario a plant that could not itself supply the missing carbohydrate would give a Fix⁻ phenotype, while the rhizobial mutant in a plant that normally synthesizes the compound, would have no effect.

2.2. Secreted proteins. Several strains of *Rhizobium* have been shown to secrete symbiotically active proteins. Amongst these, *nodO* is required for nodulation of *Vicia hirsuta* by mutants of *R. leguminosarum* bv. *viciae* deleted of *nodFELMNT* (Downie, Surin 1990), as well as to extend the host range of a *R. trifolii nodE* mutant to include *V. sativa* (Economou *et al.* 1994). *nodO* encodes a Ca^{2+}-binding protein that is thought to form cation-specific channels in membranes of leguminous plants (Economou *et al.* 1990; Sutton *et al.* 1994). Secretion of NodO is dependent on a C-terminal signal of about 24 residues (Sutton *et al.* 1996), and the *prsDE* genes which encode two type-I-like inner-membrane proteins (Finnie *et al.* 1997). In addition to NodO, at least three other proteins are secreted via this system, two of which (PlyA and PlyB) are glycanases involved in the processing of bacterial exopolysaccharides (Finnie *et al.* 1998). Although NodO and NodS have distinct biochemical functions (the *nodS* gene encodes an *N*-methyl transferase; Jabbouri *et al.* 1995), a *nodO* homolog of *Rhizobium* sp. BR816 was shown to complement a *nodS* mutant of NGR234 for nodulation of *L. leucocephala* (van Rhijn *et al.* 1996). This suggests that secreted proteins may complement or supplement some Nod factor deficiencies.

Sequencing the symbiotic plasmid of NGR234 revealed flavonoid-inducible genes encoding components of a type III secretion system (T3SS) (Freiberg *et al.* 1997). In a number of bacterial pathogens, T3SSs are induced upon contact with host cells (of plants or animals) and deliver virulence proteins directly into the eukaryotic cytosol (Lee 1997). NGR234, as well as several strains of *R. fredii*, was shown to excrete at least three to five proteins in a NodD1-, and T3SS-dependent manner (Krishnan *et al.* 1995; Viprey *et al.* 1998; Marie *et al.* 2001; M. Göttfert *et al.* this volume). In USDA257, the *nolXWBTUV* cluster (corresponding to the *nopX*, *rhcC1*, *nolB*, *rhcJ*, *nolU* and *nolV* genes of NGR234) (Viprey *et al.* 1998) regulates the nodulation of *Glycine max* in a cultivar-specific manner (Meinhardt *et al.* 1993), whereas the T3SS of NGR234 profoundly affects nodulation of various legumes such as *Crotalaria juncea*, *Pachyrhizus tuberosus* and *Tephrosia vogelii* (see below). The absence of conserved *nod*-box regulatory elements in the promoter regions of the *nol, nop* and *rhc* operons indicate that NodD1 dependent transcriptional regulation of the T3SS genes is mediated by another factor. y4xI, an HrpG homolog which is under the control of a functional *nod*-box, is thought to be the key intermediary in the regulatory cascade between flavonoids and activation of the T3SS machinery in NGR234 (Viprey *et al.* 1998; Marie *et al.* 2001). Delayed induction of this secretion pathway in comparison with loci involved in the elaboration of the Nod factors (see Perret *et al.* 1999), suggests that the T3SS machinery is assembled after Nod factors have been elaborated and protein export begins when intimate contact between bacteria and root-hairs has been established. Thus, the secreted proteins (some of which have direct effects on plants and are thus "effector" proteins) such as NopJ, NopX and y4xL of NGR234 would function after the bacteria have entered the plant, possibly during development of infection threads.

It has been suggested that bacterial invasion of plant cells triggers non-specific defense reactions, and that successful pathogens overcome these defences (Gabriel, Rolfe 1990). Similarly, invading symbionts have probably evolved different strategies to lower host-plant defences. T3SS proteins and polysaccharides may contribute to this phenomenon. Some plants would perceive these compounds as part of the infection pathway, and react to their presence by increased nodulation as in the cases of *M. sativa* inoculated with Exo⁺ *R. meliloti* and *Tephrosia vogelii* which favors strains of NGR234 with a functional T3SS. In contrast, *Crotalaria juncea* and *Pachyrhizus tuberosus*, which are poorly nodulated by wild-type NGR234, produce many effective nodules when inoculated with T3SS mutants, suggesting that secreted proteins have detrimental effects on certain hosts (Viprey *et al.* 1998; Marie *et al.* 2001). It is thus not unreasonable to suggest that this last group of plants exhibit a sort of hyper-sensitive response towards NGR234 and the proteins that it secretes.

2.3. Fine-tuning of infection-thread development. Although correlations between Nod-factor structure and rhizobial specificity in nodulation are hard to draw, it is generally agreed that Nod factors are required for deformation and curling of root-hairs as well as entry of rhizobia into the plant (see Broughton *et al.* 2000). Once within the infection threads however, other rhizobial components are necessary for nodule development. In symbioses between *Medicago sativa* and *Rhizobium meliloti*, these include the extracellular-polysaccharides called succinoglycans, EPSII, and K antigens (Pellock *et al.* 2000). Succinoglycan is highly efficient in mediating both infection thread initiation and extension. EPSII is significantly less efficient than succinoglycan in mediating both steps in the invasion process while K antigen is less efficient than succinoglycan in mediating extension of infection threads (Pellock *et al.* 2000; Walker *et al.* this volume). Moreover, the mere fact that one class of compounds may substitute for another suggests that initiation and development of infection threads probably involves redundant mechanisms. If redundancy in rhizobial control of infection thread development is a general feature of nodulated legumes, then other legume-*Rhizobium* systems should also require extracellular-polysaccharides for infection thread formation. As discussed under Polysaccharides/surface components, the available evidence suggests that extracellular-polysaccharides probably play similar roles in many symbioses.

Redundancy in the type of extracellular-polysaccharides that support growth of infection threads coupled with preliminary observations suggesting that proteins secreted by the T3SS are important in infection-thread development (N.M. Boukli, M. Saad, P. Skorpil, unpublished) raises the question of whether polysaccharides and proteins are complementary. Another way of phrasing this question is to ask if extracellular-polysaccharides could replace secreted proteins (and *vice versa*) in nodule development of specific plants? If the answer is yes, it would explain the various phenotypes found when different plants are inoculated with, e.g. a *rhcN* deletion mutant (*rhcN* encodes the ATPase that provides the energy to drive protein export). Perhaps the effector proteins play irreplaceable roles in plants such as *Tephrosia vogelii*, while in hosts like *Lotus japonicus* that are indifferent to the presence or absence of a functional T3SS (Marie *et al.* 2001), extracellular-polysaccharides functionally replace the effector proteins.

To generalize these speculations, it is necessary to find an explanation for the absence of T3SSs in certain rhizobia. Although most rhizobia probably contain a T3SS, and this includes the genome of the only *Mesorhizobium loti* strain that has been completely sequenced, that of the "symbiotic" island of a related *M. loti* strain does not (see Marie *et al.* 2001). Similarly, the sequence of *R. meliloti* lacks this type of secretion system (see Long *et al.* this volume). In both cases, components of a type four-secretion system (T4SS) including most of the *vir*-genes (from *virB*1 to *virB*11) have been found however. Although it remains to be seen if these *vir*-loci function in protein secretion into the plant, a unifying hypothesis would be that extracellular-polysaccharides and proteins are able to complement or supplement one another in infection-thread development. Either T3SS or T4SS machines could secrete the "effector" proteins. Taken together, the various combinations of known polysaccharides and proteins provide ample scope for fine-tuning infection-thread development to the large array of plant hosts.

3. References

Broughton WJ, Jabbouri S, Perret X (2000) J. Bacteriol. 182, 5641-5652
Cheng H-P, Walker GC (1998) J. Bacteriol. 180, 5183-5191
Downie JA, Surinm B (1990) Mol. Gen. Genet. 222, 81-86
Economou A *et al.* (1990) EMBO J. 9, 349-354
Economou A, Davies AE, Johnston AWB, Downie JA (1994) 140, 2341-2347

Finnie C *et al.* (1997) Mol. Microbiol. 25, 135-146
Finnie CA *et al.* (1998) J. Bacteriol. 180, 1691-1699
Freiberg CR *et al.* (1997) Nature 387, 394-401
Gabriel DW, Rolfe BG (1990) Annu. Rev. Phytopathol. 28, 365-391
Jabbouri SR *et al.* (1995) J. Biol. Chem. 270, 22968-22973
Krishnan HB, Kuo C-I, Pueppke SG (1995) Microbiol. 141, 2245-2251
Lee C (1997) Trends in Microbiol. 5, 149-156
Marie C, Broughton WJ, Deakin WJ (2001) Curr. Opinion. Plant Biol.
Meinhardt LW *et al.* (1993) Mol. Microbiol. 9, 17-29
Pellock BJ, Cheng H-P, Walker GC (2000) J. Bacteriol. 182, 4310-4318
Perret XC *et al.* (1999) Mol. Microbiol. 32, 415-425
Perret X, Staehelin C, Broughton WJ (2000) Microbiol. Mol. Biol. Rev. 64, 180-201
Reed JW, Glazebrook J, Walker GC (1991) J. Bacteriol. 173, 3789-3794
van Rhijn P *et al.* (1996) Mol. Plant-Microbe Interact. 9, 74-77
Sutton JM, Lea EJ, Downie JA (1994) Proc. Natl. Acad. Sci. USA 91, 9990-9994
Sutton JM *et al.* (1996) Mol. Plant-Microbe Interact. 9, 671-680
Viprey V *et al.* (1998) Mol. Microbiol. 28, 1381-1389

4. Acknowledgements

Dora Gerber is thanked for her unfailing support of our work. We are extremely grateful to Graham C. Walker and Brett Pellock for their comments on this manuscript. Financial assistance was provided by the Fonds National de la Recherche Scientifique and the Université de Genève.

THE TYPE III SECRETION SYSTEM OF *BRADYRHIZOBIUM JAPONICUM* 110S*PC*4

A. Krause[1], A. Doerfel[1], C. Schmid[1], H. Hennecke[2], H. Krishnan[3], M. Göttfert[1]

[1]Inst. für Genetik, Technische Universität Dresden, Dresden, Germany
[2]Inst. für Mikrobiologie, Eidgenössische Technische Hochschule, Zürich, Switzerland
[3]Dept of Agronomy, University Missouri, Columbia, MO 65211, USA

1. Introduction

Symbiosis between rhizobia and legumes is initiated by an exchange of signal molecules. Communication starts with flavonoids that are secreted by the host plant via the root system. These flavonoids lead to the activation of the transcriptional regulator NodD, a member of the LysR family. NodD binds to the *nod* box, a conserved sequence within the promoter of nodulation genes, and initiates their transcription (Schlaman *et al.* 1998). In *Bradyrhizobium japonicum* 110, a second regulatory system consisting of the genes *nodV* and *nodW* is involved in *nod* gene regulation (Göttfert *et al.* 1990; Loh *et al.* 1997). These genes are indispensable for the infection of several host plants. The Nod proteins catalyze the synthesis of Nod factors, modified lipochito-oligosaccharides, that are secreted by the bacterial cell. These signal molecules induce the first steps of nodule development (Hadri, Bisseling 1998).

Little is known about alternative signaling pathways that may eventually also influence later stages in nodule development. For *Sinorhizobium fredii,* it was shown that flavonoids stimulate the secretion of proteins (Krishnan, Pueppke 1993). Sequence analysis of the closely related *Rhizobium* sp. strain NGR234 (Freiberg *et al.* 1997) and of *Mesorhizobium loti* (Kaneko *et al.* 2000) indicated the presence of a type III secretion system. Many bacterial pathogens of plants and animals use such systems for the export of proteins (Hueck 1998). In several cases, it was demonstrated that these proteins enter the host cell. In NGR234, two secreted proteins have been identified (Viprey *et al.* 1998). Mutations in the type III secretion system affect symbiosis depending on the host (Meinhardt *et al.* 1993; Viprey *et al.* 1998).

Recently, sequence analysis of a 410- kb chromosomal region of *B. japonicum* revealed the presence of a type III secretion system (Göttfert *et al.* 2001). In this work, the gene cluster is referred to as *tts* (type three secretion). Here, we present the results of mutational, transcriptional and further nucleotide sequence analyses of this region.

2. Material and Methods

Bradyrhizobium japonicum 110*spc*4 is a spontaneous spectinomycin resistant mutant of USDA3Ib110 and is referred to as wild type (wt) in this work. Δ136 is a mutant derivative of 110*spc*4 (Figure 1). Strain 149 has a *uidA* gene integrated into the chromosome downstream of the *nod* box promoter preceding *ttsA* (Figure 1). For expression analyses, *B. japonicum* was grown in YEM medium to the stationary phase. Cells were resuspended in the same amount of fresh medium and genistein was added to a concentration of 1 µM. Activity was measured after 24 hours. Methylumbelliferyl-glucuronide was used as substrate for β-D-glucuronidase and methylumbelliferon was measured according to standard protocols.

3. Results and Discussion

3.1. A deletion within the *tts* gene cluster affects symbiosis in a host specific manner. The known *tts* gene clusters of rhizobia have at least 17 genes in common (Figure 1). To analyze their importance for symbiosis, we created the deletion derivative Δ136 that lacks several open reading frames. Compared to the wild type, nodulation with strain Δ136 was clearly delayed on *Macroptilium atropurpureum* and to a lesser extent on soybean (Figure 2). No nodulation defect was observable with *Vigna unguiculata.* Twenty days after inoculation, *M. atropurpureum* infected with strain Δ136 still had fewer nodules, however of increased size. Despite the fact that fewer nodules were induced on *M. atropurpureum,* acetylene reduction was hardly impaired on the basis of acetylene reduced per plant (data not shown). Hence, we observed an increase of nodule dry weight with the mutant strain.

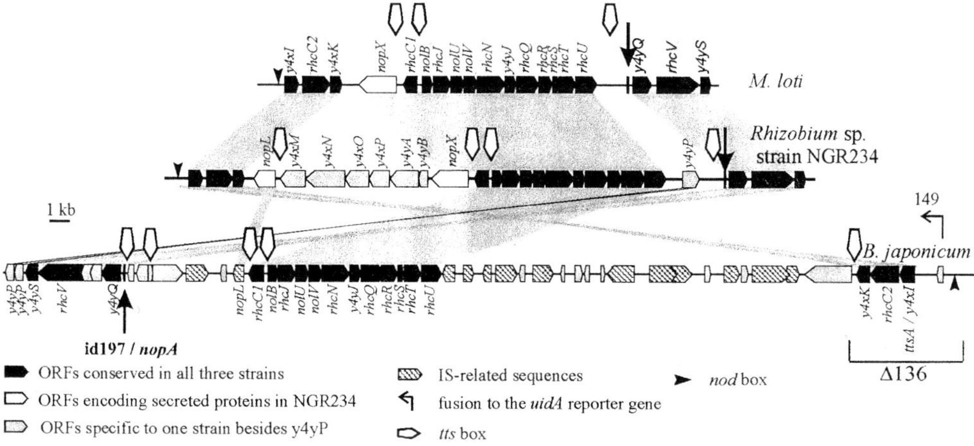

Figure 1. Comparison of type III secretion gene clusters. Conserved regions are connected by shaded areas. The arrow points to the newly identified *nopA* gene that is conserved in all three strains.

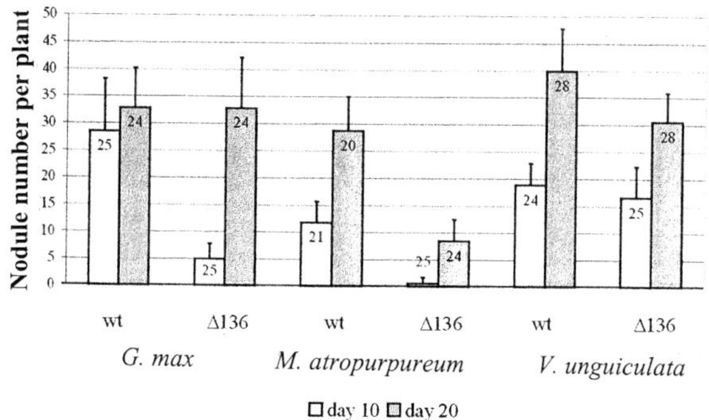

Figure 2. Nodulation phenotype of mutant Δ136 (Figure 1). Numbers inside the columns indicate the quantity of plants used in the assay.

3.2. The *nod* box preceding y4xI *(ttsA)* is regulated by *nodW* and the *nodD* region. It is known that the common *nod* genes, which are preceded by a *nod* box, are activated by *nodW* and *nodD1*. Because y4xI *(ttsA)* is also preceded by a *nod* box (Figure 1), we expect it to be regulated by *nodW* and *nodD1* as well. To test this, we constructed a transcriptional fusion with the *uidA* reporter gene. The fusion was inserted into the chromosome of the wt and different mutant strains. As predicted, expression was inducible by genistein in the wild-type background (Figure 3). The expression pattern did not change in the y4xI *(ttsA)* mutant, indicating that this gene is not involved in the regulation of the *nod* box promoter. However, no induction was observed in the *nodW* or the *nodD* mutant. Thus, either one of the regulatory systems is acting through the other system or both systems are acting at the same promoter. We favor the last idea due to the aforementioned similarity of the *nod* box with that in the common *nod* gene region. The Y4xI (TtsA) protein exhibits similarity to members of the two component regulatory systems. Furthermore, we found that it is involved in the regulation of a *nolU-lacZ* fusion. Because of this, we believe that Y4xI (TtsA) acts in a regulatory cascade downstream of NodW and NodD1 as transcriptional activator of the *tts* cluster. Therefore, we suggest to change the designation of y4xI to *ttsA* (type three secretion).

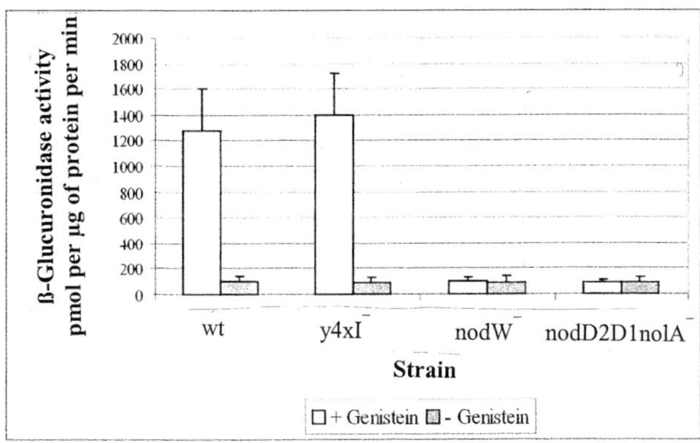

Figure 3. Activity of a transcriptional *nod* box-*uidA* fusion in the wild type and mutant strains. Values are from at least five independent cultures.

3.3. Genes in the *tts* cluster are preceded by a common motif. If TtsA is the major activator of the genes in the *tts* cluster, then one might expect to find common sequence motifs upstream of transcription units. The result of a corresponding analysis is summarized in Figure 4. Upstream regions of all three rhizobial strains have significant similarity. The *rhcC1-nolB* intergenic regions exhibit a higher similarity than the remaining upstream sequences. In the alignment, we did not include the sequence of the upstream regions of *nopX* and *nolB* of *S. fredii*. The sequence of this strain (accession number Ay034152) is almost identical to the sequence of NGR234 and is invariant in the conserved elements. In *S. fredii*, however, the transcriptional start sites of *nopX* and *nolB* have been determined (Kovács *et al.* 1995). They are located 11 bp and 12 bp downstream of the conserved elements, respectively. Therefore, it is likely that the conserved sequences serve as promoter elements. Nevertheless, the impact of the conserved nucleotides on the expression level remains to be elucidated.


```
Bjap     rhcC1 90 bp - TCCGTCAGGTTTTCGTCAGCTCGGCAGCCTA - 49 bp  nolB
M.loti   rhcC1 80 bp - TCAGTCAGCTTGTCGTCAGCTCGGCCACCTA - 71 bp  nolB
NGR234   rhcC1 85 bp - GTAGTCAGCGTGTCGTCAGCTCGCCTCGCTA - 39 bp  nolB
         Consensus         GTCAG T TCGTCAGCTCG C    CTA


         Bjap      AGGCTCGTCAGCTTTTCGAAAGCTAGCGCCCTA -  64 bp  nopA
         M.loti    AGACTCGTCAGGTTCTCGAAAGCTCCTGCTCGTA - 446 bp  nopA
         NGR234    ACAATTGTCAGCTTTTCGAAAGCTGGAGCTCATA - 410 bp  nopA
         Bjap      ATCATCGTCAGCTTTTCGACAGGTGTTCGGGCTA - 221 bp  id205
         NGR234    CTGATTGTCAGCTTCTCGAAAGGTATGTCTCTTA - 261 bp  nopL
         Bjap      CCGATCGTCAGCTTTTCGAAAGCTAAAGCCCCCA - 109 bp  nopL
         Bjap      GCGAT-GTCAGGTTTTGGAAAGCAAACGTGAGTA -  36 bp  id274
         M.loti    GAACTCGTCAGTTTACCGAAAGCTAAACCGCTCA - 100 bp  nopX
         NGR234    AGCCTCGTCAGTTTCTCGAAAGCTAAACCGCTCA - 189 bp  nopX
         Consensus         TcGTCAG TT TCGAAAGcTa   c c tA
                                                         c
tts box consensus         tcGTCAGcTT TCGaaAGcT   c c tA
                          c              tc             c
```

Figure 4. Alignment of putative promoter regions within the *tts* gene cluster. The numbers indicate the distance from the putative start codon. The upper part denotes the intergenic region between the divergently oriented *nolB* and *rhcC1* genes. Nucleotides that are conserved in all regions are underlined. Nucleotides that are conserved in all sequences except one are printed in capital letters. Lower case letters show the positions that are conserved in at least seven upstream regions.

3.4. The *tts* gene cluster contains a so-far unidentified conserved ORF that might encode a secreted protein. Reanalysis of the nucleotide sequence of all three rhizobial strains revealed the presence of a conserved ORF upstream of y4yQ (Figures 1 and 5). All are of comparable size. Similarity on the amino acid level is highest between the proteins of NGR234 and *M. loti*. In an earlier work, we determined the N-terminal amino acid sequence of an exported protein of *B. japonicum*. The obtained sequence AVDNVGG fits nicely to the N-terminal end of the deduced gene product of the new ORF. We suggest to name the gene *nopA* (nodulation outer protein) in line with the nomenclature suggested by Marie *et al.* (2001).

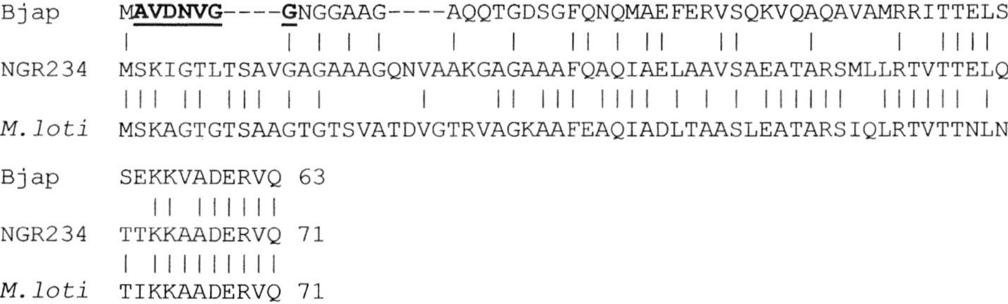

Figure 5. Alignment of NopA proteins. The amino acid sequence that was obtained by N-terminal sequencing of an extracellular protein of *B. japonicum* is underlined.

4. References

Freiberg C *et al.* (1997) Nature 387, 394-401
Göttfert M *et al.* (1990) Proc. Natl. Acad. Sci. USA 87, 2680-2684
Göttfert M *et al.* (2001) J. Bacteriol. 183, 1405-1412
Hadri AE, Bisseling T (1998) In Spaink HP, Kondorosi A, Hooykaas PJJ (eds),
    The Rhizobiaceae, pp 403-416, Kluwer Academic Publishers, Dordrecht, The Netherlands
Hueck CJ (1998) Microbiol. Mol. Biol. Rev. 62, 379-433
Kaneko T *et al.* (2000) DNA Res. 7, 331-338
Kovátcs L *et al.* (1995) Mol. Microbiol. 17, 923-933
Krishnan HB, Pueppke SG (1993) Mol. Plant-Microbe Interact. 6, 107-113
Loh J *et al.* (1997) J. Bacteriol. 179, 3013-3020
Marie M *et al.* (2001) Curr. Opin. Plant Biol. 4, 336-342
Meinhardt LW *et al.* (1993) Mol. Microbiol. 9, 17-29
Schlaman HRM *et al.* (1998) In Spaink HP, Kondorosi A, Hooykaas PJJ (eds),
    The Rhizobiaceae, pp 361-386, Kluwer Academic Publishers, Dordrecht, The Netherlands
Viprey *et al.* (1998) Mol. Microbiol. 28, 1381-1389

COMMON THEMES IN SYMBIOSIS AND PATHOGENESIS: THE CASE OF *BRUCELLA ABORTUS*

R.A. Ugalde

Instituto de Investigaciones Biotecnologicas, Universidad Nacional de General San Martin, CONICET, Avenida General Paz entre Albarellos y Constituyentes, INTI, (1650), San Martin, Buenos Aires, Argentina

1. Introduction

Brucella spp. are intracellular pathogens that, together with *Agrobacterium* and *Rhizobium*, belong to the α-2-subgroup of proteobacteria. *Brucella* spp. invade professional and non-professional phagocytic cells through the endocytic pathway exploiting the autophagic machinery and establish a unique replication niche in the endoplasmic reticulum. No cell damage is observed upon intracellular multiplication, thus indicating that *Brucella* is a refined parasite well adapted to the intracellular lifestyle. In this sense, *Brucella* resembles an endosymbiont rather than a pathogen. The presence of several virulence genes homologous to those that in *Agrobacterium* or *Rhizobiaceae* are involved in bacterial/plant interaction, suggests common pathways for intracellular adaptation in plants and mammals.

2. Gene Discovery through Genomic Sequence

A quantum leap in the study of bacterial pathogenesis is the availability of complete genome sequence. Genomics allows us to infer functions of gene products by direct comparison of nucleotide or amino acids sequences. *B. abortus* partial genome sequence revealed the presence of genes sharing homology with genes involved in *Rhizobium* symbiosis or *Agrobacterium* virulence. Some of the most significant examples are: the *Sinorhizobium meliloti ndvB*, *S. meliloti phaA*, *S. meliloti bacA*; *Bradyrhizobium japonicum nodV*, *B. japonicum nodW*, *B. japonicum hemH*; *A. tumefaciens exoC*, *A. tumefaciens virB10*, *A. tumefaciens attJ*. In order to assess if those genes are functional in *Brucella*/cell interaction, a series of null mutants were obtained and their phenotypes analyzed in cell and mouse models.

3. *Brucella abortus ndvB*

This gene codes for the cyclic β (1-2) glucan synthetase (*cgs*), an integral inner membrane protein of 2831 amino acids with 51% identity to *S. meliloti* NdvB. In *S. meliloti, ndvB* is required for effective invasion of the nodules and in *A. tumefaciens chvB* is involved in tumor induction. *B. abortus cgs* mutants have reduced virulence in mice and defective intracellular multiplication in HeLa cells. *B. abortus cgs* expressed under its own promoter complemented *S. meliloti ndvB* and *A. tumefaciens chvB* mutants. Studies on the interaction between *B. abortus cgs* mutants with mammalian cells, revealed that cyclic β (1-2) glucan is not required for cell invasion. However, the ability to prevent fusion between early endosomes with lysosomes is severely reduced. A small number of cyclic β (1-2) glucan minus mutant cells reach the endoplasmic reticulum to establish their intracellular multiplication niche, explaining their reduced virulence. The addition of purified β (1-2) glucan during cell infection partly restored the wild-type phenotype.

4. *Brucella abortus exoC*

This gene codes for the enzyme phosphoglucomutase (Pgm), a cytoplasmatic protein of 544 amino acids that is 75% identical to ExoC of *A. tumefaciens* and *Mesorhizobium loti*. ExoC (Pgm) is required in *A. tumefaciens* for virulence and in *M. loti* for nodulation. This phenotype was complemented by *B. abortus pgm* gene expressed under its own promoter. In *M. loti* and *A. tumefaciens pgm* is part of the glycogen (*glg*) operon. Surprisingly, all the *glg* genes, except for *pgm* which is highly conserved, are absent in *B. abortus*. *Pgm* null mutants are avirulent, losing the ability to persist in the spleen of mice beyond four weeks. However, the ability to infect and multiply inside HeLa cells is partially conserved.

5. *Brucella abortus hemH*

This gene codes for the ferrochelatase, the enzyme which catalyzes the incorporation of ferrous iron into protoporphyrin IX, the intermediate precursor of protoheme. *B. abortus hemH* codes for a 352 amino acid protein that is 70% and 58% identical to *M. loti* and *B. japonicum hemH*, respectively. In *B. japonicum* ferrochelatase is essential for normal nodule development. Mutants in this gene incited nodules on soybean that did not fix nitrogen, contained few viable bacteria and did not express leghemoglobin heme apoprotein. *B. abortus hemH* null mutants are avirulent in mice, conserved the ability to invade the cells but have lost the capacity to multiply inside them.

6. *Brucella abortus virB* operon

The complete *virB* operon coding for a number of proteins that form a type IV secretion system is present in *B. abortus*. In *A. tumefaciens virB* operon is involved in the conjugation of the T-DNA from bacteria to plant cell. *B. abortus virB* null mutants are avirulent in mice and do not multiply inside HeLa or macrophage cells. The mutants are unable to prevent the fusion between the early endosome with the lysosome and, as a consequence, they do not reach the endoplasmic reticulum to establish their intracellular multiplication niche.

Section 13:
Nitrification, Denitrification and the Nitrogen Cycle

CHAIR'S COMMENTS: THE NITROGEN CYCLE

D. Werner, D. Ingendahl, E. ter Haseborg, H. Kreibich, P. Vinuesa

Fachgebiet Zellbiologie und Angewandte Botanik, Fachbereich Biologie der Philipps-Universitaet, D-35032 Marburg, Germany

In contrast to general belief, the largest nitrogen pool on our planet is not in the atmosphere (4×10^{15} tons) but in the primary rocks (1.9×10^{17} tons). Compared to these pools, the N_2 dissolved in the sea is only around 2.2×10^{13} tons. The other nitrogen pools (of nitrate, ammonia, organic nitrogen in the soil and in the sea and the nitrogen fixed in plants and animals) are in the order of 2×10^{12} tons. Compared to the pools of N_2, this is a tiny fraction. However, these pools are by four orders of magnitude larger, compared to N_2, fixed annually by organisms in the soil and in the sea, which is in the range of 200 to 300 million tons per year (Burns, Hardy 1975).

The carbon pool on this planet has a similar scale with, of course, very interesting differences in the various pools and turnover rates. The total carbon on this planet is around 10^{17} tons, from which more than 70% are in the lithosphere, the second largest pool is the oceans with 3.8×10^{13} tons C. The carbonate fraction in the upper soil layers contains around 1.7×10^{12} tons C and the organic soil carbon is in the range of 1.6×10^{12} tons C. The atmosphere contains only about 0.75×10^{12} tons C and the biomass of the plants with around 0.6×10^{12} tons C has a similar size. It is also very interesting to compare the annual turnover rate produced by humans from combustion of fossil energies (6.3×10^9 tons C) with the turnover rate by photosynthesis of land plants (120×10^9 tons C), the transfer rates from plant biomass to the soil (60×10^9 tons C) and the respiration rate from the soil into the atmosphere (62×10^9 tons C) per year. Also the equilibrium between atmosphere and the ocean is apparently not complete, with a small surplus of uptake versus transfer in the atmosphere (Werner 2000).

Surveys on various aspects of the nitrogen cycle in the biochemistry and ecophysiology of the different components of the cycle have been published by Zumft (1997) on denitrification, on atmospheric trace gases such as N_2O and NO by Conrad (1996), on nitrogen fertilizer use by Kawashima *et al.* (2000), on nitrogen fixation in agricultural systems by Peoples and Herridge (2000), on nitrogen mineralization and nitrification by Van Veen (2000) and on nitrogen use efficiency by Simek and Cooper (2001).

The four contributions of this session contribute to various aspects of the nitrogen cycle. Daniel Arp and co-workers report on "A gene to genome look at *Nitrosomonas europaea*" (this volume). Hermann Bothe and co-workers study "The distribution of dinitrogen fixing and denitrifying bacteria in soils assessed by molecular biological methods" (this volume). Denitrification is studied with *Bradyrhizobium japonicum* genes by Socorro Mesa and co-workers (this volume) and the plant aspects of plant ureases is contributed by Joe Polacco *et al.* (this volume).

From our laboratory some results on "Population shift of denitrifying bacteria in interstitial and surface waters downstream of a purification plant documented by amplified 16S ribosomal DNA restriction analysis (ARDRA) patterns" are presented. Sixteen different genotypes of denitrifying bacteria have been isolated from the various river sediments before and after the purification plant. Some genotypes were only present in the free flowing water while other genotypes were isolated only from the hyporheic interstitial. Other genotypes had a ubiquitous occurrence. A further group of genotypes could be only detected in the hyporheic interstitial and in the surface waters upstream of the purification plant (ter Haseborg *et al.* 2001). Altogether 37 ARDRA types were isolated in the interstitial and in the surface water of the Lahn River in Germany. Sixteen of them were only obtained in one sampling period. This means that the variation in the various years in the structure of the bacterial community is rather high. Five ARDRA types were found in continuing years, representing stable denitrifying groups in the interstitial as well as in the surface water. From agricultural soils, 26 ARDRA types have been isolated (Cheneby *et al.* 2000) compared to the 37 groups from the aquatic environment. The temporal and spatial diversity of denitrifying bacteria in the aquatic environment may therefore be larger than in agricultural systems.

Nitrosomonas europaea and *Nitrosomonas eutropha* are able to nitrify and denitrify at the same time, growing under oxygen limitation (Bock *et al.* 1995). These simultaneous processes form large quantities of N_2O and N_2, which causes a significant nitrogen loss. Nitrite is another important intermediate, which can occur under certain temperature conditions in surface waters (von der Wiesche, Wetzel 1998) and also in agricultural soils (Burns *et al.* 1995). The concentrations in the soil were highly variable and ranged from 0 to 2.7 µg N x g^{-1} soil. An increase in soil pH produced large nitrite flushes (Burns *et al.* 1995). The known biodiversity of nitrifying bacteria has increased over the last years. Nine species of ammonia oxidizers and seven species of nitrite oxidizers have been described (Spiek, Bock 1998).

References

Bock E *et al.* (1995) Arch. Microbiol. 163, 16-20
Burns LC *et al.* (1995) Soil Biol. Biochem. 27, 47-59
Burns RC, Hardy RWF (1975) Nitrogen Fixation in Bacteria and Higher Plants, Springer-Verlag, Berlin, Germany
Cheneby DL *et al.* (2000) FEMS Microbiol. Ecol. 34, 121-128
Conrad C (1996) Microbiol. Rev. 60, 609-640
ter Haseborg E *et al.* (2001) Appl. Environ. Microbiol.
Kawashima H *et al.* (2000) In Balázs E *et al.* (eds), Biological Resource Management Connecting Science and Policy, pp.309-314, Springer-Verlag, Berlin, Germany
Peoples MB, Herridge DF (2000) In Pedrosa FO *et al.* (eds) Nitrogen Fixation: From Molecules to Crop Productivity, pp.519-524, Kluwer Academic Publishers, Dordrecht, The Netherlands
Simek M, Cooper JE (2001) In Shiyomi M and Koizumi H (eds), Structure and Function in Agroecosystem Design and Management, pp.227-251, CRC Press, Boca Raton, Florida
Spiek E, Bock E (1998) Biospektrum 4, 25-31
Van Veen JA (2000) In Balázs *et al.* (eds) Biological Resource Management Connecting Science and Policy, pp.71-80, Springer-Verlag, Berlin, Germany
Von der Wiesche M, Wetzel A (1998) Wat. Res. 32, 1653-1661
Werner D (2000) Research Report, MBR, Marburg, Germany
Zumft WG (1997) Microbiol. Mol. Biol. Rev. 61, 533-616

MOLECULAR BIOLOGY OF AMMONIA OXIDATION BY *NITROSOMONAS EUROPAEA*

D.J. Arp, L.A. Sayavedra-Soto, N.G. Hommes

Dept of Botany and Plant Pathology, Oregon State University, Corvallis, OR 97331-2902, USA

1. Introduction

Nitrification is a bacterial process in which ammonia (NH_3) is oxidized to nitrate (NO_3^-). Nitrification is carried out by two groups of bacteria (Bock *et al.* 1986). NH_3 oxidizing bacteria transform NH_3 to nitrite (NO_2^-) while NO_2^- oxidizing bacteria subsequently oxidize the NO_2^- to NO_3^-. In both cases, the oxidation of the N compound provides the reductant required by the bacteria for energy and biosynthesis, i.e. the bacteria are chemolithotrophs. The bacteria that oxidize NH_3 are obligate chemolithotrophs (Bock *et al.* 1986), i.e. they are entirely dependent upon NH_3 as a source of electrons in spite of the fact that NH_3 is not a particularly good energy source thermodynamically (Hooper 1989). These bacteria are also obligate autotrophs – they obtain most or all of their carbon for growth from CO_2 (Killham 1986).

In croplands fertilized with NH_3 or urea-based compounds, oxidation of NH_3 by these bacteria leads to the loss of available N through leaching of the oxidized forms of nitrogen. While NH_3, which is present primarily as ammonium (NH_4^+) in soils, remains bound to typical soils (i.e. having a net negative charge), NO_3^- is more mobile in the soil and is readily leached into ground and surface waters. Furthermore, when the NO_3^- subsequently serves as a substrate for denitrification (the process where NO_3^- and other forms of oxidized N are reduced to N_2), then the NH_3 fertilizer is lost to the atmosphere as N_2. Trace amounts of NO and N_2O, both greenhouse gases, are also released in the processes of nitrification and denitrification (McLaren *et al.* 1991). The industrial production of ammonia requires large inputs of energy in the form of natural gas and H_2. The loss of N through the processes initiated by nitrification corresponds to a loss of energy used in the production of NH_3. In spite of the negative role of nitrification in croplands, nitrification is beneficial to the treatment of NH_4^+ in sewage (by initiating the conversion of NH_4^+ to N_2) and may have potential in the bioremediation of polluted soils and waters (through the indiscriminate action of the monooxygenase that initiates nitrification) (Rasche *et al.* 1991; Vitousek *et al.* 1997). NH_3 oxidizing bacteria in general and *N. europaea* in particular are the focus of this review.

2. Ammonia Oxidation Pathway

The oxidation of NH_3 to NO_2^- is carried out in two steps (Figure 1) (Hooper, 1989). NH_3 is first oxidized to hydroxylamine (NH_2OH) by ammonia monooxygenase (AMO). NH_2OH is subsequently oxidized to NO_2^- by hydroxylamine oxidoreductase (HAO). In the reaction catalyzed by AMO, one O from O_2 is inserted into NH_3 while the second O is reduced to H_2O. This reaction requires two additional electrons. Because NH_3 is the only source of reductant for these bacteria, the electrons required for the formation of H_2O must come from the subsequent oxidation of NH_2OH. Of the four electrons released in the oxidation of NH_2OH, two must be directed towards the oxidation of NH_3 and the remaining two are used for other reductant-requiring cellular processes such as biosynthesis and ATP generation (Wood 1986).

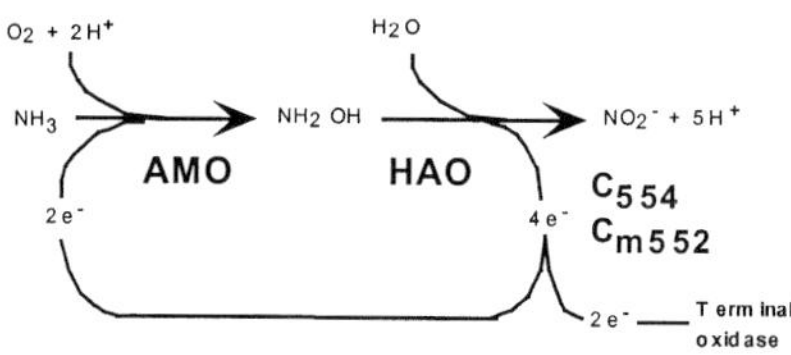

Figure 1. Ammonia oxidation pathway

3. Ammonia Monooxygenase

Although AMO has not yet been purified to homogeneity in an active form, considerable information has accumulated regarding its structure and activities. AMO is a membrane-bound enzyme which probably consists of three polypeptides. A 27 kDa polypeptide (AmoA) is labeled with ^{14}C when AMO activity is inactivated with the mechanism-based inactivator, $^{14}C_2H_2$ (Hyman, Wood 1985). A 38 kDa polypeptide (AmoB) co-purifies with the 27 kDa polypeptide (McTavish *et al.* 1993a). A 31.4 kDa polypeptide (AmoC) is coded by a gene contiguous with AmoA (Sayavedra-Soto *et al.* 1998). AMO most likely contains Cu because the inhibitor profile of the enzyme includes Cu selective chelators (Bedard, Knowles 1989) and because Cu can activate AMO activities in lysates of *N. europaea* (Ensign *et al.* 1993). The substrate range of AMO is remarkably nonspecific. In addition to the oxidation of N in NH_3, AMO can catalyze the oxidation of C-H bonds to alcohols (Hyman, Wood 1983), C=C bonds to epoxides (Hyman, Wood 1984b), C≡C bonds to oxirenes (presumably) (Hyman *et al* 1988) and sulfides to sulfoxides (Juliette *et al.* 1993a, 1993b). The substrates for these reactions include alkyl (Hyman *et al.* 1988) and aryl hydrocarbons (Hyman, Wood 1985), halogenated hydrocarbons (Hyman, Wood 1984a; Rasche *et al.* 1991; Vannelli *et al.* 1990), aromatic molecules (Keener, Arp 1994) and other compounds.

The genes *amoCAB*, coding for the structural proteins of AMO, were isolated and sequenced (Bergmann, Hooper 1994; McTavish *et al.* 1993a; Sayavedra-Soto *et al.* 1994). The genes *amoC*, *amoA* and *amoB* are contiguous (Klotz *et al.* 1997; Sayavedra-Soto *et al.* 1994) and are transcribed as a single mRNA (Sayavedra-Soto *et al.* 1994). Two nearly identical (>99%) copies of *amoCAB* exist in the genome of *N. europaea* (McTavish *et al.* 1993a, 1993b). A third, somewhat divergent copy (60% identity) of *amoC* is also present (Sayavedra-Soto *et al.* 1994). Mutants of *N. europaea* with either copy of *amoA* inactivated will grow, which indicates that both copies can be expressed (Hommes *et al.* 1998). However, mutants in copy 1 grew about 25% more slowly than wild type cells, while mutants in copy 2 grew at rates similar to wild type (Hommes *et al.* 1998). Ammonia-dependent O_2 uptake rates, [^{14}C] incorporation into AmoA, as well as the *amo* transcript levels in the mutants showed a pattern similar to their growth rates. Transcripts corresponding to *amoC* and *amoAB* as well as *amoCAB* have been identified in *N. europaea* (Sayavedra-Soto *et al.* 1994, 1996) (Figure 2). Of these the *amoC* mRNA is very stable, probably because it can form stable stem-loop structures which protect it from degradation. The roles of the different *amo* mRNAs are unknown and they may originate from *amoCAB* mRNA processing or from transcription from *amoC* and *amoA* (Sayavedra-Soto *et al.* 1994). A potential transcription start site for *amoA* was identified 114 bp upstream of the start codon in the intergenic region between *amoC* and *amoA* (Hommes *et al.* 2001). Transcript analysis for *amoC* showed two potential transcript start sites located 166 and 103 bp upstream of the *amoC* start codon (Hommes *et al.* 2001). All three transcript start sites had putative σ^{70} promoter sequences associated with them. The DNA sequence of the regions upstream of *amoC* and *amoA*, including the putative promoter sequences, were identical for the two copies of *amoCAB*.

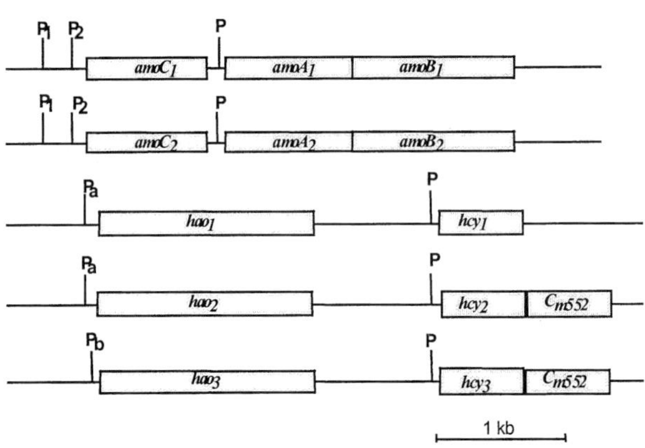

Fig. 2. Map of genes for AMO, HAO, and Cytochrome s

4. Regulation of *amo* Expression

Although AMO is essential to the growth of *N. europaea*, the specific rate of NH_3 oxidation varies. For example, when cells taken from early stationary phase were suspended in fresh medium, AMO activity increased two-fold or more in two hours (Hyman, Arp 1992). Ammonia-starved cells recovered ammonia-oxidizing activity in a process which required *de novo* protein synthesis (Gerards *et al.* 1998). Following treatment with light or C_2H_2 (which both specifically inactivate AMO), cells of *N. europaea* eventually recover the ability to oxidize NH_3. This recovery corresponds to the synthesis of a limited set of proteins, which includes the 27 kDa protein of AMO (Hyman, Arp 1992, 1995).

A question arises as to what signal the cells respond to when synthesizing new AMO proteins. NH_3 is an obvious candidate to provide this signal but NH_3 also provides both a source of energy and N for these cells. Thus, it is difficult to assign a specific role to NH_3 as a signaling molecule. We investigated the effects of NH_3 on both protein synthesis and gene transcription in *N. europaea*. AmoA (27 kDa) was synthesized when cells were incubated simultaneously with NH_4^+ (as a potential inducer of AMO synthesis and nitrogen source), C_2H_2 (so NH_3 cannot serve as an energy source), NH_2OH (as an energy source) and $^{14}CO_2$ (as a radiotracer to detect *de novo* protein synthesis in this autotrophic bacterium). The synthesis of AmoA required NH_3 and was strongly influenced by changes in the free NH_3 concentration (Hyman, Arp 1995). Other potential N sources (amino acids, nitrite) did not induce synthesis. The regulation of expression of AmoA is influenced by the concentration of NH_3 available to the cells.

We also investigated the regulation of transcription of *amoA* and *hao* by NH_4^+ (Sayavedra-Soto *et al.* 1994). In $^{14}CO_2$ labeling experiments the response to NH_3 appeared to be global (detected as a smear of RNA) at the transcription level (Sayavedra-Soto *et al.* 1996) in contrast to the few synthesized proteins (defined bands) observed at the translational level (Hyman, Arp 1995). When C_2H_2-treated cells (AMO inactivated) were subsequently exposed to NH_4^+, transcripts for *amoA* and *hao* were produced even if the incubations were carried out in the continued presence of C_2H_2 (Sayavedra-Soto *et al.* 1996). However, neither transcripts for *amoA* nor *hao* were produced in the absence of NH_4^+. Unlike *amoA*, a very stable mRNA encoding *amoC* can be found for at least 72 hours after NH_4^+ is removed (Sayavedra-Soto *et al.* 1998). Analysis of the two identified transcription start sites for *amoC* showed that they responded differently to the addition of NH_4^+. Whereas in the absence of NH_4^+ transcripts starting at both potential promoters were found (i.e. derived from the stable *amoC* mRNA), in the presence of NH_4^+ transcripts from the distal promoter greatly predominated (Hommes *et al.* 2001).

5. Hydroxylamine Oxidoreductase

This remarkable periplasmic enzyme is a homotrimer (subunit size 60 kDa), and each subunit harbors eight c-type hemes (Hooper *et al.* 1997; Igarashi *et al.* 1997). Seven of these hemes are each covalently bound to the protein by two thioether linkages typical of c-type hemes. The eighth heme has a third point of covalent attachment to the protein through a tyrosine residue. This unusual heme, designated P460, is at the site of hydroxylamine oxidation. The crystal structure of HAO has been determined and has revealed the orientation of the hemes in each subunit (Igarashi *et al.* 1997). The eight hemes group into four clusters. The structure also revealed that the subunits of HAO are covalently cross-linked through one of the hemes. All eight hemes have different mid-point potentials (Hooper *et al.* 1997), but it is not yet known which potential is associated with which heme.

The gene that codes for HAO (*hao*) is expressed as a monocistronic transcript (Sayavedra-Soto *et al.* 1994). The genome of *N. europaea* encodes three copies of *hao* (McTavish *et al.* 1993a) (Figure 2). The coding regions for the three copies are nearly identical (Bergmann *et al.* 1994). Mutants with any one copy disrupted grew with no discernible difference from the wild type

(Hommes *et al.* 1996). We investigated the regulation of transcription of *hao* by NH_4^+ (Sayavedra-Soto *et al.* 1996). HAO mRNA was induced under the same conditions as AMO mRNA but to a lower extent. HAO activity increased 5% to 20% under these conditions. DNA sequencing of the flanking regions upstream of the three copies of *hao* was done (Hommes *et al.* 2001). The sequences of *hao₁* and *hao₂* were found to be nearly identical for 160 bp upstream. The sequence of *hao₃* copy diverged from the other two copies 15 bp upstream of the start codon. Transcript analysis identified putative transcript start sites for *hao₁* and *hao₂* 71 bp upstream of the start codon, and 54 bp upstream of the start codon for *hao₃* (Hommes *et al.* 2001). All of these transcript start sites had σ^{70} promoter sequences associated with them.

The gene for HAO has also been examined in the closely related strain *Nitrosomonas* sp. ENI-11 (Yamagata *et al.* 2000). Gene mapping revealed that *hao₁* was located about 23 kb upstream of *amoCAB₁*, *hao₂* was located about 15 kb downstream from *amoCAB₂* and *hao₃* was located about 75 kb upstream of *amoCAB₂*. The two copies of *amoCAB* were more than 360 kb apart. Unlike *N. europaea*, three single *hao* mutants were created in *Nitrosomonas* sp. ENI-11 which had 68 to 75% of wild type growth rates and 58 to 89% wild type HAO activity (NH_2OH dependent NO_2^- formation) (Yamagata *et al.* 2000).

6. Cytochromes

Ammonia-oxidizing bacteria are rich in additional cytochromes (Hooper *et al.* 1997). Cytochrome C_{554} is a unique periplasmic tetraheme cytochrome that can accept electrons from HAO (Figure 2). There are three copies of the genes that code for this cytochrome (*hcy* or *cyc*) located about 1200 bp downstream of each copy of *hao* and are transcribed separately from *hao* (Bergmann *et al.* 1994; Hommes *et al.* 1994; Sayavedra-Soto *et al.* 1996). Two copies have been sequenced and were found to be identical (Bergmann *et al.* 1994; Hommes *et al.* 1994). A putative transcription start site was determined for *hcy* 97 bp upstream of the start codon and was associated with a σ^{70} promoter sequence (Hommes *et al.* 2001). Another membrane-bound cytochrome is encoded by genes contiguous with two of the three copies of *cyc* (Bergmann *et al.* 1994) (Figure 2). Cytochrome C_{m552} contains one heme and is also likely to be involved in electron transfer from HAO, albeit mediated through cytochrome C_{554} (Hooper *et al.* 1997). A soluable cytochrome C_{552} involved with ammonia oxidation has also been identified. The amino acid sequence and solution structure of this C_{552} have been determined (Fujiwara *et al.* 1995; Timkovich *et al.* 1998).

7. Genomics

The sequence of the *N. europaea* genome is currently being determined by the Joint Genome Institute at the Lawrence Livermore National Laboratory with funding from the US Dept of Energy (http://www.jgi.doe.gov/JGI_microbial/html/nitrosomonas/nitro_homepage.html). The sequencing is under the direction of Jane Lamerdin and the sequence annotation is under the direction of Frank Larimer at Oak Ridge National Laboratory. Collaborators include: Daniel J. Arp, Oregon State University; Alan B. Hooper, University of Minnesota; Martin G. Klotz, University of Louisville; and Jenny M. Norton, Utah State University. Random clones were sequenced representing more than 8x coverage of the genome and the project is currently in the finishing phase. The determination of the sequence of the genome of *N. europaea* represents the first elucidation of the genome of an obligate aerobic chemolithoautotroph and suggests prototypically basic capabilities. Similarity searches suggested much about the bacterium's metabolic capabilities. For example, all central biosynthetic pathways are essentially complete, but degradative pathways are largely absent. *N. europaea* appears well specialized for growth only on NH_3 and CO_2.

8. Summary

With the application of molecular biological techniques to *N. europaea*, rapid progress has been made in characterizing the genes involved in ammonia oxidation including their organization, sequence, and regulation. This information complements the body of biochemical data about ammonia oxidation in *N. europaea* that already exists. The regulation of *amo* expression in particular appears to be a complex event involving both different transcription patterns in conjunction with mRNA processing. Similarly, the two promoter types for the *hao* genes suggest that the copies of *hao* might respond to different environmental signals. However, the physiological significance of the presence of multiple copies of these genes remains unclear and is still under investigation. The nearly complete *N. europaea* genomic sequence has already begun to reveal a wealth of information about the physiology of *N. europaea*. With this new genetic information in hand, the relationship between ammonia oxidation and other physiological process can now be explored with a view toward understanding some of the physiological consequences of the chemolithoautotrophic lifestyle that *N. europaea* has adopted.

9. References

Bergmann DJ *et al.* (1994) J. Bacteriol. 1767, 3148-3153

Bergmann DJ, Hooper AB (1994) Biochem. Biophys. Res. Commun. 204, 759-762

Bock E *et al.* (1986) In Nitrification, Prosser JI (ed), 17-38, IRL Press Ltd, Oxford

Ensign SA *et al.* (1993) J. Bacteriol. 175, 1971-1980

Fujiwara T *et al.* (1995) Current Microbiol. 31, 1-4

Gerards S, Duyts H, Laanbroek J (1998) FEMS Microbiol. Lett. 26, 269-280

Hommes NG *et al.* (1994) Gene 146, 87-89

Hommes NG *et al.* (1996) J. Bacteriol. 178, 3710-3714

Hommes NG *et al.* (1998) J. Bacteriol. 180, 3353-3359

Hommes NG *et al.* (2001) J. Bacteriol. 183, 1096-1100

Hooper AB (1989) In Autotrophic bacteria Schlegel HG, Bowien B (eds), pp. 239-281, Science
    Tech Publishers, Madison, Wisconsin

Hooper AB *et al.* (1997) Antonie van Leeuwenhoek 71, 59-67

Hyman MR, Arp DJ (1992) J. Biol. Chem. 267, 1534-1545

Hyman MR, Arp DJ (1995) J. Bacteriol. 177, 4974-4979

Hyman MR *et al.* (1988) Appl. Environ. Microbiol. 54, 3187-3190

Hyman MR, Wood PM (1983) Biochem. J. 212, 31-37

Hyman MR, Wood PM (1984a) In Crawford RL, Hanson RS (eds), Microbial Growth on C$_1$
    Compounds, Proceedings of the 4th International Symposium, pp. 49-52, American Society
    for Microbiology, Washington, DC

Hyman MR, Wood PM (1984b) Arch. Microbiol. 137, 155-158

Hyman MR, Wood PM (1985) Biochem. J. 227, 719-725

Igarashi N *et al.* (1997) Nature Structural Biol. 4, 276-284

Juliette LY *et al.* (1993a) Appl. Environ. Microbiol. 59, 3718-3727

Juliette LY *et al.* (1993b) Appl. Environ. Microbiol. 59, 3728-3735

Keener WK, Arp DJ (1994) Appl. Environ. Microbiol. 60, 1914-1920

Killham K (1986) In Prosser JI (ed), Nitrification, pp. 117-126, IRL Press, Oxford

Klotz MG *et al.* (1997) FEMS Microbiol. Letters 150, 65-73

Klotz MG, Norton JM (1995) Gene 163, 159-160

McLaren RS *et al.* (1991) J. Mol. Biol. 221, 81-95

McTavish H *et al.* (1993a) J. Bacteriol. 175, 2436-2444

McTavish H *et al.* (1993b) J. Bacteriol. 175, 2445-2447

Norton JM *et al.* (1996) FEMS Microbiol. Letters 139, 181-188
Rasche ME *et al.* (1991) Appl. Environ. Microbiol. 57, 2986-2994
Sayavedra-Soto LA *et al.* (1998) FEMS Microbiol. Letters 167, 81-88
Sayavedra-Soto LA *et al.* (1994) J. Bacteriol. 176, 504-510
Sayavedra-Soto LA *et al.* (1996) Mol. Microbiol. 20, 541-548
Semrau JD *et al.* (1995) J. Bacteriol. 177, 3071-3079
Timkovich R *et al.* (1998) Biophysical J. 75, 1964-1972
Vannelli T *et al.* (1990) Appl. Environ. Microbiol. 56, 1169-1171
Vitousek PM *et al.* (1997) Ecol. Alications 7, 737-750
Wood PM (1986) In Prosser JI (ed) Nitrification, pp. 39-62, Soc. for Gen. Microbiol., IRL Press,
    Oxford
Yamagata A *et al.* (2000) Biosci. Biotechnol. Biochem. 64, 1754-1757

10. Acknowledgements
This work was supported by DOE grant DE-FG03-97ER20266 to DJ Arp and LA Sayavedra-Soto.

DIVERSITY OF DINITROGEN FIXING AND DENITRIFYING BACTERIA IN SOILS ASSESSED BY MOLECULAR BIOLOGICAL METHODS

H. Bothe, A. Mergel, C. Rösch

Botanical Institute, The University of Cologne, Gyrhofstr. 15, D-50923 Köln, Germany

Up until now, few analyses of the bacterial population have been performed with soils because of their complexity and variability in the chemical composition and the microbial life (for a review see Bothe *et al.* 2000). It has been stated that more than 10^4 different ribotypes exist per g of soil (Torsvik *et al.* 1990) of which only a few have been cultivated as yet. The study of microorganisms participating in the conversion of nitrogen gases in soils is of paramount importance. Soils are believed to be a significant sink and/or source for N_2, N_2O and NO, and the latter two gases have detrimental impacts on the atmosphere (Crutzen 1979). Both NO and N_2O are produced by denitrification (the conversion of nitrate to N_2 via nitrite, NO and N_2O) and nitrification (the oxidation of ammonia via hydroxylamine to nitrite and then to nitrate with the concomitant release of small amounts of NO and presumably also N_2O). Dinitrogen and also N_2O are reduced by nitrogenase in dinitrogen fixation. Little is known about the factors, which govern the uptake and evolution of the gases in soils and direct the life of bacteria of the N-cycle. Molecular biological techniques offer a new avenue for analyzing the population of N-converting bacteria in soils to get insights on their potential roles in the production and consumption of gases.

Initially, this laboratory analyzed the soil composition of denitrifying and N_2-fixing bacteria by DNA-DNA hybridizations with 0.5–0.8 kb probes recognizing part genes coding for target enzymes of dinitrogen fixation (*nifH* of nitrogenase reductase) and denitrification (*narG* of nitrate reductase, *nirS* of the cytochrome cd_1 containing nitrite reductase, *nirK* of the Cu-nitrite reductase and *nosZ* of the nitrous oxide reductase). The probes specifically hybridized with DNA from numerous microorganisms of culture collections or of bacteria isolated from soils (Linne von Berg, Bothe 1992; Kloos *et al.* 1998). These investigations showed that the highest number of culturable denitrifying and N_2-fixing bacteria occurs in the upper (5 cm) soil layer and decreases with the depth in several soils examined. The message obtained was, however, limited, as only a small percentage of soil bacteria, about 1% or even less (Amann *et al.* 1997) can be cultured in standard media. Therefore a protocol has recently been developed for the isolation of DNA from soils which allowed to assess the relative distribution of bacterial DNA hybridizing with the different probes in the soil horizons (Mergel *et al.* 2001a,b). This study also showed that the upper soil layer contained the highest content of DNA hybridizing with the probes for denitrification and N_2-fixation and that this content decreased with the depth of an acid soil of an oak-hornbeam forest in the vicinity of Cologne (Chorbusch soil; Mergel *et al.* 2001a) and of a strongly acid Norway spruce stand in the Black Forest near Villingen (Mergel *et al.* 2001b). Bacterial DNA content was consistently higher in soils taken from the vicinity of plant roots than from the bulk, plant free soil. However, there was no selective enrichment of dinitrogen fixing and denitrifying bacteria at the roots of the plants (assessed by hybridizing with the probes of genes specific for dinitrogen fixation and denitrification). The percentage of dinitrogen fixing and denitrifying bacteria was estimated to make up to 5–10% of the total population of bacteria in such soils.

The result that the highest numbers of dinitrogen fixing and denitrifying bacteria occur in the upper (5 cm) soil layer is somewhat surprising. The nitrate content did not decrease with the depth of the soil, particularly in the case of the oak-hornbeam forest in the vicinity of Cologne ("Chorbusch" forest). Thus denitrifying bacteria were expected to be enriched with the depth of the soil. Just the opposite was observed. In the upper soil layer the oxygen and nitrate contents are probably non-limiting. Therefore the counts of high numbers of nitrogen fixing and denitrifying bacteria in the upper layer raise the question about the selective advantage of the possession of denitrification and N$_2$-fixation at this location. The stands examined are never waterlogged. Therefore one can only argue that specific microsites with favorable conditions for denitrification and N$_2$-fixation to proceed exist in the upper layer of these soils. Perhaps physiological traits like denitrification and N$_2$-fixation are only retained in the organisms which may have developed at other locations with selective pressures for these processes.

The DNA isolated from the two soils Villingen and Chorbusch was pure enough for PCR amplifications using conserved oligonucleotide primers for *nifH* and for denitrification genes (*nirS, nirK, nosZ*) and also for 16S-rRNA (as a general bacterial probe). For each of these genes, 16-60 PCR-products obtained with DNA from the upper soil layer have now been cloned and sequenced. The data obtained give first insights on the composition of the bacterial community being present in the upper layer of an acid forest soil. Data are presented here for the 16S-rRNA (Figure 1) and the *nifH* (Figure 2) segments. Comparisons are given for the sequences deposited in the databanks and the own sequences from the oak-hornbeam forest Chorbusch (marked with C) and from the Norway spruce stand of Villingen, Black Forest (marked with V). The following general conclusions can be drawn from these sequence comparisons of Figure 1:

(i)   All sequences obtained are new and not identical with any deposited into the databanks.

(ii)   Many of the sequences cluster with Actinomycetes and the "*Acidobacterium*" phylum which are mainly non-culturable as yet. None of the sequences of PCR-products clustered with those of the *Bacillus/Paenibacillus* group.

(iii)   Some of the PCR-products clustered on the sequenced range (328 bp sequenced out of the about 1.5–1.6 kb of the whole gene) with proteobacteria (e.g. C6A23 next to *Azospirillum amazonense*, C6016, C6021, C6060-62 next to γ-proteobacteria).

(iv)   None of the PCR-products gave identical sequences which might reflect the high diversity of bacteria in such soils.

(v)   Culturable bacteria from the same soil depth of Chorbusch or Villingen (marked with A for grown in heterotrophic mineral medium, "*Azospirillum* medium", or with Y for YEM medium (see Kloos *et al.* 1998)) clustered within the *Bacillus-Paenibacillus* group.

When primers for *nifH* were used, only few different amplificates (Figure 2) were obtained compared to the situation with the gene probes for denitrification (not documented). However, identical sequences were obtained several fold (up to 9) from cloned PCR-products using *nifH* primers. As judged from the microscopic inspection also, it may be that only few N$_2$-fixing bacteria, however, in higher copy numbers, occur in such soils. This impression needs to be verified by more detailed investigations. Some of the DNA amplified with the *nifH* primers came from organisms related to well-known N$_2$-fixing bacteria like *Rhizobium, Bradyrhizobium japonicum* or *Azospirillum* (Figure 2).

Of course, a limited number of PCR-products only can be cloned and sequenced. The primers used in the present study have continuously been improved with the availability of more sequences in the databanks and provide PCR-products in the case of a wide range of culturable bacteria. This was the case for *nifH* and all genes assayed in denitrification (*nirS, nirK, nosZ*). However, as a large percentage of the soil bacteria is non culturable as yet, it may well be that their nitrogen fixation or denitrification genes are not amplified by the primers used under the conditions employed. The data set obtained until now may still be too small to represent the biodiversity of a bacterial soil population. However, this start of the characterization already showed that this approach allows access to a whole regime of still undiscovered bacteria in soils. Details will be published elsewhere (C. Rösch, A. Mergel, H. Bothe, in preparation).

Figure 1.

16S rDNA sequences of PCR products obtained with DNA from two different acid forest soils.

The sequence of *Natronobacterium salstagnum* was used as outgroup. 328 characters of the 16S-rRNA gene (N 357 to N 684 of the *Escherichia coli* sequence) were analyzed by the NJ algorithm (250 boot-straps).

Abbreviations: C = from the Chorbusch soil, V = from Villingen, Y = from colonies on agar plates containing YEM-medium, A = from colonies on agar plates containing heterotrophic mineral medium (= *Azospirillum* medium). The clones marked with C6… only come from PCR products obtained by amplifying segments of the Chorbusch DNA using primers for the 16S-rDNA.

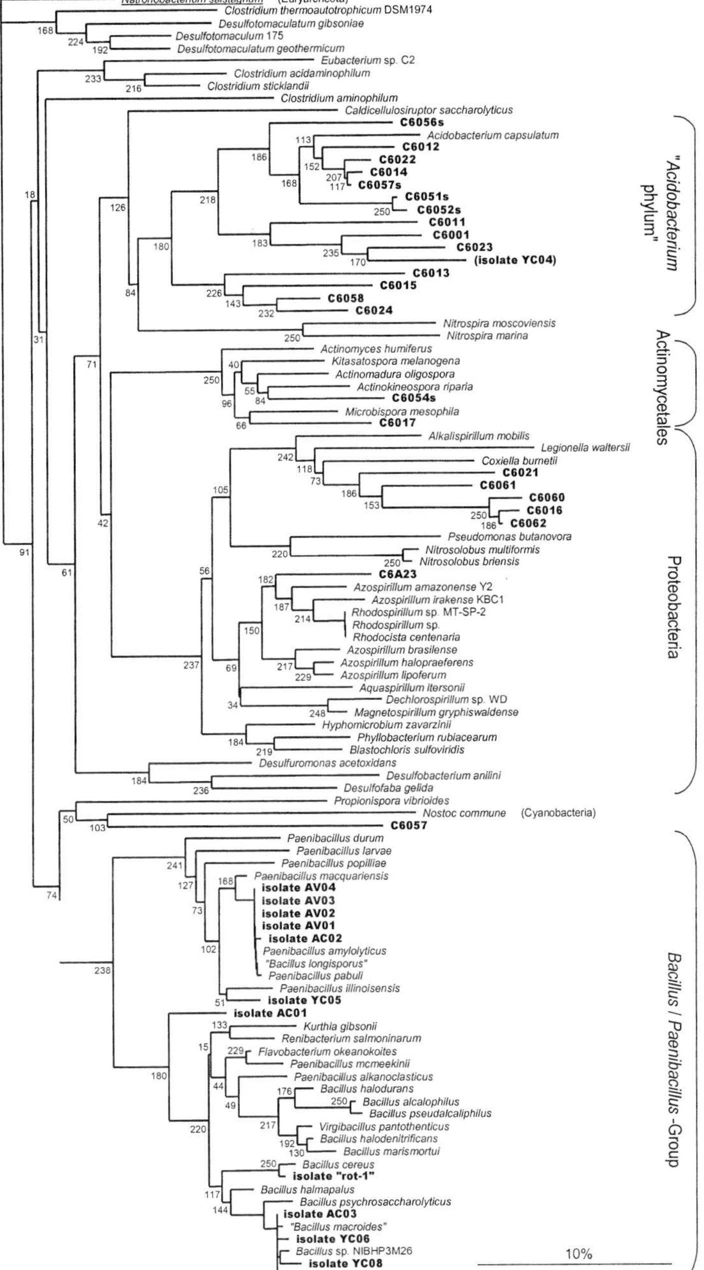

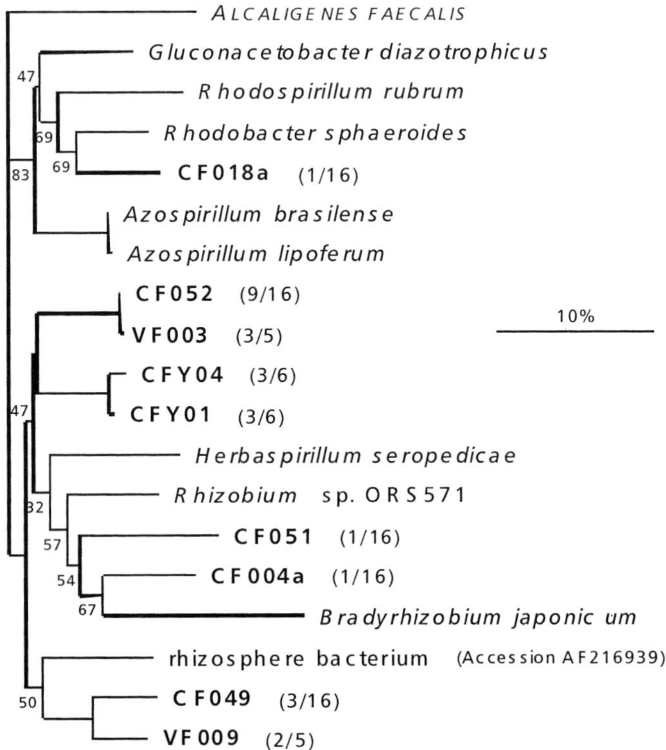

Figure 2. Sequences of PCR products obtained for *nifH* from two different acid forest soils.

First number in brackets: amount of identical sequences obtained for this clone. Second number in brackets: amount of PCR-products sequenced at one location. The sequence of *Alcaligenes faecalis* was used as outgroup. 397 characters of *nifH* gene (N60 to N465 of the *Sinorhizobium meliloti* sequence) were analyzed by the NJ algorithm (100 bootstraps). Abbreviations mean: C = from the Chorbusch soil, V = from Villingen, F = for *nifH,* Y = from an enrichment culture grown in YEM-medium.

References
Amann R, Glockner F-O, Neef A (1997) FEMS Microbiol. Rev. 20, 191-200
Bothe H, Jost G, Schloter M, Ward BB, Witzel K (2000) FEMS Microbiol. Rev. 24, 673-690
Crutzen PJ (1979) Ann. Rev. Earth Planet. Sci. 7, 443-472
Kloos K, Hüsgen UM, Bothe H (1998) Z. Naturforsch. [C] 53, 69-81
Linne von Berg KH, Bothe H (1992) FEMS Microbiol. Ecol. 86, 331-340
Mergel A, Schmitz O, Mallmann T, Bothe H (2001a) FEMS Microbiol. Ecol. 36, 33-42
Mergel A, Kloos K, Bothe H (2001b) Plant and Soil 230,145-169
Torsvik V, Goksoyr J, Daae FL (1990) Appl. Environ. Microbiol. 56, 782-787

Acknowledgements
This work was kindly supported by a grant from the Deutsche Forschungsgemeinschaft.

BRADYRHIZOBIUM JAPONICUM DENITRIFICATION GENES

S. Mesa[1], L. Velasco[1], M. Göttfert[2], E.J. Bedmar[1], M.J. Delgado[1]

[1]Depto de Microbiología del Suelo y Sistemas Simbióticos, Estación Experimental del Zaidín, CSIC, PO Box 419, 18080-Granada, Spain
[2]Institüt für Genetik, Technische Universität Dresden, Mommsenstrasse 13, D-01062 Dresden, Germany

1. Introduction

Denitrification is an alternative form of respiration in which bacteria reduce sequentially nitrate (NO_3^-) or nitrite (NO_2^-) to nitrogen gas (N_2) anaerobically. Among the enzymes involved in denitrification, nitrate reductase reduces nitrate to nitrite, and nitrite reductase catalyzes the reduction of nitrite to nitric oxide (NO); nitric oxide is further reduced to nitrous oxide (N_2O) by nitric oxide reductase; finally, N_2O is converted to N_2 by the nitrous oxide reductase enzyme. Reduction of nitrogen oxides is coupled to energy conservation and permits cell growth under oxygen-limiting conditions. Although the denitrification process is initiated by respiratory nitrate reduction, this reaction is not unique to denitrification since it also occurs in ammonification and assimilatory nitrate reduction. Thus it is considered that the defining reaction in denitrification is the reduction of nitrite to the first gaseous intermediate, NO. Comprehensive reviews covering the physiology, biochemistry and molecular genetics of denitrification have been published elsewhere (Zumft 1997; Baker *et al.* 1998; Watmough *et al.* 1999; Hendriks *et al.* 2000). *Bradyrhizobium japonicum* is a gram-negative soil bacterium with the unique ability to establish an N_2-fixing symbiosis with soybeans (*Glycine max* L. Merr.). *B. japonicum* cells have been shown to couple nitrate reduction to ATP generation (Bandhari, Nicholas 1984) and to assimilate and denitrify $^{15}NO_3^-$ simultaneously to $^{15}NH_4^+$ and $^{15}N_2$, respectively (Vairinhos *et al.* 1989). The presence of a complete denitrification system has also been demonstrated in bacteroids of *B. japonicum*; within the nodules, denitrifying activity has been shown to generate ATP and to maintain nodule integrity and nitrogenase activity (O'Hara, Daniel 1985).

2. Materials and Methods

Cells of *B. japonicum* 3I1b110 (US Department of Agriculture, Beltsville, MD, USA) and mutant derivatives *nirK* GRK13, *norC* GRC131 and *nosZ* GRZ25 (Mesa *et al.* 2001) were used. Plasmid isolations, restriction enzyme digestions, agarose gel electrophoresis, ligations and *E. coli* transformations were performed according to standard protocols (Sambrook *et al.* 1989). Seeds of *Glycine max* L. Merr. cv. Williams were surface-sterilized, germinated, planted in sterile Leonard jars, and grown in controlled environment chambers as previously described (Delgado *et al.* 1998). For nitrate treatments, the mineral nutrient solution was supplied with 4 mM KNO_3. Plants were harvested 40–45 days after planting. Plant dry weight and reduced nitrogen content (Kjeldhal analysis) were determined on plants that had been heated at 85°C for 48 h. Bacteroids were prepared by grinding 1 g of nodules with 50 mM Tris-HCl buffer (pH 7.4) containing 250 mM mannitol (Delgado *et al.* 1998).

3. Results and Discussion

The *B. japonicum* 3I1b110 *nirK*, *norCBQD* and *nosRZDFXYL* genes were identified using heterologous probes from *Alcaligenes faecalis* (Nishiyama *et al.* 1993), *Paracoccus denitrificans* (de Boer *et al.* 1996) and *Pseudomonas stutzeri* (Viebrock, Zumft 1987) encoding the Cu-containing respiratory nitrite reductase, nitric oxide reductase and nitrous oxide reductase, respectively. The deduced primary sequence of *nirK* (EMBL, accession number AJ002516) has greater than 68%

identity with translated sequences of *nirK* genes from other denitrifiers, including other members of the *Rhizobiaceae* family such as *Sinorhizobium meliloti* (http://sequence.toulouse.inra.fr/ rhime/public/Access/RhimeFormRA.html) and *Rhizobium hedysary* (Toffanin *et al.* 1996). Alignment of the deduced NorC amino acid sequence (EMBL, accession number AJ132911) revealed significant identity with NorC proteins of other denitrifiers, ranging from 52% with the deduced amino acid sequence of *norC* from *Paracoccus halodenitrificans* to 82% with that of *S. meliloti*. The deduced primary sequence of the structural *nosZ* gene (EMBL, accession number AJ002531) exhibits 63%, 64% and 77% identity with the products of *nosZ* from *Pseudomonas aeruginosa*, *Achromobacter cycloclastes* and *S. meliloti*, respectively, which provides a strong argument to consider the *B. japonicum nosZ* product as a nitrous oxide reductase enzyme (Figure 1).

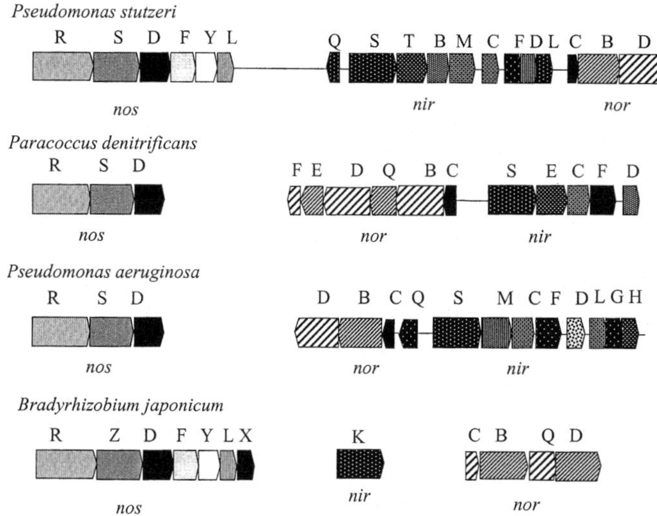

Figure 1. Denitrification gene cluster of *P. stutzeri* and *P. aeruginosa* (Zumft 1997), *P. denitrificans* (Baker *et al.* 1998) and *B. japonicum*.

Cleavage of genomic DNA from *B. japonicum* strain 3l1b110 by the restriction enzymes *Pme*I, *Pac*I and *Swa*I was used together with pulsed-field gel electrophoresis and Southern hybridization to locate the *nirK, norCBQD* and *nosRZDFYLX* denitrification genes on the chromosomal map of *B. japonicum* strain 110*spc*4 published earlier (Kündig *et al.* 1993; Göttfert *et al.* 1998). Restriction of parental and mutant strains genomic DNAs with *Pac*I, *Pme*I and *Swa*I followed by pulsed-field gel electrophoresis resulted in different fragment patterns that allowed determination of the position of the selected genes. Complementary mapping data obtained by hybridization using *B. japonicum* 3l1b110 *nirK, norBQD* and *nosZD* as gene probes revealed that *nirK, norCBQD* and *nosRZDFYLX* genes were dispersed over the entire chromosome, being located close to the *groEL₂, cycH* and *cycVWX* genes, respectively, on the strain 110*spc*4 genetic map (Mesa *et al.* 2001). In *P. denitrificans* the *nir* and *nor* genes are linked on a 17.7- kb fragment (Baker *et al.* 1998), and genes neccesary for denitrification are concentrated at 20 to 36 minutes on the *P. aeruginosa* chromosome, where they form two separate loci, the *nir-nor* and *nos* gene clusters (Vollack *et al.* 1998). In *P. stutzeri* the *nos* genes are within 14- kb of the *nir-nor* genes, forming a single denitrification cluster of about 30 kb (Zumft 1997). Organization of the denitrification genes in *B. japonicum* resembles that of *Rhodobacter sphaeroides* strains 2.4.1 and IL106 where *nirK* and the *norCB* genes are not contiguous on the chromosome (Bartnikas *et al.* 1997; Tosques *et al.* 1997), and the *nos* locus is located on a 115- kb plasmid (Schwintner *et al.* 1998) (Figure 1).

To study expression of denitrification genes, *lacZ* fusions were constructed and transferred by conjugation into *B. japonicum* 3Ι1b110. After aerobic growth, cells of strain 3Ι1b110 containing either the *nirK-lacZ*, *norC-lacZ* or the *nosZ-lacZ* fusion had basal levels of β-galactosidase activity, regardless of the presence or the absence of nitrate in the incubation medium (Figure 2a). When the cells were incubated under 1% O_2 in the absence of nitrate, values of β-galactosidase activity were 3 to 6 times greater than in cells grown in air, and there was a 12- to 27-fold increase in β-galactosidase activity after incubation of the cells under both oxygen-limiting conditions and the presence of nitrate (Figure 2a).

Bacteroids isolated from nodules of plants inoculated separately with cells of strain 3Ι1b110 containing the *nirK-lacZ*, *norC-lacZ* or the *nosZ-lacZ* fusion also showed β-galactosidase activity (Figure 2b). In plants that were N_2-dependent, β-galactosidase activity was induced 5–10 times above basal levels found in bacteroids formed by strain 3Ι1b110 transformed with plasmid pMP220 (Figure 2b). Although denitrification activity, measured as N_2O production, was induced by addition of 4 mM KNO_3 to the mineral nutrient solution, values of bacteroidal β-galactosidase activity were similar to those found in plants not treated with nitrate (Figure 2b). Similar results were obtained in bacteroids isolated from plants that were only N_2-dependent and treated with 5, 10 and 20 mM KNO_3 five days before measurements. If nitrate does not induce expression of denitrification genes in bacteroids, and nitrogen-fixing *B. japonicum* bacteroids use the high-affinity cbb_3-type oxidase to produce ATP (Preisig *et al.* 1993; Arslan *et al.* 2000), could oxygen within the nodules be a limiting factor preventing maximal expression of bacteroidal denitrification genes?

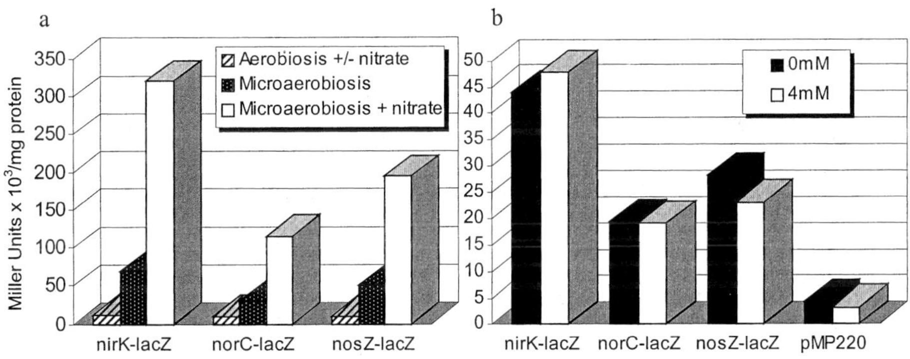

Figure 2. β-Galactosidase activity in free-living cells (a) and bacteroids (b) of *B. japonicum* 3Ι1b110 containing the *nirK-lacZ*, *norC-lacZ* or the *nosZ-lacZ* fusion. Cells were cultured aerobically and microaerobically in a medium supplemented or not with 10 mM KNO_3. Plants were grown in the absence and the presence of 4 mM KNO_3 for 45 days.

To analyze the symbiotic phenotype of the *nirK*, *norB* and *nosZ* mutants, soybean plants were inoculated separately with the *B. japonicum* parental strain 3Ι1b110 and mutant derivatives *nirK* GRK13, *norC* GRC131 and *nosZ* GRZ25 (Table 1). In plants that were only N_2-dependent, no differences in nodule number and nodule fresh weight were found among soybeans inoculated with either the wild-type *B. japonicum* 3Ι1b110 or each one of the mutant strains; moreover, values of total plant dry weight and nitrogen content were similar, regardless of the bacterial strain used for inoculation. However, the nodule number and the nodule fresh weight of plants inoculated with the *nirK* and the *norC* mutants and grown with 4 mM KNO_3 nitrate were significantly lower (P ≤ 0.05) than those of soybeans inoculated with the parental strain or the *nosZ* mutant. Similarly, the depressive effect on nodulation also affected the dry weight and N content of the plants inoculated with strains GRK13 and GRC131, which were significantly lower (P ≤ 0.05) than those of plants inoculated with the *nosZ* mutant or the parental strain 3I1b110. Whether mutation in the *nirK* or *norC* genes affected viability and persistence of the mutants in the rhizosphere or some other stage during nodule formation and development is not known. Denitrifying wild-type cells of *Pseudomonas* sp. strain RTC01 had greater advantage to colonize the rhizosphere of maize plants than those of an isogenic mutant deficient in the ability to synthesize respiratory nitrite reductase, which indicates that the presence of a functional structural gene confers higher rhizosphere competence to a microorganism (Philippot *et al.* 1995).

Table 1. Nodule number (NN), nodule fresh weight (NFW, g/plant), plant dry weight (PDW, g/plant), and total N (N, mg/plant) of soybean plants inoculated with *B. japonicum* 3I1b110 and mutant derivatives *nirK* GRK13, *norC* GRC131 and *nosZ* GRZ25. Values followed by the same letter are not significantly different at P ≤ 0.05 according to the test of Tukey.

B. japonicum strain	Nitrate treatment (mM)							
	0				4			
	NN	NFW	PDW	N	NN	NFW	PDW	N
3I1b110	38 a	0.50 a	2.04 a	49.87 a	47 a	0.87 a	6.35 a	143.67 a
GRK13	44 a	0.50 a	1.88 a	49.53 a	35 b	0.57 b	4.70 b	120.72 b
GRC131	40 a	0.43 a	1.79 a	47.64 a	33 b	0.48 b	4.81 b	108.45 b
GRZ25	42 a	0.49 a	1.98 a	52.28 a	48 a	0.75 a	5.88 a	138.37 a

4. References

Arslan *et al.* (2000) FEBS Lett. 470, 7-10
Bandhari B *et al.* (1984) FEBS Lett. 168, 321-326
Baker SC *et al.* (1998) Microb. Mol. Biol. Rev. 62,1046-1078
Bartnikas TB *et al.* (1997) J. Bacteriol. 179, 3534-3540
de Boer AP *et al.* (1996) Eur. J. Biochem 242, 592-600
Delgado MJ *et al.* (1998) Plant Physiol. Biochem. 36, 279-283
Ghiglione *et al.* (2000) Appl. Environ. Microbiol. 66, 4012-4016
Göttfert M *et al.* (1998) In de Bruijn FJ, Weinstock GM, Lupski JR (eds) Bacterial Genomes: Physical Structure and Analysis, pp. 625-628, Chapman and Hall, New York
Hendriks J *et al.* (2000) Biochim. Biophys. Acta-Bioenergetics 1459, 266 273
Kündig C *et al.* (1993) J. Bacteriol. 175, 613-622
Mesa S *et al.* (2001) Arch. Microbiol.
Nishiyama N *et al.* (1993) J. Gen. Microbiol. 139, 725-733
O'Hara GW, Daniel RM (1985) Soil Biol. Biochem. 17, 1-9
Preisig *et al.* (1993) Proc. Natl. Acad. Sci. USA 90, 3309-3313
Schwintner C *et al.* (1998) FEMS Microbiol. Lett. 165, 313-321

Sambrook J *et al.* (1989) Molecular Cloning: A Laboratory Manual, 2nd. ed., Cold Spring Harbor
Laboratory Press, Cold Spring Harbor, New York
Toffanin A *et al.* (1996) Appl. Environ. Microbiol. 62, 4019-4025
Tosques IE *et al.* (1997) J. Bacteriol. 179, 1090-1095
Vairinhos F *et al.* (1989) J. Gen. Microbiol. 135, 189-193
Viebrock A, Zumft WG (1987) J. Bacteriol. 169, 4577-4580
Vollack KI *et al.* (1998) Microbiol. 144, 441-448
Watmough NJ *et al.* (1999) Biochim. Biophys. Acta 1411, 456-474
Zumft, G (1997) Microbiol. Mol. Biol. Rev. 61, 533-616

5. Acknowledgements

This work was supported by Grant PB97-1216 from Dirección General de Enseñanza Superior e Investigación Científica y Tecnológica (DGESIC). We thank H. Hennecke and H.-M. Fischer for supplying *B. japonicum* strains, and M.L. Fernández-Sierra for technical assistance.

ACTIVATION OF PLANT UREASES: BACTERIA REVISITED IN GREEN?

M. Bacanamwo[1], S.K. Freyermuth[1], J.M. Palacios[2], C.-P. Witte[3], J.C. Polacco[1]

[1]Dept of Biochemistry, Univ. of Missouri, Columbia, MO 65211, USA
[2]Depto Biotecnología, Univ. Politécnica, ETSIA, Ciudad Universitaria, Madrid 28040,
  Spain
[3]Dept Cell. Environ. Phys., Scottish Crop Res. Inst., Invergowrie, Dundee DD2 5DA,
  Scotland

1. Importance of Urea, Urease and Ni to Plants

Most or all plants have ureolytic activity (Hogan *et al.* 1982). To produce it they must deal with potentially toxic nickel (Ni), essential for an active urease. There are no other known plant Ni metalloenzymes (Gerendás *et al.* 1999) as opposed to six (urease, hydrogenase, CO dehydrogenase, methyl-CoM reductase, acetyl-CoA synthase and a superoxide dismutase) in bacteria (Ragsdale 1998). EST databases and mutational analysis of urease expression in soybean indicate that plants have specific permeases for Ni uptake and at least as many urease accessory genes as bacteria. Accessory genes encode proteins essential for insertion of Ni into apourease *in vivo*. It is worthwhile to ask why plants maintain this armamentarium for synthesizing urease and provisioning it with active site Ni. Just how important is the hydrolysis of urea to ammonia and CO_2, and what is the source of urea?

Comparison of deduced and mature N-terminal AA sequences (Torisky *et al.* 1994) and localization analyses (Faye *et al.* 1985) indicate that ureases are cytoplasmic. Urease-negative soybean mutants grown with NH_4NO_3 or N_2 as nitrogen (N) sources accumulate urea and exhibit necrotic leaf tip "urea burn" (Stebbins *et al.* 1991). Our evidence indicates that the metabolic source of all or most of this urea is arginine (Arg) (Stebbins, Polacco 1995). Arg is a pervasive N storage/transport compound in seeds, roots, tubers, bulbs, etc. In soybean seeds, Arg constitutes 18% of storage protein N (Micallef, Shelp 1989), and germination signals a massive conversion of liberated Arg to ornithine and urea via an inducible mitochondrial arginase (Goldraij, Polacco 1999). More germane to N fixation is the fate of ureides (allantoin and allantoate) exported from fixing soybean nodules. No more than a minor portion of the ureides was reported to be converted to urea in soybean plants (Stebbins, Polacco 1995; Winkler *et al.* 1987) and suspension culture (Stahlhut, Widholm 1989). It is worth considering, however, that some varieties may generate urea from ureides, and those that do so may have a greater ability to fix N under water stress. Soybean N-fixation is sensitive to water-deficit (Sinclair, Serraj 1995) under which leaf ureides accumulate (Purcell *et al.* 1998). Ureides have been shown to inhibit nodule nitrogenase (Serraj *et al.* 1999). Purcell *et al.* (2000) have proposed that sensitive varieties build up leaf ureides because leaf allantoate <u>amido</u>hydrolase (which generates ammonia and CO_2 directly from allantoate) is less active due to reduced xylem delivery of manganese (Mn). Drought-tolerant varieties accumulate less ureide because they have an allantoate <u>amidino</u>hydrolase, which generates urea from allantoate and which does not require Mn. Mn stimulation of N fixation under water-deficit is consistent with this model as well as an earlier report (Shelp, Ireland 1985) that the drought-tolerant cultivar, Maple Arrow, produced urea from allantoate (Vadez, Sinclair 2000). While it is surprising that the presumed narrow germplasm base of North American soybean varieties could encompass two routes of ureide breakdown, urea producers (amidinohydrolases) are more common in nature. Recently, amidinohydrolases of ureidoglycolate (Piedras *et al.* 2000) and allantoate (Muñoz *et al.* 2001) were purified from *Chlamydomonas* and chickpea, respectively.

An active bacteroid hydrogenase, a Ni metalloenzyme, has been reported to improve the energetics of N fixation (Ruiz-Argüeso *et al.* 2000). Are Ni activation of hydrogenase and urease

---

related? Olson *et al.* (2001) reported that the hydrogenase accessory proteins, HypA and HypB, were required for full activity of urease in *Helicobacter pylori*. We observed reduced activities of urease and hydrogenase of the phylloplane bacterium, *Methylobacterium* spp., when it colonized urease accessory gene mutants, *eu2/eu2* and *eu3/eu3*, of soybean (Holland, Polacco 1992). Examination of urease may help define the defective step in the induction of hydrogenase activity in some *Rhizobium leguminosarum*-legume associations (López *et al.* 1983).

Table 1. Comparison of soybean urease isozymes

	Ubiquitous	Embryo
Tissue source	Embryo, seed coat, leaf, callus, roots	Embryo
Crude seed extract Sp. Act.[1]	10^{-3} µmole/min x mg protein	1 µmole/min x mg protein
Subunit size[2]	95 kd	93.5 kd
Holomeric structure[2]	α_3	α_3 (*Eu1-a*) or α_6 (*Eu1-b*)
pH optima[3,4]	5.5, 8.8	7.5
Km	0.8 mM[3]	19-476 mM[1]
Hydroxyurea sensitivity[5]	Lesser	Greater

[1]Holland *et al.* 1987; [2] Polacco *et al.* 1985; [3] Kerr *et al.* 1983; [4] Torisky *et al.* 1984; [5] Polacco, Winkler 1984; Polacco *et al.* 1982.

2. Urease: The Substrate of Urease Activation – in Plants, Bacteria, and Fungi

In soybean there are two distinct, non-allelic urease isozymes – the embryo (seed, embryo-specific) and the ubiquitous (metabolic, tissue-ubiquitous) – encoded by the Eu1 and Eu4 genes, respectively (summarized in Polacco, Holland 1994). Some distinguishing properties are presented in Table 1. The embryo has either an $\alpha3$ or $\alpha6$ structure, while the ubiquitous has been observed to be $\alpha3$ (Polacco *et al.* 1985). Seemingly quite different is the $(\alpha,\beta,\gamma)3$ structure of most bacterial ureases. However, these three distinct subunits are colinear with the single plant (and fungal [Figure 1]) subunit. *Klebsiella* and Jack bean ureases have >50% AA identity along their aligned regions (Mulrooney, Hausinger 1990).

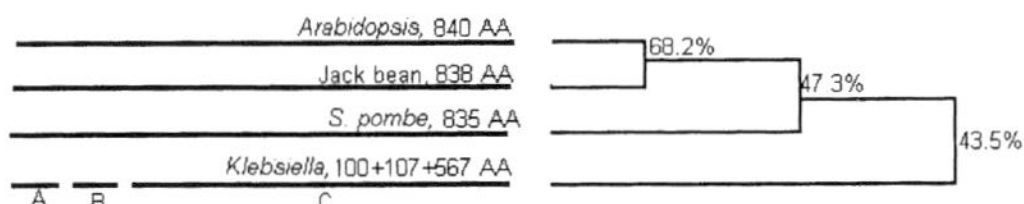

Figure 1. Subunit structure of bacterial and eukaryotic ureases and the AA identity over their co-aligned regions. *S. pombe* urease has a single 95 kDa subunit, but with sequence identity between the bacterial and plant ureases. Identities were determined with DNASIS 2.5 software (Hitachi Software Engineering Corp., Yokohama).

The relatedness of a fungal (*Schizosaccharomyces pombe*) urease (Tange, Niwa 1997) to two plant ureases is illustrated in Figure 1. Jack bean urease is the embryo isozyme. The tissue distribution and metabolic function *Arabidopsis* urease (Zonia *et al.* 1995) indicate that it is the ubiquitous type. Limited sequence data for the soybean ureases indicate a high degree of relatedness of both isozymes to the Jack bean embryo urease (Torisky *et al.* 1994; Krueger *et al.* 1987). Our working hypothesis is that correctly identified accessory proteins from soybean and *Arabidopsis* can activate the related, plant-like urease of *S. pombe*. Table 2 indicates that soybean contains two accessory genes each of which activates both urease isozymes.

3. Urease Accessory Gene Candidates in Soybean

Table 2 summarizes our knowledge of the genes controlling the urease activities of soybean. Since the structural genes are fusions of two or three bacterial genes, we reasoned that the number of accessory genes could be reduced as well if the bacterial counterparts were represented as domains on fused plant proteins. This hypothesis is appealing not only because we would have fewer genes to characterize, but also because complexes among bacterial accessory proteins have been demonstrated. Bacterial urease clusters contain, other than 2 or 3 structural genes plus Ni transporters, a maximum of 4 accessory proteins: UreD- a chaperone-like protein, UreE- a Ni-binding protein with a His-rich C-terminus, UreF- essential for proper insertion of Ni in the active site, and UreG- probably a GTPase. UreD, UreF and UreG form a complex that interacts with apo-urease. Each is essential for an active urease, *in vivo*. UreE deletions have about 50% normal urease activity, but Ni supplementation can increase this basal level. Hausinger's lab has been in the forefront of advancing our knowledge of bacterial urease activation (see Hausinger, Karplus 2001 for a review).

Table 2. Genes controlling urease production in soybean

	UREASE ACTIVITY IN MUTANT		
LOCUS	Ubiquitous	Embryo	FUNCTION
Eu1	+	-	STRUCTURAL (embryo)[1,2]
Eu4	-	+	STRUCTURAL (ubiquitous)[1]
Eu2	-	-	Accessory
Eu3	-	-	ACCESSORY (UREG) [3,4]
Eu5	-	+	Accessory

[1]Meyer-Bothling, Polacco 1987; [2]Torisky *et al.* 1994; [3]Polacco *et al.* 1999; [4]Freyermuth *et al.* 2000.

4. Soybean Eu3 Encodes a UreG-like Protein with Ni-binding Capacity

EST databases of several plants identified UreG, though we first cloned it serendipitously. This is the best conserved accessory gene in bacteria and appears to be well conserved in plants, that of *Arabidopsis* and soybean sharing 80% identity and 88% similarity. By exploiting plant UreG cDNAs, we conclude that soybean Eu3 encodes the UreG ortholog: (1) The *eu3-e1/eu3-e1* null mutant lacks UreG antigen, mRNA and a genomic EcoRV fragment with UreG homology. (2) A dominant, leaky allele, *Eu3-e3*, contains an alanine to valine substitution. (3) Anti-UreG inactivates the *Eu3* gene product in *in vitro* activation of urease in mixed *eu2/eu2 + eu3-e1/eu3-e1* developing embryo extracts (Freyermuth *et al.* 2000). Soybean UreG binds to a Ni column, as expected from a His-rich N-terminus, exclusively found in plant UreGs (Witte *et al.* 2001). This His-run led us to consider a UreG-UreE fusion at first, especially since UreE has yet to be reported in any plant database.

5. UreG Complementation Across Kingdoms

The UreG sequence is well conserved between bacteria and plants. Hence it is not too surprising that potato UreG was able to complement, albeit at low efficiency, a UreG deletion in the *Klebsiella* operon (Witte *et al.* 2001). This occurred in spite of the His-rich N-extension in the plant UreG. We observed no *Arabidopsis* UreG complementation of a UreG disruption of *R. leguminosarum*.

The greater similarities in both UreG and urease (Figure 1) between *S. pombe* and plants lead to a greater expectation for plant UreG to function in *S. pombe*.

6. Soybean UreD and UreF: Looking for a Locus

UreD, UreF and UreG were annotated by the *S. pombe* genome project (http://www.sanger.ac.uk/Projects/S_pombe/). We exploited UreD and UreF as (i) heterologous probes to recover the soybean cDNA homologs, and (ii) a source of PCR primers to confirm their presence in genomic DNA clones (generously provided by Dr N. Honey, Massey Univ., NZ) complementing urease mutants of *S. pombe* (Kinghorn, Fluri 1984). We are now attempting to correct *S. pombe* mutants, identified as *ureD*, *ureF* and *ureG*, with the soybean orthologs.

RT/PCR analysis of an *eu2* mutant allele revealed no alteration in its UreD or UreF ORF. If the sequence of a second mutant allele, and mRNA quantification, both indicate that UreD and UreF proteins are functional in *eu2/eu2* we are obliged to consider that Eu2 encodes a new accessory protein. *In vitro* activation assays (above) indicate that Eu2 interacts with Eu3 (the UreG protein) (Polacco *et al.* 1999), and *eu2/eu2* appears to be able to take up and translocate Ni (Holland, Polacco 1992). No UreE ORF has been identified in any eukaryotic genome to date.

RT/PCR analysis of mutant AJ6, tentatively assigned to new locus Eu5 (Table 2), revealed that 4 of 14 cDNA clones with UreD homology lacked a 15 AA region which defines an exon in *Arabidopsis* (Figure 2). These represent mRNA transcripts from the developing cotyledon, which is urease-positive (Table 2). Ten of ten transcripts from wild type were normal. We are now examining transcripts from leaf tissue, which is urease-negative in AJ6.

Figure 2. Alignment of the first 120 AA (out of ~295) of UreD of *Arabidopsis* (*Arab*), tomato and soybean (soy). The 15-AA run, missing in 4 of 14 AJ6 transcripts, represents the excision of an entire exon according to the splice sites of *Arabidopsis* (between introns 3 and 4).

If we posit that splicing of the UreD transcript is impaired in AJ6, the site of the AJ6 lesion still remains to be determined. Since the sequence of normally spliced variants shows no alteration it is possible that an intron junction sequence is impaired. Regardless of the role that alternative splicing plays in the AJ6 UreD transcript, alternative splicing may be a general mechanism among dicots to maintain UreD protein at a low level. In *Klebsiella*, this is achieved by a sub-optimal ribosome-binding site and an unusual start codon (Park *et al.* 1994). We have observed a high frequency of mis-spliced UreD transcripts in tomato and *Arabidopsis*, often leading to the introduction of a stop codon immediately after exon IV in the latter.

In conclusion, plant urease activation differs from bacterial: (i) Plants have to activate more than one urease, usually produced at quite different levels. (ii) Plant UreG has a His-rich N-extension, like bacterial HypB and unlike all bacterial UreGs to date. (iii) Plants don't seem to have Ni-binding UreE. (iv) Plants have alternative splicing of UreD, possibly to maintain it at low levels. (v) Plants may have another accessory protein, Eu2.

7. References

Faye L *et al.* (1986) Planta 168, 579-585

Freyermuth SK *et al.* (2000) Plant J. 21, 53-60

Gerendás J *et al.* (1999) J Plant Nutr. Soil Sci. 162, 241-256

Goldraij A, Polacco JC (1999) Plant Physiol. 119, 297-304

Hausinger RP, Karplus PA (2001) In Karplus PA *et al.* (eds), Handbook of Metalloproteins, Wiley, Chichester, UK

Hogan ME *et al.* (1982) Phytochem. 22, 663-667

Holland MA *et al.* (1987) Dev. Genet. 8, 375-387

Holland MA, Polacco JC (1992) Plant Physiol. 98, 942-948

Kerr PS *et al.* (1983) Physiol. Plant 57, 339-345

Kinghorn JR, Fluri R (1984) Curr. Genet. 8, 99-105

Krueger RW *et al.* (1987) Gene 54, 41-50

López M *et al.* (1983) Plant Sci. Lett. 29, 191-199

Meyer-Bothling LE, Polacco JC (1987) Mol. Gen. Genet. 209, 439-444

Micallef BJ, Shelp BJ (1989) Plant. Physiol. 90, 624-630

Mulrooney SB, Hausinger RP (1990) J. Bacteriol. 172, 5837-5843

Muñoz A *et al.* (2001) Plant Physiol. 125, 828-834

Olson JW *et al.* (2001) Mol. Microbiol. 39, 1786-182

Park *et al.* (1994) Proc. Natl. Acad. Sci. USA 91, 3233-3237

Piedras P *et al.* (2000) Arch. Biochem. Biophys. 378, 340-348

Polacco JC *et al.* (1982) Plant Physiol. 70, 189-194

Polacco JC, Winkler RG (1984) Plant Physiol. 74, 800-803

Polacco JC *et al.* (1985) Plant Physiol. 74, 794-800

Polacco JC, Holland MA (1994) In Setlow JK (ed), Genetic Engineering, vol 16, pp. 33-48, Plenum Press, New York

Polacco JC *et al.* (1999) J. Exp. Botany 50, 1149-1156

Purcell LC *et al.* (1998) J. Plant Nutr. 21, 949-966

Purcell LC *et al.* (2000) Crop Science 40, 1062-1070

Ragsdale SW (1998) Curr. Opin. Chem. Biol. 2, 208-15

Ruiz-Argüeso T *et al.* (2000) In Triplett EW (ed), Prokaryotic Nitrogen Fixation: A Model System for Analysis of a Biological Process, pp. 489-507, Horizon Scientific Press, Wymondham, UK

Serraj R *et al.* (1999) Plant Physiol. 119, 289-296

Shelp BJ, Ireland RJ (1985) Plant Physiol. 77, 779-783

Stahlhut RW, Widholm JM (1989) J. Plant Physiol. 134, 85-89

Stebbins N *et al.* (1991) Plant Physiol. 97, 1004-1010

Stebbins NE, Polacco JC (1995) Plant Physiol. 109: 169-175

Tange Y, Niwa O (1997) Curr. Genetics 32, 244-246

Torisky RS *et al.* (1994) Mol. Gen. Genet. 242, 404-414

Vadez V, Sinclair TR (2000) J. Exp. Bot. 51, 1459-1465

Winkler RG *et al.* (1987) Plant Physiol. 83, 585-591

Witte CP *et al.* (2001) Plant Mol. Biol. 45, 169-79

Zonia LE *et al.* (1995) Plant Physiol. 107, 1097-103

Section 14:
Novel Applications in Nitrogen Fixation

CHAIR'S COMMENTS: PLANT GROWTH PROMOTION IN LEGUMES AND CEREALS BY LUMICHROME, A RHIZOBIAL SIGNAL METABOLITE

F.D. Dakora[1], V. Matiru[1], M. King[1], D.A. Phillips[2]

[1]Botany Dept., Univ. of Cape Town, P/B Rondebosch 7701, Cape Town, South Africa
[2]Dept of Agronomy and Range Science, Univ. of California, Davis, USA

1. Introduction

Bacteria in the *Rhizobiaceae* (i.e. rhizobia) affect fundamental processes in plants through the use of powerful molecules. Some of these compounds such as the phytohormone IAA (Law, Strijdom 1989; Hirsch *et al.* 1997) have been known for a long time. These bacterial molecules include nodulation factors or lipochito-oligosaccharides (Lopez-Lara *et al.* 1995; Prome, Dermont 1996; Smith *et al.* this volume). Another signal compound that has been identified from the culture filtrate of *Sinorhizobium meliloti* is lumichrome (Phillips *et al.* 1999). This molecule enhances root respiration in alfalfa (Volpin, Phillips 1998) and can trigger an increase in net carbon assimilation and plant growth (Phillips *et al.* 1999). The aim of this study was to test the effect of lumichrome on the growth of tropical grain legumes and cereals under glasshouse conditions.

2. Materials and Methods

Legumes used in this study included cowpea (*Vigna unguiculata* (L) Walp) and soybean (*Glycine max*). The cereals sorghum (*Sorghum bicolour* (L) Moench) and maize (*Zea mays*) were also tested. In all experiments, surface-sterilized seed material was sown in the sterile 1 liter pots containing vermiculite and watered every second day with a nutrient solution containing 1 mM NH_4NO_3, and the antibiotics ampicillin (125 mg l^{-1}) and rifampicin (10 mg l^{-1}).

Lumichrome effects on plant growth were tested by treating the roots or seeds with the compound. The treatments applied included soaking sterile seeds for 2 h in 5 nM lumichrome, or applying 0, 5 or 50 nM lumichrome to emerging seedlings of the test plants three times a week. Plants developing from seeds soaked in 5 nM limichrome were watered with lumichrome-free solution. The pots were initially covered with transparent, clear plastic wraps that were removed after 6 d. The plants were harvested at various time intervals and separated into shoots and roots for dry matter determination after oven drying at 70°C for 48 h.

3. Results and Discussion

Culturing cowpea seedlings for 11 d either after soaking the seeds in 5 nM lumichrome or watering seedlings with 5 nM concentration resulted in 35–41% increase (P < 0.05) in shoot dry matter and 36–49% increase in total biomass (Figure 1). The trifoliate leaf dry matter also increased significantly by 70% with lumichrome application to cowpea due to increase in leaf size and area. However, treatment with 50 nm lumichrome resulted in decreased plant biomass. When watered with 5 nM lumichrome, soybean also showed a significant 17% increase in unifoliate leaf area, 76% increase in trifoliate and 30% increase in total leaf area per plant. This resulted in an overall 10% increase in soybean plant biomass relative to control.

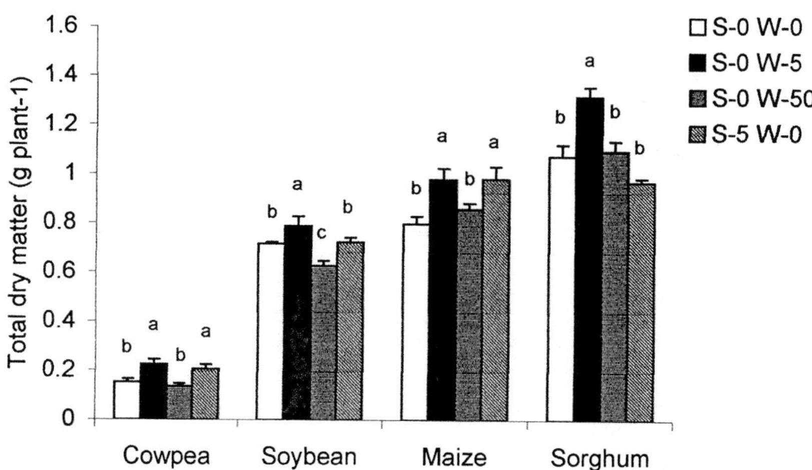

Figure 1. Effect of lumichrome on dry matter yield of selected legume and cereals harvested at 11 (cowpea), 23 (soybean and maize) or 37 (sorghum) days after planting. S-0 W-O = control; S-0 W-5 = watered with but not soaked in 5 nM lumichrome; S-0 W-50 = watered with 50 nM; S-5 W-O = soaked in lumichrome for 2 h only.

Maize and sorghum plants were similarly treated with 5 nM lumichrome. As with cowpea and soybean, the application of this compound significantly enhanced growth of the two cereal species. In maize, total leaf area per plant increased significantly by 26%, shoot dry matter by 36% and total biomass by 23% in lumichrome-treated plants compared to control (Figure 1). Soaking cowpea seeds in 5 nM lumichrome before planting caused a significant 33% and 18% increase in shoot and root biomass, respectively. Watering seedlings with 5 nM lumichrome also promoted 40% increase in sorghum shoot growth at 37 d.

Taken together, these data indicate that lumichrome has a growth-promoting effect on a wide range of plants including legumes and cereals. It is therefore likely that the robust growth of symbiotic legumes is due partly to the positive effects of lumichrome released by rhizobia in addition to improved N nutrition.

References

Hirsch AM *et al.* (1997) Plant and Soil 194, 171-184
Law IJ, Strijdom BW (1989) S. A. J. Plant and Soil 6, 161-166
Lopez-Lara IM *et al.* (1995) Plant Mol. Biol. 29, 465-477
Phillips *et al.* (1999) Proc. Nat. Acad. Sci. USA 96, 12275-12280
Prome JC *et al.* (1996) In Stacey G, Keen NT (eds) Plant Microbe Interactions, pp. 272-307, Chapman and Hall, New York
Volpin H, Phillips DA (1998) Plant Physiol. 116, 777-783

SYMBIOTIC AND DEVELOPMENTAL MUTANTS OF WHITE SWEETCLOVER (*MELILOTUS ALBA* DESR.)

A.M. Hirsch[1], M.R. Lum[1], R.S.N. Krupp[1], W. Yang[1], T.A. LaRue[2]

[1]Dept Mol., Cell and Develop. Biol., Univ. California, Los Angeles, CA 90095-1606, USA
[2]Boyce Thompson Institute, Tower Road, Ithaca, NY 14853-1801, USA

1. Introduction

White sweetclover (*Melilotus alba* Desr.) is an autogamous diploid with more than 70 years' worth of biochemical and genetic investigations that serve as the foundation for the current research. All sweetclover species studied to date are diploid with n=8 chromosomes (Smith, Gorz 1965). *Melilotus* species are easy to grow, and exhibit a short seed-to-seed cycle time (3 months).

A large number of biochemical studies have been performed on members of this genus because they produce significant quantities of coumarin, the simplest of all flavonoids. More is probably known about the biochemistry of flavonoid synthesis in sweetclover than in any other legume because of efforts to find cultivars that do not cause sweetclover-bleeding disease (Smith, Gorz 1965). Improper curing or ensiling of sweetclover hay leads to the formation of dicoumarol, a compound that interferes with vitamin K activation of prothrombin, which is necessary in blood clotting. This hemorrhagenic agent, given the trade name Dicumarol, is used to prevent blood clots after surgery. Warfarin, probably the best-known anticoagulant, is a derivative of dicoumarol.

Figure 1. *Arabidopsis* (left) and *M. alba* (right) are similar in height and flowering time.

In the process of searching for coumarin mutants, a large number of other mutants were generated by Gorz and Haskins, Goplen, Micke, and others. Sweetclover produces a broad spectrum of flavonoids and related molecules, many of which have already been extensively studied. Gorz and Haskins and co-workers identified genes important for seed color (*Y/y*; Specht *et al.* 1976) and for seed and seedling color (*C/c*; Gorz, Haskins 1975). It is very likely that plants with pale seeds or seedlings are flavonoid mutants. Goplen (1992) found that a single recessive gene controlled the white sepal trait in sweetclover, and Goplen and Micke also uncovered a large number of floral mutants (Goplen 1976; Schiebe, Micke 1965). We also have obtained a large number of mutants in our screens of EMS- and neutron bombardment-generated M_2 plants generated by Kneen and LaRue (1988) from the U389 line (Hirsch *et al.* 2000). This report describes some of these mutants as well as additional ones generated from the U390 line.

2. Materials and Methods

2.1. Isolation of mutants. Seeds from the wild-type *Melilotus alba* (Desr.) line U389 were mutagenized as described earlier (Kneen, LaRue 1988). Mutants of the U390 line were obtained by treatment with EMS (Hirsch, LaRue, unpublished). The U390 line, a wild-type dwarfed U389 (Gengenbach *et al.* 1969), is approximately the same size and has a similar flowering time as *Arabidopsis* (Figure 1). M_2 seeds were planted in sand-vermiculite (1:1), watered with a medium without N, and inoculated with the wild-type *Sinorhizobium meliloti* strain Rm1021 to look for symbiotic mutants. The Antho⁻ and non-nodulating white sweetclover mutants were derived from Nod⁺Fix⁺ U389 plants. The inflorescence mutants were derived from the U390 line.

2.2. Genetic analysis. The T280 line was crossed to the wild-type U390 line, and F_1 hybrid seeds resulted. The Antho⁻ mutants were crossed to the wild-type U389 line to obtain F_1 hybrids. The F_1 flowers were selfed to produce the F_2 generation. From the segregating F_2 populations, F_3 plants were obtained and analyzed for their phenotypes.

2.3. Plant growth. Sweetclover seeds of the non-nodulating mutants were surface-sterilized briefly in 95% ethanol, followed by immersion in commercial bleach for 45 min. The seeds were then washed five times with sterile water, placed on 0.8% Phytagar (GiboBRL), and grown in the dark for 48 h.

2.4. Inoculation studies. The wild-type *S. meliloti* strain Rm1021 was used as inoculum. Liquid cultures were grown shaking in an incubator at 30°C, washed once with sterile water, and re-suspended to approximately 10^8 cells per ml. For the mycorrhizal studies, aseptic spores of *Glomus intraradices* (Premier Tech, Quebec) were used for inoculating the roots. The plants were grown in a sand-vermiculite mixture watered with one-strength Hoagland's medium minus P.

3. Results and Discussion

3.1. Floral mutants. The inflorescence of mutant T280 differs from that of the wild-type U390 line in that the basal bracts of the inflorescence contain in their axils, a reiteration of inflorescence initiation instead of individual flowers (Figure 2). Many of the floral primordia in these basal inflorescences do not show any obvious sepal, petal, stamen or carpel primordia at the same stage of development as wild type. Mature flowers develop from some of these aberrant primordia, but many abort before elongation of the floral organs. Other

Figure 2. Inflorescence of the *coi* (compound inflorescence) mutant.

meristems produce normal-looking flowers. Nevertheless, very few seeds were obtained from selfing the *coi* mutants. F_2 progeny segregated in a ratio that approximated 3:1.

We also obtained two lines (T149 and T148) that produce inflorescences, which were even more undeveloped than those of the *coi* mutant. The inflorescence was characterized by incompletely formed, secondary inflorescences composed of green, anomalous buds (data not shown). In contrast to the cauliflower-inflorescence mutants with unifoliate leaves previously described by Goplen (1967) and Micke (unpublished), this mutant has trifoliate leaves. The T149 and T148 mutants resembled the vegetable broccoli (*brc*) (Hirsch *et al.* 2000). Because the *brc* mutant was sterile, we were unable to do a genetic analysis.

Figure 3. U389 (left) has reddish stems whereas the Antho⁻ mutant BT20 has green stems (right).

3.2. Flavonoid mutants. Fifteen independently isolated stable Antho⁻ (anthocyanin-minus) mutants were generated in the U389 background using different mutagens. One Antho⁻ mutant was found to be luteolin-minus (Hrazdina, LaRue, unpublished), but will not be described here. Three other Antho⁻ mutants are described.

BT20 and BT65 are characterized by having green stems instead of the red stems that typify the U389 and U390 lines. BT19 has pale tan-colored stems. The green stems of BT20 are compared with the wild-type U389 line in Figure 3.

The Antho⁻ mutants were crossed to U389 and F_1 seeds were obtained. Analysis of the F_2 population demonstrated a Mendelian ratio of 3 red to 1 green, indicating that the mutation segregated as a monogenic recessive. We have now intercrossed the various Antho⁻ mutants for allelism tests.

Studies on nodulation and mycorrhizal establishment in the mutant roots are in progress. Preliminary results suggest that these three Antho⁻ mutants are affected in their symbiotic interactions. Although BT19, BT20 and BT65 are nodulated by Rm1021, the nodules show some unusual phenotypes. Moreover, their interaction with *Glomus intraradices* is impaired (Lum, Hirsch, unpublished).

Table 1. Effects of *Sinorhizobium meliloti* and *Glomus intraradices* on wild-type U389 and *sym* mutant sweetclover (*Melilotus alba* Desr.).

Genotype/Mutant	Nodulation Phenotype*	Mycorrhizal Phenotype
Wildtype/U389	Hac[+a], Inf[+b], Nod[+c]	Pen[+d], Ves[+e], Arb[+f]
sym1/BT62, BT58, and BT35	Had[+g], Inf⁻; small, white nodules (12% of the plants) for BT62 but BT58 and BT35 are Nod⁻	Pen⁺, Ves⁺, Arb[±] (BT62), Pen⁻, Ves⁻, Arb⁻ (BT35, BT58)
sym2/BT59	Hac⁺, Inf⁺; similar to U389, white, ineffective nodules (25%), but 1.7% of the studied plants formed effective nodules	Pen⁺, Ves⁺, Arb⁺
sym3/BT61, BT69, and BT70	Has[+h], Had⁻, Hac⁻, Inf⁻, Nod⁻	Pen⁻, Ves⁻, Arb⁻
sym5/BT71	Had⁺, Inf⁻; occasional small, white nodules on 13% of the plants	Pen⁻, Ves⁻, Arb⁻

*Based on Utrup *et al.* (1993), Wu *et al.* (1996) and this report; [a] root hair curling, i.e. shepherd's crook formation; [b] infection thread formation; [c] nodule development; [d] penetration of hyphae; [e] vesicle formation; [f] arbuscule formation; [g] root hair deformation; [h] root hair tip swelling.

3.3. Symbiotic mutants. The responses of the *M. alba* non-nodulating mutants to inoculation with Rm1021 have already been extensively described (Utrup *et al.* 1993; Wu *et al.* 1996). We analyzed their responses to inoculation with *G. intraradices* and found that several of the non-nodulating mutants are Myc⁻ as has been described for other legumes (Lum, Li, LaRue, Schwartz, Kapulnik, Hirsch, submitted). Interestingly, one of the non-nodulating *sym* mutants, *sym2*, appears to be colonized more extensively by mycorrhizal fungi than its wild-type parent (Table 1). This mutant is also responsive to *S. meliloti*; 25% of the plants were reported to develop white, ineffective nodules (Utrup *et al.* 1993).

The *sym3* mutant and its three alleles, BT61, BT69, and BT70, showed no penetration of the root by *G. intraradices*, even months after inoculation. Two *sym1* mutants were also Myc⁻, but a third one, BT62, exhibited some mycorrhizal formation. BT62 had previously been described as a weak allele by Utrup *et al.* (1993).

In conclusion, a number of single-gene recessive *M. alba* mutants are available for studying both symbiotic and developmental processes. In particular, white sweetclover has several distinct advantages as a legume model for studying the role of flavonoids in plant development as well as in symbiosis: (1) like *Arabidopsis*, there are a large number of available mutants, which are likely to be mutated in critical steps for flavonoid synthesis; and (2) sweetclover appears to have fewer copies of genes encoding enzymes for the phenylpropanoid pathway than either *Medicago sativa* (alfalfa) or *M. truncatula* (data not shown). It is well known that many of the flavonoids serve as important signaling molecules in plant-microbe interactions. Some recent data also indicates that flavonoids also impact plant growth and development (Woo *et al.* 1999). White sweetclover may also serve as a useful model for studying flower development because similar to *Arabidopsis* and *Antirrhinum*, the sweetclover inflorescence is determinate and consists of a simple raceme. Pea, in contrast, is indeterminate, and produces secondary inflorescences that generate one or more floral meristems (Ferrándiz *et al.* 1999; Singer *et al.* 1999). The study of legume floral mutants will help expand our knowledge of the genetic mechanisms of flowering by uncovering the key genes regulating inflorescence and floral development in plants beyond the model species.

4. References

Ferrándiz C *et al.* (1999) Dev. Genet. 25, 280-290

Gengenbach BG, Haskins FA, Gorz HJ (1969) Crop Sci. 9, 607-610

Goplen BP (1967) Can. J. Genet. Cytol. 9, 136-140

Goplen, BP (1992) Can. J. Plant Sci. 72, 1259-1262

Gorz HJ, Specht JE, Haskins FA (1975) Crop Sci. 15, 235-238

Haskins FA, Gorz HJ (1965) Genetics 51, 733-738

Hirsch AM *et al.* (2000) In Triplett EW (ed.) Prokaryotic Nitrogen Fixation: A Model System for the Analysis of a Biological Process, pp. 627-642, Horizon Scientific Press

Kneen BE, LaRue TA (1988) Plant Sci. 58,177-182

Scheibe A, Micke A (1967) In Induzierte Mutationen und ihre Nutzung (Induced Mutations and their Utilization), Erwin-Baur Gedächtnisvorlesungen IV, Gatersleben 20-24 June 1966, pp. 231-236, Akademie-Verlag Berlin

Singer S *et al.* (1999) Bot. Rev. 65, 385-410

Smith WK, Gorz HJ (1965) In Advances in Agronomy, pp. 163-231, Academic Press, New York

Specht JE *et al.* (1976) Phytochem. 15, 133-134

Utrup LJ, Cary AJ, Norris JH (1993) Plant Physiol. 103, 925-932

Woo H-H, Orbach MJ, Hirsch AM, Hawes MC (1999) Plant Cell 11, 2303-2315

Wu C, Dickstein R, Cary AJ, Norris JH (1996) Plant Physiol. 110, 501-510

5. Acknowledgements

This research was supported in part by National Science Foundation grants 96-30842 and 97-23982, UC Mexus grant 017451, and a grant from the Council on Research from the UCLA Academic Senate to AMH. A USPHS National Research Service Award GM07185 and BioStar Grant S98-86 supported MRL.

RHIZOBIAL SIGNALS AND CONTROL OF PLANT GROWTH

D.L. Smith[1], B. Prithiviraj[1], F. Zhang[2]

[1]Plant Science Department, Macdonald Campus, McGill University, Montreal, QC, Canada
[2]Bios Agriculture Inc., Macdonald Campus, McGill University, Montreal, QC, Canada

1. Introduction

As our understanding of plant-microbe interactions improves it is clear that these relationships are subtle, sophisticated and complex. It seems certain that plants and microbes can, at least under some circumstances, control elements of each other's growth and development. For field grown plants, there are also the effects of various biotic and abiotic stresses and their effects on the interactions between plants and microbes. It is clear that we have come a long way in our comprehension of some of these systems. However, it is also clear that we have a long way to go. Below we present a review of over ten years work in this area.

2. The Soybean Nitrogen-fixing Symbiosis: Signal Exchange under Suboptimal Conditions

Soybean evolved under warm conditions, and is physiologically adapted for warm conditions. The optimum temperature for soybean nodulation is 25–30°C. Southwestern Quebec is on the North American northern limit for soybean production. Soybean plants emerging in the region frequently turn light green once cotylendonary N reserves are exhausted, and remain that way until nodulation is complete (Zhang *et al.* 1995). We postulated that low soil temperatures inhibited soybean nodulation. We were able to show that the time from inoculation increased by about 2 days per °C decrease in temperature between 25 and 17°C, and by approximately a week per °C between 17 and 15°C (Zhang, Smith 1994; Zhang *et al.* 1995). Further, the time that was most sensitive to low temperature was the first 12 h after inoculation (Zhang, Smith 1994), a time when signal exchange was taking place.

Signal exchange begins with the secretion of phenolic compounds, flavonoids and isoflavonoids, by host plants (reviewed by Schlaman *et al.* 1998). These compounds induce the *nod* genes, resulting in the production of the bacteria-to-plant signals (Kondorosi 1992). These are lipo-chito-oligosaccharides (LCOs). These compounds play a key role in initiating the early events of the legume-rhizobia nitrogen-fixing symbiosis (Lhuissier *et al.* 2001). In this role they are referred to as Nod factor. Extensive work on LCOs has revealed that members of the group are responsible for the host specificity of rhizobia (Perret *et al.* 2000). LCOs invoke multiple physiological responses in the host: root hair deformation (Spaink *et al.* 1991; Prithiviraj *et al.* 2000a), induction of nodulin genes essential for infection thread formation (Horvath *et al.* 1993) and cortical cell division (Schlaman *et al.* 1998). Purified Nod factors *per se* are capable of initiating complete nodule structures at submicromolar concentrations in some legume-rhizobia systems (Denarie, Cullimore 1993).

We added plant-to-bacteria signal compounds (generally genistein) to bacteria used as inocula on soybean plants and found that this accelerated the very earliest stages of nodulation (Zhang, Smith 1994), leading to earlier subsequent nodule development and the onset of nitrogen fixation. As a result, plants inoculated with genistein treated *Bradyrhizobium japonicum* cells were larger and contained more nitrogen. While the optimum genistein concentration is known to be 5 µM at 25°C (Kosslak *et al.* 1987), at 15°C it was higher, at 20 µM (Zhang, Smith 1995). At the same time we were able to show that, at lower root zone temperatures soybean roots contained less genistein (Zhang, Smith 1996a).

Zhang and Smith (1996b, 1997) showed that incubation of *B. japonicum* with genistein, prior to application as an inoculant, or directly applied into the seed furrow at planting, increases soybean nodulation, N_2 fixation and total N yield, when field conditions are such that they would normally delay or inhibit nodulation. This was caused by a shortening of the time between inoculation and when the infection thread reached the bottom of the root hairs.

Pan and Smith (2000a) showed that genistein treated *B. japonicum* 532C, with a genetic marker, had higher levels of nodule occupancy than the untreated cells under greenhouse conditions. Paau *et al.* (1990) reported that adding soybean meal to the fermentation medium can alter the nodule readiness of the rhizobia and that this has an effect on the competitiveness of the inoculant strains (McDemott, Graham 1990).

Soybean nodulation is also inhibited by mineral nitrogen (Zahran 1999). We were able to show that when N concentration in the rooting medium increased above 50 mg L^{-1}, genistein concentrations in root systems decreased, and daidzein concentration in plant root systems decreased steadily as N application level increased from 0 to 150 mg L^{-1} (Zhang *et al.* 2000). Inoculation of soybean with *B. japonicum* that has been pre-activated with genistein and daidzein also improves plant nodulation and nitrogen fixation under levels of mineral nitrogen that are inhibitory to nodulation (Pan, Smith 2000b). Pan and Smith (2000b) found that the plants receiving pre-incubated *B. japonicum* cells had more nodules, nodule weight, and plant nitrogen content, especially in a low N containing sandy soil and where 20 kg N ha^{-1} were added as mineral fertilizer.

A company, Bios Agriculture Inc., has commercialized these findings. They have accumulated data from over 200 site-years in the Canadian soybean production areas, and in the northern tier states of the USA. In general, this technology results in a yield increase on the order of 10%, when applied to early seeded soybean crops.

3. Lipochito-oligosaccharides and Controls on Crop Production

The chitin backbone of the LCO may cause it to act, in some way, like an elicitor in disease systems. LCO reduces salicylic acid levels in alfalfa roots, potentially aiding in suppression of legume host defense responses, and thereby ensuring successful infection (Martinez-Abarca *et al.* 1998). More recently, we have shown that a strong reduction in SA level in the leaf tissues occurred when soybean plants were sprayed with LCO (Prithiviraj *et al.* 2000b). Nod factors can alter phytoalexin concentrations in plant tissues. Nod Rm–IV(C16:1, S) increases medicarpin content in alfalfa (Savoure *et al.* 1994). Schmidt *et al.* (1994) reported an increase in concentrations of the flavonoids daidzen, coumestrol and genistein in root exudates of soybean (*Glycine max*) following exposure to the appropriate Nod factor (LCO) (Savoure *et al.* 1994). Nod factors have also been shown to induce the expression of chalcone synthase, chalcone reductase and PR genes (Krause *et al.* 1997). Co-inoculation of soybean plants with *B. japonicum* and the mycorrhizal fungus *Glomus mossae*, enhanced root colonization of the fungus, and similar results were observed when highly purified Nod factors (Nod NGR-V (MeFuc, Ac) from *Rhizobium* sp. NGR 234 (Xie *et al.* 1995) were applied.

LCOs have been shown to provoke a range of physiological responses in non-host plants. For instance, LCOs induce rapid and transient alkalination of tobacco (Baier *et al.* 1999) and tomato cells (Stahlein *et al.* 1994) in suspension cultures. LCOs can restore cell division and embryo development in a carrot mutant (de Jong *et al.* 1993). Finally, LCOs caused a resumption of cell division in somatic embryo cultures of Norway spruce (*Picea abies*) (Egertsdotter, von Arnold 1998), even in the absence of auxin and cytokinin (Daychok *et al.* 2000).

Induction of *nod* genes by non-host plants has been reported (Hungria, Stacey 1997; Le Strange *et al.* 1990), raising a possible explanation for the reported PGPR activity of rhizobia in crops such as rice (Biswas *et al.* 2000a,b; Prayitno *et al.* 1999). Increased growth of maize and bean in intercropping systems could be attributed to the reciprocal stimulation of *Rhizobium* and *Azospirillum* by root exudates of bean and maize (Hungria, Stacey 1997).

During the last decade we have found that treatment of soybean seeds with genistein-induced cultures of *B. japonicum* enhanced the germination and emergence of soybean and other crop plants under field conditions, when compared to uninduced cultures of *B. japonicum* or genistein. This finding was made first in the field and is a classic example of serendipity in biological research. The stimulation of seed germination only by genistein induced *B. japonicum* cells suggests that the observed effects might be due to the LCO present in the induced cultures. We have recently shown that the major LCO of *B. japonicum* enhanced the germination and early growth of a variety of crop plants and the model plant *Arabidopsis* (Prithiviraj *et al.* 2000c). Similar effects were observed with a number of synthetic LCOs (unpublished results). Presoaking of seeds in LCO solutions induced rapid emergence of soybean, maize and cotton under field conditions. LCO treatments also increased early growth of maize and soybean in pot and hydroponic experiments. Irrigation of maize seedlings with LCO solution doubled variables such as leaf area, plant height, and root and shoot dry weight. When hydroponically grown three-day-old seedlings of soybean and maize were treated with 10^{-7}, 10^{-9} or 10^{-11} M LCO the biomass of both soybean and maize was increased. At 10^{-9} M and 10^{-7} M LCO, the soybean root biomass was 7–16% larger and roots were 34–44% longer than in the control. LCO treatment also had positive effects on the root growth of soybean and maize, increasing the total length, projected area and surface area.

Enhancement of plant photosynthesis due to *B. japonicum* soybean associations has been reported. Imsande (1989a,b) reported increased net photosynthesis and grain yield in soybean inoculated with *B. japonicum* as compared with plants not inoculated but adequately supplemented with N fertilizer. Thus it seems probable that rhizobial associations enhance photosynthesis and that this might be mediated by signal molecules. To test the hypothesis that LCO is responsible for the increased photosynthesis a series of experiments were conducted in the greenhouse and in the field. Spray application of LCO at submicromolar concentrations enhanced the photosynthetic rates of soybean, maize, rice, bean, canola, apple and grapes. On average there was a 10–20% increase in the photosynthetic rate and this was concomitant with an increase in stomatal conductivity and constant or decreased leaf internal CO_2 concentration. Under field conditions, spray application of LCO at concentrations of 10^{-6}, 10^{-8} and 10^{-10} M resulted in increased soybean grain yields of up to 40%. Taken together, the results of our experiments suggest the possible use of this novel class of signal molecules in improving crop production.

4. Concluding Remarks

Under field conditions most plants are stressed for at least a part of every day. Signal exchange between plants and microorganisms is particularly susceptible to the imposition of abiotic and biotic stresses, because this system involves two genomes, each with its own environmental sensitivities, and because the two organisms may not be in direct contact with each other at the time of signal exchange, so that this process takes place through time and space, under potentially unfavorable conditions. We have shown here a practical example of environmental conditions disrupting the signaling process, leading to less efficient establishment of the soybean–*B. japonicum* nitrogen-fixing symbiosis, and some potentially practical solutions to this situation. In the course of this work we made a serendipitous discovery with regard to the potential control of plant growth by the signal molecules of microbes. This finding has potentially large applications in the understanding of the controls on plant growth and, also, in world food production.

5. References

Baier R *et al.* (1999) Planta 210, 157-164
Biswas JC *et al.* (2000a) Soil Sci. Soc. Am. J. 64, 1644-1650
Biswas JC *et al.* (2000b) Agron. J. 92, 880-886
Daychok JV *et al.* (2000) Plant Cell Rep.3, 290-297
De Jong AJ *et al.* (1993) Plant Cell 5, 615-620
Denarie J, Cullimore J (1993) Cell 74, 951-954

Egertsdotter U, von Arnold S (1998) J. Exp. Bot. 49, 155-162
Horvath B *et al.* (1993) Plant J. 4, 727-733
Hungria M, Stacey G (1997) Soil Biol. Biochem. 29, 819-830
Imsande J (1989a) J. Exp. Bot. 39, 1313-1321
Imsande J (1989b) Agron. J. 81, 549-556
Kondorosi A (1992) In Verma DPS (ed.) Molecular Signals in Plant-Microbe Communication,
    pp. 25-340, CRC Press, Boca Raton, FL
Kosslak RM *et al.* (1987) Proc. Nat. Acad. Sci. USA 84, 7428-7432
Krause A *et al.* (1997) Molec. Plant-Microbe Int. 10, 388-393
Le Strange KK *et al.* (1990) Molec. Plant-Microbe Int. 3, 214-220
Lhuissier FGP *et al.* (2001) Ann.Bot. 87, 289-302
Martinez-Abarca F *et al.* (1998) Molec. Plant-Microbe Int. 11, 153-155
McDermott TR, Graham PH (1990) App. Environ. Micro. 56, 3035-3039
Paau *et al.* (1990) In Gresshoff PG *et al.* (eds) Nitrogen Fixation: Achievements and Objective,
    pp. 617-624, Chapman and Hall, New York
Pan B, Smith DL (2000a) Plant Soil 223, 229-234
Pan B, Smith DL (2000b) Plant Soil 223, 235-242
Perret X *et al.* (2000) Microbiol. Mol. Biol. Rev. 64, 180-201
Prayitno J *et al.* (1999) Australian J. Plant Phys. 26, 521-535
Prithiviraj B *et al.* (2000a) J. Exp. Bot. 51, 2045-2051
Prithiviraj B *et al.* (2000b) Abstract PE3, 17th North American Conference on Symbiotic Nitrogen
    Fixation, 23-28 July 2000, p. 66, Quebec, Canada
Prithiviraj B *et al.* (2000c) Abstract E6, 17th North American Conference on Symbiotic Nitrogen
    Fixation, 23-28 July 2000, p. 38, Quebec, Canada
Savoure A *et al.* (1994) EMBO J. 13, 1093-1102
Schmidt J *et al.* (1994) Proc. Nat. Acad. Sci. USA 85, 8587-8582
Schlaman HR *et al.* (1998) Development 124, 4887-4895
Spaink HP *et al.* (1991) Nature 354, 125-130
Staehelin C *et al.* (1994) Proc. Nat. Acad. Sci. USA 91, 2196-2200
Xie Z-P *et al.* (1995) Plant Physiol. 108, 1519-1525
Zahran HH (1999) Microbiol. Molec. Biol. Rev. 63, 968-989
Zhang F, Smith DL (1994) J. Exp. Bot. 279, 1467-1473
Zhang F, Smith DL (1995) Plant Physiol. 108, 961-968
Zhang F, Smith DL (1996a) J. Exp. Bot. 47, 785-792
Zhang F, Smith DL (1996b) Plant Soil 179, 233-241
Zhang F, Smith DL (1997) Plant Soil 192,141-151
Zhang F *et al.* (1995) Environ. Exper. Bot. 35, 279-285
Zhang F *et al.* (2000) J. Agron. Crop Sci. 184, 197-204

6. Acknowledgements

The authors thank Dr X. Zhou for her support in maintaining the laboratory and field sites, S. Leibovitch for assistance in HPLC techniques and general cooperation from Bios Agriculture Inc., G. Stacey for provision of LCO samples to act as standards for our HPLC work, and R. Carlson for provision of artificial LCOs.

WHY DO LEGUME NODULES EVOLVE HYDROGEN GAS?

Z. Dong[1], D.B. Layzell[2]

[1]Dept of Biology, Saint Mary's University, Halifax, NS, B3H 3C3 Canada
[2]Dept of Biology, Queen's University, Kingston, ON, K7L 3N6 Canada

1. H_2 Evolution from N_2-Fixation Systems

Hydrogen (H_2) gas is an obligate byproduct of the N_2-fixing enzyme, nitrogenase, claiming about 33% of the reducing power and ATP that flows to the enzyme. In some legume symbioses, the bacteria also produce an uptake hydrogenase (HUP) that is able to oxidize the H_2 and thereby recover the reducing power used in H_2 production. However, many N_2 fixing legume nodules evolve H_2 due to the absence (HUP⁻) or low activity of the uptake hydrogenase (Arp 1992). In a HUP⁻ symbiosis, large amounts of H_2 can diffuse out of the nodule into the soil. For example, at peak growth every hectare of a N_2-fixing soybean field will produce about 5000 L H_2 d^{-1}. This hydrogen evolution represents an energy equivalent to about 5% of the crop's net photosynthetic C gain for that day (Dong, Layzell 2001). It is interesting to note that the majority (>75%) of the rhizobia strains isolated from major soybean production areas in United States are HUP⁻ (Uratsu *et al.* 1982). Also, all known clover and alfalfa symbioses are HUP⁻.

The existence of HUP has been considered to be beneficial since it makes it possible for the symbiosis to recover at least a portion of the energy used for H_2 production (Postgate 1998). Some laboratory studies have shown that HUP⁺ symbioses support a higher N content and enhanced legume growth (Albrecht *et al.* 1979; Bergersen *et al.* 1995), and many attempts have been made to introduce the HUP genes into endemic HUP⁻ strains (Uratsu *et al.* 1982; Arp 1992). However, the expected benefits from HUP⁺ were not always apparent, especially in field studies and some studies showed negative effects of HUP on yield (Arp 1992).

2. The Fate of H_2 Released from Nodules

Soil is a major sink for the H_2 produced by legume nodules (Conrad *et al.* 1980). In fact, most soils are able to remove H_2 from the atmosphere where it exists at a level of only 0.55 ppm H_2. However, if a nodulated root system of a HUP⁻ symbiosis is transplanted into non-legume soil, the H_2 produced by the nodules can be measured readily as a net evolution from the soil surface (Layzell, Atkins, Smith, Zhang, unpublished). In contrast, no H_2 evolution can be detected from the surface of a soil that has supported the growth of a HUP⁻ legume symbiosis.

La Favre *et al.* (1983) showed that H_2 production from legume nodules induced H_2 oxidation capacity of the soil, and that this capacity and the number of H_2 oxidizing bacteria decreased exponentially with distance from the nodule. Soil microorganisms within a few cm of the legume nodules rapidly oxidized the H_2. Popelier *et al.* (1985) found a highly significant positive correlation between the microbial biomass of the soil and the soil H_2 uptake rate. Despite numerous attempts (Conrad *et al.* 1979a, 1979b, 1983; Haring, Conrad 1994; Haring *et al.* 1994; Kluber *et al.* 1995; Lechner, Conrad 1997), the microorganisms responsible for H_2 oxidation in soils have yet to be identified and Conrad (1988) even questioned whether the H_2 oxidation was, in fact, biological or chemical.

The H_2 exposure rate of soil within a few cm of N_2-fixing nodules has been calculated to be 30-250 nmoles cm^{-3} h^{-1} (Dong, Layzell 2001). A similar H_2 exposure rate was used to treat a large volume of field soil (300 mL to 70 L) held within a plastic container. After 7 to 10 d of exposure, the H_2 uptake rate of soil increased rapidly. After 3 weeks of H_2 treatment, the apparent Km(H_2) of soil had increased from 40.2 to 1028 ppm H_2, and the Vmax of the treated soil increased from 4.35 to 836 nmoles H_2 cm^{-3} h^{-1} (Dong, Layzell 2001). Similar results have also been reported in field soils adjacent to H_2 evolving legume nodules (Conrad, Seiler 1979). These data suggest that the H_2 treatment enhances the growth and hydrogenase activity of certain soil microorganisms.

The reducing power from H_2 could be used in a number of biological reactions in soil, including ATP formation, lipid production, or the reduction of CO_2 or N_2. To assess the fate of the reducing power from H_2, the exchange rates of CO_2, O_2 and H_2, were measured simultaneously as a pretreated soil was provided with a range of H_2 exposure rates (Dong, Layzell 2001). When the soil received an H_2 exposure rate equivalent to that used in its pretreatment (147 nmoles H_2 cm^{-3} h^{-1}), 60% of the electrons from H_2 were passed to O_2 and 40% were used in CO_2 fixation. At H_2 exposure rates greater than this, net CO_2 fixation was measured. This observation was consistent with higher C content following cultivation of HUP^- legumes (Popelier *et al.* 1985), and in alfalfa-based rotation fields than the monoculture maize fields (Gregorich *et al.* 2001).

The additional microbial activity in the H_2 treated soil was frequently associated with a greatly increased population of springtails, small insects of the order Collembola that feed on soil bacteria and fungi. Also, in field studies in which H_2 was provided to soils, an increase in the nematode and earthworm populations were observed (Dong, Zhang, Layzell, unpublished).

3. Plant Growth Promotion in H_2-Treated Soils

Since evolution and crop breeding programs have not favored HUP^+ symbioses over HUP^- symbioses, it may be possible that H_2 evolution by nodules is beneficial to the growth and yield of the symbiosis.

To test this hypothesis, plant growth was compared in soils that were previously exposed to air or to H_2 in air at an H_2 exposure rate that increased from 63 to 250 μmoles L^{-1} soil h^{-1} over a 9 week period. This H_2 exposure rate was similar to that experienced by soils adjacent to legume nodules during plant growth (Dong, Layzell 2001). After 43 d, soybean plants grown in H_2 treated soil had 14% more dry weight than plants grown in untreated soil (Figure 1) (Wu, Dong, Layzell, unpublished).

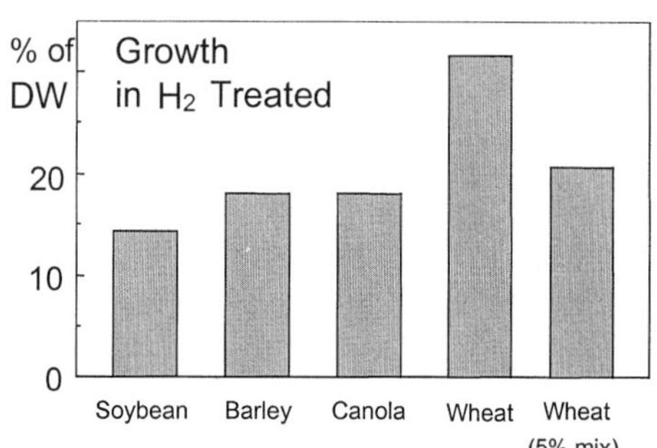

Figure 1. The effect of soil pretreatment with H_2 on the growth of soybean, barley and canola relative to growth in untreated soils. In spring wheat, a growth response was seen in soils in which 100% or 5% of the total soil volume was treated with H_2.

A plant growth response was also observed in non-legumes. For example, 44 d old barley plants grown on H_2 treated soils were 18% larger than plants grown on air treated soil (Figure 1). In similar studies with 35 d old canola, the H_2 treated soil stimulated growth by 18% compared to untreated soil (Figure 1) (Wu, Dong, Layzell, unpublished). In spring wheat, soil pretreated with H_2 for 30 d was found to enhance plant biomass of 29 d old plants by 32% compared with plants grown in air treated soil (Figure 1) (Dong, Layzell, unpublished). These results show that H_2 treatment of soil enhances fertility and promotes plant growth, not only for legumes, but also for various non-leguminous crops that may grow concurrent with or subsequent to the time of H_2 fertilization.

In spring wheat, growth promotion can also be achieved in soils in which only 5% of the soil volume was previously treated with H_2; the balance of the soil being untreated (Figure 1) (Dong Layzell, unpublished). Similarly, the plant growth promoting activity of H_2 treated soils has been shown to be extractable since aqueous extracts of the soil (3 mL per seed) enhanced the growth of 25 d old barley plants by 32% when compared with seeds watered with extracts of air treated soil (Willms, Layzell, unpublished).

To assess whether the plant growth promoting activity of H_2 treated soil was due to bacteria or fungi, spring wheat was grown in a soil:promix (1:19) mixture in the presence and absence of antibiotics (penicillin and streptomycin) and a fungicide (benomyl). The antibiotic treatment eliminated the growth response observed in the H_2 treated soils when compared with the untreated control soils (Figure 2) (Dong, McLearn, unpublished).

These studies illustrate that exposure to H_2 at levels similar to that which occurs next to legume nodules greatly enhances the ability of that soil to support the growth of both legumes and non-legumes. The growth promotion activity seems to be associated with the growth and hydrogen oxidation capacity of soil bacteria. Moreover, the activity is extractable and is present even when the soil is diluted to only 5% of the total soil volume.

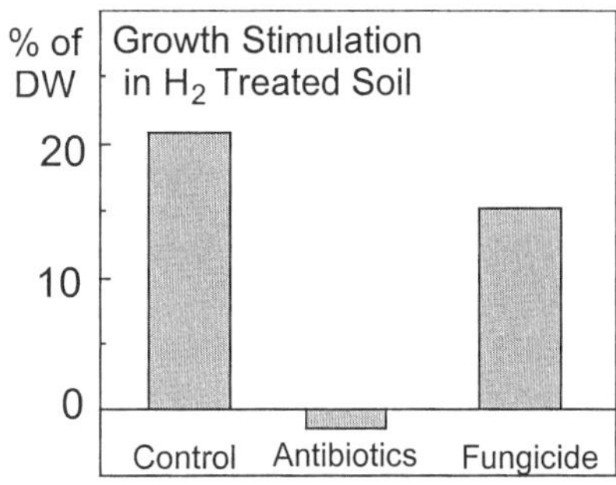

Figure 2. The effect of antibiotics (penicillin and streptomycin at 50 mg L^{-1}) and fungicides (benomyl at 36 mg L^{-1}) on the growth stimulation of spring wheat in H_2 treated soils. The growth response was expressed as a % increase in the biomass of 35 d old plants grown in H_2 treated soils compared to untreated soils.

4. Preliminary Field Trials

To assess whether H_2 fertilization of soil enhances the growth and yield of field-grown plants, trials were set up in Minto, Manitoba (MB) and Truro, Nova Scotia (NS) in 2000. Soils pretreated with H_2 were added with seeds at a rate of about 1.65 L m^{-2} in NS and 6.4 L m^{-2} in MB.

In seven-week-old barley and spring wheat plants in MB, tiller number per plant increased by 47% and 27%, respectively, compared with plants in the air-treated or control soils (Figure 3) (Kettlewell, Layzell, unpublished). Similarly, in eight-week-old barley in NS the number of heads was 48% higher when the seeds were sowed in a bed of H_2 treated soil compared with plants that were germinated with air treated soil (data not shown; Dong, Caldwell, unpublished).

The ultimate effect of the H_2 treated soil on crop yield was less than that observed on tiller or head number. Yield increases of 3% to 16% were observed relative to the air treated soil, but not all data were significantly different at the 95% confidence level (data not shown). The lower yield response than early growth response was attributed to the high seeding densities that were used in this study. Clearly, more field studies are needed.

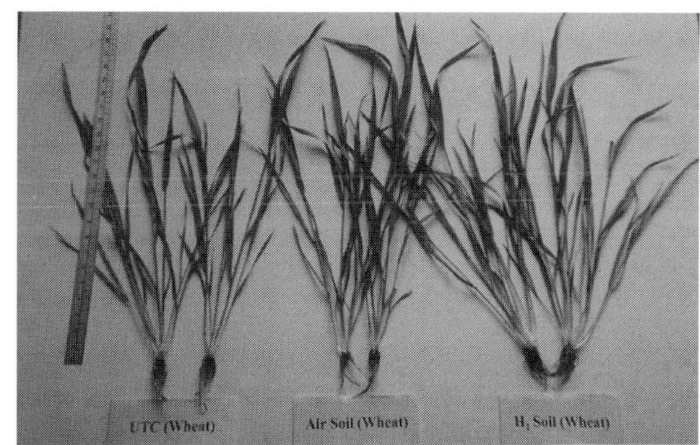

Figure 3. The effect of H_2 treated soil on the growth and tiller number in spring wheat grown in the field in Minto, MB. Field soils were pretreated with air or H_2 (peak exposure rate of 200 µmoles L^{-1} soil h^{-1}) for 5 weeks before being placed in a furrow (6.4 L m^{-2}) with wheat seeds. Note the larger number of tillers in these 7-week-old plants in H_2 soil compared to soils exposed to air and the untreated control (UTC) soils (Kettlewell, Dong, Layzell, unpublished).

5. Is Soil H₂ Fertilization Part of the Benefit of Crop Rotation?

For more than 2000 years, legumes have been used in crop rotations to enhance the growth and yield of the subsequent cereal crops. While legumes can leave residual N in the soil for the subsequent crop, this cannot account for all the rotation benefit (Bolton *et al.* 1976; Hesterman *et al.* 1986; Fyson, Oaks 1990). Consequently, a variety of mechanisms have been proposed, including legume effects on altering soil structure or nutrient balance, breaking the disease cycle or opening channels in soil for deeper rooting.

The data described in this report indicated that soil H_2 fertilization may play a significant role in contributing to the benefit of legume crops in rotation with cereals. Although the mechanism by which H_2 fertilization of soils enhances plant growth has yet to be proven, the most likely explanation involves the enhanced growth of H_2-oxidizing microorganisms in the soil. These organisms may improve the nutrient status of soil or act as plant growth promoting rhizobacteria (PGPR), enhancing the plants disease resistance or growth regulator balance (Hart *et al.* 1986; Williams, Sparling 1988; Insam *et al.* 1991; Srivastava, Singh 1991; Bankole, Adebanjo 1996; Omar, AbdAlla 1998). These bacteria population changes in the soils adjacent to H_2 releasing nodules may be associated with the enhanced growth response of plants rotated with legumes. If so, H_2 fertilization of soils could achieve some of the beneficial effects of crop rotation, without the need to implement actual crop rotation.

This study would also help to account for the evolutionary questions surrounding why HUP⁻ symbioses have thrived when there are genes (in many cases within the same genus and species) for the more energetically efficient HUP⁺ symbioses. Perhaps the plant growth advantages of the HUP⁻ symbioses offset the greater energy efficiency of the HUP⁺ symbioses.

6. References

Albrecht SL *et al.* (1979) Science 203, 1255-1257

Arp DJ (1992) In Stacey RH *et al.* (eds), Biological Nitrogen Fixation, pp. 432-460, Chapman and Hall, New York

Bankole SA, Adebanjo A (1996) Crop Protection 15, 633-636

Bergersen FJ *et al.* (1995) Soil Biol. Biochem. 27, 611-616

Bolton EF *et al.* (1976) Can. J. Soil Sci. 56, 21-25

Conrad R (1988) Adv. Microb. Ecol. 10, 231-384

Conrad R and Seiler W (1979) Soil Biol. Biochem. 11, 689-690

Conrad R *et al.* (1979a) Soil Biol. Biochem. 11, 689-690

Conrad R *et al.* (1979b) FEMS Microbiol. Lett. 6, 143-145

Conrad R *et al.* (1980) Geophys. Res. 85, 5493-5498

Conrad R *et al.* (1983) FEMS Microbiol. Lett. 18, 207-210

Dong Z, Layzell DB (2001) Plant and Soil 229, 1-12

Fyson A, Oaks A (1990) Plant and Soil 122, 259-266

Gregorich EG *et al.* (2001) Can. J. Soil Sci. 81, 21-31

Haring V, Conrad R (1994) Biol. Fertil. Soils 17, 125-128

Haring V *et al.* (1994) Biol. Fertil. Soils 18, 109-114

Hart PBS *et al.* (1986) NZ J. Agric. Res. 29, 681-686

Hesterman OB *et al.* (1986) Agron. J. 78, 19-23

Insam H *et al.* (1991) Soil Biol. Biochem. 23, 459-464

Kluber HD *et al.* (1995) FEMS Microbiol. Ecol. 16, 167-176

La Favre JS *et al.* (1983) Appl. Environ. Microbiol. 46, 304-311

Lechner S, Conrad R (1997) FEMS Microbiol. Ecol. 22, 193-206

Omar SA, AbdAlla MH (1998) Folia Microbiol. 43, 431-437

Popelier F *et al.* (1985) Plant and Soil 85, 85-96
Postgate J (1998) Nitrogen Fixation, Cambridge University Press, Cambridge
Srivastava SC, Singh JS (1991) Soil Biol. Biochem. 23, 117-124
Uratsu SL *et al.* (1982) Crop Sci. 22, 600-602
Williams BL, Sparling GP (1988) Soil Biol. Biochem. 20, 579-581

7. Acknowledgements

We appreciate the valuable technical assistance and expertise of Lishu Wu, Claude Caldwell, Feng Zhang, Bubby Kettlewell, Nicole McLearn, Patricia Irvine and Melissa Smith. Don Smith and Prof. Craig Atkins are thanked for useful discussions and insights.

Section 15:
Applied Aspects of Nitrogen Fixation

CHAIR'S COMMENTS: BIOLOGICAL NITROGEN FIXATION AND SUSTAINABLE AGRICULTURE

G. Hernández

CIFN/UNAM, Res. Program Plant Mol. Biol., Cuernavaca, Morelos, México

Nitrogen is the major limiting nutrient for most crop species. The acquisition and assimilation of nitrogen for plant growth and development is second in importance only to photosynthesis (Vance 1997). Biological nitrogen fixation constitutes the main natural input of nitrogen into the biosphere. This represents around 50% of the total fixed nitrogen, the other half is fixed through chemical processes. The chemical reduction of nitrogen for fertilizers production, mainly via the Haber-Bosch process, has been a fundamental process for mankind's development since it has made it possible to feed around 40% of the total world population. The striking rise in cereal yields in developing countries during the last half of the century is directly attributable to a ten-fold increase in nitrogen fertilizer use. The "Green Revolution" favored the selection of crops that respond favorably to chemical fertilizers. The great increase in the production and use of chemical fertilizers for agricultural production during the last decades has resulted in serious ecological alterations, such as: the volatilization of nitrogen oxides to the atmosphere that causes the depletion of ozone, the depletion of non-renewable resources, an imbalance in the global nitrogen cycle and leaching of nitrate into groundwater (Kinzing, Socolow 1994).

Sustainable agriculture is broadly defined as agriculture that is managed towards greater resources efficiency and conservation while maintaining an environment favorable for the evolution of all species. One of the driving forces behind agricultural sustainability is the effective management of nitrogen in the environment. Successful manipulation of nitrogen inputs through the use of biological fixed nitrogen result in farming practices that are economically viable and environmentally prudent.

The primary source of biologically fixed nitrogen for agricultural system is through the *Rhizobium* (and related genera) – legume symbiosis (Vance 1997). The amount of nitrogen fixed by legumes is quite amazing. Legumes provide 25–35% of the worldwide protein intake. The agronomic use of symbiotic nitrogen fixers used as inoculants or "biofertilizers" is a good alternative to chemical fertilization. An important goal for sustainable agriculture, which will result in humanitarian and economic benefits, is to enhance the use and to improve the yield of biologically fixed nitrogen by legumes. Environmental and management limitations to legume growth are the major factors regulating nitrogen fixation, although practices that either limit the presence of effective rhizobia in the soil or enhance soil nitrate can also be critical (Peoples *et al.* this volume).

In the Nitrogen Fixation Research Center (CIFN/UNAM), in Cuernavaca, Mexico, a global project on the research on biological nitrogen fixation for sustainable agriculture has been carried out for several years. An important specific project on this subject includes the production and evaluation of *Rhizobium* biofertilizers for beans. Common bean (*Phaseolus vulgaris*) is the second most important crop in Mexico; it constitutes the main protein source for Mexicans' diet. Dr Jaime Mora, is the scientist from CIFN/UNAM responsible for the *Rhizobium* biofertilizers project, which is being done in collaboration with INIFAP, the agricultural research institute from the Agriculture Ministry from the Mexican government. Native *R. etli* or *R. tropici* strains isolated from different regions of Mexico and Central America, as well as genetically engineered strains improved for symbiotic nitrogen fixation, have been tested as bean biofertilizers in different experimental fields. Also, different agricultural technologies for watering and adding the biofertilizer, such as the dripping technology, were tested. The best results obtained in the field trials gave around 80% crop yield using biofertilizer as compared to the yield obtained following the addition of chemical fertilizer. Besides the ecological benefits, the latter may represent around a 10-fold saving in agricultural costs to Mexican farmers.

The group of Mariangela Hungria and Diva de S. Andrade in Londrina, Brazil have studied the role of biological nitrogen fixation in the two most important legume crops: bean and soybean (*Glycine max*). They are characterizing the biodiversity of indigenous rhizobial populations and their effect in inoculation with introduced improved strains. In both crops selected strains usually increase grain yield. There are other limiting factors from improving crop yield such as soil conditions (temperature, moisture, acidity).

Mpepereki *et al.* (this volume) emphasize the benefits that biological nitrogen fixation can bring to the cultivation of soybean in poor and marginalized communities of Sub-Saharan Africa. They are developing a research-extension model for promoting biological nitrogen fixation among peasant farmers of Zimbabwe. Potential to improve food security and alleviate poverty among the rural poor is tremendous.

Biological nitrogen fixation for cropping systems is also important for industrialized countries. Martin H. Entz is studying this issue for the prairie provinces of Canada. Integrated agricultural systems that include both ruminant livestock and crop production, arguably provide the best opportunities for capturing biological nitrogen fixation benefits in food production. The advantages of integrated food production systems as compared to monocultures, due to the role of biological nitrogen, are being analyzed.

Another important aspect of the research of biological nitrogen fixation towards sustainable agriculture includes associative nitrogen fixers. These microorganisms, including genera such as *Azospirillum, Herbaspirillum,* and *Acetobacter*, may associate as endophytes or may colonize the rhizosphere of important cereal crops and may be advantageous for crop production. Crop growth promotion by associative microorganisms is not always provided through biological nitrogen fixation. It is well documented that the growth promotion of maize by *Azopirillum,* a root colonizer, is due to the excretion of auxins and other phytohormones that promote root growth and allow a better capacity for absortion of nutrients from the soil. At CIFN in Mexico, a project is being carried out with Dr Jesus Caballero-Mellado as responsible in collaboration with INIFAP, with the aim of developing, producing and distributing biofertilizers, based in *Azospirillum*, for cereals crops such as maize, wheat, sorghum. *Azospirillum* biofertilizer was used in around 2 million ha in crop fields from different states of Mexico during 1999 and 2000. An average increase of 26% in the production of basic grains was obtained in 75% of the fields inoculated with the biofertilizer. This also represents great saving for farmers.

References

Kinzing AP, Socolow RH (1994) Physics Today 47, 24-35
Vance CP (1997) In Legocki A, Bothe H, Pühler A (eds) Biological Fixation of Nitrogen for
    Ecology and Sustainable Agriculture, pp. 179-186, Springer-Verlag, Berlin, Germany

MAXIMIZING THE CONTRIBUTION OF BIOLOGICAL NITROGEN FIXATION IN TROPICAL LEGUME CROPS

D.S. Andrade[1], M. Hungria[2]

[1]IAPAR, Cx. Postal 481, 86001-970, Londrina, PR, Brazil
[2]Embrapa Soja, Cx. Postal 231, 86001-970, Londrina, PR, Brazil

1. Introduction

Soybean (*Glycine max* L.) and common bean (*Phaseolus vulgaris* L.) are the main legume crops grown in Brazil and in some South America countries. In Brazil, 32 million soybean grains are produced in 12.8 million hectares with average yield of 2500 kg ha^{-1}. Farmers desiring high profits employ high technology and large land areas destined for crop exportation. A similar situation is verified in neighboring countries like Argentina and Paraguay. Unlike soybean, common beans are mainly cropped for food by smallholders with a low level of technology resulting in average yield of only 700 kg ha^{-1} in 4.5 million hectares. Biological nitrogen fixation (BNF) plays an important role in the successful management of both crops.

2. Rhizobial Soil Population and Diversity

Soybean was introduced in Brazil 120 years ago and several experiments have shown that uncropped soils are void of bradyrhizobia able to establish an effective symbiosis with this legume. The crop expanded in the 1960s and has been intensively inoculated since then, so that today most soils where this legume is grown show a very high population of soybean bradyrhizobia, estimated in 103 to 106 cells g^{-1} of soil. In areas cropped to soybean for the first time, the few nodules formed were identified as *Bradyrhizobium japonicum* and *B. elkanii* strains used in commercial inoculants and dispersed from other cropped fields, as well as some fast-growing indigenous rhizobia (Ferreira *et al.* 2001). Sequencing of the 16S rRNA genes of those fast-growing strains has detected similarity with *Rhizobium tropici* and *Rhizobium* genomic species Q (unpublished data). Furthermore, several strains resembling agrobacteria that effectively nodulate soybeans were also isolated from soybean nodules in both Brazil and Paraguay (Figure 1, Chen *et al.* 2000). However, although many Brazilian soybean cultivars are effectively nodulated by those fast-growing strains, and those bacteria are usually found in high number in soils, they compete poorly with *B. japonicum* and *B. elkanii* (Hungria *et al.* 2001).

Soybean seeds, even when harvested in an area with a very high population of naturalized strains, usually carry very few viable cells. As an example, in 28 field experiments performed from 1996 to 2001 in areas cropped with soybean for the first time, nodule number varied from 0 to 3 nodules per plant, with an average of 0.15 (unpublished data).

A different situation is found with the bean crop, since almost all soils, even when they have never been cropped before with this legume, show a very high population of indigenous rhizobia, estimated in 103 to 106 cells g^{-1} soil depending on crop and soil management practices (Andrade 1999; unpublished data). Furthermore, a high level of rhizobial diversity is also found in soil. For example, in a survey of soils from seven Brazilian states, 38 different RFLP-PCR profiles were detected. When 207 strains from two of those States (Pernambuco and Paraná) were characterized, Pernambuco, with alkaline soils and semi-arid climate, and Paraná, with acid soils and tropical and subtropical climates, a very high number (90% of the isolates) of unique strains was shown, as revealed by the BOX-PCR analysis (Grange 2001). There was no effect of either the bean cultivars (black or colored seeds) used as trap plants, or of the ecosystem on the rhizobial diversity. The sequencing of the 16S rRNA genes of some of those strains has shown that bean plants had the capacity to trap several rhizobial species, as shown in Figure 2 (Grange 2001).

Another difference from soybean is that common bean seeds usually carry many rhizobial cells, probably due to the harvesting method. However, this is a questionable hypothesis since little is known of survival ability of rhizobia on seeds. For example, when non-sterilized seeds from thirty different sites in Brazil (Paraná and Minas Gerais) were used, independent of whether they came from smallholders or if they were certified seeds produced with high technology, nodules formed in 37% of the plants and the number per plant varied from 2 up to 45 (unpublished data).

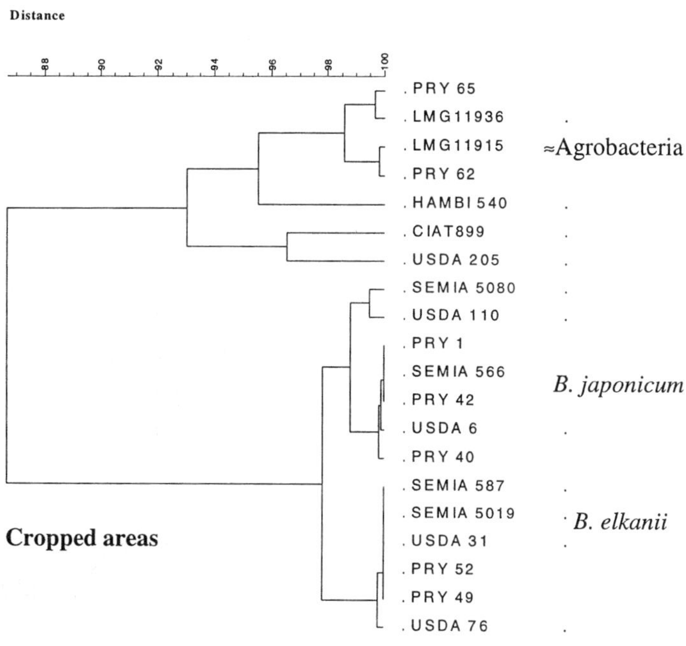

Figure 1. Dendrogram built with the UPGMA algorithm with the aligned 16S rRNA partial sequences of isolates from field-grown soybean nodules in Paraguay (PRY), from areas cropped for several years with this legume and of fourteen reference strains belonging to three genera. See Chen *et al.* (2000).

3. Selection of Efficient and Competitive Strains and Seed Inoculation

As the history and the diversity of rhizobia in the soil are quite different for both crops, the approaches of the strain selection programs have to be different. For the soybean crop, a continuous selection program is mandatory, since bacteria have to meet the increased N demand of more productive cultivars. The main approach consists of re-isolating adapted strains from areas, which have been previously inoculated, looking for variant genotypes with higher competitiveness and BNF capacity (Santos *et al.* 1999; Hungria, Vargas 2000). New strains are annually tested in national field trials, and recommendation for commercial inoculants can be changed if a more efficient and competitive strain is identified. Currently reinoculation guarantees a mean increase of 4.5% in grain yield, but can reach values as high as 25%. Therefore inoculation is practiced by 58% of the farmers, with 12.5 million doses (55% peat based and 45% liquid) of inoculants sold in the country last year, representing 99% of the Brazilian market.

A different approach is used for the common bean crop and the strain selection is based on the search for efficient and competitive strains within the diverse indigenous population. A first premise of the Brazilian program is that strains have to belong to the species *Rhizobium tropici*, which shows higher genetic stability. A successful example was the strain PRF 81 (=SEMIA 4080), isolated from a soil of Paraná State that consistently increased yield by up to 900 kg ha^{-1}, reaching yields of 3000 to 4000 kg ha^{-1} (Hungria *et al.* 2000). The sequencing of 16S rRNA genes has shown a high similarity of PRF 81 with *Rhizobium* genomic species Q strain BDV5102 isolated from *Daviesa leptophylla* in Australia (Lafay, Burdon 1998).

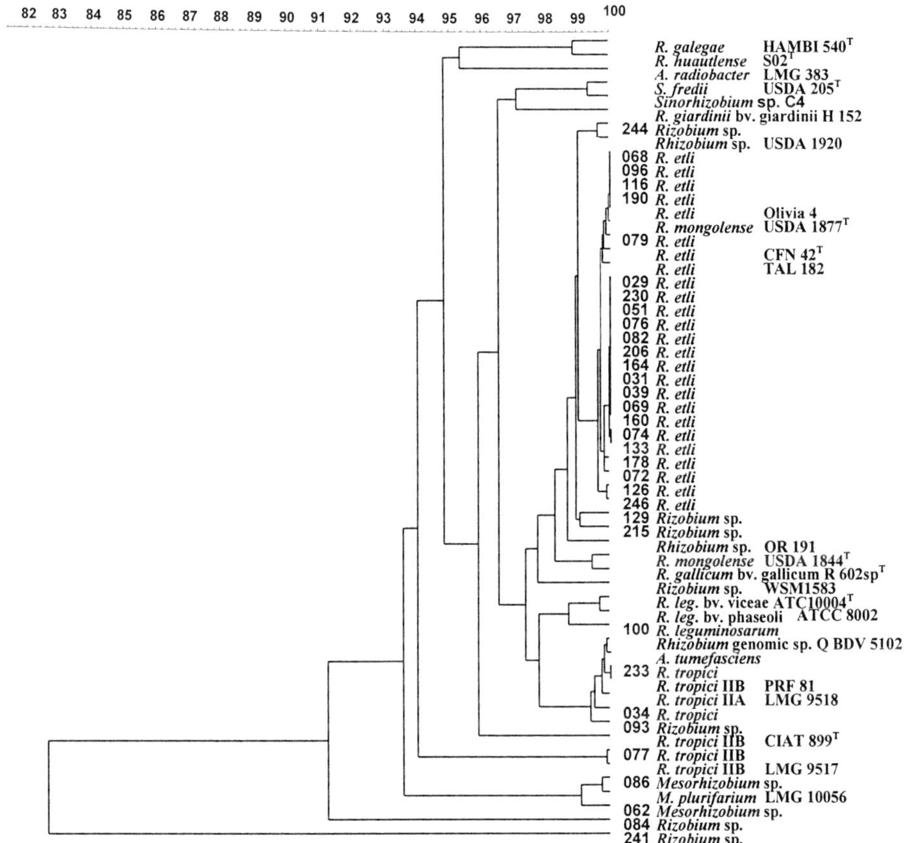

Figure 2. Dendrogram built with the UPGMA algorithm with the aligned 16S rRNA partial sequences of Brazilian common bean isolates and of several type and reference strains. After Grange (2001).

Response to inoculation also depends on soil/crop management practices. In soybean, the zero-tillage management system reduces soil temperature and increases soil moisture resulting in a higher number of bradyrhizobial cells in the soil, higher rates of BNF and higher grain yield (Hungria, Stacey 1997; Ferreira *et al.* 2000; Hungria, Vargas 2000). Rhizobial number of diversity is also affected by soil management, and when samples from eight different sites in Brazil were analyzed, using soybean promiscuous primitive cultivars as trap plants, the following rhizobia were identified: *R. tropici*, *R. huautlense*, *Rhizobium* genomic species Q, rhizobia resembling agrobacteria, *Rhizobium* OR191, *B. japonicum* and *B. elkanii*. In soils under no-tillage, species present included *B. japonicum*, *B. elkanii* and *R. tropici*, while under conventional tillage the only species detected was *R. tropici* (unpublished data). These results indicate that even when management techniques considered appropriate for the tropics are used, rhizobial diversity can be drastically reduced.

For the bean crop, PCR-RFLP analysis of the 16S-23S rRNA intergenic spacer (IGS) and the 16S rRNA gene, indicated that the rhizobial populations from bean nodules cultivated in an unlimed acidic oxisol were less diverse than those from the limed soil (Andrade 1999). The 16S rRNA gene nucleotide sequences of isolates presented similarity values ranging from 97 to 100% with *R. etli*, *R. gallicum*, *R. tropici*, *R. mongolense*, *R. leguminosarum*, *Sinorhizobium meliloti*, *Agrobacterium rhizogenes* and *A. tumefaciens*. However, isolates affiliated to *R. tropici* IIB and to *R. leguminosarum* bv. *phaseoli* were predominant independent of lime application. The richness index (number of IGS groups) increased from 2.2 to 5.7 along the soil-liming gradient and, based on species, showed an increase from 0.5 to 1.4. The Shannon index (species diversity) ranged from 0.9 in unlimed soil to 1.4 in limed soil, and based on the number of IGS groups, the diversity increased from 1.8 to 2.8 (Andrade 1999).

4. Main Limiting Factors of BNF in Tropics

Although inoculation with efficient and competitive strains can increase grain yield of soybean, common bean and other legumes in Brazil some limiting factors to the biological process are often reported under field conditions.

High temperature and low soil moisture are major causes of nodulation failure, affecting all stages of the symbiosis and limiting rhizobial growth and survival in soil. They may also contribute to undesirable changes in rhizobia, including plasmid deletions, genomic rearrangements and reduced diversity. Practices such as use of cover crops and no-tillage management decrease soil temperature and increase soil moisture, benefiting BNF, as discussed earlier.

Seed treatment with pesticides is used by most farmers and can drastically reduce nodulation and BNF. In Brazil, when soybean seeds were treated with ten fungicides recommended for the crop, decreases in nodulation varied from 8 to 88%, affecting BNF and yield (Campo, Hungria 1999). Soil acidity affects several steps in the development of the symbiosis, mainly nodulation.

Nutrients such as phosphorus and molybdenum often limit BNF, mainly under acidic conditions. For example, in an experiment performed in the State of Paraná, Brazil, supplying additional molybdenum increased soybean yield from about 3100 to 3400 kg ha^{-1} (Campo *et al.* 1999).

5. Conclusions

Biological nitrogen fixation can supply the N demand of important crops grown in South America, such as soybean and common bean. Soybean is an exotic plant in Brazil and therefore soybean bradyrhizobia diversity is restricted to changes caused by adaptation of inoculant strains to the soil. Some fast-growing indigenous strains are also able to nodulate soybean and were identified mainly as *R. tropici*, but are poor competitors against *Bradyrhizobium*. On the other hand, a high level of diversity is found among common bean rhizobia. Selection programs have been conducted to obtain more efficient and competitive strains for both crops and inoculation usually results in grain yield increases. Biological nitrogen fixation is usually limited by environmental factors, as high temperature and low soil moisture, soil fertility problems (such as acidity and deficiency of P and Mo) and agriculture practices (e.g. seed treatment with fungicides, tillage practices).

6. References

Andrade DS (1999) Ph.D Thesis, Wye College, University of London, Wye, Kent, UK
Chen LS *et al.* (2000) Appl. Environ. Microbiol. 66 (11), 5099-5103
Campo RJ, Hungria M (1999) Pesquisa em andamento 21, Embrapa Soja, Londrina, p. 1-7
Campo RJ *et al.* (1999) Pesquisa em Andamento 19, Embrapa Soja, Londrina, p. 1-7
Ferreira MC *et al.* (2000) Soil Biol. Biochem. 32, 627-637
Ferreira MC *et al.* (2001) Plant Soil
Grange L (2001) M.Sc. Thesis, Londrina, UEL
Hungria M, Stacey G (1997) Soil Biol. Biochem. 29 (5/6), 819-830
Hungria M, Vargas MT (2000) Field Crops Res. 65, 151-164
Hungria M *et al.* (2000) Soil Biol. Biochem. 32 (11-12), 1515-1528
Hungria M *et al.* (2001) Biol. Fert. Soils
Lafay B, Burdon JJ (1998) Appl. Environ. Microbiol. 64, 3989-3997
Santos MA *et al.* (1999) FEMS Microbiol. Ecol. 30, p. 261-272

7. Acknowledgements

We wish to thank CNPq (PRONEX 41.96.0884.00), INCO-EC (ERBIC18CT980321) and CNPq (520396/96-0), for financial support and Dr George G. Brown for his assistance with the English text.

SOYBEAN N₂ FIXATION AND FOOD SECURITY FOR SMALLHOLDER FARMERS: A RESEARCH-EXTENSION MODEL FOR SUB-SAHARAN AFRICA

S. Mpepereki, F. Makonese, K.E. Giller

Dept of Soil Science, Univ. of Zimbabwe, P.O.B. MP 167, Mt Pleasant, Harare, Zimbabwe

1. Introduction

Nitrogen remains the single most limiting nutrient for crop growth in most developing countries. Exploitation of biological N fixation offers a unique opportunity to harness "free" fertilizer from a relatively low-cost technology. Soybean has been shown to respond well to inoculation with appropriate strains of rhizobia and to fix large quantities of N under field conditions. In parts of Africa including Nigeria, Malawi, Zambia and Zimbabwe, promiscuous soybean has been successfully grown without inoculants demonstrating its potential as a vehicle for conveying the benefits of BNF to poor and marginalized communities (Mpepereki *et al.* 2000). Legumes however still constitute only a small proportion of planted crops, receive few or no inputs and they are often considered as minor women's crops (Svubure 2000). Soybean, despite its multiple benefits, is even less well known in most parts of sub-Saharan Africa.

A working group of the African Association for Biological Nitrogen Fixation (AABNF), meeting in Accra Ghana in February 2001, proposed a new 21st Century Paradigm for achieving impacts through BNF research that said "research in biological N fixation must be nested into larger understandings of system N dynamics and land management goals before the comparative benefits of N₂ fixation may be realistically appraised and understood by society-as-a-whole". The group identified promiscuous soybean as the best legume to provide BNF benefits to the greatest number of rural communities in terms of improving soil fertility, protein nutrition and household incomes. Research must address smallholder farmers' concerns across the diverse agro-ecological zones of Africa and recognize that process research and application of molecular techniques are useful tools to work on recognized constraints within farming systems with the final goal being food security and improved nutrition for poor and excluded communities. Experiences with promiscuous soybean in Nigeria and Zimbabwe where research-extension initiatives have led to widespread adoption of BNF technologies by smallholder farmers could provide useful lessons.

This paper describes the conceptual models developed for research and extension of soybean BNF technologies to smallholder farmers in Zimbabwe and outlines the implementation of a research and extension initiative that has aimed to bring tangible socio-economic benefits to poor and marginalized smallholder farmers.

2. Soybean in Zimbabwe Smallholder Agriculture

Soybean has been grown in Zimbabwe's large-scale commercial farms since the 1930s but due to the colonial land tenure system the crop was never promoted among smallholders who were pushed to marginal areas with sandy soils and poor rainfall considered unsuitable for soybean. A general myth was that the crop was too sophisticated for the peasant farmers who also did not have cooling facilities or the know how to handle rhizobial inoculants. Protein malnutrition, general declining soil fertility and a poor resource base against a background of increasing mineral N fertilizer prices, however also ravaged smallholders practicing maize monoculture, following World Bank/IMF-induced removal of government subsidies. Cheaper alternative sources of N inputs were urgently required and soybean BNF was a natural choice because of its multiple benefits.

It was against this background that the smallholder soybean promotion program in Zimbabwe was initiated. A National Soybean Promotion Task Force representing both private and public institutions and coordinated by the University of Zimbabwe was set up. The objectives were to promote soybean BNF among smallholders and to carry out appropriate research to support its successful integration into smallholder farming systems. Conceptual models were drawn up to guide the overall soybean BNF promotion effort.

3. Conceptual Framework

In the African context, BNF must primarily contribute to the food security of rural communities and secondarily to environmental quality through reduced N inputs to the ecosystem and C sequestration in legume biomass. The conceptual framework for soybean BNF contributions to an integrated crop-livestock system is illustrated in Figure 1.

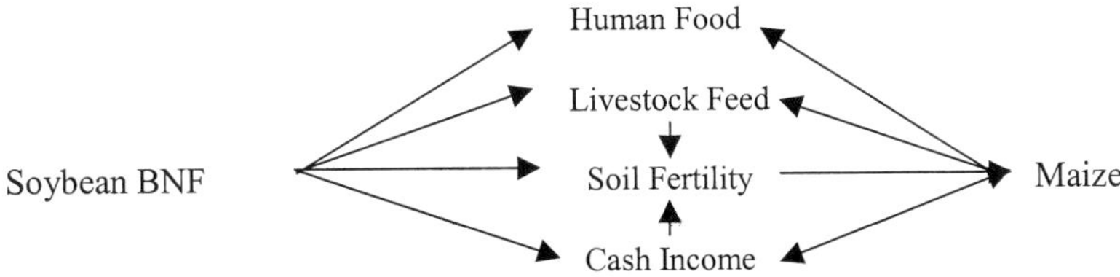

Figure 1. Conceptual framework for soybean BNF contribution to food security in a smallholder maize-based farming system. Soybean BNF generates biomass (grain, crop residues).

Each pathway presents opportunities and challenges for research and extension; constraints must be removed or reduced (research) and the benefits clearly demonstrated to the clients (promotion), the farmers. The conceptual model for the research agenda is shown in Figure 2 and identifies the research areas that were considered key to ensure that the products of soybean BNF could be translated into quantifiable benefits for adopting communities. Appropriate cultivars (genetics/breeding), compatible effective rhizobia (rhizobium ecology, inoculant technology), crop management (agronomy) and pest and disease control all contribute to higher yields which in turn directly boost food security through consumption of produce, processing and value addition and marketing to generate income.

4. Soybean BNF Research Program

The initial research phase involved a preliminary survey to establish the extent of soybean cultivation and utilization by smallholder farmers and the status of their knowledge of legume N fixation and its benefits. Indigenous rhizobial populations were estimated and their potential effectiveness on local soybean germplasm established in comparison with commercial inoculants. Isolates associating with promiscuous soybean varieties were characterized. Soybean biomass yields, amounts of N fixed and residual fertility effects on maize grown in rotation were quantified. The second phase of the research agenda is on-going and focused on improving soybean yields through improved field management of the legume-rhizobium symbiosis with emphasis on agronomic practices and exploiting promiscuous nodulation where commercial inoculants are not readily available. Pest and disease management and the integration of soybean into crop-livestock systems through utilization of grain and crop residues as feeds are also part of phase two studies. A parallel program is looking at the quality and nutritional value of soya-based foods processed by village women with a view to develop small to medium scale processing enterprises. To strengthen the socio-economic aspects, a research program has commenced to look at institutional constraints to the marketing of soybeans by smallholder farmers.

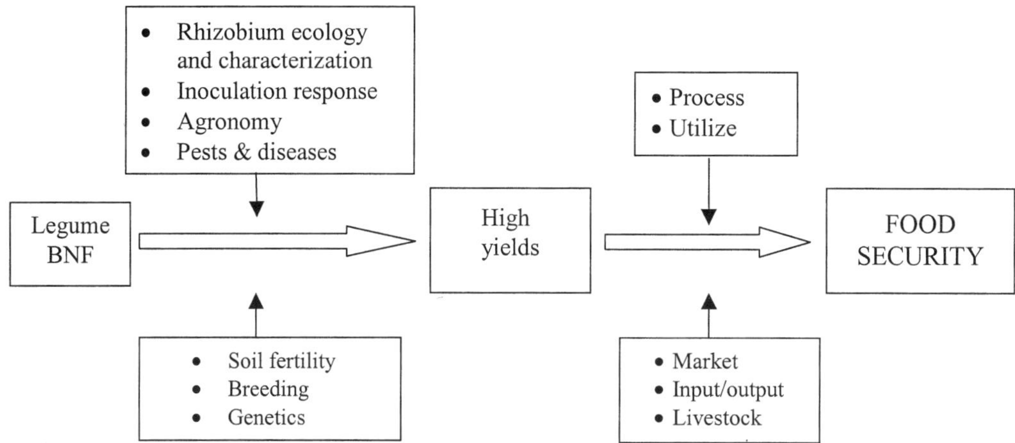

Figure 2. Conceptual framework for research activities for soybean BNF to impact on food security of smallholder farm communities.

5. Soybean BNF Promotion Program

To ensure that the socio-economic benefits of BNF reach the ordinary people, a phased promotion program was initiated from 1996 with a coordinating unit (CU) located at the University of Zimbabwe. The main strategy has been training farmers, extension agents and agro-dealers from both public and private sectors in practical BNF technology application including inoculant handling, seed dressing and agronomic aspects, soybean processing and use as food including removal of anti-nutrition factors and marketing of inputs and outputs through a train-the-trainer program. Selected community representatives are trained and sent to train others in their local areas. The Coordinating Unit (CU) works under the guidance of a National Soybean Promotion Task Force with representation from all stakeholder groups including several NGOs to ensure that information flows to all the players through training workshops, electronic (radio, television) and written (newsletters, brochures) media. The CU encourages linkages and facilitates stakeholders to play their part in ensuring that the benefits of soybean BNF reach as many farming communities as possible as illustrated in Figure 3.

Non-governmental organizations (NGOs) have played an important part in sourcing funds for BNF technology training for rural communities and local extension agents and micro-finance. The CU has set up strategically located farmer-managed and extension-supervised technology transfer demonstration plots to show proper application of BNF technology, soybean varietal performance and rotational effects with maize. After harvest farmers have been trained to grade their crop for seed and for the market and given tips on managing crop residues for soil fertility enhancement and as livestock feed. In all cases farmers undertake most of the tasks to ensure self-reliance and sustainability. Traders and agro-dealers have added soybean inputs such as seed, inoculants and fertilizers, and harvested grain to their list of trading items.

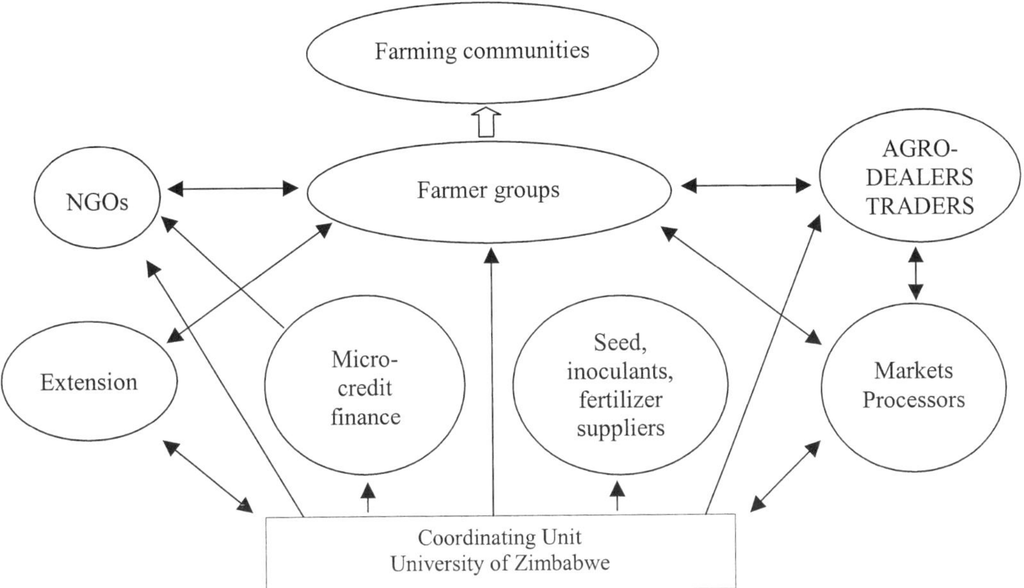

Figure 3. Coordination linkages of stakeholders in the soybean promotion program in Zimbabwe. Key activities in the links include training, information exchange, adaptive research and movement of inputs, outputs and cash.

6. Outcomes

In general the research-extension program has successfully introduced and brought benefits of soybean BNF to thousands of smallholder families, exploding the myth that the crop is too sophisticated for them to manage. The results of work to extend promiscuous soybean to smallholders in southern Africa and Nigeria have already been reviewed (Mpepereki *et al.* 2000). Promiscuous soybean has enabled farmers with no access to commercial inoculants also to adopt soybean. Up to 50% of soybean produced in Hurungwe district in northern Zimbabwe in the last three seasons (1998–2000) was promiscuous, while in Zambia promiscuous Magoye still forms the backbone of smallholder soybean production (Javaheri, 1996). Promiscuous soybean effectively nodulated with indigenous rhizobia, fixed 82-92% of their N with amounts ranging from 46–73 kg N/ha without inoculation and produced a larger biomass and contributed more N to the cropping system (Table 1).

Table 1. Maize yields for two seasons following soybean in a sandy loam soil in a smallholder farm, Hurungwe, Zimbabwe (1998/99).

Soybean variety (96/97)	Soybean biomass incorporated (t ha^{-1})	Maize yields (t ha^{-1})	
		97/98	98/99
Magoye (prom.)	5.4	2.3	1.2
Local (prom.)	4.9	2.1	1.4
Roan (spec.)	3.2	1.8	0.9
Nyala (spec.)	2.8	1.4	0.8
Maize control	Nil	0.19	0.2

Yields of maize after soybean were significantly higher than maize after maize, demonstrating significant residual fertility effects of soybean (Table 1). This is a positive contribution to sustainable food production and security as maize is the staple for many sub-Saharan communities. Residual fertility effects on maize have been consistently obtained under farmer management and boosted adoption of soybean BNF against a background of rising mineral N fertilizer prices and depreciating local currencies.

An important benefit of soybean BNF has been the boost in household incomes from grain sales by farmers, with volume sold from four districts rising from 65 tons in 1997 to over 800 tons in 1999. A critical element in the promotion program has been the consolidation of loads to achieve economies of scale that have enabled the relatively small production of each farmer to be sold on the lucrative commodity exchange as part of a large parcel. Produce marketing is a key element of the conceptual framework for promoting soybean.

A study of the economic potential of soybean showed that the crop was most profitable for the poorest farmers as it had lower input costs but gave the highest return on investment (Rusike *et al.* 2000). Poor farmers who adopted soybean for the first time between 1997 and 2001 have testified that they earned more money from soybean sales than from any other crop that they have ever grown (Table 2). The significant boost in family dietary protein availability (Table 2) is a critical element of household food security, a key benefit of BNF among poor rural communities.

Table 2. Grain, protein and cash returns from soybean for Tapera smallholder farm in Zimbabwe (1998).

Soybean variety	Total grain yield (kg ha^{-1})	Protein from 15% seed retained (kg ha^{-1})	Cash from 70% grain sold (US$ equiv.)
Magoye	2100	126	471
Local	1900	114	302
Roan	2800	168	496
Nyala	3100	186	560

Average smallholder planting: 0.4 ha; average yield: 0.8 t ha^{-1}; average price: US$360 t^{-1} (2001). Poor nutrition among the HIV-infected is contributing to the high death toll from AIDS. Significant savings from use of rhizobium inoculants also were reported in Zambia (Carr *et al.* 1998). Integration of soybean into local diets has driven adoption as soybean is processed to substitute several expensive grocery items that include milk and meat.

7. References

Carr *et al.* (1998) In Mpepereki, Makonese (eds), Harnessing Biological Nitrogen Fixation in African Agriculture, University Zimbabwe/CTA, Harare, Zimbabwe

Javaheri (1981) Mimeo Government of Zambia, Lusaka

Kasasa *et al.* (1998) In Waddington *et al.* (eds), Soil Fertility Research for Maize-based Farming Systems in Malawi and Zimbabwe, SoilFertNet/CIMMYT, Harare, Zimbabwe

Mpepereki *et al.* (2000) Field Crops Res. 65, 137-149

Svubure (2000) M.Phil. Thesis University of Zimbabwe, Harare, Zimbawe

8. Acknowledgements

We thank the Rockefeller Foundation for funding our recent BNF research and extension work.

THE IMPORTANCE OF BIOLOGICAL NITROGEN FIXATION IN CROPPING SYSTEMS OF INDUSTRIALIZED COUNTRIES

M.H. Entz

Dept of Plant Science, University of Manitoba, Winnipeg, MB, R3T 2N2 Canada

1. Introduction

The major source of biological N fixation (BNF) in crop production systems of industrialized countries is through inclusion of legume plants in cropping systems. This paper describes the historical context of why BNF with legumes is still important in modern agriculture, even though inorganic fertilizer appears quite available, inexpensive, and easy to use. The major legume systems that supply N to cropping systems are reviewed, and recent developments in enhancing BNF in modern cropping systems are discussed.

2. Historical Context

European agriculture is relatively new (<130 years) to many of the industrialized, grain-exporting countries such as the USA, Canada, Australia, Argentina and others. Grain production in these countries has relied, to a significant extent, on indigenous soil N supplies, and this "N mining" effect is well documented (e.g. Campbell *et al.* 1990). For example, in western Canada (Manitoba, Saskatchewan and Alberta), a deficit of approximately 25 million t of N was created between 1883 and 1990 through crop production (Morrison, Kraft 1994). Similar observations were made after cropping for less than 50 years in Australia (Grace *et al.* 1995). The need to "replace" some of this lost N has been recognized for some time, and it was discussed in 1924 at a special conference organized by the American Society of Agronomy entitled: "The legume problem". The goal of the conference was to identify opportunities for including legumes in USA dryland cropping systems. This goal is still important today, since rebuilding soil biological fertility with inorganic fertilizer N has been found to be uneconomical in many areas, especially those where water is limiting (Morrison, Kraft 1994).

A second important reason for emphasizing BNF in industrialized cropping systems is to reduce reliance on fossil-fuel energy, which is currently used in inorganic N production.

It must be recognized that while some crop production regions are starved for N (e.g. dryland regions of western North America and Australia), other agricultural regions often suffer from an excess amount of N. Therefore, the need for BNF in industrialized countries is not uniform across nations, or even within nations.

3. Plants for Biological N Fixation

3.1. Grain legumes. The main grain legumes in industrialized countries include soybean (*Glycine max* L.), pea (*Pisum sativa* L.), lentil (*Lens culinaris* Medikus), lupin and bean (*Phaseolus* spp.). The area dedicated to grain legume production has increased dramatically in the past 20 years. Some of the more dramatic examples include lupin production in western Australia (117,939 ha in 1980; 2,042,000 ha in 2000) and dry pea production in western Canada (49,300 ha in 1980; 1,219,000 ha in 2000) (FAOSTAT 2001). On the other hand, only modest gains in soybean and dry pea area were recorded for EU countries during this period, and soybean production in the USA has remained stable at approximately 29 million ha during the past 20 years (FAOSTAT 2001).

With the exception of *Phaseolus* spp., most of the aforementioned grain legume species can supply much of the N required for their own growth through BNF. Grain legumes also contribute N to following crops. In western Canada, it is estimated that 10 to 15 kg ha^{-1} of N are contributed for each 1000 kg ha^{-1} pea seed harvested (Wright, 1990). BNF by grain legumes is influenced by soil and edaphic factors including soil pH, moisture and indigenous soil N status. Higher BNF by field pea (31% increase) and lentil (10% increase) under no-till compared with tilled systems in western Canada was attributed to a soil environment more conducive to *Rhizobia* bacteria function (Matus *et al.* 1997). A positive effect of no-till on BNF of soybean was reported in Brazil by Andrade and Hungria (this volume).

3.2. Forage and pasture legumes. Major forage legumes in temperate regions include alfalfa (*Medicago sativa* L.) and clovers (e.g. Red clover, *Trifolium repens* L.). Most of these crops are harvested for hay, though some direct grazing by livestock also occurs (Entz *et al.* 2001a). Area under forage production is decreasing in some industrialized countries. For example, fodder maize (*Zea mays* L.) has displaced significant areas of alfalfa hay production in EU countries. In Argentina, a significant perennial pasture hectarage has been replaced with a grain legume (mainly soybean)/maize system (Panigatti 1992). In the North America Northern Great Plains, 7.8 and 3.8 million hectares are currently dedicated to perennial forage legumes for hay production and grazing, respectively (Entz *et al.* 2001a).

BNF by alfalfa and clover crops is well documented. In dry subhumid Manitoba, Canada, Kelner *et al.* (1997) established that net N additions of an alfalfa hay crop were 84, 148 and 137 kg ha^{-1} in the first, second and third years of the stand, respectively. Most of the N benefits of alfalfa to following grain crops are captured in the first two or three grain crops (Campbell *et al.* 1990), though significant additions have still been detected 13 years after alfalfa crop termination (Hoyt 1990).

Major forage legume species in humid areas of Uruguay and Argentina include red clover and *Lotus* spp. These species, often grown in combination with perennial forage grasses (e.g. *Festuca* spp. and *Lolium* spp.), are rotated with cereal and oilseed grain crops, often in an eight year rotation (half forage; half grain). In a long-term study at La Estanzuela, Uruguay, Rossello (1992) observed that four years of a legume grass mixture (grazed) increased total N in the 0 to 10 cm soil depth by 500 kg ha^{-1}. This organic N was found to be quite unstable after the forage stand was terminated. Rossello strongly suggests that a better understanding of N loss mechanisms (i.e. denitrification, leaching, selective soil erosion) is necessary. Agronomic practices that have been shown to increase grain crop utilization of the biologically-fixed N elsewhere [no-till forage termination systems (Mohr *et al.* 1999); elimination of post forage fallow periods (Campbell *et al.* 1994)] are now being implemented in Uruguay.

A unique cropping system, which has been used by farmers in southern Australia for over 40 years, involves annual, self-seeding *Medicago* ("*medic*") and *Trifolium* spp. grown in sequence with grain and oilseed crops (Grace *et al.* 1995). The single year medic and clover pasture plants have a high BNF potential and are capable of supplying most or all of the N for one or two following cereal grain crops (Grace *et al.* 1995). BNF of these medic and clover plants is sometimes limited by residual herbicides used in the grain phase of the rotation (B. Bellotti, personal communication).

Forage-based cropping systems provide benefits other than BNF, benefits that increase the overall sustainability of crop production. For example, deep roots of alfalfa can extract nitrate-N which may have leached below the rooting zone of annual grain crops (Campbell *et al.* 1994). Scientists at the University of Minnesota (J. Lamb; M. Russelle and co-workers) are currently developing and testing non-BNF alfalfa cultivars for deep nitrate extraction, though conventional (i.e. N-fixing) alfalfa cultivars are capable of significant subsoil nitrate extraction (Campbell *et al.* 1994). Perennial forage legumes are also used in salinization management (Entz *et al.* 2001a).

3.3. Green fallow and cover crop legumes. Fallow periods are often included in grain production systems in semi-arid zones in an effort to replenish soil water reserves and control weeds (Campbell *et al.* 1990). Green fallow refers to a practice where short-duration legume crops are grown during this fallow period. Suitable legume species and their soil water use characteristics have been described by Biederbeck and Bouman (1994). BNF by legumes during the green fallow phase can contribute a fertilizer N replacement value of up to 150 kg ha^{-1} (Badaruddin, Meyer 1989). A cultivar of Chickling vetch (*Lathyrus sativus* L.), 'AC Greenfix', was recently developed specifically for green fallow (Biederbeck, personal communication). AC Greenfix has high tolerance to indigenous inorganic soil N and has a high water use efficiency.

In wetter zones, where continuous grain cropping is feasible, cover crops offer an opportunity to add N to the soil system. These cover crops can be relay cropped with grain or vegetable crops, and are often used in orchards. Heat and water resources for relay cropping legume cover crops with winter wheat have recently been documented for western Canada (Thiessen Martens, Entz 2001). The fertilizer N replacement value of relay cropped legume cover crops in Manitoba range from 0 to 70 kg ha^{-1} (Hoeppner 2001). Trials are currently underway at the University of Manitoba to adapt self-seeding medics for use as cover crops in western Canadian cropping systems.

4. Future of BNF

In a paper entitled "Past, present and future BNF credits from legumes to western Canadian agriculture", Biederbeck *et al.* (1996) predicted that by 2005, BNF from legumes would amount to 550 million kg of N annually. This amount of N represents almost one-third of the total inorganic N fertilizer used in Canada in 1996. Biederbeck's prediction was based on a scenario of increased use of green fallow, expansion of grain legume production, and a modest increase in the land area dedicated to perennial forage legumes. All of these predictions are being realized. In fact, present expansion of grain legume production is greater than predicted. Biederbeck *et al.* concluded that, in the "2005 scenario", forage legumes, grain legumes, green fallow would account for 45%, 34% and 15% of the BFN, respectively.

Ecological and organic agriculture is growing rapidly in industrialized countries. This shift to a more holistic food production system, where external synthetic crop production inputs are reduced, presents an important opportunity to expand the role of BNF in modern agriculture. Evidence of the importance of legumes in organic farming systems was provided in a recent organic farm survey (Entz *et al.* 2001b). Forty-two percent of the landbase on survey farms in western Canada were found to be dedicated to legume production.

5. Conclusions

BNF from legumes increases the biological efficiency of crop production through direct additions of N to the soil system, and through many indirect benefits associated with perennial cropping and pasturing. Farm economics are improved when some N can be supplied through BNF; this is especially important in those areas of the world where indigenous soil N has been "mined". The role of BNF will increase in the future as farmers in industrialized nations adopt more ecologically-friendly farming systems.

6. References

Badaruddin M, Meyer DW (1989) Agron. J. 81, 419-424
Campbell CA *et al.* (1990) Publ. 1841, Canadian Government Publishing Centre, Ottawa, Canada
Campbell CA *et al.* (1994) J. Environ. Qual. 23, 195-201
Biederbeck VO, Bouman OT (1994) Agron. J. 86, 543-549
Biederbeck VO (1996) In Proc. Soils and Crops Workshop, University of Saskatchewan, Saskatoon, Canada
Entz MH *et al.* (2001a) Agron. J.

Entz MH *et al.* (2001b) Can. J. Plant Sci. 81, 351-354
FAOSTAT (2001) http://apps.fao.org/
Grace PR *et al.* (1995) J. Exper. Agric. 35, 857-864
Hoeppner JW (2001) M.Sc. Thesis, Dept. Plant Science, Univ. of Manitoba,
    Winnipeg, Canada
Hoyt (1990) Can. J. Soil Sci. 70, 109-113
Kelner DJ *et al.* (1997) Agric. Ecosys. and Enviro. 64, 1-10
Matus A *et al.* (1997) Can. J. Plant Sci. 77, 197-200
Mohr RM (1999) Agron J. 91, 622-630
Morrison IN, Kraft D (1994) International Institute for Sustainable Development,
    Winnipeg, Canada
Thiessen Martens J, Entz MH (2001) Can. J. Plant Sci. 81, 217-220
Wright AT (1990) Can. J. Plant Sci. 70, 1023-1032
Panigatti JL (1992) Rev. INIA Inv. Agr. No.1, Tomo II, La Estanzuela, Uruguay
Rossello (1992) Rev. INIA Inv. Agr. No.1, Tomo I, La Estanzuela, Uruguay

LIMITATIONS TO BIOLOGICAL NITROGEN FIXATION AS A RENEWABLE SOURCE OF NITROGEN FOR AGRICULTURE

M.B. Peoples[1], K.E. Giller[2], D.F. Herridge[3], J.K. Vessey[4]

[1]CSIRO Plant Industry, GPO Box 1600 Canberra, ACT 2601, Australia
[2]Dept of Soil Science & Agric. Engineering, University of Zimbabwe, MP 167 Harare, Zimbabwe
[3]NSW Agriculture, RMB 944 Tamworth, NSW 2340, Australia
[4]Dept of Plant Science, University of Manitoba, Winnipeg, MB, R3T 2N2 Canada

1. Summary

Biological systems that can fix atmospheric N_2 include a range of free-living microorganisms and associative or symbiotic relationships between microbes and plants. But although there are many potential sources of fixed N in terrestrial ecosystems, symbiotic N_2 fixation by legume-rhizobia associations provide the largest inputs of N for agriculture. Environmental and management limitations to legume growth are the major factors regulating N_2 fixation, although practices that either limit the presence of effective rhizobia in the soil, or enhance soil nitrate concentrations can also be critical. There is the potential for large increases in N_2 fixation and enhanced benefits for farmers in the short-term through the wider availability of high quality rhizobial inoculants, basic improvements in crop agronomy, the introduction of legumes to new areas, and changes in residue management. However, since much of these technologies are already known, the prospects for such increases must be considered within the context of the present constraints to the adoption by farmers of existing knowledge. Further increases in the inputs of fixed N into agro-ecosystems might come from rhizobial strain selection and plant breeding, or the use of modern molecular techniques to transfer the capacity to fix N to non-legume crops. But compared to the large potential gains that could be made via improved management, the impact of genetic changes to either the microsymbiont or host is likely to be relatively marginal.

2. Introduction

The fixation of atmospheric N_2 can occur in the free-living state with some diazotrophs or via associative relationships on the roots or within the tissues of plants, while other organisms require a symbiosis with specific host plants. Inputs of fixed N by these diverse systems provide a renewable source of N for many terrestrial ecosystems. Biological N_2 fixation (BNF) contributes directly to agricultural production where the fixed N is harvested in grain or other food for human or animal consumption, and indirectly by adding N to the soil for the benefit of companion plant species or following crops.

3. Comparative Inputs of Fixed N by Different Organisms

Experimental estimates of N_2 fixation by various organisms are presented in Table 1. Free-living N_2-fixers probably contribute only small amounts of N to farming systems (Table 1). The data tend to be inconclusive concerning the role of diazotrophs associated with non-legumes in temperate agriculture although studies have demonstrated potential for significant inputs of fixed N by some tropical grasses and crops such as sugarcane (Table 1). Symbiotic associations between *Anabaena* and the aquatic fern *Azolla*, or *Frankia* and actinorrhizal trees such as *Casuarina* and *Alnus* may also fix useful amounts of N in agro-ecosystems. However, it is the symbiosis between legumes and rhizobia (*Rhizobium*, *Bradyrhizobium*, *Allorhizobium*, *Azorhizobium*, *Mesorhizobium* and *Sinorhizobium* spp.) that is generally responsible for the largest amounts of fixed N (Table 1).

4. Contributions of Fixed N by Legumes

The ability of legumes to progressively improve the N status of soils has been utilized for thousands of years in crop rotations and traditional farming systems. The 163 million ha of legume crops grown each year, legume components of the 200 million ha under temporary pastures or forage crops, the 10–12 million ha of perennial legume cover-crops in rubber and oil-palm plantations, and the legume trees and shrubs in agroforestry systems all contribute fixed N to agriculture. Collectively experimental and on-farm data both suggest potential inputs of several hundreds of kg of fixed N/ha per year (Table 1), often with maximum rates of N_2 fixation of 3–4 kg shoot N/ha per day (Unkovich, Pate 2000). The amounts of N_2 fixed with most legumes are regulated by environmental or management constraints to plant growth associated with soil nutrients, water supply, diseases and pests. Legumes commonly fix around 20–25 kg of shoot N for every tonne of shoot dry matter accumulated across a range of environments (Figure 1) unless their capacity to fix N is restricted by local practices which either limit the presence of effective rhizobia (no inoculation, poor inoculant quality), or directly affect soil N fertility (excessive tillage, extended fallows, fertilizer N, rotations; Peoples, Herridge 2000). In addition to fixed N in shoots, legume roots may contribute up to 0.7 kg of fixed N for every kg of fixed N accumulated in shoots (Kelner *et al.* 1997; see also Peoples, Herridge 2000).

Table 1. Estimates of the annual amounts of N_2 fixed in different agricultural systems[a].

N_2-fixing organism	System	Range measured (kg shoot N/ha)	Commonly observed (kg shoot N/ha)
Free-living	Crops	0–80	0–15
Associative	Tropical grasses	10–45	10–20
	Crops	0–240	25–65
Symbiotic	Azolla	10–150	10–50
	Actinorrhizal trees	10–440	30–75
	Green manure legumes	5–325	50–150
	Forage legumes	5–680	50–250
	Crop legumes	0–450	30–200
	Tree legumes	5–470	50–300

[a] Adapted from Ledgard and Giller (1995) with additional information from Peoples and Craswell (1992); Peoples *et al.* (1996); Unkovich, Pate (2000); Maskey *et al.* (2001); and Peoples *et al.* (2001).

5. Prospects for Enhancing BNF in Farming Systems

5.1. Changes in management. To be able to fix atmospheric N_2 it is necessary for the legume to form an effective symbiotic relationship with rhizobia in the soil. In its simplest form the symbiosis is established when indigenous rhizobia infect the roots of legumes to produce nodules. But populations of appropriate rhizobia are not always adequate or may be absent entirely such as when legumes are introduced to new areas. Under these conditions the only way to introduce the required rhizobia is through the inoculation of legume seeds before, at, or soon after sowing. Many studies have reported improvements in legume growth and grain yield in response to inoculation (van Kessel, Hartley 2000). However, with a few exceptions, inoculation technology has not been widely adopted by farmers. This reflects in part inadequate demonstration and promotion of the benefits of inoculation, difficulties in applying inoculants, limited potential for inoculant production and distribution, poor quality control, and economic constraints to resource poor farmers.

Promiscuously-nodulating soybean varieties is one approach being evaluated in Africa as a potential solution to these problems (Giller *et al.* 2000).

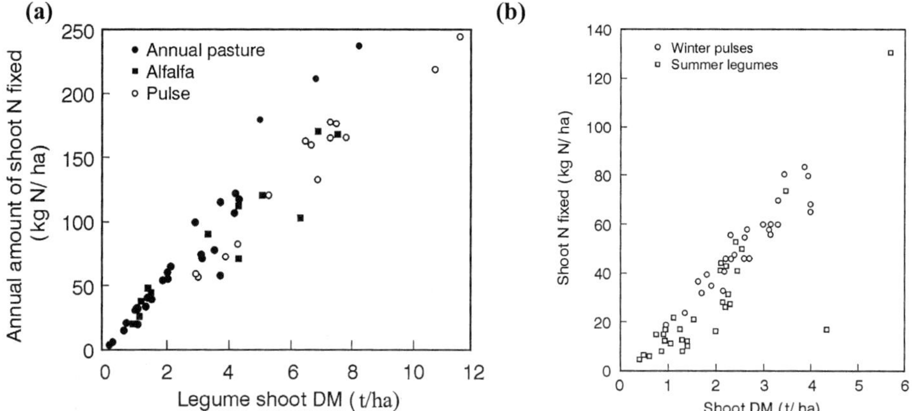

Figure 1. Examples of the relationships between amounts of shoot N fixed (kg N/ha) and shoot dry matter (DM, t/ha) for (a) annual pasture legumes, alfalfa, and pulse crops in SE Australia (Peoples *et al.* 2001), and (b) winter (lentil, chickpea) and summer (mashbean, soybean) crops growing in different locations in Nepal (Maskey *et al.* 2001).

Basic improvements in crop agronomy hold great promise for immediate increases in N_2 fixation where legumes are already grown (Figure 2). However, just as with inoculation technology, the prospects for such increases must be considered within the context of the present constraints to the adoption of existing knowledge by farmers. For example, research has demonstrated that nutritional deficiencies induced by poor P supply or soil acidity commonly restrict legume growth and BNF (Giller, Cadisch 1995; Peoples *et al.* 1995). Yet the implementation of the simple fertilizer strategies needed to ameliorate these soil limitations may require economic, transport and knowledge barriers to be removed. Other research has indicated relationships between biomass production, rates of BNF and legume population (Figure 3), and poor legume density and vigor have been implicated as important factors contributing to low inputs of fixed N in farmers' fields in South Asia (Maskey *et al.* 2001). However, access to better quality seed to ensure good germination and greater tolerance to pests and diseases is likely to be limited for resource poor farmers.

Even in the absence of access to inoculants, fertilizers or quality seed there still appears to be considerable potential for enhancing BNF inputs in farming systems through the inclusion of more legumes in farming systems (Figure 2). The small area of land cropped with legumes has been identified as a major constraint to N_2 fixation in Africa (Giller *et al.* 2000), but this can be considered to be a common factor since few countries have cropping ratios of cereal:legume less than 10:1. The ratio of land area seeded to cereals versus pulse crops has not changed in 40 years, whereas the productivity of cereals relative to pulse crops has increased significantly (Figure 4). Increased areas of legumes might be achieved by including more leguminous pastures or crops in rotations, the wider use of intercropping or cover crop strategies where legumes and non-legumes are grown together, or the introduction of legumes to areas of fallowed or degraded land. Legume green manures and agroforestry systems are other strategies that can potentially enhance inputs of fixed N (e.g. Peoples *et al.* 1996). However, they are not without limitations that restrict their wide spread use (Giller *et al.* 2000). Changes to common practices by Asian farmers so that legume residues are retained rather than removed after grain harvest would also have a huge impact on the subsequent benefits of fixed N to following crops. On-farm data from South Asia suggests that the return of up to 50 kg fixed N/ha is commonly forgone when above-ground vegetative residues are removed along with the grain.

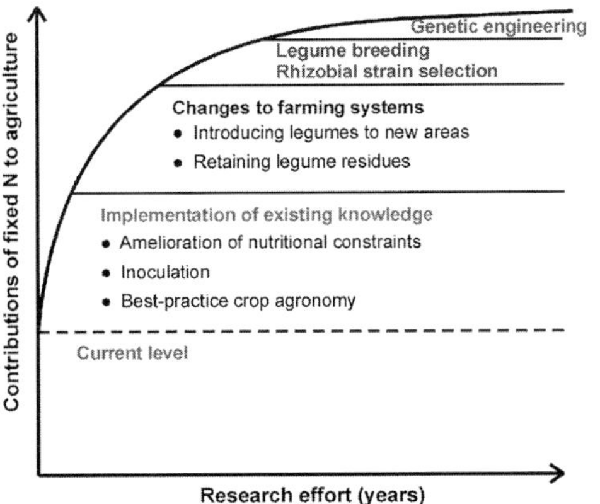

Figure 2. Conceptual representation of possible approaches to increase inputs of fixed N and the potential time scale of benefits to agriculture. Modified from Giller and Cadisch (1995).

Figure 3. On-farm data collected from Vietnam that illustrates the relationship between plant population and N_2 fixation.

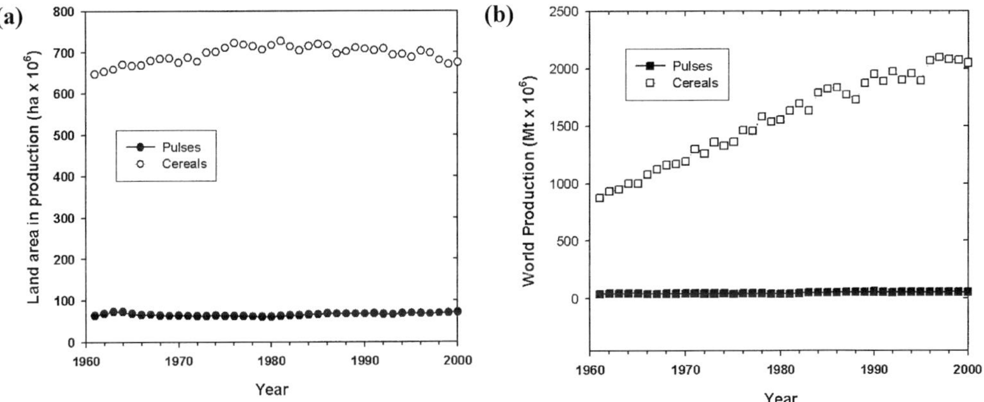

Figure 4. Global area under production (a) and production levels (b) of cereal and pulse crops (FAO 2001).

5.2. Genetic improvements. Further increases in the potential inputs of fixed N might require selection of better adapted rhizobial strains and/or genetic improvements through plant breeding (Giller, Cadisch 1995; Herridge, Rose 2000). In the long-term the use of modern molecular techniques to transfer of the capacity to fix N to non-legume crops could be imagined to have benefits to agriculture, but there are enormous physiological and biochemical obstacles that will need to be overcome before this goal can be realized (Ladha, Reddy 2000). There is also a concern that if cereals could fix their own N that this may encourage an increased reliance upon monocultures rather than rotations with legumes and other crops which may bring with it other problems for agriculture (Giller, Cadisch 1995). Ultimately the gains in amounts of N_2 fixed likely to be gained through genetics will be far more modest than the adoption of current agronomic knowledge (Figure 2).

6. Conclusions

Globally, legumes in farming systems should routinely be fixing >100 kg N/ha/year, but in reality they don't. A major reason for this is that the relevant technology is either not in the hands of the farmers, or they cannot adopt it because of economic or operational imperatives. These issues have been raised before (Giller, Cadisch 1995), but the ability to overcome constraints at the farm level, or to undertake applied BNF research that will be of direct benefit to farmers continues to deteriorate rather than improve. The poor countries become poorer, the training and support programs like NifTAL that were instrumental in advancing BNF in the past (particularly in the developing world) have been wound up or redirected, and BNF receives little attention from the CGIAR institutes. There is a need for some strong policy intervention to redress this trend. The recent initiative such as the FAO-sponsored meeting during March 2001 in Rome on BNF is an encouraging step in the right direction.

7. References

FAO (2001) FAOSTAT Agricultural Database, http://apps.fao.org/

Giller KE, Cadisch G (1995) Plant Soil 174, 255-277

Giller KE *et al.* (2000) In Pedrosa FO *et al.* (eds), Nitrogen Fixation: From Molecules to Crop
    Productivity, pp. 525-530, Kluwer Academic Publ., Dordrecht, The Netherlands

Herridge D, Rose I (2000) Field Crops Res. 65, 229-248

Kelner *et al.* (1997) Agric. Ecosystem Environ. 64,1-10

Ladha JK, Reddy P (2000) The Quest for Nitrogen Fixation in Rice, IRRI, Los Baños,
    Philippines, 354 pp.

Ledgard ST, Giller KE (1995) In Bacon PE (ed), Nitrogen Fertilization in the Environment,
    pp. 443-486, Marcel Dekker Inc., New York

Maskey SL *et al.* (2001) Field Crops Res. 70, 209-221

Peoples MB, Craswell ET (1992) Plant Soil 141, 13-39

Peoples MB, Herridge DF (2000) In Pedrosa FO *et al.* (eds), Nitrogen Fixation: From Molecules to
    Crop Productivity, pp. 519-524, Kluwer Academic Publ., Dordrecht, The Netherlands

Peoples MB *et al.* (1995) Plant Soil 174, 83-101

Peoples MB *et al.* (1996) Plant Soil 182, 125-137

Peoples MB *et al.* (2001) Plant Soil 228, 29-41

Unkovich MJ, Pate JS (2000) Field Crops Res. 65, 211-228

van Kessel C, Hartley C (2000) Field Crops Res. 65, 165-181

8. Acknowledgements

The financial support of the Australian Centre for International Agricultural Research (ACIAR) and NSERC (Canada) is gratefully acknowledged, as is the cooperation of our many collaborators throughout the world.

Poster
Papers

ISOLATED FeMo-Co CATALYTIC REACTIVITY IN NON-PROTEIN SURROUNDING: SUBSTRATE AND INHIBITOR INTERACTIONS (C₂H₂, N₂, CO)

T.A. Bazhenova, M.A. Bazhenova, G.N. Petrova

Inst. of Problems of Chem. Phys., Russ. Acad. Sci., 142432, Chernogolovka, Russia

To understand the mechanism of substrate reduction in the active site of nitrogenase the catalytic reactivity of the isolated FeMo-co with respect to C_2H_2, and/or CO, and/or N_2 reduction in non-enzymatic surroundings (DMF, Eu/Hg, PhSH) was investigated. The quality of the FeMo-co extracted (before and after catalytic reactions) was evaluated from its Fe:Mo ratio and its ability to reconstitute the activity of FeMo-co-deficient MoFe protein (NifB-Kp1) in crude extracts of *Klebsiella pneumoniae* Kp5058. It has been found that the structural integrity of FeMo-co was retained during these reactions. Study of a steady-state kinetics of C_2H_2 reduction catalyzed by FeMo-co extracted has been carried out. It has been found in particular that the FeMo-co cluster reduced by Eu/Hg demonstrates substrate induced cooperativity among two sites capable to bind and reduce acetylene (Bazhenova *et al.* 2000). The bell-shaped profile was observed for dependence of C_2H_2 reduction on PhSH concentration. The C_2D_2 reduction stereospecificity was examined by FTIR spectroscopy. *Cis*-$C_2D_2H_2$ was shown to be the main product (Bazhenova *et al.* 2001).

Carbon monoxide (CO) is not reduced in this system, but it is a potent inhibitor of C_2H_2 reduction catalyzed by isolated FeMo-co. For CO the type of inhibition was shown to be reversible and competitive rather than noncompetitive (Bazhenova *et al.* 2001). The inhibition constants for ethylene and ethane formation were found to be different: K_i (atm) = 0.004 for C_2H_4 and 0.009 for C_2H_6. Distinct C_2H_2 bonding sites are inhibited by CO differently.

We have found that at low, unsaturating, C_2H_2 pressure, dinitrogen inhibits acetylene reduction catalyzed by FeMo-co in DMF solution with Eu/Hg as a reducing agent. The type of inhibition was shown to be competitive and reversible, K_i = 0.49 atm N_2 both for C_2H_4 and C_2H_6 formation (Bazhenova 2001). The value of N_2 inhibition constant for acetylene reduction catalyzed by the isolated FeMo-co is similar to those for wild-type, α195Gln (Kim C-H *et al.* 1995) and α195Asn (Fisher *et al.* 2000) nitrogenases. We therefore conclude that it is possible to obtain in non-enzymatic conditions a state of the isolated cofactor which is capable of binding the N_2 molecule.

On the basis of these results we concluded that the isolated FeMo-co being reduced by Eu/Hg has two distinct substrate and inhibitor binding sites. One of these (site 1) has a high affinity for C_2H_2 binding and reduction (K_m = 0.006 atm C_2H_2) and can also bind reversibly N_2 or CO molecules without reduction. The other site (site 2) has a much lower affinity for C_2H_2 binding (the effective K_m for both two C_2H_2 binding sites is 0.08 atm) and can also bind CO but not N_2.

The analysis of all results obtained shows that the isolation of FeMo-co out of the protein matrix results in a loss of the ability to catalyze *the reduction of N_2*; the capability to reduce acetylene and protons, and the capability to coordinate reversibly N_2 molecule under the action of appropriate chemical reducing agent is conserved practically unchanged.

References:

Bazhenova TA *et al.* (2000) Kinetics and Catalysis, 41, 499-510
Bazhenova TA *et al.* (2001) Kinetics and Catalysis
Kim C-H *et al.* (1995) Biochem. 34, 2798-2808
Fisher K *et al.* (2000) Biochem. 39, 15570-15

We are grateful to Professor B.E. Smith and Dr C. Gormal for providing us the crude extracts of *Klebsiella pneumoniae* Kp5058. This work is supported by RFBR (Grant No.: 01-03-33278).

CHARACTERIZATION OF NITROGENASE γ PROTEIN VARIANTS GENERATED BY SITE-DIRECTED MUTAGENESIS

L.M. Rubio, P.W. Ludden

Department of Biochemistry, UW-Madison, WI, USA

Dinitrogenase is a heterotetrameric $(\alpha_2\beta_2)$ enzyme that contains the iron-molybdenum cofactor (FeMo-co) at its active site. *Azotobacter vinelandii* mutant strains unable to synthesize FeMo-co accumulate an apo form of dinitrogenase, with a subunit composition $\alpha_2\beta_2\gamma_2$, that can be activated *in vitro* by the addition of FeMo-co. The γ subunit is able to interact with both FeMo-co and apodinitrogenase, leading to the suggestion that it facilitates FeMo-co insertion into the apoenzyme.

The non-*nif* gene encoding the γ protein has been recently cloned, sequenced, and found to encode a NifY-like protein that belongs to the NifY/NifX/VnfX family of iron and molybdenum (or vanadium) cluster-binding proteins (Rubio *et al.*, manuscript in preparation). Comparison of their amino acid sequences pointed to the only conserved Cys (Cys[166] in the γ sequence) as a good candidate for cluster binding and, therefore, we have generated variants of the γ protein with Ala or Ser in place of Cys[166]. Purified preparations of wild-type and mutant variants of γ were used to compare their binding properties to FeMo-co and to apodinitrogenase. Results presented in this work are consistent with a role for Cys[166] in stabilization of the γ-FeMo-co complex.

ACETYLENE REDUCTION WITH *AZOTOBACTER VINELANDII* Mo-NITROGENASE: ROLE OF GLUTAMINE-191 IN α-SUBUNIT OF MoFe PROTEIN

K. Vichitphan, W.E. Newton

Dept of Biochemistry, The Virginia Polytechnic Institute and State University, Blacksburg, VA 24061, USA

1. Introduction

The FeMo-cofactor (FeMo-co) is one of two types of prosthetic group found in the larger of the two nitrogenase component proteins, called the MoFe protein, and it is strongly implicated as the substrate binding and reduction site. The glutamine-191 residue in the α-subunit of the MoFe protein was targeted for substitution because it is located between the P cluster (the second type of prosthetic group in the MoFe protein) and the FeMo-co. Moreover, its side chain is involved in a hydrogen-bond network from one of the terminal carboxylate of the homocitrate component of FeMo-co through to the backbone NH of αGly-61, which is adjacent to the P cluster-ligating residue, αCys-62. The effect of substitution at the position α-191 on the properties of the FeMo-co could be mediated through hydrogen bonding and the homocitrate-Mo linkage. Substitution with lysine in this position, to give the αLys-191 altered MoFe protein, produces an unusual phenotype, i.e. no nitrogen fixation, CO-sensitive H_2 evolution, low C_2H_2 binding affinity, and C_2H_6 formation from C_2H_2 reduction (Scott *et al.* 1992; Fisher *et al.* 2000).

2. Results and Discussion

A variety of altered MoFe proteins, namely the αSer-191, αHis-191, αGlu-191, and αArg-191 altered MoFe proteins, have been purified to homogeneity and have the following results compared to wild type MoFe protein. All four altered MoFe proteins have decreased catalytic activity but charged amino-acid side chains decrease catalytic activity more (37% and 18% of wild type under Ar alone for αGlu-191 and αArg-191 altered MoFe protein, respectively). Based on a pH profile study, the decrease in catalytic activity could be due to a shift in the pK_a of deprotonated and/or protonated group(s), resulting in changing pH for maximum activity. In the presence of 10% C_2H_2/90% Ar, the four altered MoFe proteins use less of the electron flux for C_2H_2 reduction (21-46%) than does wild type (89%). The decreased rate of electron distribution to C_2H_2 reduction correlates with a decrease in C_2H_2 binding affinity (higher K_m). Moreover, the αSer-191 altered MoFe protein exhibits a biphasic response for C_2H_2 reduction at pH 8.0 with two K_m values (0.007 and 0.800 atm of C_2H_2 for C_2H_4 production), which suggest at least two C_2H_2 binding sites. The αHis-191, αGlu-191, and αArg-191 altered MoFe proteins produce ethane (C_2H_6) from C_2H_2 due to both their higher K_m of C_2H_2 reduction and the pK_a shift. These results suggest that the glutamine at the position α-191 has an effect on both the C_2H_2 binding site(s) and the pK_a of responsible protonated/deprotonated group(s) during nitrogenase catalysis.

3. References

Scott DJ *et al.* (1992) J. Biol. Chem. 267, 20002-10

Fisher K *et al.* (2000) Biochem. 39, 29070-79

4. Acknowledgement

We thank Dr K. Fisher and the NIH (Grant DK-37255 to W.E.N).

AZOTOBACTOR VINELANDII NITROGENASE CONTAINING ALTERED MoFe PROTEINS WITH SUBSTITUTIONS AT α-278^{SER}: INTERACTIONS AMONG SUBSTRATES AND INHIBITORS

H. Li, K. Kloos, W.E. Newton

Dept of Biochemistry, Virginia Tech, Blacksburg, VA 24061, USA

1. Introduction

FeMo-cofactor is one of two prosthetic groups bound within the MoFe protein of nitrogenase. The FeMo-cofactor's polypeptide environment appears to be intimately involved in the delicate control of the MoFe protein's interactions with its substrates and inhibitors (Fisher *et al.* 2000). In this work, the α-subunit 278-Serine residue of the MoFe protein was targeted. Altered MoFe proteins of *Azotobacter vinelandii* Mo-nitrogenase, the α-278^{Thr}, α-278^{Cys}, α-278^{Ala} and α-278^{Leu} MoFe proteins, were used to study interactions among H^+, C_2H_2, CO and N_2.

2. Results and Discussion

All strains except the α-278^{Leu} mutant are Nif$^+$. We determined the K_m of C_2H_2 reduction for the altered MoFe proteins. The α-278^{Ala} and α-278^{Cys} MoFe proteins apparently bind C_2H_2 similarly to the wild type, whereas the α-278^{Thr} and the α-278^{Leu} MoFe proteins both have a K_m ten times higher than the wild type for C_2H_2 reduction and, unlike wild type, both produce C_2H_6. These results suggest that the C_2H_2-binding site is affected by substitution at the α-278^{Ser} position. Like the wild type, N_2 is also a competitive inhibitor for the reduction of C_2H_2 by the α-278^{Thr}, α-278^{Cys} and α-278^{Ala} MoFe proteins, but the K_i for N_2 inhibition is higher than that with the wild type MoFe protein.

When reducing C_2H_2, the α-278^{Ala} and α-278^{Cys} MoFe proteins respond to CO similarly to the wild type, whereas the α-278^{Thr} MoFe protein is much more sensitive to CO. Under a nonsaturating concentration of CO, the α-278^{Leu} MoFe protein catalyzes the reduction of C_2H_2 with sigmoidal kinetics, which is consistent with inhibitor-induced cooperativity between two C_2H_4-evolving sites. This phenomenon was previously observed with the α-277^{His} MoFe protein, which has the α-277^{Arg} substituted by histidine (Shen *et al.* 1997). These data suggest that the MoFe protein has two C_2H_2-binding sites, one of which appears to be located near the α-277-278 residues and, therefore, most likely on the Fe_4S_3 sub-cluster of the FeMo-cofactor.

3. References

Shen *et al.* (1997) Biochem. 36, 4884-4894
Fisher K *et al.* (2000) Biochem. 39, 2970-2979

4. Acknowledgement

Supported by the NIH (DK-37255).

KINETICS OF TRANSFER OF TWO ELECTRONS IN NITROGENASE WITH PHOTODONOR

L.A. Syrtsova, V.A. Nadtochenko, N.N. Denisov, E.A. Timofeeva, N.I. Shkondina V.Yu. Gak

Institute of Problems of Chemical Physics of Russian Academy of Sciences, Chernogolovka, Moscow region, 142432 Russia

The kinetics of minimum turnover of nitrogenase, transfer of two electrons from a photodonor (a system containing eosin and NADH or $4',5'$-dibromofluorescein and NADH) to Fe protein (Av2) and the kinetics of transfer of the first and second electrons from Av2 to MoFe protein (Av1) were studied by kinetic laser spectroscopy. The effects of the substrates of nitrogenase (nitrogen, acetylene, and protons) on the intramolecular electron transfer in nitrogenase were studied. Analysis of the effect of photodonor excitation radiation intensity on the rate of electron transfer was used to determine the transfer rate constants for the first (k_1) and second (k_2) electrons from Av2 to Av1. In the presence of MgATP two electrons are sequentially transferred from Av2 to Av1, and no delay between these reactions was detected. The first electron transferred from Av2 to Av1 is not targeted to the substrate; $k_1 = 154 \pm 15$ sec^{-1} at 23°C for the system $4',5'$-dibromofluorescein–NADH; $k_2 = 53 \pm 5$ sec^{-1}, 95 ± 9 sec^{-1}, and 24 ± 2 sec^{-1} at 23°C in the presence of nitrogen, acetylene, and argon, respectively. An unidentified slow step ($k_3 = 18 \pm 2$ sec^{-1} at 23°C) may be associated with electron transfer within Av1.

Steps of nitrogenase reaction with photodonor DBF-NADH presence of nitrogen:

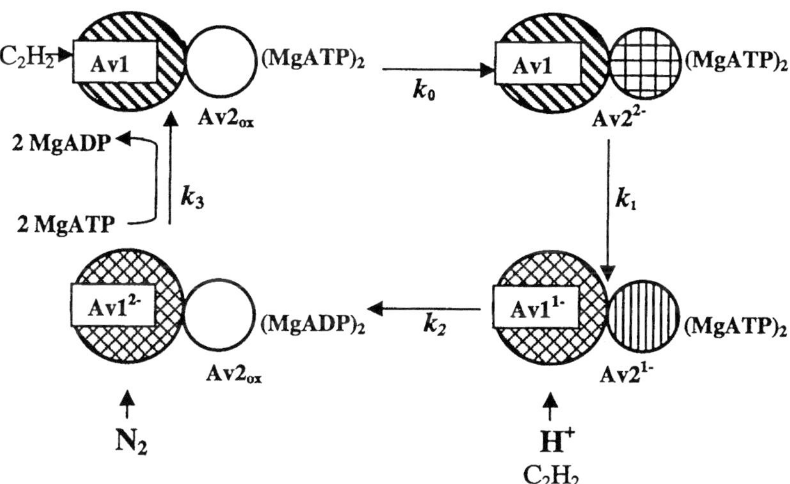

The little circle is Av2, large circle is Av1 in reduced 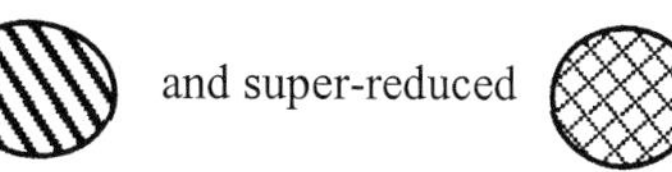 and super-reduced states.

$$k_0 = 200 \text{ c}^{-1}, \quad k_1 = 158 \text{ c}^{-1}, \quad k_2 = 54 \text{ c}^{-1}, \quad k_3 = 18 \text{ c}^{-1}$$

References

Syrtsova LA *et al.* (2000) Biochem. (Moscow) 65, 1145-1152

MECHANISTIC INSIGHTS INTO THE FUNCTIONING OF WILD-TYPE AND ALTERED *AZOTOBACTER VINELANDII* NITROGENASE MoFe-PROTEINS THROUGH THEIR INTERACTIONS WITH HCN AND CN⁻

M.J. Dilworth[1], K. Fisher[2], W.E. Newton[2]

[1]Centre for Rhizobium Studies, Murdoch University, Murdoch, W. Australia 6150
[2]Dept of Biochemistry, The Virginia Polytechnic Institute and State University, Blacksburg, VA 24061, USA

1. Introduction

At neutral pH in nitrogenase assays, cyanide is present as the substrate HCN (~98%) and the potent reversible inhibitor CN^- (~2%). HCN is reduced by six electrons to CH_4 and NH_3, and four electrons to CH_3-NH_2. Under certain conditions, additional ammonia can also be formed during nitrogenase-catalyzed HCN reduction. "Excess ammonia", i.e. the amount of NH_3 in excess of CH_4, has been suggested to arise from an initial two-electron/two-proton step converting HCN into CH_2=NH (methyleneimine). When any of this intermediate escapes from the active site, it is hydrolyzed to formaldehyde, which has never been detected, and "excess ammonia". Additional two-electron/two-proton steps yield CH_3-NH_2, some of which escapes, and finally CH_4 and NH_3 (Li *et al.* 1982).

2. Results and Discussion

We have examined the interaction of HCN and CN^- with wild type and two altered *Azotobacter vinelandii* nitrogenase MoFe proteins containing substitutions at the α-195^{His} position (α195^{Gln} and α195^{Asn}). This residue is of considerable interest because previous work has shown that the α195^{Gln} altered MoFe protein is ineffective in normal N_2 and azide reduction but effective in HCN and H^+ reduction even though the hydrogen bond to the central sulfide of the FeMo-cofactor remains intact (Dilworth *et al.* 1998). We have determined the relative rates of CH_4, NH_3, CH_3-NH_2 and H_2 product formation from HCN-reduction assays as a function of HCN concentration. It was necessary to separate the ammonia and methylamine before product determination because the phenol-hypochlorite method commonly used to measure ammonia also recovers methylamine. In particular, we have focused on the source of the "excess ammonia". Unlike the wild type and α195^{Asn} MoFe proteins, the α195^{Gln} MoFe protein catalyzes HCN reduction without the production of "excess ammonia" and without suffering significant inhibition by CN^-. We propose that the lack of production of "excess ammonia" is a consequence of a substantially lower affinity of the α195^{Gln} MoFe protein for HCN binding, i.e. HCN is less able to displace intermediates from the active site. We have developed a sensitive assay, using acetylacetone, to determine if formaldehyde is a product of HCN reduction. We were able to accurately recover (>95%) formaldehyde (10–50 nmol) from supplemented mock nitrogenase assays. However, nitrogenase assays conducted with either wild type or α195^{Asn} MoFe protein in the presence of 1 mM NaCN, emphatically showed that formaldehyde was not produced. Our search for the missing product continues.

3. References

Li *et al.* (1982) Biochem. 21, 4393-4402
Dilworth MJ *et al.* (1998) Biochem. 37, 17495-17505

4. Acknowledgements

We thank the NIH (Grant DK-37255 to W.E.N).

PRE-STEADY-STATE ANALYSIS OF REDUCED MoFe-PROTEIN INTERMEDIATES GENERATED DURING ENZYME TURNOVER FROM WILD-TYPE AND ALTERED *AZOTOBACTER VINELANDII* NITROGENASE MoFe-PROTEINS

K. Fisher[1], D.J. Lowe[2], W.E. Newton[1]

[1]Dept of Biochemistry, The Virginia Polytechnic Institute and State University, Blacksburg, VA 24061, USA
[2]Biological Chemistry Dept, John Innes Centre, Norwich, UK

1. Introduction

We have studied the pre-steady-state kinetics of wild type and altered nitrogenases during turnover in order to characterize the different MoFe-protein catalytic intermediates using rapid-freeze EPR and UV/visible stopped-flow spectroscopy. The major drawback with such experiments, in addition to the large amounts of protein required, is that the reduced states of the MoFe-protein which bind substrates, are only generated under turnover conditions and therefore the intermediates of interest are only a small proportion of the total protein present. EPR and stopped-flow spectroscopy are particularly useful techniques because they allow us to detect conformational and redox changes that are likely to control the delivery of electrons and protons to the FeMo-cofactor during substrate reduction.

2. Results and Discussion

Multiple variations of the S=3/2 EPR signal have been previously observed for wild-type MoFe protein. A signal designated as Ia occurs at neutral pH in the resting MoFe protein, signal Ib is observed at higher pH, and signal Ic is only observed during turnover (Smith *et al.* 1973). We have detected similar S=3/2 signals from the more-reduced states of the wild-type MoFe protein during catalytic turnover and provide evidence that they arise from different conformations of the FeMo-cofactor within the MoFe protein. We attribute signal Ib to MoFe protein essentially at the E_3 redox level and suggest that the conformational change results as a consequence of protons being incorporated into reduced substrate. Signal Ic is generated much more slowly than Ib and, therefore, likely arises from an even more-reduced redox state than Ib.

We have extended these studies to include the α-195^{Gln}, α-195^{Asn}, and α-191^{Lys} MoFe proteins which bind N_2 and reduce it very poorly, bind N_2 and do not reduce it, and do not bind N_2, respectively. The α-191^{Lys} MoFe protein is particularly interesting because it exhibits neither signal Ib nor signal Ic under turnover conditions. In addition, stopped-flow spectrophotometric studies with this protein do not show the oxidation after primary electron transfer that is observed with wild-type nitrogenase. This oxidation has been postulated to arise from the P-clusters and to be the step that irreversibly commits N_2 to reduction. All evidence suggests that this protein cannot attain the redox levels necessary for N_2 binding and reduction. Although the α-195^{Gln} and α-195^{Asn} MoFe proteins show an oxidation, it is short lived and the Ib and Ic EPR signals accumulate during catalysis, indicating a problem with the N_2-reduction process. Our EPR and stopped-flow spectroscopic data show a correlation between N_2 binding and the different intermediate forms of MoFe protein.

3. References

Smith *et al.* (1973) Biochem. J. 135, 331-341

4. Acknowledgement

The authors thank the National Institutes of Health (Grant DK-37255 to W.E.N.) and the UK's BBSRC (to D.J.L.) for support.

SUBSTRATE REDUCTION AND CO SUSCEPTIBILITY OF THE ΔNIFV AND α-Q191K MoFe PROTEINS OF *AZOTOBACTER VINELANDII* NITROGENASE

W.E. Newton, K. Vichitphan, K. Fisher

Dept of Biochemistry, The Virginia Polytechnic Institute and State University,
Blacksburg, VA 24061, USA

1. Introduction

Substrates and inhibitors interact at the FeMo-cofactor (FeMo-co) prosthetic group on the MoFe protein component of *Azotobacter vinelandii*. Homocitrate, the *nifV*-gene product, is an integral component of FeMo-co (Hoover *et al.* 1987) that ligates the Mo of FeMo-co, and hydrogen bonds with α-Gln191. It may direct intra-MoFe-protein electron/proton transfer, substrate reduction and inhibitor binding, all of which may involve nearby amino-acid residues (Fisher *et al.* 2000). Insight into important acid-base groups within the MoFe protein arose from the effects of changing pH in the "3-buffers" system (Pham, Burgess 1993) on the H_2-evolution rate +/-CO, using wild-type, α-Q191K and citrate-complemented (from the Δ*nifV* mutant) MoFe proteins. Unlike wild type, the last two suffer CO inhibition of H_2 evolution.

2. Results and Discussion

We found:

(a) All three MoFe proteins produce bell-like curves, indicating two (at least) acid-base groups, one of which must be deprotonated (pK_a ca. 6.0) and one protonated (pK_a ca. 8.5) for nitrogenase activity. Above pH ca. 7, all three curves are misshapen, suggesting that more than one acid-base group contribute to the pK_a of ca. 8.5.

(b) Because all six curves overlap below pH ca. 7, the group responsible for the pK_a of ca. 6 is unaffected by both substitutions and CO and so is not likely to be α-Gln191.

(c) The curves for both altered proteins are very similar (and different to wild type above pH 7.5) suggesting that α-Gln191-homocitrate is a component of the pK_a at 8.5.

(d) The wild-type curve with CO more closely resembles those of the altered proteins without CO and suggests that: (i) substitutions in the α-Gln191-homocitrate system mimic the effect of added CO; (ii) CO "masks" a group's contribution to the pK_a at 8.5; and (iii) CO binds close to α-Gln191-homocitrate.

(e) The CO-induced shift in the pH-activity curve causes CO inhibition of H_2 evolution for wild type (only above pH 7.5) and α-Q191K. No such shift occurs for ΔNifV, where inhibition occurs throughout the pH range, indicating a different mechanism of inhibition.

(f) With added CO, the α-Q191K MoFe protein produces a symmetrical bell-shaped curve, suggesting (with the above) that three groups contribute to the pK_a of ca. 8.5.

3. References

Hoover TR *et al.* (1987) Nature 329, 855-857
Fisher K *et al.* (2000) Biochem. 39, 10855-10865
Pham DN, Burgess BK (1993) Biochem. 32, 13725-13731

4. Acknowledgement

Support was from the NIH (#DK-37255).

NITROGENASE CATALYZED DINITROGEN REDUCTION – A NEW MECHANISTIC APPROACH

A.K. Sra, A. Kohen

Dept of Chemistry, University of Iowa, Iowa City, IA 52242, USA

1. Introduction

The molecular mechanism of the enzymatic reduction of the inert gas N_2 remains an open question, despite the immense amount of work that has been done in this area (Burgess, Lowe 1996; Rees, Howard 2000). We are conducting a mechanistic investigation of the dinitrogen (N_2) reduction by the enzyme. Our goal is to develop a unique methodology that will enable the examination of the N_2 reduction mechanism within the enzyme's complex kinetic cascade (Thorneley, Lowe 1983, 1996).

2. Procedure

A methodology is being established to measure competitive ^{15}N kinetic isotope effects (KIEs). Competitive KIEs are effects on the second order rate constant V/K. These KIEs are only sensitive to kinetic steps from the free N_2 binding to the first irreversible step (Cook 1991). Thus, these effects are not "masked" by the slow rate limiting steps. The enrichment of ^{15}N in the remaining dinitrogen substrate is measured at various fractional conversions (f) by isotope ratio mass spectrometry. The KIEs are calculated from: where R_t is the isotopic ratio at time t and R_0 the isotopic ratio at t = 0 (Kohen, Klinman 1999; Melander, Saunders 1987). Triple labeling ^{15}N KIE experiments ($^{15}N_2$:$^{15}N^{14}N$:$^{14}N_2$) assist in the eluci-dation of the intrinsic mechanism (Cook 1991).

$$^{15}(V/K) = \frac{\ln(1-f)}{\ln\left(1 - f\frac{R_t}{R_0}\right)}$$

3. Analysis and Discussion

KIEs are a manifestation of changes in bond order along the reduction path. The experimental findings will be used to reevaluate various theoretical models suggested and will lead to the identification of model(s) consistent with the results. We hope to provide one of the first experimental tools to shed light on this fascinating chemical process. In the future, D_2O and D_2 effects on the ^{15}N KIEs and studies with several mutants will enable examination of the reductive protonation of N_2.

4. References

Burgess BK, Lowe DJ (1996) Chem. Rev. 96, 2983-3011
Cook PF (ed.) (1991) Enzyme Mechanism From Isotope Effects, CRC Press, Boca Raton, FL
Kohen A, Klinman JP (1999) Chem. Biol. 6, R191-198
Melander L, Saunders WH (1987) In Krieger RE (ed.) Reaction Rates of Isotopic Molecules,
    Malabar, FL
Rees DC, Howard JB (2000) Curr. Opin. Chem. Biol., 559-566
Thorneley RNF, Lowe DF (1983) Biochem J. 215, 393-403
Thorneley RNF, Lowe DF (1996) J. Biol. Inorg. Chem., 576-580

5. Acknowledgements

The wild-type Mo nitrogenase from *Azotobacter vinelandii* was a generous gift from B.K. Burgess. We thank R. Hoffmann for useful discussion regarding the theoretical models involved.

CHARACTERIZATION OF BetP, AN OSMOTICALLY INDUCED BETAINES TRANSPORTER IN *SINORHIZOBIUM MELILOTI*

A. Boscari, K. Mandon, M.-C. Poggi, L. Dupont, D. Le Rudulier

Lab. Bio. Végétale et Microbiologie, CNRS FRE 2294, Université Nice-Sophia-Antipolis, Parc Valrose, 06108 Nice Cedex 2, France

1. Introduction

Variations of the osmotic environment within the rhizosphere may affect all steps of plant-rhizobia interaction, from the root colonization to nodule development and function. Betaines, mainly glycine betaine (GB) and proline betaine (PB) are common osmoprotectant in Gram-negative bacteria. In *S. meliloti* accumulation of GB and PB restores turgor pressure and growth in response to salt stress. Different GB and PB uptake systems have been detected in *S. meliloti*, but up to now only one, Hut, an ABC transporter, has been characterized (Boncompagni *et al.* 2000).

2. Results and Discussion

A PCR strategy was used to isolate BetP, a secondary transporter which belongs to the BCCT (Betaine Choline Carnitine Transporter) subfamily in *S. meliloti*. The *betP* gene was localized on the pSymb and the gene product displayed significant identities with the choline transporter BetT of *Escherichia coli* and with the GB transporters OpuD of *Bacillus subtilis* and BetP of *Corynebacterium glutamicum*. Sequence analysis suggested that the BetP protein of *S. meliloti* contained 12 transmembrane spanning segments and two hydrophile N and C terminal cytoplasmic extensions.

The transport characteristics of *S. meliloti* BetP were determined in an *E. coli* mutant strain devoided of choline, GB and proline uptake. The BetP protein was only active under high osmotic conditions and was specifically involved in GB and PB uptake at high affinity, exhibiting a Km of 16 µM towards GB and 56 µM towards PB. Using a transcriptional *betP-lacZ* fusion recombined into the genome of *S. meliloti*, the *betP* gene was shown to be constitutively expressed. Consequently, the activation of GB mediated BetP transport by high osmolarity may involve a post-transductional mechanism. Indeed, in the wild-type strain of *S. meliloti,* betaines transport activity was enhanced by 3.5-fold within five minutes following an osmotic upshock, a feature also observed in presence of chloramphenicol, while no activation could be detected in a *betP-Km* derivative mutant strain. Thus, the activity of BetP is stimulated by elevation of the osmolarity and may allow *S. meliloti* to respond rapidly to sudden changes in the osmotic pressure of the environment. The *betP* mutant showed a clear delay in its growth when transferred from low to high osmolarity medium containing betaines, presenting a much longer lag phase than the wild-type (5 h versus 12 h in media containing GB and 7 h versus 20 h using PB as osmoprotectant). For a long term adaptation in environment of elevated osmolarity, *S. meliloti* uses at least another betaines uptake system, most probably transcriptionally induced, since 16 h after transfer from low to high osmolarity medium, the *betP* mutant retains 40% of the betaine transport capacity of the wild-type strain.

The BetP protein appeared to be a major betaines transporter involved in osmoregulation. BetP may play an important role in legume-rhizobia interaction subjected to osmotic stress, as betaines are available within the rhizosphere, particulary proline betaine which is excreted by alfafa roots.

3. References

Boncompagni *et al.* (2000) J. Bacteriol. 182, 3717–3725

NtcA ACTIVATES THE *nblA* GENE OF THE CYANOBACTERIUM *SYNECHOCOCCUS* sp. PCC 7942

I. Luque[1,2], A. Contreras[2], G. Zabulon[1], J. Houmard[1]

[1]Unité des Membranes Végétales, Ecole Normale Superieure, UMR8543, Paris, France
[2]División de Genética, Universidad de Alicante, Apartado 99, E-03080 Alicante, Spain

1. Introduction

NtcA is a global transcriptional regulator for nitrogen control present in every cyanobacteria tested (Herrero *et al.* 2001) and most of its targets are genes involved on nitrogen assimilation. Non-diazotrophic cyanobacteria degrade their light-harvesting antennae, the phycobilisomes, when exposed to a variety of stress conditions, including nitrogen starvation. This phenomenon, termed chlorosis or bleaching has been shown to be dependent of the response regulator NblR in *Synechococcus* sp. PCC 7942 (Schwarz, Grossman 1998). The expression of *nblA*, a key gene in degradation of phycobilisomes is induced during nitrogen starvation, although previous work failed to show the direct involvement of NtcA on *nblA* expression (Bhaya *et al.* 2000).

2. Results and Discussion

To throw some light on transcriptional regulation of *nblA* in response to nutrient starvation, we performed Northern and primer extension analyses in wild type, NblR- and NtcA- strains from *Synechococcus* sp. PCC 7942 and gel retardation assays with purified NblR and NtcA. Results show that (i) after nitrogen depletion, *nblA* transcript accumulation was impaired in NtcA- cells; (ii) *nblA* expression was still responsive to nitrogen in the absence of NblR; (iii) purified NtcA and NblR bind to the nblA promoter region; and (iv) the NtcA-mediated increase in *nblA* transcripts is not via NblR, since *nblR* transcript levels were not affected by *ntcA* inactivation or by the nutritional conditions tested. As a whole, these results demonstrate that NtcA directly activates nblA transcription under nitrogen starvation conditions.

Primer extension and sequence analysis also indicate that *nblA* transcription can initiate at several promoters. The most active one, P*nblA*-2, is indeed directly regulated by NtcA under nitrogen starvation conditions and constitutes a novel type of NtcA activated promoter, with putative NtcA binding sites centered at −68 and −100.5 from the transcription star point.

Our results indicate that *nblA* promoter region is a complex regulatory region and that both NtcA and NblR greatly stimulate transcription from P*nblA*-2 in response to nitrogen starvation and, in the case of NblR, also in response to other signals. Lack of viability of the NblR⁻-NtcA⁻ double mutant and other observations made in the course of this work anticipate additional regulatory connections between the NblR and NtcA regulons.

3. References

Bhaya *et al.* (2000) The Ecology of Cyanobacteria, pp. 397-442, Kluwer Academic Publishers, Dordrecht, The Netherlands
Herrero A *et al.* (2001) J. Bacteriol. 183, 411-425
Schwarz R, Grossman AR (1998) Proc. Natl. Acad. Sci. USA 95, 11008-11013

STUDIES ON THE INTERACTION BETWEEN P$_{II}$ AND NifA IN *AZOSPIRILLUM BRASILENSE* BY USING A YEAST TWO-HYBRID SYSTEM

J. Du, S. Chen, Y. Zhao, J. Li

Dept of Microbiology, China Agricultural University, Beijing 100094, China

P$_{II}$ is specifically required for nitrogen fixation in *A. brasilense*. P$_{II}$ activates NifA, preventing an inhibitory effect on the N-terminal domain (Arsene *et al.* 1996). However, the mechanism of P$_{II}$ interacting with NifA directly or indirectly is unknown. Here we investigate the interaction between P$_{II}$ and NifA in *A. brasilense* Sp7 by using yeast two-hybrid system.

The *glnB* gene encoding P$_{II}$ was PCR amplified and fused to the DNA-binding domain in the pGBD-C2 plasmid vector to generate a recombinant pGBD-GlnB plasmid. Four DNA fragments corresponding respectively to the N-terminal domain, the central domain, the C-terminal domain of NifA and the complete NifA protein were PCR amplified and fused respectively to the DNA-activation domain in the pGAD-C2 plasmid vector to generate the four recombinant pGAD-NifAN, pGAD-NifAM, pGAD-NifAC and pGAD-NifA plasmids. After introducing appropriate plasmid combinations in yeast cells, the existence of direct interaction between NifA and P$_{II}$ was analyzed with the yeast two-hybrid system by testing for the expression of *lacZ, trp, leu, his* and *ade* genes. The experimental results showed that P$_{II}$ directly interacts with the N-terminal domain of NifA and that P$_{II}$ did not have any interactions with the central domain and the C-terminal domain of NifA. No interaction occurred if *glnB* was frame-shift mutated by cloning into pGBD-C3.

References

Arsene *et al.* (1996) J. Bacteriol. 178, 4830-4838

DNA MACROARRAY ANALYSIS OF *SINORHIZOBIUM MELILOTI* GENE EXPRESSION DURING INFECTION OF *MEDICAGO TRUNCATULA*

F. Ampe, E. Kiss, F. Sabourdy, J. Batut

Laboratoire de Biologie Moléculaire des Relations Plantes-Microorganismes,
CNRS INRA, BP27, 31326 Castanet-Tolosan, Cedex France

1. Introduction

The complete sequencing and annotation of the alfalfa symbiont *Sinorhizobium meliloti* (Galibert *et al.* 2001) highlighted the presence of genes potentially implicated in the infection process preceding the differentiation of bacteroids and nitrogen fixation. These genes include homologs of virulence genes from plant and mammal intracellular pathogens, cell-surface components, adhesins, outer membrane proteins, pili proteins, regulators, etc. To test the possible implication of these genes in the infection of *Medicago truncatula* and *M. sativa,* we have designed nylon gene-arrays with PCR products corresponding to the candidate genes. Arrays were hybridized with cDNAs from *S. meliloti* grown in various culture conditions and from plant nodules.

2. Results and Discussion

Transcriptome analysis using dedicated nylon gene-arrays enabled to detect the expression of ca. 75% of the genes tested. Control genes behaved as expected thus validating the procedure.

Cluster analysis as well as differential expression made it possible to identify a variety of regulatory patterns. Thanks to the use of both *S. meliloti* and a plant mutant (Bénaben *et al.* 1995) blocked at various stages of the symbiotic interaction, we could identify genes that were specifically induced or repressed during the infection of *Medicago* by *S. meliloti.* Future work using specific mutants and gene expression fusions is required to characterize their function. Altogether, data suggest that transcriptome experiments are appropriate for the study of the infection process.

Although some improvements of the technique are still needed (e.g. detection of low-expressed genes), the use of whole genome arrays should enable a better understanding of the development of the rhizobium-legume symbiosis

3. References

Bénaben, Duc, Lefebvre, Huguet (1995) Plant Physiol. 107, 53-62
Galibert *et al.* (2001) Science 293, 668-672

4. Acknowledgements

EK is supported by an INRA post-doctoral fellowship. This work was supported by the Toulouse Génopôle.

IAA INFLUENCE ON GENE REGULATION FOLLOWED BY DNA MACROARRAY AND TWO-DIMENSIONAL ELECTROPHORESIS

C. Bianco[1], E. Imperlini[1], S. Camerini[1], B. Senatore[1], R. Calogero[2], T. Pandolfini[3], A. Spena[3], R. Defez[1]

[1]International Institute of Genetics and Biophysics, Naples-Italy
[2]Dipartimento di Scienze Cliniche e Biologiche Università di Torino, Italy
[3]Dipartimento Scientifico Tecnologico, Facoltà di Scienze MM.FF.NN, Verona-Italy

Indole-3-acetic acid (IAA), known to provide auxin activity in higher plants, has been shown to regulate gene expression in prokaryotic systems. IAA is able to circumvent the necessity for cAMP in eliciting gene expression in the arabinose system of *E. coli* and has an effect on RNA and protein synthesis in this bacterium. However, why indole derivatives can have these abilities is not understood, and this situation illustrates the need for multifaceted approaches in the investigation of cellular metabolism to clarify their mechanisms of action.

We have measured the genomic and proteomic expression patterns of *E. coli* K-12 following treatment with IAA in a minimal medium using L-arabinose as sole carbon source. In this study, we applied DNA macroarray technology and two-dimensional gel electrophoresis in combination with mass spectrometric identification of selected proteins. We have introduced into *Rhizobium leguminosarum* bv. *viciae* two genes converting tryptophan into IAA and observed changes in both of root nodule development and of bacteroid survival and morphology (see Camerini *et al.* and Senatore *et al.* this volume). Our results attempt to correlate induction and repression of specific genes/gene products with the expression pattern in both *Rhizobium* and enteric bacteria.

STRUCTURE AND FUNCTION OF NEW NATURALLY GENERATED pSYMs OF *RHIZOBIUM ETLI*

S. Brom, L. Girard, C. Tun-Garrido, D. Romero

Programa de Genética Molecular de Plásmidos Bacterianos CIFN. UNAM. Apdo.
Postal 565-A, Cuernavaca, Mor., México

The pSym of *Rhizobium etli* strain CFN42 is subject to conjugative transfer, depending on the presence of p42a, another endogenous, self-transmissible plasmid (Brom *et al.* 2000). We have analyzed transconjugants generated during pSym transfer. This allowed the identification of new pSyms, which we have named pSyms*. The pSyms* contain deletions of the pSym and integrations of p42a. The pSyms* have acquired some features of p42a, such as self-transfer, and compatibility with the wild-type pSym, thus allowing the construction of derivatives containing both, a pSym* and a wild-type pSym.

Regarding their structure, we have determined that most of the pSyms* (7/10), share one of the endpoints for the recombination between the pSym and p42a. The sequence analysis of the p42a border shows homology to a fragment of pRleVF39b, a plasmid from *R. leguminosarum* strain VF39, subject to a rearrangement with the endogenous pSym (Zhang *et al.* 2001), and to a sequence encoding a prophage lambda integrase. Also, we have seen that the generation of pSym* is independent of RecA, suggesting the involvement of a very specific recombination system.

As to their symbiotic performance, derivatives containing a pSym*, showed approximately a two-fold increase in nitrogen fixation, measured through acetylene reduction. This increase could imply that the pSym region deleted in these strains contains a repressor for nitrogen fixation. Alternatively, it could be due to an increase in the plasmid's copy number, provided by the new oriV. The increased nitrogen fixation is abolished in the presence of a wild-type pSym, furthermore, in one case the effect observed is the reverse, a drastic reduction in nitrogen fixation. The strains containing a pSym*, and the derivatives containing two pSyms showed a decrease in competitivity for nodulation. The strains containing two pSyms are very stable *in vitro*, however, the derivatives isolated from nodules usually lost the wild-type pSym.

References

Brom *et al.* (2000) Plasmid 44, 34- 43
Zhang *et al.* (2001) J. Bacteriol. 183, 2141- 2144

Acknowledgements

We thank Laura Cervantes and Javier Rivera for excellent technical assistance. Work was partially supported by grant IN202599 (DGAPA, UNAM).

WHOLE GENOME MICROARRAY HYBRIDIZATION USED TO VALIDATE A PCR-BASED GENOMIC SUBTRACTION PROTOCOL

Y. Dong, E.W. Triplett

Dept of Agronomy, University of Wisconsin-Madison, 1575 Linden Drive, Madison, WI, USA

In recent work, 3000 genes in a maize endophyte, *Klebsiella pneumoniae* 342 (Chelius *et al.* 2000) were discovered by hybridization with a microarray containing nearly all ORFs from *E. coli* K12 (Dong *et al.* 2001). Those genes unique to strain 342 compared to K12 are now being identified. A PCR-based subtraction hybridization protocol from Clontech (Cat. no. K1809-1 based on work by Akopyants *et al.* 1998) was used to prepare putative 342-specific sequences following subtraction with *E. coli* K12 DNA. The putative 342-specific DNA and K12 DNA were labeled with Cy5 and Cy3, respectively. The labeled DNA samples were hybridized on *E. coli* K12 ORF microarrays (Richmond *et al.* 1999). Of the 3000 *K. pneumoniae* 342 genes identified in Dong *et al.* (2001), 521 (17.36%) of the genes in common between K12 and 342 were still present after subtraction. Among the genes of high (>75%) and intermediate identity (55–75%) between the two strains, 256 (13.52%) and 265 (23.96%) of those genes remained in the sample after subtraction. Sequence analysis of randomly selected 24 clones containing subtracted DNA showed that about 85 to 87.5% of the clones had no or very low homology (less than 50% in DNA level) to *E. coli* K12 DNA. Thus by these two measures, the subtraction protocol provided a 5–6-fold enrichment of the sequences of interest. These experiments illustrate the utility of microarray analysis for the rapid validation of genome subtraction protocols. Sequence analysis of randomly selected 67 clones containing putative 342-specific DNA after subtraction with DNA from a clinical isolate, *K. pneumoniae* MGH78578 (http://genome.wustl.edu/Projects/bacteria/klebsiella.shtml), showed that about 73% to 77% of these clones (also roughly a 5-fold enrichment) had no or low homology to the clinical strain. A subtraction library by subtracting *K. pneumoniae* MGH78578 DNA from *K. pneumoniae* 342 DNA was also constructed. *K. pneumoniae* 342 specific genes with interest that related to the interaction with maize roots will be discussed.

References:

Akopyants *et al.* (1998) Proc. Natl. Acad. Sci. USA 95, 13108-13113
Chelius *et al.* (2000) Appl. Environ. Microbiol. 66, 783-787
Dong *et al.* (2001) Appl. Environ. Microbiol. 67, 1911-1921
Richmond *et al.* (1999) Nucleic Acids Res. 27, 3821-3835

IDENTIFICATION OF *RHIZOBIUM*-SPECIFIC PROTEIN FAMILIES BY COMPARATIVE GENOMICS

J. Gouzy, J. Batut, F. de Bruijn, D. Kahn

Laboratoire de Biologie Moléculaire des Relations Plantes-Microorganismes, INRA/CNRS, BP 27, 31326 Castanet-Tolosan Cedex, France

Two complete *Rhizobium* genome sequences have now been determined, for *Sinorhizobium meliloti* (Galibert *et al.* 2001) and *Mesorhizobium loti* (Kaneko *et al.* 2000). It thus becomes possible (i) to compare the structures of the genomes; (ii) to systematically investigate which genes are shared between different rhizobia; and (iii) to identify which genes may be considered as *Rhizobium*-specific.

The comparison of the *S. meliloti* and *M. loti* genome structures showed limited macrosynteny between their chromosomes and no synteny between *S. meliloti* pSym plasmids and the *M. loti* genome. In particular no synteny was found between *S. meliloti* pSymA and the *M. loti* symbiotic island. Similarly no synteny was observed between pSymA and the symbiotic region of the *Bradyrhizobium japonicum* genome (Göttfert *et al.* 2001).

Clustering of homologous proteins coded by the *S. meliloti* and *M. loti* genomes generated 2347 protein families. Among these, 131 families were found in *S. meliloti* only, and 142 families in *M. loti* only. Two-hundred and seventeen protein families were *Rhizobium*-specific. The CyaG family of nucleotide cyclases was identified among these. *Rhizobium* nucleotide cyclases were systematically analyzed using the MKDOM program (Gouzy *et al.* 1999) in order to identify the various types of domain arrangements. Forty-three cyclase related proteins were found in both genomes, which represents a considerable expansion when compared to other genomes. Four families are shared by the two rhizobia, including the CyaF, CyaG and CyaH families which appeared *Rhizobium* specific. The CyaF class appears to have expanded remarkably in rhizobia, with seven paralogs in *S. meliloti* and four paralogs in *M. loti*. CyaF cyclases can be defined structurally by an N-terminal catalytic domain and C-terminal TPR repeats. The role of this novel, *Rhizobium*-specific cyclase family remains to be defined.

References
Galibert F *et al.* (2001) Science 293, 668-672
Göttfert M *et al.* (2001) J. Bacteriol. 183, 1405-1412
Gouzy J *et al.* (1999) Computers and Chemistry 23, 333-340
Kaneko T *et al.* (2000) DNA Res. 7, 331-338

ANALYZING THE FUNCTION OF TWO *nodT* GENES IN *RHIZOBIUM ETLI*. ARE THESE COPIES DIRECTLY INVOLVED IN THE NODULATION PROCESS?

A. Hernández-Mendoza, T. Scheublin, N. Nava, O. Santana, C. Quinto

Departamento de Biología Molecular de Plantas, Instituto de Biotecnología,
Universidad Nacional Autónoma de México Apdo. Postal 510-3 C.P. 62251, Cuernavaca, Morelos, México

In *Rhizobium etli* we have previously described the isolation of two *nodT* copies, one located on plasmid c (*nodT*plc) and the other on chromosome (*nodT*cro) (1).

The *nodT*plc mutant does not have a clear nodulation phenotype. However, two ORFs that have 30 to 50% identity with *Escherichia coli cpxAR* genes were localized upstream *nodT*plc. CpxAR form a two-component signal transduction system that in *E. coli* responds to heat shock and membrane damage. A putative σ^{24} promotor sequence was also found upstream *nodT*plc. We are analyzing now the role of this copy in heat resistance.

In order to analyze the *nodT* cro function we have been making efforts to obtain an insertion in this gene without success. However, if we first complement with the wild-type gene *in trans*, stable insertions were obtained, suggesting that this gene is essential. On the other hand, two ORFs were found upstream *nodT*cro that have 73% to 63% identity with *ameAB* from *Agrobacterium tumefaciens* and *mexAB* from *Pseudomonas aeruginosa*, respectively. NodT has 50% and 30% identity with AmeC and OprM, respectively. AmeABC and MexAB-OprM form multidrug efflux pumps.

Our results suggest that in *R. etli*, *nodT* genes are not directly involved in the nodulation process.

References

Hernández-Mendoza A (1999) M.S. Thesis, IBT-UNAM

Acknowledgement

A. H.-M. was supported by a CONACyT Student Fellowship No. 90288.

A SIMPLE SYSTEM FOR CLONING RHIZOBIUM PLASMIDS

I. Hernández-Lucas[1], T.M. Finan[2], P. Mavingui[1], E. Martínez[1]

[1]Centro de Investigación sobre Fijación de Nitrógeno, UNAM, Apdo. Postal 565-A
Cuernavaca, Morelos, México
[2]Department of Biology, McMaster University, Hamilton, Ontario, Canada

In this work, we developed a simple system to clone *Rhizobium* plasmids in *E. coli*. This method can be used to clone plasmids potentially from many gram-negative bacteria. The requirements for this method are a Tn5 inserted in the replicon to be cloned, the integrative vector pTH509 (Chain 2000), which contains 300 bp of the IS50, the *oriT* (RK2), a ΩSp cassette and the F origin of replication. The system consists of a conjugation between *E. coli* harboring pTH509 and *Rhizobium*-bearing plasmid with a Tn5 insertion. A single cross-over occurs between the IS50 of the integrative vector and that of target *Rhizobium* plasmid. *Rhizobium* transconjugants are selected and mated with *E. coli*. The resultant *E. coli* strain carries the *Rhizobium* plasmid under the control of the F origin of replication. We have applied this simple system (based on two matings) to clone plasmids of *Rhizobium etli*, *Rhizobium tropici* and *Sinorhizobium meliloti*. The average size of the replicons that has been obtained is 200 kb. In contrast, plasmids of higher molecular weight (600 kb) were recalcitrant to cloning as a whole molecule, although fragments of 200 kb of such replicons were obtained. In the case of *R. tropici* pSym we obtain three different regions that cover almost the complete symbiotic plasmid of 600 kb, one of these fragments carried all the genes necessary to nodulate and fix nitrogen. Our results showed that this technique can be used to isolate different parts of high molecular replicons or to clone directly plasmids of moderate size, from *Rhizobium* species. Currently some of the replicons that we isolated are being sequenced.

References
Chain P *et al.* (2000) J. Bacteriol. 182, 5486-5494

Acknowledgements
We thank José Augusto, Ramírez -Trujillo, Miguel Angel Gaitan, José A Gama, Victor González for technical help and discussion, and Susana Brom, who kindly supplied strains.

ANALYSIS OF *CIS*-ACTING REGIONS WITHIN A *RepABC*-TYPE PLASMID REPLICATOR REGION

S.R. MacLellan, C.D. Sibley, T.M. Finan

Department of Biology, McMaster University, Hamilton, Ontario, L8S 4K1 Canada

The 1700 kb plasmid of *Sinorhizobium meliloti* possesses a *repABC* operon that controls replication and partitioning of the plasmid. We are engaging in genetic and biochemical analyses to determine the nature of *cis* and *trans* acting elements that control replication, segregation, copy number control, and incompatibility functions in the plasmid.

We have isolated a minimal, functioning replication region of 5.4 kb and the resulting mini-pExo derivative has been used in genetic studies to deduce the functions of encoded gene products and *cis*-acting regions. The 5.4 kb fragment encodes three genes (*repA*, *repB* and *repC*) and confers replication, efficient segregation, and incompatibility against the parental pExo (pSymB) plasmid when cloned into an otherwise non-replicating vector. Interestingly, a 4.6 kb fragment that contains the *repA* promoter and approximately 500 bp downstream of *repC* is not efficiently segregated during cell division. This indicates that stability determinates lie outside these boundaries.

On the basis of genetic analyses, we conclude that the *repABC* genes are transcribed from a single promoter upstream of *repA*. This promoter exists within 100 bp upstream of the predicted translational start of RepA. An insertion in *repA* eliminates replication likely due to polar effects on the expression of RepC. A 2.3 kb subclone containing only the *repC* open reading frame (plus upstream and downstream non-coding DNA) confers replication indicating that the expression of RepC is sufficient for replication. However, replication is only manifest when this fragment is cloned downstream of a vector based *lac* promoter suggesting that a promoter does not exist immediately upstream of *repC*.

The *repABC* gene fragment exerts incompatibility against its parental plasmid. A 242 bp sequence isolated from the intergenic region between *repB* and *repC* is sufficient to confer this incompatibility. Interestingly, a fine scale genetic dissection of this region indicates that the incompatibility locus is at least 130 bp long and that the entire region is needed for incompatibility. A frameshift mutation in *repA* partially relieves the incompatibility effect suggesting that RepA is a *trans*-acting incompatibility factor.

Both the *repB-repC* intergenic region directly upstream of *repC* and a nucleotide region downstream of the *repC* gene are essential for replication. Subcloning experiments wherein each region was deleted independently abolished replication.

We are continuing to delineate the minimal genetic requirements for replication, segregation and incompatibility.

PREDICTION AND CHARACTERIZATION OF REPLICON FUSIONS IN *RHIZOBIUM* SP. NGR234

X. Guo[1], P. Mavingui[1], M. Flores[1], X. Perret[2], W.J. Broughton[2], R. Palacios[1]

[1]Centro de Inv. Sobre Fijación de Nitrógeno, UNAM, 565-A, Cuernavaca, Morelos, Mexico
[2]Lab. Biol. Mol. Plantes Supérieures, Univ. de Genève, Genève, Switzerland

1. Introduction

Repeated DNA sequences in a genome represent potential sites for homologous recombination. Recombination between reiterated sequences present in different replicons may lead to their co-integration. *Rhizobium* sp. NGR 234 contains three replicons: the chromosome, the pSym and a megaplasmid. Based on the analysis of the sequence of the pSym (Freiberg *et al.* 1997) and partial sequences from the megaplasmid and the chromosome, the repeated sequence NGRRS-1 (6 kb) shared by the three replicons was identified. Co-integrations among the three replicons should be produced by homologous recombination between pairs of NGRRS-1.

2. Materials and Methods

The detection of the different rearrangements and the obtention of subpopulations pure for a specific rearrangement were performed by a PCR-based technique, which detects the join points corresponding to the rearrangement, coupled with an artificial selection. We have referred to such experimental strategy as natural genomic design (Flores *et al.* 2000).

3. Results and Discussion

The PCR products corresponding to the joint points of predicted co-integrations among the three replicons (chromosome-megaplasmid, chromosome-pSym, megaplasmid-pSym) were detected in the culture of the wild-type strain NGR234. A derivative population, CFNX416, possessing a co-integration between the pSym and the megaplasmid was isolated by artificial selection and confirmed by different techniques: plasmid profile, pulse-field gel electrophoresis (PFGE), Southern hybridization with specific replicon probes (nifH and exoBDFLK genes). The size of this co-integrated plasmid is estimated to be more than 2500 kb. Such rearrangement has no apparent influence on the growth in both rich and minimal media. After 30 days' culture (almost 200 generations) in liquid medium, transferred every day, most of the cells (more than 90%) still contained the co-integrated structure. Reversion to the wild type genome architecture and the partition of the co-integrate into new sized replicons were also observed. Considering that no exogenous DNA was introduced and that the experiments were performed without any selective pressure, the results suggest that: (i) the natural genomic design strategy is efficient for the artificial selection of alternative genome structures; and (ii) the bacterial genome is capable of rapid evolution that can be observed in action.

4. References

Flores M *et al.* (2000) Proc. Natl. Acad. Sci. USA 97, 9138-9143
Freiberg C *et al.* (1997) Nature (London) 387, 394-401

5. Acknowledgements

We thank M. Rodríguez, A. Moreno, V. Quinto, Rosa Ma. Ocampo for technical assistance as well as G. Dávila and D. Romero for suggestions helpful to the project. This work was supported in part by grants L0013N and 0028 from CONACyT-Mexico and IN214498 fromDGAPA-UNAM-Mexico.

SYMBIOTIC BACTERIA IN ROOT NODULES OF *ACACIA MANGIUM*

A. Ngom[1], M. Abe[1], T. Uchiumi[1], A. Suzuki[1], A. Nuntagij[2], S. Kotepong[2], S. Higashi[1]

[1]Dept Chem. and BioSci., Kagoshima Univ., Kagoshima 890-0065, Japan
[2]Soil Sci. Div., Dept Agr., Rhizobium Build., Bangkok 10900, Thailand

1. Introduction

Acacia mangium, a leguminous tree, which can survive in infertile soil because its symbiotic ability with nitrogen-fixing bacteria. Acacia tree has many purposes, such as fuel-wood, timber, wind protection and livestock fodder. Moreover, the leguminous trees in dry tropical areas form thick-leaved branches, which spread widely and give good shade over the crop yard to be protective from strong sunlight. There are few reports concerning leguminous tree and their symbiont. The symbiotic performance between *Acacia mangium* and its nodulating bacteria were analyzed.

2. Procedures

Isolated nodule bacteria were cultured in TY or YM medium. Genomic DNA was isolated and used as template for PCR of 16S rRNA and a couple of *nod* genes. General biological characters, generation time, optimum pH and temperature for growth, enzyme secretion during growth were measured. Nodule structure, bacteroid existence in a symbiosome was also observed. The infection pathway was investigated using Acacia seedlings by slide culture technique (Higashi 1979).

3. Results and Discussion

Twenty-two strains were tested for nodulation to *Acacia* species and other genera. All tested strains could nodulate and showed nitrogen-fixing activity on *Acacia mangium*. No root hairs had been observed on the surface of Acacia young roots. However, a few short, deformed root hairs had emerged from the root when the seedlings were inoculated with the symbiotic bacteria. The infection thread could be recognized in the induced root hair around 2 weeks after inoculation. The nodule morphology of Acacia was typical determinate type. The existence of bacteria in the nodule cell was exhibited in two groups. That is, the symbiosome is formed by several bacteroid cells enwrapped in a peribacteroid membrane (PBM), called plural type, and by single bacteroid cell in a PBM, called singular type.

The phylogenetic analysis of 16S rRNA gene revealed that the Acacia strains fell into three genera, *Rhizobium*, *Bradyrhizobium* and *Ochrobacterum*. This result showed that the Acacia strains from Southeast Asia were very diverse. Strain DASA35030, from Thailand, showed 93% similarity to *Ochrobacterum intermedium*. This strain showed rod shape and grew faster (generation time 60 min) than other *Rhizobium* and *Bradyrhizobium* strains. The optimum temperature for growth of strain DASA35030 was 37°C though other strains were 30°C. Furthermore, pH growth range of strain DASA35030 was very wide pH 5 to pH 9.5 comparing with other strains. However, most Acacia strains preferred the alkaline condition rather than the acidic condition.

Lebunh *et al.* (2000) isolated *Ochrobacterum* strains from rhizoplane of wheat. Our report might be the first one that *Ochrobacterum* strain was isolated from root nodule of leguminous plant. The most plausible interpretation of this fact could be an example that it is a horizontal transfer of symbiotic gene set of *Rhizobium* into *Ochrobacterum*.

4. References

Higashi S, Abe M (1979) Appl. Environ. Microbiol. 39, 297-301
Lebunh *et al.* (2000) J. Syst. Evol. Microbiol. 50, 2207-2223

DEVELOPING GENETIC TOOLS FOR *FRANKIA*, THE BACTERIAL PARTNER OF THE ACTINORHIZAL SYMBIOSIS

A.K. Myers, T. Rawnsley, L.S. Tisa

Dept of Microbiology, Univ. of New Hampshire, Durham,
NH 03824-2617, USA

Genetic analysis of *Frankia* is in its infancy. There are no known systems for gene transfer. Reliable standardized mutagenesis protocols have not yet been firmly established. One major drawback for genetic studies on *Frankia* is the lack of genetic markers. To identify potential genetic markers, twelve strains of *Frankia* were screened for resistance to antimetabolites, antibiotics and heavy metals by the use of a growth inhibition assay. Several strains had distinctive patterns of resistance that are potentially useful as genetic markers. The physical properties of the *Frankia* genome are being investigated by the use of PFGE. To develop a collection of genetically marked *Frankia* strains, chemical and physical mutagens were used to generate axotrophic, antimetabolite-resistant and antibiotic-resistant mutants. In the case of strain EuI1c, EMS induced tetracycline-resistant and lincomycin-resistant mutants occurred at frequencies of 3.2 x 10^{-3} and 4.7 x 10^{-4}, respectively. With strain Cc1.17, EMS induced lincomycin-resistant and 5-fluorouracil-resistant mutants occurred at frequencies of 1 x 10^{-5} and 4 x 10^{-5}, respectively. The conjugative transposon Tn916 was successfully introduced into *Frankia* by filter mating with *Enterococcus faecalis* as the donor. These transconjugates were stable and have been maintained in culture. PCR and DNA hybridization experiments confirmed the insertion of the *tetM* gene. This is the first successful transfer and expression of foreign DNA into *Frankia*.

GENOMIC ANALYSIS AND GENETIC STUDY OF PHOSPHATE REGULATION GENES (*pho* REGULONS) IN *SINORHIZOBIUM MELILOTI* 1021

Z. Yuan, R. Morton, T.M. Finan

Department of Biology, McMaster University, Hamilton, Ontario, L8S 4K1 Canada

In bacteria, the transport and assimilation of phosphorus compounds from the environment is regulated by the PhoR-PhoB two component system. PhoR is a transmembrane sensor histidine kinase and PhoB is a transcriptional activator. In phosphorus limiting conditions PhoR activates PhoB by phosphorylation. The Pho regulon consistes of genes whose expression is regulated by PhoB. Phosphorylated PhoB binds to an 18 base pair (*pho* Box) containing two 7 bp repeats of 5'---CT{T/G}TCAT---3', which are present in the promoter region of phosphate regulated genes.

In *S. meliloti* we have previously identified two phosphate transport systems. The PhoCDET system is an ABC-type high affinity phosphate transport system. The OrfA-Pit system is a low affinity system which is expressed in cells growing in excess phosphate but repressed in cells growing under conditions of Pi limitation. We have cloned a 7.5 kb *Hind*III gene fragment which includes the *phoR-pstS-pstC-pstA-pstB* and partial *phoU* genes. PstSCAB forms another ABC type phosphate transporter in *E. coli*. By Tn*5*-B20 and *lacZ-aacC1* cassette gene disruptions/fusions, we mutated the *pstA, pstB and phoR* genes respectively. We found: (a) *pstA-pstB-phoU-phoB* are in one operon, (b) *pstB* expression is not regulated by the media phosphate concentration and is independent of *phoB*, (c) in free-living cells, *pstB* mutants behave in the same manner as *phoU* or *phoB* regulatory mutants, i.e. they exhibit an alkaline phosphatase negative phenotype, (d) in plant tests, a *pstB* mutant had normal nitrogen fixation ability and like *phoB* mutations, the *pstB* mutation suppressed the Fix⁻ phenotype of *phoCDET* mutants, (e) *phoB* expression is neither regulated by phosphate concentration nor does its expression appear to be autoregulated. Sequence analysis showed that there is a putative *pho* box upstream of *pstS* genes.

We collected all the available *pho* boxes from *S. meliloti* and *E. coli*, then built a matrix with the frequency of the four nucleotides at each position of the *pho* box and generated the *S. meliloti* *pho* box consensus motif CTGTCAT AAAT CTGTCAT. Employing this matrix we analyzed the complete *S. meliloti* genome sequence. Approximately 110 putative *pho* boxes were found to be located around different promoter regions: 71 *pho* boxes belong to known genes, and 40 to unknown genes. To further analyze the Pho regulon, we cloned about 20 genes or promoters including *nadE₂, Smc01907, Smb20427, Smc00801, sra, rhbF, CRP, recF, katA, ppk, afuA* which contain putative *pho* boxes and are studying their promoter activity in low and high phosphate condition.

DNA BINDING PROPERTIES OF RepA, A PROTEIN INVOLVED IN PARTITION AND AUTOREGULATION OF THE REPLICATOR REGION OF THE SYMBIOTIC PLASMID OF *RHIZOBIUM ETLI (repABC* FAMILY)

A. Pérez-Oseguera, M.A. Ramírez-Romero, M.A. Cevallos

Centro de Investigación sobre Fijación de Nitrógeno, A. P. A.P. 565-A., UNAM
Cuernavaca, Morelos, México

As a member of the *repABC* plasmid family, the replicator region of the *R. etli* symbiotic plasmid, contains three genes (*repA, repB* and *repC*) and a conserved region sequence between *repB* and *repC* (Ramírez-Romero *et al.* 1997).

A genetic analysis of the *repABC* genes showed that they are organized in an operon. The *repA* and *repB* genes encode proteins required in plasmid stability, and in plasmid copy-number control. Also, it has been shown that RepA acts as a *trans*-acting incompatibility factor (Ramírez-Romero *et al.* 2000).

Little is known about the regulation of the *repABC* genes. To understand the transcriptional characteristics of this operon: the transcriptional start site was identified. The –10 and –35 hexameric elements of the repABC promoter were localized by mutational analysis, and the negative transcriptional regulator and its target site was identified by genetic and biochemical analysis.

Several *repA::gusA* fusions were constructed to identify the minimal region of DNA necessary for the expression of the operon. The results indicate that the promoter region is contained within 82 bp upstream of *repA*. The transcription start site of the *repABC* operon was found 57 bp upstream of the initiation codon of *repA*. Upstream of this site, a sequence similar to the –35 and –10 boxes of the *Escherichia coli* σ70 promoter consensus was recognized. A mutagenesis analysis of the putative –35 and –10 boxes showed that they were correctly identified.

The β-glucuronidase activity of the *repA::gusA* fusion was repressed in the presence of RepA but not with RepB and RepC, indicating that RepA is the negative transcription regulator.

To demonstrate that RepA interacts with the sequences located upstream of the *repA* gene, two strategies were followed. First, a DMS footprinting analysis in vitro of *R. etli* strains containing the *repA::gusA*, in the presence or absence of RepA was performed. The results of these experiments suggest that the putative binding site for RepA is one inverted repeat sequence of 14 nucleotides located upstream of *repA*. Second, gel mobility shift assays were performed with a purified RepA hexahistidine derivative and with a 160 bp region upstream of *repA*, as target DNA. These experiments showed that RepA is able to bind to this DNA, and that this activity has an absolute requirement of ATP or ADP, suggesting that ATP hydrolysis is not necessary for DNA binding.

References

Ramírez-Romero *et al.* (1997) Microbiol. 134, 2825-2831
Ramírez-Romero *et al.* (2000) J. Bacteriol. 182, 3117-3124

CONSTRUCTION OF A MICROARRAY COVERING THE *SINORHIZOBIUM MELILOTI* 1021 GENOME

S. Rüberg[1], B. Linke[1], E. Krol[1], S. Weidner[1], U.D. Schiller[2], S. Kurtz[2], R. Giegerich[2], A. Pühler[1], A. Becker[1]

[1]Lehrstuhl für Genetik, Universität Bielefeld, Germany
[2]AG Praktische Informatik, Technische Fakultät, Bielefeld, Germany

Sequencing of the genome of the symbiotic soil bacterium *S. meliloti* 1021 was recently finished by an international consortium (Galibert *et al.* 2001). The genome consists of three replicons: the 3.5 Mb chromosome, the 1.35 Mb megaplasmid pSymA and the 1.7 Mb megaplasmid pSymB. Annotation of the genome sequence predicted 6204 protein-encoding genes (http://sequence.toulouse.inra.fr/rhime/Consortium/home.html). To construct a microarray on glass slides as well as a macroarray on nylon membranes comprising DNA fragments representing the 6204 protein-encoding ORFs, primer pairs for amplification of 6163 internal fragments of these ORFs by PCR and 41 70mer oligonucleotides were designed using the Primer-Lib software (B. Linke, Bielefeld, unpublished) that creates a primer database and allows for the design as well as the verification of primer sequences on the whole genome scale. Most of the primer pairs amplify 300 to 350 bp fragments, whereas the remaining primer pairs amplify 80 to 299 bp fragments that represent short genes. The Tm of all primers varies less than 1°C. All primers carry 5' extensions that allow the reamplification of the amplified fragments by PCR using standard primers. Currently, the amplification of fragments for 6163 different *S. meliloti* 1021 ORFs is under way.

*Hin*dIII and *Bsr*GI restriction sites in the 5' extensions that are not present in the amplified sequences of 6051 ORFs can be used for directed cloning into the mobilizable suicide vector pK19mob2HMB. For cloning 4992 fragments longer than 220 bp were selected to allow for integration of the resulting plasmids in the *S. meliloti* genome by homologous recombination. Since internal fragments of the predicted protein-coding ORFs were chosen, in most cases plasmid integration will result in a loss of gene function. Currently, cloning of these fragments and conjugation is carried out.

A 183 bp linker cassette and 1498 30-bp tags were designed for the construction of a set of miniTn5 transposons each containing two of these tags. The linker cassette and the tag sequences were designed using the DNA sequence compiler software (Feldkamp *et al.* http://ls11-www.cs.uni-dortmund.de/molcomp). All sequences were designed to be as different as possible from the *S. meliloti* 1021 genome sequence and from each other. The Tm of all tags that contain 24 bp variable sequence and are flanked by sequence overhangs compatible to *Hin*dIII and *Kpn*I restriction sites varies less than one 1°C. These tags are suitable to be inserted into the *Kpn*I and *Hin*dIII sites of transposon mTn5-GNm-L derived from transposon mTn5-GNm (Reeve *et al.* 1999) by insertion of the linker cassette into the *Sfi*I site. Tag sequences inserted into the linker cassette can be amplified by quantitative PCR using the two standard primer pairs individually or together. Tags were linked to a coated glass surface after ligation of an amino modified linker and denaturation. A hybridization was carried out using Cy3- and Cy5-labeled tags that were amplified from a mixture containing 96 different tag-containing fragments in different ratios. Cross-hybridization experiments showed that the tags can be differentiated by hybridization. Mutants carrying individually tagged transposons are suitable for the analysis of mixtures of these mutants in competition experiments.

References:

Reeve WG *et al.* (1999) Microbiol. 145, 1307-1316
Galibert F *et al.* (2001) Science 293, 668-672

A *FadD* MUTANT OF *SINORHIZOBIUM MELILOTI* SHOWS MULTICELLULAR SWARMING MIGRATION AND IS IMPAIRED IN NODULATION EFFICIENCY ON ALFALFA ROOTS

M.J. Soto[1], M. Fernández-Pascual[2], J. Sanjuan[1], J. Olivares[1]

[1]Dept de Microbiología del Suelo y Sistemas Simbióticos, Estación Experimental del Zaidín CSIC, Granada, Spain
[2]Centro de Ciencias Medioambientales CSIC, Madrid, Spain

1. Introduction

Motility and chemotaxis play a crucial role in the ability of *Sinorhizobium meliloti* to nodulate its host, alfalfa. Swarming is a form of bacterial translocation, generally dependent on flagella, that involves cell differentiation and is characterized by a rapid and coordinated population migration across solid surfaces. This kind of active motility, mostly found in pathogenic bacteria in which it plays an important role in the colonization of natural environments, has not yet been described for any member of the family *Rhizobiaceae*. Although the pathways of signal integration are still poorly understood, there is evidence showing that extracellular chemical signals, physiological parameters as well as surface contact provide stimuli triggering swarm cell differentiation (Fraser, Hughes 1999).

2. Results

We have isolated a Tn*5* mutant of *S. meliloti* GR4 showing conditional swarming. Multicellular migration in the mutant strain QS77 seems to be a cell density-dependent phenomenon induced on semisolid minimal medium in response to certain amino acids, influenced by the viscosity of the medium, and abolished by glucose. The mutation in QS77 lies within a gene encoding a homolog of the FadD protein (long chain fatty acyl CoA ligase) of several microorganisms, including the RpfB protein of the phytopathogenic *Xanthomonas campestris*, a putative FadD protein mediating regulation of virulence factor synthesis (Barber *et al.* 1997). Like an *Escherichia coli fadD* mutant and in contrast to the *S. meliloti* wild type strain GR4, QS77 cannot grow using oleate as sole carbon source. Interestingly, the *S. meliloti fadD* mutant shows a defective symbiotic phenotype. Although QS77 is able to induce nitrogen-fixing root nodules on alfalfa plants, it exhibits a less efficient nodulation phenotype than the wild-type, together with a significant reduction in competitiveness. Furthermore, the expression of genes involved in motility (*che*, *mofli*, and *flaA* belonging to the flagellar regulon), and symbiosis (*nodC*) are significantly reduced in the *fadD* mutant. In *trans* expression of multicopy *fadD* restored growth ability of an *E. coli fadD* mutant as well as growth, control of motility and symbiotic phenotype of QS77.

3. Conclusions

We have identified conditional swarming motility in *S. meliloti*. FadD plays an important role in the control of this multicellular migration. Our results indicate that in *S. meliloti*, fatty acid derivatives may act as intracellular signals controlling motility and symbiotic performance through gene expression.

4. References

Barber *et al.* (1997) Mol. Microbiol. 24, 555-566
Fraser H (1999) Curr. Opin. Microbiol. 2, 630-635

THE CHROMOSOMAL REPLICATION INITIATION PROTEIN DnaA, AND THE IDENTIFICATION OF THE CHROMOSOMAL ORIGIN OF REPLICATION FROM *SINORHIZOBIUM MELILOTI*

C. Sibley, S.R. MacLellan, T.M. Finan

Department of Biology, McMaster University, Hamilton, Ontario, L8S 4K1 Canada

The organization of the *Sinorhizobium meliloti* tripartite genome, consisting of one chromosome of 3500kb and two large plasmids (1700kb and 1400kb), has stimulated interest in the mechanisms of replication initiation and replicon segregation during cellular division. Prokaryotic replication initiation has been an intense area of investigation, and has led to the characterization of some essential proteins involved in the process. The DnaA protein has been implicated as one such essential component in the formation of an active complex at the chromosomal origin of replication. The DnaA protein binds to 9 base pair elements in the oriC, leading to the melting of the duplex DNA and the initiation of the replication fork.

The goal of this work was to purify *S. meliloti* DnaA and identify the origin of replication on the *S. meliloti* chromosome. The *S. meliloti dnaA* gene was originally cloned into pBADHisA and *S. meliloti* DnaA(His$_6$) was successfully over-expressed in *E. coli*. However, virtually all of the over-expressed protein was represented as insoluble inclusion bodies. Changes in the induction procedure as well as the *E. coli* expression strain did not result in an increased expression of soluble *S. meliloti* DnaA(His$_6$). Subsequently, the *S. meliloti dnaA* gene was cloned as a PCR product in frame into the pGEX5X-1 plasmid, which is designed for high level intracellular expression of proteins as fusion to glutathione S-transferase. Over-expression of soluble *S. meliloti* DnaA as a fusion protein was successful in *E. coli*. The over-expression of soluble fusion protein was optimized, by inducing protein synthesis once cells had grown to an OD = 0.6 with 0.1 mM IPTG. Cells were then incubated overnight at room temperature.

S. meliloti DnaAGST has been successfully purified. Induced culture (100 mL) was pelleted and resuspended in 10 mL ice-cold STE buffer. Lyozyme was added, incubated on ice for 15 min and just prior to sonication 100 µL 1M DTT and 1.4 mL 10% Sarkosyl was added. Cells were sonicated and the crude lysate was spun (16,000 rpm) to pellet insoluble material. The effective concentrations of Sarkosyl and Triton X-100 were then brought to 0.7% and 2%, respectively and incubated at room temperature for 45 minutes. The soluble lysate was then incubated with 1 mL of 50% glutathione sepharose in PBS at room temperature for 1 hour with agitation. The binding mixture was then added to a column, washed with PBS and *S. meliloti* DnaAGST was eluted with 1 mL volumes of elution buffer containing 20 mM GSH. The 25kDa GST domain was effectively cleaved by treatment with Factor Xa to liberate the 57 kDa *S. meliloti* DnaA.

The origin of replication on the *S. meliloti* chromosome has not previously been experimentally identified. Analysis of the *S. meliloti* chromosome sequence revealed a region similar to the oriC locus of the closely related organism *Caulobacter crescentus* which contains a significant A/T rich region upstream of the *hemE* gene. A 3 kb fragment containing the entire *S. meliloti hemE* ORF, the A/T rich region as well as both up and downstream DNA was PCR amplified. To test if the PCR product contained the *S. meliloti* chromosomal origin of replication it was cloned into pUCP3OT, a plasmid unhable to replicate in *S. meliloti* cells. Conjugal transfer of the recombinant plasmid into *S. meliloti* showed that the 3 kb *S. meliloti hemE* region was able to confer autonomous replication to the pUCP30T plasmid. A genetic analysis of the *S. meliloti* origin of replication is currently underway.

HORIZONTAL TRANSFER OF THE *SINORHIZOBIUM MELILOTI* pExo MEGAPLASMID

L.B. Koski, T.M. Finan, G.B. Golding

Department of Biology, McMaster University, Hamilton, Ontario, L8S 4K1 Canada

Plasmids play one of the strongest roles in prokaryotic evolution, moving from cell to cell, among bacterial strains, related bacterial species and even domains of life. We have phylogenetically analyzed the pExo (pSymB) megaplasmid of the gram-negative soil bacterium *Sinorhizobium meliloti* and estimate that 60% of the megaplasmid has been acquired through horizontal gene transfer. Previous studies have suggested that horizontal transfer in the rhizobial lineage has occurred only between closely related strains. Our results suggest otherwise; horizontal transfer seems to have taken place between pExo and many distantly related species. A large portion of the horizontally transferred genes appear to be related to *Streptomyces coelicolor*, *Escherichia coli*, and *Pseudomonas aeruginosa*, all of which, like *S. meliloti*, are found in the soil. Interestingly, the highest number of ORFs related to any one species is a gram-positive species, *Streptomyces coelicolor*. Analysis of the functional distribution of pExo ORFs reveal that most of the horizontally transferred genes (43%) are involved in cellular processes, which comprise proteins important for the transport of small molecules, DNA uptake, protection responses and nodulation.

THE cAMP RECEPTOR PROTEIN (CRP)-MEDIATED REPRESSION EFFECT ON *KLEBSIELLA PNEUMONIAE nif* PROMOTERS IN *ESCHERICHIA COLI*

Z.-T. Li, Y.-C. Sun, J. Li, Z.-X. Tian, F.-M. Li, Y.-P. Wang

National Laboratory of Protein Engineering and Plant Genetic Engineering, College of Life Sciences, Peking University, Beijing 100871, PR China

1. Introduction

The quality and quantity of carbohydrate available to the cell is the major limiting-factor for the capacity of energy-consuming nitrogen fixation. In enteric-bacteria, in response to the PTS system, the cAMP receptor protein (CRP) mediates the glucose effect at transcriptional level. Recently, we have observed that CRP can repress a σ^{54}-dependent promoter in a cAMP dependent, binding site independent fashion. Direct interaction between CRP and promoter bound σ^{54} RNA polymerase plays an important role in repression (Wang *et al.* 1998), which may suggest that the CRP-mediated repression effect on a σ^{54}-dependent promoter is general.

2. Results and Discussion

In order to investigate the above hypothesis, a series of *nif* promoters from *Klebsiella pneumoniae* (i.e. *nifB*, *nifE*, *nifF*, *nifH*, *nifJ* and *nifU* promoters) were cloned and fused in frame with the *lacZ* reporter gene on plasmid pGD926. When these constructs were introduced into *Escherichia coli cya crp* mutants respectively, their activities were measured. The results show that: first, these promoters can all be activated by their cognate activator NifA, expressed by a constitutive promoter; secondly, their activity can all be repressed by the CRP protein in a cAMP dependent fashion, ranging from 5- to 48-fold. Sequence analysis of the above promoters indicates that some CRP-binding sites may exist on the upstream control sequences, and several of them overlap with previously identified NifA-binding sites. However, little relationship has been found between the location and the conservation of these putative CRP-binding sites and the fold of repression.

The *crp* genes from *K. oxytoca* and *K. pneumoniae* have been isolated and sequenced. At the nucleotide sequence level, they are about 85% homologous to *E. coli crp* gene. At the deduced amino acid sequence level, they are virtually identical to that of *E. coli*. The only difference is a serine at position 118 of *K. oxytoca* and *K. pneumoniae* rather than the alanine in *E. coli*. This difference has also been observed in *Salmonella typhimurium* and *Klebsiella aerogenes* CRP proteins. Previous studies demonstrated that it did not cause significant differences in function.

Taken together, these results indicate that although our experiments were conducted in a heterogeneous genetic background, the same repression phenomenon could also exist in homogeneous strain. It suggests that CRP-cAMP-mediated repression on *nif* promoters has physiological significance.

3. Reference

Wang *et al.* (1998) EMBO J. 17, 786-796

4. Acknowledgement

This work is supported in part by grants from National Natural Science Foundation of China (grants 39925017 and 39970168), the Ministry of Education of China and the 863 program from the Ministry of Science and Technology of China.

GENETIC AND FUNCTIONAL ANALYSIS OF NOVEL *RHIZOBIUM TROPICI* CIAT899 SYMBIOSIS-RELEVANT LOCI IDENTIFIED THROUGH Tn*5* INSERTION MUTAGENESIS AND SELECTION OF ACID-SENSITIVE MUTANTS ON ACIDIFIED SOLID MEDIA

P. Vinuesa[1,3], F. Neumann-Silkow[1], A. Aranda-Rickert[1], C. Pacios-Bras[2], H. Steele[1], H.P. Spaink[2], E. Mörschel, D. Werner[1]

[1]FB Biologie, Dept. of Cell Bio and Applied Botany, Philipps-University, Germany
[2]Leiden University, Institute for Mol. Plant Sci., The Netherlands
[3]Centro de Investigación sobre Fijación de Nitrógeno, UNAM, Mexico

Research on the genetic and physiological basis of acid tolerance in rhizobia is both of practical and fundamental significance. Firstly, acid soils constrain nodulation and N_2-fixation in many legume-*Rhizobium* symbioses of agronomic interest (Glenn *et al.* 1999). Since both the micro- and the macrosymbiont can represent the acid-sensitive component of the symbiosis, investigations aimed at finding acid-tolerant plant germplasm and compatible rhizobial strains are of utmost agronomic and ecological relevance (Zahran *et al.* 1999). Secondly, rhizobia, as intracellular microsymbionts, have been postulated to inhabit an acidic lytic compartment, the symbiosome, which resembles in several aspects the phagosomes inhabited by mammalian intracellular pathogens (Mellor *et al.* 1989; Parniske 2000). As a corollary, it can be hypothesized that the capacity of the microsymbiont to adapt to an acidic environment may be of fundamental importance for host cell invasion and the consequent establishment of a N_2-fixing *Rhizobium*-legume symbiosis. The aim of this study was to identify novel rhizobial genes required for acid tolerance, and eventually also for symbiosis. For this purpose *R. tropici* CIAT899 was chosen as the model organism, since this strain is well known to be highly tolerant to several environmental stress factors, including acidity, being able to grow on media acidified down to pH 4.0 (Graham *et al.* 1994). Four prototrophic acid-sensitive Tn*5* insertion mutants of *Rhizobium tropici* CIAT899 were identified by plating a mutant bank on acidified minimal medium (pH 4.2). These mutants (899-PV1, 899-PV2, 899-PV4, 899-PV9) elicit Ndv⁻ and Fix⁻ nodules on bean plants. Light and electron microscopy of the nodules elicited by strains 899-PV4 and 899-PV9 showed that they can invade nodules. However, bacteroids are not efficiently released from infection threads, and do not elongate. Very few symbiosomes are found within highly vacuolated host cells, lysing prematurely, indicating that the mutated loci affect bacteroid intracellular accommodation and nodule differentiation. Cosmids complementing the defects displayed by the 4 strains could be isolated. The sequence of cosmid subclones complementing 899-PV4 and 899-PV9 was determined, resulting in the identification of two novel symbiosis-relevant loci. The Tn*5* insertions in strains 899-PV4 and 899-PV9 were mapped by sequencing. The former disrupted a gene with high sequence identity to the *Agrobacterium tumefaciens* chromosomal virulence locus *acvB* (Vinuesa *et al.* submitted). We demonstrated by means of reciprocal complementation studies, that the agrobacterial and rhizobial *acvB* genes are orthologs. Biological basis of the acid tolerance displayed by these mutants is presently not understood.

References

Glenn A *et al.* (1999) In Booth RI (ed), Bacterial Responses to pH, pp. 112-130, Wiley, Chichester
Graham PH *et al.* (1994) Can. J. Microbiol. 40, 189-207
Mellor R (1989) J. Exp. Bot. 40, 831-839
Parniske M (2000) Curr. Opin. Plant Sci. 3, 320-328
Vinuesa P *et al.* (2001) Mol. Plant-Microbe Interact., (in review)
Zahran HH (1999) Microbiol. Mol. Biol. Rev. 63, 968-989

A FLAVOPROTEIN MODULATES INHIBITION BY AICAR OF RESPIRATORY GENE EXPRESSION IN *SINORHIZOBIUM MELILOTI*

C. Cosseau, J. Batut

Lab. Biol. Mol. des Relations Plantes-Microorganismes, CNRS-INRA BP27 31326, Castanet, France

1. Introduction

The interaction between rhizobia and their host plants culminates in the formation of specialized organs called nodules in which differentiated bacteria fix nitrogen to the benefit of the host plant. The main focus of our group is the identification of the physiological signals that control the expression of bacterial genes important for nitrogen fixation. Symbiotic expression of most of the nitrogen fixation genes identified so far in *Sinorhizobium meliloti* is under the control of the two component regulatory system fixLJ that responds to the microaerobic environment in the nodule (David *et al.* 1989). Soberon *et al.* (2001) reported that "AICAR" a purine intermediate down-regulates *fixN* expression in *S. meliloti*. Recently, we showed that a 151 amino acid protein of previously unknown function (Batut *et al.* 1989) modulates inhibition by "AICAR".

2. Results and Discussion

We showed that addition of "AICAR" to the culture medium reduces the microaerobic level of expression of *fixN-lacZ*, *fixT-lacZ*, *fixK-lacZ* and *nifA-lacZ* gene fusion by 10-, 3.7-, 7.5- and 1.6-fold respectively. "AICAR" has no effect when *fixK* is constitutively expressed. These results suggest that "AICAR" affects *fix* gene expression upstream of *fixK* and *nifA*. We speculate that the metabolite interferes with the activity of the two components FixLJ system either directly or indirectly via an unknown protein intermediate. This is consistent with the proposed role of AICAR in eukaryotic cells (Kaiser *et al.* 1996).

It is to point out that the repression by "AICAR" occurs in a strain mutated for a gene known as the *ORF151*, located downstream of *fixK*. *ORF151* gene product is homologous to Nim proteins of *Bacteroides fragilis*, to an unkown protein from *Methanobacterium thermoautotrophicum* and to 4 homologous genes found in *Mezorhizobium loti* genome. Nim proteins are 5-nitroimidazole reductase (Carlier *et al.* 1997). We purified the ORF151 protein as an inteine fusion and further biochemical analysis reveals that it encodes a flavoprotein. These properties of the ORF151 suggest that, as Nim proteins do on 5-nitroimidazole drugs, it could have a reductase activity on AICAR.

In the symbiosis, an *ORF151* mutant is Nod$^+$ and fix$^+$ and symbiotic activity of a *fixN-lacZ* fusion is only reduced by 2-fold in the *ORF151* mutant. However when AICAR is endogenously accumulated (in a *purH* mutant), *fixK* is repressed by 10-fold in the nodule. We speculate that AICAR may accumulate under particular symbiotic conditions as a result of either bacteroids or plant cell metabolism and inhibit *fixK* and *nifA* expression. ORF151 would prevent AICAR effect and preserve nitrogen fixation and respiration.

3. References

Batut J *et al.* (1989) EMBO 8, 1279-86
Carlier *et al.* (1997) Antimicrob. Agent Chemotherapy, 1495-1499
David J *et al.* (1998) Cell 54, 671-683
Kaiser *et al.* (1996) Physiol. Plant. 98, 833-837
Soberon *et al.* (2001) Molec. Plant-Microbe. Int. 14, 572-576

EFFECT OF OVEREXPRESSION OF A SOYBEAN CYTOSOLIC GLUTAMINE SYNTHETASE GENE ON NODULATION AND GROWTH OF TRANSGENIC PEA GROWN AT DIFFERENT LEVELS OF NITRATE

H. Fei[1], J.K. Vessey[1], B. Luit[1], S. Chaillou[2], B. Hirel[2], E. Carrayol[2], J.D. Mahon[3], P.L. Polowick[3]

[1]Dept of Plant Science, U. Manitoba, Winnipeg, MB, Canada
[2]Laboratoire du Métabolisme et de la Nutrition des Plantes, INRA, Route de St-Cyr, Versailles Cedex, France
[3]Plant Biotechnology Institute, Saskatoon, SK, Canada

The impact of overexpression of a soybean cytosolic glutamine synthetase gene (*GS15*) on nodulation and growth of pea (*Pisum sativum* L. cv. Green feast) was studied. Three constructs of the *GS15* fused with a constitutive 35S CaMV promoter, a nodule-specific promoter (LBC$_3$) and a root-specific promoter (rolD), respectively, were introduced into pea plants via *Agrobacterium rhizogenes*. The transgenic lines homozygous for *GS15* were isolated from 16 plants in each line by PCR analysis, and analyzed by Southern blot hybridization. Four lines with only one copy of *GS15*, i.e. 35SGS15-DB917, LBC$_3$GS15-PLP225, rolDGS15-DB681 and rolDGS15-DB779, were selected for Western blots and hydroponic culture.

Western blot analysis revealed that the *35SGS15* construct containing a constitutive promoter was overexpressed in the leaves and roots. The *rolDGS15* with a root-specific promoter was strongly overexpressed in roots and nodules, but also was detectable in leaves. The *LBC$_3$GS15* protein was relatively abundant in the nodules and roots compared to control.

The hydroponic experiment showed that the overexpression of *35SGS15* and *LBC$_3$GS15* constructs did not affect nodulation in the N free treatment, while *rolDGS15* suppressed nodulation in the rolDGS15-DB681 line, but not in rolDGS15-DB779 line. In the treatment with 0.1 mM NO$_3^-$, nodulation was inhibited in the 35SGS15-DB917 and rolDGS15-DB779 lines. Although nodule numbers for transgenic lines tended to be higher at the 1.0 mM NO$_3^-$ level, these values were not significantly different (P = 0.05). Overexpression of *rolDGS15* also suppressed nodulation in DB681line in the 10.0 mM NO$_3^-$ treatment.

Overexpression of the *35SGS15* did not influence biomass accumulation in all NO$_3^-$ treatments, but the *LBC3GS15* tended to enhance biomass production, particularly in the treatments with 0.1 and 10.0 mM NO$_3^-$ (biomass increased by about 20% in both treatments). Overexpression of the *rolDGS15* significantly inhibited biomass in DB681 line in all NO$_3^-$ treatments, but stimulated biomass accumulation in DB779 line at 0.1 mM NO$_3^-$ level and did not affect biomass in other treatments.

These results showed that the constructs of soybean cytosolic *GS15* fused with different promoters were mainly overexpressed in specific tissues. The impact of overexpression of the *GS15* on biomass accumulation of pea at different NO3$^-$ levels was different from that on nodulation, implying that overexpression of *GS15* may affect biomass accumulation by means other than nodulation. We are currently analyzing GS activity and nitrogen fixation rates to further characterize the impact of overexpression of *GS15* on biomass production.

WHOLE GENOME STUDY OF FixLJ TARGETS IN *SINORHIZOBIUM MELILOTI*

L. Ferrières, S. Rouillé, D. Kahn

Laboratoire de Biologie Moléculaire des Relations Plantes-Microorganismes
INRA/CNRS, BP 27, 31326 Castanet-Tolosan Cedex, France

In *Sinorhizobium meliloti*, expression of nitrogen fixation genes is controlled by a hierarchically organized regulatory cascade (David *et al.* 1988). The two component FixLJ system operates at the head of this cascade, which suggests that it plays a global regulatory role in response to the microaerobic conditions found in alfalfa root nodules. However, up to now only two genes were known to be directly activated by the FixJ transcriptional activator – *nifA* and *fixK*.

In order to identify potential new FixJ targets in the genome of *S. meliloti* we have developed an *in vitro* cyclic selection (SELEX) procedure using a fusion protein between GST and the DNA binding domain of FixJ. In a first phase, the procedure was applied to synthetic oligonucleotides randomized over 20 bp in order to select DNA fragments with good affinity for FixJ. Protein/DNA complexes were adsorbed on glutathione-sepharose beads, bound DNA was eluted, amplified and protein/DNA complexes were selected again. After six rounds of such a selection the randomized DNA appeared highly enriched in FixJ binding fragments. These fragments were cloned, sequenced and their complexes with phosphorylated FixJ (FixJ~P) were characterized by gel retardation and DNase I footprinting experiments. This analysis led to the definition of two recognition patterns for FixJ~P. The first pattern CTAAGTAGTTCCC (14 bp) is similar to the high affinity FixJ~P binding site mapped in the *fixK* promoter (Galinier *et al.* 1994). The second pattern CTACGTAG (8 bp) lies in the middle of a 40 bp region protected by FixJ~P against DNase I attack.

In a second phase, the SELEX procedure was applied to a pool of DNA fragments covering the entire *S. meliloti* genome, generated by random priming as described by Singer *et al.* (1997). This resulted in the identification of 20 new FixJ binding sites in addition to the known *fixK* and *fixK'* promoters. These sites appear unequally distributed over the three replicons constituting the *S. meliloti* genome (Galibert *et al.* 2001); while the chromosome and the pSymA megaplasmid contain 9 and 10 sites respectively, only 3 sites are located on pSymB. Two of the new FixJ targets appear to result from a duplication of the *fixK* promoter, evident from the presence of a downstream truncated *fixK* ORF. This duplication of the *fixK* promoter confers FixJ-dependent microaerobic induction to the downstream gene, as evidenced by RT-PCR and reporter gene experiments. Similar promoter duplications, including a truncated *nifH* ORF, were previously observed for the *nifH* promoter and were found to confer *nifA* regulation (Better *et al.* 1983; Murphy *et al.* 1993). Such a 'promoter hijacking' may therefore be a more common phenomenon in the *S. meliloti* genome than originally thought, allowing for the recruitment of new genes under particular physiological conditions.

References
Better *et al.* (1983) Cell 35, 479-485
David M *et al.* (1988) Cell 54, 671-683
Galibert *et al.* (2001) Science 293, 668-672
Galinier *et al.* (1994) J. Biol. Chem. 269, 23784-23789
Murphy P *et al.* (1993) J. Bacteriol. 175, 5193-5204
Singer B *et al.* (1997) Nucleic Acids Res. 25, 781-786

ADDITIONAL REGULATORY ELEMENTS OF THE *R. ETLI fix* GENES ARE LOCATED ON DIFFERENT PLASMIDS

M. Granados, D. Romero, L. Girard

Programa de Genética Molecular de Plásmidos Bacterianos, Centro de Investigación sobre Fijación de Nitrógeno, UNAM, Cuernavaca, Morelos, México

1. Introduction

Nitrogen fixation in Rhizobia is regulated via control of transcription of *fix* and *nif* genes, under low oxygen conditions (Soupene *et al.* 1995). The regulatory elements identified so far (FixL, FixJ, FixK, FnrN, NifA and RpoN) participate in regulatory cascades whose architecture is species-, or even strain-specific. In one cascade, NifA and RpoN control the expression of genes coding for nitrogenase. Another cascade involves the two-component system formed by FixL-FixJ and FixK (Fischer 1994) and is specific for symbiotic diazotrophs. This cascade controls the expression of the *fixNOQP* operon, which codes for a terminal oxidase with a high affinity for oxygen, necessary for bacteroid survival in the nodule (Preisig *et al.* 1993). In *Rhizobium etli* CFN42, two copies of the *fixN* and *fixK* genes are present, one located on the pSym and the other on a cryptic plasmid (pCFN42f); this plasmid also harbors a *fixL* gene coding for an atypical FixL protein: sequence analysis of the predicted FixL polypeptide suggests that it is a probable hybrid histidine-kinase, two-component sensor protein. An interesting characteristic of this strain is the absence of a structurally similar *fixJ* gene (Girard *et al.* 2000). This led us to hypothesize the existence of another protein, forming with FixL a two-component regulatory system.

2. Results and Discussion

The objective of this work is the identification of additional regulatory elements required for optimal microaerobic expression of *R. etli* CFN42 *fixK*f and *fixN*d genes. The experimental strategies for the identification of these elements included random mutagenesis and complementation experiments. The results showed that *R. etli* CFN42 FixL is unable to activate *fixK*f expression directly. Different derivatives of *R. etli* CFN42 cured of different plasmids or with deletions on pSym were analyzed. The results obtained suggest the existence of different genetic elements required for an optimal microaerobic *fixK*f expression, and that should be encoded on pSym and pCFN42f. On the other hand, the expression of *fixNOQP*d is under the control of a complex regulatory system, involving the participation of FixL, FixKf and FnrN (López *et al.* submitted); by a mutagenesis strategy we isolated an interesting mutant that affects drastically *fixN*d expression. However, this mutation does not affect *fixK*f expression, suggesting that is affecting a novel element, acting downstream of *fixK*f in the cascade for *fixN*d expression.

3. References

Fischer H (1994) Microbiol. Rev. 58, 352-386
Girard L *et al.* (2000) Mol. Plant-Microbe Interact. 13, 1283-1292
Preisig O *et al.* (1993) Proc. Natl. Acad. Sci. USA 90, 3309-3313
Soupene E *et al.* (1995) Proc. Natl. Acad. Sci. USA 92, 3759-3763

A MUTATION IN THE CONSERVED CYSTEINE MOTIF IMPAIRS THE ACTIVITY OF AnfA

B. Jepson, S. Austin

Dept of Molecular Microbiology, John Innes Centre, Colney Lane, Norwich, Norfolk, UK

1. Introduction

Azotobacter vinelandii is able to synthesize three genetically different nitrogenases, which are characterized by the presence of either a molybdenum, vanadium or iron atom in their respective co-factors. Each of the nitrogenases requires an activator to express the structural genes required for the enzyme. AnfA is the activator protein for the iron only nitrogenase. The *in vivo* activity of the protein requires the presence of the nitrogenase Fe-protein (Joerger *et al.* 1991). AnfA has the characteristic three-domain structure of a σ^N-dependent transcriptional activator. Previous work has shown that an amino-terminally truncated ΔNAnfA has been shown to remain constitutively active *in vivo* (Frise *et al.* 1994) and *in vitro* (Austin, Lambert 1994), implying regulation of AnfA activity is via the N-terminus. A conserved cysteine motif is present at the N-terminal domain. The cysteine residues 21 and 26 have been previously shown to be required for *in vivo* activity (Premakumar *et al.* 1994). This motif could provide potential ligands for a metallocluster, which may regulate the activity of the protein in a redox sensitive manner.

We have shown previously that the isolated N-terminal domain of AnfA (NAnfA) was a red/brown color when purified and displayed spectra characteristic of oxidized 2Fe-2S clusters identified in other proteins. A mutant version of the N-terminal domain where one of the conserved cysteines was changed to an alanine (C26A) did not display the spectral features observed with the native protein. This implicates the N-terminal cysteine motif in the formation of the Fe-S cluster. In this work we describe the effect of the C26A mutation on the activity of the full-length protein.

2. Results and Conclusions

Wild-type and C26A mutant versions of full-length AnfA were overexpressed in *E.coli* and purified. Neither protein displayed the spectral features observed in the isolated N-terminal domain, indicating the apo-forms of both proteins had been purified. The wild-type protein was shown to be active *in vitro* as judged by its ability to activate transcription from the *anfH* promoter. However C26A AnfA had a 4–5-fold reduced ability to form open complexes at *anfH* compared to the wild-type protein. A similar reduction in ATPase activity was also observed in the mutant protein while DNA binding to an *anfH* promoter was the same for both wild-type and mutant proteins.

Thus the apo-form of AnfA is competent to activate transcription indicating that the presence of the Fe-S cluster is not required for activity of the protein. However even in the apo-form of AnfA the cysteine 26 residue is still important for the activity of the protein, implying that the cysteine motif may have a role in regulation of AnfA activity other than as a potential ligand to the Fe-S cluster.

3. References

Austin S, Lambert J (1994) J. Biol. Chem. 269, 18141-18148
Frise E, Green A, Drummond M (1994) J. Bacteriol. 176, 6545-6549
Joerger RD, Wolfinger ED, Bishop PE (1991) J. Bacteriol. 173, 4440-4446
Premakumar R, Loveless TM, Bishop PE (1994) J. Bacteriol. 176, 6139-6142

CHARACTERIZATION OF DEFENSE RESPONSES OF *SESBANIA ROSTRATA*

K. Norihito[1], S. Takehisa[1], L. Sam[2], S. Satoshi[1], M. Hiroki[1], M. Yukihiro[1], H. Marcelle[2], O. Hiroshi[1]

[1]Dept Global Agri. Sci., U. Tokyo, Tokyo, Japan
[2]Dept Plantengenetica, VIB, U. Gent, Gent, Belgium

Sesbania rostrata is an annual, fast-growing legume from the Sahel region of West Africa and parts of Angora, Mozambique and Madagascar with tropical climates and flooded soils, where it engages in symbiotic nitrogen fixation with *Azorhizobium caulinodans*. Nitrogen-fixing nodules are formed at lateral root bases and at the bases of adventitious rootlets that are located on the stem. We want to examine the potential overlap between symbiosis- and pathogenesis-related plant responses and collected a few genes commonly implicated in defense responses for use as molecular markers.

Chalcone synthase (CHS) and phenylalanine ammonia-lyase (PAL) are two key enzymes in the biosynthesis of isoflavonoid phytoalexins. In general, these antimicrobial plant compounds accumulate strongly upon pathogen attack. Other important plant proteins in the response to pathogen attack are hydrolytic enzymes such as β-1,3-glucanase (GLU), that inhibit growth of fungal infection.

To isolate homologs from *S. rostrata*, DNA sequences for these genes were aligned and degenerate primers were designed corresponding with conserved regions. DNA fragments were amplified on a cDNA template prepared from a mixture of *S. rostrata* root primordial samples harvested at different time points after inoculation with *Ralstonia solanacearum*. This wide host range pathogen causes a nonhost defense response upon infection of the adventitious rootlets. Sequencing resulted in the identification of an *S. rostrata* clone for each of the three defense genes. Subsequently gene specific primers were designed for RT-PCR detection of the transcripts corresponding to SrPAL, SrGLU and SrCHS.

To check whether these genes are expressed in *S. rostrata* following pathogen attack, RT-PCR analysis was applied to *Botrytis cinerea* infected leaf material. Uninfected leaves were compared with leaves harvested at 4, 8, 17, 30 and 48 hours after inoculation with the fungus. PAL, CHS and GLU genes were induced already at 4 hours after infection (Lievens 2001). Strongest induction was observed for β-1,3-glucanase. To isolate the full length β-1,3-glucanase sequence, 3' and 5' RACE were performed. The deduced amino acid sequence contained an ORF of 372 amino acids with 84% identity with *Phaseolus vulgaris* β-1,3-glucanase and 83% identity with *Medicago sativa* acidic glucanase. In wounded leaves, SrGLU transcripts were expressed from 2 h to 2 days. At present the expression of these genes upon *Ralstonia* infection and during nodule development is being investigated.

Reference
Lievens, S (2001) Ph.D Thesis, Dept Plantengenetica, VIB, U. Gent

Acknowledgements
We thank Tomoko Kanamori for critical advice. This work was supported in part by Belgium government fellowship.

SITE DIRECTED MUTAGENESIS OF THE *AZOTOBACTER VINELANDII* REGULATORY FLAVOPROTEIN NifL

S. Perry[1], R. Little[1], F. Reyes-Ramirez[2], R. Dixon[1]

[1]Dept of Molecular Microbiology, John Innes Centre, Norwich, NR4 7UH, UK
[2]School of Biological Sciences, University of East Anglia, Norwich, NR4 7TJ, UK

The nitrogen fixation-specific regulatory flavoprotein, NifL, modulates the activity of the *nif* gene transcriptional activator NifA in response to redox, carbon and nitrogen status. Sequence analysis of *Azotobacter vinelandii* NifL has revealed a conserved N-terminal PAS domain and a conserved C-terminal kinase-like domain (HATPase), although NifL has no detectable kinase activity. We have performed site directed mutagenesis to determine the role of critical residues in NifL function.

Previous work has demonstrated that NifL binds FAD located in its PAS domain (Soderback *et al.* 1998). Mutation of three conserved residues in the PAS domain (Y60, Y83 and Y129) eliminated the redox response but did not affect the nitrogen response *in vivo* (Figure 1). All these mutations appeared to impair FAD binding. NifLY83A retained the response to ADP but the oxidized form of this protein was unable to inhibit NifA activity *in vitro*.

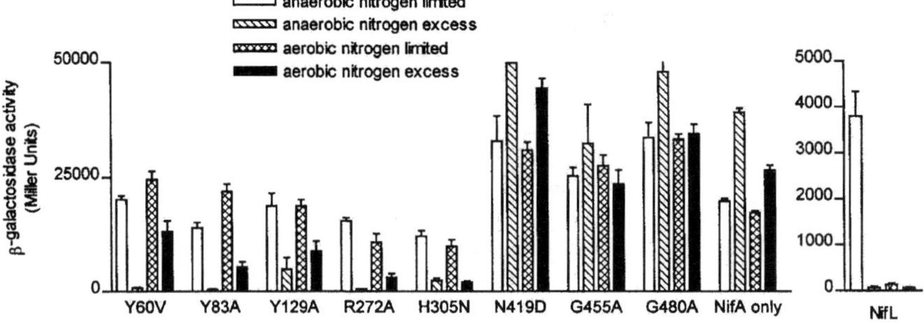

Figure 1. Influence of mutations on NifL activity *in vivo*

The central region of NifL has no significant homology to other proteins and its function is unknown. Mutation of two residues in this region (R272, and H305) eliminated the redox response but did not affect the nitrogen response *in vivo* (Figure 1). On purification NifL R272A was yellow in color and had FAD-specific spectral features comparable to wild type NifL. We propose that redox signaling to NifA involves the central region of NifL and that signaling of the redox and nitrogen status occurs independently.

NifL interacts with NifA by direct protein-protein interaction and this interaction is increased in the presence of ADP (Money *et al.* 1999), although the precise role of adenosine nucleotide binding is not understood. Three mutations in conserved residues (N419D, G455A and G480A), which have been well characterized in various histidine protein kinases, eliminated the response of NifL to both nitrogen and redox status (Figure 1). Hence, nucleotide binding is a major determinant of NifL activity.

References

Money T *et al.* (1999) J. Bacteriol. 181, 4461-4468
Söderbäck E *et al.* (1998) Mol. Microbiol. 28, 179-192

NifA ACTIVATION OF TARGET PROMOTERS BY BINDING TO NON-CANONICAL UPSTREAM ACTIVATING SEQUENCES (UAS)

M. Martinez, M.V. Colombo, J. Palacios, J. Imperial, T. Ruiz-Argüeso

Dept of Biotecnología, ÉTSIA, Universidad Politécnica, 28040-Madrid, Spain

1. Introduction

The prokaryotic enhancer-binding protein NifA stimulates transcription by binding to a conserved sequence (5'-TGT-N10-ACA-3') located upstream of -24/-12 promoters such as those of nitrogen fixation (*nif* and *fix*) genes. Recycling of hydrogen generated by the nitrogenase complex in legume nodules is mediated by a nickel-containing hydrogenase and contributes to increase the energy efficiency of the nitrogen-fixation process. The hydrogenase structural genes (*hupSL*) of *Rhizobium leguminosarum* bv. *viciae* strain UPM791 are temporally and spatially co-expressed in pea nodules with *nif* genes (Brito *et al.* 1995). The symbiotic transcription of *hupSL* takes place from a -24/-12 type promoter (P1), which is activated by NifA and requires IHF (Brito *et al.* 1997).

2. Results and Discussion

Promoter deletion assays identified an enhancer region spanning from position -173 to -140 that was essential for NifA-dependent P1 activation but that contains no NifA-binding canonical UAS. Band shift experiments using purified NifA from *Azotobacter vinelandii* and the C-terminal half of NifA from UPM791 demonstrated that NifA binds to this region. Extensive site directed mutagenesis analysis identified three extended half-boxes (ACAA n(5) ACAA n(12) TTGT) within this region that were required for full NifA-dependent P1 activation. Random mutagenesis of P1 promoter identified no additional elements other than the extended half-boxes, IHF and RpoN binding sites that were required for P1 promoter activity. The relative contribution of each extended half-box to activation was examined by comparing the activity associated to mutant promoters altered in one, two or the three half-boxes. The most important one was the TTGT half-box, but a strong cooperative effect among them was observed. A mechanism of activation based on a weak, cooperative binding of NifA to these half-boxes with formation of an oligomeric NifA-complex is proposed.

3. References

Brito B *et al.* (1995) Mol. Plant-Microbe Interact. 8, 235-240
Brito B *et al.* (1997) Proc. Natl. Acad. Sci USA 94, 6019-6024

4. Acknowledgements

Work was supported by grants PB95-0232 from DGICYT, BIO96-0503 from CICYT, and by the Programa de Grupos Estratégicos de la Comunidad Autónoma de Madrid.

IDENTIFICATION OF A SOYBEAN NODULE PROTEIN THAT BINDS TO THE GAGA ELEMENT OF THE HEME BIOSYNTHESIS GENE *Gsa1*.

I. Sangwan, M.R. O'Brian

Department of Biochemistry, State University of New York at Buffalo, Buffalo, New York 14214, USA

The soybean *Gsa1* gene encoding the heme and chlorophyll synthesis enzyme glutamate 1-semialdehyde aminotransferase (GSAT) is strongly expressed in symbiotic root nodules. The *Gsa1* promoter contains a perfect dinucleotide repeat GAGA element that binds a nuclear factor to positively affect transcription. Using a yeast one-hybrid system, we isolated soybean nodule cDNA that encodes a protein that binds to the *Gsa1* GAGA element. A peptide fusion comprising the yeast GAL4 activation domain and the GAGA binding protein (GBP) strongly activated a reporter *HIS3* or *lacZ* gene under the control of the GAGA element. Furthermore, gel mobility shift assays show that GBP binds the GAGA element *in vitro*. GBP is a basic protein and contains a nuclear localization signal. Interestingly, the N-terminal portion is homologous to histones H1 from numerous organisms. In *Drosophila*, binding of GAGA factor to GAGA elements relieves transcriptional repression exerted by histone HI. We speculate that soybean GBP may compete with histone HI for the activation of the *Gsa1* gene. Northern blot analyses show that mRNA encoding GBP was expressed highly in nodules but not in uninfected roots, and thus *Gbp* is a regulated gene. Furthermore, the *Gbp* gene is expressed in leaves of dark-grown etiolated plantlets, but not after exposure to light for 24 hours. We are investigating the hypothesis that GBP compensates for light for high levels of expression of *Gsa1* in the dark.

DISPARATE OXYGEN RESPONSIVENESS OF TWO REGULATORY CASCADES CONTROLLING THE EXPRESSION OF NITROGEN FIXATION GENES IN *BRADYRHIZOBIUM JAPONICUM*

M.A. Sciotti, H.M. Fischer, H. Hennecke

Institute of Microbiology, Eidgenössische Technische Hochschule, Schmelzbergstrasse 7, CH-8092 Zürich, Switzerland

The expression of nitrogen fixation genes in *Bradyrhizobium japonicum* depends on two parallel regulatory cascades (see also article by H.M. Fischer *et al.*, this volume). The first one involves the products of the *fixL*, *fixJ* and *fixK₂* genes. The oxygen-responsive two-component regulatory system FixLJ induces the expression of *fixK₂*, which is also negatively regulated by its own product. The second cascade consists to the *regS* and *regR* genes and the *fixR-nifA* operon. This operon harbors two overlapping promoters P1 and P2. The RegSR two-component system activates the transcription of the *fixR-nifA* operon from P2, while transcription from P1 is activated by NifA under microaerobic conditions.

We have investigated the modulation by oxygen of the induction by FixJ of *fixK₂* expression, the induction by NifA of a specific target gene, *nifH*, and the induction by RegSR of *fixR-nifA*. To this aim, we translationally fused the specific target genes with a *lacZ* reporter gene and chromosomally integrated the fusions. The *fixR-nifA* promoter fused with the reporter gene harbors a TG>CT mutation in the –24 region of the σ^{54}-/NifA-dependent P1 promoter. This fusion can therefore not be activated by NifA and depends exclusively on RegSR for its induction. The fusion-carrying strains were grown under different atmospheric oxygen concentrations, namely 0.1, 0.5, 2, 5, and 21%. β-Galactosidase activity was assayed after 24, 48 and 72 hours.

Expression of *nifH-lacZ* is strongly induced at ≤ 0.5 % oxygen and almost undetectable at ≥ 5% oxygen. This further documents the redox sensitivity of *B. japonicum* NifA. In contrast, expression of *fixK₂* shows significant basal activity under aerobic conditions and a three-fold increase in the range of ≥ 2% to ≤ 5% oxygen. Expression of P2-*fixR-lacZ* was also found to be regulated by the oxygen conditions. RegSR-dependent expression shows basal level of activity under aerobic conditions and a four-fold increase in the range of ≥ 0.5% to ≤ 2% oxygen. In a *regR⁻* background, the P2 activity is almost completely abolished. In the *regS⁻* background, similar results were obtained, except at ≤ 0.5% oxygen, where low but significant induction was maintained.

We conclude that the expression of the *fixR-nifA* operon under the control of the P2 promoter is strictly RegR-dependent. RegS is required for maximal expression from P2. Activation by RegSR is modulated by oxygen. Since disruption of the *regS* gene does not completely abolish the microaerobic induction and taking into account the lack of an obvious structural basis for oxygen sensing in RegS, we propose that the RegSR system senses the cellular oxygen status indirectly via an alternative mechanism.

Disparate oxygen responsiveness of the FixLJ-, NifA-, and RegSR-regulated genes may result in ordered temporal activation of the symbiotic genes during invasion of the host plant by the bacteria and formation of a functional nodule.

REGULATION OF AMINO ACID TRANSPORTER AAP UNDER ACIDIC CONDITIONS

H.L. Steele, B.U. Becker, P. Müller, P. Vinuesa, D. Werner

FB Biologie, FG Zellbiologie und Angewandte Botanik, Philipps Universität Marburg, Karl-von-Frisch-Strasse, 35032 Marburg, Germany

2D SDS-PAGE of proteins from acid tolerant *R. tropici* CIAT899 wild-type and its acid-sensitive derivative PV4, grown under neutral and acidic conditions, revealed a differentially expressed protein with high homology (79% identity over 24 amino acids) to AapJ from *R. leguminosarum* bv. *viciae* (Walshaw, Poole 1996). The mutant had less AapJ only under acidic conditions, indicating that acidity had disrupted regulation of expression.

The *R. leguminosarum* bv. *viciae* ABC transporter, encoded by the *aapJQMP* operon, is involved in the exchange of various nutrients between rhizobia and their environment (Walshaw, Poole 1996). This ABC transporter has an optimum efficiency between pH 6.0 and pH 7.0 (Poole *et al.* 1985). To compensate for the reduced efficiency under acidic conditions, rhizobia may produce higher numbers of transporters. It is possible that highly acid tolerant strains, such as *Rhizobium tropici* CIAT899, are better able to regulate their transporters under acidic conditions than less acid tolerant strains, thus giving them a competitive advantage. The aim of this research is to construct a reporter gene fusion to the intact *aapJQMP* operon in *R. tropici* CIAT899 and in the less acid tolerant *R. leguminosarum* TAL1400. This should allow analysis of *aap* regulation, as well as comparison between the two strains in their ability to regulate expression of the *aap* operon under acidic conditions. We present here the strategy we are using to achieve this.

A 556 bp fragment of *aapJ* from *R. tropici* CIAT899 was PCR amplified, cloned and sequenced to confirm its identity (100% identity over 182 amino acids). The 556 bp piece is located directly downstream of the signal sequence which guides the protein to its periplasmic location. Southern blot hybridization analysis with DIG labeled *aapJ* indicated that it is present as a single copy. To determine whether *aap* genes are influenced by acidity, a reporter gene ´*phoA* (Rodríguez-Quiñones *et al.* 1994) was inserted directly downstream of the signal sequence of *aapJ* so that the intact ABC transporter is produced as well as the reporter-gene fusion product. To achieve this the vector pJQ200 (Quandt, Hynes 1993) was modified to include ´*phoA* (Rodríguez-Quiñones *et al.* 1994).

The region upstream of *aapJ* including the *aapJQMP* promoter and signal sequence has been PCR amplified from the two strains, ligated in the vector pJQ200/´*phoA*, transformed into *E. coli* strain DH5α, and plated onto media containing X-phosphate. Blue clones have been obtained indicating that the DNA fragment was inserted in the correct orientation and reading frame into the vector. The hybrid plasmid now has to be sequenced to confirm that the correct insert is present. The vector will then be transformed into an *E. coli* donor strain and used to mutagenize the wild-type in a conjugation experiment. The transconjugants will be grown at different pHs and the expression measured by determining alkaline phosphatase activity. Our aim is to show that this operon is differently regulated under acidic conditions, and that the ability to regulate the operon under acidic conditions is one of the factors influencing acid tolerance.

References

Poole PS *et al.* (1985) J. Gen. Microbiol. 131, 1441-1448
Quandt J, Hynes MF (1993) Gene 127, 15-21
Walshaw DL, Poole PS (1996) Mol. Microbiol. 21, 1239-1252

The authors thank the DFG for financing this research through the SFB395.

EnodDR1 GENE OF WHITE CLOVER RESPONDS TO *nod* GENES DERIVED FROM *RHIZOBIUM LEGUMINOSARUM* BV. *TRIFOLII* STRAIN 4S

A. Suzuki, S. Ishiba, T. Shitaohta, M. Abe, T. Uchiumi, S. Higashi

Dept of Chemistry and Bioscience, Kagoshima University, Kagoshima 890-0065, Japan

The plasmid pC4S8, which possesses the *nod* genes of *Rhizobium leguminosarum* bv. *trifolii* strain 4S (Had$^+$, Hac$^+$, Inf$^+$, Nod$^+$, Nif$^+$, Fix$^+$) as an insert sequence, was constructed and transformed into the pSym-cured strain H1. All the symbiotic features (Had$^+$, Hac$^+$, Inf$^+$, Nod$^+$) except for nitrogen fixation, of strain 4S were restored in the transconjugant H1(pC4S8).

We attempted to analyze the gene(s) of the host plant (white clover) that responded to *Rhizobium nod* signal(s) at a very early step of infection. The genes of host plant responsive to the *Rhizobium nod* genes were differentially screened. As a result, 20 cDNA clones were isolated as up-regulated genes and 44 cDNA clones were isolated as down-regulated genes. After partial sequencing, 27 cDNA clones among the down-regulated genes were represented by A3, designated *TrEnodDR1*. From the results of full sequence, *TrEnodDR1* gene is composed of 1755 bp and the number of deduced amino acid residues is 487. There is no similar gene with this nucleotide and amino acid sequences reported in database. A computer search showed that there are two possible transmembrane helices and a proline-rich domain in the N-terminal region.

The expression level of *TrEnodDR1* in nodules was reduced to 60% compared with that of the nodule detached root. It was also suppressed in 48 hour-seedlings inoculated with the strain H1(pC4S8) and strain H1. It was at 20 and 60%, respectively, compared with the signal of un-inoculated seedlings. These results predicted that *TrEnodDR1* gene may be involved with the nodulation process.

To analyze the function of TrEnodDR1 protein, white clover and *Lotus japonicus* cv. Gifu (a model leguminous plant) were transformed via *Agrobacterium tumefaciens* using both sense- and antisense-*TrEnodDR1* gene driven by CaMV35S promoter. The germination efficiency of the seeds of the T2 generation from transgenic *L. japonicus* was analyzed. In the case of the seeds from antisense transgenic *L. japonicus*, about 95% of seeds were normally germinated, whereas only about 45% germinated with the sense seeds. Moreover, the existence of the transgene of T2 germinated plants was confirmed by PCR using specific primers. The germination efficiency of genuine transgenic T2 sense plants is only 13%. These results suggest that TrEnodDR1 protein might influence the germination process negatively. We are now using the inoculation test of *Rhizobium* on the regenerated sense- and antisense-*TrEnodDR1* transgenic plants.

ANAEROBIC INDUCTION AND NODULE INFECTED-CELL SPECIFIC EXPRESSION OF *L_jNOD12* GENE FROM *LOTUS JAPONICUS*

K. Szczyglowski[1], S. Bandyopadhyay[2], J. Jorgensen[3], L. Amyot[1]

[1]Agriculture and Agri-Food Canada, SCPFRC, Ontario, Canada
[2]DOE-Plant Research Laboratory, Michigan State University, MI, USA
[3]University of Aarhus, Denmark

Plants respond to hypoxic or anoxic conditions by various anatomical changes and physiological adaptations that include regulation of gene expression. Oxygen concentration in the infected cells of nitrogen-fixing nodules (10 to 50 nM) is among the lowest that has been measured in aerobic cells of any organism. However, the mechanism by which nodule infected cells adapt to microaerobic conditions is poorly understood. Here, we describe a novel late nodulin gene, *LjNOD12*, from *Lotus japonicus* and show that its expression is regulated by oxygen concentration. The cDNA corresponding to *LjNOD12* gene encodes a polypeptide with a predicted molecular mass of 11.5 kDa (nodulin Nlj12). *LjNOD12* mRNA accumulates in the infected cells of *L. japonicus* nodules concomitantly with the commencement of nitrogen fixation. We show that *LjNOD12* mRNA is unstable under atmospheric oxygen concentration and that expression of the *LjNOD12* gene can be induced in uninoculated *L. japonicus* roots within 0.5 h after their exposure to anaerobic conditions. These results suggest that the microaerobic conditions regulate expression of the *LjNOD12* gene, and that nodulin Nlj12 may constitute part of a mechanism required for the adaptation of plant cells to the hypoxic milieu of nodules.

CRYPTIC *dct* GENES OF *MESORHIZOBIUM* SP. STRAINS R7A AND CJ1

J.E. Weaver, J.T. Sullivan, C.W. Ronson

Dept of Microbiology, Otago University, PO Box 56, Dunedin, New Zealand

1. Introduction

The *dct* operon is required for the transport and utilization of the C4-dicarboxylates succinate and malate. Symbiotic strains of *Mesorhizobium* sp. contain two copies of the *dct* operon, one on the symbiosis island and one elsewhere on the chromosome. Non-symbiotic mesorhizobia contain the chromosomal operon but most are unable to utilize succinate. In addition, *M. loti* strain R7A symbiosis island *dctB* and *dctD* mutants are Dct⁻ suggesting that the chromosomal *dct* operon is not normally expressed. Non-symbiont strains CJ1 and N18 were found to mutate to Dct⁺ after extended periods on succinate media, suggesting that these strains undergo some form of activating mutation. These observations indicate that the chromosomal operon is cryptic. The aims of this work were to determine why the chromosomal operon is cryptic and to identify the adaptive mutation occurring in CJ1.

2. Materials and Methods

Mesorhizobium strains R7A and CJ1 have been described (Sullivan *et al.* 1996; Sullivan and Ronson 1998). Strain N18 is a newly-isolated non-symbiont. Standard molecular biology methods were used.

3. Results and Discussion

The chromosomal *dct* operons of R7A and *Mesorhizobium* sp. strain CJ1 complemented *Rhizobium leguminosarum* bv. *viciae dctA*, *dctB* and *dctD* mutants, indicating they are functionally intact. This was confirmed by sequencing. However the operons did not complement non-symbiotic mesorhizobia. Sequence analysis of the *dct* operon from a CJ1Dct⁺ variant indicated that the adaptive mutation did not occur in the operon. A DNA library of CJ1Dct⁺ was introduced into a Dct⁻ non-symbiont and a cosmid pJW1 isolated that conferred the ability to utilize succinate. pJW1 was used as a probe against genomic DNA from several mesorhizobia and only hybridized to CJ1 and N18, the strains that give rise to Dct⁺ variants. pJW1 hybridized to a plasmid in these strains, suggesting that there is some form of communication between the plasmid-encoded activating mutation and the chromosomally-encoded *dct* operon. Sequence obtained from the plasmid suggested it encodes the ability to synthesize nicotinamide and to catabolize rhizopine, in addition to the ability to utilize succinate. Tn*5* mutagenesis indicated three genes may be involved in the switch to succinate utilization, a *lacI* family repressor and two genes of unknown function. Further work is aimed at characterizing the role of these genes and identifying the actual adaptive mutation.

4. References

Sullivan JT *et al.* (1996) Appl. Environ. Microbiol. 62, 2818-2825
Sullivan JT, Ronson CW (1998) Proc. Natl. Acad. Sci. USA 95, 5145-5149

REGULATION OF THE *BRADYRHIZOBIUM JAPONICUM sdh* OPERON ENCODING SUCCINATE DEHYDROGENASE

V. Gulley, D.J. Westenberg

Department of Biological Sciences, University of Missouri - Rolla, MO, USA

Succinate is one of the primary carbon and energy sources within the nitrogen-fixing nodule. The rhizobia depend on energy from these plant-supplied substrates for nitrogen fixation. Succinate is oxidized by succinate dehydrogenase (Sdh) which participates directly in the electron transport chain. A critical role for Sdh in nitrogen fixation is supported by a *R. meliloti* strain lacking succinate dehydrogenase activity that forms ineffective nodules. Also, a correlation exists between succinate dehydrogenase activity and increased efficiency of nitrogen fixation (Emerich *et al.* 1978; Katznelson *et al.* 1957; Soberón *et al.* 1989; Tajima 1990). We propose that it would be possible to enhance nitrogen fixation by increasing the level of respiratory enzymes such as succinate dehydrogenase. Therefore, we are investigating regulatory factors influencing *B. japonicum sdhCDAB* expression. These regulatory factors would serve as targets for increasing *sdhCDBA* expression in *B. japonicum*.

Expression of *sdhCDBA* was measured using an operon fusion linking approximately 500 bases upstream of the *sdh* promoter and 100 bases downstream (including the entire untranslated leader region) to a promoterless *lacZ* gene. The *sdh::lacZ* fusion construct was introduced into *B. japonicum* strain USDA 110d. A low level of β-galactosidase activity was measured in xylose grown cells. However, the same strain grown in minimal medium containing malate and succinate, had 10-fold higher levels of β-galactosidase activity. Pyruvate and arabinose/gluconate resulted in intermediate levels of β-galactosidase activity.

B. japonicum containing the *sdh::lacZ* fusion construct was also grown on minimal medium containing each carbon source and incubated with either high aeration (aerobic) or in sealed test tubes (microaerobic) and assayed for *sdh:lacZ* expression. On all carbon sources, an increased level of β-galactosidase activity is detected under microaerobic conditions compared to aerobic conditions. Higher *sdh* expression under microaerobic conditions is unique to *B. japonicum* but would be consistent with the need for higher Sdh activity in the microaerobic nitrogen-fixing nodule.

References

Emerich DW *et al.* (1978) J. Bacteriol. 137, 153-160

Gardiol A *et al.* (1982) J. Bacteriol. 151, 1621-1623

Katznelson H *et al.* (1957) Can. J. Microbiol. 3, 879-884

Soberón M *et al.* (1989) J. Bacteriol. 171, 465-472

Tajima S (1990) In PM Gresshoff *et al.* (eds), Nitrogen Fixation: Achievements and Objectives, pp. 309-314, Chapman and Hall, New York

EVIDENCE FOR AHL AUTOINDUCER PRODUCTION BY THE SOYBEAN SYMBIONT *BRADYRHIZOBIUM JAPONICUM*

D.J. Westenberg

Department of Biological Sciences, University of Missouri – Rolla, MO, USA

During the rhizobium/legume symbiosis, the bacterial partner must make the transition from a free-living organism to an intracellular symbiont (bacteroid) capable of nitrogen fixation. Many bacterial genes are regulated in response to the transition from free-living bacterium to bacteroid. For example, attachment proteins would no longer be needed but nitrogenase would be needed. In addition, it would not be wise for the nitrogen-fixing bacteroid to begin fixing nitrogen until a sufficient cell density is achieved.

A number of symbiotic and pathogenic bacteria regulate the expression of symbiosis or virulence specific genes in response to cell density (quorum sensing). In a quorum sensing regulatory system, the bacterium produces an autoinducer molecule (AI) that is secreted to the surrounding medium. Once the AI reaches a high concentration, the AI interacts with regulatory proteins that either activate or repress specific genes. In gram-negative bacteria, two types of AIs have been observed (AI-1 and AI-2). AI-1 molecules are N-acyl-homoserine lactones (AHL) and the structure of AI-2 has not yet been determined. Recently, evidence for a peptide AI molecule in *B. japonicum* has been presented. AHL AIs have been detected in *Rhizobium leguminosarum* and *Rhizobium meliloti*. However, to date, AHL autoinducers have not been detected in the soybean symbiont, *B. japonicum*.

Using the NTL4/pZLR4 indicator strain described by Piper *et al.*, we screened twelve strains of *B. japonicum*. Three of the twelve strains (61A1186, 61A224, and 61A227) produce AHLs (Figure 1). To our knowledge, this is the first evidence of AHL autoinducer production in *B. japonicum*.

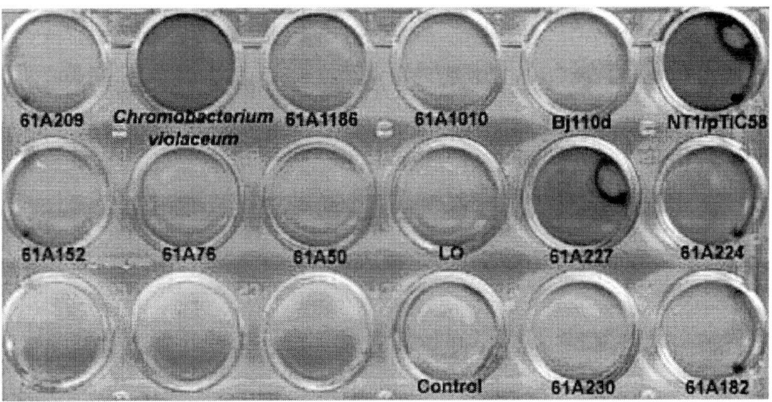

Figure 1. Induction of *lacZ* expression by NTL4/pZLR4 in response to culture supernatants from various *B. japonicum* and control strains.

The number and type of autoinducer molecule(s) is currently being pursued along with experiments to measure the time course of autoinducer production and optimization of autoinducer production. Autoinducer production by *B. japonicum* is a potential target for improving the competitiveness of inoculum strains and we are using the indicator strain to screen gene libraries for the genes responsible for autoinducer production and the genes that may be regulated by these molecules.

HEME-DEPENDENT DEGRADATION OF THE Irr PROTEIN FROM *BRADYRHIZOBIUM JAPONICUM*

J. Yang, M.R. O'Brian

Department of Biochemistry, State University of New York at Buffalo, NY 14214, USA

1. Introduction

The bacterium, *Bradyrhizobium japonicum*, can establish a symbiotic relationship with soybean plants that is manifested as root nodules. Heme is essential for respiratory metabolism and nitrogen fixation and, therefore, we are interested in the regulation of heme biosynthetic pathway. In *B japonicum*, the heme biosynthetic pathway is regulated by iron via the iron response regulator (Irr) protein. Under iron limitation, Irr accumulates and negatively regulates hemB, the gene encoding the heme biosynthesis enzyme ALA dehydratase. However, when iron is sufficient, Irr degrades to derepress the pathway (Hamza *et al.* 1998). This degradation is mediated by heme, which binds directly to the protein at the heme regulatory motif (HRM) (Qi *et al.* 1999). Here, we address the mechanism of heme-mediated Irr degradation *in vitro*.

2. Results and Discussion

Purified Irr was degraded by hemin in the presence of oxygen and a reducing agent (DTT). The redox inactive heme analog zinc protoporphyrin and free ferric iron did not catalyze degradation. In addition, Irr was oxidatively carbonylated during the degradation process. These data indicate that degradation of Irr involves oxidative carbonylation of the protein. Carbonylated products both smaller and larger than the Irr monomer were observed, suggesting both oxidative cleavage and peptide cross-linking, respectively. The Irr mutant IrrC29A does not bind heme with high affinity, and is stable *in vivo* in the presence of iron (Qi *et al.* 1999). IrrC29A was carbonylated to a much lesser extent than the wild type protein. Furthermore, oxygen was required for iron-dependent degradation *in vivo*. We suggest that heme binding to Irr at the HRM catalyzes the localized generation of reactive oxygen species leading to Irr oxidation and degradation.

3. References

Hamza I *et al.* (1998) J. Biol. Chem. 273, 21669-21674
Qi Z *et al.* (1999) Proc. Natl. Acad. Sci. USA 96, 13056-13061

FUNCTIONAL CHARACTERIZATION OF THREE P_{II} HOMOLOGS IN *RHODOSPIRILLUM RUBRUM*

Y. Zhang[1,2,3], E.L. Pohlmann[1,3], P.W. Ludden[2,3], G.P. Roberts[1,3]

[1]Department of Bacteriology
[2]Department of Biochemistry
[3]The Center for the Study of Nitrogen Fixation, University of Wisconsin - Madison,
   Madison, WI 53706, USA

The GlnB (P_{II}) protein, the gene product of *glnB*, has been characterized previously in the photosynthetic bacterium *Rhodospirillum rubrum*. We recently have identified two other P_{II} homologs in that organism, GlnK and GlnJ. Although the sequences of these three homologs are very similar, they have both distinct and overlapping functions in the cell. While GlnB is required for the activation of NifA activity in *R. rubrum*, GlnK and GlnJ do not appear to be involved in that process. In contrast, either GlnB or GlnJ can serve as a critical element in the regulation of the reversible ADP-ribosylation of dinitrogenase reductase, catalyzed by the DRAT/DRAG regulatory system. Similarly, either GlnB or GlnJ is necessary for normal growth on a variety of minimal and rich media, and any of these proteins is sufficient for normal posttranslational regulation of glutamine synthetase. Surprisingly, in their regulation of DRAT/DRAG system, GlnB and GlnJ appear to be responsive to not only changes in nitrogen status, but also changes in energy status, revealing a new role for this family of regulators in central metabolic regulation.

INTERDOMAIN CROSS-TALK LEADS TO CONTROL OF NifA PROTEIN ACTIVITY IN RESPONSE TO AMMONIUM IONS IN *HERBASPIRILLUM SEROPEDICAE*

R.A. Monteiro, E.M. Souza, R. Wassem, M.G. Yates, F.O. Pedrosa, L.S. Chubatsu

Department of Biochemistry, Universidade Federal do Paraná, Curitiba, PR, Brazil

1. Introduction

Transcriptional activity of the NifA protein of *H. seropedicae* is controlled by oxygen and ammonium. The mechanism by which this occurs is not yet well understood. The NifA protein comprises three domains; the N-terminal domain that has a putative regulatory function, the central domain has ATPase activity and the C-terminal domain is involved in DNA binding. In this work, we present evidence that the N-terminal domain of the NifA protein of *H. seropedicae* interacts *in vitro* with an N-truncated form of the NifA protein, supporting a regulatory NH_4^+ dependent role for the N-terminal domain.

2. Materials and Methods

Purified proteins (> 98% pure) were used. DNA-binding and ATPase activities of the N-truncated NifA protein were assayed as described by Monteiro *et al.* (1999a) and Weiss *et al.* (1992), respectively. Proteolysis by trypsin and proteinase K was according to Soderback *et al.* (1998).

3. Results and Discussion

Increasing concentrations of the N-terminal domain inhibited DNA-binding (Figure 1) and ATPase activity (Figure 2) of the N-truncated NifA protein. The N-terminal domain also inhibited proteolysis of the N-truncated NifA. These results indicated a direct contact of the N-terminal domain with the Central and/or C-terminal domains of the N-truncated NifA, and support. These results confirm the observation that the N-terminal domain negatively controls NifA activity *in vivo* (Monteiro *et al.* 1999b). The N-terminal domain of *H. seropedicae* NifA also strongly inhibited ATP hydrolysis by *Azotobacter vinelandii* NifA but not by the *K. pneumoniae* NtrC, indicating the specific nature of the binding to the NifA protein.

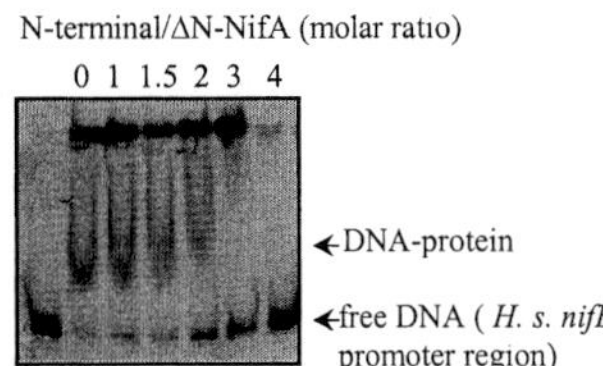

Figure 1. Inhibition of DNA-binding activity by the N-terminal domain.

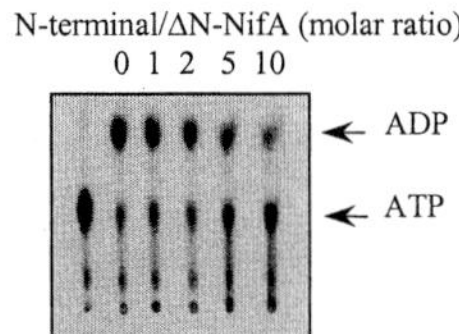

Figure 2. Inhibition of ATPase activity by the N-terminal domain.

4. References

Monteiro RA *et al.* (1999a) FEBS Lett. 447, 283-286
Monteiro RA *et al.* (1999b) FEMS Microbiol. Lett. 180, 157-161
Soderback E *et al.* (1998) Mol. Microbiol. 28, 179-192
Weiss V *et al.* (1992) Proc. Natl. Acad. Sci. 89, 5088-5092

COMPLEX GENOME ORGANIZATION OF THE GENUS *AZOSPIRILLUM*

C.C.G. Martin-Didonet, L.S. Chubatsu, E.M. Souza, F.G.M. Rego, L.U. Rigo,
F.O. Pedrosa

Depto de Bioquímica e Biologia Molecular, Universidade Federal do Paraná, CP 19046,
CEP 81531-990, Curitiba, PR, Brazil

1. Introduction

Several members of the alpha-proteobacteria have a highly complex genome organization (Martin-Didonet *et al.* 2000; Jumas-Bilak *et al.* 1998). Some bacteria have more than one chromosome in addition to several plasmids. *Agrobacterium tumefaciens*, *Ochrobactrum intermedium* and *Rhodobacter sphaeroides* have four replicons and the largest DNA molecules were considered to be chromosomes (Jumas-Bilak *et al.* 1998; Suwanto *et al.* 1989). One chromosome of *A. tumefaciens* is linear and another circular (Allardet-Servent 1998). These results altered the definition of the bacterial genome and highlighted its dynamic nature and plasticity. In this study, we examined the presence of multiple replicons in five *Azospirillum* species using PFGE and hybridization with 16S and 23S rDNA probes. The results obtained suggested the presence of two to five chromosomes in *Azospirillum*. The restriction profile of *A. brasilense* FP2 with endonuclease indicated the presence at least five *rrn* operons.

2. Materials and Methods

Bacterial strains used: *Azospirillum brasilense* strains FP2, Sp7, Cd and 245; *A. lipoferum* Sp59 and JA25; *A. amazonense* strains Y2 and Y6; *A. irakense* and *A. halopraeferens*. Media and growth conditions: strains of *A. lipoferum, A. brasilense, A. iraquense* and *A. halopraeferens* were grown in NFb medium and the strains of *A. amazonense* in LGI medium. Pulsed-field gel electrophoresis (PFGE) of chromosomal DNA was done using a Gene Navigator pulsed-field system (Pharmacia). The chromosomes of *S. cerevisiae* and *S. pombe* were used as molecular mass markers. DNA hybridization: *A. brasilense* 16S and 23S rDNA genes labeled with ^{32}P were used as probes for DNA hybridization. Restriction endonucleases: the enzymes I-*Ceu*I, *Xba*I and *Spe*I were used.

3. Results and Discussion

The analysis of PFGE patterns allowed determination of the genome size of the *Azospirillum* strains studied which varied from 4.8 to 9.7 Mbp. *Azospirillum lipoferum* has the largest and *A. iraquense* the smallest genome size in the genus. At least two megareplicons of each strain hybridized with 16S and 23S rDNA probes, suggesting the presence of multiple chromosomes in *Azospirillum*, some of which were linear. The profiles were clearly different among the strains. The *rrn* copy number of *A. brasilense* FP2 was estimated to be at least five by hybridization of restricted total DNA, suggesting that the chromosomes may have more than one copy of the 16S rRNA gene. Our results indicate that *Azospirillum* genome topology is highly plastic and variable.

4. References

Allardet-Servent A (1998) J. Bacteriol. 175, 7869-7874
Jumas-Bilak E *et al.* (1998) J. Bacteriol. 180, 2749-2755
Martin-Didonet CCG *et al.* (2000) J. Bacteriol. 184, 4113-4116
Suwanto A *et al.* (1989) J. Bacteriol.171, 5850-5859

DOES PHOSPHORUS DEFICIENCY AFFECT THE EXPRESSION OF AQUAPORIN GENES IN *PHASEOLUS VULGARIS* NODULES?

J.J. Drevon[1], J. Escout[2], H. Gherbi[1], S. Herrmann[3], H. Payré[1], M. Soussi[1], D. Tousch[3]

[1]INRA Sols Symbioses Environnement, Place P. Viala, Montpellier 34060 Cedex 01, France
[2]CIRAD Biotrop, Avenue Agropolis, Montpellier, 34000, France
[3]Univ. Montpellier II, Biochimie et Physiologie Moléculaires des Plantes, Montpellier, France

1. Introduction

Aquaporins are abundant in legume-nodule inner-cortex (Serraj *et al.* 1999) where they are postulated to have a major control of the O_2 diffusion for N_2 fixation (Minchin 1997). Since P deficiency increases nodule permeability (Ribet, Drevon 1995), we examined whether P deficiency affects the expression of aquaporin genes in N_2-fixing nodules.

2. Procedure

P. vulgaris was inoculated with *Rhizobium tropici* CIAT 899 and grown hydroaeroponically under P sufficiency versus P deficiency as described by Vadez *et al.* (1999). Nodules were harvested during vegetative stage, and immediately fixed and embedded in methacrylic resin for *in situ* hybridization (ISH) during four weeks according to Schumpp (1998). Sense and antisense ^{35}S-UTP-labeled RNA were synthesized from a putative aquaporin-1 (*Mip*-1) cloned in pGEMT (Promega), using SP6 and T7 polymerase (Promega).

3. Results and Discussion

The *in situ* hybridization signal was found in (i) the nodule cortex, with the notorious exception of the middle cortex, i.e. the cell layers in between the vascular traces and the external cortex, (ii) the vascular traces, and (iii) the non-infected cells of the infected zone, but not in the infected cells. The intensity of the signal varied with tissues and with treatments: it was higher in inner cortex and non-infected cells than in vascular traces and external cortex; under P deficiency, it was significantly increased in the inner cortex and in the non-infected cells.

Thus aquaporins of inner-cortex would contribute to the water uptake for the turgidity of inner-cortex cells in agreement with the osmoregulation hypothesis for nodule-permeability (Drevon *et al.* 1997). The aquaporin in non-infected cells of the infected zone suggest that aquaporin may be involved in solute transport from vascular traces towards the infected zone and reversely. The role of aquaporin in external cortex is more elusive.

4. References

Drevon *et al.* (1997) In Elmerich C, Kondorosi A, Newton WE (eds), Biological Nitrogen Fixation for the 21st Century, pp. 465-466, Kluwer Academic Publishers, The Hague, The Netherlands
Minchin FR (1997) Soil Biol. Biochem. 29, 881-888
Ribet, Drevon (1995) J. Exp. Bot. 46, 1479-1486
Serraj *et al.* (1998) Planta. 206, 684-686
Schumpp (1998) DEA Report Université Montepllier II, France
Vadez V *et al.* (1999) Euphytica 12, 645-648

5. Acknowledgements

This work was supported by the EC project FYSAME (ERBIC18CT960081) and by AUPELF and INRA postdoctoral fellowships for M. Soussi.

USE OF PEA (*PISUM SATIVUM* L.) MUTANTS IMPAIRED IN ROOT FORMATION TO STUDY THE ROLE OF AUXIN IN NODULE DEVELOPMENT

Z.B. Pavlova[1], N.V. Ischenko[1], V.A. Voroshilova[2], V.E. Tsyganov[2], A.Y. Borisov[2], I.A. Tikhonovich[2], L.A. Lutova[1]

[1]Dept of Genetics, St.-Petersburg State University, Universitetskaya nab. 7/9, 199034, St-Petersburg, Russia
[2]Lab. of Genetics of Plant-Microbe Interactions, All-Russia Research Institute for Agricultural Microbiology, Podbelsky h . 3, 196608, St.-Petersburg, Pushkin-8, Russia

Studying symbiotic traits of plant hormonal mutants has been proposed as an approach to investigate the functions of phytohormones in the legume-*Rhizobium* symbiosis. It is well known that root development and especially the response of roots to environmental stimuli are regulated mainly by the plant hormone auxin. Thus, mutants of leguminous plants impaired in root development are the very convenient tools for studying the role of auxin in the symbiotic root nodule formation.

Two allelic mutant lines JI819 and JI1743 (*age*) characterized by root agravitropic response (Blixt, 1970) and recently obtained mutant line SGEcrt (*crt*) characterized by curly roots (Tsyganov *et al.* 2000) were used in the present research. These mutants were compared with parental lines in *in vitro* culture for exogenous auxin sensitivity. As a result the increased sensitivity to auxins was revealed for all analyzed mutants. Also mutant line SGEcrt has been characterized by two-fold increased content of free IAA (Tsyganov *et al.* 2000). Thus the significant role of auxin in accomplishment of mutant phenotypes for all characterized mutants has been demonstrated.

The comparative analysis of nodulation ability and nodule histological differentiation of mutants and parental lines was performed. All mutants had significantly decreased (1.8-2.5 times) number of nodules in comparison with parental genotypes after inoculation with effective *R. leguminosarum* bv. *viciae* strain CIAM 1026 (Safronova, Novikova 1996). At the same time the nodules of line SGEcrt formed after inoculation by strain *R. l.* bv. *viciae* VF39 with constitutive expression of reporter gene *gusA* did not differ from initial line SGE by their histological differentiation.

Thus, auxin influences nodulation ability but it is unlikely to be involved in control of nodule histological differentiation.

References

Blixt S (1970) Pisum Newslett. 2, 11-12
Safronova, Novikova (1996) J. Microbiol. Methods 24, 231-237
Tsyganov VE *et al.* (2000) Ann. Bot. 86, 975-981

Acknowledgements

The seeds of *age* lines were kindly provided by Dr M. Ambrose (JI Centre, UK) and strain *R. l.* bv. *viciae* VF39 with constitutive expression of reporter gene *gusA* by Prof. U.B. Prifier (Oekologie des Bodens, RWTH-Aahen, Germany). This work was financially supported by the Netherlands Organization for Scientific Research (NWO) grant 0147-007.017 and Russian Ministry of Education grant E00-6.0-248.

ROLE OF A CATALASE-PEROXIDASE IN *RHIZOBIUM LEGUMINOSARUM*

K.M. Gray, J.-Y. Chen, A.J. Crockford, A.M. Carey, H.D. Williams

Dept of Bio. Sciences, Imperial College of Science, Tech. and Medicine, London, UK

1. Introduction

Catalase levels in *Rhizobium leguminosarum* biovar *phaseoli* strain 4292 are cell-density regulated and are controlled by the accumulation of extracellular molecules (Crockford *et al.* 1995). Catalase activity reaches its peak in mid-exponential phase before declining on approach to stationary phase. This control is different to enteric bacteria where catalase activity reaches its peak in stationary phase. We were interested in investigating the role of catalases in *R. leguminosarum*.

2. Results

Southern blotting indicated that *R. leguminosarum* has potential homologs to the *E. coli katG* and *katE* genes, which encode a catalase-peroxidase and a monofunctional catalase respectively. A 4.7kb insert, containing a potential *R. leguminosarum katG* homolog, was cloned and sequenced. Sequencing identified three open reading frames (ORFs), including a divergently arranged unit consisting of an *oxyR* homolog gene (0.9 kb) and a *katG* homolog gene (2.2 kb) plus a partial *panB* homolog (0.8 kb). The *R. leguminosarum* KatG showed strong homology to *Escherichia coli* KatG (60.3% similiarity) but strongest homology to catalase-peroxidase from *Streptomyces reticuli* (66.8% similarity). The ORF adjacent to the *katG* gene is most similar to the oxidative stress sensing transcription factor OxyR.

A *katG*::Ω mutant was constructed. Two potential mutants were identified as they showed weak bubbling when colonies were flooded with 3% H_2O_2 compared to the wild type, which is indicative of loss of catalase activity. These two mutants have an 80% decrease in catalase activity compared to the wild type in exponential phase and activity stained gels indicated that they were defective in the production of a catalase-peroxidase (KatG).

Using a *katG* mutant we investigated the role of KatG in protecting *R. leguminosarum* against H_2O_2. We showed that KatG protects against H_2O_2 in exponential phase but not stationary phase cultures. KatG also appears to have a critical role in the H_2O_2 induced adaptive response as, unlike the wild type, pre-treatment of the *katG* mutant with a sub-lethal dose of H_2O_2 does not result in resistance to subsequent lethal doses. Interestingly KatG does not have a role in symbiosis as the mutant nodulated and promoted plant growth as efficiently as the wild type.

To investigate the regulation of *katG* in *R. leguminosarum* we constructed a plasmid-borne *katG-lacZ* transcriptional gene-fusion. During growth *katG* expression was shown to parallel the pattern of catalase activity, with maximal expression in mid-exponential phase and minimal levels throughout stationary phase. This suggests that KatG is regulated, at least in part, at the level of transcription.

3. Conclusions

We have cloned and sequenced a catalase-peroxidase, KatG, in *R. leguminosarum*. KatG has a role in oxidative stress protection, particularly in exponential phase. However the catalase-peroxidase does not have a role in symbiosis. Preliminary regulatory studies, using a *katG-lacZ* gene-fusion, suggest that the catalase-peroxidase is controlled at the level of transcription.

PEA LATE NODULIN GENES CHARACTERIZED BY A SIGNAL PEPTIDE AND CONSERVED CYSTEINE RESIDUES

T. Kato[1], N. Suganuma[1], K. Kawashima[1], M. Miwa[1], Y. Mimura[1], M. Tamaoki[1], H. Kouchi[2]

[1]Dept Life Science, Aichi University of Education, Kariya, Aichi 448-8542, Japan
[2]Dept Appl. Physiology, Nat'l Inst. Agrobiological Resources, Tsukuba, Ibaraki 305-8602, Japan

1. Introduction

The legume-*Rhizobium* symbiosis is established by coordinated expression of genes of both partners. Dozens of plant genes, which are exclusively expressed in nodules and termed nodulin genes, have been isolated and are presumed to be involved in symbiosis. A few, such as leghemoglobin (Lb), uricase and glutamine synthetase, have been characterized, though functions of most nodulin genes remain unclear. We previously isolated a set of nodulin genes from pea (*Pisum sativum*) and compared their expression in effective Sparkle nodules and ineffective nodules induced on the pea mutant E135 (*sym13*) (Suganuma *et al.* 1995). The E135 mutant forms ineffective nodules in which nitrogenase proteins are synthesized, but lack activity. In this type of nodule, most nodulin genes are expressed. However, expression of six of them is decreased in the E135 nodules, indicating that the functions of those genes may be related to nitrogen fixation. One of them, *PsN5*, is the Lb gene, and further characterization showed that this type of Lb has a higher oxygen-binding affinity and is expressed in the central tissue of the nodules (Kawashima *et al.* 2001). These results suggested that this Lb is more effective for nitrogen fixation. Here, we investigated the primary structures of the remaining five nodulin genes whose expression is reduced in E135 nodules, *PsN1*, *PsN6*, *PsN314*, *PsN335* and *PsN466*, and their temporal and spatial expression patterns during nodule development.

2. Results and Discussion

Determination of the nucleotide sequences showed their deduced amino acid sequences were homologous to the early nodulins PsENOD3/14 and the late nodulin PsNOD6. They all encoded small polypeptides containing a putative signal peptide and two pairs of cysteine clusters, such as Cys-X$_5$-Cys and Cys-X$_4$-Cys. Hydropathy analysis showed that the amino N-terminal regions are hydrophobic. A small gene family encoding small polypeptides of late nodulins containing a putative signal peptide and two pairs of cysteine clusters has been recently reported in *Vicia faba* (Frühling *et al.* 2000). Five nodulin genes characterized in this study appeared to belong to the same gene family. Multiple bands were detected in each gene by genomic Southern analysis. Northern analysis showed that they all were exclusively expressed in nodules and were in a similar way to the Lb gene during nodule development. They were expressed from the interzone II-III to the distal part of nitrogen-fixing zone in effective nodules, like the Lb gene. However, in ineffective E135 nodules, their expression regions were narrower than those in effective nodules. These results indicated that these nodulins are abundant in nodules and that their continuous expression during nodule development is closely associated with nitrogen-fixing activity.

3. References

Frühling M *et al.* (2000) Plant Sci. 152, 67-77
Kawashima K *et al.* (2001) Plant Physiol. 125, 641-651
Suganuma N *et al.* (1995) Plant Mol. Biol. 28, 1027-1038

4. Acknowledgements

This work was supported by Special Coordination Funds of the Science and Technology Agency of the Japanese government.

CHARACTERIZATION OF THE FLAGELLAR BIOSYNTHESIS REGULATORY GENE *flbD* IN *AZOSPIRILLUM BRASILENSE* YU62

J. Wang[1], D. Yan[1], S. Chen[2], J. Li[2]

[1]National Laboratory for Agrobiotechnology, China Agricultural University, Beijing 100094, China
[2]Department of Microbiology, China Agricultural University, Beijing 100094, China

A positive clone of a 2.6 kb *Sal*I fragment was isolated by screening the *Azospirillum brasilense*Yu62 genomic library by using a probe of PCR product of conserved central region of *nifA* gene. Sequencing of the 2.6 kb *Sal*I fragment revealed three ORFs. One ORF (from 713 bp to 2227 bp) shares 59% identity with FlbD of *Caulobacter crescentus*, which is a global transcriptional regulator of many σ^{54}-dependent flagellar genes. Two incomplete ORFs were found in the flanking regions of *flbD*. The upstream one shows homology with *motA* in many bacteria, e.g. *Rhodospirillum centenum*. MotA is a membrane protein that enables flagellar motor rotation. The downstream incomplete ORF is homologous to *flhA* which encodes another membrane protein suggested to be involved in regulation or secretion of some flagellar proteins. The overall homology suggests this 2.6 kb fragment is a portion of a flagellar gene cluster in *A. brasilense*.

The *flbD* mutant designated as *A. brasilense*YF2 was obtained by inserting a 1.2 kb kanamycin resistance cassette (Kan^r) at the *Aat*II site of *flbD*. This mutant strain can neither swim in liquid medium (observed under light microscopy), nor swarm on semisolid plates. Electron microscopy showed that YF2 (the *flbD* mutant) grown in semisolid medium lost both polar and lateral flagella (Fla⁻Laf).

A *flbD* complementation plasmid pFV5 was constructed by inserting the 2.6 kb *flbD* fragment into the *Xho*I site of plasmid pVK100. Plasmid pFV5 was introduced into Yu62 (wild-type) and YF2 (*flbD*⁻ mutant), respectively. Motility was restored after introducing pFV5 into *flbD* mutant YF2, with a larger swarming circle than wild type Yu62.

Further electron microscopic studies indicated that the complemented mutant strain YF2/pFV5 possesses a single polar flagellum and 2-3 lateral flagella (fewer than wild type Yu62), whereas the wild type strain Yu62 carrying pFV5 has multiple polar flagella but loses lateral flagella. It appears that different flagellation gave rise to differences in motility of the above strains. In summary, the mutation and complementation analysis clearly indicated that *flbD* is essential for both polar and lateral flagellar biosynthesis in *A. brasilense*.

REGULATORY CIRCUITS CONTROLLING BOTH NITROGENASES IN *RHODOBACTER CAPSULATUS*

S. Groß[1], T. Drepper[1], B. Masepohl[1], T. Rehmann[1], A.F. Yakunin[2], P.C. Hallenbeck[2], W. Klipp[1]

[1]Ruhr-Universität Bochum, LS Biologie der Mikroorganismen, D-44780 Bochum, Germany
[2]Université de Montréal, Dépt. de Microbiologie et Immunologie, Montréal, Canada

The phototrophic purple bacterium *R. capsulatus* is able to reduce atmospheric dinitrogen to ammonia either via a molybdenum (*nif*-encoded) or an alternative heterometal-free (*anf*-encoded) nitrogenase. Expression and activity of both nitrogenase systems is controlled by ammonium on at least three different levels. At the first level, transcription of the *nifA1*, *nifA2* and *anfA* genes – coding for the transcriptional activators of the other *nif* and *anf* genes – is controlled by the Ntr system in dependence on ammonium availability. As in other bacteria the *R. capsulatus* Ntr system consists of the two-component regulatory system NtrB/NtrC, two P_{II}-like proteins (GlnB and GlnK) and the P_{II}-modifying enzyme GlnD. In addition, two ammonium transporters (AmtB and AmtY) might play a role in ammonium-dependent signal transduction. In a *glnK* mutant (constitutively expressing *amtB*) NtrC-dependent gene expression is derepressed in the presence of ammonium, whereas in an *amtY* mutant NtrC-dependent gene activation is abolished in the absence of ammonium. Both the *glnK-amtB* operon and *amtY* are part of the Ntr regulon. In contrast to most NtrC-dependent genes, which are constitutively expressed in an *R. capsulatus glnB* mutant, *glnK* expression is still repressed by ammonium. However, in a *glnB/glnK* double mutant ammonium suppression of *glnK* expression is relieved.

Besides NtrB/NtrC the *R. capsulatus ntr* gene region codes for a second two-component system (NtrY/NtrX). The sensor kinase NtrY contains two membrane spanning elements and includes a PAS domain characteristic for redox sensing proteins. Although NtrB effectively phosphorylates and thereby activates NtrC, an *ntrB* mutant still exhibits nitrogenase activity albeit at a reduced level compared to the wild type. In contrast, an *ntrB/ntrY* double mutant no longer synthesizes nitrogenase suggesting that NtrB and NtrY can (partially) substitute for each other. A *glnK-ntrB* double mutant constitutively expresses both NifA and NifH, and exhibits nitrogenase activity even in the presence of ammonium. Therefore the double mutant mimics a *glnB-glnK* double mutant. Complementation of the *glnK-ntrB* double mutant with either *glnK* or *glnB* abolishes NifH synthesis due to GlnK-/GlnB-dependent ammonium regulation at the level of NifA activity.

At the second level of control, the activity of the transcriptional activators NifA1, NifA2 and AnfA is inhibited in an NtrC-independent manner. This post-translational ammonium control of NifA activity is partially released in the absence of GlnK, and completely abolished in a *glnB-glnK* double mutant, whereas AnfA activity is still repressed by ammonium in the *glnB-glnK* mutant background.

At the third level of regulation, both nitrogenase reductases (NifH and AnfH) are controlled by DraT/DraG-mediated reversible ADP-ribosylation. In the presence of ammonium, oxygen or in the dark "switch-off" occurs. Both GlnB and GlnK as well as AmtB are involved in ammonium control of the DraT/DraG system. While GlnB might affect the ADP-ribosylation reaction in response to increasing ammonium concentrations, GlnK seems to be involved in the demodification of nitrogenase under N_2-fixing conditions when *glnK* expression is highly induced. An *amtB* mutant is unable to carry out the short-term ADP-ribosylation.

CHARACTERIZATION OF BRADYRHIZOBIUM JAPONICUM MUTANTS WITH INCREASED SENSITIVITY TO GENISTEIN FOR *nod* GENE INDUCTION

F. D'Aoust[1], H. Ip[2], A.A. Begum[1], T.C. Charles[2], D.L. Smith[3], B.T. Driscoll[1]

[1]Dept Nat. Res. Sci.
[3]Dept Plant Sci., McGill U, Ste-Anne-de-Bellevue, Quebec, Canada
[2]Dept of Biol., U Waterloo, Waterloo, Ontario, Canada

The establishment of symbiosis between soybeans [*Glycine max* (L.) Merr.] and *Bradyrhizobium japonicum* has been shown to be elongated by suboptimal soil temperatures, and this has a negative impact on yield. The delays appear to be partially due to reduced plant-microbe signaling. *B. japonicum* mutants with increased sensitivity to genistein were isolated by UV mutagenesis and transformed with plasmid pZB32 (carrying a *nodY::lacZ* gene fusion). The mutants were found to have higher *nodY* expression than the wild type in the presence of genistein. The increased sensitivity of all mutants to genistein was more apparent under suboptimal inducer concentration (0.1µM) and/or temperature (15°C). The kinetics of *nodY* gene induction were determined for five strains (Bj30050, 53, 56, 57, 58) under different temperature and inducer conditions. These five strains were also found to produce more lipochito-oligosaccharide than the wild type, at both 25°C and 15°C. Three of the ten mutant strains (including Bj30056 and 57) were unable to fix nitrogen with soybeans grown at optimal temperatures. We are continuing the investigation of *nod* gene expression in these mutants, including the effects of cell-density, by both β-galactosidase and semi-quantitative mRNA analyses.

THE *SINORHIZOBIUM MELILOTI lpiA* GENE IS TRANSCRIPTIONALLY ACTIVATED BY LOW pH

W.G. Reeve[1], R.P. Tiwari[1], J. Castelli[1], M.J. Dilworth[1], A.R. Glenn[2], J.G. Howieson[1]

[1]Centre for *Rhizobium* Studies, School Biological Sciences and Biotechnology, Murdoch University, Murdoch, WA 6150, Australia

[2]Office of the Pro Vice-Chancellor (Research), University of Tasmania, Hobart, Australia

1. Introduction

One important environmental stress encountered by root nodule bacteria is that of unfavorably low pH. The isolation of acid-tolerant *Sinorhizobium meliloti* strains from Sardinia and Greece (Howieson, Ewing 1986) made possible the establishment of medic pastures on over 400,000 ha of acid soils in Western Australia. An understanding of the genetics of pH response is being sought to identify the mechanisms required for successful persistence of these acid-tolerant strains of *S. meliloti*. One approach has been to create *gusA* fusions (Reeve *et al.* 1998, 1999) to identify genes that are low-pH activated.

2. Results and Discussion

Two such genes are *phrR* (pH regulated regulator) and *lpiA* (low pH inducible gene). The expression of *phrR* responds to copper, zinc, H_2O_2 and ethanol besides low pH. The *lpiA* gene encodes a transmembrane protein similar to hypothetical membrane proteins from *S. meliloti*, *Synechocystis* sp. and *Escherichia coli*. The *fsrR* (fused sensor-regulator) gene upstream to *lpiA* encodes a protein similar to *Methanobacterium thermoautotrophicum* sensory transduction regulatory proteins.

The *lpiA* mutant is no more sensitive to stress than the wild-type. Transcriptional activation of *lpiA* is specific to low pH; expression increasing at least 20-fold between pH 7.0 and 5.7. High concentrations of calcium reduced expression significantly at low pH. We have previously shown that high concentrations of calcium promote cell growth and survival at low pH (Reeve *et al.* 1993); the influence of calcium on the expression of a gene at low pH is an effect observed in *S. meliloti* at the genetic level for the first time.

Mobilization of a plasmid-borne *lpiA-gusA* fusion into different genetic backgrounds revealed that the low pH-specific activation was not regulated by ActR (a regulator essential for low pH tolerance of *Sinorhizobium*). The low-pH responsive promoter has been pinpointed within a 372 bp region upstream to the *lpiA* start codon. The *lpiA* gene is not required for symbiotic effectiveness although it is expressed in the nodule.

3. References

Howieson JG, Ewing MA (1986) Austral. J. Agric. Res. 37, 55-64

Reeve WG *et al.* (1993) Soil Biol. Biochem. 25, 581-586

Reeve WG *et al.* (1998) Microbiol. 144, 3335-3342

Reeve WG *et al.* (1999) Microbiol. 145, 1507-1516

PROBING FOR pH-REGULATED GENES AND PROTEINS IN *SINORHIZOBIUM MELILOTI* USING TRANSCRIPTIONAL AND PROTEOMIC ANALYSIS

W.G. Reeve[1], N. Guerreiro[2], R.P. Tiwari[1], J. Stubbs[2], B.J. Fenner[1], M.J. Dilworth[1], A.R. Glenn[3], B.G. Rolfe[2], M.A. Djordjevic[2], J.G. Howieson[1]

[1]Centre for *Rhizobium* Studies, School of Biological Sciences and Biotechnology, Murdoch University, Murdoch, W.A. 6150, Australia
[2]Genomic Interactions Group, RSBS, Australian National University, Canberra, Australia
[3]Office of the Pro Vice-Chancellor (Research), University of Tasmania, Hobart, Australia

1. Introduction

Knowledge of the stress-activated genetic circuits in *S. meliloti* is required to understand how acid-tolerant strains can survive in acidic soils to form an effective symbiosis. To identify global changes in *S. meliloti* in response to pH stress we have used two approaches. With the first approach, minitransposon-induced transcriptional fusions to a promoterless *gusA* (Reeve *et al.* 1998, 1999) were used to identify pH-responsive sinorhizobial genes. With the second approach, the pH-responsive protein complement was identified using proteomic analysis (Guerreiro *et al.* 1999; Natera *et al.* 2000; Chen *et al.* 2000).

2. Results and Discussion

Sequencing the rhizobial DNA flanking the minitransposon insertions revealed that genes required for cytochrome biosynthesis, DNA modification, lipid metabolism, membrane transport, and regulation were transcriptionally activated by low pH. Other insertions disrupted membrane proteins of unknown function.

To detect low pH-inducible proteins possibly essential for growth under acidic conditions, we compared protein profiles of cells exposed to a persistent or transient acid-stress. We found approximately 50 proteins that are pH-regulated; sixteen of which yielded N-terminal sequences after Edman degradation. Transient acid exposure down-regulated proteins involved in nitrogen metabolism and up-regulated a hypothetical protein. Continuing acid exposure down-regulated proteins required for amino-acid transport, proteolysis, lipid metabolism, and unknown proteins. Persistent acid-stress up-regulated chaperone proteins and others involved in proteolysis and carbon metabolism. The non-overlap of the proteins identified by the two methods is probably due to: loss of proteins of IEP>7 or <4, loss of membrane proteins or greater sensitivity of the fusion technique for proteins of low copy number. Overall, the total number of proteins involved in pH-response and acid-tolerance may be as high as 100.

3. References

Chen H *et al.* (2000) Electrophoresis 21, 3833-3842
Guerreiro N *et al.* (1999) Electrophoresis 20, 818-825
Natera SHA *et al.* (2000) Mol. Plant Microbe Interact. 13, 995-1009
Reeve WG *et al.* (1998) Microbiol. 144, 3335-3342
Reeve WG *et al.* (1999) Microbiol. 145, 1507-1516

IDENTIFICATION AND CHARACTERIZATION OF *regR* AND *regM*, TWO PUTATIVE SENSOR KINASE GENES IN *RHIZOBIUM LEGUMINOSARUM*

J.E. Macintosh, H.L. Whelan, M.F. Hynes

Dept of Biological Sciences, University of Calgary, Calgary, AB, Canada

1. Introduction

Bacteria use two-component signal transduction systems to sense and respond to environmental signals. Two-component signal transduction systems generally consist of a histidine protein kinase and a cognate response regulator that participate in histidine-aspartate phosphorelays. The histidine protein kinase senses the environmental signal, autophosphorylates, signaling the response regulator which modulates an output response, typically transcriptional regulation (Appleby *et al.* 1996). Two novel sensor kinase genes, *regR* and *regM* have been found in *Rhizobium leguminosarum*. They are being characterized by sequence analysis, mutagenesis, phenotypic assays, and *gusA*-promoter fusion analysis by comparing expression in wild-type and mutant backgrounds.

2. Procedures

Sequence analysis was performed using BLAST programs. *regR* and *regM* were mutated with the insertion of a kanamycin cassette within the gene and introduced into the genome using a suicide vector (Quandt, Hynes 1993). Mutant phenotype was analyzed using competition, motility, osmolarity, biofilm, and nodulation assays. To identify genes regulated by RegR, the mutant background was mutagenized with mTn5SS*gusA40* (Wilson *et al.* 1995). This created random promoter-*gusA* fusions within the genome. Wild-type *regR* was introduced *in trans* on a broad host range plasmid. Differential *gusA* expression in complemented vs. uncomplemented with wild-type *regR* suggests a role in transcriptional regulation.

3. Results and Discussion

regM appears to be a histidine protein kinase that is reiterated in the genome, thus a double mutant must be constructed before further analysis. *regR* appears to be a hybrid histidine kinase-response regulator. The phenotype of *regR* mutants displayed no discernible difference from wild-type phenotype. However, RegR appears to play a role in transcription regulation. Of the 1380 mutants screened in the *gusA* screen, 37 mutants appear to be positively regulated and 87 mutants appear to be negatively regulated.

4. References

Appleby JL, Parkinson JS, Bourret RB (1996) Cell 86, 845-848

Quandt J, Hynes MF (1993) Gene 127, 15-21

Wilson KJ, Sessitch A, Corbo JC, Giller KE, Akkermans ADL, Jefferson RA (1995) Microbiol. 141, 1691-1705

PROTEIN-PROTEIN INTERACTIONS OF REGULATORY AND OTHER GENE PRODUCTS INVOLVED IN NITROGEN FIXATION IN *RHODOBACTER CAPSULATUS*: IDENTIFICATION OF NEW *anf* GENES

B. Masepohl, P. Dreiskemper, T. Drepper, S. Groß, N. Isakovic, A. Pawlowski, K. Raabe, K.-U. Riedel, C. Sicking, W. Klipp

Ruhr-Universität Bochum, LS Biologie der Mikroorganismen, D-44780 Bochum, Germany

The photosynthetic purple bacterium *Rhodobacter capsulatus* can fix atmospheric dinitrogen either via a molybdenum (*nif*-encoded) or an alternative heterometal-free (*anf*-encoded) nitrogenase. Synthesis and activity of both nitrogenases is tightly controlled by ammonium on at least three different levels. At the first level, transcription of the *nifA1*, *nifA2*, and *anfA* genes – coding for the transcriptional activators of the other *nif* and *anf* genes – is controlled by the Ntr system consisting of the two-component regulatory system NtrB/NtrC and the signal transduction protein GlnB (PII). In addition, expression of *glnK-amtB* and *amtY* – coding for a PII-paralog and two putative (methyl)-ammonium transporters – is also under control of the Ntr system. At the second level of regulation, activity of NifA1, NifA2, and AnfA is inhibited in an NtrC-independent manner. This post-translational ammonium control of NifA activity is partially released in the absence of GlnK, and completely abolished in a *glnB-glnK* double mutant, whereas AnfA activity is still repressed by ammonium in the *glnB-glnK* mutant background. At the third level of regulation, both GlnB and GlnK as well as AmtB are involved in ammonium control of the DraT/DraG system, which mediates reversible ADP-ribosylation of both nitrogenase reductases in response to changes in ammonium availability. Remarkably, in a *glnB-/glnK* double mutant ammonium control of the molybdenum (but not of the alternative) nitrogenase is completely relieved, leading to synthesis of active nitrogenase in the presence of high concentrations of ammonium.

To identify proteins interacting with the above-mentioned regulatory proteins, yeast two-hybrid (Y2H) studies were carried out. For this purpose, an *R. capsulatus* DNA library was constructed using a low-copy GAL4-based Y2H system. This library covered the *R. capsulatus* genome several-fold with in-frame fusions to the activating domain of the prey plasmid every 12 bp along the coding strand. To test the library, GlnB was used as a bait. As expected, the by far strongest interaction was found for NtrB confirming the current regulatory model. Weaker interactions were found for NifA2, DraT, and the ATP-dependent helicase PcrA. Using GlnK as a bait, interaction with the Ras-like protein Era could be demonstrated, whereas no interaction was found with NtrB. However, the roles of the interactions between GlnB/PcrA and GlnK/Era, respectively, remain to be elucidated. In addition, analysis of defined protein pairs demonstrated interaction of GlnB with NifA1, GlnK, and GlnB itself. Furthermore, protein interactions were found for NifA1-NifA1, NifA1-NifA2, NifA1-GlnK, NifA2-NifA2, and NifA2-GlnK. These results suggest that (i) NifA1 and NifA2 may form both homodimers as well as heterodimers, and (ii) post-translational regulation of NifA activity in response to ammonium is mediated via direct interaction of the transcriptional activator with the PII-like signal transduction proteins.

Using Anf1 (encoded by a gene located downstream and co-transcribed with *anfHDGK*) as a bait, two new putative cytoplasmatic Anf proteins were identified, called Orf349 and Orf1065 (TetR). The role of Orf349 and Orf1065 as Anf proteins was corroborated by analyses of *in vivo* nitrogenase activity of corresponding mutant strains. Both *orf349* and *orf1065* mutants did not reduce acetylene via the alternative nitrogenase, whereas the activity of the molybdenum nitrogenase was not affected. Surprisingly, the *orf349* mutant was still able to grow diazotrophically with the Anf system albeit at a reduced level compared to the parental strain.

GLYOXYLATE CYCLE IN *R. LEGUMINOSARUM*: CHARACTERIZATION OF A MUTANT CARRYING A Tn*5*B22 INSERTION DISRUPTING THE MALATE SYNTHASE-ENCODING GENE.

A. Garcia-de los Santos[1], A. Morales[1], I. Hernandez-Lucas[1], C.Y. Yost[2], S. Brom[1], M.F. Hynes[2]

[1]CIFN/UNAM, Apdo. Postal 565-A, Cuernavaca, Mor. México
[2]Dept of Biological Sciences, University of Calgary, Calgary, AB, Canada

1. Introduction

We have been interested in isolating rhizobial genes which are turned on by environmental signals. Sugars commonly found as components of plant cell walls may be among these molecular signals. Oresnik *et al.* (1998) reported the isolation of a plant-inducible rhamnose locus involved in competition for nodulation. These data encouraged us to search for other genes whose expression is induced by sugars that form part of plant polysaccharides.

2. Procedures

We screened a collection of 3000 *R. leguminosarum* mutants carrying random insertions of the *lacZ* transposon Tn*5*B22 for induction by arabinose.

3. Results and Discussion

We identified a chromosomal locus inducible by arabinose, a component of plant polysaccharides. BLASTX analysis of the ORF disrupted by Tn*5*B22 showed high similarity to different bacterial malate synthase G (MSG) enzymes. MSG participates in the metabolism of glyoxylate, by catalyzing the irreversible aldol condensation of glyoxylate and acetyl CoA to form malate. The similarity with MSG sequences is reinforced by the presence of 21 residues which are strictly conserved in the 23 MSG sequences previously published (Howard *et al.* 2000). Furthermore, a *lacZ* fusion in the putative malate synthase encoding gene was also induced by glycolate, a compound which activates the transcription of MSG in *E. coli* (Molina *et al.* 1994). Pea plants inoculated with the MSG mutant were shorter than those inoculated with the wt strain and their leaves presented a marked chlorosis. The nodules induced by the mutant were white, indicating that this mutation has a negative effect on nitrogen fixation.

4. References

Howard BR *et al.* (2000) Biochem. 39, 3156-3168
Molina I *et al.* (1994) Eur. J. Biochem. 224, 541-548
Oresnik IJ *et al.* (1998) MPMI 11, 1175-1185

5. Acknowledgements

We thank CONACYT and DGAPA for supporting the postdoctoral stay of Alejandro García de los Santos in the laboratory of Michael Hynes at the University of Calgary.

IDENTIFICATION OF GENES REGULATED BY THE ALTERNATIVE SIGMA FACTOR σ^{54} IN *RHIZOBIUM LEGUMINOSARUM* BV. *VICIAE* VF39SM

S.R.D. Clark, M.L. Tabche, B.T. Steven, A. Cieslak, M.F. Hynes

Dept of Biological Sciences, University of Calgary, Calgary, AB, Canada

1. Introduction

σ^{54} is an 'alternative' sigma factor encoded by *rpoN*; recognized promoters display a highly conserved consensus -12/-24 sequence, 5'- CTGGCAC-N5-TTGCA -3' (Beynon *et al.* 1983). σ^{54}-dependent systems in *Rhizobium* include dicarboxylic acid transport and nitrogen regulation, and the *fixGHIS*, *fixNOQP*, and *fnrN* genes of *Rhizobium leguminosarum* bv. *viciae* VF39SM are dependent on σ^{54} for full induction under microaerobic conditions (Clark *et al.* 2001). σ^{54} can bind DNA in the absence of core polymerase, allowing the sigma factor to be involved in negative as well as positive regulation, depending on the spacing of promoter structures (Merrick 1993).

2. Procedure

The strategy used to identify putative targets for σ^{54} is similar to that used by Bittinger *et al.* (2000). A VF39SM *rpoN*::ΩSp mutant, Rlv1046, was mutagenized with the transposon, Tn*5*-B30 (Simon *et al.* 1989). This element carries a promoterless neomycin gene, generating a pool of random reporter fusions. Each mutant was replica plated, and one copy received pCO28 by conjugation. This plasmid carries the wild type *rpoN* gene, complementing the *rpoN*::ΩSp allele. Both the complemented and uncomplemented forms of each mutant were screened on TY containing increasing concentrations of neomycin; differences in resistance levels were predictive for the nature of each insert under the conditions tested (i.e. Nm^{r} only in the presence of wt *rpoN* suggests σ^{54}-dependence). Inserts cloned from total genomic DNA were identified by homology.

3. Results and Discussion

Of approximately 2400 mutants screened so far, 54 display putative, positive regulation and 21 display putative, negative regulation. Sequences obtained show homology to transposases including ISRm3 and ISRle39 (Rochepeau *et al.* 1997); leucine-response protein transcriptional regulators; and nuclease inhibitors. One of the transposases, as well as the LRP-type regulator, carry the insert in the opposite orientation to the host genes. Upstream flanking sequences will be obtained using a primer to the Nm^{r} gene.

4. References

Beynon J, Cannon M, Buchanan-Wollaston V, Cannon F (1983) Cell 34, 665-671
Bittinger MA, Handelsman J (2000) J. Bacteriol. 182, 1706-1713
Clark SRD, Oresnik IJ, Hynes MF (2001) Mol. Gen. Genet. 264, 623-633
Merrick MJ (1993) Mol. Microbiol. 10, 903-909
Simon R, Quandt J, Klipp W (1989) Gene 1989 80, 161-169
Rochepeau P, Selinger LB, Hynes MF (1997) Mol. Gen. Genet. 256, 387-396

A NOVEL REGULATORY REGION RESPONSIBLE FOR SUPPRESSION OF AEROBIC EXPRESSION OF *fixN*d IN *RHIZOBIUM ETLI*

L. Girard, L. Díaz, D. Romero

Progr. de Genet. Mol. de Plásmidos Bact. Centro de Investigación sobre Fijación de Nitrógeno, UNAM. Av. Universidad s/n. C. P. 62210, Cuernavaca, Morelos, México

Recently, we reported the presence of two *fix* regions in *Rhizobium etli* CFN42 (Girard *et al.* 2000; Soberón *et al.* 1999). Gene fusion analysis coupled with mutation of each regulatory element was used to study the mechanisms employed in *R. etli* CFN42 for the regulation of nitrogen fixation genes. This circuit integrates the participation of genes located in two different plasmids and shows several novel characteristics when compared with the conventional systems as *R. meliloti* and *B. japonicum*. Among these, microaerobic expression of the *fixN* reiterations exhibits a differential dependence for FixL, where transcription of *fixN*f was suppressed in the absence of FixL; expression of *fixN*d, however, is reduced only 50% in this mutant (Girard *et al.* 2000). Moreover, new data indicate the participation of *fnrN* homologs in the *fixN*d microaerobic expression (López *et al.* submitted).

To localize which sequences in the *fixN* regulatory regions are responsible for this differential expression, derivatives containing deletions of the upstream and 5' coding sequence were generated by PCR, and fused with a promoterless β-glucuronidase gene. The expression patterns were analyzed under aerobic and microaerobic conditions. Plasmids containing transcriptional fusions were introduced into different *R. etli* CFN42 derivatives and cultured under aerobic and microaerobic conditions to study its transcriptional regulation.

Differential regulation of *fixN* reiterated genes is maintained. However, a very interesting behavior was observed for a new *fixN*d deleted derivative. Deletion of codons 17 to 95 increases the *fixN*d microaerobic expression and allows its aerobic expression. Surprisingly, the aerobic expression observed in derivatives harboring this deletion is still dependent on the same regulators operating under microaerobic conditions, FixKf, FnrNd and FnrNchr. These results suggest the existence of a regulatory region within the *fixN*d coding region that diminish its expression in both aerobiosis and microaerobiosis. To define the minimal region involved in this new transcriptional pattern for a cytochrome oxidase, new derivatives carrying small deletions will be generated.

References

Girard L *et al.* (2000) Mol. Plant-Microbe Int. 13, 1283-1292
Soberón *et al.* (1999) Appl. Environ. Microbiol. 66, 2015-2019

Acknowledgements

We thank Laura Cervantes and Javier Rivera for excellent technical assistance. Work partially supported by grant IN202599 (DGAPA, UNAM).

UNMODIFIED GlnK INTERACTS WITH NifL *IN VIVO*, STIMULATING NifL-MEDIATED INHIBITION OF NITROGENASE EXPRESSION IN *AZOTOBACTER VINELANDII*

P. Rudnick, C. Kunz, M. Gunitilaka, C. Kennedy

Dept of Plant Pathology, The University of Arizona, Forbes 204, Tucson, AZ 85721 USA

In the γ-subgroup of Proteobacteria, represented by *A. vinelandii* and the enteric bacterium *Klebsiella pneumoniae*, Mo-dependent nitrogenase, encoded by *nif* genes, is positively regulated by the σ^{54}-dependent activator, NifA, whose activity is controlled by a cognate inhibitor, NifL. In response to inhibitory signals, such as oxygen and fixed nitrogen, a non-activating NifL:NifA complex is formed (Dixon 1998) To elucidate the genetic basis of the fixed nitrogen response in *A. vinelandii* has been an ongoing project in our lab.

In *K. pneumoniae*, expression of *nifLA* is tightly controlled by the *ntr* system, and hence responsive to fixed nitrogen. However, in *A. vinelandii nifLA* expression is constitutive, suggesting that control of NifA activity by NifL is the main mechanism controlling the response. In *A. vinelandii*, mutations that affect the activity of PII-uridylyltransferase (GlnD) prevent NifA activity and these mutations can be rescued by deletion of *nifL* (Contreras *et al.* 1991; Colnaghi *et al.* 2001), indicating that GlnD controls an aspect of the NifL-NifA interaction. We sought to determine the role of GlnK, the only PII-like protein in *A. vinelandii*, in regulating NifA activity. These experiments had been hampered because *glnK* is essential in *A. vinelandii* (Meletzus *et al.* 1998).

In this work, we made use of a *glnD* null suppressor strain MV72 (*gln-71*), which carries an ammonium insensitive glutamine synthetase (GS) activity (Colnaghi *et al.* 2001), to construct a mutation at the site of covalent uridylylation on GlnK (Y51→F). This allele was only stable in a *gln-71* background, indicating that GlnK-UMP is required for GS activity probably by reversing adenylylation of GS by ATase. The *glnKY51F* mutation was crossed into the chromosome, selecting for a Tet^r cassette in the 5' end of the adjacent *amtB* gene. Transformant *glnKY51F* strains exhibited severely decreased NifA activity as measured by growth in N-free medium and expression of a *nifH-lacZ* reporter, when compared to *glnK$^+$* strains constructed in the same background. To determine if NifL might be a target for GlnK, a yeast two-hybrid system was used to test protein interactions. In this system, GlnK and GlnKY51F interacted with NifL, indicating that NifL controls NifA in response to the uridylylation state of GlnK; under conditions of nitrogen excess, NifL inhibits NifA in response to the binding of unmodified GlnK. This model contrasts to one proposed for regulation of NifA activity in *K. pneumoniae* where GlnK relieves NifL inhibition of NifA in any uridylylation state (He *et al.* 1998) and suggests that these organisms have evolved separate mechanisms to control the same regulatory pathway.

References

Colnaghi *et al.* (2001) Microbiol. 147, 1267-1276
Contreras *et al.* (1991) J. Bacteriol. 173, 7741-7749
Dixon R (1998) Arch. Microbiol. 169, 371-380
He *et al.* (1998) J. Bacteriol. 180, 6661-6667
Meletzus *et al.* (1998) J. Bacteriol. 180, 3260-3264

Acknowledgements

This work was funded by a USDA-NRI grant to C. Kennedy. We thank L. He, S. Kustu and W. vanHeeswijk for strains and PII-antiserum, and R. Dixon for communication of unpublished results.

EXPRESSION OF SYMBIOTIC AND NONSYMBIOTIC GLOBIN GENES RESPONDING TO THE MICROSYMBIONTS INFECTION ON *LOTUS JAPONICUS*

T. Uchiumi[1], Y. Shimoda[1], T. Tsuruta[1], Y. Mukoyoshi[1], A. Suzuki[1], S. Sato[2], K. Senoo[3], S. Tabata[2], M. Abe[1], S. Higashi[1]

[1]Dept of Chemistry and Bioscience, Kagoshima University, Kagoshima 890-0065, Japan
[2]Kazusa DNA Research Institute, Chiba 292-0812, Japan
[3]Dept of Sustainable Resource Science, Mie University, Mie 514-8507, Japan

All plants are expected to have hemoglobins; nodule specific globins are called 'symbiotic globins' and globins not involved in symbiosis with rhizobia and *Frankia* are called 'nonsymbiotic globins'. Leguminous plants have both of the globin genes. However, symbiotic globin (leghemoglobin) accumulates specifically and abundantly in root nodules. The mechanism of controlling the expression of both the globin genes and the function of nonsymbiotic globins in leguminous plants have not been revealed.

Lotus japonicus has three symbiotic (*LjLb1, 2, 3*) genes and one nonsymbiotic (*LjNSG1*) globin gene. RT-PCR with specific primer sets for each globin genes showed that all the symbiotic globin genes were expressed specifically and strongly in root nodules. The expression of *LjLb1, 2, 3* could not be detected in root, leaf, and stem of a mature plant, however, the low-level expression was detected in young seedlings. This suggests that symbiotic globin may have unknown function other than oxygen transporter for microsymbiont. *LjNSG1* was expressed at low level in root, leaf, and stem. The expression of *LjNSG* 1 was enhanced in root nodules, whereas, it was strongly repressed in the roots colonized by mycorrhizal fungi *Glomus* sp. R10. The expression of the nonsymbiotic globin gene (*Mhb1*) was also repressed in the mycorrhiza of *Medicago sativa*.

Nitric oxide (NO) is now known to be an "on signal" activating plant defenses against a pathogen (Klessig *et al.* 2000). Considering that heme can serve as a target of NO and that nonsymbiotic hemoglobin is localized in nuclei as same as NO synthase (Seregélyes *et al.* 2000), nonsymbiotic hemoglobin might play a critical role in establishing symbiosis between plants and mycorrhizal fungi.

References

Klessig DF *et al.* (2000) PNAS 97(16), 8849-8855
Seregélyes C *et al.* (2000) FEBS Lett. 482, 125-130

A GLUTAMINE-AMIDOTRANSFERASE-LIKE PROTEIN MODULATES FixT ANTI-KINASE ACTIVITY IN *SINORHIZOBIUM MELILOTI*

H. Bergès, C. Checroun, S. Guiral, A.M. Garnerone, P. Boistard, J. Batut

Lab. de Bio. Mol. des Relations Plantes-Microorganismes, CNRS-INRA, Castanet, France

1. Introduction

S. meliloti forms N_2-fixing nodules on the roots of alfalfa and closely related plants. Expression of nitrogen fixation genes is under both positive and negative control. This regulation depends on a regulatory cascade, on top of which the two-component regulatory system FixLJ activates expression of nitrogen fixation genes in response to microoxic conditions, such as those that prevail inside the nodule. Under microoxic conditions, the sensor histidine kinase FixL autophosphorylates and transfers its phosphate to the FixJ transcriptional regulator protein. Phosphorylated FixJ then activates transcription of two intermediate regulatory genes, *nifA* and *fixK,* that both encode transcriptional regulators. *FixK* is also indirectly responsible for negative regulation of the cascade since it controls expression of a gene, *fixT,* that negatively affects expression of FixLJ dependent genes (Foussard *et al.* 1997). We have shown recently that the FixT protein negatively affects the expression of *nifA* and *fixK* by inhibiting phosphorylation of the sensor hemoprotein kinase FixL and, by consequence, phosphorylation of FixJ (Garnerone *et al.* 1999). Whether FixT serves a mere homeostatic function in *S. meliloti* or whether FixT allows integration of a physiological signal by the FixLJ system was so far unknown. We addressed this question by looking for *S. meliloti* mutants in which the FixT protein would not be active in repression.

2. Results and Discussion

We have isolated an *S. meliloti* mutant strain that phenotypically escapes the repressor activity exerted by FixT. The mutation lies in a gene named *asnO*. This gene encodes a protein homologous to glutamine-dependent asparagine synthetases. The *asnO* gene did not appear to affect asparagine biosynthesis and may serve a regulatory function in *S. meliloti*. We provide evidence that *asnO* is active during symbiosis and that its expression is induced in microoxic conditions (Bergès *et al.* 2001).

The present work argues in favor of a physiological function associated with *fixT*. This finding brings support to the previous suggestion that FixT may allow integration of an additional signal by the FixLJ two-component regulatory system whose activity is primarily regulated by oxygen. We propose as a working model that the absence of AsnO may result in an imbalance in the pool of a metabolite (e.g. a substrate or a product of AsnO), that would affect the intrinsic repressing activity of FixT or, equally, the interaction between FixT and FixL. Because glutamine, a likely by-product of nitrogen fixation in symbiotic rhizobia, is a predicted substrate of the AsnO protein, it is tempting to speculate that *asnO* and *fixT* may provide a link between the nitrogen status of bacteria – or of the plant cell – and nitrogen fixation activity and reducing power generation. Possibly, such a genetic device may connect the nitrogen needs of the plant to the nitrogen fixation activity of the microsymbiont.

3. References

Bergès H *et al.* (2001) BMC Microbiol. 1, 6
Foussard M *et al.* (1997) Mol. Microbiol. 25, 27-37
Garnerone AM *et al.* (1999) J. Biol. Chem. 274, 32500-32506

DOES POLY-β-HYDROXYBUTYRATE INTERFERE WITH NITROGEN FIXATION IN BACTEROIDS?

S. Povolo, S. Casella

Dept Biotecnologie Agrarie, University of Padova, 35020 Legnaro, Padova, Italy

Most rhizobia accumulate poly-β-hydroxybutyrate (PHB) during free-living growth but not all are able to store the polymer in symbiotic life. The role of PHB during nitrogen fixation in bacteroid is unclear. A PHB-synthase (*phbC*) mutant of *Rhizobium etli* has more nitrogenase activity and higher seed content (Cevallos *et al.* 1996). In contrast, bacteroids of *Sinorhizobium meliloti* do not accumulate PHB and nitrogen fixation is not affected by the lack of PHB-synthase (Povolo 1994). In *S. meliloti* a mutation in a gene involved in carbon flux regulation (*aniA*) reduces nitrogen fixation suggesting a complex interplay among different metabolic pathways, namely PHB, glycogen and nitrogenase activity (Povolo, Casella 2000). Recently a Tn5-mutant of *Rhizobium tropici* with enhanced symbiotic nitrogen fixation was isolated. The mutation mapped in a gene of glycogen synthesis (Marroqui 2001). The aim of this work was to isolate and characterize *R. tropici* mutants unable to accumulate PHB.

Mutants deficient in *phbC* were obtained by chromosomal integration of a kanamycin resistance cassette. DNA isolation and handling were performed as described (Sambrook *et al.* 1989). PHB was assayed by gas chromatography. Glycogen was extracted from cells and analyzed according to Ruà *et al.* (1993). For microscopy, dry smears of formalin-treated cells or bacteroids were stained with 0.05 mM Nile red and examined by epifluorescence microscopy. Assay for symbiotic performance on *Phaseolus vulgaris* was carried out in Leonard jars using a nitrogen-free medium. Not inoculated controls were used. Protein concentration was determined by the method of Bradford.

R. tropici CIAT899 accumulated PHB during free-living growth but not much in the symbiotic state. The PHB-mutants of *R. tropici* (strains 900 and 901) showed nitrogen fixation activities similar to wild type (Table 1). Another effect of the *phbC* mutation was an increase of glycogen content (Table 1), confirming what was previously observed for *R. etli*. If PHB in *R. tropici* plays a role in storage/regeneration of reducing equivalents, thus interacting with the nitrogen-fixing apparatus during symbiotic life, glycogen can take the place of PHB when the *phbC* gene is knocked out.

Table 1. Poly-β-hydroxybutyrate, glycogen content and symbiotic traits of *R. tropici* strains.

Strain	PHB mg (mg protein)$^{-1}$	Glycogen μg(mg protein)$^{-1}$	Nodule per plant number	Nodule per plant mg dw	C_2H_4 production mg nod^{-1} h^{-1}
CIAT899	2.16	36.6 ± 1.1	86	55.3	466.8 ± 25
900	0.00	71.1 ± 1.3	73	41.7	428.2 ± 32
901	0.11	62.2 ± 2.2	81	32.1	431.2 ± 46

References

Cevallos MA *et al.* (1996) J. Bacteriol. 178, 1646-1654

Marroqui S *et al.* (2001) J. Bacteriol. 183, 854-864

Povolo S, Casella S (2000) Arch. Microbiol. 174, 42-49

Povolo S *et al.* (1994) Can. J. Microbiol. 40, 823-829

Ruà J *et al.* (1993) J. Gen. Microbiol. 139, 217-222

Sambrook J *et al.* (1989) Molecular Cloning: A Laboratory Manual. 2nd edn, Cold Spring Harbor Laboratory Press, Cold Spring Harbor, NY

INVESTIGATION OF THE ADENYLTRANSFERASE ACTIVITY IN *RHODOSPIRILLUM RUBRUM*

A. Jonsson, S. Nordlund

Dept of Biochemistry and Biophysics, Stockholm University, S-106 91 Stockholm, Sweden

1. Introduction

Adenyltransferase (ATase) is the bifunctional effector enzyme that controls the activity of glutamine synthetase (GS) by adenylation. GS in *R. rubrum* can be ADP-ribosylated without losing the γ-glutamyl transferase activity (Woehle *et al.* 1990). The deadenylation reaction of GS is not affected by the effectors PII-UMP, glutamine or α-ketoglutarate in *R. rubrum*, as it is in *Escherichia coli*, i.e. deadenylation is probably regulated differently in this diazotroph (Johansson *et al.* 1999).

2. Procedure

We are purifying ATase from *R. rubrum* by monitoring ATase activity by incorporation of [α-^{32}P] AMP into GS. We are also using Western blotting techniques and the γ-glutamyl transferase assay to detect the modification status of GS.

3. Results and Discussion

The high-speed supernatant and two different ion exchange fractions (100 and 250 mM NaCl) from *R. rubrum* cells grown under nitrogen-fixing conditions show adenylation activity with ATP, MgCl$_2$ and glutamine added. ADP can substitute for ATP giving almost the same decrease in GS activity. Thus, ATase may use either ADP or ATP as substrate for adenylation of GS. The rate of the adenylation is decreased by α-ketoglutarate (Jiang *et al.* 1998). Adenylated GS treated with snake venom phosphodiesterase shows an increase in activity, indicating conversion to the unmodified form. Deadenylation of GS in the supernatant is stimulated by Pi and α-ketoglutarate. We cannot observe deadenylation activity in the ion-exchange fractions even if PII-UMP, Pi and α-ketoglutarate are added. Thus, a component required for the deadenylation may be absent or the added PII-UMP demodified. The 250 mM NaCl fraction requires PII for modification of GS with either ATP or ADP. Western blotting with polyclonal antibodies against *E. coli* ATase shows that *R. rubrum* ATase is approximately 115 kDa the same as *E. coli* (Caban *et al.* 1976). A strong immunoreaction around 55-60 kDa in the 100 mM fraction suggest that ATase might be degraded within its Q-linker (Jaggi *et al.* 1997). The 100 mM fraction also contains most of the adenylation activity but no deadenylation activity.

4. References

Caban CE *et al.* (1976) Biochem. 15, 1569-1579
Jaggi R *et al.* (1997) EMBO J. 16, 5562-5571
Jiang P *et al.* (1998) Biochem. 37, 12782-12794
Johansson M *et al.* (1999) J. Bacteriol. 181, 6524-6529
Woehle DL *et al.* (1990) J. Biol. Chem. 265, 13741-13749

5. Acknowledgements

Thanks to authors and we thank Dr W.C. van Heeswijk for antibodies against ATase from *E. coli*, and *E. coli* strains RB 9017 *glnE⁻* and och RB9040 *glnD⁻*. Supported by the Swedish Natural Science Research Council.

NITROGEN FEEDBACK REGULATION IN PHOSPHORUS DEFICIENT *LOTUS JAPONICUS* PLANTS

S. Keller[1], G. Colebatch[2], M. Udvardi[2], C. Sautter[1], U.A. Hartwig[1]

[1] Institute of Plant Sciences, Swiss Federal Institute of Technology, 8092 Zürich, Switzerland
[2] Max Planck Institute of Molecular Plant Physiology, 14424 Potsdam, Germany

1. Introduction

A major limiting factor for legume growth and symbiotic N_2 fixation is phosphorus (P). In extreme cases P deficiency prevents nodulation and symbiotic N_2 fixation. The legume *Lotus japonicus* and its microsymbiont *Mesorhizobium loti* serve as models to study the mechanisms of the inhibitory effect of low P supply on the symbiosis. The aim of this experiment was to evaluate the difference in gene expression of *Lotus japonicus* grown under a sufficient and a very low phosphorus supply. This may in turn help elucidate the inhibitory mechanism of phosphorus on symbiotic N_2 fixation.

2. Material and Methods

One hundred and forty-four seedlings of *Lotus japonicus*, wt were grown for 42 days in sandculture. The nutrient solution contained N 1.5 mM and P 0.075 mM. After day 35, the P supply was changed to 0.5 mM and 0.0005 mM to half of the plants. At day 42 plants were harvested and frozen in liquid N_2. Total RNA was extracted and ^{33}P labeled ss cDNA was hybridized with filters containing 1580 different clones of a cDNA library (nodule tissue of *Lotus japonicus*). Six micro array filters per treatment have been processed. Data were normalized with the *Haruspex* expression software.

3. Results and Discussion

Under low P supply three different clones encoding for asparagine synthetase LJAS2 have been down-regulated significantly. In earlier studies an accumulation of asparagine under P deficiency has been observed (Steward, Larher 1980; Almeida *et al.* 2000). Repression by nitrogen-transport amino acids of the expression of the asparagine synthetase gene ASN2 in Arabidopsis has been shown by Lam *et al.* 1998 and Wang *et al.* 2000. Therefore, we suggest a feedback mechanism of asparagine on the expression of asparagine synthetase.

4. References

Almeida JPF *et al.* (2000) J. Exp. Bot. 51, 1289-1297
Lam HM *et al.* (1998) Plant J. 16, 345-353
Steward GR, Larher F (1980) Biochem. of Plants 5, 609-635
Wang RC *et al.* (2000) Plant Cell 12, 1491-1509

RHIZOBITOXINE BIOSYNTHETIC GENES IN *BRADYRHIZOBIUM ELKANII*

T. Yasuta [1], S. Okazaki[1], K. Yuhashi[1], H. Ezura[2], K. Minamisawa[1]

[1]Institute of Genetic Ecology, Tohoku University, Sendai 980-8577, Japan
[2]Plant Biotechnology Institute, Nishi-Ibaraki, 319-0292, Japan

We cloned and sequenced a cluster of genes involved in the biosynthesis of rhizobitoxine, a nodulation enhancer produced by *Bradyrhizobium elkanii*. The nucleotide sequence of the cloned 28.4-kb DNA region encompassing *rtxA* showed that several open reading frames (ORFs) were located downstream of *rtxA*. A large deletion mutant of *B. elkanii*, USDA94Δ*rtx*::Ω1, which lacks *rtxA*, ORF1 (*rtxC*), ORF2 and ORF3, did not produce rhizobitoxine, dihydrorhizobitoxine, or serinol. The broad host-range cosmid pLAFR1, which contains *rtxA* and these ORFs, complemented rhizobitoxine production in USDA94Δ*rtx*::Ω1. Further complementation experiments involving cosmid derivatives randomly inserted by using a kanamycin cassette revealed that at least *rtxA* and *rtxC* are necessary for rhizobitoxine production. Insertional mutagenesis of the N-terminal and C-terminal regions of *rtxA* indicated that *rtxA* is responsible for two crucial steps, serinol formation and dihydrorhizobitoxine biosynthesis. An insertional mutant of *rtxC* produced serinol and dihydrorhizobitoxine, but no rhizobitoxine. Moreover, *rtxC* was highly homologous to the fatty acid desaturase of *Pseudomonas syringae* and included the copper-binding signature and eight histidine residues conserved in membrane-bound desaturase. This result suggested that *rtxC* encodes dihydrorhizobitoxine desaturase for the final step of rhizobitoxine production.

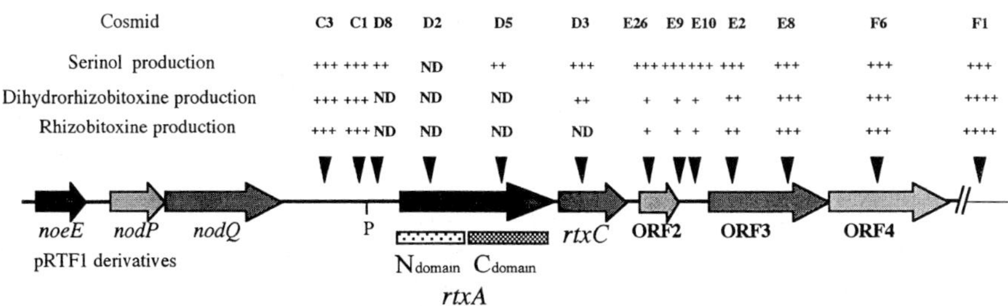

Figure 1. [AT1]Serinol, dihydrorhizobitoxine and rhizobitoxine production of *Bradyrhizobium elkanii* USDA94Δ*rtx*::Ω1 as complemented with pRTF1 cosmid derivatives created by insertion of a kanamycin cassette. The insertion point of the kanamycin cassette is indicated by the arrowhead [AT2] on pRTF1. N-domain and C-domain of the *rtxA* gene show the regions where their deduced amino acid sequences are homologous to aminotransferase and O-acetylhomoserine sulfhydrylase, respectively. P, putative promoter (according to the sequence).

RELATIONSHIP BETWEEN REDUCTIVE METABOLISM AND THE EXCHANGES OF CO_2, O_2, H_2 AND N_2 IN LEGUME NODULES

J. Mykoo, Y.-P. Cen, A. Guse, D.B. Layzell

Dept of Biology, Queen's University, Kingston, ON, K7L 3N6 Canada

1. Introduction

Legume nodules are involved in the simultaneous exchange of four gases: CO_2, O_2, H_2 and N_2. These exchanges are linked integrally through the metabolism of C, N and O within the nodule tissues. In theory, it should be possible to use a measure of nitrogenase activity and our understanding of nodule biochemistry to predict, with accuracy, the relative exchanges of CO_2 and O_2. This report describes preliminary efforts to make these predictions and measurements.

2. Methods

A detailed mathematical model was constructed to describe the current understanding of the complex quantitative relationships that exist among bacteroid and plant metabolism in nodules. By extending an earlier study (Cen *et al.* 2001), the model calculated CO_2 and O_2 exchanges associated with N_2 fixation, ATP, carbohydrate, polyhydroxybutyrate and glycogen synthesis in bacteroids, as well as the metabolism of sucrose and the synthesis of malate, ATP, carbohydrate, asparagine and ureides in the plant fraction.

Measurements of H_2 production in excised soybean (*Glycine max* cv. Maple Glenn) nodules were combined with assumptions regarding nodule growth rate and the substrates and end products of metabolism to predict the CO_2 and O_2 exchanges associated with metabolism. To test the model predictions, a gas exchange system was built to measure simultaneously the production of H_2 and CO_2 and the uptake of O_2, the latter involving a novel differential oxygen analyzer (Willms *et al.* 1997). Respiratory quotient (RQ) was calculated as $-CO_2/O_2$ exchange.

3. Results and Discussion

The model predicted that the bacteroid fraction of nodules is highly reductive (RQ = 1.96) whereas the plant fraction is highly oxidative (RQ = 0.13). Whole nodules were predicted to have a respiratory quotient (RQ = $-CO_2/O_2$) of 1.08, a value significantly lower than that measured in excised soybean nodules (1.25).

To explain the major discrepancy between measured and theoretical RQ values, a sensitivity analysis of the model was conducted, based on the possibility that the model's assumptions may be incorrect. The only reasonable changes in model assumptions that were able to generate predicted RQ values similar to measured values involved the complete elimination of the ureide biosynthetic pathway. Therefore, the model predicted that short term (<7 min) exposure of a nodule to Ar:O_2 should result in an increase in RQ of about 0.18 units. If this does not occur, there are likely to be major problems with our current understanding of the pathway of ureide synthesis in nodules. Experiments are currently underway to test this hypothesis.

4. References

Cen Y-P, Turpin DH, Layzell DB (2001) Plant Physiol. 126
Willms J, Dowling N, Dong Z, Hunt S, Shelp B, Layzell D (1997) Anal. Biochem. 254, 272-282

EVIDENCE OF A CONFORMATIONAL PROTECTION MECHANISM OF NITROGENASE IN *ACETOBACTER DIAZOTROPHICUS*

A. Ureta, S. Nordlund

Dept of Biochemistry and Biophysics, Stockholm University, S-106 91 Stockholm, Sweden

1. Introduction

Acetobacter diazotrophicus is a gram-negative diazotroph, endophytic in sugarcane, where it has been shown to play an important role in the nitrogen supply required for growth without nitrogen fertilizer. To understand its diazotrophic physiology under endophytic conditions, we need to establish how it functions diazotrophically in the presence of O_2 as well as the source of reductant and energy. To address this question we have studied nitrogenase activity in the presence of glucose or pyruvate as energy/reductant source under low (5%) or high (20%) O_2 levels.

2. Materials and Methods

A. diazotrophicus, strain PAL5, was grown as described (Stephan *et al.* 1991) and diazotrophic conditions were established by growing cells with a starter of 1 mM NH_4Cl. Nitrogenase activity was determined as acetylene reduction. SDS-PAGE, Western blotting and protein concentration were determined by standard protocols.

3. Results and Discussion

Nitrogenase activity was supported by 0.1 M glucose under 20% and 5% O_2 whereas, with pyruvate, nitrogenase activity occurred only under 5% O_2. When cells were transferred to 20% O_2, with pyruvate, nitrogenase activity was lost and could not be recovered at 5% O_2 concentration. However, glucose addition restored nitrogenase activity. Together, these results indicate that the diversion of electrons to O_2 cannot account for this inactivation and that reactivation requires a reductant that can supply electrons for nitrogenase and for respiration, generating ATP as well as lowering the intracellular O_2 concentration.

Western blot analysis showed that the Fe protein is initially protected from degradation but after longer exposure to 20% O_2 with pyruvate, it was degraded. This indicated that a putative conformational mechanism protecting nitrogenase is present in *A. diazotrophicus*. Western blot using antibodies raised against the FeSII (Shetna) protein of *Azotobacter vinelandii* showed cross reaction with a protein of approximately 15Kd present in both N-limited or N+ extracts of *A. diazotrophicus* grown either under 5% or 20% O_2. The interaction of the *A. diazotrophicus* FeSII-like protein and both components of nitrogenase was further confirmed by co-immunoprecipitation of the Fe or MoFe protein under aerobic but not anaerobic/reducing conditions. These results show that a transient conformational protection mechanism of nitrogenase operates in *A. diazotrophicus* when nitrogen-fixing cells experience a sudden change in redox-state.

4. References

Moshiri F *et al.* (1994) Mol. Microbiol. 14, 101-114
Stephan MP *et al.* (1991) FEMS Microbiol. Lett. 77, 67-72

5. Acknowledgements

We are grateful to F. Moshiri for providing antibodies against the FeSII protein and an expression plasmid pAVFESII. The work was supported by grants to SN from SJFR and CTS.

CHARACTERIZATION OF IRON-REGULATED OUTER MEMBRANE PROTEINS IN *BRADYRHIZOBIUM JAPONICUM*

H.L. Prince[1], A.N. Gray[1], E. Boncompagni[2], M.L. Guerinot[1]

[1]Dartmouth College, Hanover, NH, USA
[2]Université de Nice - Sophia Antipolis, Nice, France

1. Introduction

Iron-containing proteins figure prominently in the nitrogen-fixing symbiosis between soybeans and *Bradyrhizobium japonicum*. Iron acquisition, however, presents unique challenges both in soil and *in planta*. Many rhizobial species, including *B. japonicum*, produce and utilize iron-chelating siderophores as a means of solubilizing and acquiring iron. Siderophore utilization requires a specific uptake system, including a high-affinity outer membrane receptor. Purification of *B. japonicum* strain 61A152 iron-regulated outer membrane proteins has led to the cloning of two genes with similarity to known siderophore receptors. The *fegA* gene is predicted to encode a protein with similarity to hydroxamate siderophore receptors (LeVier, Guerinot 1996).

2. Results and Discussion

A *fegA* knockout, constructed by insertion of an omega spectinomycin/streptomycin resistance cassette into the coding region, is incapable of utilizing the hydroxamate siderophore ferrichrome and produces a dramatic phenotype *in planta*, failing to develop a nitrogen-fixing symbiosis. Ten putative *fegA* mutant suppressors were isolated from wild type-like nodules infrequently observed on soybeans inoculated with the *fegA* knockout strain. These suppressor strains have not regained the ability to use ferrichrome but are able to establish a nitrogen-fixing symbiosis. We are currently creating genomic libraries from the suppressor strains to complement the *fegA* knockout.

The gene encoding the second iron-regulated OMP in *B. japonicum* 61A152, *hemR*, is predicted to encode a protein with significant similarity to heme receptors from other gram-negative bacteria (Genco, Dixon 2001). We have also sequenced two ORFs downstream of *hemR*. Recently, sequences encoding a heme uptake system in *B. japonicum* strain USDA 110 were deposited in Genbank. The heme receptor genes from USDA110 and 61A152 are 60% identical and 66% similar. The two downstream ORFs, *orf110* and *orf165*, are 84% and 92% identical and 85% and 95% similar, respectively. There are 19 transmembrane domains predicted in the 61A152 heme receptor.

Using a simple plate assay, we have established that *B. japonicum* is able to utilize heme and leghemoglobin as iron sources. We are currently constructing a stable knockout of the *hemR* gene and the resulting mutant will be characterized in both free-living conditions and *in planta*. We have also cloned a predicted heme receptor from *S. meliloti* strain Rm 1021, and currently we are making a knockout of the heme receptor gene. We have shown that Rm1021 cells are able to utilize heme and leghemoglobin as iron sources. A third iron-regulated OMP fron *B. japonicum* 61A152 is currently being isolated using 2D-gel SDS-PAGE analysis to facilitate the cloning of the gene.

3. References

Genco CA *et al.* (2001) Mol. Microbiol. 39, 1-11
LeVier K *et al.* (1996) J. Bacteriol. 178, 7265-7275

CONTROL OF NITROGENASE REACTIVATION BY GlnZ PROTEIN IN *AZOSPIRILLUM BRASILENSE*

E.M. Souza[1], G. Klassen[1,2], M.B. Steffens[1], J. Inaba[1], M.G. Yates[1], L.U. Rigo[1], F.O. Pedrosa[1]

[1]Dept of Biochem. and Mol. Biol.
[2]Dept of Basic Path., UFPR, Curitiba-PR, Brazil

1. Introduction

Nitrogenase switch-off by ADP ribosylation in some diazotrophs is due to the inactivation of the Fe-protein (Lowery 1986) by the transfer of an ADP-ribosyl residue to one of the subunits, catalyzed by dinitrogenase reductase ADP-ribosyl transferase (DRAT), and the activity is recovered by dinitrogenase reductase-activating glycohydrolase (DRAG) (Lowery 1986). *ntrC* mutants of *A. brasilense* were impaired in nitrogenase switch-off (Pedrosa, Yates 1984) suggesting that NtrC may participate in a signaling circuit including DRAT/DRAG, involving a signal protein. The PII protein and its paralog, GlnZ in *A. brasilense* are candidates for such signal molecules. Here we show that the nitrogenase activity of a *glnZ* mutant is switch-on deficient and that GlnZ could be involved in the control of the DRAT/DRAG system.

2. Material and Methods

A. brasilense strains FP2 (wild type) (Pedrosa, Yates 1984), 7611 (*glnZ::Ω*) (de Zamaroczy 1998) and 7611 pJI1 (*glnZ::Ω* carrying the plasmid-borne wild-type *A. brasilense glnZ* gene Tc^R) were grown in liquid NFbHPN medium (Pedrosa, Yates 1984) at 120 rpm and 30°C. Plasmid pJI1 was constructed by cloning a DNA fragment containing the *A. brasilense glnZ* gene into the vector pLAFR3.18. Nitrogenase activity was determined by the acetylene reduction method (Scholhorn, Burris 1966).

3. Results and Discussion

Addition of ammonium chloride (0.2 mmol L^{-1}) to nitrogenase derepressed *A. brasilense* FP2 (wild type) cultures caused almost complete inhibition of nitrogenase activity. The activity was fully recovered after approximately 15 minutes, following exhaustion of ammonium ions from the medium. The *A. brasilense* strain 7611 (*glnZ* mutant) was also fully derepressed for nitrogenase activity and the addition of ammonium ions caused a similar level of inhibition. However, recovery of the initial nitrogenase activity was partial (20–40%) even after 80 min of ammonium depletion. Since added ammonium was completely taken up after 10 min in both the wild type and *glnZ* strains this result suggests that the GlnZ protein is involved in the mechanism of reactivation of the ADP-ribosylated iron protein, but is not essential for nitrogenase switch-off by ammonium ions. Furthermore, full complementation of the *glnZ* mutant by the *A. brasilense glnZ* gene strongly supports this role for the GlnZ protein. In contrast, nitrogenase switch-off by anaerobic conditions was reversed completely by oxygenation and at the same rate in both the wild type and the *glnZ* mutant. This result indicates that GlnZ is neither involved in the mechanism of anaerobic inactivation nor in the aerobic reactivation of nitrogenase.

4. References

Lowery RG *et al.* (1986) J. Bacteriol. 166, 513-518
Pedrosa FO, Yates MG (1984) FEMS Microbiol. Let. 23, 95-101
Scholhorn R, Burris RH (1967) Proc. Natl. Acad. Sci. 58, 213-216

5. Acknowledgements

We thank CNPq/MCT/PRONEX for financial support.

THE *nifX* GENE PRODUCT IS ESSENTIAL FOR NITROGENASE ACTIVITY UNDER IRON OR MOLYBDENUM LIMITATION IN *HERBASPIRILLUM SEROPEDICAE*

G. Klassen[1,2], F.O. Pedrosa[1], E.M. Souza[1], M.G. Yates[1], L.U. Rigo[1]

[1]Department of Biochemistry and Molecular Biology
[2]Department of Basic Pathology, UFPR, 19046, 81531-970, Curitiba-PR, Brazil

1. Introduction

In *Klebsiella pneumoniae* the products of six genes are essential for the biosynthesis of the nitrogenase cofactor (FeMo-co): *nifH, nifQ, nifB, nifV, nifE* and *nifN* (Kim, Rees 1992). Shah *et al.* (1999) demonstrated that the product of the *nifX* gene, enhanced the yield of FeMo-co *in vitro*. The *nifENXorf1orf2* gene cluster was found downstream from the *nifHDK* in *Herbaspirillum seropedicae* (Klassen 1999). Here we show that *nifX* and/or *orf1orf2* of *H. seropedicae* are required for nitrogenase activity under iron or molybdenum deficiency, suggesting the involvement of these genes in FeMo-co biosynthesis.

2. Materials and Method

H. seropedicae mutants *nifX* and *orf1* were produced by Km cassette insertions (Klassen 1999). The strains were grown in NFbHP medium (Pedrosa, Yates 1984) supplemented with 20 mmol/L NH₄Cl (NFbHPN). Ferrous sulfate was omitted from the medium and sodium molybdate free-medium was prepared as described (Schneider 1991). Nitrogenase activity of *H. seropedicae* strains was measured in NFbHP semi-solid medium containing sodium glutamate (0.5 mmol/L).

3. Results

The *nifX* and *orf1* mutants were capable of fixing nitrogen in the presence of iron or molybdenum sufficiency. However, when iron was omitted from the medium those mutants failed to reduce acetylene whereas the wild type strain had normal activity. Addition of 0.16 and 16 µg/mL of Fe.EDTA recovered 50% and full acetylene reduction activity respectively of the mutants. Both mutants were fully complemented by the *nifNXorf1orf2* genes. Equal expression of β-galactosidase activity in the both mutants with *lacZ* fusion was observed under low and high iron conditions supporting the suggestion that *nifXorf1orf2* are required for nitrogenase activity rather than *nif* gene expression. The *nifX* and *orf1* mutants were tested in molybdenum-free medium. The results show that both mutants had two-fold reduced nitrogenase activity under molybdenum deficiency compared to the wild type strain. The results indicate that NifX and Orf1 may be involved in both iron and molybdenum mobilization or incorporation into FeMo-co in *H. seropedicae*.

4. References

Kim J, Rees DC (1992) Nature 360, 553-560
Klassen G *et al.* (1999) FEMS Microbiol. Lett. 181, 165-170
Pedrosa FO, Yates MG (1984) FEMS Microbiol. Lett. 23, 95-101
Schneider K *et al.* (1991) Anal. Biochem. 193, 292-298
Shah VK *et al.* (1999) J. Bacteriol. 181, 2797-2801

5. Acknowledgements

We thank CNPq/MCT/PRONEX for financial support.

MOLECULAR EVOLUTION OF URICASE II GENES AND CHARACTERIZATION OF TRANSGENIC *LOTUS JAPONICUS* CONTAINING ANTI-SENSE URICASE GENES

S. Tajima[1], K. Shimomura[1], M. Nomura[1], K. Takane[2], H. Kouchi[2]

[1]Dept Life Sci., Fac. Agri, Kagawa University, Miki-cho, Ikenobe, Kagawa, Japan
[2]Natl Inst. Agrobiol. Res., Kannondai, Tsukuba 305-8602, Japan

1. Introduction

Uricase (Nodulin 35) gene expression in uninfected cells of root nodules of tropical legumes like *Glycine max*, *Phaseolus* and *Vigna* has been studied as a typical case of symbiotically enhanced gene expression contributing to assimilation of fixed nitrogen. Uricase expression itself can be observed in various legume tissues, because the reaction introduces irreversible degradation of purine bases and the activity of very low intensity was detected in all senescent or dividing plant cells for recruiting nucleic acid-N. However, the molecular mechanism and physiological role of uricase over-expression that may be specific in tropical legumes is still obscure. In this report, we tried to identify the profile of the significance of uricase gene activation in legumes.

2. Materials and Methods

To analyze the mechanism, we compared the uricase gene structure and the expression profiles (*in situ* hybridization) after cDNA and genomic cloning of uricase genes from *Glycine max* (Takane *et al.* 1997), *Medicago sativa* (Cheng *et al.* 2000) and *L. japonicus* (Takane *et al.* MPMI 2000).

In order to analyze the metabolic role of uricase gene in *Lotus japonicus* GIFU, we cloned the uricase gene from a cDNA library of *Lotus japonicus* nodules and introduced an anti-sense uricase gene for decreasing uricase content in uninfected cell of the nodules. Using an *Agrobacterium tumefaciens*-mediated gene transfer system, we have obtained over 20 independent transgenic plants. Southern blot analysis showed that these plants had introduced anti-sense uricase construct.

3. Results and Discussion

The result shows that legume uricases are divided into two major groups. Group-I is comprised of soybean, *Vigna*, and *Phaseolus*, which are well demonstrated to bear ureide-transporting determinate legumes (Streeter 1991). Canavalia bearing determinate nodules, which is also described as an ureide transporter (Kim and An 1993), is close to but distinct from group-I. Group-II is comprised of alfalfa, pea, *Astragalus*, and *Lotus japonicus*. Alfalfa and pea are well known as amide-transport legumes (Streeter 1991).

When we surveyed the gene expression by *in situ* hybridization technique, the data revealed that alfalfa uricase genes were expressed in uninfected cells of nodule tissue in small extent, and in *Lotus japonicus* nodules the magnitude was higher. The data suggested that cell type specific expression was observed with identical manners to that of Nod-35 in soybean nodules in various legumes, although the magnitude of the expression was much higher in Nod-35 in soybean nodules. The data also showed that enhancement of uricase gene expression was observed not only in tropical legumes but also in various non-ureide determinate type nodules.

An anti-sense RNA experiment using transgenic *Lotus japonicus* plants suggested uricase gene expression would influence to carbon metabolism of the plants because the transgenic plants were dwarfed under low light and grew normally under natural light. We also found that all of these transgenic plants grow more slowly than non-transgenic plants. After infection with *M. loti*, all of these transgenic plants were nodulated well.

A MODEL OF ADENYLATE GRADIENTS IN THE INFECTED CELLS OF LEGUME NODULES

H. Wei[1], P.P. Thumfort[2], C.A. Atkins[2], D.B. Layzell[1]

[1]Biology, Queen's U, Kingston, ON, Canada K7L 3N6
[2]Botany, U Western Australia, Nedlands, WA 6009, Australia

1. Introduction

In metabolically active legume nodules, the adenylate energy charge (AEC = [ATP + 0.5ADP] / [ATP + ADP + AMP]) is low (0.70) compared with other aerobic tissues (typically $\geq$ 0.80). Non-aqueous fractionation of soybean nodules showed that the low AEC was due to the plant fraction (0.66), not the bacteroids (0.76) (Kuzma *et al.* 1999). However, an O_2 limitation of nodule metabolism reduced bacteroid AEC to 0.56 but had no effect on that of the plant fraction. Therefore, the bacteroids were O_2 limited while the plant fraction was not. To explain why the plant fraction AEC was so low (0.66) when the oxygen supply to it was sufficient, we hypothesized that there were steep ATP gradients from the mitochondrial region near the gas space (i.e. the site of ATP synthesis) to the cytosol surrounding the symbiosomes. A mathematical model was developed to test this hypothesis.

2. Methods

The model assumed that the infected cells were tightly packed rhombic dodecahedra with intercellular gas spaces along the edges, from which O_2 diffused to the center of the cell. The Thumfort *et al.* (2000) model was modified to predict the diffusion and gradients of O_2 and ATP across the cell, assuming cytosolic ATP demand for NH_4^+ assimilation, malate transport and plant growth + maintenance. The model was also used to simulate the effects of an Ar:O_2 atmosphere (5–7 min exposure), where the demand for NH_4^+ assimilation was assumed to be zero.

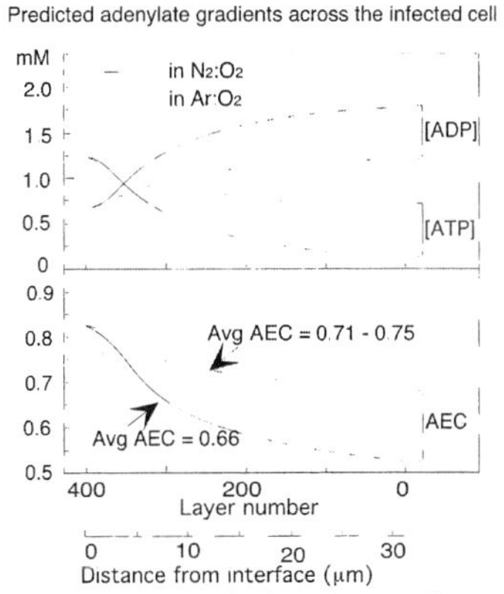

3. Results and Discussion

The model supported the hypotheses, predicting that glutamine synthetase activity in the cytosol was largely responsible for generating a gradient in plant AEC of 0.84 near the space to 0.55 near the center of the cell to achieve an average AEC of 0.66 (Figure 1). Under Ar:O_2, the adenylate gradients were predicted to be significantly reduced, resulting in an average plant fraction AEC of 0.71–0.75 (Figure 1). This prediction was consistent with previous studies of adenylates in whole nodules (deLima *et al.* 1994), and will be tested using non-aqueous fractionation of control and Ar:O_2 treated nodules.

Figure 1. Predicted adenylate gradients in the bacteria infected cells.

4. References

de Lima *et al.* (1994) Physiol. Plant. 91, 687-695
Kuzma *et al.* (1999) Plant Physiol. 119, 399-407
Thumfort *et al.* (2000) J. Theor. Biol. 204, 47-65

VOLTAGE-ACTIVATED CATION-SELECTIVE CHANNEL PERMEABLE TO NH$_4^+$ ON THE SYMBIOSOME MEMBRANE OF THE MODEL LEGUME *LOTUS JAPONICUS*

D.M. Roberts[1], S.D. Tyerman[2]

[1]Department of Biochemistry, Cellular and Molecular Biology, Univ. Tennessee, Knoxville, TN, USA

[2]Department of Biology, Flinders University of South Australia, SA, Australia

The symbiosome membrane is responsible for the metabolite exchange that represents the crux of nitrogen-fixing symbioses: the efflux of fixed nitrogen (NH$_4^+$) and the uptake of dicarboxylates as an energy source for the symbiont (Udvardi, Day 1997). In the present study we have examined the transport activities of the symbiosome membrane of *Lotus japonicus* by patch clamp recording. Excised *L. japonicus* symbiosome membrane patches show a rectified, inward (towards the cytosolic compartment) current that is activated by negative (with respect to the cytosol) voltage potentials. A predominant selectivity for monovalent cations was observed, although evidence for a low calcium permeability was also obtained. The channel is permeable to both K$^+$ and NH$_4^+$, but NH$_4^+$ showed a slightly higher conductance (apparent K$_m$=17 mM). The channel shows no intrinsic rectification, but depends upon the presence of divalent cations. In the absence of divalent cations (either Mg^{2+} or Ca^{2+}) the channel is largely open and symmetrical, but the addition of divalent cations results in inhibition of the current with the direction of rectification dependent upon which side of the membrane possesses a higher divalent cation activity. However, based on measurements with intact symbiosomes, the direction of current is proposed to be towards the cytosol and Mg^{2+} on the cytosolic side is proposed to mediate the rectification and voltage-dependence of the current. Overall, the data suggest that *L. japonicus* possesses an ammonium-permeable cation channel that is similar to the cation channel previously described on the soybean symbiosome (Tyerman *et al.* 1995; Whitehead *et al.* 1998) and has the properties necessary to transport fixed NH$_4^+$ from the symbiosome space to the plant cytosol. In addition, coordinate regulation of this channel and the electrogenic H$^+$-ATPase may also play a role in symbiosome homeostasis by regulation of the transmembrane potential and pH of the symbiosome space.

References
Udvardi MK, Day DA (1997) Annu. Rev. Plant. Physiol. Plant. Mol. Biol., 48493-48523
Tyerman SD, Whitehead LF, Day DA (1995) Nature 378, 629-632
Whitehead LF, Day DA, Tyerman SD (1998) Plant J. 16, 313-324

rhrA-, *rhtA-*, *sitB-* AND *sitD*-LIKE GENES ARE INVOLVED IN IRON ACQUISITION IN *SINORHIZOBIUM MELILOTI* 242

R. Platero, F. Battistoni, A. Arias, E. Fabiano

Laboratorio de Ecología Microbiana, Departamento de Bioquímica, Unidad Asociada a la Facultad de Ciencias, Instituto de Investigaciones Biológicas Clemente Estable, Ministerio de Educación y Cultura, Av. Italia 3318, Montevideo 11600, Uruguay

Rhizobia are able to acquire iron from different ferri-siderophores, hemin, hemoglobin and leghemoglobin in iron-depleted conditions. Four Tn*5*-1063a mutants of *Sinorhizobium meliloti* 242 defective on iron-acquisition were isolated and characterized. Mutant H38, unable to grow on iron-chelated medium, did not produce detectable amount of siderophores. Furthermore, the tagged locus of H38 was almost identical to an *araC* like gene found in the rhizobactin regulon of *S. meliloti* 2011. Mutants H26, H36 and H21 presented impaired growth on iron-chelated medium despite that they were still able to produce siderophores. The interrupted DNA region of H21 mapped in *rhtA*, a gene that encodes the outer membrane receptor of rhizobactin 2011. Tagged loci of H26 and H36 were found to share significant similarities with ABC (ATP-binding cassette) iron transporter genes of several gram-negative bacteria and with two ORFs present in the sequenced chromosome of *S. meliloti* SU47 predicted to encode ABC iron transport proteins. Despite the relevant role of iron in the rhizobia-legume symbiosis and the predictable importance of possessing high affinity iron acquisition systems, none of these loci were essential for N_2-fixation.

STUDIES OF THE INTERACTION BETWEEN THE AMMONIUM TRANSPORT PROTEIN AmtB AND THE SIGNAL TRANSDUCTION PROTEIN GlnK

G. Coutts, M. Merrick

Dept of Molecular Microbiology, John Innes Centre, Norwich, NR4 7UH, UK

In *E. coli* there are two members of the PII signal transduction protein family, GlnB and GlnK. GlnB facilitates the sensing of the intracellular nitrogen status and regulates the activities of the NtrBC system and adenyltransferase. While GlnK can partially substitute for some of the functions of GlnB, the primary function of GlnK is unknown. In both bacteria and archaea the *glnK* gene is highly conserved in an operon along with *amtB*, which encodes a high affinity ammonium transporter, and we have proposed that this conservation reflects the fact that GlnK and AmtB interact (Thomas *et al.* 2000a). If this is the case then we expect to find GlnK associated with the inner membrane in wild type or Δ*glnB* strains of *E. coli* but not in an Δ*amtB* strain. We have fractionated cells grown under nitrogen-limiting conditions and probed these fractions using Western blots with an antibody against PII (GlnK or GlnB) and we do indeed observe the predicted association.

To investigate the possible physiological role of the GlnK-AmtB association we examined the effects of ammonia shock on the interaction. It is known that ammonia shock results in the de-uridylation of GlnK in *E. coli*, and when strains were fractionated pre- and post-ammonia shock we observed a significant increase in the membrane-association of GlnK post-ammonia shock. We then used native PAGE to confirm that the non-uridylated form of GlnK has a markedly higher affinity for AmtB than uridylylated GlnK. A His-tagged form of AmtB was used to show that the level of AmtB in the membrane does not change post-ammonia shock.

The major part of AmtB that is located in the cytoplasm is the 32 amino acid C-terminal domain and this is therefore a potential site of interaction between AmtB and GlnK (Thomas *et al.* 2000b). Deletion of this region does not inactivate the transport activity of AmtB but does prevent GlnK association to the membrane. We therefore propose that the C-terminus is either required directly for interaction with GlnK or is necessary to allow AmtB to adopt a conformation that allows GlnK to interact.

In conclusion we propose the following model – that as a response to an increase in the intracellular nitrogen status the GlnK protein associates to AmtB thereby switching off ammonium transport, by post-translational control. The sequestration of GlnK to the membrane would also result in a rapid decrease in the intracellular pool of GlnK, which could in turn affect the activities of any cytoplasmic targets of GlnK.

References
Thomas G *et al.* (2000a) Trends Genet. 16, 11-14
Thomas G *et al.* (2000b) Mol. Microbiol. 37, 331-344

Acknowledgment
Graham Coutts acknowledges a Ph.D studentship from the BBSRC.

TRANSCRIPTIONAL REGULATION OF NITROGEN ASSIMILATION BY THE PRODUCTS OF NITROGEN FIXATION

J. Bussell[1], M. Ferguson[1], H. Winter[2], A. Mann[1], P. Storer, C. Atkins[1], P. Smith[1]

[1]Department of Botany, The University of Western Australia, Crawley, Australia
[2]Biologie, Universitat Osnabruck, Osnabruck, Germany

De novo purine nucleotide synthesis is the major assimilation pathway for fixed N in species of the Tribes *Phaseoleae*, *Desmodieae* and *Indigofereae* within the Phaseoloid Group of papilionoid legumes (Atkins, Smith 2000). The amide group of glutamine provides half of the N for the purine ring while amino-N, as the amino group of aspartate and the intact glycine molecule, provides the other two. The product, IMP, is oxidized to ureides (allantoin and allantoic acid) and exported from nodules in xylem. The ureides are the major form of N for the nutrition of the host. Nine *pur* genes encode the 10 enzymes in the purine pathway. Regulation of *pur* genes by products of N_2 fixation was investigated by growing inoculated cowpea plants with their root systems in 80% Ar: 20% O_2. Although after 16–18 days under these conditions the plants begin to suffer N deficiency up to ca. 16 days nodule development is normal. Normal levels of nitrogenase activity are expressed and, because high rates of H_2 evolution occur, the nodules utilize incoming phloem-delivered sugar and do not accumulate starch. In nodules grown in Ar:O_2 *Vupur1* (PRPP amidotransferase), *4* (FGAR amidotransferase), *5* (AIR synthetase), *6* (AIR carboxylase), *7* (SAICAR synthetase) and *8* (adenylosuccinate AMP lyase) are expressed at levels required to maintain basic cellular processes, while in air, expression increases dramatically as N_2 fixation begins, consistent with regulation of these genes by products of fixation. *Vupur2* (GAR synthetase), *3* (GAR transformylase), *9* (AICAR transformylase/IMP cyclohydrolase), IMPDH (IMP dehydrogenase) and uricase transcripts showed little change in level of expression over time. However, expression of *Vupur2*, *9* and IMPDH were reduced in Ar:O_2 compared to air while *Vupur3* and uricase was the same in both treatments. These data indicate that unlike the situation in some prokaryotes the purine pathway in legume nodules is not coordinately regulated and only some of the encoding genes are transcribed in response to the flux of fixed N entering the infected cell.

Transient exposure of an established symbiosis to Ar:O_2 causes the activity of AIRS and uricase, but not GART, to decline rapidly. However, only AIRS protein rapidly disappeared in Ar and only the gene encoding AIRS (*Vupur5*) appeared to be transcriptionally regulated. Within 3 h, expression of *Vupur5* was markedly reduced, returning to its initial level once plants were transferred back into air. The activity of uricase, the ultimate step in purine oxidation, was rapidly lost in Ar:O_2 but the protein was stable, consistent with post-translational/allosteric regulation. The prompt loss of *de novo* purine synthesis that occurs when N_2 fixation is inhibited could be accounted for by rapid turnover of AIRS protein and by regulation of transcription and translation of *Vupur5*.

Taken together these data indicate that the ureide pathway is regulated closely by the flux of fixed N, that not all genes involved are equally sensitive, and that a number of different levels of regulation are involved. Current research is examining the nature of *pur* gene promoters with a view to identifying the transcription factors and mechanisms involved.

References

Atkins CA, Smith PMC (2000) In Triplett E (ed), Prokaryotic Nitrogen Fixation, pp 559-587, Horizon Scientific Press, Wymondham, UK

ALFALFA NODULATION BY A PQQ-LINKED GLUCOSE DEHYDROGENASE DEFECTIVE MUTANT OF *SINORHIZOBIUM MELILOTI* RCR2011

C.E. Bernardelli, M.F. Luna, M.L. Galar and J.L. Boiardi

CINDEFI, Facultad de Cs. Exactas (UNLP)-CONICET, Argentina

We have previously reported extracellular oxidation of glucose, via a periplasmic PQQ-linked glucose dehydrogenase (GDH), in cultures and in bacteroids of different rhizobia. It has been previously shown that the expression of GDH provides different bacteria with energetic and competitive advantages. In this study we have investigated the possible involvement of GDH during the symbiotic association between *S. meliloti* and alfalfa roots. Inoculation of alfalfa with a *gdh* mutant of *S. meliloti* RCR2011 (RmH580, from T. Finan) resulted in a significant decrease in number of nodules per plant with fewer nodules above the root tip position at the time of inoculation compared with control plants inoculated with the wild-type strain. These differences were maintained irrespective of the inoculum dosage (Figures A and B).

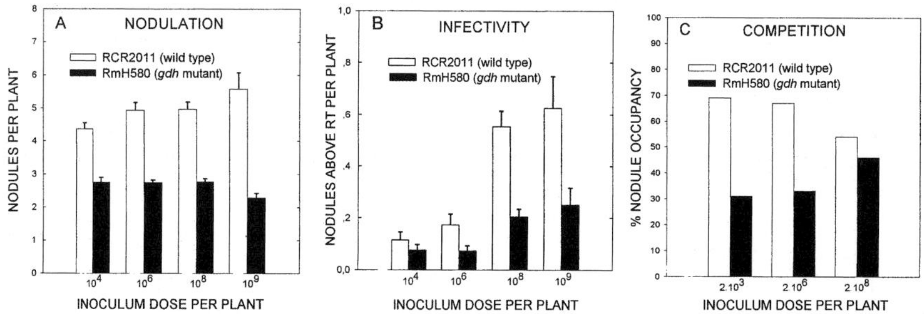

In co-inoculation experiments, nodulation and infectivity increased with the ratio wild-type/mutant strain in the inoculum. Nodule occupancy by wild-type and mutant was proportional to the relative concentration of each strain in the inoculum. However, at equal concentrations of wild-type and mutant in the inoculum, significantly more nodules were formed by the wild type, particularly at low inoculum rates (Figure C). Therefore, the lack of GDH in *S. meliloti* seems to be associated with an impaired symbiotic phenotype indicating that expression of this metabolic route may be important during some stage(s) of nodule formation. Using an *S. meliloti* strain carrying a *gdh::lacZ* fusion (RmG212, from T. Finan) we investigated the expression of *gdh* at different stages of the symbiotic association. *In situ* expression of β-galactosidase was observed in root-colonizing rhizobia, in bacteria growing inside the infection threads, and in bacteroids at different developmental stages of alfalfa nodules (between 3 to 12 weeks after inoculation). In young nodules, the nodule core was stained but not the apical meristem. In mature nodules, staining was observed in the central tissue but not in the senescent zone nor in the apical meristem. These results show that GDH is expressed from the early root colonization and mantained thereafter all over the symbiosis.

The symbiotic defects reported here for the *gdh* mutant, the transcription of *gdh* during different symbiotic stages, and the already reported GDH activity in bacteroids from alfalfa nodules indicate a role for the extracellular oxidation of aldoses during the symbiotic association.

Supported in part by ANPCyT (PICT 97 No. 1196).

CHARACTERIZATION OF THE UREASE GENE CLUSTER FROM *RHIZOBIUM LEGUMINOSARUM* BV. *VICIAE*

A. Toffanin[1,2], E. Cadahia[1], H. Manyani[1], J. Imperial[3], T. Ruiz-Argüeso[1], J.C. Polacco[4], J. Palacios[1]

[1]Dept Biotecnología, Universidad Politécnica de Madrid, 28040-Madrid, Spain
[2]Dept Chimica e Biotecnologie Agrarie, Università di Pisa, 56124 Pisa, Italia
[3]Consejo Superior de Investigaciones Científicas, Dept Biotecnología, 28040-Madrid, Spain
[4]Dept Biochemistry, University of Missouri, Columbia, MO 65211, USA

1. Introduction

Urease catalyzes the hydrolysis of urea into ammonia and CO_2. This enzyme enables the utilization of other nitrogenous compounds, such as ureides, whose N can be used exclusively as urea. Moderate levels of urease activity (ca. 300 nmol min.mg protein^{-1}) were detected in *Rhizobium leguminosarum* bv. *viciae* UPM791 vegetative cells. This activity did not require urea for induction, and was partially repressed by the addition of ammonium. Also, this bacterium was able to use other urea-rendering compounds (inosine, hypoxanthine, xanthine, urate) as nitrogen source. This ability was dependent on the presence of urease activity. A role for urease in internal urea metabolism rather than in use of external urea is inferred because urease deficiency led to the excretion of urea when cells were grown with arginine as N source. Lower urease activities (ca. 100 nmol min.mg protein^{-1}) were measured in bacteroids obtained from pea nodules. Urease-deficient derivatives of this strain were not impaired in either nitrogenase or hydrogenase in bacteroids.

2. Results and Discussion

Urease synthesis requires the concerted action of at least seven proteins, the products of the *ure* genes (Mobley *et al.* 1995). An *R. leguminosarum* bv. *viciae* DNA region of ca. 9 kb containing the urease structural genes (*ureA*, *ureB* and *ureC*), accessory genes (*ureD*, *ureE*, *ureF* and *ureG*), and 5 additional orfs (*orf83*, *orf135*, *orf207*, *orf223* and *orf287*) encoding proteins of unknown function has been sequenced. Three of these orfs (*orf83*, *orf135* and *orf207*) were conserved in the *Sinorhizobium meliloti* urease gene cluster (Miksch *et al.* 1994). *orf287* encodes a potential transmembrane protein with a C-terminal GGDEF domain, for which a nucleotidyl cyclase activity has been proposed (Pei, Grishin 2001). A mutant affected in *orf287* exhibited normal levels of urease activity in vegetative cells. *R. leguminosarum* UreE contains a long histidine-rich C-terminus (24 of the last 57 residues). A mutant strain affected in *ureE* gene was partially complemented by nickel addition to the medium, suggesting the involvement of this protein in nickel provision for urease.

3. References

Miksch G *et al.* (1994) Mol. Gen. Genet. 242, 539-550
Mobley HL *et al.* (1995) Microbiol. Rev. 59, 451-480
Pei J, Grishin NV (2001) Proteins 42, 210-216

4. Acknowledgements

This work has been funded by Projects PB95-0232 and BIO96-0503 from Spain's DGICYT and CICYT, respectively, and by Programa de Grupos Estratégicos from Comunidad de Madrid.

ALTERATION OF GLUTAMINE SYNTHETASE AND ARGINASE ACTIVITIES IN *LOTUS JAPONICUS* AND THE CONCOMITANT EFFECTS ON NITROGEN METABOLISM

R.J. Sunley, R. Parsons

Division of Environmental and Applied Biology, University of Dundee, UK

1. Introduction

Sensing and regulating N status is crucial for the efficient use of the metabolically expensive N_2 fixing apparatus. It has been hypothesized that leguminous nitrogen fixation and indeed more general nitrogen acquisition in plants, is regulated by a feedback mechanism involving cycled, nitrogen-rich organic compounds (Cooper, Clarkson 1989; Parsons, Sunley 2001). Using *Lotus japonicus* as our model legume, we have observed some interesting changes in N_2 fixation and N metabolism in plants with altered glutamine synthetase (GS) and arginase activities.

2. Procedure

A novel system of applying low levels of the herbicide glufosinate (commercially known as BASTA/Challenge/Rely) was used to inhibit GS in the root tissues of *Lotus japonicus*. Nitrogenase activity was assayed using a closed system acetylene reduction assay (Parsons, Baker 1996). Transgenic plants were obtained through the development of a transformation protocol in which seedling cotyledon attachment sites were wounded and infected with *Agrobacterium* (Chang *et al.* 1994). Regenerant shoots were selected and screened for transgene insertion by kanamycin resistance and PCR. Plant tissues were profiled for changes in N metabolism by MTBSTFA derivatization and analysis by capillary GC-MS.

3. Results and Discussion

Plants treated bi-weekly with 112–188 µg glufosinate to the root systems for three weeks had significantly greater specific nodulation. Nitrogenase activities were consistent with control plants. GC-MS analysis identified significant changes in leaf alanine, serine, glutamine and particularly asparagine. Transgenic *Lotus* plants positively screened for a transgene designed to constitutively suppress arginase activity consistently showed substantial accumulation of asparagine in the leaves. The results from this work show that N_2 fixing activity can be significantly altered by modifying N metabolism by changing crucial enzyme activities such as GS. Further metabolic profiling associated with a complement of N metabolism mutants such as our asparagine accumulator will provide an excellent strategy for furthering our understanding of the specific aspects of N metabolism involved in N status regulation. There may also be potential for improving leguminous N_2 fixing phenotypes, tailored for more efficient and sustainable agricultural systems.

4. References

Chang *et al.* (1994) Plant J. 5, 551-558
Cooper HD, Clarkson DT (1989) J. Exp. Bot. 40, 753-762
Parsons R, Baker A (1996) J. Exp. Bot. 47, 421-427
Parsons R, Sunley RJ (2001) J. Exp. Bot. 52 (Sp. Iss.), 435-44

POLY-β-HYDROXYBUTYRATE AND GLYCOGEN METABOLISM IN *RHIZOBIUM LEGUMINOSARUM*

E.M. Lodwig[1], K. Findlay[2], A.J. Downie[2], P.S. Poole[1]

[1]School of AMS, University of Reading, Whiteknights, PO Box 228, Reading, RG6 6AJ, UK
[2]John Innes Centre, Norwich Research Park, Colney, Norwich, NR4 7UH, UK

The role of the carbon storage compounds poly-β-hydroxybutyrate and glycogen in bacteroids is not fully understood. Both compounds are synthesized in bacteroids however the overall pool sizes vary considerably. Furthermore, mutation of the biosynthetic pathways has been shown to affect the efficiency of symbiotic nitrogen fixation in beans (Cevallos *et al.* 1996; Marroqui *et al.* 2001). Therefore the pathways appear to be involved in the regulation of bacteroid metabolism. We have constructed single mutants in *phaC* (PHB synthase) and *glgA* (glycogen synthase), and a double mutant, to assess their impact on the *R. leguminosarum* bv. *viciae* symbiosis. A mutant in *phaC* in *R. leguminosarum* strain A34 (mutant RU1328) which is unable to accumulate PHB in free-living cells exhibits a 50% reduction in nitrogenase activity and a 40% reduction in the dry weight of plants. However, the efficiency of the RU1328 symbiosis is not consistent as in a second experiment there were no differences in nitrogenase activity of RU1328 or dry weight of peas. A mutant in *glgA* (RU1448) which is unable to accumulate glycogen in free-living cells exhibits no significant difference in nitrogenase activity or dry weight when inoculated onto peas. Both mutants in *phaC* and *glgA* alter the carbon balance in infected plant cells therefore carbon metabolism of the plant may have a strong influence on the effects of the mutations. The mutant in both genes (RU1478) nodulates pea but forms bacteroids that senesce prematurely and appear to give a Fix⁻ phenotype (nitrogenase activity is not yet available). Together these data suggest that at least one pathway is required for nitrogen fixation but that carbon metabolism via either pathway is relatively plastic.

References

Cevallos *et al.* (1996) J. Bacteriol. 178, 1646-1654
Marroqui *et al.* (2001) J. Bacteriol. 183, 854-864

DISCOVERY OF A HEME UPTAKE SYSTEM IN THE SOIL BACTERIUM *BRADYRHIZOBIUM JAPONICUM*

A. Nienaber, H. Hennecke, H.M. Fischer

Institut für Mikrobiologie, Eidgenössische Technische Hochschule Zürich, Schmelzbergstrasse 7, CH 8092 Zürich, Switzerland

In *Bradyrhizobium japonicum*, the nitrogen-fixing symbiont of soybeans, we have identified a heme uptake system, Hmu, which comprises a cluster of nine open reading frames, (Figure 1; Nienaber *et al.* 2001). Predicted products of these genes include HmuR, a TonB-dependent heme receptor in the outer membrane, HmuT, a periplasmic heme-binding protein, and HmuUV, an ABC transporter in the inner membrane. Furthermore, we identified homologs of ExbBD and TonB, which are required for energy transduction from the inner to the outer membrane. Mutant analysis and complementation tests indicated that HmuR and the ExbBD-TonB system but not the HmuTUV transporter are essential for heme uptake or heme acquisition from hemoglobin and leghemoglobin. The TonB system seems to be specific for heme uptake since it is dispensable for the uptake of FeCitrate and ferrichrome. We therefore propose the existence of a second TonB homolog functioning in the uptake of siderophores. When tested on soybean host plants, *hmuT-hmuR* and *exbD-tonB* mutants exhibited wild-type symbiotic properties. Thus, heme uptake is not essential for symbiotic nitrogen fixation but it may enable *B. japonicum* to have access to alternative iron sources in its non-symbiotic state. In fact, growth on heme iron had been demonstrated for various rhizobial species (Noya *et al.* 1997), and a heme uptake system similar to that described here was found most recently in *Rhizobium leguminosarum* bv. *viciae* (Wexler *et al.* 2001).

Transcript analyses and studies with *lacZ* fusions showed that expression of *hmuT* and *hmuR* is induced under low-iron conditions. In *B. japonicum*, two regulatory proteins are known to be involved in iron homeostasis, the ferric uptake regulator (Fur) and the iron response regulator (Irr) (Hamza *et al.* 1998, 1999). Iron-dependent expression of *hmuT* and *hmuR* was preserved in *fur* or *irr* mutant backgrounds although maximal induction levels were decreased. We conclude either that both regulators, Fur and Irr, independently contribute to the iron control of *hmuT* and *hmuR* or that a yet unknown iron regulatory system is exerting this type of regulation. An A/T-rich element (ICE, for iron control element), located in the promoter region of the divergently transcribed *hmuTUV* and *hmuR* genes, is involved in iron control as indicated by the loss of *hmu* induction upon insertion of 2 nucleotides into ICE. We speculate that ICE represents a binding site for a novel iron regulatory protein.

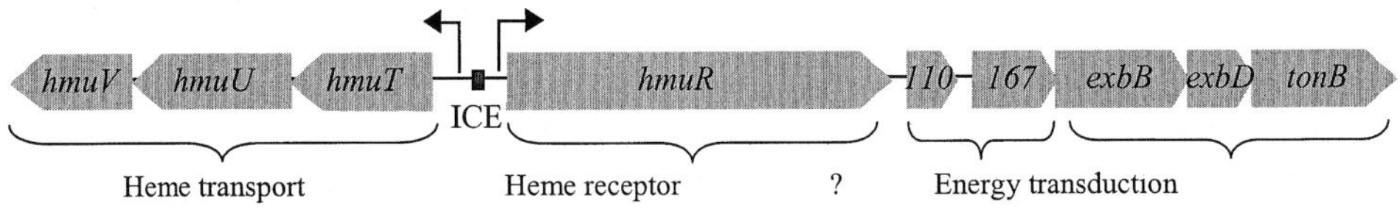

Figure 1. The *B. japonicum* locus encoding a heme uptake system.

References

Hamza I *et al.* (1998) J. Biol. Chem. 273, 21669-21674
Hamza I *et al.* (1999) J. Bacteriol. 181, 5843-5846
Nienaber A *et al.* (2001) Mol. Microbiol.
Noya F *et al.* (1997) J. Bacteriol. 179, 3076-3078

NITRATE APPLICATION FROM LOWER ROOTS CAN PROMOTE NODULATION AND N₂ FIXATION IN SOYBEAN

H. Yashima[1], H. Fujikake[1], T.I. Sato[2], N. Ohtake[1], K. Sueyoshi[1], Y. Takahashi[3], T. Ohyama[1]

[1]Faculty of Agriculture, Niigata University, Niigata 950-2181, Japan
[2]Faculty of Bioresource Sciences, Akita Pref. University, Akita 010-0195, Japan
[3]Niigata Agricultural Experiment Station, Niigata 940-0826, Japan

It is well known that combined nitrogen, especially nitrate, strongly inhibits nodulation and N_2 fixation activity in soybean. However, we have observed that a basal dressing of deep placement of controlled release nitrogen fertilizer (coated urea) did not inhibit N_2 fixation, consequently, the plant growth and the seed yield were promoted compared with those in a conventional cultivation (Takahashi *et al.* 1991). The result indicated that a continuous supply of low level of nitrogen from lower part of roots might not inhibit nodulation and N_2 fixation in the nodules on upper roots. In this paper, the effects of nitrate supply from the lower roots on the nodulation and N_2 fixation in the upper nodules were investigated using two-layered pots separating the upper roots in vermiculite medium and the lower roots in hydroponics. The 0-0, 1-1, 5-5, 0-5, and 5-0 treatments were imposed where a solution containing 0 mM, 1 mM or 5 mM nitrate was supplied to the roots in the lower pot from transplanting at V1 stage to R3 stage (the former number) and R3 to R7 stages (the latter number). Nitrogen-free solution was supplied to the roots in the upper pot through the experimental period. The plants were harvested at R1, R3 and R7 stages.

At R7 stage the total plant dry weight was highest in 5-5 treatment (127 g plant^{-1}), moderate in 1-1 (94 g), 5-0 (95 g) and 0-5 (74 g) treatments, and lowest in 0-0 treatment (51 g). At R3 stage the upper nodule dry weight was higher (0.88 g plant^{-1}) in the plants with continuous supply of 1mM nitrate (1-1 treatment) and exceeded 0-0 treatment (0.61 g) in the plants which were solely depended on N_2 fixation. Similarly total ARA (acetylene reduction activity) per plant at R3 stage was higher in 1-1 treatment than in 0-0 treatment (69 and 38 µmole C_2H_4 h^{-1} plant^{-1} respectively). These results confirmed that a continuous supply of low level of nitrate from the lower roots could promote nodulation and N_2 fixation in the upper nodules that were separated from the lower roots. Although the number of upper nodules were higher in 5-5 treatment than in 0-0 treatment, the nodule dry weight and ARA were depressed by continuous 5 mM nitrate supply from the lower roots (5-5 treatment). A withdrawal of 5 mM nitrate after R3 stage (5-0 treatment) enhanced greatly the nodule growth (1.44 g plant^{-1}) and ARA per plant (40µmole C_2H_4 h^{-1} plant^{-1}) at R7 stage where ARA declined in other treatments. The nitrate concentrations in the upper nodules were very low (less than 0.1 mg N gDW^{-1}) irrespective of the treatments, although the concentration in the lower roots were increased at 2.7 and 6.3 mg N gDW^{-1} by 1 mM and 5 mM nitrate supply.

The concentration and period of nitrate supply from the lower roots affected both the number and dry weight as well as N_2 fixation activities (ARA) of the upper nodules. Continuous supply of low level of nitrate (1mM) from the lower roots promoted the growth of shoot and roots. Therefore the acceleration of photosynthesis and nutrient uptake might increase nodule growth and N_2 fixation activity. A high level of nitrate supply to the lower roots depressed the upper nodule growth and ARA. These responses of the nodules in the upper roots may be regulated by some systemic control such as photosynthetic supply and N demand. These results support the concept that maximum soybean yield can be obtained by compatible utilization of N_2 fixation in nodules and nitrate absorption from roots. The separation of N_2 fixation site and nitrate absorption site, e.g. by deep placement of N fertilizer can promote nodulation, N_2 fixation and seed yield of soybean.

References

Takahashi *et al.* (1991) Soil Sci. Plant Nutr. 37, 223-231

RAPID AND REVERSIBLE NITRATE INHIBITION ON NODULE GROWTH AND N$_2$ FIXATION IN SOYBEAN

H. Fujikake[1], H.I. Yashima[1], T. Suganuma[1], T. Sato[2], N. Ohtake[1], K. Sueyoshi[1], T. Ohyama[1]

[1]Faculty of Agriculture, Niigata University, Niigata 950-2181, Japan
[2]Faculty of Bioresource Sciences, Akita Pref. University, Akita 010-0195, Japan

In soybean plants, the development and function of root nodules are markedly depressed when the nodulated roots are in direct contact with a high level of NO_3^- a main form of inorganic nitrogen in field. A short-term effect of presence or absence of 5 mM NO_3^- in culture solution on the nodule growth of hydroponically grown soybean was observed by a computer microscope. In addition, a split roots $^{14}CO_2$ tracer experiment was carried out to investigate the involvement of photosynthate partitioning to nodules and roots in relation to the NO_3^- inhibition of nodule growth and N$_2$ fixation.

Under N-free conditions the nodules on the primary roots grew continuously from 11 to 25 d after planting (DAP) and the diameter of nodules became about 5 mm. After 5 mM NO_3^- addition to the nutrient solution, the nodule growth was quickly depressed from the next day as shown in Figure 1. On the other hand, the depression of the nodule growth was rapidly recovered by withdrawal of NO_3^- from the medium. It appears that the inhibitory effect of 5 mM NO_3^- on the nodule growth is rapid and reversible under the experimental conditions. All the nodule tissue cells (infected, uninfected, sclerenchyma and inner cortex cells) enlarged in the N-free medium, especially the infected cell growth was prominent. The 5 mM NO_3^- decreased the cell expansion in any tissue. The acetylene reduction activity (ARA) of the plants supplied with 5 mM NO_3^- for the last week were significantly lower than those of the control plants, however the ARA were almost recovered during the cultivation with the N-free solution for a week after 5 mM NO_3^- cultivation.

The lateral roots were split into two compartments, and 5 mM NO_3^- was supplied to one side and N-free solution to the other for 3 d. Then after the exposure of 370 kBq $^{14}CO_2$ to a whole shoot for 2 h, the effect of 5 mM NO_3^- supply on the ^{14}C partitioning to each organ was measured. The percentage distribution of ^{14}C in the underground parts was about 8.2% in the nodules and about 17.4% in the lateral roots in N-free compartment. Whereas in 5 mM NO_3^- compartment ^{14}C distribution was much lower in the nodules (2.3%) and higher in the roots (29.5%) than those in N-free side. This result indicates that the photosynthate transported from shoot might be predominantly utilized in the nitrate-fed roots and consequently the photosynthetic supply to the nodules was decreased, then it retarded the nodule growth and depressed N$_2$ fixation.

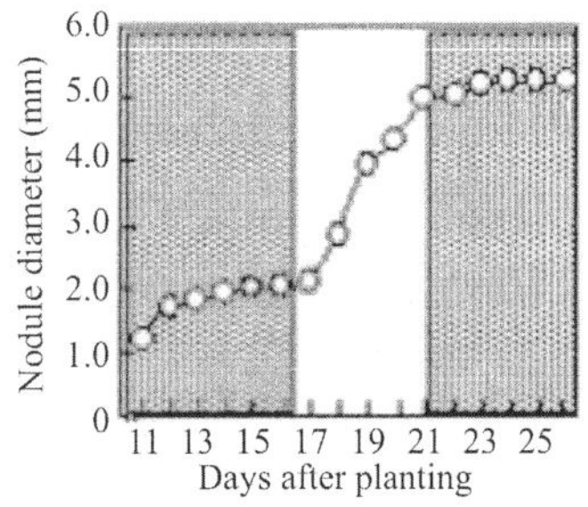
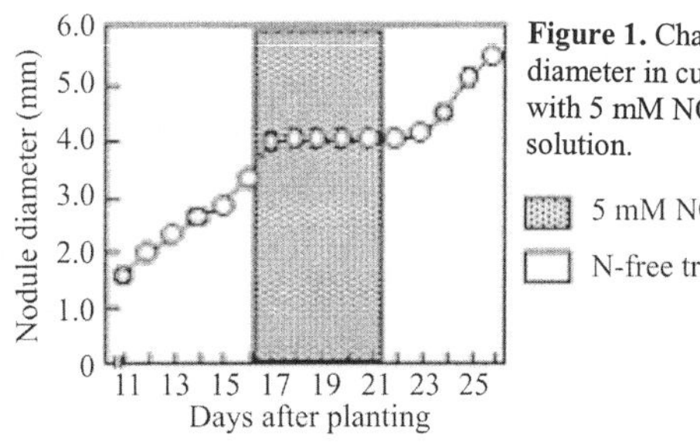

Figure 1. Changes in nodule diameter in culture solution with 5 mM NO_3^- or N-free solution.

LOCALIZATION OF CARBONIC ANHYDRASE IN LEGUME NODULES

C.A. Atkins [1], P.M.C. Smith [1], A.J. Mann [1], P.P. Thumfort [2]

[1]Dept Botany, Univ. Western Australia, Crawley WA 6009 Australia
[2]Dept Biology, MIT, Cambridge, MA 02319, USA

The plant tissue fraction of nodules from a range of legume species shows significant rates of carbonic anhydrase (CA) activity compared to the supporting root. The partially purified native CA protein from bean (*Phaseolus vulgaris* L.) nodules differed from the leaf isozyme in size, and in sensitivity to inhibitors. This evidence led to the suggestion that the smaller isozyme in legumes was a nodule-specific form, associated in some functional way with nitrogen fixation (Atkins 1974). Separate extracts of the central infected zone (CZ) and the cortex of nodules from *Lupinus angustifolius* L., *Vigna unguiculata* L. [Walp], *Pisum sativum* L., *Vicia faba* L. and *Medicago sativa* L. as well as those of *Phaseolus vulgaris* L. both contained significant CA activities (Atkins *et al.* 2001). Immunoassay using antisera to a putative nodule CA (*Mscal*) cloned from *M. sativa* (Coba de la Pena *et al.* 1997) also indicated expression in both tissue types. Quantitative confocal microscopy using laser scanning imaging and a fluorescent CA-specific probe (5-dimethylaminonaphthalene-1-sulfonamide [DNSA]) localized expression to the infected cells in the CZ tissue and to a narrow band of 2-3 files of cells in the cortical tissue that corresponded to the inner cortex. In the infected cells the enzyme activity was evenly distributed in the cytosol but in the inner cortical cells was restricted to the periphery.

It is reasonable to suppose that in the infected cells CA will be coordinately expressed with PEPC in the cytosol, to ensure an adequate rate of CO_2 hydration in a neutral or slightly acidic environment for C4 acid synthesis (cf. mesophyll cells of the C4-leaf; Burnell, Hatch 1988). Despite this apparently obvious role for CA, Raven and Newman (1994) have calculated that the uncatalyzed rate of CO_2 hydration was adequate for organic acid synthesis in soybean nodules and other possibilities, such as the short term (transient) buffering of intracellular pH change or ion transport, should not be ignored.

With the realization that nitrogenase activity is regulated by the flux of O_2, and that this may be varied as a result of reversible changes to diffusive resistance in the cortex (Hunt, Layzell 1993), there has been renewed interest in events in this tissue. While there is general acceptance that the mechanism probably involves a variable "aqueous barrier" to the diffusion of gases, and especially O_2, the underlying processes that cause reversible changes to the water relations of these cells are yet to be described. However, a number of observations have led to the idea of a mechanism(s) closely related to that in stomatal guard cells relying on synthesis of malate through PEPC. An alternative role, unrelated to malate synthesis, is plausible. Convection, as well as diffusion, may contribute to the aeration of plant tissues faced with a significant aqueous barrier (Beckett *et al.* 1988).

References
Atkins CA (1974) Phytochem. 13, 93-98
Atkins CA *et al.* (2001) Plant Cell Environ. 24, 317-326
Beckett PM *et al.* (1988) New Phytologist 110, 463-468
Burnell JN, Hatch MD (1988) Plant Physiol. 86, 1252-1256
Coba de la Pena T *et al.* (1997) Plant J. 11, 407-420
Hunt S, Layzell D (1993) Ann. Rev. Plant Physiol. Plant Molec. Biol. 44, 483-511
Raven JA, Newman JR (1994) Plant Cell Environ. 17, 123-130

GATING OF THE NODULIN 26 AQUAGLYCEROPORIN BY pH, CALCIUM AND PHOSPHORYLATION

D.M. Roberts, N. Chanmanivone

Dept of Biochem. Cell. and Molec. Biol., Univ. of Tennessee, Knoxville, TN 37996, USA

Nodulin 26 is a member of the Major Intrinsic Protein (MIP) superfamily of membrane channel proteins that includes water and solute transporters. The protein is expressed in nitrogen-fixing nodules on soybean roots and is targeted to the symbiosome membrane that encloses the endosymbiotic rhizobia bacterium within infected cells of the nodule. It constitutes the predominant protein component of this membrane (Weaver *et al.* 1991). In previous work we have shown that soybean nodulin 26 is an aquaglyceroporin that mediates the flux of water and uncharged solutes across the symbiosome membrane (Dean *et al.* 1999). However, compared to most aquaporins, nodulin 26 shows a substantially lower rate of water transport. For example, compared to the prototypical aquaporin, mammalian AQP 1, nodulin 26 has a 30- to 50-fold lower unitary conductance (Dean *et al.* 1999). An examination of various aquaporins shows that the rate and selectivity of transport can be modulated by a variety of exogenous factors including pH and phosphorylation (Németh-Cahalan, Hall 2000; Zeuthen, Klaerke 1999; Yasui *et al.* 1999). To address the factors that regulate the permeability of nodulin 26, we have investigated the effects of pH, Ca^{2+} and phosphorylation on the regulation of nodulin 26 transport.

The transport properties of nodulin 26 were evaluated upon expression in *Xenopus* oocytes as described previously (Dean *et al.* 1999). The intrinsic transport rate of nodulin 26 was elevated three-fold by either lowering the pH in the recording bath or by phosphorylation of serine 262 within the carboxyl-terminal region of the protein. In contrast, it was found that nodulin 26 transport was substantially inhibited (5- to 6-fold) by the elevation of intracellular calcium within the oocyte.

Overall, the results suggest that the aquaporin activity of nodulin 26 is modulated by pH and calcium, factors that may impact water flow and osmoregulation of the symbiosome membrane. With respect to pH effects, the regulation of nodulin 26 aquaglyceroporin activity may be coupled to the electrogenic proton pump that acidifies the lumen of the symbiosome, particularly under conditions of high metabolic activity (Uvardi, Day 1997). The effects of calcium are more complex. *In vivo*, nodulin 26 is phosphorylated on a unique serine residue (S262 within the carboxyl terminal domain) by a calcium-dependent protein kinase that is co-localized on the symbiosome membrane (Weaver *et al.* 1991). The present results suggest that this event would enhance the transport rate of the protein. However, assay in *Xenopus* oocytes also suggests that Ca^{2+} ion alone inhibits the activity, perhaps through a separate mechanism (e.g. calmodulin binding as has been reported for other MIP proteins; Németh-Cahalan, Hall 2000). The opposing effects of calcium and the possible interplay of phosphorylation in regulating these effects, as well as the larger question of the symbiotic role of up and down regulation of nodulin 26 transport, needs to be investigated further. (Supported by USDA NRICGP 9703548 and NSF MCB-9904978.)

References
Dean RM, Rivers RL, Zeidel ML, Roberts DM (1999) Biochem. 38, 347-353
Németh-Cahalan KL, Hall JE (2000) J. Biol. Chem. 275, 6777-6782
Weaver CD, Crombie B, Stacey G, Roberts DM (1991) Plant Physiol. 95, 222-227
Yasui M, Hazama A, Kwon TH, Nielsen S, Guggino WB, Agre P (1999) Nature 402, 184-187
Zeuthen T, Klaerke DA (1999) J. Biol. Chem. 274, 21631-21636

PROTEOME ANALYSIS OF NODULATION-RELATED PROTEINS IN THE WILD TYPE AND A SUPERNODULATION MUTANT (*sunn*) OF THE MODEL LEGUME, *MEDICAGO TRUNCATULA*

U. Mathesius, G. Keijzers, J.J. Weinman, B.G. Rolfe

Research School Biol. Sci., Australian National University, Canberra, Australia

We studied differential protein expression in the wild type and the supernodulating mutant, *sunn*, of *Medicago truncatula* in response to *Rhizobium* infection to elucidate the nature of the *sunn* mutation. *Sunn* is a supernodulating mutant with a proposed defect in the regulation of auxin transport and/or sensitivity, suggesting that an auxin defect is underlying its supernodulating phenotype.

We have previously established a proteome reference map for *Medicago truncatula* using two-dimensional gel electrophoresis combined with peptide mass fingerprinting to separate and identify differentially expressed proteins. The *M. truncatula* EST database was used as an essential tool for the identification of the proteins because it provided a database for the generation of theoretical peptide mass fingerprints and prediction of protein function using BLAST searching.

Wild type and *sunn* showed only minor differences in the untreated state, but differential protein expression after *Rhizobium* inoculation, including auxin-induced and ripening-related proteins, two nodulins, and S-adenosyl methionine synthase, involved in ethylene synthesis. The differential expression of these proteins supports the hypothesis that hormonal regulation is affected in *sunn*. The number of hormonally regulated proteins changing in amount during nodule development might reflect the importance of plant hormones in nodule organogenesis.

GENETIC TRANSFORMATION OF *PHASEOLUS VULGARIS*

E. Diego, O. Avendaño, G. Hernández, M. Lara

CIFN/UNAM, Res. Program Plant Mol. Biol., Cuernavaca, Morelos, México

Beans (*Phaseolus vulgaris*) are the most widely consumed legumes in the world. This crop is the principal protein source for people in Mexico and other American countries. *P. vulgaris* establishes symbiosis with *Rhizobium etli*. Both the macro and micro-symbionts originated in America and have co-evolved for a long time. Our main research interest is towards *P. vulgaris* functional genomics; this effort is part of the global project on *Phaseolus* genomics from an international consortium recently formed. An efficient and reliable genetic transformation procedure is a critical issue that needs to be satisfied in order to propose a certain legume species for genomic projects. In this regard, we now report a genetic transformation system of *P. vulgaris* var. Negro Jamapa 81, mediated by *Agrobacterium tumefaciens*.

First we established a regeneration protocol for three different bean cultivars, widely consumed in Mexico: Negro Jamapa 81, Flor de Mayo and Peruano. The explants used were cotyledonary nodes, from germinated seedling. Shoots were induced, through direct organogenesis, in MS media with BAP (5 mg/L) and sucrose. Around two shoots per explant were formed in the three varieties tested. The use of liquid media for shoot induction and development was a key element for an efficient regeneration.

Based in the efficient regeneration procedure, a reliable protocol for genetic transformation mediated by *Agrobacterium tumefaciens* has been developed for the Negro Jamapa 81 variety. Cotyledonary nodes were cocultivated with *A. tumefaciens* LBA4404, bearing pBI121 plasmid, followed by *in vitro* regeneration using kanamycin (kan) for selection. The kan concentration (50 mg/L) used only for a short period (10 days) were critical factors for efficient regeneration of putative (kan^R) tranformants. Kan^R shoots were generated from the apical and lateral meristematic regions of the explant. Stable integration of the transgenes in the kan^R putative primary transformants (T$_0$) were evidenced by PCR and *gus* activity analyses. Near 60% of the plants analyzed were PCR$^+$ or were stably transformed; while 40% were false positives (PCR$^-$). The frequency for obtaining PCR$^+$ plants was reproducible among different experiments. The transformation efficiency was 7%, meaning that 7 transgenic plants are obtained from 100 explants. The T$_0$ plants were fertile and showed a normal symbiotic phenotype with *R. etli*. The T$_0$ plants were self-fertilized and the transfer of the transgenes to the progeny is being confirmed by PCR and Southern analysis. All the T$_1$ plants checked, so far, are PCR$^+$ indicating the stable heredity of the transgenes.

PEA (*PISUM SATIVUM* L.) GENE *Sym35* IS HOMOLOGOUS TO THE GENE *Nin* OF *LOTUS JAPONICUS* (REGEL.)

K. Larsen, A.Y. Borisov[1], V.E. Tsyganov[1], V.A. Voroshilova[1], A.O. Batagov[1],
L.H. Madsen[2], Y. Umehara[2], N. Sandal[2], L. Schauser[2], N. Ellis[3], I.A. Tikhonovich[1],
J. Stougaard[2]

[1]All-Russia Research Institute for Agricultural Microbiology, St.-Petersburg, Pushkin 8,
  Podbelsky chaussee, 3, 168608, Russia
[2]Aaarhus University, Laboratory of Gene Expression, Gustav Wieds Vej 10, DK-8000,
  Aarhus C, Denmark
[3]John Innes Centre, Norwich, NR4 7UH, UK

Despite the fact that garden pea (*Pisum sativum* L.) is an old and well-developed system for wide-range genetic and physiological research including studying plant-microbe interactions it is, unfortunately, not suitable for molecular genetics and gene manipulation due to large size of the genome in physical sense. A major advance in the study of pea symbiotic loci (to date more than 40) at the molecular level can therefore be achieved by taking advantage of other legumes more suitable for molecular genetics. One of those model legumes is *Lotus japonicus* (Regel.) K. Larsen. Recently the first genetically defined legume symbiotic gene, the *Nin* controlling early stages of nodule formation in *L. japonicus*, was cloned and sequenced. Using comparative phenotypic analysis the pea gene *Sym35* was found as the most probable candidate to be a homologous gene for *Nin*. It opened up a possibility to clone the gene *Sym35* using molecular probes developed for *Nin*. Clones containing different fragments of cDNA and genomic copies of the pea *Nin* homolog were identified in "Root Hair Enriched" and "Genomic" pea gene libraries kindly provided by Dr H. Franssen (Wageningen University, The Netherlands), respectively. Subsequent sequencing of the fragments gave information about primary structure of this gene in pea and showed high homology (44%). Based on the sequence information, the molecular markers (primers allowing specifically amplify the fragments of allelic variants of pea *Nin* in various lines) for pea gene homologous to the gene *Nin* of *L. japonicus* were developed. Those markers allowed establishing the fact that the genes *Sym35* and pea *Nin* are in the same position on pea genetic map using co-segregation analysis and therefore they are the same gene. Detailed sequencing of three independent mutant alleles (lines SGENod⁻-1 (*sym35*), SGENod⁻-3 (*sym35*) and RisNod8 (*sym35*)) and corresponding wild type ones (lines SGE and Finale) revealed specific mutations in all cases: early stop codons and substitution of amino acid, respectively.

Thus, the first pea symbiotic gene identified by means of chemical mutagenesis was cloned with the use of achievements of molecular genetics in model legumes that opens up a possibility for further comparative analysis of the gene *Sym35/Nin* function and regulation of expression in these two legume plant species.

EXPRESSION PROFILING IN THE EUROPEAN UNION GENOME PROJECT MEDICAGO: CONSTRUCTION OF ARRAYS REPRESENTING THE *M. TRUNCATULA* ROOT INTERACTION TRANSCRIPTOME

P. Gamas[1], J. Denarié[1], L. Brechenmacher[2], D. van Tuinen[2], V. Gianinazzi-Pearson[2], M. Crespi[3], P. Mergaert[3], A. Kondorosi[3], F. Krajinski[4], P. Franken[5], A. Pühler[6], J. Kalinowski[8], A. Becker[7], H. Küster[6]

[1]LBMPRM, CNRS-INRA, Toulouse, France; [2]UMR BBCE-IPM, CMSE-INRA, Dijon, France; [3]ISV, CNRS, Gif-sur-Yvette, France; [4]LG Molekulargenetik, Universität Hannover, Germany; [5]MPI, Marburg, Germany; [6]Lehrstuhl für Genetik, Universität Bielefeld, Germany; [7]Zentrum für Genomforschung, Bielefeld, Germany; [8]IIT GmbH, Bielefeld, Germany

During the past years, *Medicago truncatula* emerged as a model legume to study plant-microbe interactions as well as general aspects of plant biology (Cook, Denarie 2000). Thus, this legume is currently analyzed in detail in several international genome projects in the US as well as in Europe (Harris, Frugoli 2001). In order to identify genes expressed in specific symbiotic or pathogenic interactions of this legume as well as genes relevant for the development of plant organs, e.g. flowers, seeds and leaves, expression profiling experiments are beginning to be performed on a larger scale. These experiments capitalize on the ever-increasing collection of ESTs available in public databases, e.g. the TIGR *Medicago truncatula* Gene Index (http://www.tigr.org/tdb/mtgi/).

Eight European labs participate in the expression profiling part of the European Union genome project MEDICAGO. This project is based on a collection of ~15,000 ESTs obtained in the Toulouse and Dijon labs (http://sequence.toulouse.inra.fr/Mtruncatula.html) that cluster into ~6000 different groups. This collection representing roots, young nodules and mycorrhiza was enriched by ESTs from roots interacting with pathogenic fungi and by ESTs from an SSH library of mycorrhizal roots. We are currently generating a 6k macroarray (nylon filter) of this root interaction transcriptome (Mt6k-RIT) that also includes well-studied *M. truncatula* marker genes and non-*Medicago* control ESTs. Preceding the Mt6k-RIT production, a small 96 clone array containing a representative collection of the root interaction transcriptome was constructed (MtSTAMP) that will serve for controlling probe labeling and for setting up experimental conditions. Both the MtSTAMP and the Mt6k-RIT array are to be made available to the *Medicago* community after initial experiments have been performed within the MEDICAGO Consortium.

Using the MtSTAMP and the Mt6k-RIT array, we will carry out expression profiling experiments in order to identify genes differentially regulated during symbiotic or pathogenic root interactions as well as during different physiological conditions. We will thus set up a database linking EST sequence with expression data. In the meantime, we started to use this EST collection to establish experimental conditions for transcriptome analysis on microarrays (glass slides). During the project, the Mt6k-RIT set will be enriched with ESTs obtained from different *M. truncatula* organs or tissues, e.g. roots, nodules under drought and saline stress, roots challenged with pathogenic fungi or treated with Nod factors, flowers and seeds. These ESTs either stem from conventional or from SSH cDNA libraries. After normalization, microarrays (glass slides) will be produced carrying a non-redundant set of ESTs representing the *M. truncatula* transcriptome from all organs or conditions studied. The results of expression profiling will be entered into a database for a global comparison of regulation patterns of the genes identified in different organs or conditions analyzed.

References

Cook D, Denarié J (2000) Grain Legumes 28, 12-13
Harris J, Frugoli J (2001) Plant Cell 13, 458-463

ANALYSIS OF EARLY EVENTS IN NON-NODULATING CHICKPEA LINES

W. Hopper[1], F. Tilton[1], O.P. Rupela[2], J. Thomas[1]

[1]Centre for Biotechnology, SPIC Science Foundation, Chennai 600032, India
[2]Agronomy Division, ICRISAT, Patencheru 502 324, India

1. Introduction

Non-nodulation (Nod⁻) mutants are useful tools to understand the mechanism of nodulation in legumes. The chickpea non-nodulation lines ICC435M, ICC4918M, ICC4993M (Rupela 1992) and ICCV2M (O.P. Rupela, unpublished) were developed by pure line selection from their normal nodulating parent lines. Flavonoid synthesis/secretion may not be the limiting factor for nodulation in these Nod⁻ lines (Hopper *et al.* 1998). To identify the nodulation block in Nod⁻ lines, the present study examined their phenotypes and expression of certain early nodulin genes such as ENOD40, ENOD2 and ENOD12.

2. Results and Discussion

The Nod⁻ selections showed poor root hair deformation except for ICCV2M (equal to its parent line). Cortical cell divisions were completely absent. The phenotypes suggested that the nodulation process is probably blocked at the early stages.

ENOD40 from *Sesbania* and ENOD12 from pea were used as probes for chickpea for Southern and Northern hybridizations but the heterologous probes gave faint signals. RT-PCR was performed on total RNA of nodulation variants to study the expression of ENOD40, ENOD2 and ENOD12 using primers designed from soybean and pea. They were also amplified by PCR from chickpea genomic DNA to be used as homologous probes. The expression pattern of these genes in nodulation variants ICC4918M and ICCV2M and their normal parent lines were compared. ENOD40 and ENOD12 were expressed both in normal and non-nodulating chickpea lines.

The strategy of differential display of mRNA by PCR (DDRT-PCR) is being adopted to analyze the expression of nodulation related genes.

3. References

Hopper W *et al.* (1998) In Elmerich C, Kondorosi A, Newton WE (eds) Biological Nitrogen Fixation
for the 21st Century, pp. 240, Kluwer Academic Publishers, Dordrecht, The Netherlands
Rupela OP (1992) Crop Sci. 32, 349-352

4. Acknowledgements

This work was carried with funding from Department of Biotechnology, Government of India.

FUNCTIONAL CHARACTERIZATION OF NODULIN 26 AND GAMMA TIP ORTHOLOGS FROM THE NODULES OF *LOTUS JAPONICUS*

J.F. Guenther, D.M. Roberts

Dept Biochem. Cell. Molecular. Biol., Univ. of Tennessee, Knoxville, TN 37996, USA

During the establishment of rhizobia/legume symbioses a subset of nodulins are targeted to the symbiosome membrane that encloses the rhizobia bacteroid. Among these proteins is nodulin 26 (Fortin *et al.* 1987), which constitutes 10–15% of the total symbiosome membrane protein (Weaver *et al.* 1991). Nodulin 26 is a member of the MIP superfamily, an ancient family of conserved integral membrane proteins. These proteins have been shown to mediate the flux of water, small uncharged solutes or both (multifunctional aquaglyceroporins) (Borgnia *et al.* 1999).

The fact that nodulin 26 is the major MIP in the symbiosome membrane argues for a selective role in symbiosis. Many roles have been suggested for nodulin 26 ranging from osmoregulation (via water or solute transport; Dean *et al.* 1999; Guenther, Roberts 2000) to metabolite transport (e.g. NH_3; Niemietz, Tyerman 2000). Though much is known regarding nodulin 26 transport, its biological function remains unresolved. To address the biological role of nodulin 26 and other MIPs in the nodule, we have investigated these proteins in the genetically tractable model legume *Lotus japonicus*. Using RT-PCR, cDNAs encoding two MIPs were identified that are expressed in mature nitrogen-fixing nodules of *L. japonicus*. These proteins were designated LIMP 1 and LIMP 2 (Lotus Intrinsic Membrane Protein). LIMP 1 is a member of the TIP (Tonoplast Intrinsic Protein) subclass of plant MIPs and shows high sequence similarity to gamma TIP (78% identity) (Guenther, Roberts 2000). It is expressed at high levels in nodules and roots. Functional analysis by expression in *Xenopus* oocytes shows that LIMP 1 is a water-selective aquaporin.

In contrast, LIMP 2 shows high similarity to soybean nodulin 26 (68% identity). LIMP 2 shows nodule specific expression based on Northern blot analysis. Western blot with an anti-LIMP 2 antibody shows that it is localized to the *L. japonicus* symbiosome membrane. Further, similar to nodulin 26 (Weaver *et al.* 1991), LIMP 2 is phosphorylated on its carboxyl-terminal domain by a calcium-dependent protein kinase (CDPK) and is likely a regulatory target of this enzyme. Finally, LIMP 2, similar to nodulin 26 (Dean *et al.* 1999), is multifunctional since it is able to transport water as well as uncharged solutes such as glycerol upon expression in oocytes. Thus, two functionally distinct MIPs are expressed in *L. japonicus* nodules: LIMP2 which is an ortholog of soybean nodulin 26; and LIMP 1 which is a water-selective TIP present both in roots and nodules that likely plays a separate osmoregulatory role. Interestingly, TIP proteins have been localized to regions of the nodule cortex involved in the regulation of gas diffusion, raising the possibility that they serve in this role (Serraj *et al.* 1998). The identification of these two nodule MIPs in a genetically tractable legume will allow the investigation of their roles in nodule physiology. (Supported by USDA NRICGP 9703548 and NSF MCB-9904978.)

References

Borgnia M, Nielsen S, Engel A, Agre P (1999) Annu. Rev. Biochem. 68, 425-458

Dean RM, Rivers RL, Zeidel ML, Roberts DM (1999) Biochem. 38, 347-353

Fortin MG, Morrison NA, Verma DP (1987) Nucl. Acids Res. 15, 813-824

Guenther JF, Roberts DM (2000) Planta 210, 741-748

Niemietz CM, Tyerman SD (2000) FEBS Lett. 465, 110-114

Serraj R, Frange N, Maeshima M, Fleurat-Lessard P, Drevon JJ (1998) Planta 206, 681-684

Weaver CD, Crombie B, Stacey G, Roberts DM (1991) Plant Physiol. 95, 222-227

NITROGEN FIXATION ASSOCIATED WITH TUBERCULATE ECTOMYCORRHIZAE

L. Paul[1], B. Chapman[2], F.B. Holl[1], C. Chanway[1,3]

[1]Faculty of Agricultural Sciences
[2]B.C. Ministry of Forests, Williams Lake, BC
[3]Faculty of Forestry, UBC, Vancouver, BC

Tuberculate ectomycorrhizae (TEM) are a unique form of mycorrhizae where the ectotrophic fungus forms tuber-like structures (tubercles) on roots of the host plant somewhat similar in appearance to legume root nodules. The function of tubercles has yet to be precisely determined. This study focuses on the possibility that TEM on lodgepole pine roots supply fixed nitrogen from diazotrophic bacteria known to reside within tubercles. The acetylene reduction assay (ARA) was used to quantify nitrogen fixation associated with pine TEM *in situ*. Nitrogen fixation ranged from 0–4.27 $g/m^3/yr$ depending on soil type and pine stand age. Three bacterial species were isolated from the inner matrix of tubercles using N-free media and identified using GC Fame and BIOLOG as *Bacillus* spp., *Micrococcus luteus* and *Methylobacterium extorquens*. *In vitro* ARA results were positive for *Bacillus* spp. and *M. luteus*, but not *M. extorquens*. Because TEM often have a pink cast, we looked for hemoglobin in extracts of TEM tissues, initially using Drabkin's Reagent and then by cellulose acetate electrophoresis with Ponceau S and o-dianisidine staining for protein and heme, respectively. Human hemoglobin was used as a comparative standard. Both assays indicated the presence of hemoglobin or a hemoglobin-like compound within TEM tissues.

ROLE OF ARGININE IN ROOT NODULE NITROGEN METABOLISM OF *DATISCA GLOMERATA*

A.M. Berry[1], T.M. Murphy[2], K. Pawlowski[3], P.A. Okubara[4], K.R. Jacobsen[5]

[1]Dept Environmental Horticulture, U Calif, Davis, CA, USA
[2]Sect Plant Biology, U Calif, Davis, CA, USA
[3]Albrecht v. Haller Inst, Univ Goettingen, Germany
[4]USDA-ARS, Wash St U, Pullman, WA, USA
[5]Dept Agronomy, U Calif, Davis, CA, USA

In root nodules of the actinorhizal host *Datisca glomerata*, glutamine synthetase (GS) is expressed at high levels in uninfected tissue, but not in infected tissue, as shown by *in situ* hybridization and immunolocalization. The absence of GS in the infected tissue excludes the pathway of nitrogen assimilation via ammonia excretion from the microsymbiont and plant glutamine synthesis in the infected tissue, which is common to legumes and to other actinorhizal species. We further show that arginine is highly elevated in *D. glomerata* nodule amino acid profiles, but not in root, xylem sap or leaf extracts. Glutamine and glutamate levels are elevated in both nodule and xylem sap, however, indicating their importance in whole-plant nitrogen transport. We hypothesize that arginine is the major storage form of nitrogen within the nodule; it is synthesized in the infected tissue and exported to the uninfected tissue, where it is further catabolized by arginase and urease to yield free ammonium. The observation that the nodules are positive for urease activity supports this hypothesis. These data explain the expression pattern of glutamine synthetase in the uninfected tissue and suggest a novel role for arginine in nodule nitrogen metabolism.

EFFECTS OF SUCROSE, pH, TEMPERATURE AND MINERAL NITROGEN ON GROWTH AND NITROGENASE ACTIVITY OF *GLUCONACETOBACTER DIAZOTROPHICUS* ON SOLID MEDIUM

B. Pan, J.K. Vessey

Department of Plant Science, University of Manitoba, Winnipeg, MB, R3T 2N2, Canada

1. Introduction

The development of cereal crops that can supply their own mineral nitrogen requirements has been the goal of many researchers. The discovery of *G. diazotrophicus* (Cavalcante, Döbereiner 1988) offers new promise on achieving significant levels of N_2 fixation in cereal crops. The first step in the process of having *G. diazotrophicus* fix N_2 in cereals would be to understand the environmental conditions that affect the growth and nitrogenase activity of *G. diazotrophicus*. The objectives of this study were to characterize the effects of sucrose, pH, mineral nitrogen and temperature on the cell growth and nitrogenase activity of *G. diazotrophicus* on solid media and to determine if the bacterium has hydrogenase uptake (Hup) activity.

2. Procedure

G. diazotrophicus was grown on LGI-P medium. Combinations of citric acid and potassium phosphates were used to obtain media with different pH. *G. diazotrophicus* strain PAL5 (ATCC 49037) and MAD3A (a *nifD⁻* mutant for Hup⁺ determination) were used. Nitrogenase activity (H_2 evolution under $Ar:O_2$) was measured in real time at ambient O_2 concentrations (Vessey 1992). Hup activity of *G. diazotrophicus* was assessed by measuring H_2 consumption by MAD3A in a closed chamber with wild type PAL5 and the Hup⁺ *Ralstonia eutropha* (ATCC 17699) as controls.

3. Results

The results show that sucrose concentration had a great effect on both cell growth and nitrogenase activity of *G. diazotrophicus*. Nitrogenase activity was very low, but still measurable at 0.1% sucrose. The lowest nitrogenase activity was found when cells of *G. diazotrophicus* were grown at a medium pH of 4.5. The optimum pH of the medium for nitrogenase activity was 5.5. Nitrogenase activities decreased as medium KNO_3 or $(NH_4)_2SO_4$ concentration increased from 0.1 to 10 mM. At 10 mM KNO_3, nitrogenase activity was 15 times less than of the control. Nitrogenase activity increased as temperature increased from 15 to 30°C. Nitrogenase activity of the 5-day-old *G. diazotrophicus* colonies grown at 30°C was almost three times that of the 14-day-old colonies grown at 15°C. Our results also indicate that *G. diazotrophicus* has Hup capacity, but its Hup activity was relatively lower under the culture conditions (i.e. H_2 consumption by the *nif⁻* mutant was approximately 7% of the rate of H_2 evolution by the wild-type *G. diazotrophicus*).

In conclusion, our study showed that *G. diazotrophicus* could grow and fix N_2 at different concentrations of sucrose, mineral nitrogen, pH and temperature. The flexibility of *G. diazotrophicus* to fix N_2 under a wide range of conditions increases the possibility of adapting *G. diazotrophicus* to fix N_2 in new host species.

References

Cavalcante VA, Döbereiner J (1988) Plant Soil 108, 23-31
Vessey JK (1992) Plant Soil 139, 185-194

COLONY STRUCTURE PROTECTS NITROGENASE FROM O₂ INACTIVATION IN THE SUGARCANE ENDOPHYTE *GLUCONACETOBACTER DIAZOTROPHICUS*

Z. Dong[1], C.D. Zelmer[2], J.K. Vessey[2], M.J. Canny[3], M.E. McCully[3], B. Luit[2], B. Pan[2], R.S. Faustino[4], G.N. Pierce[4]

[1]Dept of Biology, St. Mary's University, Halifax, NS, Canada
[2]Dept of Plant Science, Univ. Manitoba, Winnipeg, MB Canada
[3]Division of Plant Industry, CSIRO, Canberra, ACT 2601, Australia
[4]Dept of Physiology, Univ. Manitoba, Winnipeg, MB, R3E 3J7

1. Introduction

Nitrogenase activity requires large amounts of ATP and reductant from aerobic respiration (Hunt 1993), but nitrogenase is sensitive to inactivation by oxygen. *Gluconacetobacter diazotrophicus* (Yamada 1997), an aerobic, N_2-fixing bacterium that inhabits intercellular spaces of sugarcane (Dong 1994), has the unusual ability to fix nitrogen on solid media under relatively high (21 kPa) partial pressures of atmospheric oxygen. This bacterium does not form special structure with its host (e.g. nodules) to limit exposure to oxygen. On solid agar media *G. diazotrophicus* produces lens shaped, mucilagenous colonies. In a series of three experiments, we tested the hypothesis that colony structure provides a suitable O_2 environment for nitrogenase activity in *G. diazotrophicus*.

2. Results and Discussion

G. diazotrophicus colonies were raised on solid LGI-P medium (Cavalcante 1988) for five days under ambient (21 kPa) and low (2 kPa) partial pressures of atmospheric oxygen. Whole colonies were observed live using laser scanning confocal microscopy to quantify differences in cell positioning or freeze-substituted, fixed, embedded and sectioned for light microscopy. We found that the uppermost bacterial cells were located at a greater depth below the surface of the colony mucilage under the ambient O_2 treatment (85–110 µm) than in the low O_2 treatment (45–60 µm).

Colonies raised at 21 kPa O_2 for 5 days were assessed for nitrogenase activity (measured by H_2 evolution) then gently spread over the surface of the agar with a bent glass rod. After disruption the nitrogenase activity of the colonies was re-assessed. This showed that disruption of the colony structure resulted in a decrease in H_2 evolution to 3.5% that of the intact colonies.

As *G. diazotrophicus* colonies age, their structure begins to break down; the mucilage slumps to one side of the colony. By the 8th day post-inoculation, 75% of the colonies on a petri plate are collapsed. Nitrogenase activity of the colonies was tracked from day 3 to day 8 post-inoculation to determine the effect of colony structure disruption on nitrogenase activity. As the colonies increased in size and then collapsed over the time period, nitrogenase activity of *G. diazotrophicus* peaked at day 6, then decreased to 76% of maximum at day 8.

These data support a role for colony structure in the protection of nitrogenase by *G. diazotrophicus*.

3. References

Cavalcante VA, Döberiener, J (1988) Plant and Soil 108, 23-31
Dong Z *et al.* (1994) Plant Physiol. 105, 1139-1147
Hunt S, Layzell DB (1993) Ann. Rev. Plant Physiol. Plant Mol. Biol. 44, 483-511
Yamada Y (1997) Biosci. Biotechnol. Biochem. 61, 1244-1251

Acknowledgements

This research was funded by Cargill, Ltd. and the AAFC/NSERC Research Partnership Program.

ENDOPHYTIC COLONIZATION AND *IN PLANTA*-NITROGEN FIXATION BY *HERBASPIRILLUM* SP. ISOLATED FROM WILD RICE

A. Elbeltagy[1], K. Nishioka[1], T. Sato[1], H. Suzuki[1], B. Ye[1], T. Hamada[2], T. Isawa[1], H. Mitsui[1], K. Minamisawa[1]

[1]Institute of Genetic Ecology, Tohoku University, Sendai, Japan
[2]Marine Biotechnology Institute, Kamaishi, Japan

From surface-sterilized culms of wild and cultivated rice, we obtained 11 gram-negative isolates of nitrogen-fixing bacteria by a modified Rennie medium. Based on 16S rDNA sequences, those diazotrophic isolates were phylogenetically close to *Herbaspirillum*, *Ideonella*, *Enterobacter* and *Azospirillum*. Isolates B65, B501 and B512 from wild rice phylogenetically belong within *Herbaspirillum*, however, the result of basic properties, carbon source utilization and the diagnostic probe sequences indicated that those three isolates cannot be classified into any known species of this genus. The isolates B65, B501 and B512 are possibly new species of *Herbaspirillum*.

To examine endophytic behavior of *Herbaspirillum* sp. B501 isolated from *Oryza officinalis* W0012 in the original wild rice and cultivated rice plants, the *gfp* gene encoding green fluorescent protein (GFP) was introduced into the bacteria. Observations by fluorescence stereomicroscopy showed that the GFP-tagged bacteria colonized shoots and seeds of aseptically grown seedlings of the original wild rice after seed inoculation. Conversely, no GFP fluorescence was observed for shoots, and only weak signals for seeds, of cultivated rice *O. sativa*.

Transmission electron microscopy demonstrated that *Herbaspirillum* sp. B501 colonized intercellular spaces in young leaves and in coleoptiles that were easily detected in the young seeding stage. Bacteria were often found outside the exodermis of roots, but rarely seen within the tissues. According to our observations, the majority of cells of *Herbaspirillum* sp. B501 invaded and colonized intercellular spaces and also found in intracellular spaces of rice shoots. Colony counts of surface-sterilized rice seedlings inoculated with the GFP-tagged bacteria indicated significantly more bacterial populations inside the original wild rice than in cultivated rice varieties.

Moreover, after bacterial inoculation, *in planta*-nitrogen fixation in young seedlings of wild rice, *O. officinalis*, was detected with the acetylene reduction and $^{15}N_2$ gas incorporation assays. Therefore, we conclude that *Herbaspirillum* sp. B501 is a diazotrophic endophyte compatible with wild rice, particularly *O. officinalis*.

POLYHYDROXYALKANOATES PRODUCTION BY *GLUCONACETOBACTER DIAZOTROPHICUS*

M. Rojas, J. Martínez, M. Heydrich

Dept of Microbiology, Faculty of Biology, Univ. Havana, CP 10400, La Habana, Cuba

1. Introduction

Polyhydroxyalkanoates (PHA) are biocompatible, biodegradable and thermoplastic polymers that can be applied in multiple biotechnological branches, industry and agriculture. Some microbial genera are able to produce PHAs, in recent studies it was demonstrated that diazotrophic microorganisms are also strong producers of PHA. In this study the capacity to produce PHA of *Gluconacetobacter diazotrophicus,* a nitrogen-fixing endophyte of sugarcane, in different conditions was demonstrated.

2. Materials and Methods

Type strain PAl5 (ATCC 49037) of *Gluconacetobacter diazotrophicus* and two indigenous strains isolated from sugarcane in Cuba (4-02 and 1-05) were tested. Screening of colonies was carried out in LGI-P (Reis *et al.* 1994) and SYP media (Caballero and Martínez 1994) supplemented with Nile Red. PHA extraction-quantification was tested by our modification of Lan and Slepecky (1960) spectrophotometric method.

3. Results and Discussion

These results demonstrated that *G. diazotrophicus* can produce the polymer. Some strains were selected using a plate method with Nile Red as fluorogenic indicator of PHA accumulation. Microscopic examination of strains grown in media plus Nile Red showed fluorescence in some cells. *G. diazotrophicus* accumulate efficiently PHA, in nitrogen-fixing and not nitrogen-fixing conditions, establishing the capacity for production of high levels of PHA (Table 1).

Table 1. Comparison of PHA accumulation, yield (product/biomass), carbon:nitrogen and carbon:phosphorous ratio among strains of *G. diazotrophicus* at 48 h of culture in different media.

Strains	LGI-P Not nitrogen fixation			LGI-P Nitrogen fixation			SYP		
	Pal 5	4-02	1-05	Pal 5	4-02	1-05	Pal 5	4-02	1-05
PHB g/L	0.235	0.063	0.226	0.138	0.014	.0.101	0.982	1.265	0.998
Yield (P/B)	3.95	15.36	3.58	2.89	0.80	2.60	48.37	36.12	35.01
C:N		351:1			5264:1			13.58:1	
C:P		243.4:1			243.4:1			4.38:1	

Bergensen *et al.* (1991) propose that PHA reserves can support N_2 fixation during prolonged periods of N_2 fixation. Bacteria can also use PHA as a source of energy and reductive power.

4. References

Bergensen *et al.* (1991) Proc. R. Soc. London B 245, 59-64

Caballero M, Romero M (1994) Appl. Environ. Microbiol. 60, 1532-1537

Reis *et al.* (1994) World J. Microbiol. Biotechnol. 10, 101-104

5. Acknowledgements

This work was partially supported by the Canadian International Development Agency (CIDA).

INTERACTIONS IN THE MICROBIAL ENDOPHYTIC COMMUNITY IN SUGARCANE

M. Heydrich, M. Rojas, J. Manzano, A. Rodríguez

Dept of Microbiology, Faculty of Biology, University of Havana, La Habana, Cuba

1. Introduction

The endophytic microbial community is a dynamic structure, this colonization of the plant by bacteria raises questions about their role (Reinhold-Hurek and Hurek 1998). They are influenced by biotic and abiotic factors, so in addition to inoculation studies another approach is to determine the bacterial physiology and functions. It also seems crucial to study the conditions and interactions to which the bacterial endophytic are exposed in their microenvironment.

2. Materials and Methods

Strains of *Gluconacetobacter diazotrophicus* (PAl5 ATCC 49037, 4-02 and 1-05), *Enterobacter agglomerans* and *Bacillus licheniformis* isolated from sugarcane in Cuba were studied in their capacity of growth and N_2 fixation in eight carbon sources. Cells were grown at 30°C and 150 rpm in LGI-P medium (Reis *et al.* 1994) with 1% of C sources. The effect of six nitrogen sources (5mM) on the growth was also proved. Concentrations of 0.025, 0.25 and 2.5 of indole 3acetic acid (AIA) and giberellic acid (GA) were used to determine the growth and nitrogen fixation capacity of the strains. Interactions among nine selected isolates from the endophytic community were determined by the Bauer and Kirby method.

3. Results and Discussion

From the different carbon sources tested cane juice shows the best growth and the highest rate of nitrogenase activity for the studied strains. Also their capacity of growth and N_2 fixation at low concentrations of hormones demonstrate that they are able to improve the benefits to the plant. The analysis of positive and negative interactions among endophytic community members in sugarcane contributes to enhance their effect in this important graminaceous crop.

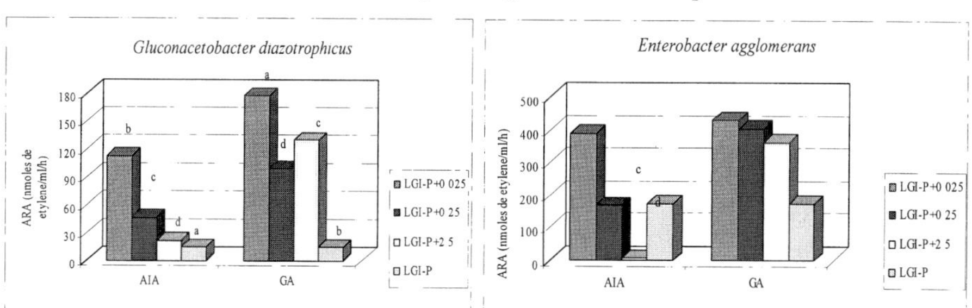

Figure 1. Effect of AIA and GA on nitrogenase activity in *G. diazotrophicus* and *E. agglomerans*.

4. References

Reinhold-Hurek B, Hurek T (1998) Crit. Rev. Plant Sci. 17, 29-54
Reis *et al.* (1994) World J. Microbiol. Biotechnol. 10, 101-104

5. Acknowledgements

This work was supported by the Canadian International Development Agency (CIDA).

RESPIRATION AND OTHER PHYSIOLOGICAL RESPONSES AS OXYGEN PROTECTION IN PLANT-NITROGEN FIXING ENDOPHYTE INTERACTIONS IN SUGARCANE

R. Rodés, E. Ortega, P. Ortega-Rodés, L. Fernández, A Ma. Pérez

Department of Plant Biology, Plant Physiology Lab., University of Havana, Cuba

1. Introduction

The stem of sugarcane harbors nitrogen-fixing endophytic bacteria which could contribute to the plant's nitrogen nutrition (Boddey *et al.* 1995). To enhance the plant-microorganism interaction, the system should provide an adequate habitat for the protection of the endophytes against O_2. Respiratory protection has been considered (Flores-Encarnación *et al.* 1999) to be present in different associative interactions to avoid high O_2 concentration and its negative effects on nitrogenase.

2. Material and Methods

Three experiments were conducted. Two nitrogen-fixing endophytes from Cuban sugarcane, *Gluconacetobacter diazotrophicus* (strain T2) and isolate 9°C (an unidentified bacterium) were grown in liquid and/or semisolid LGI-P medium. Nitrogenase activity (ARA) and respiratory rate (RR) (CO_2 evolution measured by IRGA, Qubit System Inc.) were determined at different O_2 levels. Roots from tissue-cultured sugarcane plants and from plants growing in pots (10 months old), were introduced by different periods in liquid cultures of *G. diazotrophicus*. Respiratory rate, guiacol-dependent-peroxidase (GdP) activity and superoxide dismutase (SOD) (Misra and Fridovich 1977), were measured in the roots. Respiration rate was also measured in the cultures of T2 that were in contact with the roots.

3. Results and Discussion

Both diazotrophs fixed N_2 in a wide range of O_2 levels. The nitrogenase activity of T2 is higher than 9°C and its optimum activity was at 10–15% O_2. The respiration patterns were different in semisolid and liquid medium; this probably indicates different tolerance and protection mechanisms against O_2.

The early stages (24 hours) of the interaction between roots and T2 cultures were characterized by an increased respiration rate of the roots and of the *G. diazotrophicus* cultures and by an increase in the activity of the enzymes SOD and GdP in the roots.

4. References

Boddey RM *et al.* (1995) Plant and Soil 174, 195-209
Flores-Encarnación M *et al.* (1999) J. Bacteriol. 181, 6987-6995
Misra H, Fridovich I (1977) J. Biol. Chem. 252, 6421-6423

5. Acknowledgements

This research was supported by CIDA/(Havana U./Carleton U. Project). Technical assistance of M. Diez Cabezas and V. García is gratefully acknowledged.

HYDROGEN AND NITROGEN-FIXING ENDOPHYTES IN SUGARCANE

E. Ortega, L. Fernández, R. Rodés, P. Ortega-Rodés

Department of Plant Biology, Plant Physiology Lab., University of Havana, Cuba

1. Introduction

Hydrogen (H_2) evolution has been used to measure nitrogenase activity in legume nodules (Hunt, Layzell 1993). Dong (1995) proposed to use the method of hydrogen evolution for measuring nitrogenase in *Gluconacetobacter diazotrophicus* cultures. For several years, we have been using hydrogen evolution to determine the nitrogenase activity for N_2-fixing microorganisms from sugarcane with good results.

2. Material and Methods

Nitrogen-fixing endophytes, isolated from sugarcane (var. ML 318) growing in the field in Fontanar, La Habana, were used for the research. T2 is a strain of *G. diazotrophicus*, and 9C is an unidentified sugarcane endophyte. Both microorganisms have the capacity to fix nitrogen. Ethylene production and H_2 evolution kinetics were measured with by gas chromatography (GC) from Qubit System Inc., in cultures with LGI-P medium (Reis *et al.* 1994) at 30°C. For Total Nitrogenase Activity (TNA), we replaced the N_2 with Ar. The Apparent Nitrogenase Activity (ANA) was determined in N_2. All measurements occur at 2% O_2.

3. Results and Discussion

TNA and ARA had similar values in both strains. Relative efficiency was 0.65 for T2 and 0.56 for 9C. Both are in the range of 0.4 to 0.7 reported for *Rhizobium* (Hunt, Layzell 1993). H_2 evolution from 9C is higher than ARA in solid LGI-P medium in a range from 5 to 20% acetylene.

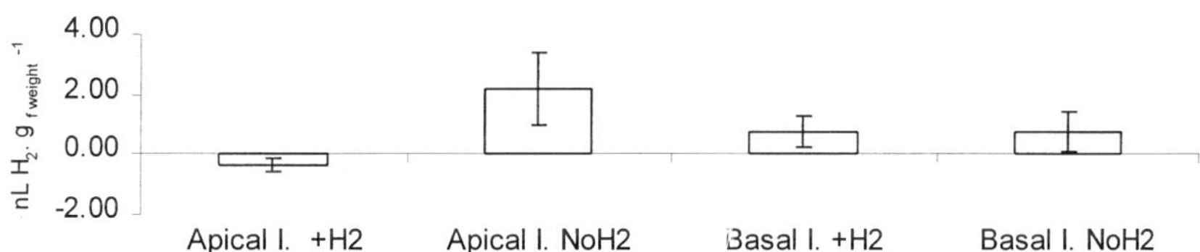

Figure 1. Hydrogen exchange by internode tissue in sugarcane. Treated tissues incubated 21 h at 30°C, with 0.5 mL, 5000 ppm per tube of 64 mL.

Apical tissues showed the capacity for net H_2 uptake. We hypothesized that some other microorganism should be using the H_2 evolved by N_2-fixing bacteria.

4. References

Hunt S, Layzell D (1993) Ann. Rev. Plant Physiol. Plant Mol. Biol. 44, 483-511
Dong Z (1995) Ph.D. Thesis, Carleton University, Ottawa
Reis V *et al.* (1994) World J. Microbiol. Biotechnol. 10, 101-104

5. Acknowledgements

This research was supported by CIDA/(Havana U./Carleton U. Project). Technical assistance of M. Diez Cabezas and V. García is gratefully acknowledged.

BIOLOGICAL NITROGEN FIXATION IN SUGARCANE: A GENOME PERSPECTIVE

E.M. Nogueira[1], F. Vinagre[1], H.P. Masuda[1], C. Vargas[1], V.L.M. de Padua[1],
F.R. da Silva[2], R.V. dos Santos [2], J.I. Baldani[3], P.C.G. Ferreira[1], A.S. Hemerly[1]

[1]Lab. Biologia Molecular, Depto. Bioquímica Médica, ICB, UFRJ, 21941-590,
  Rio de Janeiro, RJ
[2]CBMEG, UNICAMP, Campinas, SP, Brazil
[3]EMBRAPA/Agrobiologia, Seropedica, RJ, Brazil

1. Introduction

Distinct endophytic diazotrophic bacteria have been isolated from sugarcane organs, including *Gluconacetobacter diazotrophicus, Herbaspirillum seropedicae* and *H. rubrisubalbicans*. In this unique type of association, the bacteria live in the intercellular spaces and vascular tissues of most plant organs without causing disease symptoms (Baldani *et al.* 2000). Besides fixing nitrogen, the endophytic diazotrophs produce plant growth-regulating hormones, such as auxin (Fuentes-Ramirez *et al.* 1993). It is still not clear which mechanisms are involved in the establishment of this particular type of interaction and which kinds of molecules are mediating signaling between plant and bacteria. To address these questions, we have investigated gene expression profiles in sugarcane plants colonized by the endophytic diazotrophs *G. diazotrophicus* and *Herbaspirillum* spp., using the SUCEST (Sugar Cane EST Sequencing Project) database.

2. Material and Methods

For the annotation, the database (http://sucest.led.ic.unicamp.br) containing 81,223 clusters was searched. Transcriptional profiles were constructed by electronic northern. Nineteen cDNA libraries that represent distinct tissues/organs of sugarcane plants and the two that represent *in vitro* growing plants co-cultivated for seven days with *G. diazotrophicus* and *H. rubrisubalbicans*, named AD1 and HR1 respectively, were used in our analysis. EST representation (frequency in AD1 or HR1 library/frequency in the second best represented library of non-infected tissues) > 2 was classified as *Preferential*. ESTs represented only in AD1 and/or HR1 cDNA libraries were classified as Exclusive.

3. Results and Discussion

Using the EST database of SUCEST, genes encoding proteins that might function in processes involved in the association with the endophytic diazotrophs, such as nitrogen metabolism, plant growth, plant-microbe signaling and early nodulin homologs, were annotated and their expression profile was analyzed by electronic northern. The selected dataset comprised 1827 ESTs. In all the processes investigated, various ESTs preferentially or exclusively expressed in the AD1 and /or HR1 cDNA libraries were identified. An inventory of sugarcane genes, whose expression was modulated by the association, was generated. These preliminary data suggest that the plant might be actively involved in the interaction, responding to distinct processes during the association.

4. References

Baldani J *et al.* (2000) In Pedrosa F *et al.* (eds), Nitrogen Fixation: From Molecules to Crop
    Productivity, pp. 397-400, Kluwer Academic Publishers, Dordrecht, The Netherlands
Fuentes-Ramirez LE *et al.* (1993) Plant Soil 154, 145-150

5. Acknowledgement

Supported by FAPERJ (in collaboration with ONSA – FAPESP), FINEP, CNPq, CAPES.

AZOSPIRILLUM DOEBEREINERAE AND *HERBASPIRILLUM FRISINGENSE*: TWO NEW DIAZOTROPHIC SPECIES FROM C4-FIBER PLANTS

B. Eckert[1], G. Kirchhof[1], O.B. Weber[2], V.M. Reis[2], F. Olivares[2], M. Stoffels[1], M. Schloter[1], I.J. Baldani[2], A. Hartmann[1]

[1]GSF-Institute of Soil Ecology, D-86764 Neuherberg/Munich, Germany
[2]EMBRAPA, CNPAB, Seropedica, CEP 23851-970, Rio de Janeiro, Brazil

The C4-plants *Miscanthus sinensis*, *Spartina pectinata* and *Pennisetum purpureum* are becoming increasingly important as industrial crops for alternative agricultural production of renewable resources for energy and fiber as well as charcoal production in Europe and Brazil on set-aside land. Since it is known that these plants have only a low requirement for additional nitrogen fertilization, the presence of nitrogen-fixing bacteria associated with these plants was investigated. Using washed pieces of roots and stems of *M. sinensis* cv. Giganteus, *Miscanthus sacchariflorus*, and *Spartina pectinata* growing in Freising, Bavaria, Germany as well as *Pennisetum purpureum* cvs. grown in Brazil, nitrogen-fixing bacteria were isolated using NFb- and JNFb-semisolid media. Pure cultures of diazotrophs were characterized using a polyphasic approach (Eckert *et al.* 2001; Kirchhof *et al.* 2001). Besides known diazotrophic bacteria like *Herbaspirillum seropedicae* and *Azospirillum lipoferum* two new bacterial species were found.

Azospirillum doebereinerae sp. nov.: These bacteria are closely related to *A. lipoferum* and *A. brasilense* with 96.6 and 95.9% 16S rDNA sequence similarity. Two 16S rDNA targeting diagnostic oligonucleotide probes were developed to rapidly identify *A. doebereinerae* with fluorescence *in situ* hybridization. The bacteria differ from *A. lipoferum* by their inability to use N-acetylglucosamine and D-ribose and their ability to grow without supplemented biotin and from *A. brasilense* by their growth with D-mannitol and D-sorbitol. The new bacterial species was named in honor of Dr. Johanna Döbereiner for her great achievements.

Herbaspirillum frisingense sp. nov.: These bacteria form a homogenous group within the *Herbaspirillum* genus with only 1–34% similarity to *H. seropedicae* and *H. rubrisubalbicans* in DNA-DNA hybridization. The 16S rDNA similarity is 98.5-99.1%. Species-specific oligonucleotide probes were designed for diagnostic fluorescence *in situ* hybridization. On the basis of utilization of adipate (-), N-acetyl-D-glucosamine (+), meso-erythritol (-), L-rhamnose (-) and meso-inositol (-), *H. frisingense* can be distinguished from the other *Herbaspirillum* spp. Two days after inoculation of *H. frisingense* to axenically grown *Miscanthus* seedlings, bacteria could be demonstrated invading intracellular spaces in the rhizodermis; seven days after inoculation, they abundantly colonized xylem vessels in the vascular tissue.

References

Eckert B *et al.* (2001) Int. J. System. Evolut. Microbiol. 51, 17-26
Kirchhof G *et al.* (2001) Int. J. System. Evolut. Microbiol. 51, 157-168

ORGANIZATION OF NITROGEN FIXATION AND DICARBOXYLATE TRANSPORT GENES IN *PSEUDOMONAS STUTZERI*

N. Desnoues[1], M. Lin[2], R. Carreño-López [1,3], X. Guo[1], L. Ma[1], L. Gao [2], C. Elmerich[1]

[1]Microbiologie et Environnement, CNRS URA D2171, Institut Pasteur, Paris, France
[2]IAAE CAAS, Beijing, PO Box 5109, P.R. China
[3] Centro Microbiología, University of Puebla, México

1. Introduction

The A15 strain, identified first as *Alcaligenes faecalis*, was isolated from rice paddies in China. The strain can fix nitrogen under microaerobic conditions in the free-living state. It was found to colonize the root surface of rice and to invade the intercellular spaces of the root tissues. It can be considered as a facultative endophyte of rice and it appears as a particularly interesting strain to promote rice growth. Based on 16S rDNA analysis the strain was reassigned to *Pseudomonas stutzeri*.

2. Nitrogen Fixation Genes and Electron Transfer

Two clusters containing nitrogen fixation genes have been cloned. The general organization is very similar to that of *Azotobacter vinelandii*. A cluster carries nitrogenase structural genes (*nifHDK*), *nifTY* and *nifNE* (AJ313205). A second cluster carries *nifLA* and *nifBQ* (AJ307878). *nifA* is located 7 kb upstream of *nifB*. Inactivation of *nifA*, *nifB*, *nifH* and *nifE* led to a Nif negative phenotype while a *nifY* mutant remained Nif[+]. A second copy of *nifY* was found upstream of *nifH* as well as a set of genes sharing identity with the *rnfCDGEFH* operon (AJ297529), involved in electron transport to nitrogenase in *Rhodobacter capsulatus*. Different mutants have been constructed and it was found that the *rnf* cluster is required for nitrogen fixation in *P. stutzeri*.

3. Dicarboxylic Acid Transport

Several DNA regions encoding two-component sensor-response regulator systems have been isolated. The nucleotide sequence of a 9 kb region carrying *rmlACD* (involved in dTDP rhamnose synthesis) and *dctBDPQM* (involved in dicarboxylic acid transport) was determined. This region shared high identify with the locus AE004929 of *Pseudomonas aeruginosa* PAO1, except that the order of *rml* genes was changed. Inactivation of *dctBD* genes led a mutant strain impaired in utilization of succinate and fumarate, but growth with malate was not impaired. The efficiency of colonization of the rice root system by the *dct* mutant was decreased.

COLONIZATION OF RICE ROOTS BY *RHIZOBIUM LEGUMINOSARUM* BV. *TRIFOLII* ANALYZED BY CMEIAS[©] COMPUTER-ASSISTED MICROSCOPY

F.B. Dazzo, Y.G. Yanni, B.G. Rolfe

[1]Dept of Microbiology, Michigan State University, East Lansing, MI 48824, USA
[2]Sakha Agricultural Research Station, Kafr El-Sheikh, 33717 Egypt
[3]RSBS, Australian National University, Canberra, Australia

1. Introduction

Rhizobium leguminosarum bv. *trifolii* forms a natural, endophytic association with rice roots that can promote the growth of this important cereal plant, resulting in higher grain production with less N fertilizer inputs (Yanni *et al.* 1997; Prayitno *et al.* 1999; Biswas *et al.* 2000; Yanni *et al.* 2001). Earlier work with rice in gnotobiotic culture indicated that the rhizobia accumulate and penetrate sites where lateral roots emerge, multiply within dead cortical cells and intercellularly within the lateral rootlets, and can disseminate further inside of growing roots (Reddy *et al.* 1997; Yanni *et al.* 1997; Prayitno *et al.* 1999). Here, we acquired 140 georeferenced SEM images of rice roots colonized by strain E11, and used CMEIAS[©] (Center for Microbial Ecology Image Analysis Software; Dazzo *et al.* 2001; Liu *et al.* 2001) to analyze the spatial distribution of ~100,000 cells *in situ* on the root surface.

2. Results and Discussion

SEM showed that E11 colonized the rice rhizoplane with a slightly clustered distribution, entered into small crevices at epidermal junctions, and developed eroded epidermal pits (Yanni *et al.* 2001; Mateos *et al.* 2001). Geostatistics produced an exponential isotropic model that accurately defined the distribution of the surface-colonized bacteria, indicating a spatial dependence of interacting cells that influences colonization by neighboring bacteria within a separation distance of up to ~9μm. Block kriging analysis provided vivid, continuous interpolation maps of the intensity of bacterial colonization, even in areas of the root that could not be examined. This study established statistical proof of the existence and the relevant spatial scale of bacterial cell–cell interactions that influence their colonization of plant root surfaces, and provides a foundation of information upon which colonization of rice by different endophytic strains of rhizobia can be compared quantitatively.

3. References

Biswas J *et al.* (2000) Agronomy J. 92, 880-886
Biswas J *et al.* (2000) Soil Sci. Soc. Amer. J. 64, 1644-1650
Dazzo F *et al.* (2001) http://lter.kbs.msu.edu/Meetings/2001_Symposium/dazzo_etal.htm
Liu J *et al.* (2001) Microb. Ecol. 41, 173-194
Mateos P *et al.* (2001) Can. J. Microbiol. 47, 475-487
Prayitno J *et al.* (1999) Austr. J. Plant Physiol. 26, 521-535
Reddy P *et al.* (1997) Plant and Soil 194, 81-98
Yanni Y *et al.* (1997) Plant and Soil 194, 99-114
Yanni Y *et al.* (2001) Austr. J. Plant Physiol. 28

CHARACTERIZATION OF GENES THAT PLAY A ROLE IN SWARMING MOTILITY OF *AZOSPIRILLUM BRASILENSE*

R. Carreño-López [1,2], L. Pereg-Gerk [1], B.E. Baca [2], C. Elmerich [1]

[1]Microbiologie et Environnement, Institut Pasteur, 25-28 rue Dr. Roux, 75724 Paris Cedex, France
[2]Centro de Microbiologìa, Universidad Autónoma de Puebla, Apdo. Postal 1622, C.P. 72000, Puebla, Pue. México

Azospirillum bacteria live in close association with plants roots. The attachment of the bacterium to the root surface is essential for the establishment of an efficient colonization. Bacteria are highly motile in liquid media by way of a single polar flagellum responsible for swimming. Motility by swarming on semisolid media is made possible by numerous lateral flagella (Michiels *et al.* 1991). The aim of this work was to characterize a locus affecting swarming motility of strain Sp7S, a derivative of *A. brasilense* Sp7, Sp7S which was also shown to carry a mutation in capsular polysaccharide production (gene *flcA*) (Pereg-Gerk *et al.* 1998) and was impaired in surface root colonization. A gene bank of the strain Sp7 was introduced by conjugation into Sp7-S and a plasmid clone termed pAB7115 that restored swarming was isolated. The physical map of the plasmid was established and the locus responsible for the complementation was identified after mutagenesis with Tn5. Determination of the nucleotide sequence of the complementing region resulted in the identification of two genes.

The deduced translation product of the first one was highly similar to a phosphoribosyl-aminoimidazol carboxylase, encoded by *purK*, a gene involved in purine biosynthesis. Inactivation of *purK* did not lead to auxotrophy as described in *Escherichia coli* and *Salmonella typhimurium* (Tiedeman *et al.* 1989). The second one (ORF1) showed similarity with proteins of unknown function described in *E. coli*, *Aqualifex aeolicus* and *Synechocystis* spp., and a low similarity was found with a diguanylate cyclase motif from *Acetobacter xylinus* (24%). Disruption of both genes, in wild type Sp7, led to prototrophic mutant strains with decreased swarming properties, although both mutants showed lateral flagella. It is hypothesized that ORF1 and *purK* genes play a role in controlling the swarming motility possibly in the functioning of the lateral flagella. Root surface colonization of wheat was studied. No major difference was found between the wild type and the two mutants, although an effect on long-term colonization cannot be excluded, because lateral flagella enable bacterial movement along the root surface.

References

Michiels KW *et al.* (1991) J. Gen. Microbiol. 137, 241-2246
Pereg-Gerk L *et al.* (1998) MPMI. 11, 177-187
Tiedeman AA *et al.* (1989) J. Bacteriol. 171, 205-212

Acknowledgements

This work was partially supported by the ECOS-ANUIES-SEP CONACyT program (France-México). R. Carreño-López was a predoctoral fellow of the Secretaria de Educacion Publica (SEP) and was supported by the Institut Pasteur.

SUGARCANE GROWTH PROMOTION DEPENDS ON THE PLANT CULTIVAR AND GENOTYPE OF *GLUCONACETOBACTER DIAZOTROPHICUS*

J. Muñoz-Rojas, J. Caballero-Mellado

Centro de investigación sobre fijación de nitrógeno, UNAM. Apo. Postal 565-A, Cuernavaca, Mor. México

Experiments carried out in Brazil estimate that the contribution of biological nitrogen fixation (BNF) in some sugarcane cultivars ranged from 50 to 80% of total plant nitrogen. The bacteria that are most important in sugarcane-associated BNF remains unknown. However, *G. diazotrophicus* has been suggested as a strong candidate responsible for the BNF observed in field experiments. In the present study, micropropagated plantlets of five sugarcane cultivars (MEX 57-473; MY 5514; CP 72-2086; SP 70-1143; MEX 69-290) were inoculated with three *G. diazotrophicus* strains corresponding to different electrophoretical genotypes (UAP 5560-ET 1; PAl 5[T]-ET3; PAl 3-ET 5). Plant growth and bacterial colonization were monitored at different stages. The cell number of *G. diazotrophicus* decreased drastically in relation to plant age, regardless of the bacterial genotype or the sugarcane cultivar. However, the persistence of the three strains was significantly higher in a cultivar (MEX 57-473) than in the others. In addition, some strains (e.g. PAl 5[T]) persist for longer periods in higher numbers than other strains (e.g. PAl 3) inside the plant in all the cultivars tested. Sugarcane growth promotion was observed only in the cultivar MEX 57-473 inoculated with strain PAl 5[T]. This positive effect on the sugarcane plants was apparently caused by mechanisms other than nitrogen fixation.

THE ROLE OF ENDOPHYTIC DIAZOTROPHS IN SUGARCANE ROOT MORPHOGENESIS AND DEVELOPMENT

F.L. Olivares[1], V.M. Reis[2], J.I. Baldani[2], A.R. Façanha[1], V.L.D. Baldani[2], F.P. Ferreira[1], Q.R. Batista[1], B.R. Barreto[1], A.P. Simões[1], L.G. da Silva[1], I.G. de Azevedo[1], F.C. Brasil[3], R.O.P. Rossielo[3], F.C. Miguens[1], E.K. James[4]

[1]Centro de Biociências e Biotecnologia - UENF, Av. Alberto Lamego 2000, 28015-620, Campos dos Goytacazes, Rio de Janeiro, Brazil
[2]Embrapa Agrobiologia, Km 47, 23851-970, Seropédica, Rio de Janeiro, Brazil
[3]Dept de Solos, UFRRJ, 23890-000, Seropédica-RJ, Brazil
[4]Dept of Biological Sciences, University of Dundee, Dundee, DD1 4HN, UK

1. Introduction

The interactions between sugarcane and diazotrophic bacteria have been receiving much attention as an alternative means to extend nitrogen fixation to grasses. However, other mechanisms aside from N fixation may be involved in growth promotion. One of the most remarkable effects in inoculation experiments is the increase of root growth and morphological changes followed by endophytic colonization of sugarcane roots. Despite this, there are no studies emphasizing molecular aspects and the cell biology of root growth and development. The purpose of this study was to evaluate the effects of endophytic establishment on root morphogenesis and development.

2. Material and Methods

Sugarcane plant var. RB 72-454 was inoculated with *Herbaspirillum seropedicae* strain HRC54 and *Gluconacetobacter diazotrophicus* strain PAL5 under axenic conditions and after seven days transferred to pots in greenhouse conditions. Two harvests, 35 and 63 days after inoculation (d.a.i.), were performed and the inoculation effect was measured by height and fresh/dry matter accumulation, total length and area of the root system by digital image analysis. To examine more carefully root morphology and development, histochemical detection and quantification of lateral root mitotic sites and emerged lateral roots were performed at the whole root system.

Table 1. Effect of the endophytic establishment of *H. seropedicae* strain HRC54 and *G. diazotrophicus* strain PAL5 on sugarcane plants variety RB 72-454.

	PAL 5	HRC54	CONTROL
1. Plant Growth Parameters (35 d.a.i.)			
Height (cm)	48.0 a	43.4 a	33.1 b
Fresh Matter Weight (roots)	4.50 a	3,79 a	2.01 b
Fresh Matter Weight (aerial part)	1.79 a	1.31 ab	0.69 b
2. Root Morphology/Development			
Total length (cm)	41.2	30.6	12.6
Total area (cm^2)	197.3	159.2	68.2
Lateral Root Mitotic Sites *in vitro* (7 d.a.i.)	31	62	4
Emerged Lateral Roots *in vitro* (7 d.a.i.)	110	159	27
Lateral Root Mitotic Sites (63 d.a.i.)	70	69	91
Emerged Lateral Roots (63 d.a.i.)	1651	2553	1108
3. Endophytic Establishment (roots)	2.5×10^4	p2.5×10^5	---
4. Total Nutrient Accumulation (aerial part)			
Nitrogen (mg)	7.30	3.73	2.79
Phosphorus (mg)	4.84	2.91	1.80
Potassium (mg)	16.34	10.30	6.72

SOME PROPERTIES AND EFFECTIVENESS OF NITROGEN-FIXING RHIZOBACTERIA ISOLATED FROM RICE ROOTS

V.K. Chebotar[1], U.G. Kang[2], H.M. Park[2], N.M. Makarova[1], T.A. Popova[1], S. Akao[3]

[1]All-Russia Research Institute for Agricultural Microbiology, Shosse Podbelskogo 3,
St. Petersburg – Pushkin 8, 196609, Russia
[2]National Yeongnam Agricultural Experiment Station, P.O. Box 6, Milyang 627-130, Korea
[3]National Institute of Agrobiological Resources, Tsukuba, Ibaraki, 305, Japan

Concern about possible health and environmental consequences of using increasing amounts of mineral fertilizers and chemical pesticides have led to strong interest in alternative strategies to ensure competitive yields and protection of crops. This new approach to farming, often referred as sustainable agriculture, seeks to introduce agricultural practices that are more friendly to the environment and that maintain the long-term ecological balance of the soil ecosystem. The main objective of our study was to isolate strains with high root colonization activity possessing beneficial properties: biological nitrogen fixation, biocontrol activity and plant growth promotion. An original method for isolation of rhizobacteria inhabiting roots of rice grown in South Korea and Russia has been used. Among 256 isolated strains, 8 promising strains with high root colonization activity possessing beneficial properties (biological nitrogen fixation, biocontrol activity and plant growth promotion) were selected. These strains showed high motility and enzyme activity compared with other tested isolates. Acetylene reduction activity in pure cultures showed that tested strains demonstrated higher activity than reference strain of *Herbaspirillum seropedicae* BR11175 after 24 and 48 hours of incubation. Promising isolates demonstrated strong plant growth stimulation as was shown by two methods. It is interesting to note that in cell suspension at 1:1 dilution inhibition of growth of maize seedling roots was observed while a dilution 1:50 showed significant stimulation. Tested strains produce growth-promotion substances as was shown by HPLC analysis (data not shown). KR076 and KR083 strains, but not *Herbaspirillum seropedicae* showed very high biocontrol activity against major phytopathogens *Fusarium culmorum*, *Fusarium solani*, *Pythium* spp., *Phytophtora capsici*, *Rhizoctonia solani* in plate tests. Colonization of the plant roots by soil-borne or introduced bacteria is a very important step in the establishment of effective plant-bacteria interactions. Success in introduction of beneficial bacteria to the plant seeds or seedlings usually depends on the colonization potential of introduced strains. We studied colonization of rice roots cv. Keumobyeo by promising isolates in competition with reference *gus*-marked strain *Pseudomonas fluorescens* WCS365 known as active root colonizer. Some tested strains demonstrated high colonization ability. Pot experiments with promising isolates showed that only strain KR181 has increased yield of rice by 26.5% comparing with uninoculated control. The effect of this strain on number of panicles per hill, chlorophyll content and rice yield was the same as in variant with NPK 100%. Other tested strains did not show any positive effect on growth parameters of rice even though they demonstrated some beneficial properties. However the pot experiments with lettuce and radish have shown high efficiency of KR076 and KR083 strains. Seed inoculation of barley and sorghum in field tests with strain KR076 has been increased yield by 11% and 99% respectively. So, it is not clear what mode of action is responsible for success of inoculation and further pot and field experiments with rice and other crops should be done.

IDENTIFICATION OF INDOLE-3-BUTYRIC ACID IN *AZOSPIRILLUM* CULTURES

L.J. Martínez Morales[1], A. Sánchez Ahedo[2]

[1]Microbiological Research Center, University Autonomous of Puebla, Puebla, Pue. México
[2]Mexican Institute of Petroleum, México D.F. México

Azospirillum is a metabolically versatile organism with the ability to associate with the agronomically important plants; it produces several metabolites involved in beneficial effects to the plants; for instance, for the production of phytohormones by the microorganism there is more than one pathway for the synthesis of indole-3-acetic acid, however, there has not been described yet a metabolic pathway for the biosynthesis of indole-3-butyric acid by bacteria.

In this work, we show the presence of a substance with auxine activity identified as indole-3-butyric acid. *Azospirillum brasilense*, was grown at 30°C in a minimal medium for 72 hours and indoles were extracted from the supernatant. This sample was analyzed by high performance liquid chromatography (HPLC), gas chromatography (GC) and gas chromatography-mass spectrometry (GC-MS). The retention times matched those of the authentic IBA standard.

Large scale production of IBA was achieved in cultures grown in fermenters. Identification of the compound was carried out by HPLC and indole butyric acid was collected from the HPLC to perform assays on roots of maize seedlings (*Zea mays*), inducing the formation of root hair proliferation as well as adventitious roots, the positive biologic activity was observed when compared with authentic IBA standard. It must be noted that this is a report for the actual production of IBA by *Azospirillum brasilense*.

ENDOPHYTIC BACTERIA IN INDUCED ROOT NODULES ON EGGPLANT AND TOMATO

G. Yongjun

G-Laboratory, Beijing Liuhe Newstar Biological Technology Co. Ltd.
Beijing, 100832 China

Through optical microscopes, transmitting electron microscopes and scanning electron microscopes, the ultrastructure of and endophytic bacteria in artificially-induced nodules on the roots of eggplant (*Solanum melongena* L.) and tomato (*Lycopersicon esculentum* Mill.), plants were observed and studied. The results show that a large number of nodules formed on the root system due to intrusion of autogenous azotobacters. Within the root nodule cells, two strains of azotobacters were found. One was bacillus-brevis-type azotobacter of genus *Berjerinckia* Derx 1950 and the other *Azotobacter chroococcum* Beijerinck 1901. The latter was found in three different development forms, i.e. thallus, sporangium and sporangiospores, in the root nodule. They and their host plant formed a new type of symbiosis, which is quite rare in nature. This new type of symbiotic relationship between them will be elaborated in this paper.

ENDOPHYTIC ASSOCIATION BETWEEN *RHIZOBIUM LEGUMINOSARUM* BV. *VICIAE* AND ROOTS OF NONLEGUME CROPS

N.Z. Lupwayi[1], G.W. Clayton[1], V.O. Biederbeck[2], W.A. Rice[1]

[1]Agriculture and Agri-Food Canada, Lacombe-Beaverlodge Research Centre, AB, Canada
[2]Agriculture and Agri-Food Canada, SPARC, Swift Current, SK, Canada

1. Introduction

Endophytic rhizobia have been observed in nonlegumes and it is speculated that they improve the nutrition of these crops. The objective of this work was to investigate the effects of pea-based crop rotations on endophytic rhizobia and other bacteria in barley, wheat and canola roots.

2. Procedures

Barley, wheat and canola were each grown (a) continuously, (b) following uninoculated peas, or (c) following inoculated peas. Plants were sampled at flag-leaf stage (barley and wheat) or flowering stage (canola), and roots were surface-sterilized and macerated to extract endophytic bacteria. The endophytes were plate-counted on nutrient agar (NA), and their diversity was evaluated using the BIOLOG method. Endophytic rhizobia in the macerate were enumerated by the most probable number (MPN) method. At crop maturity, grain yields and N were measured.

3. Results and Discussion

Table 1. Effects of previous crop on populations of wheat endophytic bacteria and yield.

Previous crop	Endophytic rhizobia	Endophytic bacteria	Diversity of bacteria	Wheat grain yield	Wheat grain N
	Log (cells g^{-1} soil)		H'	Mg ha^{-1}	kg ha^{-1}
Wheat	0.69b	6.64a	2.72b	1.03b	19.5b
Uninoculated peas	3.27a	7.00a	2.83b	1.70a	34.0a
Inoculated peas	2.00c	6.49a	3.12a	1.64a	32.7a

Wheat results are presented (Table 1). Endophytic rhizobium populations were in the order: crop following uninoculated peas > crop following inoculated peas > continuously-grown crop. This means either (a) that indigenous rhizobia, stimulated by growing a legume crop, are better endophytic colonizers than inoculant rhizobia, or (b) that the inoculant rhizobium strain used in this experiment was a poor endophytic colonizer. The diversity, but not the populations, of all bacteria was greatest after inoculated peas. Grain yields and N uptake of crops grown after peas were significantly greater than yields of continuously-grown crops. Yield was correlated more with populations of endophytic rhizobia than with those of all endophytic bacteria, implying that endophytic rhizobia probably contribute to the rotational benefits of legumes to nonlegumes.

ACETOBACTER DIAZOTROPHICUS AND *HERBASPIRILLUM SEROPEDICEAE* NITROGENASE ACTIVITY IN THE SANDY-LOAM SOIL UNDER ITALIAN RYEGRASS *(LOLIUM MULTIFLORUM* LAM.) CULTIVATION

A. Sawicka, J. Klama

Department of Agricultural Microbiology, Agricultural University of Poznan, Wolynska 35, 60-635 Poznań, Poland

1. Introduction

Acetobacter diazotrophicus and *Herbaspirillum seropediceae* were discovered in crop plants characteristic for tropical climate such as sugarcane, sorghum, sweet potatoes and grasses. Until now, very little is known about the occurrence of these bacteria in the climate of the temperate zone. Therefore, the objective of this experiment was to recognize possibilities of the influence of the inoculation of Italian ryegrass with *Acetobacter* and *Herbaspirillum* on N_2 fixation activity.

2. Material and Methods

Italian ryegrass (*Lolium multiflorum* Lam.) cultivar Kroto was cultivated in an experimental box with loamy sand soil in a cold greenhouse. Grass seeds were inoculated with *Acetobacter diazotrophicus* and *Herbaspirillum seropediceae* directly before sowing. Non-inoculated seeds were treated as control. Nitrogenase activity in the soil under grass was measured after 3, 7, 9 and 12 weeks after sowing date using the acetylene to ethylene reduction method (ARA).

3. Results

The activity of N_2 fixation increased up to the 9th week of grass development (Table 1).

Table 1. Nitrogenase activity in the soil under Italian rye grass.

Experimental combinations	C_2H_4 (nanomoles h^{-1} plant^{-1})			
	After 3 weeks from sowing	After 7 weeks from sowing	After 9 weeks from sowing	After 12 weeks from sowing
Grass without inoculation (control)	0.8	7.2	9.3	3.5
Grass inoculated with *Acetobacter*	1.8	11.5	18.6	7.1
Grass inoculated with *Herbaspirillum*	1.2	7.4	10.2	4.3
Grass inoculated with *Acetobacter* and *Herbaspirillum*	1.5	9.6	14.7	5.6

The highest nitrogenase activity was observed 7 and 9 weeks after sowing of inoculated seeds irrespective of the bacteria used for inoculation. The best effect was recorded when grass seeds were inoculated with *Acetobacter*.

EXPRESSION OF THE DIRECT GLUCOSE OXIDATIVE PATHWAY IN *ACETOBACTER DIAZOTROPHICUS* PAL 3

M.F. Luna, C.E. Bernardelli, J.L. Boiardi

CINDEFI, Facultad de Cs. Exactas (UNLP)-CONICET, Argentina

A. diazotrophicus oxidizes glucose extracellularly to gluconate via a PQQ-linked glucose dehydrogenase (GDH). This is considered the main route for glucose catabolism in this bacterium. Moreover, aldose oxidation through GDH allows N_2-fixing *A. diazotrophicus* to direct the electron flux through a more efficient branch of its respiratory chain generating extra utilizable energy (Luna *et al.* 2000). In this study the expression of GDH by *A. diazotrophicus* PAL3 under different environmental conditions was tested.

A. diazotrophicus PAL 3 was grown fixing N_2 in continuous cultures at different glucose concentrations. GDH activity increased with the concentration of glucose in the medium. Growth yield was maximum for 10 g/L of glucose and decreased for concentrations above 20 g/L. Cultures with 10 and 20 g/L glucose were C-limited whereas at higher sugar concentrations cultures were under carbon-excess conditions. Nevertheless glucose was almost entirely consumed. Carbon-excess cultures accumulated gluconate and ketogluconates. This is a typical overflow metabolism behavior. It is commonly observed that bacteria under carbon-excess excrete partially oxidized intermediates, capsular material and protein. In the case of *A. diazotrophicus*, although some polysaccharides and protein could be detected, gluconate was the main overflow product. Therefore, within the plant, where this organism lives in a sugar rich environment (likely at very low growth rates), *A. diazotrophicus* seems to be able to oxidize glucose at high rates provided that oxygen is not limiting. The extracellular glucose oxidation yields biologically utilizable energy that can be used for N_2-fixation and, probably could also function as a mechanism of respiratory protection of nitrogenase.

The pH of the culture medium had a profound influence on the steady state biomass concentration and growth yield. The range of optimum pH was between 5.5 and 6.0 and yields decreased towards acidic or alkaline pHs. The culture became unstable and washed-out (no growth) at pH values over 7.7. These data are in accordance with those obtained in batch cultures: at an initial pH of 3.5, biomass yield was half of that obtained at 5.5 and no growth was observed at 7.5. At these pH value no GDH activity could be detected. Since this enzyme is on the inner membrane oriented towards the periplasm, the pH of the growth environment could influence either its synthesis or activity, as already reported for other microorganisms. To study the pH-dependence of GDH activity and synthesis, cells grown at different pHs were incubated with glucose at the optimum pH (6.0) and at the culture pH, and the gluconate production rates examined. At pHs ranging from 4.5 to 6.5 no significant differences between both activities were measured. But at pH values distant from the optimum, and particularly under relative alkaline environments, although GDH is actively synthesized (gluconate production at pH 6.0 was high), very low activities could be detected at the culture pHs. From these results it could be concluded, in accordance with previous speculations (Luna *et al.* 2000), that glucose metabolism in *A. diazotrophicus* proceeds mainly via GDH and, under conditions where extracellular glucose oxidation is somehow impeded, growth is profoundly affected. In accordance with this proposal a *gdh* mutant of *A. diazotrophicus* PAL 3 could be grown in glucose containing media (indicating that the direct oxidative pathway is not the only route for glucose catabolism in this bacterium), but biomass yields were significantly lower than those obtained with the parental strain under N_2-fixation.

References

Luna MF, Mignone CF, Boiardi JL (2000) Appl. Microbiol. Biotechnol. 54, 564-569

Partly supported by ANPCyT (PICT 97 No. 1196)

INVOLVEMENT OF CYTOCHROME c IN THE BIOSYNTHESIS OF IAA IN *GLUCONACETOBACTER DIAZOTROPHICUS* AND ASSESSMENT OF A ROLE FOR IAA PRODUCTION IN SUGARCANE GROWTH ENHANCEMENT

S. Lee[1], E. Pierson[1], E. Escamilla[2], C. Kennedy[1]

[1]Department of Plant Pathology, University of Arizona, Tucson, AZ 85721, USA
[2]Instituto de Fisiologia Celular, Universidad Nacional Autonoma de Mexico,
  Mexico City, Mexico

Gluconacetobacter diazotrophicus, an endophyte of sugarcane, is beneficial to sugarcane growth possibly by two mechanisms, one dependent, and one not, on nitrogen fixation by the bacterial partner (Sevilla *et al.* 2001). To test the hypothesis that IAA production is a factor leading to sugarcane growth enhancement (Fuentes-Ramirez *et al.* 1993), genes known to be involved in IAA biosynthesis in other organisms were sought by PCR and complementation strategies. These were not successful. Screening of Tn*5* mutants of *G. diazotrophicus* strain PAl5 led to the isolation of strain MAd10, which produced very little IAA (~6% of wild-type levels). The mutation which led to decreased IAA production was not associated with the insertion site of Tn*5* in MAd10, determined by cloning Tn*5* and flanking regions in a suicide vector and reinsertion of this DNA into the *G. diazotrophicus* genome. To determine the site of the mutation leading to decreased levels of IAA in MAd10, a pLAFR3 library carrying *G. diazotrophicus* DNA inserts was transferred by conjugation into MAd10, followed by screening of transconjugants for IAA production. In two IAA$^+$ transconjugants, the cosmids isolated shared identical regions in the insert fragments; analysis by subcloning, complementation, and DNA sequencing, indicated that the mutation in MAd10 was located in the *ccmC* gene, involved in cytochrome c maturation. The *ccm* operon was sequenced and found to encode Ccm proteins of ~50% identity to those of the corresponding operon in *Bradyrhizobium japonicum*. Insertion of kan cassettes into the *G. diazotrophisuc ccm* genes cloned on suicide vectors, followed by their reintroduction into the *G. diazotrophicus* genome, led to the construction of several *ccm* mutants. The mutations in *ccmC, ccmD,* or *ccmE* genes led to the IAA$^-$ phenotype, and each produced ~4–6% levels of IAA compared to strain PAl5. Mobilization of the entire *ccm* operon into these mutants restored IAA production. Therefore cytochrome c is likely to be an essential component of an IAA biosynthetic enzyme in *G. diazotrophicus*.

Spectral analysis of cytochrome c, heme-associated peroxidase activities, and membrane-associated respiratory activities in the wild-type and mutant strains showed that the Ccm proteins of *G. diazotrophicus* are involved in cytochrome c biogenesis. Growth on several media and ability to fix nitrogen were not influenced in the *ccm* mutant strains; the only phenotype observed was a decrease in IAA production. The effect of a *ccmC* mutation on plant growth enhancement was examined, either singly or in combination with a mutation in *nifD*. Regardless of nitrogen supply, plants inoculated with wild-type PAl5 were larger than uninoculated plants. Plants inoculated with a *ccmC* mutant, a *nifD* mutant, or a *ccmC-nifD* double mutant were no larger than uninoculated plants. These results are consistent with the hypothesis that both nitrogen fixation and IAA production are factors that allow *G. diazotrophicus* to benefit plant growth, and that a threshold of input by either factor must be attained before growth enhancement is achieved.

References

Fuentes-Ramirez LE *et al.* (1993) Plant Soil 154, 145-150
Sevilla M *et al.* (2001) Mol. Plant Microbe Interact. 14, 358-366
This work was funded by NSF Integrative Plant Biology Grant, IBN-9728184. We thank M. Flores-Encarnacion, M. Contreras-Zentella, and L. Garcia-Flores for help with *ccm* mutant analysis.

EVIDENCE TO ILLUSTRATE THE BENEFICIAL YIELD RESPONSE OF SPRING WHEAT AND RICE TO INOCULATION WITH NITROGEN-FIXING BACTERIAL ENDOPHYTES

P.J. Riggs, R. Mortiz, Y. Dong, M. Chelius, E.W. Triplett

Dept of Agronomy, University of Wisconsin, Madison, WI 53706, USA

Two experiments were performed to examine the potential of various endophytic nitrogen-fixing bacteria to increase yield on spring wheat and dry land rice. The long-term goal of these experiments is to locate bacteria and grass genotypes that can form biological nitrogen-fixing symbioses, and thereby decrease the level of industrial fertilizer required by modern agriculture. The strains P101 (*Panteoa* sp.), P102 (*Panteoa* sp.), K342 (*Klebsiella* sp.), K342 nif-h11 (*Klebsiella* sp. nif h-11 mutant), and PAL5 (*Acetobacter diazotrophicas*) were chosen as inoculants. Previous experiments have illustrated these strains ability to positively affect yield of maize, sugarcane and, to a lesser extent, spring wheat in the field and greenhouse. These bacterial strains were isolated from a variety of plant sources including maize, switch grass, and sugarcane.

Strain K342 wild type produced a significant increase in dry weight (25%) in spring wheat (cv. Trenton) compared to both the uninoculated control and the K342 nif-h11 mutant. Nitrogen fixation may be the mechanism underlying this stimulation in yield. Nitrogen yield data illustrate that plants inoculated with the K342 wild type had 20% more N in their shoot tissues than the plants inoculated with K342 nif-h11 mutant and the uninoculated controls. These data suggest that there is an interaction between the bacteria and the wheat that is providing nitrogen to the wheat plants inoculated with the K342 wild type, relative to the control and nif-h11 mutant. The next steps will be to perform $^{15}N_2$ reduction assays to directly measure the amount of biological nitrogen fixation that is occurring in this system and to place a fluorescent tag on the nitrogen-fixing genetic apparatus within the bacterium to illustrate the expression of the enzymes of nitrogen fixation *in vivo*. This will allow us to state confidently that there is biological nitrogen fixation occurring in this system.

Results from the rice (cv. Nipponbare) experiment illustrate several points. First, that strain P101 produces a large increase in the dry weight of rice shoot tissue (50%) compared to the uninoculated control. The mechanism for this yield increase may be due to a biological nitrogen fixation symbiosis between P101 and the host rice plants. This hypothesis is supported by the total nitrogen yield data, which indicated that compared to the uninoculated control the rice plants inoculated with P101 contained 70% more nitrogen on a dry weight basis. Strain PAL5 also produced a significant increase in total nitrogen per unit dry weight of shoot tissue (55%) when compared to the uninoculated control. This supports the idea that these two bacteria are potentially fixing nitrogen for the rice plant. The next steps will be to produce and test P101 and PAL5 mutant strains against an uninoculated control and the wild type. This will help to rule out other plausible reasons for the observed yield increase such as the production of plant growth promoting compounds. Finally, $^{15}N_2$ reduction assays will be completed to directly measure the amount of biological nitrogen fixation that is occurring in this system, and the use of GFP-labeled bacteria will allow use to verify that the inoculants are pursuing an endophytic lifestyle.

ASSOCIATION BETWEEN NITROGEN-FIXING BACTERIA AND *CONZATTIA MULTIFLORA*

E.T. Wang[1], S. Sarabia[1], E. Martinez-Romero[2], O. Dorado[3], G. Boll[1], A. Lemos[1]

[1]Dept Microbiol. ENCB, LPN, Mexico
[2]Program Ecologia Molecular y Microbiana, CIEN, UNAM, Mexico
[3]Dept Biol., UAEM, Mexico

1. Introduction

Among different nitrogen fixation systems, the symbiosis of legumes-rhizobia is the most important one and has been studied thoroughly. Compared with the nodulating species, nitrogen nutrient of the non-nodulating leguminous plants is almost ignored, although they are also important in the natural ecosystems. *Conzattia multiflora* is a leguminous tree belonging to the subfamily of *Caesalpinioideae*. It grows only in Mexico. There is no record about its symbiotic or pathogenic microbes and no information about its economic value. In this study, we isolated and characterized some endophytic bacteria from this tree.

2. Isolation and Inoculation of Bacteria

Twelve isolates with yellow or colorless colonies were isolated from 27 pieces of surface sterilized cortex using semisolid nitrogen-free medium (Tapia-Hernandez *et al.* 2000). All were gram-negative, facultative anaerobic rods. They grow in LB medium at 37°C. In McConkey medium, the isolates with colorless colonies in PY could absorb the dye and had rose to purple colors (similar to those of *Klebsiella)*. The four isolates with yellow colonies did not absorb the dye in McConkey medium. Four rDNA types were identified among the 12 isolates from the analysis of PCR-RFLP of 16S rDNAs. Representative isolates for the 4 rDNA types were inoculated onto the seedlings and were reisolated. The number of bacteria varied from 10^5 to 10^6 cells per gram of plant tissue in the inoculated plants while no bacteria were counted from the control plants when they were grown in cotton-sealed tubes. Two isolates, NF5 and NF9, could significantly improve the growth of seedlings as shown by the increased height of the plants. No nodules were observed on seedlings in the laboratory or on trees in fields and in pots. We also found these bacteria inside the *C. multiflora* seeds and these bacteria could be eliminated by germinating the seeds in a mixture of streptomycin (100 g mL^{-1}) and tetracycline (5 g mL^{-1}).

3. Nitrogen Fixation by the Bacteria

Low, but stable, acetylene reduction activity was detected in the four representative isolates when the bacteria were grown in semisolid nitrogen-free media. Modification of the carbon source or the addition of low concentration of yeast-extract did not enhance the acetylene reduction activity. In PCR-amplification of *nifH* using the primers nifH-1 and nifH-2, a band with molecular size similar to the *nifH* fragment of *R. etli* CFN42 was obtained from all the 4 isolates. The sequence of this 500 bp PCR-fragment from NF1 and NF9 was 67.7 and 73.8% similar to the *nifH* gene of *R. leguminosarum* and was 91.1% similar to each other. However, acetylene reduction in the cortex was not detected.

We concluded that: (i) this tree had no nodules, but contained a large number of endophytic bacteria in the cortex; (ii) these bacteria are facultative anaerobic nitrogen fixers; and (iii) these bacteria contained a *nifH* gene most similar to that of *R. leguminosarum*.

4. Reference

Tapia-Hernandez A *et al.* (2000) Micro. Ecol. 39, 49-55

ACTIN FILAMENTS ARE REQUIRED FOR SUCCESSFUL INFECTION THREAD DEVELOPMENT DURING NODULATION OF ALFALFA BY *SINORHIZOBIUM MELILOTI*

V. Imbruce[1], G.C. Walker[2], H.-P. Cheng[1]

[1]Dept Biology, Lehman College, CUNY, Bronx, NY 10468, USA
[2]Dept Biology, MIT, Cambridge, MA 02139, USA

1. Introduction

The *Sinorhizobium meliloti* and *Medicago sativa* (alfalfa) establish symbiosis after a series of chemical signal exchanges. The *S. meliloti* cells enter alfalfa root nodules through plant produced infection threads in the middle root hairs and convert atmospheric nitrogen into ammonia inside the nodules. The process of root hair invasion that results in the formation of infection thread is a key step in the establishment of this symbiosis. While the development of infection threads can now be visualized using green fluorescence protein (GFP) expressing *S. meliloti* cells, the mechanism of infection thread development remains unclear. We explored the possibility that the development of infection threads is an inward growth of root hair and thus shares some of the same mechanism as root hair growth.

Root hairs grow at their tips and require cytoplasmic streaming to transport cell wall and membrane materials from the base of cells. The cytoplasmic streaming is driven by myosin traveling along actin filaments, which are part of cell cytoskeleton. Actin filaments are long, contractile proteins that are supported by a framework of microtubules and they are depolymerized in the presence of cytochalasin B (CB), a mold metabolite. Depolymerization of actin filaments using CB stops cytoplasmic stream in plant root hair cells.

2. Results and Discussion

The effect of CB on alfalfa root hair cytoplasmic streaming was examined by directly exposing root hairs to CB under phase contrast microscope. Alfalfa seedlings were put on microscope slides in clear liquid growth media. Root hair cytoplasmic streaming is clearly visible under a phase contrast microscope and it was stopped immediately after the addition of CB.

To test the effect of CB on the development of infection threads, alfalfa seedlings were grown on microscope slides covered with a layer of solid growth media to facilitate live viewing. Slides were inoculated with GFP expressing *S. meliloti* cells that enables the real time monitoring of the development of infection threads using fluorescence microscopy. Root hairs with infection threads were exposed to CB through the application of CB containing agarose gel disks. The development of infection threads was blocked completely by the presence of CB at 0.1 μg/mL concentration. In the presence of lower concentration of CB, some infection threads formed spherical balls at the end of the infection thread and then formed new branches of infection threads.

Our findings that CB stops the cytoplasmic streaming and infection thread development suggest that the development of infection thread requires the presence of actin filaments. These findings and our early findings that the formation of infection thread depends on the presence of bacterial polysaccharide raise the possibility the formation of infection threads is the result of the communication between the bacterial cells and alfalfa root hair cells.

3. Acknowledgements

We thank corresponding author, Dr Hai-Ping Cheng.
This work is supported by an NIH Grant 41353-09-16 to HC and an NIH grant GM 31030 to GCW.

CHEMICAL CHARACTERIZATION OF ROOT EXUDATES FROM RICE (*ORIZA SATIVA*) AND THEIR EFFECTS ON THE CHEMOTACTIC CAPACITY OF ENDOPHYTIC BACTERIA AND PLANT GROWTH-PROMOTING BACTERIA (*AZOSPIRILLUM* AND *BACILLUS* SPP.)

M. Bacilio-Jiménez[1,2], S. Aguilar-Flores[3], E. Zenteno[4]

[1]Dept of Microb. Center for Biol., Res. of the Northwest (CIB), La Paz, BCS, Mexico
[2]Dept of Biochem., Nat. Inst. of Resp., Disease. Tlalpan, 14080 D.F. Mexico
[3]Dept of Botany, Nat. School of Biol., Sciences. IPN, 45873 Mexico
[4]Lab. Immunol. Dept of Biochem., Faculty of Medicine, UNAM, 04510 Mexico

Root exudates are an important source of nutrients for the microorganisms present in the rhizosphere and participate in the colonization process through chemotaxis caused on soil microorganisms. The colonizing capacity of plant growth-promoting bacteria (PGPB) could be favored by the chemotaxis exerted by root exudates; however, competition for nutrients and colonization sites between native strains present in the rhizosphere and endophytic bacteria seems to be an important factor, which could determine their colonizing capacity. In this work, we performed the chemical characterization of amino acids and sugars released by rice plantlets. We also evaluated the effect of these exudates on the chemotactic response of PGPB such as *Azospirillum brasilense* and *Bacillus* spp., and rice endophytes such as *Bacillus pumilus* and *Corynebacterium flavescens*.

Rice seeds (*Oryza sativa* L.) were surface-disinfected, soaked in a solution of 150 mg nalidixic acid/L and rinsed with sterile distilled water. After germination seeds were aseptically transferred to sterile hydroponic systems. Roots were kept in half-strength complete Hoagland's solution containing 50 mg/L nalidixic acid. Plants were kept for 7, 14, 21 and 28 days. The nutrient solution was lyophilized and used for identification of amino acids in an automatic amino acid analyzer and sugars by gas-chromatography. We evaluated the chemotactic capacity towards root exudates of rice plant using an acrylic chamber provided with two holes and a well. In the holes, two capillary tubes were introduced, one containing the sample of root exudates and the other, as a control. In the well, 0.5 mL of adjusted bacterial suspension (10^4 cells/mL) was placed.

The root exudates contained higher concentrations and a larger variety of carbohydrates and amino acids during the first two weeks of culture. The main amino acids identified were histidine, proline, valine, alanine, and glycine. The main carbohydrates identified were glucose, arabinose, mannose, galactose and glucuronic acid. Glucose was the main sugar derivative identified.

The four bacterial strains investigated showed positive chemoattractant capacity toward the root exudates from rice, however endophytes are better fit to respond to these attractants and can be explained based on their origin: *A. brasilense* was isolated from maize rhizosphere and *Bacillus* spp. from soil in which rice had been grown. This effect of root exudates also gives them a clear ecological advantage during the first colonization stages over *Azospirillum* and *Bacillus* spp., allowing them to compete better with applied PGPB, such as the rice soil bacterium *Bacillus* sp. These results strongly suggest that the attractant characteristics favor endophytes and, therefore, could induce exclusion of other colonizing microorganisms from the ecological niche, explaining in part why only endophytes could colonize the rhizosphere (Bacilio-Jimenez *et al.* 2001).

References

Bacilio-Jiménez M *et al.* (2001) Soil Biol. Biochem. 33, 167-172

RESPONSE OF WHEAT TO INOCULATION WITH DIFFERENT NITROGEN FIXING BACTERIA

S.R.S. Sabry[1], S.A. Saleh[2]

[1]Field Crops Research Institute, ARC, Egypt
[2]Soil, Water, and Environment Research Institute, ARC, Egypt

The objective of this work was to study the effect of chemical N fertilizer 50% reduction combined with different nitrogen-fixing bacteria on wheat grain yield compared to applying the recommended N fertilizer (100%). One field and three pot experiments were conducted in sandy and clay-loamy soil. The results of pot experiment I indicated that seed inoculation with *Herbaspirillum* plus 50% N fertilizer gave the highest grain yield and did not significantly differ from the 100% N fertilizer treatment. The results of pot experiment II showed that seed inoculation with certain double inoculants in addition to 50% N fertilizer did not significantly differ from the 100% N fertilizer control treatment. Pot experiment III results depicted that under either sandy or clay soil conditions certain wheat seed double inoculation treatments gave similar results as mentioned above. The field experiment showed that the highest significant seed inoculation treatment in grain yield was 50% N fertilizer + *Azospirillum* SP 7 + *Herbaspirillum* in sandy soil while 50% N fertilizer + *Azorhizobium* ORS 571 + *Herbaspirillum* in clay soil. It could be concluded from this study that seed inoculation treatment with nitrogen-fixing bacteria could satisfy about 50% of N requirement for wheat.

INFORMATIVE MODEL OF BIOLOGICAL PROCESS OF SOYBEAN SYMBIOTIC NITROGEN FIXATION

A.V. Kalinichenko

Dept of Agricultural Management, Poltava State Agricultural Institute, Poltava, Ukraine

The informative model of the biological process of soybean symbiotic nitrogen fixation was created on the basis of an informative-logical method of approach. It takes into account the presence of fortuitousness in natural phenomena. The usage of this model allows to improve the quality of a soybean seed and products of its processing. Received under this technology products of processing, extrudate and oilcake, contain all usually present amino acids and also the increased quantity (by 1.5% and more) of protein.

The model allows to solve such problems as: receipt of the prognosis of the expected values of a crop; of the quantity of the nitrogen and protein accumulated by a plant; of the nitrogen quantity that remained in soil for the given bacteria culture and soybean sort. It gives a solution of a direct goal (a short-term prediction of the expected values of informative parameters) as well as of an inverse task (the choice of the best culture of tuberous bacteria or soybean sort with the aim to receive the biggest values of the determined informative parameters). In addition, in case of need, it is possible to determine the probable interim values of informative parameters and to conduct a dynamic research of the process according to the determined moment of time or according to the periods of plants growth.

The research of the efficiency of the symbiotic nitrogen fixation was conducted on the crops of the soybean of the Crimean Scientific Industrial Society "Elite", Selection and Genetics Institute of the Ukrainian Academy of Agricultural Sciences, and of some farms in the Autonomous Republic of Crimea and Odesa region. These soybean seeds were processed with rhizotorfin produced on the basis of tuberous bacteria cultures 629a, 634b, 646. While studying the issues of influence of climatic factors and soybean genotype on nitrogen fixation efficiency, 122 collection soybean sorts both from home collection and produced in the USA, Canada, China and countries of Eastern and Western Europe, were taken. The study of the efficiency of the pre-sowing processing of soybeans was conducted on the harvests of the primary crops as well as post hay crops and reaping crops.

On the basis of the model, the process of selection of the perspective material for the symbiosis partners was improved, and culture *B. japonicum* 634b and soybean sorts of Krepysh, Odes'ka 124, Arkadia Odes'ka, Yug 40, Kharosoy, were recommended as the most effective.

Solution of numerous tasks and field researches were conducted on the large amount of soybean sorts and tuberous bacteria cultures. A short-term prediction of symbiosis system behavior was presented which allows us to determine the most perspective directions of the field research for realization and receipt of the high quality production for subsequent processing. Conclusions and recommendations on usage of the given model were made and offered according to the results of the research.

Application of the given research is proposed to determine the behavior of a biological system according to different factors (external as well as internal), to determine the most real final system conditions with interdependent output factors, and in addition, it is recommended to be used when choosing the most rational cropping strategy of soybean according to the practical needs of agricultural farms; in purposeful planning of soybean agricultural producing; in selection of the most effective cultures of nitrogen fixation bacteria and soybean plants in accordance with various external conditions when determining the conditions that allow to receive a plant crop of full value; and in selecting a complementary material among macro- and microsymbionts.

OPTIMIZATION OF INOCULATION OF *CALLIANDRA CALOTHYRSUS*, *LEUCAENA LEUCOCEPHALA* AND *ACACIA MANGIUM* WITH RHIZOBIA

D. Diouf[1,2], S. Forestier[2,3], M. Neyra[2], D. Lesueur[2,3]

[1]Département de Biologie Végétale, UCAD, BP 5005 Dakar, Sénégal
[2]Laboratoire de Microbiologie des Sols, IRD Bel Air, BP 1386 Dakar, Sénégal
[3]Programme Arbres et Plantations du CIRAD-Forêt, Dakar, Sénégal

1. Introduction

Several legume inoculation techniques are well known and used within the framework of many experiments. However, until now these techniques were often used without including several factors that can optimize inoculation. The purpose of our investigations was to study such factors as the physiological stage of the bacterial culture, the size of the inoculum and the mode of inoculation on the nodulation and biomass production for three agroforestry species: *Calliandra calothyrsus*, *Leucaena leucocephala* and *Acacia mangium* inoculated respectively with CCK13 (Lesueur *et al.* 1996), Ldk4 (Lesueur *et al.* 1998) and Aust. 13c (Galiana *et al.* 1990) strains.

2. Results and Discussion

The physiological stage of the bacterial culture had no significant effects on nodule biomass and seedling growth. However, inoculation with Ldk4 culture in the plate phase significantly improved nodulation of *L. leucocephala*. For all the three species, no significant differences in shoot biomass were noted.

Nodulation and growth of inoculated seedlings were less variable depending on the size of the inoculum. The greatest number of nodules was recorded on *A. mangium* for a dilution containing 10^{10} bacteria mL^{-1}. Alternatively for *C. calothyrsus* and *L. leucocephala*, 10^9 bacteria mL^{-1} corresponds to the optimal dilution. Similar results were noted for *C. calothyrsus* (Lesueur *et al.* 1996) and for *L. leucocephala* (Lemkine, Lesueur 1998).

Inoculation method had significant effects on nodulation and seedling growth. For *C. calothyrsus* and *A. mangium*, inoculation with a liquid culture one week after sowing was more favorable for growth of the seedlings. On the other hand, inoculation of *L. leucocephala* with a bacterial culture mixed with arabic gum significantly improved seedling growth.

3. Conclusion

Our work aims to determine the types of inoculum and the best formulation that are favorable to ensure optimal growth of three agroforest species. Results showed that the improvement of the growth of *C. calothyrsus*, *L. leucocephala* and *A. mangium* by inoculation in nurseries with efficient rhizobium strains was dependent on the mode of inoculation. Our work could be used to help in the large-scale production of forest seedlings.

4. References

Galiana *et al.* (1990) Appl. Environ. Microbiol. 60, 3974-3980
Lemkine G, Lesueur D (1998) Aciar Proc. 86, 168-171
Lesueur D *et al.* (1996) Forest Farm Comm. Tree Research Reports, 62-76
Lesueur D *et al.* (1998) Aciar Proc. 86, 86-95

5. Acknowledgements

This work was supported by the European Commission (INCO Contract No. IC18-CT97-0194).

COMBINED EFFECTS OF RHIZOBIA INOCULATION AND HOST PLANT ORIGIN ON GROWTH AND NODULATION OF *CALLIANDRA CALOTHYRSUS*

D. Lesueur[1,2], D. Diouf[2,3]

[1]Programme Arbres et Plantations du CIRAD-Forêt, Dakar
[2]Laboratoire de Microbiologie des Sols, IRD Bel Air, BP 1386 Dakar
[3]Département de Biologie Végétale, UCAD, BP 5005 Dakar, Sénégal

1. Introduction

Several authors have described a significant variation in nodulation and nitrogen fixation between and within ecotypes of woody legumes such as *Gliricidia sepium* and *Leucaena* (Sanginga *et al.* 1991; Mureithi *et al.* 1994), but no data are available on growth and nodulation of *Calliandra calothyrsus*. *C. calothyrsus* is an important species in agroforestry because of its value as a forage. A greenhouse experiment was carried out in order to evaluate the performances of several ecotypes of *C. calothyrsus* inoculated with two types of rhizobia (Lesueur *et al.* 1996). Results obtained are useful for the identification of a standard inoculum for improving forage production for *C. calothyrsus* in the field.

2. Results and Discussion

Growth, nodulation and nitrogen fixation of *C. calothyrsus* trees were significantly stimulated by the inoculation with both rhizobium strains. These results confirmed those obtained by Lesueur *et al.* (1996). This positive effect of the inoculation is variable according to the *C. calothyrsus* ecotype. The occurrence of an interaction between both factors for the different parameters. Comparable interaction to those found in this study have been reported for *Casuarina* (Sanginga *et al.* 1990) and *A. cyanophylla* (Nasr *et al.* 1999).

3. Conclusion

The main objective of this work was to determine if the inoculation with one rhizobium strain had an effect on the growth and shoot total N content of several ecotypes. Results show that seed origin seemed to affect the growth, nodulation and nitrogen fixation of inoculated *C. calothyrsus*. For improving symbiotic nitrogen fixation the interaction between strain and *C. calothyrsus* seedlot will be taken into account.

4. References

Lesueur *et al.* (1996) Bois For. Trop. 248, 215-230
Mureithi *et al.* (1994) Trop. Agric. 71, 83-87
Nasr *et al.* (1999) Can. J. Bot. 77, 1-10
Sanginga *et al.* (1990) Soil Biol. Biochem. 22, 539-547
Sanginga *et al.* (1991) Biol. Fertil. Soils 11, 273-278

5. Acknowledgements

This work was supported by the European Commission (INCO Contract No. IC18-CT97-0194).

EXPLOITING THE ADAPTIVE ACID-TOLERANCE OF COMMERCIAL LEGUME INOCULANTS DESTINED FOR ACIDIC SOILS

M.E. Dunne[1], G.W. O'Hara[1], G.K. Bullard[2], M.J. Dilworth[1]

[1]Ctr for *Rhizobium* Studies, Murdoch U, Western Australia
[2]Biocare Technology PTY LTD, Somersby, NSW, Australia

Legume inoculants for acidic soils are usually grown and stored in peat under non-acid conditions. These bacteria show an adaptive acid-tolerance response (ATR); cultures grown under moderately acidic conditions withstand exposure to highly acidic conditions much better than those grown under non-acid conditions.

Rhizobium leguminosarum bv. *viciae* (WSM1455) grown at pH 7.0 and then transferred to acidic defined medium of pH 3.5 displayed a decimal reduction time (D) of 10 h. When grown at pH 4.8 and exposed to pH 3.5, the D value was 16 h. *R. leguminosarum* bv. *trifolii* (WSM409) behaved similarly; the corresponding D values were 7.2 h and 9 h when grown at pH 7.0 and 5.5. *Sinorhizobium meliloti* strains also displayed an ATR; with D values of 10 min (pH 7) and 25 min (pH 5.5) for WSM419, and 5.8 h (pH 7) and 7.9 h (pH 5.5) for WSM688.

These strains acidify the commercial culture medium, the pH fell from 7.2 to 6.1 for both *R. leguminosarum* strains and to pH 5.7 (WSM419) and 5.8 (WSM688). Cultures of WSM409 inoculated into gamma-irradiated limed peat increased 100-fold while the pH of the peat remained neutral. Further studies are necessary to assess the field benefits of inoculants grown in acidic peat.

ENHANCING NODULATION COMPETITIVENESS OF AN INOCULUM STRAIN BY CO-INOCULATION WITH A NON-NODULATING ANTIBIOTIC-PRODUCING STRAIN IN FIELD-GROWN SOYBEAN

A.L. Iniguez[1], E.A. Robleto[2], M.K. Chelius[1], E.W. Triplett[1]

[1]Dept of Agronomy, University of Wisconsin, Madison, WI, USA
[2]Tufts University, Boston, MA, USA

Commercially available *Bradyrhizobium japonicum* inoculants do not prevent nodulation by indigenous strains on field-grown soybeans. Thus, legume productivity is limited to the N_2 fixation capability of indigenous strains. In previous work, we successfully employed the use of an antibiotic-producing *Rhizobium etli* strain to reduce nodulation of *Phaseolus vulgaris* by indigenous rhizobia in the field (Robleto *et al.* 1998). The reported antibiotic, Trifolitoxin (TFX) is a small peptide antibiotic that is post-translationally modified. However, TFX does not inhibit bradyrhizobia. Here we summarize our efforts to identify bacterial strains that inhibit most strains of *Bradyrhizobium* for use in new soybean inoculum systems. *Bradyrhizobium* sp. IRj2179a was found to have inhibitory properties to a large number of bradyrhizobia. Soybean inoculum strains were tested for resistance or sensitivity to IRj2179a through inhibition assays. Only 3 of the 290 soybean-nodulating strains (USDA strains 4, 54 and 61) in our collection showed zones of inhibition (sensitive to IRj2179a). Although IRj2179a does not nodulate soybean, in this study it was used to co-inoculate soybean seeds with a resistant strain USDA 61 and a sensitive strain USDA 26. This experiment was conducted in the field in West Madison research station of the University of Wisconsin-Madison. The results showed that adding IRj2179a to the USDA 61 inoculum enhanced the competitiveness of USDA 61 versus indigenous bradyrhizobia. In addition, IRj2179a improved the ability of USDA 61 to compete for nodulation versus an inoculum strain, USDA 26, which is sensitive to IRj2179a. This system may be of broad application due to the wide range of bradyrhizobia that are inhibited by IRj2179a. Nevertheless, we need to definitively determine whether antibiotic production is required for the effects on competition observed with IRj2179a. In order to determine the role of the antibiotic, we will need to develop and test antibiotic-minus mutants of IRj2179a. Finally, if the antibiotic is involved in competitiveness, the antibiotic needs to be isolated and characterized, and genes necessary for antibiotic production should be cloned and sequenced.

References

Robleto *et al.* (1998) Appl. Environ. Microbiol. 64, 2630-2633

INFLUENCE OF RHIZOBIOPHAGE ON COMPETITION BETWEEN ROOT-NODULE BACTERIA

S.H. Salem, F.I. El-Zamik, H.B. Abd El-Fattah, M.I. Radwan

Agric. Botany Dept, Faculty of Agriculture, Zagazig University, Zagazig, Egypt

1. Introduction

Rhizobiophages maybe one of the most essential factors influencing the function of rhizobia in the competition between the different strains to nodulate the host plant. The aim of this study was to examine the occurrence of temperate and lytic rhizobiophages in the Egyptian soils and nodules of faba bean and their effect on the ability of inoculated rhizobia to form nodules and fix nitrogen.

2. Materials and Methods

Twenty-six isolates of *Rhizobium leguminosarum* bv. *viceae* in addition to twenty-three rhizobiophage isolates were isolated from rhizosphere soil samples and from broad-bean nodules obtained from soils of different locations in Egypt. Isolations were done as described by Vincent (1970) and Adams (1959). The rhizobia isolates were tested against the isolated phages (Radwan 2000). The effect of inoculation with sensitive, lysogenic or resistant isolates of *R. leguminosarum* bv. *viceae* on nodulation and nitrogen fixation of *Vicia faba* plants was also studied using Leonard jars.

3. Results and Discussion

Results showed that rhizobiophages were commonly associated with nodules of *Vicia faba* plants and in their rhizosphere regions. Temperate phages were observed to have a wider distribution than lytic isolates. They exhibited different dimensions and morphological characteristics. The presence of temperate or lytic phages with the rhizobial-inoculum in general, decreased the nodulation, biomass dry weight, nitrogenase activity and nitrogen content of the host plant. Inoculation of *Vicia faba* plants with sensitive isolates of *R. leguminosarum* bv. *viceae* alone gave the highest values of symbiotic N_2-fixation parameters.

4. References

Adams MH (1959) Bacteriophages, Wiley-Interscience Pub. Inc., New York

Radwan MA (2000) Influence of Rhizobiophage on Competition Between Root-Nodule Bacteria, Ph.D Thesis, Fac. Agric. Zagazig University, Egypt

Vincent JM (1970) A Manual for the Practical Study of Root-Nodule Bacteria, IBP Handbook No. 15, Blackwell Scientific Publications, Oxford, UK

TAGTEAM : THE DEVELOPMENT OF A COMMERCIAL DUAL INOCULANT TO ENHANCE N–FIXATION AND INCREASE AVAILABILITY OF SOIL PHOSPHATE

C.B. Kuiack, S.C. Gleddie, L. Blais

Philom Bios Inc., 318–111 Research Drive, Saskatoon, SK, S7N 3R2 Canada

Phosphorus plays an important role in symbiotic N_2 fixation, and legumes generally require more P for optimal growth and development than non-legumes. The benefit of inoculating seed with both a nitrogen-fixing and a phosphate-solublizing inoculant was first demonstrated in Philom Bios small plot field trials in 1989 (Gleddie 1992). In 1994, Wendell Rice and Perry Olsen successfully co-cultured the phosphate solubilizing fungus *Penicillium bilaii* with *Rhizobium meliloti* in a sterile peat carrier (Rice *et al.* 1994). This work instigated the development of TagTeam for pea/lentil, the world's first commercial dual inoculant.

The product development process began with a pre-project analysis of the potential of the concept. This led to the definition of the key steps and major milestones which would become the project plan. The initial research involved the development of a solid state fermentation protocol for the co-culture of *Rhizobium leguminosarum* and *P. bilaii* using sterile peat as the substrate. The challenge was to identify the optimal growth conditions that would result in the appropriate titers and proportion of both organisms in a 28 g bag. The specific requirements for nutrition, pH, available oxygen, incubation temperature, moisture level, inoculum type and carrier type were identified. The stability of the resulting product was then determined under a number of different storage conditions and the packaging was designed for optimal shelf life. Collaboration with the manufacturing sector of the company resulted in the scale up of the product to 2200 g units (bags) and the development of a standard operating procedure to supply introductory market volumes. Philom Bios internal product specifications were defined and a Quality Assurance program was developed to confirm these, as well as the specifications set by the Canadian Food Inspection Agency (CFIA). The product prototypes were tested for efficacy in small plot field research. Inoculation with both organisms significantly increased grain yield as compared to inoculation with *R. leguminosarum* or *P. bilaii* alone. TagTeam was registered with the CFIA in 1995.

Further characterization of the product led to the development of a set of use guidelines which included inoculant compatibility with seed applied pesticides. Ongoing optimization of the product continues in terms of testing for compatibility with new chemicals and new strains and improving manufacturing efficiencies to lower the cost of production and increase volumes. Additionally, the platform technology is being used to develop dual inoculants for new crops and using new formulations.

The development of a high quality peat based dual inoculant was made possible only by the effective collaboration among all key areas in the product development process.

References

Gleddie SG (1992) M.Sc. Thesis in the Department of Soil Science, University of Saskatchewan, Saskatoon, SK, Canada

Rice WA *et al.* (1994) Soil Biol. and Biochem. 27, 703-705

THIRTEEN YEARS OF HEAD-TO-HEAD COMPARISONS OF SOYBEAN INOCULANTS

J.A. Omielan, D.J. Hume, G. Chu

Department of Plant Agriculture, University of Guelph, Guelph, ON, N1G 2W1, Canada

1. Introduction

Since 1988, our research group has been conducting "Head-to-Head" comparisons of commercial and pre-commercial soybean inoculants in Ontario. The first trials had six inoculants from four manufacturers. In 2000, the trials had 19 inoculant/application method combinations from five manufacturers. This poster presents some of what we have learned from 13 years of trials.

2. Materials and Methods

The Head-to-Head trials have been conducted on land which had not previously grown soybeans and so did not have existing populations of *Bradyrhizobium japonicum*. Bicentennial was the soybean cultivar used from 1988 to 1989 while Maple Glen was used from 1990 to 1995. The current variety, Bayfield, was first used in the trials in 1996.

In recent years, conventional tillage trials have been planted using Planet Junior units on a toolbar. The plots contained 7 rows and were 2.5 meters wide and 6 meters long. Compressed CO_2 was used to apply in-furrow inoculant treatments. Since 1997, the trials have also been planted using no-till techniques with plots containing 8 rows that were 3 meters wide and 12 meters long. Nodules were sampled from 5 adjacent plants in row #2 at the start of flowering and at the start of the pod-filling period. Nodules were counted and then dried for determination of mass. Color differences were often evident in the uninoculated plots by the second nodule sampling. Plots were machine harvested and seed samples were analyzed for oil and protein content.

3. Results

Average yield responses range from over 30% to about 15% for the various inoculants. Until about 1995, sterile-carrier, powdered-peat inoculants were the top performers. Since then, performance has been in the order: liquid in-furrow sprays > seed-applied liquids > sterile-carrier powdered peats > granular inoculants > non-sterile powdered peats.

In general, high quality inoculants performed well under both conventional and no-till conditions. There are however, differences in yield levels and yield responses among locations. Seed protein concentrations are a good indicator of inoculant performance. Using a number of indicators such as nodule numbers, nodule mass, seed yield, and seed protein is necessary to comprehensively interpret the effectiveness of the different inoculants.

The magnitude of the yield responses was related to the environmental conditions as well as nodule numbers and mass. Plotting nodule numbers vs. yield response generates curves with maximums at different positions. In 1998, there was a severe drought which limited yields and yield responses to inoculants. The season in 1999 was closer to average but the responses at the no-till location were greater than at the conventional tillage location. In 2000, a wet spring resulted in late planting and early frost damaged the plants before full maturity. This again resulted in lower yields and smaller yield responses. There is no single measure of inoculant quality and performance that can predict or explain a crop's response. Multi-year and multi-location field testing will continue to be an important tool in assessing inoculant performance.

4. Acknowledgements:

Funding from the Ontario Ministry of Agriculture, Food, and Rural Affairs is gratefully acknowledged.

RESPONSE OF THE COFFEE PLANT TO THE INOCULATION WITH *AZOSPIRILLUM* SP.

T. Jiménez-Salgado, L. Vázquez-Chávez, A. Tapia-Hernández, M.A. Mascarúa-Esparza, M. Rosas-Morales, L.E. Fuentes-Ramírez

Centro de Investigaciones en Ciencias Microbiológicas, Instituto de Ciencias, Universidad Autónoma de Puebla, Apdo. Postal 1622, CP 72000, Puebla, Pue., México

The coffee plant is a crop that is cultivated in large areas of the tropics. It produces one of the greatest incomes of the non-developed countries. Competitive production of grain demands the application of large quantities of fertilizers. Nevertheless, it also produces pollution and additional impact on the finance of the farmers.

Strains of the genus *Azospirillum* have been used in agriculture. Several of those experiments showed plant-growth promoting activity by inoculation with *Azospirillum*. Beneficial effects of those strains have been reported for different crops like maize, wheat and oak. We show experiments of inoculation of coffee plantlets with different homologous *Azospirillum* isolates.

Samples of coffee plants from 2 to 4 years old, cv. Caturra, as well as rhizosphere were obtained from cultivated fields from Puebla and Veracruz states, México. Surface and inner tissues of coffee roots, leaves, and rhizospheric soil were inoculated in N-free semisolid NFb. After incubation of 3–5 d at 30°C, the growth was streaked in Congo Red plates and incubated for 3–5 d at the same temperature. Isolated red colonies were tested for acetylene reduction activity (ARA) and for phenoptypic identification. Four coffee *Azospirillum* sp. isolates were grown in N-rich NFb broth and inoculated in coffee seeds, cv. Caturra (ca. 10^7 CFU seed^{-1}). The controls were non-inoculated seeds. We measured the percentage of germination at day 30, and after 180 d registered dry weight, N content and number of leaves.

The isolation frequencies of *Azospirillum* ranged from 20 to 40% in samples of rhizosphere, rhizoplane and inner tissues. The samples of coffee soils were acidic or slightly acidic. *Azospirillum* was more abundant in slightly acidic soils. On day 30, control seeds showed germination percentage of 42% while inoculated ones increased to 67–88%. Inoculation also induced increases in biomass and N-content (Table 1), depending on the *Azospirillum* isolate. As has been observed in inoculation experiments with other plants, different strains can show different growth-promotion activity. We only tested cv. Caturra, but other plant genotypes could show effects different to those reported in this work.

Table 1. Effect of inoculation of four *Azospirillum* isolates from coffee on growth and N-content of coffee plantlets, cv. Caturra [a, b].

	Dry weight (g)	Number of leaves	N-content (%)
Control	0.857 ± 0.118^1	6.9 ± 0.30^1	1.226 ± 0.043^1
CaBuap60	1.361 ± 0.110^2	8.3 ± 0.18^2	1.846 ± 0.056^2
CaBuap120	$1.617 \pm 0.045^{2,3}$	8.9 ± 0.30^2	1.962 ± 0.067^2
CaBuap290	$1.886 \pm 0.082^{3,4}$	$9.3 \pm 0.26^{2,3}$	2.279 ± 0.102^3
CaBuap38	$2.021 + 0.075^4$	$10.3 + 0.41^3$	$2.867 + 0.093^4$

[a] The measures were registered 180 days after inoculation.

[b] Mean of 10 plants $\pm$ standard error

THE EFFECT OF MINERAL AND BIOLOGICAL NITROGEN ON MICROBIOLOGICAL TRAITS OF SMONITZA AND MAIZE YIELD

D. Djuki, L. Mandi

The Faculty of Agronomy, University of Montenegro, Cacak, Serbia

This paper presents a study on the effect of inoculation of maize seed (NSSC-640) by asymbiotic nitrogen-fixing bacteria (*Azotobacter chroococcum* strain 84) and increasing rates of mineral nitrogen on quantitative composition of microorganisms (total number of bacteria, number of actionomycetes and azotobacter) in smonitza under maize and on yield.

The following fertilization treatments were studied: N_1PK (90:75:60 kg ha^{-1}); N_2PK (120:75:60 kg ha^{-1}); N_3PK (150:75:60 kg ha^{-1}), as well as the treatment with presowing inoculation of maize seed.

Microbiological analysis included the assessment of total number of microorganisms, actinomycetes and azotobacters in the rhizosphere and edaphosphere during maize vegetation. Total numbers of microorganisms were determined by growth on the medium for the total number with an appropriate amount (0.5 mL) of 10^{-6} soil dilution, numbers of actinomycetes - by growing on synthetic agar according to Krasil'nikov with 10^{-4} of soil dilution and azotobacters - by growing on Fyodorov's agar with 10^{-2} of soil dilution.

The study results showed that the numbers of microorganism groups were affected by the type and rate of fertilizers applied, as well as by the time and sampling zone. The application of *Azotobacter chroococcum*, strain 84, resulted in an increase of the total numbers of microorganisms, actinomycetes and azotobacters, especially in the rhizosphere soil at the onset, and even in the middle of maize vegetation. Lower rates of nitrogen fertilizers (90 and 120 kg ha^{-1}) led to a significant increase of total bacteria numbers, as well as an insignificant change in azotobacter numbers, whereas their high rate (150 kg ha^{-1}) had a depressive effect on the mentioned microorganisms, particularly in the edaphosphere of maize. In contrast, actinomycete numbers were not reduced even with this nitrogen treatment.

Under the studied agroecological conditions, the highest maize yield, but not economically justified one, was obtained with the highest nitrogen rate. Seed inoculation with *Azotobacter chroococcum*, strain 84, caused an insignificant rise in maize yield, which can be associated with acid reaction of the soil studied and slower release of nitrogen accumulated in their cells.

CAB *International* 2002 *Nitrogen Fixation: Global Perspectives* (eds. T Finan, M O'Brian, D Layzell, K Vessey, W Newton)

EFFECT OF INCREASING PHOSPHORUS CONCENTRATIONS ON THE GROWTH AND PHYSIOLOGY OF *RHIZOBIUM*-INOCULATED AND NITROGEN-SUPPLIED COWPEA (*VIGNA UNGUICULATA* L. WALP) PLANTS

M.L. Izaguirre-Mayoral, O. Carballo

Instituto Venezolano de Investigaciones Científicas, Centro de Microbiología y Biología Celular, Laboratorio de Biotecnología y Virología Vegetal, Apdo 21827, Caracas 1020-A, Venezuela

Cowpea is an important grain crop in Venezuela. However, local varieties do not respond to *Rhizobium* inoculation. Nodulated plants always produce a smaller biomass and lower yields. Thus, farmers do not rely upon inoculation for better yields, even though the price of fertilizers has increased over the years. Since cowpea crops are planted mainly in savanna soils, characterized by low phosphorus (P) content, it could be hypothesized that P is the limiting factor for the physiological performance of cowpea plants. The aim of this investigation was, therefore, to analyze the effect of increasing P concentrations on the growth of *Rhizobium*-inoculated and nitrogen-supplied cowpea plants. For this purpose, plants inoculated with *Bradyrhizobium* I-125 (R^+) or provided with 15 mM nitrogen (N^+) were grown in nutrient solutions supplied with the following P concentrations (Pc): 0.05 mM (P1), 0.1 mM (P2), 0.25 mM (P3), 0.5 mM (P4), 1 mM (P5), 2 mM (P6), 3 mM (P7) and 4 mM (P8). At the flowering stage, plants were divided into roots, shoots, aerial and nodule biomass. Concentration of chlorophyll, N, P, total reducing sugars (TRS) and N-compounds were determined in the individual plant components following standard protocols. The experiment was conducted under controlled conditions and the pH of the solution was maintained at 6.2. Increasing Pc up to P6 enhanced the aerial mass of R^+ and N^+ plants but reduced their root mass regardless of the mode of N uptake. Concomitantly, increasing P up to P4 enhanced the nodule mass in R^+ plants. The ureides in R^+ plants as well as the nitrate concentration in N^+ plants remained constant between P1 and P6. Nevertheless, the aerial mass in R^+ plants was always smaller than that of N^+ plants in spite of the higher TRS, α-amino-N, chlorophyll and P concentrations in the leaf tissues. Metabolic constraints for ureide degradation in leaves seem to be the mechanism underlying the poor response of plants to rhizobial inoculation at high Pc. The P7 and P8 concentrations proved to be toxic for R^+ and N^+ plants. From present results we may conclude that P is not the key factor for the improvement of the symbiotic process and growth of cowpea plants under savanna conditions.

POTENTIAL OF NITROGEN-FIXING SYMBIOSIS SYSTEMS FOR REVEGETATION STRATEGIES IN MEDITERRANEAN ENVIRONMENTAL CONDITIONS

J. Cleyet-Marel, O. Domergue, L. Maure

Laboratoire des Symbioses Tropicales et Méditerranéennes INRA/IRD/CIRAD/AGRO-M, Montpellier, France

Anthropogenic degradation activities (overgrazing, non-regulated cultivation techniques, deforestation, quarry exploitation, etc.), together with a long dry and hot summer, with scarce, erratic, but torrential rainfalls, is a major threat to the sustainability of Mediterranean ecosystems. The removal of native trees and soil disturbance result in a lack of sufficient resident soil microflora and in many instances a number of nutrients, including N, P and K are not present in sufficient quantities to promote an acceptable plant growth.

Revegetation of degraded ecosystems requires selection of suitable plant species and adaptation of the plants to the unusual soil conditions. Therefore, nitrogen-fixing legumes are key components of the natural succession because their associated rhizobial symbioses constitute a source of N input to the ecosystem. Several Mediterranean native legumes have been selected from several areas in the South of France, some of them like *Colutea, Coronilla, Genista, Medicago* and *Spartium* belong to the shrub community and other like *Astragalus, Dorycnium, Hippocrepis, Lotus, Medicago, Onobrychis* and *Ononis* are classical herbaceous legumes.

Seeds of these plant species were collected and the rhizobial partners were isolated, characterized and the most efficient strains were selected for inoculum preparation. Legume plant seeds were germinated before seedlings were transferred in plastic containers used in forestry practice. All seedlings were inoculated with a previously selected rhizobial culture and allowed to grow for 6–9 months in an experimental nursery.

Plants were transferred in the chosen degraded ecosystem and experimental variables including survival rates, plant growth, seed production were tested. The most interesting legume species for reclamation and rehabilitation programs were identified to restore sustainable ecosystems.

DIVERSITY OF POPULATIONS OF RHIZOBIA THAT NODULATE *PHASEOLUS VULGARIS* IN FRANCE

N. Amarger, M. Bours, D. Depoil, K. Groud, F. Revoy

Laboratoire de Microbiologie des Sols, INRA, BP86510, 21065, Dijon, France

1. Introduction

Phaseolus vulgaris is native to the Americas and was imported to Western Europe in the XVI century. The crop, being considered weak in nitrogen fixation and with a highly variable response to inoculation, usually receives N-fertilizer. *Phaseolus* rhizobia belong to several species and several biovars which possess different symbiotic properties. The strains that belong to the biovar (bv.) *phaseoli* of *R. leguminosarum*, *R. etli*, *R. gallicum*, *R. giardinii*, have a narrow host-range, they nodulate only *Phaseolus* spp. Strains of *R. gallicum* bv. *gallicum*, *R. giardinii* bv. *giardinii*. *R. tropici* have wider host ranges, not fully described yet, but that include at least *Leucaena* spp. in addition to *Phaseolus* spp. *R. giardinii* isolates, whether they belong to bv. *phaseoli* or bv. *giardinii*, induce ineffective nodules. The objectives of the study were to investigate the diversity of rhizobia at the origin of nodules in field grown *Phaseolus*, and to see whether this diversity fluctuates with geographical situation and plant genotype and could have agricultural implications.

2. Material and Methods

Populations of 40 to 105 rhizobia were isolated from nodules of field grown common beans collected from seven locations in bean production areas in France. There were one or two years of sampling and one or several bean varieties per location. Nodule isolates were characterized by RFLP analysis of PCR amplified DNA fragments: 16S rDNA (Laguerre *et al.* 1994), 16S-23S ITS (Laguerre *et al.* 1996) and *nif* or *nod* gene (Laguerre *et al.* 2001) to determine the species, the intraspecies polymorphism, and the biovar, respectively.

3. Results and Discussion

The populations were found to be diverse at the species and/or at the intraspecies (biovar and 16S-23S ITS type) levels. Although some populations were composed of a single species, *R. leguminosarum* bv. *phaseoli,* two different species, *R. leguminosarum* bv. *phaseoli* and *R. giardinii*, *R. etli* or *R. tropici* were detected in the other populations but one. This latter population was dominated by *R. giardinii* bv. *giardinii* but a few isolates of *R. etli*, *R. gallicum* bv. *gallicum*, *R. gallicum* bv. *phaseoli* and *R. giardinii* bv. *phaseoli* were also identified. The composition varied among locations. Within locations, it could vary between years and between bean varieties. The proportion of *R. giardinii*, which has a Fix⁻ phenotype and was present in half of the locations, could be rather high, which should have a negative effect on the amount of nitrogen fixed by the crop. The variability observed in the composition of the populations between years and plant genoypes shows that only a part of the rhizobial diversity present in a soil is revealed by sampling a given crop at a given time. This variability could be at the origin, of at least a part, of the variability observed in the crop response to inoculation.

4. References

Laguerre G *et al.* (1994) Appl. Environ. Microbiol. 60, 56-63
Laguerre G *et al.* (1996) Appl. Environ. Microbiol. 62, 2029-2036
Laguerre G *et al.* (2001) Microbiol. 147, 981-993

DIFFERENT RESPONSES TO OSMOTIC AND SALINE STRESS OF *BRADYRHIZOBIUM JAPONICUM* E109 IN CULTURE MEDIUM AND SOIL MICROCOSMS

O.S. Correa, M.A. Soria, F. Pagliero, N.L. Kerber

Cátedra de Microbiología Agrícola, Facultad de Agronomía, Universidad de Buenos Aires, Av. San Martín 4453 (C1417DSE), Ciudad Autónoma de Buenos Aires, República Argentina

1. Introduction

Fast growing species of the *Rhizobiaceae* are, in general, more salt tolerant than slow growing species, such as soybean nodulating *Bradyrhizobium japonicum*. The aim of this work was to compare the response of *B. japonicum* E109 to different types of osmotic stress in exponential- or stationary-phase cells in liquid culture medium and soil microcosms.

2. Materials and Methods

B. japonicum E109 was obtained from the INTA Culture Collection. Shock stress experiments were carried out in HPM medium (g/L): K_2HPO_4 0.18; $MgSO_4.7H_2O$ 0.18; NaCl 0.12; $(NH_4)_2SO_4$ 0.39; glycerol 2.4; yeast extract 1.2. Exponential-phase cultures (3–4 x10^6 cells/mL) or stationary-phase cultures (1.5 10^9 cell/mL) were shocked by increasing the NaCl concentration up to 0.1 M or 0.4 M by addition of sterile 4 M NaCl. Soil microcosms were prepared in flat-bottomed glass vials (28 mm diameter x 62 mm length) containing 4 g of dried soil (electrical conductivity (EC) = 1.0 mmhos/cm). The vials with soil were sterilized by autoclaving and sterile distilled water was added to 30% or 65% of field capacity. The soil salinity was increased up to EC = 35 mmhos/cm by addition of sterile 4 M NaCl. Microcosms were inoculated with exponential- or stationary-phase bacteria and incubated at 30ºC. CFU counts in samples withdrawn from cultures and microcosms were done in Yeast Mannitol Agar (YMA), with mannitol 1 g/L, incubated for 6 days at 30ºC.

3. Results and Discussion

Survival of *B. japonicum* in HPM medium depended on its physiological state at the time of shock: stationary-phase cells were more resistant to NaCl than exponentially growing cells. Exponential-phase cells shocked with 0.1 or 0.4 M NaCl began to die after shock, whereas stationary-phase cells shocked with 0.1 M had a better survival than control cells, and cells shocked with 0.4 M NaCl showed a decrease in cell count similar to control cells. When these concentrations of NaCl are present in culture media from inoculation, growth is delayed or inhibited.

The number of exponential-phase cells counts increased from 2.2x10^6 cells/g soil up to 2x10^8 cells/g during the few days following inoculation. After 25 days, cell counts were 1.2–3.0x10^8 cell/g soil either at 30% or 65% field capacity. In NaCl-supplemented microcosms inoculated with exponential cells microcosms, cell counts decreased to 7.0x10^3 and 3.9x10^2 cells/g for 65% or 30% field capacity respectively. In microcosms without NaCl addition and inoculated with stationary-phase bacteria, cell counts did not change, irrespective of field capacity; but when NaCl was added, cell number decreased, and this decay was more pronounced at 30% field capacity.

These results suggest that different kinds of mechanisms of salt tolerance exist in *B. japonicum*. Some of them are needed to grow in the presence of salt and others allow for survival upon a sudden increase in NaCl or drought, but their operation is influenced by the environment. Most of the techniques used for physiological, biochemical and genetic studies in bacteria were designed for liquid media and many cannot be reproduced in microcosms. However, major differences in cell behavior in both environments are observed, and since microcosms would better resemble natural conditions, it is worthwhile to analyze also in microcosms the responses to stress.

EFFECT OF TWO VINEYARD GROUND COVERS ON *RHIZOBIUM LEGUMINOSARUM* BV. *TRIFOLII* POPULATIONS

O. Domergue, J. Colette, V. Chareyron, L. Maure, J. Cleyet-Marel

Laboratoire des Symbioses Tropicales et Méditerranéennes INRA/IRD/CIRAD/AGRO-M, Montpellier, France

During the past 50 years, herbicides use in vineyard has induced major disturbances in soil with a dramatic decrease in microbial activity and organic matter content. Actually, to reduce erosion and to increase organic matter rate in vineyard soils, ground cover with herbaceous plants is more and more practiced. Among the plant species available, subterraneum clover (*Trifolium subterraneum*) with its nitrogen-fixing capacity is a very useful legume species as biological soil activator. However, the absence of *Rhizobium leguminosarum* bv. *trifolii,* the subterraneum clover symbiotic partner, could limit the plant development. Therefore, the relative abundance of rhizobial populations associated to subterraneum clover in soils from three different plots was evaluated in three different situations: (i) soil without plant cover; (ii) plots with perennial grass cover during three years; (iii) plots with subterraneum clover during three years.

The determination of microbial diversity based on genetic and phenotypic characterization of isolated and cultivated strains after trapping on *Trifolium subterraneum* as a plant host was performed by PCR restriction analysis of 16S-IGS and *nifKD* genes. Marked differences were observed between rhizobial isolates trapped from soil without grass cover and isolates from soil cropped with perennial grass or with subterraneum clover.

PHYTOLOGICAL ASPECTS OF HEAVY METAL TOXICTY IN CLOVER (*TRIFOLIUM ALEXANDRINUM*)

T.M. El-Mokadem, H.S. El-Toukhy, K.I.R. Badawy

Bot. Dep. Women's College Ain Shams Univ., Heliopolis, Cairo, Egypt
Pollution Unit, Desert Research Center (DRC), Cairo, Egypt

Studies were conducted to present evidence that nodulation, biomass, chlorophyll and metal content of a forage legume (clover) may serve for ecotoxicological evaluation of contaminated areas. Substances affecting the macro- and/or microsymbionts, such as certain heavy metals were examined. The experiments were carried out as pot experiments using loamy sand soils. Unpolluted soils were first mixed with four levels of soluble salts of four tested heavy metals (Pb 50–450 mg/kg, Cd 25–200 mg/kg, Zn and Cu 250–1000 mg/kg), appropriate controls were run in parallel. Clover (*Trifolium alexandrinum* cv. Compound sids) was planted in each treated soil and subsequently inoculated with *Rhizobium leguminosarum* bv. *trifolii ARC300*. Plants were collected at 48, 80 and 140 days of growth and analyzed for nodulation, dry matter, leaf area, chlorophyll content and metal uptake.

Applying trace element resulted in a significant (P<0.05) decrease in number of nodules and nitrogen content of plants. Nodule formation was inhibited by 60, 56, 48 and 21% at higher concentrations of Pb, Cd, Cu and Zn, respectively, in plants grown for 140 days. A closer look at the nodule tissue of heavy metal treated clover was less than that of the control nodule. In the central zone, most of the infected cells contain less numbers of fully developed bacteroids. The reduction in the number of infected cells within nodules suggested that these nodules fixed little nitrogen. (McGrath, Brookes 1988). This effect of nodulation was reflected in plant nitrogen content. The comparative toxicity of the tested metals to symbiotic N_2-fixation indicates that the overall order of decreasing toxicity is Cd > Cu > Zn > Pb. Heavy metal accumulated in various parts of plants resulted in retardation of growth and reduction of chlorophyll content, which expressed itself as the visual symptoms of chlorosis. The reduction in chlorophyll content in turn, at least partly, would lead to a decrease in shoot length and biomass, this could explain the positive correlation between chlorophyll content and biomass as well as metal toxicity. The mechanisms responsible for heavy metal phytotoxicity might be related to membrane impairment (Kabata-Pendias, Pendias 1984). Trends in metal concentration in clover as a function of metal content in soil was linear; uptake of Cu, Pb and Cd by tops does not occur in a linear concentrations of the metal in soil; Zn is mobile inside the plant and can be highly translocated from root to shoot parts, causing a significant increase in Zn concentrations in clover above ground part. This means that clover is especially useful where there is a need for a sensitive indicator of differences between sites or sampling occasions. In conclusion, clover plants represent a wide array of behavior toward metal toxicity symptoms.

References

Kabata-Pendias, Pendias (1984) CRC Press Inc., Boca Raton, FL
Mc-Grath, Brookes (1988) Soil Bio. Biochem. 20, 4, 415-424

THE DELAY EFFECT OF RHIZOBITOXINE PRODUCTION ON NODULATION COMPETITIVENESS OF *BRADYRHIZOBIUM ELKANII* USDA94 IN SIRATRO

S. Okazaki[1], H. Ezura[2], K. Yuhashi[1], K. Minamisawa[1]

[1]Institute of Genetic Ecology, Tohoku University, Sendai 980-8577, Japan
[2]Plant Biotechnology Institute, Nishi-Ibaraki, 319-0292, Japan

1. Introduction

Bacterial production of rhizobitoxine, an ethylene synthesis inhibitor (Yasuta *et al.* 1999), plays a role in competitiveness of *Bradyrhizobium elkanii* for siratro (*Macroptilium atropurpureum*) nodulation (Yuhashi *et al.* 2000; Duodu *et al.* 1999) since siratro nodulation is partially restricted by endogenous ethylene (Nukui *et al.* 2000). However, the mechanism of increasing nodulation competitiveness by rhizobitoxine production is still unclear. Here, we report the competitive nodulation, kinetics of nodule occupancy and ethylene evolution rate in siratro roots using *B. elkanii* USDA94 and its rhizobitoxine production-deficient mutant RTS2.

2. Results and Discussion

Competitive nodulation enhancement of rhizobitoxine was reconfirmed in various inoculum ratio of USDA94 and RTS2. Time course of nodulation showed that nodule occupancy between USDA94 and the mutant RTS2 was almost the same before 13 days after inoculation, however the occupancy of the USDA94 continuously increased after 13 days. The results suggest a delay effect of rhizobitoxine production on competitiveness of USDA94 in siratro nodulation. Ethylene evolution rate in siratro roots was at a lower level in single-inoculation with the USDA94, while it was higher in inoculation with the mutant during nodulation. Ethylene evolution rate in roots co-inoculated with both the wild type and the mutant was decreased after 13 days after inoculation. The change in ethylene evolution rate in the co-inoculated roots might reflect an increase in nodule occupancy of the rhizobitoxine producer.

3. References

Duodu *et al.* (1999) Mol. Plant-Microbe Interact. 12, 1082-1089
Nukui *et al.* (2000) Plant Cell Physiol. 41, 893-897
Yasuta *et al.* (1999) Appl. Environ. Microbiol. 65, 849-852
Yuhashi *et al.* (2000) Appl. Environ. Microbiol. 66, 2658-2663

EFFECT OF ANAEROBIOSIS ON SYMBIOTIC EFFICIENCY OF THREE STRAINS OF *SINORHIZOBIUM MELILOTI*

D. Prévost, G. Lévesque, A. Bertrand

Soils and Crops Research and Development Centre, Agriculture and Agri-Food Canada, 2560 Hochelaga Blvd, Sainte-Foy, QC, Canada G1V 2J3

1. Introduction

During winter, climatic conditions in temperate areas can create an ice layer over agricultural soils. This situation causes an anaerobic stress that may affect winter survival and regrowth of plants (Andrews, Pomeroy 1990). The production of toxic metabolites and the reduction of C and N reserves might be responsible for a poor winter survival of perennial plant species. For instance, the freezing tolerance of alfalfa cultivars is closely related to their capacity to accumulate the raffinose family oligosaccharides (Castonguay *et al.* 1995). Furthermore, the strain of rhizobium may play an important role in the adaptation of legumes to stresses (Layzell *et al.* 1984). Our aim was to evaluate the effect of anaerobiosis during winter on the symbiotic efficiency of three strains of *S. meliloti*.

2. Material and Methods

Three strains of *S. meliloti* with the same symbiotic efficiency were used: two commercial (BALSAC and NRG-34) and one indigenous (Rm-1521). Alfalfa was grown in a mixture of topsoil/peatmoss (11:1 v/v) for three months in a greenhouse (two cuttings). Potted plants were then acclimated in the field during fall. In late fall, half of the plants were subjected to a progressively developing anaerobic stress by enclosing them in gas-tight bags and exposing them to simulated winter conditions in an unheated greenhouse. At three sampling dates, gases were sampled and gas-tight bags were removed. Plants were then placed in growth chambers at optimal temperature for a two-weeks regrowth during which nitrogenase activity was determined. At the end of regrowth, nodulation indexes and shoot dry weights were measured. Populations of rhizobia were determined by colony hybridization technique at the beginning and at the end of the simulated winter period.

3. Results and Discussion

Shoot and root dry weight, nitrogenase activity and nodulation index of alfalfa were considerably reduced after 78 days of anaerobiosis under winter conditions. None of the rhizobial strain allowed an optimal regrowth after the extreme period of anaerobic stress of 101 days. After the first 78 days of winter, plants nodulated by strain NRG-34 maintained the highest nodulation index and shoot dry weight under anaerobiosis (1% O_2) and this strain was also the most efficient under control conditions. Anaerobic stress did not affect survival of rhizobia. Populations decreased at the end of the overwintering period but each strain maintained a level of 10^5 cells/g soil. Strain NRG-34 was shown to be adapted for growth and nodulation at low temperatures (Rice *et al.* 1995) and our results show that this strain can also preserve its symbiotic efficiency after overwintering. It is then possible to select rhizobia for improving regrowth of alfalfa exposed to a wide range of winter conditions.

4. References

Andrews CJ, Pomeroy MK (1990) In Jackson MB, Davies DD, Lambers H (eds) Plant Life Under Oxygen Deprivation, pp. 85-89, SPS Academic Publishing, The Hague
Castonguay Y (1995) Crop Sci. 35, 509-516
Layzell DB *et al.* (1984) Can. J. Bot. 62, 965-971
Rice WA *et al.* (1995) Plant Soil 170, 351-358

A MUTATION IN THE *nolL* GENE OF *RHIZOBIUM ETLI* AFFECTS COMPETITIVENESS AND BACTEROID DEVELOPMENT IN *PHASEOLUS VULGARIS*

A. Corvera, D. Romero

Programa de Genética Molecular de Plásmidos Bacterianos, Centro de Investigación sobre Fijación de Nitrógeno, UNAM, Cuernavaca, México

1. Introduction

The *nolL* gene product is responsible for the acetylation of the fucosyl residue of the Nod factors of *Rhizobium* sp. NGR234, *R. etli* and *Mesorhizobium loti*. Our previous work has shown that a *R. etli nolL::*Km strain (CFNX289) displays a delayed nodulation phenotype on some bean cultivars and in *Vigna umbellata*, but nodulates *Phaseolus vulgaris* cv. Negro Jamapa with the same kinetics as wild-type CFN42.

2. Results and Discussion

Recently we have found, using more sensitive assays based on competition, that CFNX289 occupies only 10% of the nodules formed in *P. vulgaris* cv. Negro Jamapa roots when co-inoculated with CFN42. The reduced competitive ability of CFNX289 strain was partially restored (31% occupancy) upon introduction of a wild-type *nolL* gene. These results suggest the existence of subtle effects on the infection process, undetectable by measurements of nodulation kinetics. Transmission electron microscopy analysis of the nodules formed on *P. vulgaris* cv. Negro Jamapa roots revealed that a *nolL::*Km mutation also affects bacteroid development. Strain CFNX289 infects plant cells later than CFN42; additionally, the bacteroids formed have a reduced accumulation of PHB, and there are a larger number of bacteroids per symbiosome in cells infected by CFNX289 than in cells infected by CFN42. These alterations provoke differences in symbiotic effectiveness; *P. vulgaris* cv. Negro Jamapa plants inoculated with CFNX289 displayed reduced nodule dry weight and total nitrogenase activity (evaluated by ARA) at 30 days after inoculation, compared with plants inoculated with the wild type strain. The alterations found here are different from those previously described for *Bradyrhizobium-Lupinus* interaction. Our results indicate that the *nolL* gene participates not only in the structuring of the Nod factor, but also in bacteroid development.

3. References

Berck *et al.* (1999) J. Bacteriol. 181, 957-964
Corvera *et al.* (1999) Mol. Plant-Microbe Interact. 12, 236-246
Lotocka *et al.* (2000) Acta Biol. Cracovien. Ser. Bot. 42, 155-163

4. Acknowledgements

We thank Javier Rivera, Eduardo Jiménez and Mark West for their technical assistance. This work was partially supported by Grant IN202599 from DGAPA-UNAM.

TEMPORAL CHROMOSOMAL AND PLASMIDIC POPULATION GENETIC STRUCTURE OF *RHIZOBIUM ETLI* BIOVAR *PHASEOLI*

C. Silva[1], L.E. Eguiarte[1], E. Martínez-Romero[2], V. Souza[1]

[1]Departamento de Ecología Evolutiva, Instituto de Ecología
[2]Centro de Investigación sobre Fijación de Nitrógeno,
  Universidad Nacional Autónoma de México, México

1. Introduction

The frequency and significance of recombination in the evolution of bacteria is a subject of intense debate. *Rhizobium* population genetic studies show that there is a wide range of genetic structures in *Rhizobium* species, spanning from strictly clonal to essentially panmictic. We recently analyzed the genetic structure of a local population of *R. etli* bv. *phaseoli* (Silva *et al.* 1999). MLEE data revealed two genetic groups, linkage disequilibrium analysis showed equilibrium within each group, when only the different genotypes were taken into account, indicating a reticulated and epidemic genetic structure (Maynard Smith *et al.* 1993). The lateral transfer of the *pSym* in natural populations of *Rhizobium* is shown by the disparity of the groups formed by the chromosomal and the symbiotic loci. It is of interest to know if transfer of plasmids in the field involves only *pSym* or if the other (cryptic) plasmids are also exchanged. The present study employed a population genetics framework to elucidate the genetic structure of a local *R. etli* bv. *phaseoli* population in Puebla, Mexico, both at the chromosomal and plasmidic level over three consecutive years.

2. Procedures, Results and Discussion

Chromosomal variation was assessed by multilocus enzyme electrophoresis. Plasmid content was visualized via a modified Eckhardt procedure. Clustering from a matrix of pairwise mismatched distances was performed by the UPGMA method, both for the chromosomal variation and the plasmid content. The degree of non-random association of chromosomal alleles or plasmids was estimated from the V_o/V_e ratio. A high genetic diversity (H=0.501) was detected and among the 126 isolates were 43 distinctive multilocus genotypes (ETs). Fifty-six percent of the isolates belonged to three ETs, indicating a high degree of genotype dominance. Of 33 different plasmid profiles, three were found in 49% of the isolates, indicating also a dominant plasmidic structure. The MLEE and plasmid profile dendrograms grouped the isolates into the same two genetic divisions. None of the ETs or plasmid profiles was shared by the divisions. Within the genetic divisions, isolates of the same ET could have different plasmid profiles. Also, a plasmid profile could be found in different ETs, showing lateral transfer of plasmids within divisions. The linkage disequilibrium analysis showed the random association of chromosomal alleles within genetic groups, but not for the entire population. The linkage disequilibrium analyses confirmed the random association of plasmids both within genetic groups and for the entire population, indicating enough plasmid flow to break the linkage disequilibrium, although exclusive combinations exist for either chromosomal division. These results point toward a reticulated and epidemic genetic structure at the chromosomal level, and an epidemic plasmidic structure for this *R. etli* bv. *phaseoli* population.

4. References

Maynard Smith J *et al.* (1993) Proc. Natl. Acad. Sci. USA 90, 4384-4388
Silva C *et al.* (1999) Mol. Ecol. 8, 277-287

DIVERSITY OF *BRADYRHIZOBIUM* NODULATING THE COWPEA (*VIGNA UNGUICULATA* (L.) WALP.) IN SENEGAL (WEST AFRICA)

T. Wade[1,2], I.N. Doye[1,2], S. Braconnier[3], M. Neyra[1]

[1]IRD, Dakar, Sénégal
[2]Departement de Biologie Végétale, UCAD, Dakar, Sénégal
[3]CERAAS, Thiès Escale, Sénégal

1. Introduction

Cowpea (*Vigna unguiculata* (L.) Walp.) is a one of the older plants cultivated by man. In Senegal it is widely cropped and it plays an important role in human nutrition and the economy. The cultivated areas of this legume span the country, with 93% in the North and the Center-North, and 7% in the South. They are situated in sahelian to guineo-soudanian ecological regions with 300 to 800 mm rainfall per year, respectively. In this work, we present a study of diversity of indigenous *Bradyrhizobium* nodulating the cowpea.

2. Materials and Methods

Nodules were sampled from 41 sites covering a span of 600 km from the North to the Southeast of Senegal. The diversity of rhizobial strains was analyzed directly on DNA extracted from crushed nodules (Dupuy *et al.* 1992) by PCR-RFLP and the sequence analysis of 16S-23S rDNA IGS (Normand *et al.* 1996; Ponsonnet, Nesme 1994; Willems *et al.* 2001).

3. Results and Discussion

Total DNA was extracted and amplified from 202 crushed nodules. All yielded single IGS PCR products ranging from 910 pb to 1035 pb. After digestion with restriction enzymes *Hae*III and *Msp*I, 202 profiles were organized into nine distinct restriction patterns, named IGS types I to IX.

The IGS sequence analysis of the nine patterns showed that the strains belong to the *Bradyrhizobium* genus. IGS types II, III, IV and VIII are closely related to some strains isolated from *Faidherbia albida* in Senegal (Dupuy *et al.* 1992). IGS types III and IV could be grouped within *Bradyrhizobium* genospecies VII of Willems *et al.* (2001), corresponding to *Bradyrhizobium japonicum* species. IGS type II and VIII showed more than 96% similarity with genospecies IV. The five other IGS types, with lower IGS similarity, could correspond to new genospecies.

The IGS type I was found in 175 nodules collected through all prospected sites. The IGS types II, III and VIII were found in more than one site, located in the Center-North and the Southeast for the IGS type II and VIII, and in the Center-North and the North for the IGS type III. The five other IGS types were revealed only once, in sites grouped principally in the Center-North. However, due to the low frequency at which strains from IGS types II to IX are found, no clear relationship could be determined between the geographical origins of strains and their grouping by IGS sequence analysis.

4. References

Dupuy *et al.* (1992) Appl. Environ. Microbiol. 58, 2415-2419
Normand *et al.* (1996) Mol. Microbial Ecol. Manual 3.4.5, 1-12
Ponsonnet, Nesme (1994) Arch. Microbiol. 161, 300-309
Willems *et al.* (2001) Int. J. Syst. Microbiol. 51, 623-632

CONSTITUTIVE EXPRESSION OF A NODULE ENHANCED MALATE DEHYDROGENASE GENE IN ALFALFA ALTERS ORGANIC ACID SYNTHESIS AND IMPROVES GROWTH IN ALUMINUM-CONTAINING ACID CULTURE

M. Tesfaye[1], D.L. Allan[2], C.P. Vance[3,4], D.A. Samac[1,4]

[1]Dept Plant Pathology; [2]Soil, Water and Climate
[3]Agronomy and Plant Genetics; [4]Plant Science Research Unit, USDA-ARS
  Univ. of Minnesota, St. Paul, MN 55108, USA

1. Introduction

Alfalfa is very sensitive to aluminum (Al) toxicity in acid soils, which make up about 40% of the world's arable land. Addition of organic acids to nutrient solutions alleviates phytotoxic Al effects. Malate dehydrogenase catalyzes the reversible reduction of oxaloacetate to malate. A unique form of malate dehydrogenase (neMDH), expressed 5–15-fold higher in alfalfa nodules, was identified previously. This non-photosynthetic isoform plays a crucial role in providing malate to bacteroids to fix N_2 and by the plant for assimilation of N_2. This research aimed to constitutively over-express neMDH enzyme in alfalfa, to test whether such transgenic plants had enhanced organic acid synthesis and excretion, and to evaluate if selected transgenic plants had increased Al tolerance.

2. Procedures and Results

Leaf pieces from a highly regenerable alfalfa cultivar Regen-SY were transformed with *A. tumefaciens* that contains a chimeric gene consisting of a full-length neMDH eDNA under the CaMV 35S promoter. Transgene integration and expression were evaluated by DNA-blot, PCR, RNA-blot, RT-PCR, Western blot and specific enzyme analysis. PCR analysis and DNA blot hybridization confirmed that at least one copy of the transgene was present in the transgenic lines. Transgenic lines showed up to 1.6-fold increase in specific MDH enzyme activity. Gel-blot analysis of total RNA clearly showed increased levels of neMDH transcripts in both root-tips and leaf samples of transgenic lines, compared to untransformed RegenSY plants. Amounts of neMDH polypeptides were also considerably higher in root-tip samples of selected neMDH transgenic lines than in root-tips of untransformed plants. Transgenic alfalfa showed reduced biomass accumulation during the first year of growth in field plots of neutral soil at pH 7.3. When plants were grown in 50 µM Al-containing acid culture at pH 4.3 however, transgenic plants had at least 3.4-fold greater root elongation than untransformed plants, indicating the set plants may be suited to cultivation in Al-containing acid soils. Although we have no direct evidence for the mechanism of Al tolerance by transgenic alfalfa, the enhanced Al tolerance by transgenic lines and the poor growth of un-transformed Regen-SY plants in Al-containing hydroponic solution or soil culture coincided with the organic acid synthesis and exudation patterns of these plants. Transgenic plants showed up to 5-fold higher leaf and root concentrations and 7-fold root exudation of oxalate, citrate, succinate, malate and acetate than Regen SY plants. One transgenic alfalfa over-expressing organic acids was crossed with GA-AT, an acid soil tolerant alfalfa previously selected in Georgia, USA. Cosegregation of the neMDH transgene and enhanced neMDH mRNA expression was observed in Fl crosses. Transgenic alfalfa will be evaluated for nodulation and biological N-fixation ability in acid soils.

3. References

Bouton JH (1996) Crop. Sci. 31, 198-200
Kochian LV (1995) Annu. Rev. Plant. Physiol. Plant. Mol. Biol. 46, 237-260
Miller *et al.* (1998) Plant J. 15, 173-184

COMMON MECHANISMS FOR THE INTRACELLULAR SURVIVAL OF BACTERIA WITHIN PLANTS AND ANIMALS

L. Barra[1,2], G.P. Ferguson[1], K. LeVier[1], G.R. Campbell[1], C. Blanco[2], G.C. Walker[1]

[1]Dept of Biology, Massachusetts Institute of Technology, Cambridge, MA 02139, USA
[2]Lab UMR CNRS-6026, Université de Rennes-1, 35042 Rennes Cedex, France

1. Introduction

Sinorhizobium meliloti can establish a symbiotic relationship with alfalfa plants. Once inside the plant cells, *S. meliloti* differentiate into bacteroids and persist in a membrane-bound acidic compartment. The bacteria need to adapt to a variety of stresses during this association. The symbiosis can be disrupted at the stage of the initial root hair colonization, infection thread formation, release into the plant cell, and finally differentiation of the bacteria into nitrogen-fixing bacteroids.

S. meliloti mutants have been isolated that are affected at different stages in the alfalfa symbiosis (Glazebrook *et al.* 1993). Some of these mutants were released into the plant cell but were impaired in the differentiation into bacteroids. One of these mutants lacks BacA activity, a 420 amino acid, putative inner membrane protein, which is apparently essential for *S. meliloti* bacteroid development and/or intracellular survival. A homolog of the *S. meliloti* BacA protein was found to be essential for the intracellular survival of the mammalian pathogen, *Brucella abortus*, which is phylogenetically close to *S. meliloti*. Both *bacA* mutants exhibit increased resistance to bleomycin, suggesting that BacA could take part in the uptake of bleomycin (Ichige *et al.* 1997; Glazebrook *et al.* 1993).

2. Results and Discussion

New *bacA* phenotypes have been highlighted by recent physiological studies (Ferguson, unpublished) and genetic analysis (LeVier 2000), suggesting that BacA has the ability to carry out more than one function. BacA could be involved in host signal, or compound importer or exporter. These signal or compound transporters essential to the survival within the cell could not be transduced to their target in a BacA- background, leading to the lysis or early senescence of the bacteria. Recent data, using membrane destabilizing agents, suggest that BacA could also be involved in the cell envelope homeostasis. We are currently investigating, by physiological and genetic analysis, the mechanisms and pathways by which BacA could modulate the intracellular survival.

3. References

Glazebrook JIA *et al.* (1993) Genes Dev. 7, 1485-97
Ichige A *et al.* (1997) J. Bacteriol. 179, 209-16
LeVier K *et al.* (2000) Science 287, 2492-3

4. Acknowledgements

This work was supported by Public Health Services Grant GM31030 from the National Institute of Health (NIH) to GCW. LB acknowledges the receipt of a doctoral fellowship from the "Region Bretagne - Conseil Regional de Bretagne - France".

APPLICATION OF TREE LEGUME RHIZOBIA FOR REFORESTATION PROGRAMME IN THAILAND: I. SELECTION AND MANAGEMENT OF RHIZOBIA FOR TREE LEGUMES

S. Kotepong[1], A. Nuntagij[1], S. Jitacksorn[1], N. Teaumroong[2], N. Boonkerd[2]

[1]Soil Microbiology Research Group, Division of Soil Science, Department of Agriculture, Bangkok, Thailand
[2]School of Biotechnology, Suranaree University of Technology, Nakhon Ratchasima, Thailand

Leguminous trees such as *Acacia*, *Pterocartpus*, and *Leuceana* have been recommended for reforesting in denuded and degraded lands in Thailand. These leguminous trees successfully compete with native grasses and adapted to adverse conditions, i.e. drought, fires, acid, saline and low fertility soils. The advantage of leguminous trees is their association with *Rhizobium*, a symbiotic microorganism. Since the legume-rhizobial symbiosis can fix atmospheric nitrogen, thus nitrogen status of soil is improved even though no chemical nitrogen fertilizer is applied. Root nodules of five tropical leguminous trees (*Acacia mangium, Acacia auricultiformis, Pterocarpus macrocarpus, Xylia kerii* and *Millettia leucantha*) were collected from reforested sites around the country. Five trees were examined and nodules were collected for each of the species at each sites. About 300 rhizobial strains were isolated, purified and subcultured. The rhizobial strains were inoculated into their host plant and evaluated for their ability to increase host plant biomass, nodule forming and acetylene reduction assay. The data indicated that effective rhizobial strain resulted in higher plant biomass than ineffective and uninoculated treatments. Further study is carried out with the ten highest efficiency strains per each host plant.

NITROGEN FIXATION AND SALT TOLERANCE OF SOME NATURALLY-GROWING HERB LEGUMES IN EGYPT

H.H. Zahran, A.M. Atteya, E.A. Al-Sherif

Department of Botany, Faculty of Science, Beni-Suef 62511, Egypt

1. Introduction

Leguminous plants have played an important role in human development and civilizations. The legumes increase the level of available soil nitrogen and they are adapted to wide range of environmental conditions in arid regions (Miller, Jastrow 1996; Zaharn 1999). The naturally-growing herb legumes attracted the attention of scientists in the last decade, they are nodulated by diverse strains of rhizobia and actively-fixing nitrogen at levels comparable to that of crop legumes (Zahran 1998, 2001). The wild herb legumes are adapted to severe conditions such as salinity and drought encountered in the arid regions (Zahran 1999).

2. Materials and Methods

Nine species of naturally-growing herb legumes were collected from different localities at Beni-Suef Governorate. The legumes are *Melilotus indicus, Melilotus siculus, Vicia sativa, V. monantha, Medicago polymorpha, M. intertexta, Trifolium resupinatum, Trigonella hamosa* and *Lotus corniculatus*. The nitrogen-fixing ability of root nodules collected from these legumes was determined by the acetylene reduction technique (Hardy *et al.* 1973). Protein content in plants was determined using the traditional method of Lowry *et al.* (1951). The contents of N, H and C were determined in dry matter from these legumes by using an elemental analyzer. Mineral content of plants and soils was also determined by using atomic absorption spectrophotometry. The anions Cl⁻, SO_4 and HCO_3) were measured in soil extract following the standard techniques of analyses.

3. Results and Discussions

The wild herb legume (*M. indicus*) is widely-distributed. The most effective nitrogen-fixing herb legumes were: *Melilotus indicus, Melilotus siculus, Medicago intertexta* and *Trifolium resupinatum*. The nitrogen-fixing ability of these legumes ranged between 1.4–3.3 μmol C_2H_4 pl^{-1}h^{-1} compared to about 0.6 for the cultivated legume *Vicia faba*. The legumes *M. siculus, M. intertexta* and *L. corniculatus* were observed growing in the salt-affected soils in association with some halophytes such as *Spergularia marina, Carex divisia* and *Hordeum leporinum*. In salt-affected soils, Na is accumulated in higher amounts, other minerals such as Mg, Fe, Cu and Mn were also available in salt-affected soils. The herb legumes accumulated higher content of protein (about 103 mg/g dry weight at *M. indicus*). The wild herb legumes (*L. corniculatus* and *M. indicus*) kept higher levels of nitrogenase activity (about 12 and 6 μmol C_2H_4 pl^{-1} h^{-1}) under salt stress. Nodule structure of *L. corniculatus* was resistant to salt stress up to 100 mM NaCl, but less tolerant to 200 mM NaCl, infected cells were distorted and some of them were disrupted. The wild herb legumes established successful (effective) symbiotic nitrogen-fixing systems under the conditions prevailing in Egypt.

4. References

Hardy RWF *et al.* (1973) Soil Biol. Biochem. 5, 47-81

Lowry OH *et al.* (1951) J. Biol. Chem. 193, 265-275

Miller RN, Jastrow JD (1996) British Grassland Soc. Reading, pp. 105-112

Zahran HH (1998) Biol. Plant. 41, 575-585

Zaharn HH (1999) Microbiol. Mol. Rev. 63, 968-989

Zahran HH (2001) J. Biotech.

NodA PROTEIN SEQUENCES AS PREDICTORS OF THE LIPOCHITO-OLIGOSACCHARIDE Nod FACTOR TYPE: A TOOL FOR STUDYING THE MOLECULAR ECOLOGY AND EVOLUTION OF RHIZOBIUM-LEGUME SYMBIOSIS

L. Moulin[1], B. Mangin[2], F. Debellé[3], E. Giraud[1], S. Ba[1], J. Dénarié[3], C. Boivin-Masson[1]

[1]LSTM, IRD-INRA-CIRAD-ENSAM, Baillarguet, BP 5035- 34398 Montpellier Cedex 5, France
[2]BIA, INRA, BP 27, 31326 Castanet-Tolosan Cedex, France
[3]LBMRPM, CNRS-INRA, BP 27, 31326 Castanet-Tolosan Cedex, France

The nodulation gene *nodA*, which is present in a single copy in all rhizobia, is involved in the specific transfer of an acyl chain to the chito-oligosaccharide backbone of Nod factors (NFs) (Debellé *et al.* 1996; Ritsema *et al.* 1996; Quinto *et al.* 1997; Schultze *et al.* 1995).

We sequenced the entire *nodA* gene of rhizobial strains, whose NFs have been characterized and which belong to several different genera. Phylogenetic analysis of all the available sequences, indicated that the NodA proteins form clusters correlating with some NF structural features such as fucosylation and/or arabinosylation at the reducing end, and the presence of polyunsaturated fatty acids at the non-reducing end. We proposed two complementary methods to predict the NF type of a given strain. One based on the complete NodA sequences phylogenetical analysis and the other based on the analysis of amino acids present at NF-type informative positions.

These tools were used to obtain clues on the NF structure of 30 strains including novel species or symbionts with a peculiar host range. Twenty-four predictions were in accordance with available biological or biochemical data. In some cases, including new described rhizobia *Methylobacterium nodulans* and *Burkholderia* sp. STM678 (Sy *et al.* 2001; Moulin *et al.* 2001) no safe prediction could be proposed, suggesting that the strains produced new NF structures.

These results indicate that the *nodA* gene could be used as a tool to search for novel NF structures and as markers for studying rhizobium-legume coevolution.

References

Debellé F *et al.* (1996) Mol. Microbiol. 22, 303-314
Moulin L *et al.* (2001) Nature 411, 948-950
Quinto C *et al.* (1997) Proc. Natl. Acad. Sci. USA 94, 4336-4341
Ritsema T *et al.* (1996) Mol. Gen. Genet. 251, 44-51
Schultze M *et al.* (1995) Proc. Natl. Acad. Sci. USA 92, 2706-2709
Sy A *et al.* (2001) J. Bacteriol. 183, 214-220

DIVERSITY OF *SINORHIZOBIUM MELILOTI* ISOLATES NATIVE TO CENTRAL ASIAN AND SIBERIAN GENE CENTERS OF ALFALFA

M.L. Roumiantseva, E.E. Andronov, B.V. Simarov

Research Institute for Agricultural Microbiology, Ch. Podbelskogo 3,
196608 St.-Petersburg-Pushkin, Russia

Central Asian and Siberian centers of diversity located on the territory of Russia and Tadjikistan according to discovery of N.I. Vavilov (1926) and A.I. Ivanov (1988) comprise perennial species of *Medicago, Melilotus* and *Trigonella* (alfalfa cross-inoculation group) and were never studied for diversity of indigenous rhizobia. Twenty-seven isolates of *Sinorhizobium meliloti* were recovered from nodules of host plants belonged to 7 species from alfalfa cross-inoculation group and from soil samples by using trap host plants at 8 sites/15 locations in the Central Asian center of diversity. Fifty-six *S. meliloti* isolates were obtained from soil samples at 2 sites/18 locations in Siberian centre of diversity by using trap host plants. Two megaplasmids bands with a size of more than 1400 kb were revealed in all isolates with a surprising variety of sizes. From 8 to 11 plasmid profiling groups containing from 0 to 3 plasmids in addition to megaplasmids were identified in Central Asian and Siberian populations. Plasmids with a size at about 200 kb represented by similar percentage of isolates in both populations occurred to be more specific for nodule than for soil Central Asian isolates. The number of IS*Rm*2011-2 elements per genome was ranged up to 23 copies among tested isolates, while three soil isolates native to both populations were free from this IS element. Both populations did not show significant difference in average copies of IS*Rm*2011-2 per genome, while a sharp contrast was revealed between nodule and soil Central Asian isolates, as the latter possessed half as many IS*Rm*2011-2 copies per genome. RFLP analysis revealed from 7 to 8 distinct chromosomal types generated by probes on *leu* and *rec*A loci in each population, but only two types were shared by both populations. One of these two types was the type similar to that of Rm2011, which was more frequently detected among both populations. Among p*Sym*B types developed by probes on *exo/exs* and *exp* gene clusters (Becker *et al.* 1997) type Rm2011 was strongly dominant in both populations. Other types were highly specific for each population. p*Sym*A types were generated by PCR amplified 224 bp inner fragment of *nod*D1 of Rm2011 and by *nif*KDH probe (Roumiantseva *et al.* 1999). Among 14 or 19 p*Sym*A types identified among isolates of both populations type Rm2011 was predominant among Central Asian population, but that was not the case in Siberian population. The majority of the remaining types were highly specific for each population. Mean genetic diversity (H) confirms a slight difference between tested populations and between nodule and soil Central Asian isolates at chromosomal loci, and significant difference at p*Sym*B loci. A highest value of diversity was observed at p*Sym*A loci among all groups of isolates. Applied disequlibrium statistics for all pairs of studied loci showed a strict link between chromosomal and p*Sym*A pairs of loci in isolates native to both gene centers of alfalfa.

References

Becker A (1997) J. Bacteriol. 179, 1375-1384
Ivanov AI (1988) In Brezhnev DD (ed), Alfalfa (Lyutserna), pp. 1-318, Amerind Publishing Co. Pvt. Ltd., New Delhi
Roumiantseva ML (1999) Russian J. Genetics 35, 159-135
Vavilov NI (1926) Trends in Practical Botany, Genetics and Selection (in Russian) 16, 3-248

Acknowledgements

This work was financed by INTAS694.

USE OF AUTOMATED SYSTEMS TO STUDY THE BIODIVERSITY OF RHIZOBIA

K. Lindström, A. Dresler-Nurmi, L.A. Räsänen, Z. Terefework

Dept Appl. Chem. Microbiol., Viikki Biocenter, University of Helsinki, Finland

To evaluate the biodiversity of 320 rhizobial strains isolated from the nodules of *Calliandra calothyrsus* growing in humid tropics in Central America, Africa and New Caledonia, PCR-RFLP of 16S rDNA was used. RFLP of 150 strains with two enzymes revealed that the diversity of the strains was very large. A direct PCR reaction to amplify ribosomal genes was set up using intact cells without prior DNA isolation. The process was automated by using 96 Deep Well Plates for growing the bacteria and 0.2 mL Skirted 96 Tube Plates for PCR reactions. To distinguish bacterial isolates, amplified 16S rDNA genes were digested with three different enzymes and run in a 96-sample tank. 16S rDNA was successfully amplified from all tested strains without prior isolation of the DNA. Different, distinct restriction patterns were detected with each of endonucleases used. Next, the IGS regions will be analyzed by a traditional RFLP approach (agarose gel electrophoresis) as well as by basepair-accurate fluorescence based T-RFLP analysis (capillary electrophoresis). The latter will also be tried for 16S rDNA to create a reliable database of patterns representing both recognized species and *Calliandra* rhizobia.

The genetic diversity of *Rhizobium galegae* strains and taxonomically diverse rhizobia representing the recognized species was assessed using a fluorescent AFLP method. Unlike other PCR-based fingerprinting techniques such as RAPD, the AFLP reactions are not as much dependent on minor changes in PCR conditions. The AFLP fingerprints are highly reproducible, and changing the selective bases of the primers can regulate the number of bands obtained. The detection methods, fluorescent based reactions on ABI 310 and ABI 377 genetic analyzers and silver staining gave consistent fingerprint patterns from all the strains studied. However, a relatively lower number of bands were observed on the polyacrylamide gels due to their smaller resolution power and due to the size of the gel when compared with the capillary and slab gel electrophoresis separation of fragments from ABI310 and ABI377. Using capillary electrophoresis with a proper polymer and standard conditions it is possible to obtain data within the range of 50–2000 bp. Pairwise comparison of the fluorescent samples from the ABI310 and ABI377 runs revealed that more distinguished peak are resolved by capillary electrophoresis (CE) than the slab gel electrophoresis. We used the recognized species of rhizobia in this study to demonstrate the broad use and advantages of the AFLP method over other fingerprinting techniques, and show the applicability of the technique in rhizobial diversity studies. The groupings in the clusters, however, do not accord with the anticipated relatedness of most of the species supported by other rhizobial systematic data. *Rhizobium galegae*, a species that makes successful symbiosis only with *Galega officinalis* and *Galega orientalis* plants, shows an interesting host-microsymbiont specificity. The *Rhizobium galegae* strains are able to infect both plants, however, strains isolated from *Galega officinalis* form effective nodules on the respective plant, but ineffective nodules on *Galega orientalis* and vice versa. All methods delineated the *G. orientalis* strains from *G. officinalis* strains, the *G. orientalis* strains formed a tight cluster whereas the *G. officinalis* strains seem to show a greater level of genetic diversity. The AFLP results warrant a subspecies status to the two biovars that are previously delineated by their symbiotic genotypic and phenotypic characters.

THE TAXONOMY AND DIVERSIFICATION OF ROOT NODULE BACTERIA FROM THE PASTURE LEGUME *BISERRULA PELECINUS* L.

K.G. Nandasena, J.G. Howieson, G.W. O'Hara, R.P. Tiwari

Centre for Rhizobium Studies, Murdoch University, Western Australia

1. Introduction

Diversification of rhizobia in the soil is a major issue faced by both farmers and soil microbiologists (Sullivan *et al.* 1996). An opportunity to investigate this arose with the introduction into Australian soils in 1994 of a new group of root nodule bacteria for the pasture legume *Biserrula pelecinus* L. (Howieson *et al.* 1995). *B. pelecinus* is a mono-specific genus endemic to the Mediterranean region with many agricultural attributes. This study investigated the taxonomic position of root nodule bacteria from *B. pelecinus* and it is also examining the diversification of these strains following their introduction to soil.

2. Materials and Methods

Three bacterial strains (WSM1283, WSM1284, WSM1497) isolated from *B. pelecinus* growing in Morocco, Italy and Greece respectively were studied to determine their taxonomic position. A polyphasic approach, which included morphological and physiological characteristics, plasmid profiles, symbiotic performance and 16S rRNA gene sequencing was used. The genetic distances were calculated using Kimura II parameter (Kimura 1980). For the diversification study, root nodule bacteria were isolated from *B. pelecinus* from a trial field in 2000 and RAPD-PCR using the primer RPO1 was used as described by Richardson *et al.* (1995) to investigate the genetic variation among the new isolates.

3. Results and Discussion

All three strains possess a polar or subpolar flagellum and one approximately 500 kb plasmid. A BLAST similarity search revealed that all three isolates shared 99% identity with *M. loti* and *M. ciceri* according to their 16S rRNA gene. Distance matrix obtained using the nucleotide substitution rates (Knuc values) displayed the following values between the biserrula strains and representative members of the other genera: *Bradyrhizobium* > 11, *Azorhizobium* > 10, *Agrobacterium* > 7, *Rhizobium* > 7, *Sinorhizobium* > 4, *Mesorhizobium* < 2.6. Therefore the biserrula isolates were placed among the members of *Mesorhizobium* in the phylogenetic tree. The results of the physiological experiments and the symbiotic performance are also suggestive of a close relationship between biserrula bacteria and members of *Mesorhizobium*.

We have obtained five strains out of 88 field isolates that are completely different from the original inoculum strain according to their RPO1 banding profiles. Further studies, involving DNA-hybridization, whole cell protein profiles and localization of the symbiotic genes, are being carried out to investigate the probable method of diversification of these bacteria in soil.

4. References

Howieson *et al.* (1995) Aust. J. Agric. Res. 46, 997-1009
Kimura (1980) J. Mol. Evol. 16, 111-120
Richardson *et al.* (1995) Soil. Biol. Biochem. 27, 515-524
Sullivan *et al.* (1996) Appl. Environ. Microbiol. 62, 2818-2825

GENOSPECIES IN *BRADYRHIZOBIUM* – IN SEARCH OF PHENOTYPIC DIFFERENTIATION

A. Willems, J. Schelkens, M. Gillis

Laboratory of Microbiology, Dept of Biochemistry, Physiology & Microbiology, University of Gent, K.L. Ledeganckstraat 35, B-9000 Gent, Belgium

Bradyrhizobium contains three named species (*B. japonicum, B. elkanii, B. liaoningense*) and a number of unnamed groups from various host plants. Genospecies I-XI (Willems *et al.* 2001a, 2001b) were revealed by DNA:DNA hybridization and genospecies I-VII confirmed by sequence analysis of the internal transcribed spacer (ITS) between the 16S and 23S rRNA genes. We have extended the ITS sequence analysis to 11 genospecies to show that it is a good indicator of close genetic relationship, gives valuable information on the genospecies grouping, and predicts where DNA:DNA hybridization will be useful. Such data permit easy comparison between laboratories. Three of the 11 genospecies represent the currently recognized species. To aid in naming the other eight genospecies, we used the following techniques on a subset of strains to find differentiating features: API ZYM and API 50CH tests, antibiotic sensitivity tests and fatty acid analysis.

API ZYM tests were not useful as just 3 of 19 tests were positive and only alkaline phosphatase could be used to differentiate genospecies VI and VIII from the others. API 50CH gave 27/49 positive results. Because of the slow growth, the procedure was modified and growth scored on days 2, 4, 7 and 9. Eleven strains were tested but more are needed to assess the system's potential.

Antibiotic susceptibility testing used the disc diffusion method on modified YMA (containing 5 g/L mannitol and $MgCl_2.6H_2O$ instead of $MgSO_4.7H_2O$). Plates were inoculated with a cell suspension (density McFarland 0.5) and incubated with antibiotic disks for 4 days at 28°C before inhibition zone diameters were read using a digital caliper. The following antibiotics were useful to differentiate between some of the genospecies: kanamycin (K30), carbenicilin (CAR100), tetracycline (TE30), cefuroxime (CXM30), minocycline (MH30), rifampicin (RD5/30) and novobiocin (NV30). Overall, the genospecies VI and VIII were most susceptible and the genospecies II (*B. elkanii*), X and XI were resistant to many of the antibiotics tested.

For fatty acid analysis, each strain was grown on a plate of modified YMA medium as described above and from these cultures a cell suspension was prepared (density McFarland 5) to uniformly inoculate 1 to 6 plates of the same medium. The number of plates to be used was assessed from the growth of the first culture. Plates were incubated for 4 days at 28°C. Fatty acid methyl esters were extracted and analyzed using the MIDI system (Microbial ID Inc.). The major fatty acids found were 18:1 w7c and 16:0, in line with literature data. Some of the smaller constituents allowed differentiation of the genospecies, but not all genospecies could be differentiated.

So far, we can conclude that phenotypic data for the differentiation of the different *Bradyrhizobium* genospecies are difficult to obtain. Most procedures have to be modified because of the slow growth of bradyrhizobia. Some groups, such as the genospecies containing photosynthetic isolates (VI and VIII), may be recognized using fatty acid composition and antibiotic susceptibilities, but more strains from the different groups need to be included to confirm this. The API 50CH system may provide further differentiation, but also here, more strains need to be included.

References

Willems A *et al.* (2001a) Int. J. Syst. Evol. Microbiol. 51, 623-632

Willems A *et al.* (2001b) Int. J. Syst. Evol. Microbiol. 51, 1315-1322

APPLICATION OF TREE LEGUME RHIZOBIA FOR REFORESTATION PROGRAM IN THAILAND: II. BIODIVERSITY OF TREE LEGUME RHIZOBIA IN THAILAND

N. Teaumroong[1], M. Manassila[1], S. Kotepong[2], A. Nuntagij[2], N. Boonkerd[1]

[1]School of Biotechnology, Suranaree University of Technology, Nakhon Ratchasima, Thailand
[2]Soil Microbiology Research Group, Division of Soil Science, Department of Agriculture, Thailand

Rhizobial strains from tree legumes were isolated from five varieties of native tree leguminous plants in all parts of Thailand, altogether with selected the highest N_2-fixing efficiency strains. The 17 strains were isolated from *Acacia auriculaformis*, 52 strains from *A. mangium*, 16 strains from *Millettia leucantha* Kurz, 21 strains from *Pterocarpus indicus* Willd, and 42 strains from *Xylia kerii* Taub. Determination of growth rate and acid-base production on YM agar demonstrated only six strains were fast grower (two strains from *A. auriculaformis*, and four strains from *A. mangium*) and other strains were slow grower (base production). Indole Acetic Acid (IAA) production showed that nine strains could produced IAA (four strains from *A. auriculaformis*, three strains from *A. mangium* and two strains from *X. kerii*). Nodulation with soybean, mungbean and cowpea were observed. Soybean was nodulated by two strains of *X. kerii* and *A. auriculaformis* rhizobia. Furthermore, mungbean and cowpea could be nodulated by 38 and 48 rhizobial strains, respectively. DNA from selected strains were amplified by BoxAIR primer. The strains which had differences in DNA fingerprint patterns were selected prior to determine the relationship by 16S rRNA and NodA-PCR primer. The size of PCR product obtain from 16S rRNA and NodA-PCA were 1500 bp and 700-900 bp, respectively. Determination between selected strains and reference strains by PCR-RFLP of NodA and sequence of 16S rRNA gene found that some of the tree legume rhizobia were closely related to *Bradyrhizobium elkanii* strain USDA 94 and different from other reference rhizobial strains.

THE AQUATIC BUDDING BACTERIUM *BLASTOBACTER DENITRIFICANS* IS A NITROGEN-FIXING SYMBIONT OF *AESCHYNOMENE INDICA*

P. van Berkum[1], B.D. Eardly[2]

[1]Soybean and Alfalfa Research Laboratory, USDA, ARS, Beltsville, MD 20705, USA
[2]Penn State Berks-Lehigh Valley, Berks Campus, PO Box 7009, Reading, PA 19610, USA

1. Introduction

The central focus of this study initially was to confirm the phylogenetic placement of bradyrhizobial isolates of *Aeschynomene indica* since they are more closely related to *B. japonicum* and other non-bradyrhizobial members of subgroup 2b than they are to *B. elkanii*. From our analysis, which included type strains representing the named genera in the alpha subdivision subgroup 2b, we concluded that some isolates were closely related to *Bl. denitrificans*. Therefore, we examined the possibility that the type strain of *Bl. denitrificans* would form a symbiotic relationship with *A. indica*.

2. Procedure

The strains used were *Blastobacter denitrificans* type strain IFAM 1005 (LMG 8443) kindly provided by the Belgian Culture Collection of Microorganisms, BTAi1 and USDA 4424 (van Berkum *et al.* 1995). Seeds of *A. indica* were surface sterilized and the plants were grown in a greenhouse without supplemental lighting for 34 days during July/August. Each of the cultures were tested for symbiosis. Each treatment was prepared in five replications and five jars without inoculated bacteria served as controls. Determination of nitrogenase activities was as described by van Berkum *et al.* 1995. The plant tops were dried at 60°C for two days to determine dry matter and total nitrogen contents. Nodules were used to isolate bacterial occupants in culture as described (van Berkum *et al.* 1995). Nodulation of *A. indica* by *Bl. denitrificans* in growth pouches was investigated in triplicate on two separate occasions as described by van Berkum *et al.* 1995.

3. Results and Discussion

All three cultures nodulated *A. indica* while uninoculated control plants formed no nodules. Nitrogen fixation in the inoculated plants was evident by their increased growth and by the presence of acetylene reduction activity in their roots. The symbiotic response of *Bl. denitrificans* was similar to that of USDA 4424 and was superior to that of BTAi1. Isolates from nodules obtained from the *Bl. denitrificans* and the inoculated culture had identical ITS region sequences confirming that *Bl. denitrificans* formed the symbiosis. From this result the natural tendency might be to view the bradyrhizobia of *A. indica* as *Blastobacter* species or to consider changing the genus *Blastobacter* to *Bradyrhizobium*. However, it is important to examine both the taxonomic status of *Blastobacter* and the bradyrhizobia that include the phototrophs first.

4. References

van Berkum P *et al.* (1995) Appl. Environ. Microbiol. 61, 623-629

DIVERSITY OF RHIZOBIA IN THE MEDITERRANEAN BASIN

F. Zakhia[1], A. Bekki[2], H. Jeder[3], O. Domergue[1], M. Gillis[4], J. Cleyet-Marel[1], B. Dreyfus[1], P. de Lajudie[1]

[1]LSTM., IRD/CIRAD/INRA/AGRO-M, Montpellier, France
[2]Univ. Oran, Algeria
[3]IRA, Gabès, Tunisia
[4]Univ. Gent, Belgium

Demographic pressure and socio-economic changes in countries around the Mediterranean sea have major consequences on the ecosystems. Plant resources are decreasing due to overgrazing, soil erosion, aridity, salinity and desertification. To restore degraded lands and to face increased forage and agronomic needs, leguminous plants are good candidates: they are good colonizers, they establish nitrogen-fixing symbioses with rhizobial bacteria, they are resistant to aridity and salinity, and many produce human and animal food. We investigated nodulation of spontaneous legumes in North Africa (Tunisia and Algeria), belonging to several genera such as *Acacia, Acmispon, Anthyllis, Argyrolobium, Astragalus, Callicotom, Colutea, Ebenus, Genista, Hedysarum, Hippocrepis, Lathyrus, Lotus, Medicago, Onobrychis, Ononis, Prosopis, Retama, Trigonella* and *Vicia*. We isolated and characterized ca. 100 rhizobial bacteria from nodules of these plants sampled in different ecological regions. Diversity was estimated using SDS-PAGE of total cellular proteins and PCR-RFLP of rRNA operon. Some rhizobial isolates were identified as known *Sinorhizobium* and *Mesorhizobium* species, but others, from saline and/or arid zones, constituted potential new groups.

GENOTYPIC ANALYSIS OF NOVEL RHIZOBIAL ISOLATES FROM DIVERSE LEGUME SPECIES OF AGRONOMIC AND ECOLOGICAL INTEREST NATIVE TO SPAIN, INDIA, NEPAL AND MYANMAR

P. Vinuesa[1,3], A. Willems[2], E. Martínez-Romero[3], M. León-Barrios[5], M. Thynn[1,4], S.K. Mahna[6], B.N. Prasad[7], D. Werner[1]

[1]FB Biologie, Dept of Cell Biology and Applied Botany, Philipps-Universität Marburg, D-35032, Germany; [2]Dept of Microbiology, Ghent University, Belgium; [3]Centro de Investigación sobre Fijación de Nitrógeno, UNAM, Cuernavaca, Morelos, Mexico; [4]Dept of Microbiology and Cellular Biology, Univ. of La Laguna, Tenerife, Spain; [6]Dept of Botany, Univ. of Mandalay, Myanmar; [7]Dept of Botany, Maharshi Dayanand Saraswati University, Ajmer, India; [4]Dept of Botany, Tribhuvan University, Kiritipur, Kathmandu, Nepal

In the frame of a new INCO-DEV Cooperation Project, we are using molecular DNA fingerprinting techniques and computer-assisted pattern analysis to characterize the diversity of rhizobial strains isolated from the nodules of plants of notable agronomic and ecological interest native to the Canary Islands, India, Myanmar (former Burma) and Nepal. Our database currently contains over 150 isolates from these countries and diverse host genera such as *Abrus, Acacia, Adenocarpus, Calopogonium, Chamaecytisus, Crotalaria, Erythrina, Glycine, Lupinus, Prosopis, Sesbania, Spartocytisus, Teline* and *Vigna*. All isolates were first subjected to rep-PCR genomic fingerprinting, a technique suited to distinguish strains of a particular species. Selected strains were then subjected to PCR-RFLP analysis of the *rrs*, *rrl* and rDNA IGS regions (Vinuesa *et al.* 1998, 1999). The resulting patterns were compared with those of most of the currently described rhizobial species, allowing a high-resolution phylogenetic placement of the new isolates to be performed. We are particularly advanced in the elucidation of the genetic and taxonomic affinities of the strains isolated from endemic woody legumes (*Papilionoideae:Genisteae*) of the Canary Islands. As previously reported (Jarabo-Lorenzo *et al.* 2000; Vinuesa *et al.* 1998, 1999), most of these strains form a clade closely related to, but distinct from *B. japonicum* and *B. liaoningense* strains. Based on our more recent DNA-DNA hybridization data, MLEE analysis of seven enzyme loci, ITS and partial *nodC* sequencing, the strains within this clade constitute a distinct genomic species, which is present also in Moroccan soils. These strains do not nodulate *Glycine max*, nor do *B. japonicum* strains nodulate the Canarian shrub legumes (*Genisteae*). The poor correlation found between the *Bradyrhizobium* strain groupings obtained based on the analysis phenotypic and genotypic traits suggests that a revision of the minimal standards proposed by Graham *et al.* (1991) for rhizobial species descriptions should be made, giving stronger support to the genotypic data, particularly due to the wealth of sequence data that has accumulated in recent years for many rhizobial genes and strains. Several *Acacia* and *Prosopis* isolates from the arid Rajastan region were found to be highly related to *S. saheli*, others to *R. etli* and *R. hainanense*. The Nepalese *Glycine max* isolates were all closely related, having ITS sequences very similar (> 98%) to those of *B. japonicum* USDA62 and USDA122. The most divergent isolates were those from tropical legumes growing in SE Myanmar, which represent new species of uncertain taxonomic status.

References

Graham *et al.* (1991) Int. J. Syst. Bacteriol. 41, 582-587

Jarabo-Lorenzo *et al.* (2000) Syst. Appl. Microbiol. 23, 418-425

Vinuesa *et al.* (1998) Appl. Environ. Microbiol. 64, 2096-2104

Vinuesa *et al.* (1999) In Martínez-Romero E, Hernández, G (eds) Highlights on Nitrogen Fixation Research, pp. 275-279, Plenum Publishing Corporation, NY

IMPORTANT CELL SURFACE POLYSACCHARIDES IN THE *SINORHIZOBIUM MELILOTI*-ALFALFA SYMBIOSIS

J. Lloret, B. Pellock, G.R.O. Campbell, T. Shcherban, G.C. Walker

Dept of Biology, Massachusetts Institute of Technology, Cambridge, MA 02139, USA

1. Introduction

Sinorhizobium meliloti polysaccharides play an important role in the establishment of a functional symbiosis with its host plant alfalfa. Three different polysaccharides (succinoglycan, EPS II and K-antigen) are each capable of promoting infection thread initiation and extension. Though the three polysaccharides mediated infection thread growth, EPS II and K-antigen does less efficiently than succinoglycan. EPS II is a polymer of a glucose-galactose disaccharide unit modified with one acetyl and one pyruvyl substituent per repeating unit. As the case for succinoglycan, low molecular weight (LMW) forms of EPS II are the active forms in symbiosis. Small amounts of LMW EPS II, as low as 7 pmol, allow nodule invasion when added to non-infective strains defective in exopolysaccharide production. This strongly suggests that exopolysaccharides may act as a signal during the infection process, necessary for correct infection thread development.

2. Results and Discussion

S. meliloti Rm1021 usually produces succinoglycan but not EPS II. When growing in low phosphate media or carrying a null allele of the regulatory gene called *mucR*, Rm1021 produces EPS II, but only high molecular weight (HMW) forms that are not active in symbiosis. A spontaneous mutation originally called *expR101* induces Rm1021 to produce EPS II ranging from LMW to HMW, and so Rm1021 *expR101* derivatives strains induce the formation of functional nodules even when they are deficient in succinoglycan production. We have found that *expR101* mutation is actually an insertion sequence excision, which restores the function of a gene, *expR*. The predicted product of *expR* shares homology with the LuxR transcriptional regulator family.

We have also studied the possible roles of another type of surface polysaccharide, the lipopolysaccharide, in symbiosis. We have identified a mutant named *fix389* that presented a deficient phenotype after its release from the infection threads. The *fix389* mutant, which carries a null allele of a putative glycosyl transferase encoded by the *lpsB* gene, has an altered LPS core and is more sensitive to several cationic peptides. The nodules formed by the *lpsB* mutant show several abnormalities, including plant cells containing large vacuoles and bacteroids with PHB granules. The plants infected by *fix389* were much smaller than the control plants and exhibit nitrogen deficiency symptoms. We suggest that the mutant might be highly sensitive to plant's defense responses and is thus unable to establish a proper chronic infection within the plant cells.

3. References

Gonzalez JE *et al.* (1996) Proc. Natl. Acad. Sci. USA 93, 8636-8641

Pellock BJ *et al.* (2000) J. Bacteriol. 182, 4310-4318

RESPONSE OF LEGUME ROOTS TO HOG GASTRIC MUCIN BLOOD GROUP A+H SUBSTANCE, A CARBOHYDRATE LIGAND FOR LEGUME ROOT LNP

G. Kalsi, J.B. Murphy, M.E. Etzler

Section Mol. Cell. Biol., U. Calif., Davis, CA, USA

Previous studies from this laboratory established that Db-LNP, a lectin nucleotide phosphohydrolase present in the roots of the legume, *Dolichos biflorus*, is a Nod factor-binding protein present on the surface of root hair and epidermal cells in the nodulation zone of roots and that this protein most probably plays a role in the initiation of the rhizobium-legume symbiosis. This LNP is isolated from roots by affinity chromatography on hog gastric mucin blood group A+H substance (A+H BGS) and binds well to this carbohydrate ligand. Treatment of roots with A+H BGS was found to mimic the effect of rhizobial symbionts in causing a redistribution of LNP to the tips of the root hairs. We now show that treatment of *D. biflorus* roots with low concentrations of this ligand results in root hair deformations that include swelling, branching and curling as well as in the formation of pseudonodules. The ability of this carbohydrate ligand of LNP to mimic the effect of rhizobia on the roots provides further evidence that LNP is involved in the signaling events that initiate the plant response to rhizobia.

SINORHIZOBIUM SP. STRAIN NGR234 LIPOPOLYSACCHARIDES: STRUCTURAL FEATURES AND COMPARISON OF LPS FROM UNINDUCED AND APIGENIN-INDUCED CELLS

S.K. Gudlavalleti, S.B. Stevens, B.R. Reuhs, L.S. Forsberg

Complex Carbohydrate Research Center, U Georgia, Athens, GA, USA

Sinorhizobium sp. NGR234 is a wide host range endosymbiont with a prominent agricultural role; for this reason it is a subject of study in several laboratories, and the nucleotide sequence of the symbiosis plasmid has recently been determined (Freiberg *et al.* 1997). However, there is little detailed structural information on the cell surface macromolecules produced by these bacteria, or the structural alterations that occur in these molecules during differentiation and bacteroid development. We describe an initial structural investigation of the cell surface lipopolysaccharides (LPS), including a detailed structural analysis of the lipid A moiety, and an investigation into the effects of apigenin on structural alterations in the total LPS. Complementary structural information on this LPS will prove to be useful for the ongoing genetic study of this organism.

The NGR234 strain is a member of the Rf205 serogroup (group B) (Bhat *et al.* 1994), as defined by the core region carbohydrate epitopes of the rough LPS (R-LPS, lacking O-antigen). Unlike *Rhizobium etli*, strain NGR234 produces very little smooth LPS (S-LPS, containing O-antigen), when grown in culture. The major biosynthetic product is the R-LPS, and the core region of this LPS yields approximately 20 different oligosaccharide components upon mild acid hydrolysis. This contrasts with *R. etli* LPS which essentially contains two major oligosaccharide components in the core region. We have found that exposure of NGR234 cells to apigenin results in a marked alteration in the chromatographic properties and carbohydrate composition of the R-LPS fraction. The R-LPS fraction from apigenin-induced cultures shows high levels of rhamnose and altered fatty acids in addition to the normal residues. These results extend those of an earlier study (Forsberg, Carlson 1998) and provide further evidence that changes in LPS structure can be induced by plant flavonoids.

The lipid A portion of the NGR234 R-LPS has structural features in common with both the enterobacterial lipid As and the lipid As of *Rhizobium etli* and *R. leguminosarum* (Jabbouri S *et al.* 1996; Reuhs BL *et al.* 1998). In contrast to *R. etli*, the NGR234 lipid A consists of a bisphosphorylated GlcNAc disaccharide backbone, similar to those of many enteric bacteria. However, the NGR234 lipid A displays a fatty acylation profile similar to that of *R. etli* and *R. leguminosarum*, characterized by extensive amide-linked fatty acyl heterogeneity, the absence of ester-linked acyloxyacyl residues, and the occurrence of very long chain fatty acids (29-hydroxy-30-carbon acids), attached as acyloxyacyl residues to amide-linked 3-OH-18-carbon fatty acid.

References

Bhat UR, Forsberg LS, Carlson RW (1994) J. Biol. Chem. 269, 14402-14410

Freiberg, CR *et al.* (1997) Nature 387, 394-401

Forsberg LS, Carlson RW (1998) J. Biol. Chem. 273, 2747-2757

Jabbouri S *et al.* (1996) In Stacey G, Mullin B, Gresshoff PM (eds) Biology of Plant Microbe
    Interactions, pp. 319-324, ISMPMI, St. Paul, USA

Reuhs BL *et al.* (1998) Appl. Environ. Microbiol. 64, 4930-4938

Acknowledgement

Supported by NRI/USDA CSREES Grant 99-35305-8143 (to LSF) and by DOE Grant DE-FG09-93ER20097 (to the CCRC).

GROWTH AND SYMBIOTIC RESPONSE OF COWPEA (*VIGNA UNGUCULATA*) TO ELEVATED LEVELS OF UV-B RADIATION

S.B.M. Chimphango[1], C. Musil[2], F.D. Dakora[1]

[1]Botany Dept, UCT, P/B Rondebosch 7701, Cape Town
[2]Ecology and Conservation, Claremont 7735, South Africa

Depletion of the stratospheric ozone layer has led to increased levels of solar ultraviolet-B radiation (UV-B, 290-315 nm) with uncertain consequences on crop plants. Among other effects, UV-B radiation is reported to damage the photosynthetic machinery (Teramura and Sullivan 1994) and induce accumulation of flavonoids and anthocyanins in leaves of plant species as defense mechanisms (Jansen *et al.* 1998). These compounds are however known to serve as plant signals to symbiotic bacteria in the Rhizobiaceae (Hungria *et al.* 1991; Phillips 1992). We therefore hypothesized that a rise in UV-B radiation will (1) increase nodulation and N_2 fixation in legumes if the increase in flavonoid concentration in parent plants and in seeds includes node gene inducers, or (2) reduce nodulation and N_2 fixation if the damaged photosynthetic machinery results in reduced synthesis and release of root exudate compounds into the rhizosphere. The aim of this study was to assess the symbiotic performance of cowpea plants exposed to UV-B radiation.

Cowpea (*Vigna unguculata*) plants were grown outdoors in potted sand under ambient and two levels of elevated UV-B at Kistenbosch Botanical Institute, Cape Town. Elevation in UV-B above ambient conditions was achieved by artificial supplementation with filtered UV fluorescent sun lamps. Weighted dose in the ambient UV-B treatment was 6.70 kJ m^{-2} d^{-1} (average amounts for the experimental time covering January to March in Cape Town). The two levels of elevated UV-B were 8.87 kJ m^{-2} d^{-1} (UV-B1) and 10.90 kJ m^{-2} d^{-1} (UV-B2). The elevated UV-B amounts simulated the levels that would be received in the southern latitudes of the tropical region. The plants were harvested at 65 DAP and assessed for nodulation, biomass accumulation, and concentration of metabolites.

Elevated UVB1 significantly reduced nodule numbers, nodule mass and total biomass, but increased the concentration of ureides in xylem. Flavonoid concentration in tissue was not altered by UV-B1. By contrast, the higher level of elevated UV-B (UV-B2) markedly increased nodule mass, tissue concentration of flavonoids, and %N in leaves and stems. The reduction in symbiotic parameters with UV-B1 was possibly due to the unaltered tissue flavonoid concentrations, in contrast to UV-B2 where increased nodulation and tissue nitrogen concentration was accompanied by increased plant flavonoids levels. These accumulated flavonoids could protect plant cells by absorbing UV-B radiation, but whether UV-B induced flavonoids are involved in signaling rhizobia for nodule formation is still being investigated.

References

Hungria M *et al.* (1991) Plant Physiol. 97, 751-758
Jansen MAK *et al.* (1998) Trends in Plant Sci. 3, 131-135
Phillips DA (1992) Recent Adv. Phytochem. 26, 201-231
Teramura AH, Sullivan JH (1994) Photosyn. Res. 39, 463-473

Acknowledgements

The authors thank AAU and DAAD for a fellowship awarded to SBMC, NBI for financial support to CM, the NRF and URC for research grant to FDD.

THE *nod*C GENE OF *SINORHIZOBIUM MELILOTI* 102L4 IS IMPLICATED IN THE SPECIFIC NODULATION OF *MEDICAGO LACINIATA*

L.R. Barran[1], E.S.P. Bromfield[1], D.C.W. Brown[2]

[1]Agriculture and Agrifood Canada, Soils and Crops Development Centre,
  2560 Blvd Hochelaga, Ste-Foy, PQ, Canada
[2]Pest Management Centre, 1391 Sandford Street, London, ON, Canada

Medicago laciniata (cut-leaf medic) is an annual that is nodulated by a restricted range of *S. meliloti* (Brockwell, Hely 1966), e.g. strain 102L4 (=USDA1170) but not by many strains which nodulate alfalfa, e.g. RCR2011 (=1021) and 41. Our aim was to characterize the *S. meliloti* 102L4 gene that is responsible for nodulation of *M. laciniata*.

An 11 kb DNA fragment was isolated from *S. meliloti* 102L4 on the basis of its ability to complement strain RCR2011 for nodulation of *M. laciniata*. This DNA fragment contained the common nodulation genes, *nod*ABCIJ, in a single operon and the overall arrangement of genes was similar to that of strains RCR2011 and 41 (Baev *et al.* 1991; Torok *et al.* 1984; S. Long *et al.* http://www.ncbi.nlm.nih). Mutation of the *nod* IJ genes of strain 102L4 resulted in a nodulation delay on *M. laciniata, M. sativa* and on *M. alba*. A combination of Tn*5* mutagenesis and complementation studies with sub-clones of the 11 kb fragment indicated that the *nod*C gene of strain 102L4 was responsible for the specific nodulation of *M. laciniata*. The amino acid sequence of the NodC protein from strain 102L4 differed significantly from that of strains RCR 2011 and 41. The rhizobial *nod*C gene is known to influence the length of the oligosaccharide backbone of lipo-chito-oligosaaccharides (LCOs) thereby influencing the host range for nodulation (Kamst *et al.* 2000; Schultze *et al.* 1992; Roche *et al.* 1991). Our data therefore suggest that *S. meliloti* 102L4 is able to nodulate *M. laciniata* because it synthesizes LCOs in which the ratio of C4 to C5 oligosaccharides differ significantly from those produced by strains RCR2011 and 41.

References

Baev N *et al.* (1991) Mol. Gen. Genet 228, 113-124
Brockwell, Hely (1966) Aus. J. Agric. 17, 885-899
Kamst E *et al.* (2000) Mol. Gen. Genet. 264, 75-81
Roche P *et al.* (1991) Cell 67, 131-1143
Schultze M *et al.* (1992) Proc. Natl. Acad. Sci. USA 89, 192-196
Torok I *et al.* (1984) Nucleic Acids Res. 12, 9509-9524

Acknowledgements

We thank Stacey Charlton, Jeff Skeene and Nathalie Ritchot for technical assistance. The financial support of Grow-Tec Ltd, Nisku, Alberta is gratefully acknowledged.

KNOCKOUT OF AN AZORHIZOBIAL dTDP-L-RHAMNOSE SYNTHASE AFFECTS LIPO- AND EXOPOLYSACCHARIDE SYNTHESIS AND SYMBIOSIS WITH *SESBANIA ROSTRATA*

W. D'Haeze, D. Vereecke, M. Gao, R. De Rycke, M. Holsters

Dept of Plant Genetics, V.I.B., K.L. Ledeganckstraat 35, B-9000 Gent, Belgium

The micro-symbiont, *Azorhizobium caulinodans* ORS571, induces nitrogen-fixing nodules on both stem and roots of the tropical leguminous host plant, *Sesbania rostrata*. Besides azorhizobial lipochito-oligosaccharidic Nod factors (D'Haeze *et al.* 1998), surface polysaccharides (SPSs), comprising lipo- (LPSs), capsular (CPSs), and exopolysaccharides (EPSs), play a pivotal role during successful nodule invasion and development. To better understand the role of SPSs, a non-polar mutation was made in the *oac2* gene (Gao *et al.* 2001). *oac2* is part of a locus containing *oac0*, *oac1*, *oac2*, and *oac3*, supposedly involved in the synthesis of dTDP-L-rhamnose, and encoding a dTDP-L-rhamnose synthase, which catalyzes the last step of this pathway. Compared to wild-type LPSs, those synthesized by the *oac2* mutant, ORS571-oac2, partitioned in the water phase upon hot phenol extraction, showed a lower degree of polymerization based on faster migration in detergent gel electrophoresis, and had a reduced rhamnose content. Using a hydrocarbon adherence method, ORS571-oac2 was shown to be slightly more hydrophobic than the parental strain.

Inoculation of *S. rostrata* stems or roots with ORS571-oac2 produced normal nodule initiation and early stages, but further nodule development was arrested, leading to defective organs with no nitrogen fixation. *In situ* localization of ORS571-oac2 indicated no bacterial internalization and bacteroid formation in the nodule-like structures. Sections showed abnormal tissue organization, without clear demarcation between central and peripheral tissues, or any sign of bacteria internalized into plant cells. Instead, many broad infection threads were present. In contrast to wild-type nodules, a severe blue auto-fluorescence was observed in a region resembling the infection center. Electron microscopy showed that intracellular infection threads, present in nodule-like structures, were 3 to 5 times wider than those in wild-type nodules. They were surrounded by a thick layer of cell wall material. Internalization events were not seen. Auto-fluorescence and the thick infection thread walls suggest the induction of a plant defense response, either to altered ORS571-oac2 SPSs, or because the *oac2* mutant is unable to suppress a plant defense response induced by other unknown factors.

In some stem nodules, induced by co-inoculation in a 1:1 ratio with ORS571-oac2 and ORS571-V44 (an azorhizobial *nodA* mutant that is unable to produce Nod factors), ORS571-oac2 could enter plant cells. Normal central tissues were formed consisting of many plant cells occupied by the oac2 mutant. This observation suggests that normal SPSs, produced by ORS571-V44, are capable to complement *in trans* defects induced by ORS571-oac2.

To better understand the role of specific SPSs, electron microscopy showed wild-type bacteria surrounded by a thick layer of low electron-dense material, suggesting a massive amount of most likely EPSs. A similar, but thinner, layer also was present around ORS571-oac2.

Quantification of EPSs produced by ORS571 and ORS571-oac2 cultivated on nitrogen fixation medium together with the determination of global bacterial hydrophobicity suggest that the latter layer surrounding the *oac2* mutant may resemble CPSs. Accordingly, CPSs may be involved in the transition from infection pockets to infection threads, whereas LPSs and/or EPSs may be required for the internalization into plant cells.

References

D'Haeze W *et al.* (1998) Mol. Plant-Microbe Interact. 11, 999-1008
Gao M *et al.* (2001) Mol. Plant-Microbe Interact. 14, 857-866

RHIZOBIUM LEGUMINOSARUM BIOVARS SECRETE SEVERAL HOMOLOGOUS EPS-BINDING PROTEINS INTO THE ENVIRONMENT

N. Ausmees, M. Lindberg

Dept of Microbiology, Swedish University of Agricultural Sciences, SLU, Uppsala, Sweden

1. Introduction

Using phage display technique we identified a family of secreted proteins in *R. leguminosarum*, which interact with the bacterial cell surface (Ausmees *et al.* 2001). These proteins (RapA1, -A2, -A3, RapB and RapC) and some others (PlyA and PlyB, Finnie *et al.* 1998) contain one or several so-called Ra-domains, which are responsible for this binding (Figure 1). Here we show that the Ra-domain specifically recognizes a carbohydrate epitope of EPS and CPS of *R. leguminosarum*.

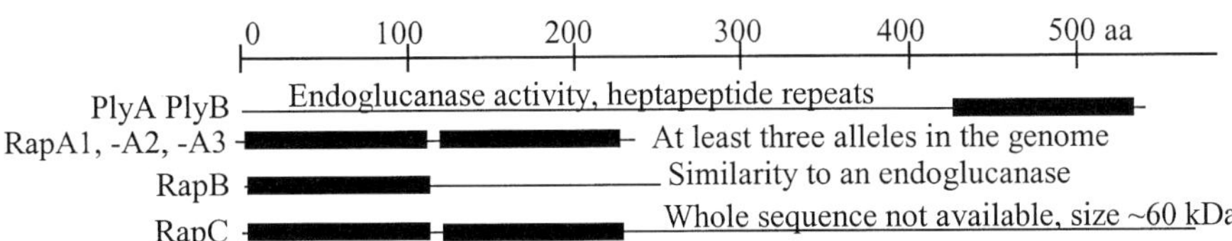

Figure 1. Domain architecture of proteins containing Ra-domains (boxes).

2. Procedures

Particle agglutination assay (PAA) was used to characterize the binding properties of the Ra-domain. RapA1-coated latex beads were mixed with bacterial cells and observed for visible aggregation. Purified EPS and CPS from different species were used to inhibit the aggregation.

3. Results and Discussion

All strains of *R. leguminosarum* bv. *trifolii* tested (R200, TA1, ANU 843, 0403) aggregated with RapA-coated beads, so did strains of *R. leguminosarum* bvs. *viciae* and *phaseoli*. Other rhizobia, e.g. *S. meliloti* and *S. fredii* USDA 257 were negative in PAA, so was *R. leguminosarum* bv. *trifolii* strain TA1-133 (*pssD*::Tn5, EPS⁻), non-mucoid and defective in EPS biosynthesis. This indicates that RapA binds to an EPS-structure present on the bacterial cell-surface. Purified EPS and CPS from several *R. leguminosarum* bv. *trifolii* strains and from *R. leguminosarum* bv. *viciae* 300, completely inhibited the aggregation between *R. leguminosarum* strains and RapA-coated beads. This shows that RapA binds to both cell-associated and culture medium EPS.

PlyA and PlyB (Finnie *et al.* 1998) and most likely RapB are involved in processing EPS. The Ra-domains in these proteins are probably involved in specific binding to the substrate. The function of the RapA proteins, however, seems to be only binding of EPS/CPS. Bearing in mind that the multicellular behavior of several gram-negative pathogens is dependent on proteinaceous fimbriae and even cellulose, it seems possible that also in rhizobia different extracellular components (proteins, EPS, cellulose) interact with each other to build an organized matrix facilitating communication between individual bacteria and contacts with the surroundings.

4. References

Ausmees N *et al.* (2001) Microbiol. 147, 549-559

Finnie C *et al.* (1998) J. Bacteriol. 180, 1691-1699

5. Acknowledgements:

Purified EPS and CPS were obtained from Dr Frank Dazzo, Michigan State University.

EFFECT OF TRUNCATIONS AND CONSTITUTIVE NtrB MUTATIONS IN MOLECULAR INTERACTIONS WITHIN THE *KLEBSIELLA PNEUMONIAE* NITROGEN-SIGNAL TRANSDUCTION PATHWAY

I. Martínez-Argudo, P. Salinas, R. Maldonado, A. Contreras

División de Genética, Universidad de Alicante, Apartado 99, E-03080 Alicante, Spain

1. Introduction

In *Klebsiella pneumoniae*, signal transduction in response to nitrogen availability is mediated by the two-component regulators NtrB and NtrC. NtrB, a bifunctional histidine-kinase, modulates the activity of the response regulator NtrC by phosphorylation. The ability of NtrB of switching between opposing kinase and phosphatase activities is regulated by PII, a trimeric protein interacting with NtrB in conditions of nitrogen sufficiency. In a previous work, we showed the usefulness of the yeast two-hybrid system to probe NtrB and NtrC homodimerization and also specific interactions between NtrC receiver and NtrB transmitter domains (Martínez-Argudo *et al.* 2001). Here we use the same *in vivo* strategy to further analyze interactions of NtrB within the nitrogen signal pathway, probing NtrB determinants for interactions with itself, NtrC and PII, and testing the effect of NtrB constitutive point mutations on these interactions.

2. Results and Discussion

Protein fusions of GAL4-AD and GAL4-BD to NtrB and derived polypeptides (including truncations and a constitutive mutation at A129T) were analyzed for their ability to interact with themselves and with equivalent fusions of signal transduction proteins PII and NtrC, in appropriate pairs, using the yeast two-hybrid system. To determine the ability of two given polypeptides to interact, we determined expression of both *GAL1:lacZ* and *GAL1:HIS3* reporters in strains of *Saccharomyces cerevisiae* Y190.

Protein-protein interactions were only detected between components of the nitrogen signal pathway. Lack of signals between heterologous two-component regulators provides evidence for specific recognition between the transmitter module of NtrB and both PII and NtrC. Contacts of NtrB with NtrC and PII are mapped to the H phosphotransfer domain and to the G kinase domain, respectively. In the latter case, the integrity of the transmitter module appears important for two-hybrid signals. Taken together, our results agree with previous data on homologous two-component systems (Park *et al.* 1998) and with recent work on PII (Piozak *et al.* 2000).

In spite of the multiple evidences for dimerization of phosphotransfer domains (Tomomori *et al.* 1999; Jiang *et al.* 2000), transmitter modules from NtrB and EnvZ do not interact when paired with themselves, a result that may reflect that contacts between H domains are not very strong. This would be compatible with a model in which the helix bundle forms and dissociates during the phosphorylation circle.

Mutation A129T affects some of the interactions tested amongst NtrB derivatives, thus supporting its effect on NtrB conformation and the sensitivity of the two-hybrid system used here.

3. References

Jiang P *et al.* (2000) Biochem. 39, 13433-13449

Martínez-Argudo I *et al.* (2001) Molec. Microbiol. 40, 169-178

Park H *et al.* (1998) Proc. Natl. Acad. Sci. USA 95, 6728-6732

Pioszak A *et al.* (2000) Biochem. 39, 13450-13461

Tomomori C *et al.* (1999) Nature Struct. Biol. 6, 729-734

NO OR NH OR BOTH: TEMPORAL EFFECTS ON NODULE INITIATION IN THE *PISUM SATIVUM/RHIZOBIUM LEGUMINOSARUM* SYMBIOSIS

M.I. Bollman, J.K. Vessey

Dept of Plant Sci., University of Manitoba, Winnipeg, MB, Canada R3T 2N2

1. Introduction

It is generally accepted that mineral N (NO or NH) inhibits nodulation in the legume-rhizobia symbioses. It has been shown that low continuous concentrations of NH (< 1.0 mM), in hydroponic culture, will stimulate nodulation in pea (*Pisum sativum* L.) (Waterer *et al.* 1992). In contrast, low concentrations of NO (as low as 0.1 mM) inhibit all aspects of nodulation (Waterer, Vessey 1993).

2. Procedure

Pea plants, grown in a continuous flow hydroponic system, were exposed to one of four sources of mineral N: Zero N; 0.5 mM NO; 0.5 mM NH; or 0.25 mM NO plus 0.25 mM NH for a period of 21 days after inoculation (DAI). N concentrations in the nutrient solutions were monitored daily and kept at a relatively constant level with additions of stock solutions by a peristaltic pump. Six plants were harvested at 7, 14 and 21 DAI. Nodule numbers and dry weights were determined. A microscopy study on two secondary roots [the third distal secondary root from the crown (#1) and the secondary root 75% of the distance from the crown to the last distal secondary root with nodules (#2)] was undertaken. Nodules were rated as to stage of development using a scale developed in our lab.

3. Results

Dry matter accumulation was greatest in the combined treatment at 7 DAI, while at 14 DAI the combined treatment was greater than the two single N sources, which were greater than the control. At 7 DAI there were no nodules in the NO treatment, and by 14 DAI, the NH treatment had more nodules than any other. The number of nodules per g root DM at 7 DAI was similar in the control, NH and combined treatment, but at 14 and 21 DAI, the control and NH treated plants were higher than either treatment containing NO.

On the secondary root #1, there were fewer nodules and they developed more slowly on the NO treated plants than any other. At 14 DAI there were more nodules, and these were more mature, on the control and NH treated plants than the other two treatments. On the younger root #2, there were fewer nodules and these were at earlier developmental stages than on root #1 at each harvest date. The relative distribution in number and development stage was similar to root #1.

4. Conclusions

From the results of this study we concluded that NO inhibits nodulation in field pea, both on an absolute and specific basis. NH, on the other hand, stimulates nodulation on an absolute basis, and does not inhibit it on a specific basis. In any of the treatments containing NO, the nodules matured slowly. The nodule proliferation in the NH treatment was greatest in the younger plants, and on those roots that formed earlier in development. The combined treatment of NO plus NH resulted in nodulation patterns early in development that were more similar to the NH treatment, and later in development that were more similar to the NO treatment.

5. References

Waterer JG *et al.* (1992) Physiol. Plant. 86, 215-220
Waterer JG, Vessey JK (1993) J. Plant Nutr. 16, 1775-1789

CHARACTERIZATION OF A HOMOGLUTATHIONE SYNTHETASE IN *MEDICAGO TRUNCATULA*

P. Frendo[1], M.J. Hernández Jiménez[1], C. Mathieu[1], L. Duret[2], D. Gallesi[1], G. Van de Sype[1], D. Hérouart[1], A. Puppo[1]

[1]Laboratoire de Biologie Végétale et Microbiologie, CNRS FRE 2294, U Nice-Sophia Antipolis, Nice, France
[2]Laboratoire de Biométrie Génétique et Biologie des Populations, CNRS UMR 5558, U Claude Bernard-Lyon 1, Villeurbanne, France

The tripeptide glutathione (γ-glutamylcysteinylglycine, GSH) is a low molecular weight thiol which is synthesized in an ATP-dependent two-step reaction. In the first step, catalyzed by γ-glutamylcysteine synthetase (γ-ECS), γ-glutamylcysteine is produced from L-glutamic acid and L-cysteine. In the second step, catalyzed by glutathione synthetase (GSHS), glycine is added to the γ-glutamylcysteine to form GSH. In plants, GSH is a storage and transport form of reduced sulfur. It has a crucial role in protecting the plant against xenobiotics, heavy metals and oxidative stress. In addition to GSH, homoglutathione (γ-glutamylcysteinyl-β-alanine, hGSH), has been detected in leguminous plants. Homoglutathione synthetase (hGSHS), an enzyme that has a higher affinity for β-alanine than for glycine, catalyzes the second step of hGSH synthesis. We are working on the characterization of GSH and hGSH metabolism in *Medicago truncatula,* a model plant for the study of legume-*Rhizobium* interactions.

Two *Medicago truncatula* cDNAs (*gshs1* and *gshs2*) corresponding to a putative glutathione synthetase (GSHS) and a putative homoglutathione synthetase (hGSHS) were characterized by heterologous expression in an *Escherichia coli* strain deficient in GSHS activity. The recombinant GSHS1 is a GSHS that does not accept β-alanine as a substrate. The recombinant GSHS2 showed a 10-fold higher affinity for β-alanine than for glycine indicating that GSHS2 is a hGSHS. Leucine-534 and proline-535 present in hGSHS were substituted in GSHS2 by two alanines that are conserved in plant GSHS. These substitutions resulted in a strongly stimulated GSH accumulation in the transformed *E. coli* strain showing that these residues play a crucial role in the differential recognition of β-alanine and glycine by hGSHS. Phylogenetic analysis of GSHS2 and GSHS1 indicated that gshs2 and gshs1 are the result of a gene duplication that occurred after the divergence between *Fabales, Solanales* and *Brassicales.* Analysis of the structure of *gshs1* and *gshs2* genes shows they are both present in a cluster and in the same orientation in the *M. truncatula* genome, suggesting that the gene duplication occurred via a tandem duplication.

ANALYZING THE FUNCTION OF TWO *nodT* GENES IN *RHIZOBIUM ETLI*. ARE THESE COPIES DIRECTLY INVOLVED IN THE NODULATION PROCESS?

A. Hernández-Mendoza, T. Scheublin, N. Nava, O. Santana, C. Quinto

Departamento de Biología Molecular de Plantas. Instituto de Biotecnología, Universidad Nacional Autónoma de México Apdo. Postal 510-3 C.P. 62251, Cuernavaca, Morelos, México

In *Rhizobium etli* we have previously described the isolation of two *nodT* copies, one located on plasmid c (*nodT*plc) and the other on chromosome (*nodT*cro) (Hernández-Mendoza 1999)

In order to analyze the *nodT*cro function we have been making efforts to obtain an insertion in this gene without success. However, if we first complement with the wild type gene *in trans*, stable insertions were obtained. Since this procedure had been successful for site-directed mutagenesis of the non-essential gene *nodT*plc in *R. etli,* we propose that *nodT*cro is necessary for viability in *R. etli* under these conditions, or may encode an essential function. We are now isolating double recombinants cured of the complementing plasmid to analyze the *nodT*cro mutant phenotype. On the other hand, two ORFs were found upstream of *nodT*cro that have 73% to 63% identity with *ameAB* from *Agrobacterium tumefaciens* and *mexAB* from *Pseudomonas aeruginosa*, respectively. NodT has 50% and 30% identity with AmeC and OprM, respectively. AmeABC and MexAB-OprM form multidrug efflux pumps.

To gain insight into the possible function of *nodT*plc gene, this copy has been mutagenized by inserting a *spectinomycin* cassette. *nodT*plc mutant does not have a clear nodulation phenotype. However, two ORFs that have 30 to 50% identity with *Escherichia coli cpxAR* genes were localized upstream of *nodT*plc. CpxAR form a two-component signal transduction system that in *E. coli* responds to heat shock and membrane damage. A putative σ24 promotor sequence was also found upstream of *nodT*plc. In order to analyze the role of this gene in response to outer membrane damage stress produced by heat shock and thermotolerance, we analyzed the growth behavior of the mutant with respect to the wild type at different temperatures. However, our results indicate that *nodT*plc is not involved in bacterial growth at different temperatures neither in basal thermotolerance under the conditions that we have analyzed.

The putative promoter sequence was subcloned into the expression vector pBB53::*gusA* in both orientations, upstream of the reporter gene *uidA* (β-glucoronidase activity). This plasmid was mobilized to wild type *R. etli* by triparental mating. GUS assays indicated that the promoter in the correct orientation has a 2–4-fold higher expression with respect to that of the promoter in opposite orientation in the presence of ethanol, heat shock conditions or in standard conditions. This suggests that the former promoter orientation has a basal activity but it does not respond to changes in temperature or presence of ethanol in the conditions that we have tested.

Taken together, our results suggest that *nodT*cro and *nodT*plc have different, but not complementary roles in the *R. etli* physiology. However, biochemical, genetical and physiological studies must be still carried out to fully characterize these genes.

References

Hernández-Mendoza A (1999) M.S. Thesis IBT-UNAM

Author Index

Index

Nif and *nif* 20, 177, 214, 229, 231, 238, 242, 392, 428
Kp2 25, 28–29
KPK2 170
kpnI 388
Krüppel transcription factor 181

lacZ 130, 374, 386, 392, 402, 403
    fusions 156, 205, 245, 252, 290, 311, 372, 408, 416, 425, 446, 530
      with *fix* genes 214, 394
      with *nif* genes 394, 428, 439
      with *nod* genes 119–120, 420
Lathyrus spp. 354, 521
Leaching 299
Lectin, DB46 121
Lectin nucleotide phosphohydrolase 115, 121, 524
Leghemoglobin 111, 184, 185, 196, 197, 198, 417, 429, 437
Legume crops, BNF in tropical 341–344
Legumes
    cover crop 354
    forage and pasture 353
    grain 352–353
    non-nodulating 485
    tree/woody 3, 384, 485, 490, 491, 512, 519, 522
    tropical 522
    wild herb 513
Lens culinaris 352
Leptochloa fusca 271
Lettuce 187
leu 374, 515
Leucaena spp. 285, 490, 491, 501, 512
Leucine regulator protein 254
Liming 344
Linkage groups 100
Lipochito-oligosaccharide 61, 99, 121, 279, 284, 288, 321, 327, 328–329, 514
Lipopolysaccharides 525, 528
Lj genes 108, 406, 429
Lodgepole pine 461
Lolium spp. 353, 481
Lotus EST project 109
Lotus intrinsic membrane protein 460
Lotus spp. 60, 66, 67, 82, 108, 353, 500, 513, 521
Lotus japonicus 122, 148–149, 286, 433, 442, 448, 460
    genes 116, 406, 429, 457
    genomics xix, 103–105, 107–108, 109–112
    'Gifu' 97, 103–105, 107–108, 405
    MG-20 strain 103–105, 108
    as model 39, 75, 95–96, 284
Lowe-Thornley model 25, 28
lpi 207, 208, 421
lps 281, 523
Lumichrome 321–322
Lupinus spp. 185–187, 352, 453, 507, 522

luxAB 60, 77
Lycopersicon esculentum 479
LysR 118

Machaerium lunatum 61
Macroptilium spp. 61, 63, 126, 289, 505
Macrozamia 88
Maize 186, 261, 263–265, 321–322, 332, 348, 353, 378, 498; *see also Zea mays*
Major intrinsic protein family 454
    Malate 153, 156, 245, 247–249, 407, 425, 453, 510
MALDI-TOF mass spectrometry 52–53
Manganese 314
Massively parallel signature sequencing 109
Mastoporan 149
Matrix glycoprotein 277–278
mcc 160
mcp 153, 157
mdh 245–246
Medic *see Medicago* spp.
Medicago genome project, EU 458
Medicago spp. 46, 126, 129, 165, 204, 353, 354, 421, 515, 527
    in Mediterranean ecosystems 500, 521
Medicago sativa 39, 75, 142, 171, 285–286, 353, 486, 511, 515
    enzymes 331, 332, 440, 446, 453, 510
    ferritin 185
    freezing, tolerance of 506
    genes 149, 181, 375, 527
    nodules 197–198, 245, 247–249, 328
    non-nodulating mutant 116, 123–126
    phenylpropanoid pathway 326
    succinoglycans 284, 286
Medicago truncatula 121, 124, 142–144, 198, 326, 375, 455, 457
    enzymes 247–248, 399, 532
    genomics xix, 99–102, 284
    as model 39, 75, 78
    mutants 116, 125
    nodules 179–182, 278
Mediterranean ecosystems, degraded 500, 521
Melilotus spp. 46, 126, 323–326, 513, 515
Melletia leucantha 512
mep 233
MerR 206
Mesorhizobium 60, 73, 284, 356, 407, 517, 521
Mesorhizobium loti 62, 112, 158–160, 291, 293–294, 394, 433
    genomics xix, 50, 55–58, 80, 87, 169, 379
    MAFF 303099 strain 67, 68, 69, 82–84
    as model 39, 75, 76
    Nod factors 148–149, 507
    secretion systems 286, 288
met 40
Metal tolerance 204–208